THE BASAL GANGLIA IV

New Ideas and Data on Structure and Function

ADVANCES IN BEHAVIORAL BIOLOGY

Recent Volumes in this Series

THE BASAL GANGLIA IV

New Ideas and Data on Structure and Function

Edited by
Gérard Percheron
INSERM
Paris, France

John S. McKenzie
University of Melbourne
Melbourne, Australia

and
Jean Féger
Université R. Descartes
Paris, France

PLENUM PRESS • NEW YORK AND LONDON

Library of Congress Cataloging-in-Publication Data

The Basal ganglia IV : new ideas and data on structure and function /
  edited by Gérard Percheron, John S. McKenzie, and Jean Féger.
      p.   cm. -- (Advances in behavioral biology ; v. 41)
  "Proceedings of the Fourth Triennial Meeting of the International
Basal Ganglia Society, held October 5-9, 1992, in Giens, Var,
France"--T.p. verso.
  Includes bibliographical references and index.
  ISBN 978-1-4612-7591-6
  1. Basal ganglia--Physiology--Congresses. 2. Basal ganglia-
-Diseases--Congresses. 3. Extrapyramidal disorders--Congresses.
I. Percheron, Gérard. II. McKenzie, John S. III. Férger, Jean.
IV. International Basal Ganglia Society. Symposium (4th : 1992 :
Giens, France) V. Title: Basal ganglia 4. VI. Title: Basal ganglia
four. VII. Series.
  [DNLM: 1. Basal Ganglia--anatomy & histology--congresses.
2. Basal Ganglia--physiology--congresses. W3 AD215 v.41 1993 / WL
307 B29716 1993]
QP383.3.B35 1993
599'.0188--dc20
DNLM/DLC
for Library of Congress   93-50823
                                                        CIP

Proceedings of the Fourth Triennial Meeting of the International Basal Ganglia Society,
held October 5–9, 1992, in Giens, Var, France

ISBN-13: 978-1-4612-7591-6   e-ISBN-13: 978-1-4613-0485-2
DOI: 10.1007/978-1-4612-0485-2

© 1994 Plenum Press, New York
Softcover reprint of the hardcover 1st edition 1994

A Division of Plenum Publishing Corporation
233 Spring Street, New York, N.Y. 10013

IBAGS IV

The 4th triennial meeting of the International Basal Ganglia Society was held at Giens,
Var, France
Monday 5th to Friday 9th October 1992

President of the meeting Mme Denise Albe-Fessard

Organiser Gérard Percheron, IBAGS President

Program Committee

Alexander R. Cools (Netherlands)
Alan R. Crossman (UK)
Jean-Michel Deniau (France)
Pierre Rondot (France)
Wolfram Schultz (Switzerland)

Organising committee
Michèle Fabre-Thorpe
Gilles Fénelon
Simone Heyner
Elisabeth Trouche
François Viallet

The meeting was organised with the kind assistance of the
GDR Neurosciences CNRS Marseille
(E. Legallet, J-C Pons and M. Alamy)
and the **VVF "La Badine" Giens**

The following Organisations and institutions are gratefully acknowledged

Centre National de la Recherche Scientifique (CNRS)
Fondation de France
Fondation IPSEN
Association pour la Neuropsychopharmacologie (ANPP)
Berlex laboratories
Charles River France
Institut de recherches internationales Servier (IRIS)
Roche
Sandoz France
SANOFI
Schering SA.
Unicet/Schering-Plough
Unimécanique

Région Provence-Alpes-Côte d'Azur

PREFACE

This volume is the fourth of an edited series based upon meetings initiated by the International Basal Ganglia Society (IBAGS), whose aims are to promote research on the system of the basal ganglia and to present the result of its activities to the scientific and general public. The present volume was generated from papers presented at the Fourth Triennial Meeting (IBAGS IV), held from 5-9 October 1993, at the Giens Peninsula, on the Mediterranean coast of France.

The policy of the IBAGS Council and of the editors was to avoid producing a mere book of abstracts. IBAGS is a multidisciplinary society. Contributors were asked first to make an effort to be understood by colleagues from different disciplines. Another request was to place their own data in a more general context, and to contribute to establishing the current state of knowledge in the field—as has been done every three years since the first IBAGS conference ten years ago. Another aim of IBAGS' books is to stimulate new approaches, ideas and hypotheses. These are the responsibility of the individual authors, with no editorial intervention.

The reader will find new questions posed and new interests revealed. Over a long period, the corticostriate connection was underrated, and the favoured picture of the system of the basal ganglia was of a fuzzy set of deep, subcortical, "centres" connected with one another to form so-called "loops". This did not take into account the fact that a loop just turns upon itself unless it is provided with input information, as well as output, if the system is to express motor effects. In this respect it is worth noting the current, almost explosive, interest in the corticostriate connection at every level of analysis, from anatomical to pharmacological. Many studies also further explore the discovery that the striatum is considerably more heterogeneous than indicated by the simple dichotomy between striosomes and matrix, which has dominated attention for more than one decade. Its minute, patchy organisation now opens up a new world of investigation and understanding. Among the traditional loops, after so much work on the dopaminergic nigrostriatal system, there are obvious strategic changes. The complex combinations of peptidergic transmission provoke much work and many questions. The nigrostriatal disconnection is no longer seen as the unique pathophysiological explanation of the parkinsonian symptoms. The role of the pallidum, particularly of its medial nucleus, is increasingly evoked. The subthalamic nucleus, long seen as the seat of turbulent hemiballism, now appears to be involved in much more subtle and complex regulations. One important and promising point is the recent interest in information processing through the system; significant progress, practical as well as theoretical, can be expected in this field.

The volume will thus signal the changes that have occurred in the last three years and the trends that are emerging. It is the hope of the editors that it will facilitate the work of other researchers, clinicians and students, and stimulate new ideas and findings.

Gérard Percheron
John S. McKenzie
Jean Féger

SPECIAL ACKNOWLEDGEMENTS

The International Basal Ganglia Society, the Organisers of the fourth meeting, would like to express particular thanks to Mrs. Michèle Fabre-Thorpe and Mrs. Simone Heyner whose commitment made the meeting and the book possible.

CONTENTS

STRUCTURE AND CONNECTIONS OF THE BASAL GANGLIA

DEVELOPMENT, DEGENERATION AND TRANSPLANTS IN THE BASAL GANGLIA

NEUROPHYSIOLOGY OF THE BASAL GANGLIA

PATHOPHYSIOLOGY OF THE BASAL GANGLIA
AND RELATED DISORDERS

NEUROCHEMISTRY AND NEUROPHARMACOLOGY
OF THE BASAL GANGLIA.

CLINICAL DISORDERS AND PHARMACOLOGY
OF THE BASAL GANGLIA

THEORIES ON BASAL GANGLIA FUNCTIONING

STRUCTURE AND CONNECTIONS

OF THE BASAL GANGLIA

THE BASAL GANGLIA RELATED SYSTEM OF PRIMATES:
DEFINITION, DESCRIPTION AND INFORMATIONAL ANALYSIS

Gérard Percheron, Chantal François, Jérôme Yelnik, Gilles Fénelon
and Boualam Talbi

Laboratoire de neuromorphologie informationnelle et de neurologie
expérimentale du mouvement. U106 INSERM, Pavillon Inserm, Hôpital de
la Salpêtrière, 47 Bd de l'Hôpital, 75651 Paris Cedex 13, France

INTRODUCTION

In the first IBAGS book we tried to define the basal ganglia and to determine their components using rational criteria. Eight years later, an answer, acceptable by most specialists, may be given. It now appears advisable to place the set of the basal ganglia in a more general system also including their inputs and outputs. Despite numerous attempts at finding direct descending connections to motoneurons, the motor action of the basal ganglia, as classically stressed, is exerted through the pyramidal system. The "basal ganglia related system" might then be seen as a cortico-baso-thalamo-cortical circuit comprising: (1) the cortico-striatal connection, (2) the striato-pallidonigral connection of the "basal ganglia core" made up of the striatum and its targets, the two pallidal nuclei and the substantia nigra (pars reticulata and lateralis), (3) the regulation of the core by the pars compacta of the substantia nigra, the subthalamic nucleus, the central complex (centre médian-parafasciculaire) and the pedunculo-pontine complex, (4) the output of the core: the pallido-thalamic and nigro-thalamic connections, (5) the pallidal thalamo-cortical and the nigral thalamo-cortical connections, (6) cortico-cortical connections, and (7) the source of the cortico-spinal connection (see Fig. 1 and Table I). Our analysis, almost exclusively based on data obtained in primates, will follow these steps successively.

THEORETICAL BASES OF ANALYSIS

The box-and-arrow diagram (Fig. 2 - 1) implicitly postulates massive communications with poor combinatory power, incompatible with elaborate information processing. Knowledge about the mediator does not suffice and spatial features must be considered. "Boxes", in fact, are metric cerebral spaces occupied by various neuronal sets (see "dictionary" of informational neuronal set theory terms at the end of the paper). The fundamental set is the neuronal species (Percheron et al., 1987, 1991b; Yelnik et al., 1987, 1991) made up of statistically similar neurons. A purely quantitative, objective neuronal typology can be obtained by measuring topologic (quantity and pattern of branching) and metric parameters on dendritic arborisations observed in three dimensions after complete reconstruction (Yelnik et al., 1984, 1987, 1991). Statistical analyses make it possible to isolate homogeneous neuronal groups and to establish a hierarchical neuronal typology and classification (Yelnik et al., 1987). Axonal typology (usually linked to the dendritic morphology) is not primarily considered in the classification. Analysis of the transfer of

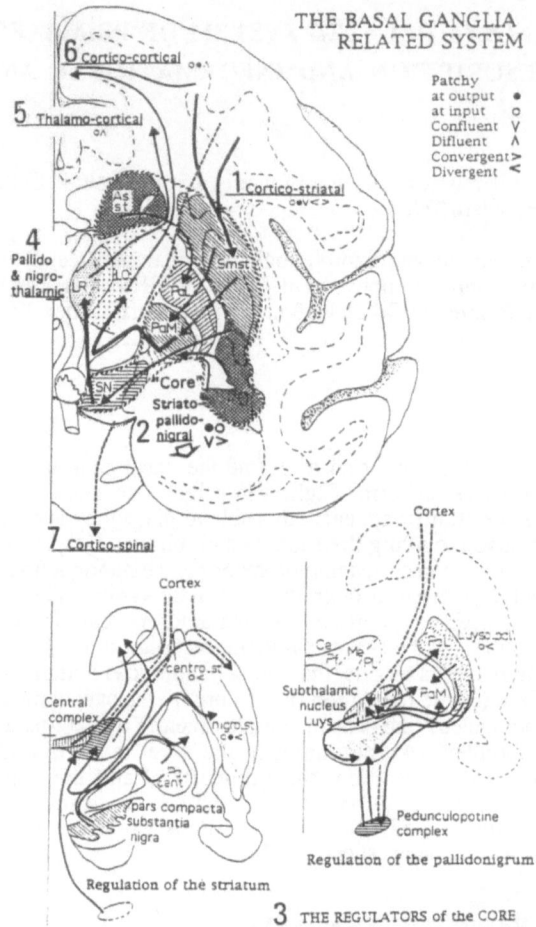

Figure 1. Semi-schematic representation of the basal ganglia system with its different stages. The numbers refers to those of the article. The symbols indicate the place where patchy maps are observed (as input or output clusters), where the communication is confluent or difluent (with regards to different sources) and where it is convergent or divergent (depending on the relative geometry of receiving dendritic arborisations and emitting cladonal arborisations in a domain of communication).

Table I. Components of the Basal Ganglia and of the Basal ganglia related systems

I - Basal ganglia system

A - Basal ganglia core

1 - Striatum
    1a - sensorimotor
    1b - "associative"
2 - Pallidonigral set (Pallidonigrum)
    2a - Pallidum nucleus lateralis (Pal. laterale)
    2b - Pallidum nucleus medialis (Pal. mediale)
    2c - Substantia nigra pars lateralis
    2d - Substantia nigra pars reticulata
3 - Striato-pallidonigral fascicle and connection

B - Basal ganglia core regulators

a - **Regulators of the striatum**
    1 - Pars compacta, substantia nigra ("locus niger")
    2 - Regio centralis thalami ("centre médian- parafasciculaire")
b - **Regulators of the pallidonigrum**
    3 - Nucleus subthalamicus (corpus Luysii)
    4 - "Nucleus" pedunculopontinus
c - **Internal regulators** (2a - Pallidum nucleus lateralis)
d - **Regulating circuits**
    Pallido-subthalamic, pallido-central, pallido-pedunculopontine,
    pallido-pallidal, pallido-nigral connections
    Nigro (compacto)-striatal and -pallidal, centro-striatal fascicles and connections
    Subthalamo- and pedunculopontino-pallidonigral fascicles and connections, etc.

II - Basal ganglia related system

A - Basal ganglia system (see I)

B - Inputs of the Basal ganglia system
1 - Cortex
    1a - **Cortico-striatal**   sensorimotor
                                "associative"
    1b - **Cortical regulators:** Cortico-central (towards central complex)
                                Cortico-subthalamic
                                Cortico-pedunculopontine connections

2 - **Raphe**

C - Ouputs of the Basal ganglia system
1 - **Pallidonigro-thalamic** (then to cortex)
    1a - **Pallido-thalamic connection.** Pallidal territory (**LO**)
        Cortical targets Supplementary motor
                            Premotor
                            Motor cortex

    1b - **Nigro-thalamic connection.** Nigral territory (**LR**)
        Cortical targets Oculomotor
                            Supplementary oculomotor
                            Frontal
                            Supplementary motor cortex
2 - **Nigro-tectal** (then to brain stem and cervical spinal cord)

neuronal information between separate "sources" and "targets" leads to the necessity of formally distinguishing two functionally different axonal components : hodons which only convey information from one source to a target and cladons (forming cladonal arborisations) which communicate with dendritic arborisations in a target (Fig. 2 - 2 and Percheron et al., 1987, 1991b). Sources and targets are connected by a fascicle or bundle of hodons. The box-and-arrow diagram represents a "macrodescription" (Smolensky, 1987) drawing up the hodonal graph (Percheron et al., 1987) of the connections between sources and targets. "Neuronal set-spaces" (the "boxes") are filled by one part (or the whole) of all neurons belonging to the same neuronal species. They may be nuclei (spaces occupied by all their somata), dendritic patriae (Fig 2 - 2) (by all their dendritic arborisations), territories (by all the cladonal arborisations from one source) and source-spaces (by all the dendritic arborisations, and somata, sending their axons to a single target) (Percheron et al., 1991b). Such neuronal set-spaces do not necessarily coincide, which stresses the necessity of precisely indicating which of them is being considered. For a given communication, first a formal distinction must be made between the intervention of information from different sources and that of different pieces of information from one single source. The first case (Fig. 2 - 3) combines information of different "modalities" (for instance sensory and motor) while the second only deals with different items of the same kind of information. Different combinations can exist beween sources and targets: "systemic separation", "unifluence" (when one source projects to one target), "difluence" (when one source projects to several targets) or "confluence" (when several sources project to the same target) (Fig. 2 - 3). Today, it appears necessary to make the "microdescription" of connections, showing how elements of one source are connected to elements of one target. Neuronal communication involves emitting by cladonal arborisations and receiving by dendritic arborisations. The relative orientation, shape and dimensions of the two sets of arborisations and of their communication domain lead to various spatial processing of information, among which are convergence or divergence. This has been previously extensively analysed and illustrated (Percheron et al., 1984a and b, 1987, 1991b; Percheron and Filion, 1991). When some elements of a neuronal set segregate and form "clusters", "patches" or "islands", there is a third level of combinatorial organisation of neuronal communication which involves neuronal subsets and neuronal subset-spaces. Description demands new conceptual tools. Three kinds of spaces must be distinguished (Fig 2 - 2). Spaces containing pools of trigger zones of close neurons elaborating coherent messages (exerting parallel processing) for the same target or subtarget (close to the "output-cell clusters" of Graybiel et al. (1991) or to the "microexcitable zones" of Crutcher and DeLong (1984) and Liles and Updyke (1985)) may be called "syllogues" (group, speaking together). They usually, but not necessarily, coincide with sets of dendritic arborisations grouped together and constituting reception subspaces that may be called dendrite-groves or "dendralses" (from alsos, grove). Patches of cladonal arborisations from the same source grouped together in a target ("input-fiber cluster") delivering a coherent message to a local part of a target may be called cladon-groves or "cladalses" (Fig.2 - 2). Subsources projecting to subtargets introduce extensive combinatory possibilities. The numbers of subtargets to which one subsource emits or of subsources received by one subtarget can lead to equi-insulation, div(ergent)-insulation, conv(ergent)-insulation, (divergent and convergent) ambi-insulation (Fig. 2- 4, column A). The pattern of connectivity, linear or branched, offers further combinations. The spatial relation between the emitting cladalses and the receiving dendralses can lead to coincidence, plurincidence, excidence or aleaincidence (Fig. 2- 4, column B) (see the patterns of Graybiel et al. 1991). All this provides considerable and varied combinatory power. The imprecise use of the term and concept of "parallel processing" biases the debate about the kinds of information processing at work in the system of the basal ganglia (Percheron and Filion, 1991). Parallel processing implies the distribution of the computing work necessary for the realisation of the same single task. The term should not be used, as is often the case, for the description of systemic separation. Parallel processing is also usually thought to be restricted to a point-to-point parallel system of channels. There are in fact many other kinds of parallel processing. Nelson and Bower (1990) described three kinds of mappings : "continuous", "patchy", and "scattered". In the first, the image in the target is a fair picture of the organisation and representation in the source. This is essentially found in sensory systems. This may be also the case in patchy connections (Fig. 2, 3-1). In this paper, we would like to stress the importance of inhomogeneous patchy connections in the basal ganglia related system.

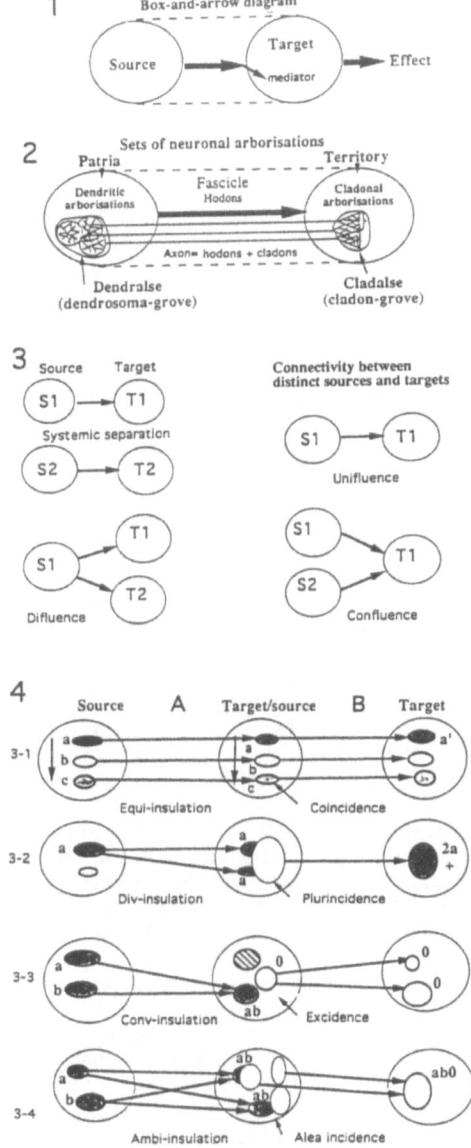

Figure 2. Successively, the box-and-arrow diagram, some neuronal set-spaces and subspaces and some combinatory possibilities provided by the existence of patchy maps. The combinations in column A depending on the relative numbers of emitting and receiving islands are independant from, but can be combined with, those of column B linked to the coincidence or not of afferent and efferent islands.

ANALYSIS OF THE SYSTEM

The cortico-striatal connection

The basal ganglia related system can be understood only if it is seen as one of the major outputs from the cortex. From the uppermost layer of the infragranular stratum ("layer Va") millions of cortico-striatal pyramidal neurons give rise to rather thin, faintly myelinated (glutamatergic hence excitatory) axons. Their messages are quite different from those sent by the cortico-spinal neurons (Bauswein et al., 1989). Almost every part of the cortex contributes to the cortico-striatal connection. The only exception is for primary sensory cortices (visual and probably auditory) which means that the system is not directly informed about the outer world. Motor maps are not continuous but patchy and iterative. In the motor cortex the body representation is ordered but made up of several separate islands for the same movement and set of muscle representations (Lemon, 1988). The fact that iterated cortical islands (like a keyboard) communicate with striatal islands, should be carefully considered.

The cortico-striatal connection is the entry to the system. Considered as a source to the striatum, the cortical surface must be subdivided into two sources sts(the sensorimotor and associative cortex). The transfer of data from various experiments on topographically precise striatal maps in macaques (Percheron et al., 1984a and b, 1987) established that the cladons from cortical sources occupy two topographically distinct intrastriatal territories: the sensorimotor and associative territories. The consideration of the two "sensorimotor" and "associative striatum" should be preferred to the historical, purely anatomical, subdivision into the caudate nucleus and the putamen. The sensorimotor territory occupied by the cladons from the somatosensory cortex, motor cortex, the motor cortex, and accessory motor cortex is located in the posterior and dorsal part of the putamen and in a thin rim of the lateral border of the caudate nucleus just overhanging the putaminal part. The connection is only grossly somatotopically organised with 3 oblique strips for lower limb, the upper limb and the face. This is not point to point continuous, but a patchy connection with a div-insulation since one spot of the sensorimotor cortex sends axons to several cladon-groves in the striatum. These are located within the matrix compartment. The cladon-groves are the place of confluence of somatosensory and motor information (Fotuhi et al., 1989), but of a separation (excidence) from the cladon-grove of the supplementary motor cortex (Alexander et al., 1988), which are reversely confluent with the post-arcuate area (Parthasarathy et al., 1992). There is a complex transformation of somatosensory information with a "remapping" made up of a concomitant divergence and a somatotopic and modality convergence (Flaherty and Graybiel, 1991).

The adjectives "associative" and "limbic" are linked to historical references. Broca's (1878) "limbic" cortex, in addition to the olfactory bulb and hippocampus, was made up of the cingulate and parahippocampal cortex. Flechsig's (1898) "association" cortex included frontal, insular, parietal, temporal and preoccipital cortex. The inclusion of a part of the "limbic" cortex in the "associative" source, also made by Gimenez-Amaya (1991), is due to the fact that the striatal "associative" territory, which comprises the rest of the striatum, including not only most of the caudate nucleus but the sizeable medial and anterior part of the putamen, receives intricated cladon-groves from historical "associative" (frontal, insular, parietal, temporal, preoccipital cortex), "limbic" (cingular, parahippocampic) and, oddly, oculomotor and accessory oculomotor cortex. There are complex intrastriatal confluences from oculomotor areas (Parthasarathy et al., 1992). There is no separate territory made by the cladon-groves from the cingulate cortex (compare Figs. 11, 13 and 14 of Selemon and Goldman-Rakic, 1985) nor from the parahippocampal cortex (Fig. 7 of Van Hoesen et al., 1981), both "interdigitated" at least with frontal and temporal islands. The cladalses of "limbic" origin are observed even in the most dorsal part of the associative striatum. This is also the case for the territory from the amygdala. Alike cortical territories, it entirely fills the associative territory in a patchy manner (see Russchen et al., 1985). In terms of cortical and amygdalar territories there is thus no "limbic striatum" nor "ventral striatum" in monkeys. The distribution of the cladalses from these different cortical "associative" source is intricated. The axons which leave the corresponding syllogues are close one to another, creating a mixed output system. The first step of the system, the cortico-striatal connection, leads to a complex transformation of cortical information at the entrance to the core of the basal ganglia.

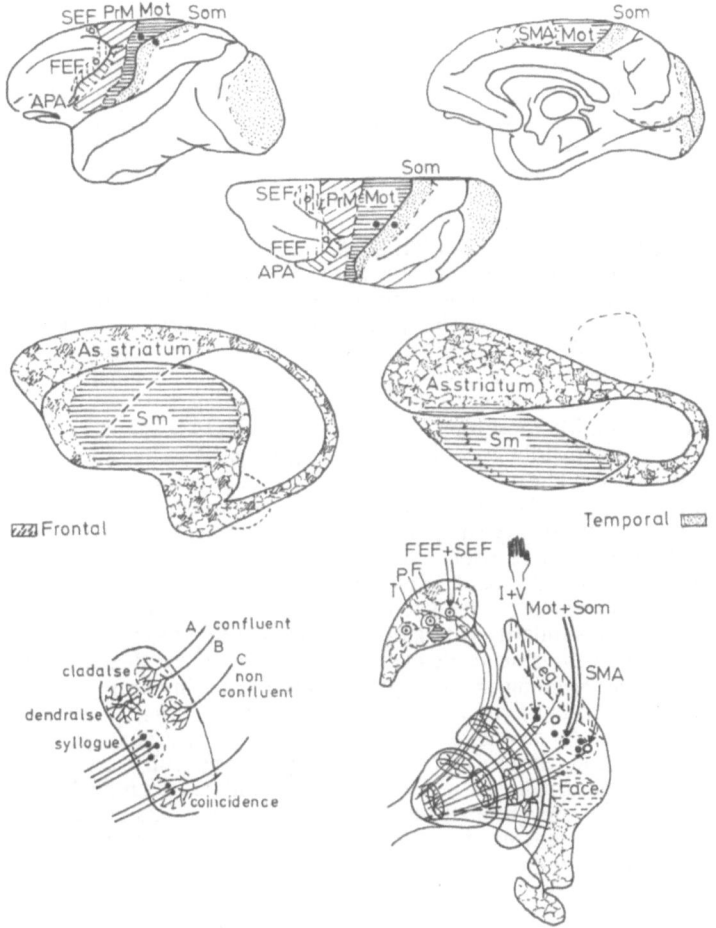

Figure 3. Corticostriatal connection. Upper row: Cortical subdivisions of the macaque brain. APA, postarcuate area. FEF, frontal eye field. F, frontal cortex. Mot, motor cortex. P, parietal cortex. PrM, premotor cortex. SEF, Supplementary eye field. SMA, supplementary motor area. Som, somatosensory cortex. T temporal cortex. Second row: the two subdivisions of the striatum. **Sm,** sensorimotor striatum..**As,** Associative striatum with cladalses from frontal and temporal origin. Lower row: Different types of clusters and of their combinatory relations. Organisation in a transverse section. I and V refer to first and last digit. See text for description and comments.

The core of the basal ganglia (the striato-pallido-nigral system)

The core of the basal ganglia (Percheron et al., 1984a and b) is made up of the striatum and its direct targets, the two pallidal nuclei and the substantia nigra.

In addition to the associative-sensorimotor subdivision and the further partition into cortical cladon-groves, analysis of the **striatum**, must consider several other levels: neuronal types, compartmentation and source-spaces. The neuronal species of the striatum have recently been reexamined on quantitative criteria by Yelnik et al. (1991). There are four sympatric species. Spiny neurons represent 96% of striatal neurons. They have rather small dendritic dwellings (430x330x230 μm), which gives a fine grain and the possibility of a large diversity of information. Spiny neurons have highly branched, very dense, axonal proximal collateral arborisations which must be taken into consideration in the analysis of the inner organisation of the striatum. Spiny neurons are the output neurons whose axons constitute the radial fibers and "pencils", the general radial organisation of which gives the striatum its name. The main mediators synthesized by spiny neurons are GABA and Substance P or enkephalin. These neurons are inhibitory. Leptodendritic neurons, with very large and thick dendritic arborisations (1200x830x530 μm) are very sparse (2%). They are also probably output neurons with unknown physiological properties, mediator and target. Spidery neurons (1%) are specific to primates. They are local circuit neurons with a very large cell body and very numerous, short, thin and beaded dendrites. The dimensions of their dendritic dwelling (610x540x400 μm), and of their concentric cladonal parcel, are also small. They are the primate cholinergic neurons. Their size, number and distribution are such that some groups of spiny neurons could be very densely innervated by cholinergic neurons while others could not. Microneurons (1%) are very small GABAergic local circuit neurons. Their size, number and distribution raise the same question as for the spidery neurons.

In spite of many studies, the compartmentation into matrix and striosomes still poses unsolved questions, particularly in primates. Both compartments are made up of the same neuronal species: spiny, spidery and microneurons (Yelnik et al., 1991). The leptodendritic neurons probably remain restricted to the matrix. The separation between the two compartments is shown in this book (Yelnik et al., 1993) to be made by a dendritic closure. The matrix is now recognized to be heterogeneous. The term "matrisome" (Graybiel et al., 1991) which evokes a similarity with striosomes is probably too strong to describe matrix islands which do not have a dendritic closure. Furthermore (see theoretical bases) a single term does not separate emitting and receiving islands. Each spiny neuron projects to only one of these targets (Féger and Crossman, 1984; Parent et al., 1984, 1989b). The number of spiny neurons devoted to each target is not known. The intrastriatal spatial distribution of clusters of spiny neurons projecting to specific targets is not fully established in primates. Neurons projecting to the pallidum and to the substantia nigra pars reticulata are restricted to the matrix where they are grouped into clusters (Gimenez-Amaya and Graybiel, 1990; Graybiel et al., 1991; François et al., 1992b). The localisation in striosomes of neurons projecting to the pars compacta is not clearly documented in primates. The patchiness of the striatal source-spaces makes possible different combinations (Graybiel et al., 1991) with different informational consequences.

The axons of the spiny neurons projecting to outside targets constitute a very dense bundle: the **striato-pallido-nigral bundle** (Percheron et al., 1984a). The axons successively cross the lateral and medial nucleus of the pallidum, the internal capsule (as the "comb system") and the substantia nigra.

The **pallidum** is made up of two neuronal species: microneurons and projection neurons. Projection neurons, when entirely reconstructed using computer assisted microscopy (Yelnik et al., 1984), have very large and flat (1500x1000x250 μm) dendritic dwellings directed parallel to one another and the lateral border of the pallidum i.e. perpendicular to the afferent axons. This arrangement was demonstrated confer a strong convergence (Percheron et al., 1984a and b, 1987). A recent debate on this point (see Percheron and Filion, 1991) calls for a detailed analysis of the striato-pallidal connection.

The electron microscopical description of this connection (Fox et al., 1966; DiFiglia et al., 1982; Cano et al., 1989; synthesized in Percheron et al., 1991b), revealed a unique feature. The very long pallidal dendrites are entirely covered by synaptic boutons. The continuous synaptic sheet, with no glial interposition, is surrounded by a cladonal sheet made up of parallel striatal cladons. In our reconstructed material, we measured the whole surface of an average pallidal dendritic arborisation. From published electron microscopic material, we measured the average surface of a striato-pallidal synapse. The two values gave an average of 40000 synapses of striatal origin per pallidal neuron.

In spite of intensive studies, the exact size and shape of the cladonal arborisations of spiny neuron axons in the pallidum are not known yet. Golgi studies (Fox and Rafols, 1975; DiFiglia et al., 1982; François et al., 1984a; Cano et al., 1989) showed few, rather short (about 100 μm) axonal branches. The intrapallidal distribution of striatal cladons is not homogeneous but laminated latero-medially. Dense, curved cladonal bands, parallel to the lateral pallidal borders are observed close to the external and internal borders of the two nuclei and in intermediate positions (Parent et al., 1984; Smith and Parent, 1986; Hedreen and DeLong, 1991; Percheron et al., 1991b; Hazrati and Parent, 1992b). Such a laminated pattern of parallel chip-like islands is a peculiar type of patchy map whose functioning is unknown. Parent's (1990) drawing suggests that these dense cladonal bands could be made up of large cladonal arborisations. They could, as well, be obtained by the juxtaposition of many short branches from many afferent axons emitted at the same level. This question is the subject of our current research. Despite Fox and Rafols' (1975) warning that a comparison with the climbing fibers of the cerebellum, making also "longitudinal axodendritic connections", shows strong difference between them, a model favouring selective pauci-innervation was recently presented (Hazrati and Parent, 1992 a and b). There are many arguments against it. The first comes from the observation that afferent axons that we were able to follow for some distance, were seen to be in "parallel contact" with a dendritic segment only for a short distance (about 55 μm, François et al., 1984a). Then they run, radially but apparently at random, in the direction of another dendritic segment from the same or another pallidal neuron, forming what we called "random cascades of striatal cladons" (Percheron et al., 1991b). The second argument is that "parallel contacts" intermittently emit only a few true contacts (DiFiglia et al., 1982; François et al., 1984a; Cano et al., 1989). They are thus unable to furnish alone the continuous sheet of synaptic boutons. The third argument is the massive cardinal convergence (Percheron and Filion, 1991) (the emitting striatal neurons being much more numerous by far than the receiving pallidal neurons). If, in the macaque, one divides up the estimated number of 31 million spiny neurons (Percheron et al., 1989) according to the number of neurons of the different targets (166000 lateral pallidal, 63000 medial pallidal and 54000 nigral neurons) (leading to 18 million spiny neurons to the lateral pallidal nucleus, 6.8 to the medial and 5.9 to the substantia nigra), there would be about 110 spiny neurons for every pallidonigral neuron, each emittting about 370 synapses to every target neuron. All these features are far from those of selective pauci-innervation.

In the two pallidal nuclei, the axons coming from the sensorimotor and the associative striatum do not mix. The sensorimotor pallidum is posterior and central. The associative pallidum is anterior, dorsal and ventral to it (Percheron et al., 1991b; François et al., 1992a), which fits with physiological data. In the sensorimotor pallidum, the somatotopic representation is patchy with "clusters" of about 1000 μm (Hamada et al., 1990). This patchiness is also observable in figures from DeLong (1971) and DeLong et al. (1985). The relation to movement is not simple and may be bidirectional (Mink and Thach, 1991). A somatotopic convergence between distal and proximal parts of the same limb or different parts of the body (arm and orofacial, superior and inferior limb, even leg, arm and face) is observable in figures from DeLong et al. (1985), Filion et al. (1988), Hamada et al. (1990), Yoshida et al. (1993). This is greatly increased after MPTP administration (Filion et al., 1988). In the sensorimotor pallidum, there is a confluence of the motor cortex and the postarcuate area (APA) after cortical stimulation (Nambu et al., 1990; Yoshida et al. 1993). There is a confluence of the sensorimotor and associative components after striatal stimulation (Tremblay and Filion, 1989). In the associative pallidum, there is a confluence of prefrontal and cingulate cortex after cortical stimulation (Yoshida et al., 1993).

Data about the sharpness and separation of representations in the pallidum are conflicting. The opposition between parallelism and convergence misses what could be the role of the core of the basal ganglia: the appropriate selection of convergent combinations of pieces of informations from different striatal subsources completing an upstream "sculpting of convergence patterns" in the corticostriate communication (Parthasarathy et al., 1992). The histologically observed pattern probably corresponds to the maximal aperture of the system, capable of a wide convergence. In given conditions, the combined actions of some clusters of spiny striatal neurons could lead to a more focused response. This ability of the system to adapt the degree of convergence to the needed task, named "dynamically focussed convergence" (Percheron and Filion, 1991) is diminished or lost in the case of dopaminergic denervation.

The neurons of the substantia nigra pars reticulata and lateralis belong to the same neuronal species as the pallidal neurons (Yelnik et al., 1987) with the same peculiar synaptology. The geometry of their dendritic arborisations is different even though nigral dendrites are also mainly perpendicular to striatal cladons (François et al., 1987). Due to the size and the flat shape of the substantia nigra, the convergence is still higher than in the pallidum. Nigral neurons essentially receive associative information (François et al., 1992a). The small sensorimotor territory overlaps with the associative territory in the intermediate part of the pars reticulata (François et al., 1992a). There is no clear somatotopic representation (DeLong et al., 1983, Schultz, 1986). The activity of nigral neurons is linked no so much to limb than to orofacial movements, licking and chewing. In the most lateral part (the pars lateralis), the activity of the neurons, which project to the superior colliculus (François et al.1984b), is linked to ocular saccades (Hikosaka and Wurtz, 1981).

The regulators of the core

Four elements of the basal ganglia system influence the core, three of which receive direct cortical information. The regulating "loops" are often complex with patchy maps. Regulation may be exerted primarily on the striatum or on the pallidonigrum (Fig. 1)

The nigro-striatal dopaminergic connection stemming from the pars compacta of the substantia nigra has been the subject of so many studies that it might hardly be seen as a subsidiary system. Its major action on the striatum, i.e. at the entrance of the core, restores its status. The pars compacta neurons belong to a single neuronal species with very large dendritic arborisations, and a few interneurons (Yelnik et al., 1987). The main input to the pars compacta neurons comes from the axons of striatal spiny neurons which contact their ventral dendrites in the pars reticulata (François et al., 1987). The trajectory of the nigro-striatal axons electively follows the pallidal laminae. The striato-nigro-striatal circuit is not a mere loop (Percheron et al., 1989) as there are many more striato-nigral than nigro-striatal neurons (about 100/1 in macaques). One connection is strongly convergent and the other divergent. The striatal innervation is not homogeneous but patchy. Dopaminergic neurons, distinct from nigro-striatal neurons (Smith et al., 1989), contribute to the innervation of the two pallidal nuclei (Lavoie et al., 1989) and mainly to the medial nucleus.

The insertion of the central complex (or centre médian-parafascicular complex) within the system of the basal ganglia is not a classical concept. In the preceding IBAGS book we demonstrated (Percheron et al., 1991a) that this should now be the case. The main input to the complex (to its pars media, François et al., 1988) comes from the sensorimotor part of the medial nucleus of the pallidum (Fénelon et al., 1990). There is only a restricted projection from the substantia nigra (personal observation). Another major input comes from the motor and premotor cortex (Akert and Hartmann-von Monakow, 1980; see Fénelon et al., 1990). The pars parafascicularis and media (and not the paralateralis) receive cholinergic input from the pedunculopontine complex (Olivier et al., 1970; Hirai and Jones, 1989). The neurons of the complex project essentially to the striatum (Fénelon et al., 1991). The projection of the pars media is restricted to the sensorimotor striatum (closing "Nauta and Mehler's loop", 1966; François et al., 1991; Percheron et al., 1991a). The centro-striatal connection is strongly patchy with oblique bands of superposed islands (François et al., 1991). The parafascicular part of the complex send its axons to the associative striatum (Fénelon et al., 1991) also in a patchy manner. The complex receives direct connections from the cortex (see Fénelon et al., 1991 for references). The paralateral part sends axons to the motor cortex (Leichnetz, 1986) and thus makes a cortico-centro-cortical circuit (Percheron et al., 1991a).

The subthalamic nucleus, or corpus Luysii, is essentially a regulator of the pallidonigral set. This closed patria is made up of a single neuronal species with rather large dendritic arborisations (Yelnik and Percheron, 1979). Its main input comes from the lateral nucleus of the pallidum through the ansa lenticularis. The pallido-subthalamic connection is GABAergic hence inhibitory. Other major inputs are from the motor, premotor, oculomotor and prefrontal cortex (Von Monakow et al., 1978). The main targets of the nucleus are the substantia nigra pars reticulata and the pallidum (Nauta and Cole, 1978; Smith et al., 1990). The lateral pallidal nucleus is involved in a pallido-subthalamo-pallidal circuit. The subthalamic territory in both pallidal nuclei is made up of several bands parallel to the lateral border of the pallidum (Nauta and Cole, 1978; Parent et al., 1989a; Carpenter and

Jayaraman, 1990; Smith et al., 1990), in register with those made by striatal afferents (Hazrati and Parent, 1992a). The functional meaning of this pattern is not known. Other connections towards the pedunculopontine complex (Nauta and Cole, 1978) and the striatum have been described. The subthalamo-pallidonigral connection is excitatory and glutamatergic (Kitai and Kita, 1987; Féger et al., 1989; Parent et al., 1989a; Robledo and Féger, 1990). The subthalamic nucleus has only a rough somatotopic organisation. The regulation made by the subthalamic nucleus is today seen as of major importance in the pathophysiology of basal ganglia disturbances.

The pedunculopontine complex, a differentiated part of the reticular formation, reciprocally linked to elements of the basal ganglia related system, should be included in this system. It receives a fair input from the medial nucleus of the pallidum (Nauta and Mehler, 1966 and personal observation), (which is apparently not the case in lower orders), from the substantia nigra (Becksteadt and Frankfurter, 1982) and from the subthalamic nucleus (Nauta and Cole, 1978; Smith et al., 1990). The phylogenetic increase of the complex and interspecific differences in connectivity was stressed by Moon Edley and Graybiel (1983). The pallidonigral-pedunculopontine connection is GABAergic hence inhibitory. The complex also receives cortical input from the motor (but not the premotor) cortex (Hartmann-von Monakow et al., 1979). The efferent projections of the complex are not well documented in primates. In rats and cats (see Moon Edley and Graybiel, 1983) it sends axons to the substantia nigra, entopeduncular nucleus and subthalamic nucleus. In addition to outputs to the core, the complex sends axons to the central complex and to the pallidal thalamic territory (Hirai and Jones, 1989). Pedunculopontine neurons are, at least in part, cholinergic (Mesulam et al., 1984) and the ascending connection could be another excitatory regulator of the core.

Internal regulators within the core. Some axons from the lateral nucleus of the pallidum appear to project to the somata essentially of the medial nucleus neurons (Hazrati et al., 1990) and to the substantia nigra (Parent and de Bellefeuille, 1983).

Another input participating in the regulation of the core does not stem from any element of the basal ganglia: the serotoninergic innervation of the striatum, pallidum (with parallel bands), substantia nigra and subthalamic nucleus (Lavoie and Parent, 1990).

The output from the core

The major output from the core of the basal ganglia projects to the motor thalamus. It is dual. It starts from two sources, the medial nucleus of the pallidum and the pars reticulata of the substantia nigra, and follows two distinct routes, the pallido-thalamic and the nigro-thalamic connections. These have been the subject of recent studies from our laboratory (Percheron et al., 1993 a and b, where references may be found). The two thalamic territories occupied by the cladons from the core of the basal ganglia are located in front of the cerebellar territory with absolutely no overlap. The thalamic neurons are of two types: "bushy" projection neurons (about 75%) and local circuit microneurons. Projection neurons have spheric dwelings of relatively small diameter which offers the possibility of a fair diversity of information. The shape and dimensions of the cladons from the pallidum and substantia nigra are still unknown. No spatial analysis of information processing can thus be made for this connection. The VA-VL system of nomenclature has been found inadequate for the description of the territories from the basal ganglia. Their borders do not correspond to those of the cytoarchitectonic nuclei traced by Olszewski (1952) for instance. A simpler and more appropriate subdivision of the lateral region has been proposed (Percheron et al., 1993b), the portion receiving pallidal input being called subregio lateralis oralis and that receiving nigral afferent subregio lateralis rostralis.

The pallido-thalamic connection is massive. From the medial nucleus of the pallidum, the bundle crosses the internal capsule, follows the V-shaped Forel' fields, and enters the thalamus ventrally in front of cerebellar bundle. The pallidal territory is extended along the whole ventrodorsal extent of the lateral region of which it occupies the lateral half. Its three-dimensional geometry is complex, curved along the lateral border in the transverse dimension and banana-shaped in the sagittal dimension. The pallidum-related subregio lateralis oralis (LO) corresponds grossly to VLo plus parts of VA, VLm and VLc of Olszewski (1952). The pallidal territory in the thalamus is very dense but inhomogeneous with denser oblique islands. There is only a rough somatotopic organisation with important overlaps, e.g. between face and upper limb, in man (Schaltenbrand et al., 1971) and in the monkey

(Anderson and Turner, 1991). The pallido-thalamic connection is GABAergic (Penney and Young, 1981; Aizawa et al. 1991) and inhibitory (Jinnai et al., 1989). The pallidal territory receives an additional cholinergic innervation from the pedunculopontine nucleus (Olivier et al., 1970; Hirai and Jones, 1989).

The nigro-thalamic connection is also massive. The thalamic nigral territory is much more extended than usually admitted, over the whole ventrodorsal extent of the anterior part of the lateral region of which it occupies the medial half. Its lateral border is irregular but there is no overlap with the pallidal territory. Its anterior border constitutes the polar border of the lateral region (Percheron et al., 1993a and b). There is thus no thalamic subdivision located in front of the pallidal and nigral territories. The nigral related subregio lateralis rostralis (LR) grossly corresponds to VAmc and parts of VA and VLm. The territory is dense with a decreasing gradient lateralwards. The inner organisation and physiology of this territory are almost unknown in primates. The nigral territory also includes the paralaminar region, which receives additional collicular input and small scattered islands in the nucleus medialis, already observed by Ilinsky et al. (1985). The nigro-thalamic connection is also GABAergic (in rats, Di Chiara et al., 1979), hence inhibitory.

The thalamo-cortical connection

The thalamo cortical connection will not be examined in detail.

The motor cortex is primarily the target of the cerebellar receiving part of the thalamus (subregio lateralis intermedius LI).

The pallidal thalamo-cortical connection (from the lateral oral subregion) is essentially directed towards the supplementary motor area (Schell and Strick, 1984), and more sparsely to the motor cortex and to the superior and inferior premotor cortex (Nakano et al., 1990).

The nigral thalamo-cortical connection (from the lateral rostral subregion) is directed towards the frontal eye field, the supplementary motor cortex, the prefrontal cortex (Ilinsky et al., 1985), the supplementary eye field (Shook et al., 1991) and the post-arcuate area (Nakano et al., 1990) but not the motor cortex. The paralaminar part of the nigral thalamus also projects to the frontal eye field. The islands included in the medial nucleus project to the frontal cortex. The nigral thalamus thus projects through different channels to the same cortical targets. Except for the supplementary motor area, these are all located in front of the cortex receiving pallidal information.

The shape and dimensions of the cladonal arborisations from the two portions of the thalamus processing information from the basal ganglia are not known in primates but could be patchy and explain the iterative pattern of cortical maps (see 1 above).

The cortico-cortical connections

The cortical targets of the motor thalamus receiving basal ganglia information are not isolated stations but have many interrelations. There are reciprocal ipsi and controlateral connections beween the motor and somatosensory, the motor (cerebellar) and the supplementary motor (pallidal), the post-arcuate (cerebellar) and the premotor (cerebellar and pallidal) cortex (Leichnetz, 1986; Gosh et al., 1987). In these areas there is thus a direct or indirect combination, with unequal proportions, of cerebellar and pallidal information. It is worthy of note that, except for the supplementary motor area, the contribution of nigral information to the cortex is separate. There are for instance no connection between the prefrontal cortex, the cortical eye field (nigral) and the motor cortex (Leichnetz, 1986). The same two cortical regions however send axons to the post-arcuate area (Arikuni et al., 1987), to the posterior part of the superior precentral cortex and to the inferior precentral cortex.

The descending output of the system

There are two different output systems of the basal ganglia: one passing through the tectum and the other through the thalamus to the cortex. The projection from the pars lateralis of the substantia nigra to the tectum has been deliberately neglected in the present account.

The main part of the pallidal related cortex is included in the cortical source of the striatal sensorimotor territory, which validates the existence of the cortico-baso-thalamo-cortical circuit. This is the case for only the frontal portion of the nigral related cortex, a part of the source of the associative territory, which also makes a circuit (Schultz, 1986). The widest remaining part of the associative cortex contributing to the striatal input is not included in corticocortical circuits passing through the basal ganglia.

The topography of the basal ganglia related cortex is close to that of the source of the corticospinal, pyramidal tract as delineated by Toyoshima and Sakai (1982) and Martino and Strick (1987). The main differences are a greater contribution of the frontal cortex and an absence of parietal contribution in the basal ganglia related cortex. This can directly influence motoneurons through the pyramidal tract, from the supplementary motor area, the premotor cortex and the motor cortex, and indirectly through corticocortical connections.

SYNTHESIS

The basal ganglia related system may be viewed as a system collecting and processing information from almost the whole extent of the cortex in order to finally furnish adequate information to the only cortex involved in movement preparation and execution. The different stages of the system correspond to different ways of processing information with successive transformations of "maps". Future discussions should take into account the fact that communication abundantly uses complex, predominanly patchy, spatial patterns.

REFERENCES

Aizawa, H, Kwak, S., Shimizu, T., and Mannen, T., 1991, Determination of GABAergic pallidothalamic termination in human brain, *J. Neurol. Sci.* 105:124-125.

Akert, K., and Hartmann-von Monakow, K., 1980, Relationships of precentral, premotor and prefrontal cortex to the mediodorsal and intralaminar nuclei of the monkey thalamus, *Acta Neurobiol. Exp.* 40:7-25.

Alexander, G.E., Koliatsos, V.E., Martin, L.J., Hedreen, J., Hamada, I., and DeLong, M.R., 1988, Organisation of primate basal ganglia "motor circuit".1. Motor cortex (MC) and supplementary motor area (SMA) project to complementary regions within matrix compartment of putamen, *Soc. Neurosci. Abstr.* 14:720.

Anderson, M.E., and Turner, R.S., 1991, Activity of neurons in cerebellar-receiving and pallidal-receiving areas of the thalamus of the behaving monkey, *J. Neurophysiol.* 66:879-893.

Arikuni, T., Sakai, M., Hamda, I., and Kubota, I.,1987, Topographical projections from the prefrontal cortex to the post-arcuate area in the rhesus monkey, studied by retrograde axonal transport of horseradish peroxidase, *Neurosci. Let,* 19:155-160.

Bauswein, E., Fromm, C., and Preuss, A., 1989, Corticostriatal cells in comparison with pyramidal tract neurons: contrasting properties in the behaving monkey, *Brain Res.* 493:198-203.

Becksteadt, R.M., and Frankfurter, A., 1982, The distribution and some morphological features of substantia nigra neurons that project to the thalamus, superior colliculus and pedunculopontine nucleus in the monkey, *Neuroscience* 7:2377-2388.

Cano, J., Pasik, P., and Pasik, T., 1989, Early postnatal development of the monkey globus pallidus: A Golgi and electron microscopic study, *J. Comp. Neurol.* 279:353-367.

Carpenter, M.B., and Jayaraman, J., 1990, Subthalamic nucleus of the monkey: connections and immunocytochemical features of afferents, *J. Hirnforsch.* 31:653-668.

Crutcher, M.D., and DeLong, M.R.,1984, Single cell studies of the primate putamen. I; Functional organisation, *Exp. Brain Res.* 53:233-244

DeLong, M.R., 1971, Activity of pallidal neurons during movement, *J. Neurophysiol.* 34:414-427.

DeLong, M.R., Crutcher, M.D., and Georgopoulos, A.P., 1983, Relation between movement and single cell discharge in the substantia nigra of the behaving monkey, *J. Neurosci.* 3:1599-1606.

DeLong, M.R., Crutcher, M.D., and Georgopoulos, A.P., 1985, Primate globus pallidus and subthalamic nucleus: functional organization, *J. Neurophysiol.* 53:530-543.

Di Chiara, G., Porceddu, M.L., Morelli, M., Mulas, M.L., and Gessa, G.L., 1979, Evidence for a GABAergic projection from the substantia nigra to the ventromedial thalamus and to the superior colliculus in the rat, *Brain Res.* 176:273-284.

DiFiglia, M., Pasik, P., and Pasik, T., 1982, A Golgi and ultrastructural study of the monkey globus pallidus, *J. Comp. Neurol.* 212:53-75.

Féger, J., and Crossman, A.R., 1984, Identification of different subpopulations of neostriatal neurons projecting to globus pallidus and substantia nigra in the monkey: a retrograde fluorescence double-labelling study, *Neurosci. Lett.* 49:7-12.

Féger, J., Vezole, I., Rentwart, N., and Robledo, P., 1989, The rat subthalamic nucleus: electrophysiological and behavioral data, *in*: "Neural Mechanisms in Disorders of movements," A.R. Crossman and M.A. Sambrook, eds., J. Libbey, London, pp. 37-43.

15

Fénelon, G., François, C., Percheron, G., and Yelnik, J., 1990, Topographic distribution of pallidal neurons projecting to the thalamus in macaques, *Brain Res.* 520:27-35.

Fénelon, G., François, C., Percheron, G., and Yelnik, J., 1991, Topographic distribution of the neurons of the central complex (centre médian-parafascicular complex) and of other thalamic neurons projecting to the striatum in macaque, *Neuroscience* 45:495-510.

Filion, M., Tremblay, L., and Bédard, P.J., 1988, Abnormal influences of passive limb movement on the activity of globus pallidus: neurons in parkinsonian monkeys, *Brain Res.* 444:165-176.

Flaherty, A.W., and Graybiel, A.M., 1991, Corticostriatal transformations in the primate somatosensory system. Projections from physiologically mapped body-part representations, *J. Neurosci.* 66:1249-1263.

Fotuhi, M., Koliatsos, V.E., Alexander, G.E., and DeLong, M.R., 1989, Patterns of sensorimotor integration in the primate neostriatum: primary somatosensory cortex (SC) and motor cortex (MC) project to coextensive territories in the putamen, *Soc. Neurosci. Abstr.* 15:285.

Fox, C.A., Hillman, D.E., Siegesmund, K.A., and Sether, L.A., 1966, The primate globus pallidus and its feline and avian homologues: a Golgi and electronmicroscopic study, *in*: "Evolution of the forebrain," R. Hassler and H. Stephan, eds., G. Thieme, Stuttgart, pp. 237-248.

Fox, C.A., and Rafols, J.A., 1975, The radial fibers in the globus pallidus, *J. Comp. Neurol.* 159:177-200.

François, C., Percheron, G., Parent, A., Sadikot, A.F., Fénelon, G., and Yelnik, J., 1991, Topography of the projection from the central complex of the thalamus to the sensorimotor striatal territory in monkeys, *J. Comp. Neurol.* 305:17-34.

François, C., Percheron, G., and Yelnik, J., 1984a, A Golgi analysis of the primate globus pallidus. I. Inconstant processes of large neurons. Other neuronal types. Afferent axons., *J. Comp. Neurol.* 227:182-199.

François, C., Percheron, G., and Yelnik, J., 1984b, Localization of nigrostriatal, nigrothalamic and nigrotectal neurons in ventricular coordinates in macaques, *Neurosci.* 13:61-76.

François, C., Percheron, G., and Yelnik, J., 1992a, Associative and sensorimotor parts of the pallidum and substantia nigra in macaques, IBAGS IV, Giens, abstr., p. 30.

François, C., Percheron, G., and Yelnik, J., 1992b, Intrastriatal localization of neurons which project to the two pallidal nuclei and to the substantia nigra in macaques, IBAGS IV, Giens, abstr., p. 30.

François, C., Percheron, G., Yelnik, J., and Tandé, D., 1988, A topographic study of the course of nigral axons and the distribution of pallidal axonal endings in the centre médian-parafascicular complex of macaques, *Brain Res.* 473:181-186.

François, C., Yelnik, J., and Percheron, G., 1987, Golgi study of the primate substantia nigra. II. Spatial organization of dendritic arborizations in relation to the cyto-architectonic boundaries and to the striatonigral bundle, *J. Comp. Neurol.* 265:473-493.

Gimenez-Amaya, J.M., 1991, The association cortex and the basal ganglia: a neuroanatomical view upon their relationship based on hodological studies, *J. Hirnforsch.* 4:501-510.

Gimenez-Amaya, J.M., and A.M. Graybiel, 1990, Compartmental origins of the striatopallidal projection in the primate, *Neuroscience*, 34:11-126.

Gosh, S., Brinkman, C., and Porter, R., 1987, A quantitative study of the distribution of neurons projecting to the precentral motor cortex in the monkey (*M. fascicularis*), *J. Comp. Neurol.* 259:424-444.

Graybiel, A.M., Flaherty, A.W., and Gimenez-Amaya, J.M., 1991, Striosomes and matrisomes, *in*: "The Basal Ganglia III," G. Bernardi, M.B. Carpenter, G. Di Chiara, M. Morelli and P. Stanzione, eds., Plenum Press, New York, pp. 3-12.

Hamada, I., DeLong, M.R., and Mano, N-I., 1990, Activity of identified wrist-related pallidal neurons during step and ramp movements in the monkey, *J. Neurophysiol.* 64:1892-1906.

Hartmann-von Monakow, K., Akert, K., and Künzle, H., 1979, Projections of precentral and premotor cortex to the red nucleus and other midbrain areas in *Macaca fascicularis*, *Exp. Brain Res.* 34:91-105.

Hazrati, L.-N., Parent, A., Mitchell, S., and Faber, S.N., 1990, Evidence for interconnections between the two segments of the globus pallidus in primates: a PHA-L anterograde tracing study, *Brain Res.* 533:171-175.

Hazrati, L.-N., and Parent, A., 1992a, Convergence of subthalamic and striatal efferents at pallidal level in primates: an anterograde double-labeling study with biocytin and PHA-L, *Brain Res.* 569:336-340.

Hazrati, L.-N., and Parent, A., 1992b, The striatopallidal projection displays a high degree of anatomical specificity in the primate, *Brain Res.* 592:213-227.

Hedreen, J.C., and DeLong, M.R., 1991, Organization of striatopallidal, striatonigral, and nigrostriatal projections in the macaque, *J. Comp. Neurol.* 304:569-595.

Hirai, T., and Jones, E.G., 1989, A new parcellation of the human thalamus on the basis of histochemical staining, *Brain Res. Rev.* 14:1-34.

Hikosaka, O., and Wurtz, R.H., 1981, The role of the substantia nigra in the initiation of saccadic eye movements, *in*: "Progress in oculomotor research," A. Fuchs and W. Becker, eds., Elsevier, Amsterdam, pp. 145-152.

Ilinsky, I.A., Jouandet, M.L., Goldman-Rakic, P.S., 1985, Organization of the nigrothalamocortical system in the Rhesus monkey, *J. Comp. Neurol.* 263:315-330.

Jinnai, K., Nambu, A., and Yoshida, S.-I., 1989, Activity of thalamic neurons conveying the basal ganglia output to the motor cortex, *in*: "Neural programming," M. Ito, ed., Karger, Basel, pp. 111-121.

Kitai, S., and Kita, H., 1987, Anatomy and physiology of the subthalamic nucleus: a driving force of the basal ganglia, in : "The Basal Ganglia II," M.B. Carpenter and A. Jayaraman, eds., Plenum Press, New York, pp. 357-373.

Lavoie, B., and Parent, A., 1990, Immunochemical study of the serotoninergic innervation of the basal ganglia in the squirrel monkey, J. Comp. Neurol. 299:1-16.

Lavoie, B., Smith, Y., and Parent, A., 1989, Dopaminergic innervation of the basal ganglia in the squirrel monkey as revealed by tyrosine hydroxylase immunochemistry, J. Comp. Neurol. 289:36-52.

Leichnetz, G.R., 1986, Afferent and efferent connections of the dorsolateral precentral gyrus (area 4, hand/arm region) in the macaque monkey, with comparisons to area 8, J. Comp. Neurol. 254:460-492.

Lemon, R, 1988, The output map of the primate motor cortex, Trends Neurosci., 11:501-506.

Liles, S.L., and Updyke, B.V., 1985, Projection from the digit and wrist area of precentral gyrus to the putamen: relation between topography and physiological properties of neurons in the putamen, Brain Res. 339:245-255

Martino, A.M., and Strick, P.L., 1987, Corticospinal projections originate from the arcuate premotor area, Brain Res. 404:307-312.

Mesulam, M.M., Mufson, E.J., Levey, A.I., and Wainer, B.H., 1984, Atlas of cholinergic neurons in the forebrain and upper brain stem of the macaque based on monoclonal choline acetyl transferase immunochemistry, Neuroscience 12:669-686.

Mink, J.W., and Thach, W.T., 1991, Basal ganglia motor control. II. Late timing relative to movement onset and inconsistent pallidal coding of movement parameters, J. Neurophysiol. 65:301-328.

Moon Edley, S., and Graybiel, A.M., 1983, The afferent and efferent connections of the nucleus tegmenti pedunculopontinus, pars compacta, J. Comp. Neurol. 217:187-215.

Nakano, K., Tokushige, A., Khono, M., Hasegawa, Y., Kayahara, T., and Sasaki, K., 1990, An autoradiographic study of cortical projections from motor thalamic nuclei in the macaque monkey, Neurosci. Res. 13:119-137.

Nambu, A., Yoshida, S.-I., and Jinnai, K., 1990, Discharge patterns of pallidal neurons with input from various cortical areas during movement in the monkey, Brain Res. 519:183-191.

Nauta, H.J.W., and Cole, C., 1978, Efferent projections of the subthalamic nucleus: an autoradiographic study in monkey and cat, J. Comp. Neurol. 188:1-16.

Nauta, W.J.H., and Mehler, W.R., 1966, Projections from the lentiform nucleus in the monkey, Brain Res. 1:3-42.

Nelson, M.E, and Bower, J.M., 1990, Brain maps and parallel computers, Trends Neurosci. 13:403-408.

Olivier, A., Parent, A., and Poirier, L.J., 1970, Identification of the thalamic nuclei on the basis of their cholinesterase content in monkey, J. Anat. 106:37-50.

Olszewski, J., 1952, "The Thalamus of the Macaca mulatta. An Atlas for Use with the stereotaxic Instrument," Basel, Karger.

Parent, A., Bouchard, C., and Smith, Y., 1984, The striatopallidal and striatonigral projections: two distinct fiber systems in primate, Brain Res. 303:385-390.

Parent, A., and de Bellefeuille, L., 1983, The pallidointralaminar and pallidonigral projections in primate as studied by retrograde double-labeling method, Brain Res. 278:11-27.

Parent, A., Hazrati, L.-N., and Smith, Y., 1989a, The subthalamic nucleus in primates. A neuroanatomical and immunohistochemical study, in: "Neural Mechanisms in Disorders of movements," A.R. Crossman and M.A. Sambrook, eds., J. Libbey, London, pp. 29-35.

Parent, A., Smith, Y., Filion, M., and Dumas, J., 1989b, Distinct afferents to internal and external pallidal segments in the squirrel monkey, Neurosci. Lett. 96:140-144.

Parthasarathy, H.B., Schall, J.D., and Graybiel, A.M., 1992, Distributed but convergent ordering of corticostriatal projections: Analysis of the frontal eye field and the supplementary eye field in the macaque monkey, J. Neurosci. 12: 4468-4488.

Penney, J.B., and Young, A.B., 1981, GABA as the pallidothalamic neurotransmitter: implications for basal ganglia function, Brain Res. 207:195-199.

Percheron, G., 1982, Principles and methods of the graph-theoretical analysis of natural binary arborescences, J. Theor. Biol. 99:509-552

Percheron, G., and Filion, M., 1991, Parallel processing in the basal ganglia: up to a point, Trends Neurosci. 14: 55-56.

Percheron, G., François, C., Parent, A., Sadikot, A.F., Fénelon, G., and Yelnik, J., 1991a, The primate central complex as one of the basal ganglia, in: "The Basal Ganglia III," G. Bernardi, M.B. Carpenter, G. Di Chiara, M. Morelli and P. Stanzione, eds., Plenum Press, New York, pp. 177-186.

Percheron, G., François, C., Talbi, B., Meder, J.-F., Fénelon, G., and Yelnik, J., 1993a, The primate motor thalamus analyzed with reference to subcortical afferent territories, Stereotact. Funct. Neurosurg. 93:32-41.

Percheron, G., François, C., and Yelnik, J., 1987, Spatial organization and information processing in the core of the basal ganglia, in: "The Basal Ganglia II," M.B. Carpenter and A. Jayaraman, eds., Plenum Press, New York, pp. 205-226.

Percheron, G., François, C., Yelnik, J., and Fénelon, G., 1989, The primate nigro-striato-pallido-nigral system. Not a mere loop, in: "Neural Mechanisms in Disorders of movements," A.R. Crossman and M.A. Sambrook, eds., J. Libbey, London, pp. 103-109.

Percheron, G., François, C., Yelnik, J., Talbi, B., Meder, J.-F., and Fénelon, G., 1993b, The pallidal and nigral thalamic territories and the problem of the anterior part of the lateral region in primates, *in*: "Thalamic networks for relay and modulation," D. Minciacchi, M. Molinari, G. Macchi and E.G. Jones, eds., Pergamon Press, New York, in press.

Percheron, G., Yelnik, J., and François, C., 1984a, A Golgi analysis of the primate globus pallidus. III. Spatial organization of the striato-pallidal complex, *J. Comp. Neurol.* 227:214-227.

Percheron, G., Yelnik, J., and François, C., 1984b, The primate striato-pallido-nigral system. An integrative system for cortical information, *in*: "Basal ganglia: structure and function," J.S. McKenzie, R.E. Kemm and L.N. Wilcock, eds., Plenum Press, New York, pp. 87-105.

Percheron, G., Yelnik, J., François, C., Fénelon, G., and Talbi, B., 1991b, The spatial organisation of information processing in the striato-pallido-nigral system, *in*: "Basal ganglia and movement disorders", A. Bigami, ed., *New Issues in Neurosci.* Vol III, N°2:211-234.

Robledo, P., and Féger, J., 1990, Excitatory influence of rat subthalamic nucleus to substantia nigra pars reticulata and the pallidal complex: electrophysiological data, *Brain Res.* 518:47-54.

Russchen, F.T., Bakst, I., Amaral, D.G., and Price, J.L., 1985, The amygdalostriatal projectios in the monkey. An anterograde tracing study, *Brain Res.* 329:241-257.

Schaltenbrand, G., Spuler, H., Wahren, W., and Rümler, B., 1971, Electroanatomy of the thalamic ventro-oral nucleus based on stereotaxic stimulation in man, *Z. Neurol.* 199:259-276.

Schell, E.R., and Strick, P.L., 1984, The origin of thalamic inputs to the arcuate premotor and supplementary motor areas, *J. Neurosci.* 4:539-560.

Schultz, W., 1986, Activity of pars reticulata neurons of monkey substantia nigra in relation to motor, sensory, and complex events, *J. Neurophysiol.* 55:660-677.

Selemon, L.D., and Goldman-Rakic, P.S., 1985, Longitudinal topography and interdigitation of corticostriatal projections in the rhesus monkey, *J. Neurosci.* 5:776-794.

Shook, B.L., Schlag-Rey, M., and Schlag, J., 1991, Primate supplementary eye field. II. Comparative aspects of connections with the thalamus, corpus striatum, and related nuclei, *J. Comp. Neurol.* 307:562-583.

Smith, Y., Hazrati, L.-N., and Parent, A., 1990, Efferent projections of the subthalamic nucleus in the squirrel monkey as studied by the PHA-L anterograde tracing method, *J. Comp. Neurol.* 294:303-323.

Smith, Y., Lavoie, B., Dumas, J., and Parent, A., 1989, Evidence for a distinct nigropallidal dopaminergic projection in the squirrel monkey, *Brain Res.* 482:381-386.

Smith, Y., and Parent, A., 1986, Differential connections of caudate nucleus and putamen in the squirrel monkey (*Saimiri sciureus*), *Neuroscience* 18:347-371.

Smolensky, P., 1987, Information processing in dynamical systems: Foundation of Harmony theory, *in*: "Parallel distributed processing," D.E. Rumelhart, J.L. McClelland and the PDP Research group, eds., MIT Press, Cambridge, Vol 1, Chap 6:194- 281.

Toyoshima, K., and Sakai, H., 1982, Exact origin of the cortico-spinal tract and the quantitative contributions to the CST in different cytoarchitectonic areas. A study with horseradish peroxidase in the monkey, *J. Hirnforsch.* 23:257-261.

Tremblay, L., and Filion, M., 1989, Responses of pallidal neurons to striatal stimulation in intact waking monkeys, *Brain Res.* 498:1-16.

Van Hoesen, G.W., Yeterian, E.H., and Lavizzo-Mourey, R., 1981, Widespread corticostriate projections from temporal cortex of the Rhesus monkey, *J. Comp. Neurol.* 199:205-219.

Von Monakow, K.H., Akert, K., and Künzle, H., 1978, Projections from the precentral motor cortex and other areas of the frontal lobe to the subthalamic nucleus in the monkey, *Exp. Brain Res.* 33:395-403.

Yelnik, J., François, C., Percheron, G., and Heyner, S., 1987, Golgi study of the primate substantia nigra. I. Quantitative morphology and typology of nigral neurons, *J. Comp. Neurol.* 265:455-472.

Yelnik, J., François, C., Percheron, G., and Tandé, D., 1991, Morphological taxonomy of the neurons of the primate striatum, *J. Comp. Neurol.* 313:273-294.

Yelnik, J., and Percheron, G., 1979, Subthalamic neurons in primates: a quantitative and comparative analysis, *Neuroscience* 4:1717-1743.

Yelnik, J., Percheron, G., and François, C., 1984, A Golgi analysis of the primate globus pallidus. II. Quantitative morphology and spatial orientation of dendritic arborizations, *J. Comp. Neurol.* 227:200-213.

Yelnik, J., François, C., and Percheron, G., 1993, Three-dimensional morphology of striatal neurons in relation to compartmental organization of the striatum, *in*: "The Basal Ganglia IV. New ideas and data on structure and function," G. Percheron, J.S. McKenzie and J. Féger, eds., Plenum Press, New York, This volume.

Yoshida, S.-I., Nambu, A., and Jinnai, K., 1993, The distribution of the globus pallidus neurons with inputs from various cortical areas in the monkeys, submitted, personal communication.

Terms and definitions of informational neuronal set theory
(G.Percheron)

I - Neuronal level

IA - Parts of neurons.

Soma   trigger zone

dendrites   hodon : part of axon conveying information between a source and a target

                    nemon : part of axon conveying information within a target

                    cladon : part of axon transmitting information within a target

IB - Neuronal arborisations. Topological metered graphs (binary arborisations with simple mathematical properties made up of 3 kinds of segments - stem, internodes, twigs - , 3 kinds of points - root, nodes or bifurcation points, tips - with own topological parameters - numbers of stems S and tips F, Percheron, 1982) and metrical parameters (lengths and diameters)

Dendritic arborisation : bush (with several stems) of dendritic trees (binary graphs with one stem) of one individual neuron.

Cladonal arborisation : cladonal tree forming a spatially separate subgraph. A single neuron can have several cladonal arborisations, then connected by nemons.

Hodonal graph : binary graph of the connection with outer targets (not to be in fact considered individually, see below)

IC - Neuronal spaces. (Fractal) metric spaces (whose border is their "outer enveloppe") occupied by neuronal arborisations, whith three dimensons - length, height and width - , shape and orientation (see Percheron et al., 1984, 1987)

Dendritic dwelling : space occupied by one dendritic arborisation.

Cladonal parcel : space occupied by one cladonal arborisation.

II - Neuronal sets level

IIA - Neuronal typology. Based on measured parameters on dendritic arborisations (with no consideration of axonal or biochemical features, which often fit)

Neuronal species : Set of neurons having statistically similar (fully reconstructed) dendritic arborisations (see Yelnik et al., 1987)

Neuronal sub-species, order, family.: statistical subdivisions or regrouping of neuronal species

IIB - Neuronal set-spaces. (Fractal) metric spaces (whose border is the "outer enveloppe" of the set) occupied by neuronal sets (see Percheron et al., 1987, 1991b)

Nucleus : space occupied the somata of the neurons belonging to the same neuronal species

Patria : space occupied by the dendritic dwellings of one neuronal species. Several neuronal species in the same patria are sympatric.

Territory (territorium) : space occupied, in one target, by the cladonal parcels of one neuronal species. Cladonal parcels from different neuronal species sharing the same territory are coterritorial

Fons : metric space occupied by the trigger zones (and usually, coincidently, by the soma and dendritic dwellings) of neurons from one neuronal species projecting to a given target

Communication domain : metric space corresponding to the intersection between a patria and a territory and where neuronal communication occurs.

IIC - Neuronal topological spaces and connectivity. Spaces with no metric dimensions but connected and labelled.

Source : neuronal space from where axons of one or several neuronal species projecting to a given (or several) target arise

Target : neuronal space receiving axons from one or several neuronal species. There are outer and inner (within the source) targets.

Hodonal graph : Labelled graph of the connections between sources and targets whose identity as well as the direction and type of the connection are important, not their metric dimensions. Hodonal graphs involving several targets may have different topological types: linear, in series or branched. There may be systemic separation. (see Fig. 2 - 3 and Percheron and Filion, 1991).

Depending on the relative number of sources and targets there can be unifluence (equal numbers of sources and targets), difluence (more targets than sources) or confluence (more sources than targets).

IID - Neuronal set maps. Inner organisation in a neuronal set-space. Can be homogeneous, "continuous", "scattered" or "patchy" (Nelson and Bower, 1990) with subsets and subspaces, see below.

III - Neuronal subsets levels

IIIA - Neuronal subset-spaces. (Fractal) metric subspaces (whose border is the "outer enveloppe" of the subset) occupied by neuronal subsets

Syllogue: Subspace in a fons occupied by the trigger-zones of a group of neurons of the same neuronal species together sending messages to the same target or subtarget

Dendralse, dendrite-grove : subspace occupied by the dendritic arborisations of a group of neurons receiving information from the same subsource (generally coincident with a syllogue)

Cladalse, cladon-grove : subspace in a target occupied by the cladonal arborisations of a group of neurons emmitting information from the same source or subsource

IIIB - Combination of connections between neuronal subspaces. (Fig. 2 - 4) Connections between subspaces can have different topological types, see IIc. Parallel separation leads to a lineary (continuous) ordered map. Depending on the relative numbers of subsources and subtargets there may be equiinsulation (same numbers), divinsulation (one subsource projecting to several subtargets), convinsulation (one subtarget receiving from several subsources), ambiinsulation (div- and conv-insulation together)

IIC - Spatial coincidence. between emitting and receiving subspaces. There can be coincidence, excidence (no coincidence between given cladalses and given dendralse-syllogues), plurincidence (one cladalse being coincident with several dendralses), aleaincidence (random distribution) (see also Graybiel et al., 1991).

IV - Neuronal communication informational parameters

Deduced from comparisons between the relative metric dimensions, shape and orientation of individual neuronal spaces (dendritic dwellings and cladonal parcels) and of the domain of communication. (see Percheron et al., 1984, 1987) in relation to the communication axis

IVA - Grain. Communication is fine grain when the size of the dwellings is small compared to that of the domain or rough grain in the reverse case

IVB - Lattice density. Volume of dendrites or cladons in a given volume of a domain of communication (Yelnik et al., 1991).

IVC - Diversity of information. Measured, in Nits, the number of non overlapping dwellings or parcels that can be placed perpendicular to the communication axis (Percheron et al., 1984, 1987).

IVD - Cardinals. The cardinal of a neuronal set is the number of its elements. There can be a divergence of cardinal if the number of elements of the target is larger than that of the source and a convergence of cardinal in the reverse case.

IVE - Informational powers. Redundancy may be measured by the number of somata that may be counted in an average dendritic dwelling. The capacity of fascicles is the number of their constituing hodons multiplied by the quantity of information that can be carried by one of them.

IVF - Types of communications. There is a divergence, when the dimensions of the emitting parcel are larger than those of the receiving dwellings (One parcel reaches more than one dwelling) and a convergence, when the dimensions of the receiving dwellings are larger than those of the emitting parcels. Convergence or divergence ratios can be computed from the ratio of the relative dimensions of average dwellings and parcels.

LOCALISATION OF CALCIUM BINDING PROTEINS IN THE NEOSTRIATUM

Ben D. Bennett and J. Paul Bolam

MRC Anatomical Neuropharmacology Unit
Mansfield Road
Oxford, OX1 3TH
U.K.

INTRODUCTION

The neostriatum represents the major division of the basal ganglia and is the primary recipient of cortical information flowing into the basal ganglia. In order to understand how extrinsic information is processed and integrated with other extrinsic inputs and with local neurones to produce signals that are transmitted to the output structures of the basal ganglia i.e. the substantia nigra and the pallidal complex, it is necessary to have detailed plans of the microcircuitry in which the morphological and chemical features, as well as the afferent and efferent connections, of each component is known. To this end, a great deal of work has already been carried out in establishing the microcircuitry of the neostriatum and the basal ganglia in general (see Smith and Bolam, 1990 for review). This work has been based on the analysis of Golgi-impregnated neurones, and more recently, on the analysis of structures immunostained to reveal either the neurotransmitter content of neurones or a neurotransmitter-related marker, for instance the synthetic enzyme or catabolic enzyme of a neurotransmitter. It is becoming increasingly apparent that antibodies against markers other than the neurotransmitter-related substances are of value in the elucidation of neural microcircuits. One such group of substances are the calcium binding proteins that are widely distributed within the central nervous system and although little is known about their functions, antibodies directed against them have been instrumental in the discovery of subpopulations of neurones within many regions of the brain (Gulyás et al., 1991; Jacobowitz and Winsky, 1991; Seress et al., 1991; Résibois and Rogers, 1992), and in the elucidation of the neuronal microcircuits to which these cells contribute (Freund and Gulyás, 1991; Gulyás et al., 1992; Miettinen et al., 1992). Within the neostriatum the calcium-binding proteins parvalbumin, calretinin and calbindin D28k have been localised in distinct subpopulations of neurones. Parvalbumin is present in a population of neurones that have been identified as the GABAergic interneurones of the striatum (Cowan et al., 1990; Kita et al, 1990) and immunoreactivity for calretinin is localised in a small population of striatal neurones (Jacobowitz and Winsky, 1991; Résibois and Rogers, 1992; Bennett and Bolam, 1993a). In contrast, the calcium-binding protein, calbindin D28k, is present in a much larger population of neurones that have the characteristics of the most common class of striatal neurone, the medium-size densely spiny neurone (Gerfen et al., 1985; Gerfen, 1992).

The object of this paper is to review the data relating to the morphology, distribution and in particular the ultrastructure and synaptic connections of neurones that display immunoreactivity for these three calcium binding proteins, parvalbumin, calretinin and calbindin D28k. Most of the data is derived from the striatum of the rat but some information comes from the analysis of the caudate-putamen of primates. The approach to these analyses has been to carry out immunocytochemistry using antibodies directed against the calcium binding proteins at both the light and electron microscopic levels.

PARVALBUMIN

General Observations in the Rat

Sections of the rat striatum incubated to reveal parvalbumin-immunoreactive structures contain a population of neurones that represent only a small proportion of the total population (3-5%, estimated) (Kita et al., 1990). The perikarya of the immunoreactive neurones are of medium to large size (147 ± 45 μm^2), generally polygonal in shape and give rise to several primary dendrites that branch close to the perikaryon and thereafter branch frequently and often at an angle close to or greater than 90°. The intensity of the staining that can be obtained with the antibody preparations against parvalbumin (PV) is often sufficiently strong to give very fine detail of the dendrites. In these well-stained sections it is evident that the dendrites of the PV-immunoreactive neurones are spine-free but sometimes have a varicose appearance. In the electron microscope it is evident, as it is in some light microscopic preparations, that the nucleus possesses indentations and is surrounded by a relatively large area of cytoplasm that is rich in organelles.

Double labelling studies with antibodies against PV and the synthetic enzyme for GABA, glutamate decarboxylase (GAD) (Kita et al., 1990) or GABA itself (Cowan et al., 1990) have revealed that, in the rat at least, there is 96-100% correspondence between PV-immunoreactive neurones and the markers for GABAergic neurones. This feature and the morphological characteristics of these neurones implies that the PV antibodies selectively mark the GABA interneurones of the striatum. GABA interneurones, although identified in the early studies of GABAergic neurones in the striatum (Iversen and Schon, 1973; Ribak et al., 1979), were recognised as GABAergic interneurones on the basis of the uptake of radiolabelled GABA combined with Golgi impregnation (Bolam et al., 1983) and immunocytochemistry for GAD (Bolam et al., 1985; Kita and Kitai, 1988). They are medium-to-large size neurones, account for only a small proportion of the total population of neurones, selectively accumulate exogenous GABA and stain more intensely for GABA markers than do other striatal GABAergic neurones. The somato-dendritic morphology, at both the light and electron microscopic levels, of PV-immunoreactive neurones and GABAergic interneurones are indistinguishable. It is thus likely that in the rat at least, PV is a faithful marker of the GABAergic interneurones. However, the reverse is not true, markers of GABA function are not faithful markers of PV neurones as in addition to GABAergic interneurones, the spiny projection neurones contain and utilise GABA (for references see Graybiel, 1990; Smith and Bolam, 1990).

In addition to the stained neurones, sections of PV-immunostained striatum contain a dense network of immunostained axons and terminals. When examined in the electron microscope the PV-immunoreactive boutons are seen to form predominantly symmetrical synapses with perikarya and dendrites (Kita et al., 1990; Bennett and Bolam, unpublished). The distribution of postsynaptic targets, both in the neostriatum and nucleus accumbens is weighted towards the proximal regions of the neurones and in the region of 50% of the terminals make synaptic contact with neuronal perikarya (Kita et al., 1990; Bennett and Bolam, unpublished). The morphology and type of synaptic specialisation formed by the PV-immunoreactive terminals is similar to that of the identified terminals of Golgi-impregnated neurones presumed to be of the same class of neurone (Takagi et al., 1984).

Parvalbumin-immunoreactive structures within the neuropil are distributed heterogeneously, although their distribution does not follow the patch/striosome-matrix organisation of the striatum (Cowan et al., 1990; Kita et al., 1990). There is a considerably greater density of PV-immunoreactive structures in the neuropil in more lateral locations of the striatum, such that medial areas can appear to be almost devoid of immunostaining whereas lateral areas of the same sections possess an intricate meshwork of PV-immunoreactive axons and dendrites.

Synaptology of Parvalbumin-Containing Structures in the Rat

Input. Ultrastructural analyses of neurones that display immunoreactivity for PV (Kita et al., 1990; Bennett and Bolam, unpublished) or the analyses of identified GABA interneurones (Bolam et al., 1983; Bolam et al., 1985) have revealed that they receive a dense synaptic input from terminals of varying morphological types. Both symmetrical and asymmetrical synapses have been identified and these occur on all parts of the neurones so far examined. Several types of afferent terminals in synaptic contact with the PV-positive neurones have been characterised on the basis of neurochemistry or origin. First, PV-positive neurones have been shown to receive symmetrical synaptic input from terminals that are themselves immunoreactive for PV (Kita et al., 1990; Chang and Kita, 1992; Bennett and Bolam, unpublished). Furthermore, the same class of neurone identified in GAD-immunostained material has been shown to receive input from many terminals that display immunoreactivity for GAD (Bolam et al., 1985). There are several possible sources of the PV- or GAD-positive terminals in contact with these cells. The most obvious is the PV-positive neurones themselves, it is thus likely that the GABA interneurones are interconnected with one another. However, in addition, these terminals could be derived from the globus pallidus (GP), as all GP neurones are PV-positive (Celio, 1990) and the GP is known to innervate the striatum (Staines et al., 1981; Arbuthnott et al., 1982). In the primate (see below) the situation is more complex as several regions that project to the striatum possess PV-positive neurones. A second class of terminals that form symmetrical synapses with PV-positive neurones are immunoreactive for choline acetyltransferase (ChAT) (Chang and Kita, 1992). The major source of ChAT-positive terminals in the rat striatum are the cholinergic interneurones that provide a dense network of ChAT-positive terminals (Bolam et al., 1984). By combining anterograde labelling with immunocytochemistry for PV, it has been shown both in the rat (Kita, 1991; Bennett and Bolam, unpublished) and the primate (Lapper at al., 1992) striatum that PV-positive neurones receive direct synaptic input from the cortex. The corticostriatal input accounts for a high proportion of those terminals that form asymmetrical synaptic contact with the PV-positive neurones. This input is located on all parts of the neurones examined including large diameter proximal dendrites, small diameter distal dendrites and perikarya. The functional significance of this input is discussed below.

Output. The output of PV-positive neurones, and hence GABAergic interneurones, has been analysed on the basis of examination of PV-positive terminals. This type of analysis can result in problems of interpretation as the PV-positive neurones in the striatum are not the *only* source of PV-immunoreactive terminals. This problem is particularly acute in the primate (see below) in which a high proportion of PV-positive terminals are derived from sources other than local PV-positive neurones. Nevertheless, in the rat the major source of PV-positive terminals are the local PV-immunoreactive neurones whereas only a small proportion are derived from extrinsic sources, mainly the GP. Analyses in this species have shown that greater than 50% of PV-positive terminals make synaptic contact with the cell body or most proximal dendrites of neurones that have the ultrastructural features of medium-size spiny neurones (Kita et al., 1990; Bennett and Bolam, unpublished). Furthermore, the combination of immunocytochemistry for PV combined with retrograde transport of horseradish peroxidase conjugated to wheatgerm agglutinin or neurobiotin, from the substantia nigra, has revealed that a high proportion of PV-positive terminals make synaptic contact with identified output neurones of the striatum (Fig. 1A). The same is true for the nucleus accumbens. Large injections of retrograde tracers, that included the medial aspects of the substantia nigra and the ventral tegmental area, retrogradely labelled neurones in the nucleus accumbens which were found to receive multiple synaptic inputs from terminals that display immunoreactivity for PV (Bennett and Bolam, unpublished). A further class of striatal neurone has been shown to be post-synaptic to PV-positive terminals. Analysis of rat striatum histochemically stained to reveal the presence of NADPH-diaphorase (Vincent, 1992) and immunocytochemically stained to reveal PV, has demonstrated that at the light microscopic level, PV-positive boutons are frequently apposed to the perikarya of diaphorase-positive cells (Fig. 1B inset). Examination of these appositions in the electron microscope has revealed that PV-postive boutons often form symmetrical synaptic contacts with the diaphorase-positive cells (Fig. 1B). However, the morphological features of PV-positive boutons in contace with NADPH-neurones is not uniform. Some PV-positive

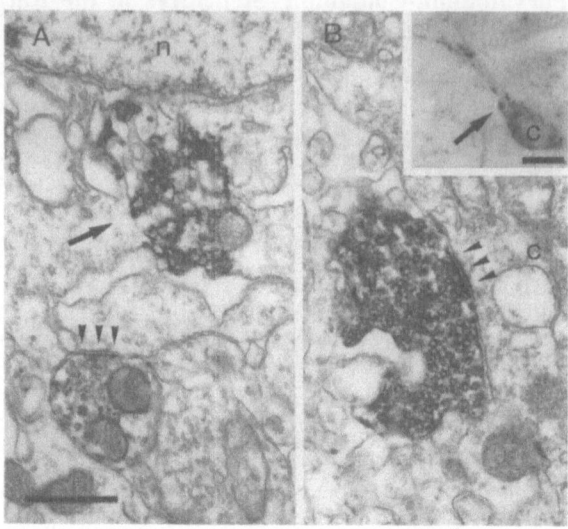

Figure 1. Parvalbumin-immunoreactive structures in the striatum of the rat.
A. This figure illustrates a parvalbumin-immunoreactive terminal forming symmetrical synaptic contact (arrowheads) with the cell body of a neurone identified as a striatonigral neurone by the retrograde transport of neurobiotin. The reaction product formed as a result of the presence of retrogradely transported neurobiotin is indicated by the large arrow. This micrograph demonstrates that parvalbumin-immunoreactive terminals are in synaptic contact with the perikarya of striatal output neurones. n, nucleus.
B. These micrographs illustrate the association of parvalbumin-positive terminals with NADPH-diaphorase reactive neurones in the striatum of the rat. The inset shows a diaphorase positive cell (c) in the light microscope. Apposed to the proximal dendrite of the cell are several parvalbumin positive boutons, one of which is indicated by a large arrow. Examination of this same neurone in the electron microscope revealed that the parvalbumin immunoreactive bouton is a large vesicle-filled axon terminal that forms symmetrical synaptic contact (arrowheads) with the diaphorase-positive cell (c).
Scale: 0.5 μm for A and B and 10 μm for B, inset.

boutons have the ultrastructural features of terminals derived from striatal interneurones whereas others are larger, often contain many mitochondria and fewer vesicles and are reminiscent of pallidal terminals that have been previously identified in other structures (Smith and Bolam, 1991). It is postulated therefore, that at least some of the PV-positive terminals within the striatum are pallidal in origin: the analysis of postsynaptic targets of anterogradely labelled pallidal terminals in the striatum is currently under investigation (Bennett and Bolam, unpublished).

Parvalbumin-Immunoreactive Structures in the Primate

In contrast to the rat, PV immunoreactivity in the neuropil of the primate caudate-putamen is heterogeneously distributed, such that areas containing low levels of PV-immunopositive structures are embedded in a surrounding matrix that is dense in PV-immunoreactive structures. This pattern of immunoreactivity forms a complex mosaic-like pattern and is in register with the striosome/matrix organisation of the caudate-putamen as revealed by examination of adjacent sections stained to reveal substance P or met-enkephalin immunoreactivity (Bennett and Bolam, unpublished).

Parvalbumin-immunoreactive neurones are present throughout the caudate-putamen and are located within both the striosome and matrix compartments. In the matrix, these cells possess morphological features in common with PV-positive cells found in the rat striatum (Cowan et al., 1990; Kita et al., 1990) i.e. PV-immunoreactive neurones have round, oval or polygonal somata which give rise to primary dendrites which branch and ramify close to the parent cell body. However, within areas corresponding to striosomes, PV-positive neurones exhibit a greater variation in their somatodendritic morphology, such that a continuum of morphological types are present. These range from neurones which are bipolar in appearance and give rise to primary, secondary and higher order dendrites which branch infrequently to become long, thin, tapering processes in their most distal regions, to neurones which possess the morphological features of PV-positive cells described in the rat (see above). Sub-sets of PV-immunoreactive neurones are present which possess both long, infrequently branching, tapering dendrites and dendrites which ramify extensively near the parent soma. The dendrites of PV-positive cells located within both the matrix and striosomal compartments occasionally cross the boundaries between the two compartments. The PV-immunoreactive structures within the neuropil of the matrix compartment are not distributed evenly. Hence, the matrix compartment possesses a higher density of PV-immunoreactive structures at more lateral locations when compared to the matrix located more medially.

In addition to the PV-immunopositive cells described above the sections of the caudate-putamen also contain a small population of large PV-immunopositive neurones. These neurones are located in the ventral parts of the putamen in regions close to or continuous with cells of the ventral pallidum or pallidal cells embedded in the internal capsule. They are not found in the caudate nucleus. The perikarya are 20-60 μm in diameter and give rise to 3-6 primary dendrites which are of large diameter (approx. 5 μm) and smooth with few if any varicosities. The dendrites course through the putamen, sometimes for distances in excess of 800 μm and occasionally cross the borders between the striosome and matrix compartments. The dendrites often have a large number of PV-immunoreactive boutons apposed to them.

Synaptology of Parvalbumin-Containing Structures in the Primate

Input. In order to identify the origin of some of the asymmetric terminals in contact with the PV-positive cells, anterograde labelling with biocytin produced by large injections in the motor cortex, has been combined with immunocytochemistry for PV. In these experiments it was demonstrated that the PV-positive cells in the putamen receive direct synaptic input from cortico-striatal terminals (Lapper et al., 1992), which provides a presumably glutamatergic, excitatory input to these putative GABAergic interneurones.

Output. Unlike the rat, PV-immunopositive terminals in the primate striatum are heterogeneous forming both symmetrical and asymmetrical synaptic specialisations with dendrites and spines. This is the case in both the caudate nucleus and putamen and in both the striosome and matrix compartments. The reason for this difference between primate and rat is that in the primate, areas which are known to project to the caudate-putamen e.g. centre-median and parafascicular (CM/Pf) nuclei of the thalamus (Sadikot et al., 1992) possess PV-immunopositive cells (Jones and Hendry, 1989). These cells represent a source of PV-immunoreactive terminals which form asymmetrical synaptic specialisations. Hence, analysis of the targets of PV-immunoreactive terminals is limited in the primate due to the localisation of PV in many afferent fibres as well as in the terminals of local PV-positive cells. Furthermore, afferents arising from the CM/Pf terminate preferentially in the matrix compartment (Sadikot et al., 1992) and some neurones in the CM/Pf complex are PV-positive (Jones and Hendry, 1989). It is therefore likely that it is the PV-staining of afferent terminals that gives rise to the patch/striosome-matrix staining of the neuropil in the caudate-putamen.

Feed-forward Inhibition in the Rat Striatum

Electrophysiological studies and the anatomical studies described above imply that GABAergic interneurones subserve a feed-forward inhibitory role in the circuitry of the striatum and nucleus accumbens (Wilson et al., 1989; Pennartz and Kitai, 1991). Consistent with the electrophysiological data, anatomical studies in the rat have shown that GABAergic interneurones, identified by PV immunocytochemistry, receive input from the cortex (Kita, 1991) and in turn make contact with the somata of spiny neurones (Kita et al., 1990), some of which have been identified as projection neurones (Bennett and Bolam, unpublished). Hence, ultrastructural analyses performed in combination with PV immunocytochemistry have allowed the elucidation of the anatomical basis for the electrophysiological phenomenon of feed-forward inhibition.

Usefulness as a Marker for GABAergic Neurones

It is apparent from foregoing discussion, that within the rat striatum, PV immunocytochemistry is extremely useful for selectively staining the dendrites, somata, axons and terminal boutons of GABAergic interneurones which allows a comprehensive analysis of the synaptology of these elements in the striatum to be conducted. One technical problem in the rat is that PV immunoreactivity is present in all neurones of the GP (Celio, 1990) and the GP projects to the striatum (Staines et al., 1981; Arbuthnott et al., 1982). However, this projection is relatively small (Bennett and Bolam, unpublished) and pallidal terminals possess morphological features that are readily identified in the electron microscope. On the other hand, the primate caudate-putamen possesses PV-immunoreactive neurones which are extensively stained by immunocytochemistry directed against this calcium-binding protein. However, it has not yet been demonstrated directly that the PV-positive cells in the primate are GABAergic. Furthermore, antibodies directed against PV, as well as disclosing the axons and terminal boutons of local putative GABAergic interneurones, also stain at least one population of afferent fibres (arising from the CM/Pf thalamic complex) and hence, have limited use in the elucidation of the microcircuitry in which putative GABAergic interneurones of the primate caudate-putamen reside. Although PV immunocytochemistry has certain limitations, it is useful for selectively staining the histochemical compartment of the caudate-putamen referred to as the matrix (Giménez-Amaya and Graybiel, 1990).

CALRETININ

Sections of the striatum of both the rat and primate stained to reveal calretinin immunoreactivity contain a moderate density of stained profiles in the neuropil as well as a small population of immunoreactive neurones that are present throughout this structure (Jacobowitz and Winsky, 1991; Résibois and Rogers, 1992; Bennett and Bolam, 1993a).

Light Microscopic Observations in the Rat

Calretinin-immunoreactive neurones are present in all regions of the striatum. There is no marked variation in their distribution in the dorsoventral and mediolateral planes. However, there is a very marked rostrocaudal gradient, with many more calretinin-immunoreactive neurones located in the rostral portions of the striatum and very few calretinin-immunoreactive cells found caudally (Bennett and Bolam, 1993a). The somata of the calretinin-immunoreactive neurones are round, oval or fusiform with a diameter of 9-17 μm (longest axis) and mean cross-sectional area of 81.3 μm^2 (range 21.2-152.5 μm^2). The somata give rise to one-three primary dendrites which in turn give rise to two or three secondary dendrites. The dendrites are long, predominantly smooth, occasionally varicose and infrequently branching. Distal dendrites taper to become very thin processes in their most distal portions. Dendritic varicosities are occasionally present. In the electron microscope the nucleus of the immunoreactive neurones is seen to characteristically possess marked indentations and a prominent nucleolus. The nucleus is generally centrally placed in the perikaryon

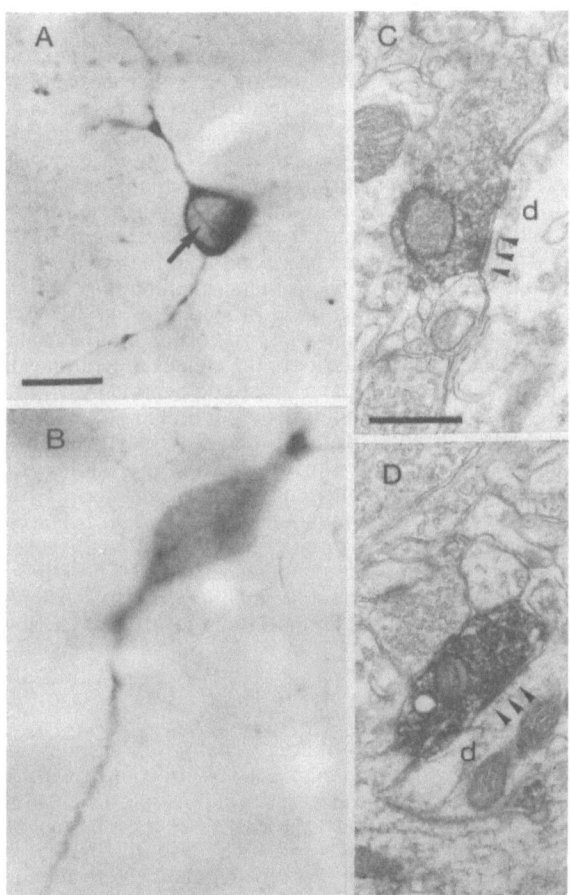

Figure 2. Calretinin-immunoreactive structures in the primate caudate-putamen. A and B illustrate the two types of calretinin-positive neurones. In A, a medium-sized neurone is illustrated. Several smooth dendrites emerge from the perikaryon of this cell that branch frequently. The nucleus of this neurone displays indentations (arrow). In B a giant calretinin-positive neurone is illustrated. This cell type is very similar to the giant parvalbumin-positive cells which are only localised in the ventral part of the putamen. These cells are considered to be misplaced ventral pallidal neurones.

Calretinin-immunoreactive structures in the rat neostriatum.C and D illustrate calretinin immunoreactive axon terminals in the rat neostriatum. In both cases the terminals are forming synaptic contacts with dendritic shafts (d). The bouton in C is in symmetrical synaptic contact with a dendrite (arrowheads) whereas that in D is forming asymmetrical synaptic contact with a dendrite (arrowheads).

Scale: A and B 20 μm, C and D, 0.5 μm.

and is surrounded by a moderate volume of cytoplasm which contains many organelles, including mitochondria, Golgi apparatus and many ribosomes arranged as polysomes. On occasions, filamentous nuclear inclusions have been observed in immunoreactive somata.

Double immunolabelling experiments have been carried out in order to determine the characteristics of calretinin-immunoreactive neurones in relation to other classes of aspiny neurones in the striatum (Bennett and Bolam, 1993a). From these experiments it is clear that in the main, calretinin is localised in a separate population of neurones to the choline acetyltransferase-containing cells i.e. cholinergic interneurones and the PV-containing neurones (i.e. GABA interneurones, see above). Furthermore, immunocytochemistry for calretinin combined with histochemistry for NADPH-diaphorase again reveals two separate populations, indicating that the calretinin-immunoreactive neurones are distinct from the class of striatal interneurones that are immunoreactive for somatostatin and neuropeptide Y (both of which co-localise with diaphorase activity) (Vincent and Johansson, 1983). Thus, based on both morphological and chemical grounds it appears that the calretinin-immunoreactive neurones are distinct from the major type of striatal neurone i.e. the medium-size densely spiny neurones and from the major types of striatal interneurones. Since neurones of this morphological type have never been identified in studies of striatal output neurones it is reasonable to suggest that they represent a population of interneurones. The calretinin-immunoreactive cells do bear some similarity to the classes of neurones in the striatum that display immunoreactivity for vasoactive intestinal polypeptide (VIP) (Takagi et al., 1984; Theirault and Landis, 1987) and those that display immunoreactivity for cholecystokinin (CCK) (Adams and Fisher, 1990). Both of these classes are similar to calretinin-immunoreactive neurones, not only in somatodendritic morphology, ultrastructure and afferent synaptic input but also in their density in the striatum. However, the overall (macroscopic) distribution of the CCK-immunoreactive neurones is markedly different from that of the calretinin-immunoreactive neurones and the VIP-positive neurones, unlike the calretinin-immunoreactive cells are preferentially associated with fibre fascicles traversing the striatum. It remains to be established, in double-labelling studies, whether there is an overlap between the populations of neurones containing calretinin, VIP and CCK.

Electron Microscopic Observations in the Rat

Input. The synaptic input to the perikarya of calretinin-immunoreactive neurones is fairly sparse, no more than one afferent synapse is seen in contact with the perikaryon in a single ultrathin section. The density of afferent synapses is greater on the dendrites and increases with distance from the cell body. At least three morphological types of afferent synaptic boutons have been identified in contact with both the perikarya and dendrites of calretinin-immunoreactive neurones. The first type is relatively small, contains spherical or ellipsoidal vesicles and forms symmetrical synapses. The second is larger, contains many mitochondria, spherical and ellipsoidal vesicles and forms symmetrical synaptic contact. The third type is similar in morphology to the first but forms asymmetrical synapses. The possible sources of the terminals forming asymmetrical synapses are the cortex, thalamus and possibly the amygdala and dorsal raphe (Smith and Bolam, 1990). Those terminals forming symmetrical synapses could be derived from either local striatal neurones or from extrinsic sources, such as the terminals of neurones in regions known to project to the striatum which form symmetrical synapses e.g. the terminals of dopaminergic nigrostriatal neurones. It is interesting to note that the large type of bouton that forms symmetrical synaptic contact with the calretinin-positive neurones is similar in morphology to one type of PV-containing terminal (unpublished observations) and therefore may be derived from the globus pallidus. It is thus likely that the calretinin-immunoreactive neurones receive afferent synaptic input from neurones located in other brain regions that project to the striatum and from local striatal neurones.

Output. In addition to the calretinin-immunoreactive neurones, the striatum contains dense neuropil staining, which under high magnification in the light microscope can be identified as a network of axons and terminal boutons. Electron microscopic analysis has confirmed that the structures are axons and boutons which form typical synaptic contacts (Fig. 2C, D). It is evident, based both on morphology and the type of synaptic specialisation, that the calretinin-immunoreactive terminals are heterogeneous. Although the majority of terminals have a very similar morphology, containing small spherical or ellipsoidal vesicles, at least one population possesses much larger vesicles. The immunoreactive terminals are seen to form both symmetrical (38%) and asymmetrical synaptic specialisations (56%). Most of those forming asymmetrical synapses are in contact with dendritic spines (52% of total) whereas those forming symmetrical synapses are mainly in contact with dendritic shafts (30% of total). This observation is not surprising in view of the fact that regions known to project to the striatum contain populations of calretinin-positive output neurones. These include, the thalamus, dorsal raphe and substantia nigra pars compacta (Jacobowitz and Winsky, 1991; Résibois and Rogers, 1992). Thus, calretinin-immunoreactive terminals that form asymmetrical synapses could be derived from the thalamus, the dorsal raphe nucleus and possibly the amygdala, since each of these regions give rise to terminals in the striatum that form asymmetrical synapses (see Smith and Bolam, 1990). Although corticostriatal terminals form asymmetrical synapses, it is unlikely to be a source of asymmetrical calretinin-positive terminals as calretinin-immunoreactivity in the cortex is not localised in pyramidal neurones i.e. the type that give rise to the striatal projection but in a population of non-pyramidal neurones (Jacobowitz and Winsky, 1991; Résibois and Rogers, 1992). It is possible, of course that the striatal calretinin-positive neurones themselves give rise to the terminals which form asymmetrical synapses but, because it was not possible to identify the axons arising from the immunoreactive cells it was not possible to test this directly (Bennett and Bolam, 1993a). In addition, the terminals of interneurones within the striatum have never previously been found to form asymmetrical synapses (Smith and Bolam, 1990). The origin of those terminals forming symmetrical synapses is also an open question, as some of the calretinin-positive neurones projecting to the striatum (e.g. substantia nigra pars compacta neurones), also form symmetrical synapses.

Light Microscopic Observations in the Primate

Sections of primate striatum incubated to reveal calretinin immunoreactivity contain a population of neurones that are similar in morphology to those identified in the rat (Bennett and Bolam, 1993a). Thus the neurones are of medium size and possess several dendrites that do not bear spines. The immunostaining in the primate is usually more extensive than that observed in the rat, such that a greater extent of the dendritic arborisation of the calretinin-positive cells is revealed. The dendrites branch frequently, close to the perikaryon (Fig. 2A) which contain nuclei with prominent indentations (Fig. 2A). The neurones in the primate are more regularly distributed throughout the caudate-putamen in comparison to the rat. In addition to the medium size neurones, the primate caudate-putamen contains a population of giant neurones that display immunoreactivity for calretinin (Fig. 2B). These cells have perikarya that are very large (approx. 30-60 μm diameter) and give rise to thick dendrites that extend for several hundred microns, usually in a dorsal-ventral plane. These giant calretinin-immunoreactive neurones have only been found in the ventral aspects of the putamen close to the ventral pallidal cell groups that are below the anterior commissure or embedded in the internal capsule. The morphology and distribution of these neurones suggest that they are the same population of cells as the large neurones identified in the PV-immunostained primate putamen (see above) and are probably not true striatal neurones but rather misplaced ventral pallidal cells (see above). It is also interesting to note that although the giant cells are identified in both PV and calretinin immunostained sections, the PV-immunoreactive boutons which surround the giant PV-immunoreactive cells are not stained in sections treated to reveal calretinin-immunoreactivity. Hence, although PV- and calretinin-immunoreactivity may be co-localised in these giant neurones, calretinin does not appear to be co-localised with PV in the PV-immunopositive boutons that innervate these large neurones.

CALBINDIN

Light and Electron Microscopic Observations in the Rat

Sections of the striatum incubated to reveal calbindin immunoreactive structures contain large numbers of stained perikarya and dendrites. Their morphological and distributional characteristics have been described extensively (Gerfen et al., 1985; DiFiglia et al., 1989; Gerfen, 1992). Thus, calbindin-positive neurones are found throughout the rostro-caudal extent of the striatum, although the intensity of staining is greatest in the most ventral regions and at its weakest or absent in the most dorsal aspects. The relatively homogeneous distribution of stained neurones is interspersed by islands that are virtually free of immunostained cells, referred to as patches or striosomes. The immunostained neurones possess both the light and electron microscopic characteristics of medium-size spiny neurones in that they are of medium-size, possess a round nucleus that occupies most of the perikaryon and does not display any indentations. The combination of calbindin immunocytochemistry and Golgi-impregnation has confirmed this (Bennett and Bolam, 1993b); Golgi-impregnated medium-sized densely spiny neurones were identified that also display immunoreactivity for calbindin (Fig. 3A).

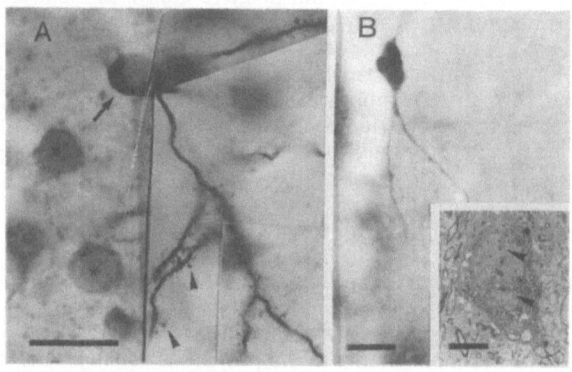

Figure 3. Calbindin D28K-immunoreactive neurones in the rat neostriatum.
A is from a section of the rat striatum that was incubated to reveal calbindin immunoreactivity and then Golgi-impregnated. On the left of the micrograph are two immunoreactive perikarya (*) that are medium sized and have a large nucleus filling most of the cytoplasm. On the right is an immunoreactive neurone that is also Golgi impregnated (arrow). This cell gives rise to several dendrites which become densely laden with spines (some of which are indicated by arrowheads). This illustrates that the major type of calbindin-immunoreactive neurone is of the medium size spiny class. In B, an immunoreactive neurone that displays the characteristics of an aspiny neurone is illustrated. These cells have irregular shaped perikarya and give rise to long aspiny dendrites. The inset shows an electron micrograph of the same neurone and illustrates that the nucleus of this cell type has prominent indentations (arrowheads). Note that the magnification of the micrograph in B is much lower than that in A.
Scale: A and B 20 μm, C, 4 μm.

In addition to the spiny calbindin-positive neurones, a second, much smaller population, of calbindin-immunoreactive neurones has also been identified (Fig. 3B) (Kiyama, et al., 1990; Roberts and DiFiglia, 1990; Bennett and Bolam, 1993b). In human post-mortem tissue a population of calbindin-positive neurones has been described that are larger and more intensely stained than the majority (Kiyama et al., 1990). Unlike the medium-size spiny calbindin-positive neurones, the larger type is spared in Huntington's disease. In grafts of rat foetal striatal tissue deposited into the quinolinic acid-lesioned caudate nucleus, a significant proportion of calbindin-positive neurones were observed that possess an indented nucleus, a feature that does not normally occur in spiny neurones (Roberts and DiFiglia, 1990). The second class of calbindin-immunoreactive neurones has also been identified in normal rat striatum (Bennett and Bolam, 1993b). They are generally more extensively stained than the spiny neurones and they possess long, essentially smooth dendrites which taper to become thin processes in their most distal portions (Fig. 3B). Correlated light and electron microscopic analysis of the second class of calbindin-immunoreactive neurones has revealed that, unlike spiny neurones, they possess indented nuclei that are surrounded by relatively large volumes of cytoplasm containing many organelles (Fig. 3B inset). Due to the difficulty of identifying a small sub-class of neurones amongst a large population of stained neurones, it has not been possible to examine the distribution of the non-spiny type nor to determine the relative proportions of the two neuronal types. However, it is clear that the second class represents only a small proportion of the total population and that they are distributed in both the patch/striosome and matrix compartments, identified by the distribution of the spiny calbindin-immunoreactive neurones.

The combination of calbindin immunocytochemistry and NADPH-diaphorase histochemistry (Bennett and Bolam, 1993b) has revealed that at least a small population of neurones which display calbindin-immunoreactivity and have the features of the aspiny neurones are also positive for NADPH-diaphorase.

The possibility that the staining of the second class of neurones by the antibodies to calbindin is due to cross-reactivity with some other antigen or calcium-binding protein should be considered. The calcium-binding protein, calretinin, shares a 60% sequence homology with calbindin D28k (Rogers, 1987; Parmentier, 1990; Rogers et al., 1990) and it is possible that the present population is in fact the population of neurones that have been identified in studies of calretinin immunoreactivity in the striatum (see above). Indeed, the morphological characteristics and the sparsity of the aspiny calbindin-positive neurones is similar to those of the calretinin-immunoreactive cells. However, for several reasons it seems unlikely that the two populations are identical. First, the antibodies used in the studies, in the normal rat at least, do not cross-react with calretinin (Celio, 1990). Secondly, the same class of neurones has been identified with three different antibody preparations (Kiyama et al., 1990; Roberts and DiFiglia, 1990; Bennett and Bolam, 1993b). Thirdly, the calretinin-immunoreactive neurones do not display reactivity for NADPH-diaphorase whereas at least a sub-population of the second class of calbindin-immunoreactive neurones are NADPH-diaphorase-positive (Bennett and Bolam, 1993b).

By definition, the second class of calbindin-positive neurones are different from the medium spiny neurones which are the major striatal projection neurones; this fact together with the failure to detect this class of neurones in retrograde tracing studies in the striatum suggests that they are a population of striatal interneurones. Although double labelling studies have not been performed, on morphological grounds it is clear that they are distinct from the GABAergic interneurones of the striatum and from the cholinergic interneurones (Bolam et al., 1984; Bolam et al., 1985; Cowan et al., 1990; Kita et al., 1990). The fact that at least a population of these neurones display reactivity for NADPH-diaphorase, suggests that there is a least a partial overlap with the third major class of striatal interneurones i.e. the population that displays immunoreactivity for somatostatin, neuropeptide Y and possesses NADPH-diaphorase activity (DiFiglia and Aronin, 1982; Takagi et al., 1983; Vincent and Johansson, 1983; Vincent et al., 1983; Smith and Parent, 1986). It is clear that only a small proportion of diaphorase-positive cells are immunoreactive for calbindin, a finding that is not due to technical problems associated with double staining experiments as many diaphorase-positive cells were identified in the same study that were negative for calbindin even when they occurred at superficial levels of the sections where immunostaining for calbindin was at its strongest and most consistent (Bennett and Bolam, 1993b). It is probable therefore that the calbindin-positive neurones represent a subpopulation of the somatostatin/NPY/diaphorase interneurones of the striatum.

SUMMARY

From the data described in the present paper it is evident that there are at least four morphologically distinct populations of neurones in the neostriatum that possess immunoreactivity for different calcium binding proteins. In addition, at least one population of giant neurones are present but these are probably not truly striatal in origin. On morphological grounds it is evident that at least three of the populations are striatal interneurones. In order to understand how information is processed in the striatum it is important to know how these neurones 'fit into' striatal microcircuits. To this end analyses of the PV-positive neurones, in the rat at least, have enabled details of the microcircuitry of these neurones to be partially elucidated. For the other classes of neurones and PV-positive neurones in the primate caudate-putamen the situation is less clear. Although further analyses will enable the afferent inputs to these cells to be established, the fact that there are populations of neurones extrinsic to the striatum that give rise to terminals in the striatum that are positive for the calcium binding proteins, means that the elucidation of the *synaptic output* of calcium binding protein-positive neurones will be more difficult.

Acknowledgements

Thanks are due to C. Francis, E. Norman, F. Kennedy and P. Jays for the technical help with the authors' work described in this paper. We would also like to thank Dr. J. Rogers, University of Cambridge, for the antiserum against calretinin. B.D.B. is supported by an MRC Research Studentship.

REFERENCES

Arbuthnott, G.W., Wright, A.K., Hamilton, M.H., and Brown, J.R., 1982, Orthograde transport of nuclear yellow: a problem and its solution, *J. Neurosci. Meth.* 6:365-368.

Adams, C.E., and Fisher, R.S., 1990, Sources of neostriatal cholecystokinin in the cat, *J. Comp. Neurol.* 292:563-574.

Bennett, B.D., and Bolam, J.P., 1993a, Characterisation of calretinin-immunoreactive structures in the striatum of the rat, *Brain Res.* 609:137-148.

Bennett, B.D., and Bolam, J.P., 1993b, Two populations of calbindin D28k-immunoreactive neurones in the striatum of the rat, *Brain Res.* 610:305-310.

Bolam, J.P., Clarke, D.J., Smith, A.D., and Somogyi, P., 1983, A type of aspiny neuron in the rat neostriatum accumulates (^{3}H)- -aminobutyric acid: combination of Golgi-staining, autoradiography and electron microscopy, *J. Comp. Neurol.* 213:121-134.

Bolam, J.P., Powell, J.P., Wu, J-Y., and Smith, A.D., 1985, Glutamate decarboxylase-immunoreactive structures in the rat neostriatum. A correlated light and electron microscopic study including a combination of Golgi-impregnation with immunocytochemistry, *J. Comp. Neurol.* 237:1-20.

Bolam, J.P., Wainer, B.H., and Smith, A.D., 1984, Characterisation of cholinergic neurons in the rat neostriatum. A combination of choline acetyltransferase immunocytochemistry, Golgi-impregnation and electron microscopy, *Neuroscience* 12:711-718.

Celio, M.R., 1990, Calbindin D-28k and parvalbumin in the rat nervous system, *Neuroscience* 35:375-475.

Chang, H,T, and Kita, H., 1992, Interneurons in the rat striatum: relationships between parvalbumin neurons and cholinergic neurons, *Brain Res.* 574:307-311.

Cowan, R.L., Wilson, C.J., Emson, P.C., and Heizmann, C.W., 1990, Parvalbumin-containing GABAergic interneurons in the rat neostriatum, *J. Comp. Neurol.* 302:197-205.

DiFiglia, M., and Aronin, N., 1982, Ultrastructural features of immunoreactive somatostatin neurons in the rat caudate nucleus, *J. Neurosci.* 2:1267-1274.

DiFiglia, M., Christakos, S., and Aronin, N., 1989, Ultrastructural localization of calbindin-D28k in the rat and monkey basal ganglia, including subcellular distribution with colloidal gold labeling, *J. Comp. Neurol.* 279:653-665.

Freund, T.F., and Gulyás, A.I., 1991, GABAergic interneurons containing calbindin D28k or somatostatin are major targets of GABAergic basal forebrain afferents in the rat neocortex, *J. Comp. Neurol.* 314:187-199.

Gerfen, C.R., 1992, The neostriatal mosaic: multiple levels of compartmental organization, *Trends in Neurosciences* 15:133-139.

Gerfen, C.R., Baimbridge, K.G., and Miller, J.J., 1985, The neostriatal mosaic: compartmental distribution of calcium binding protein and parvalbumin in the basal ganglia of the rat and monkey, *Proc. Natl. Acad. Sci.* 82:8780-8784.

Giménez-Amaya, J.M., and Graybiel, A.M., 1990, Compartmental origins of the striatopallidal projection in the primate, *Neuroscience* 34:111-126.

Graybiel, A.M., 1990, Neurotransmitters and neuromodulators in the basal ganglia, *Trends in Neurosciences* 13:244-254.

Gulyás, A.I., Miettinen, R., Jacobowitz, D.M., and Freund, T.F., 1992, Calretinin is present in non-pyramidal cells of the rat hippocampus-I. A new type of neuron specifically associated with the mossy fibre system, *Neuroscience* 48:1-27.

Gulyás, A.I., Tóth, K., Dános, P., and Freund, T.F., 1991, Subpopulations of GABAergic neurons containing parvalbumin, calbindin D28k, and cholecystokinin in the rat hippocampus, *J. Comp. Neurol.* 312:371-378.

Iversen, L.L., and Schon, F.E., 1973, The use of autoradiographic techniques for the identification and mapping of transmitter-specific neurons in the CNS, *in*: "New Concepts in Neurotransmitter Regulation," A.J. Mandell, ed., Plenum, New York, pp. 153-193.

Jacobowitz, D.M., and Winsky, L., 1991, Immunocytochemical localization of calretinin in the forebrain of the rat, *J. Comp. Neurol.* 304:198-218.

Jones, E.G., and Hendry, S.H.C., 1989, Differential calcium binding protein immunoreactivity distinguishes classes of relay neurons in monkey thalamic nuclei, *Eur. J. Neurosci.* 1:222-246.

Kita, H., 1991, GABAergic interneurones of the neostriatum, *Proc Symp. Chemical Signalling in the Basal Ganglia, Cambridge,* C3.

Kita, H., and Kitai, S.T., 1988, Glutamate decarboxylase immunoreactive neurons in the rat neostriatum: their morphological types and populations, *Brain Res.* 447:346-352.

Kita, H., Kosaka, T., and Heizmann, C.W., 1990, Parvalbumin-immunoreactive neurons in the rat neostriatum: a light and electron microscopic study, *Brain Res.* 536:1-15.

Kiyama, H., Seto-Ohshima, A., and Emson, P.C., 1990, Calbindin D_{28k} as a marker for the degeneration of the striatonigral pathway in Huntington's disease, *Brain Res.* 525:209-214.

Lapper, S.R., Smith, Y., Sadikot, A.F., Parent, A., and Bolam, J.P., 1992, Cortical input to parvalbumin-immunoreactive neurones in the putamen of the squirrel monkey, *Brain Res.* 580:215-224.

Miettinen, R., Gulyás, A.I., Baimbridge, K.G., Jacobowitz, D.M., and Freund, T.F., 1992, Calretinin is present in non-pyramidal cells of the rat hippocampus-II. co-existence with other calcium binding proteins and GABA, *Neuroscience* 48:29-43.

Parmentier, M., 1990, Structure of the human cDNAs and genes coding for calbindin D28k and calretinin, *in*: "Calcium-binding proteins in normal and transformed cells," R. Pochet, D.E.M Lawson, and C.W Heizmann, eds., Plenum, New York, pp. 27-34.

Pennartz, C.M.A., and Kitai, S.T., 1991, Hippocampal inputs to identified neurons in an *in vitro* slice preparation of the rat nucleus accumbens: evidence for feed-forward inhibition, *J. Neurosci.* 11:2838-2847.

Résibois, A., and Rogers, J.H., 1992, Calretinin in rat brain: an immunohistochemical study, *Neuroscience* 46:101-134.

Ribak, C.E., Vaughn, J.E., and Roberts, E., 1979, The GABA neurons and their axon terminals in rat corpus striatum as demonstrated by GAD immunocytochemistry, *J. Comp. Neurol.* 187:261-283.

Roberts, R. C., and DiFiglia M., 1990, Long-term survival of GABA-, enkephalin-, NADPH-diaphorase- and calbindin-d28k-containing neurons in fetal striatal grafts, *Brain Res.* 532:151-159.

Rogers, J.H., 1987, Calretinin: a gene for a novel calcium-binding protein expressed principally by neurons, *J. Cell Biol.* 105:1343-1353.

Rogers, J.H., Khan, M., and Ellis, J., 1990, Calretinin and other CaBPs in the nervous system, *in*: "Calcium-binding proteins in normal and transformed cells," R. Pochet, D.E.M Lawson, and C.W Heizmann, ed., Plenum, New York, pp. 195-203.

Sadikot, A.F., Parent, A., Smith, Y., and Bolam, J.P., 1992, Efferent connections of the centromedian and parafascicular thalamic nuclei in the squirrel monkey: a light and electron microscopic study of the thalamostriatal projection in relation to striatal heterogeneity, *J. Comp. Neurol.* 320:228-242.

Seress, L., Gulyás, A.I., and Freund, T.F., 1991, Parvalbumin- and calbindin D_{28k}-immunoreactive neurons in the hippocampal formation of the macaque monkey, *J. Comp. Neurol.* 313:162-177.

Smith, A.D., and Bolam, J.P., 1990, The neural network of the basal ganglia as revealed by the study of synaptic connections of identified neurons, *Trends Neurosciences* 13:259-265.

Smith, Y., and Bolam, J.P., 1991, Convergence of synaptic inputs from the striatum and the globus pallidus onto identified nigrocollicular cells in the rat: a double anterograde labelling study, *Neuroscience* 44:45-73.

Smith, Y., and Parent, A., 1986, Neuropeptide Y-immunoreactive neurons in the striatum of the cat and monkey: morphological characteristics, intrinsic organization and co-localization with somatostatin, *Brain Res.* 372:241-252.

Staines, W.A., Atmadja, S., and Fibiger, H.C., 1981, Demonstration of a pallidostriatal pathway by retrograde transport of HRP-labelled lectin, *Brain Res.* 206:446-450.

Takagi, H., Mizuta, H., Matsuda, T., Inagaki, S., Tataeishi, K., and Hamaoka Y., 1984, The occurrence of cholecystokinin-like immunoreactive neurons in the rat neostriatum: light and electron microscopic analysis, *Brain Res.* 309:346-349.

Takagi, H., Somogyi, P., Somogyi, J., and Smith, A.D., 1983, Fine structural studies on a type of somatostatin-immunoreactive neuron and its synaptic connections in the rat neostriatum: A correlated light and electron microscopic study, *J. Comp. Neurol.* 214:1-16.

Theirault, E., and Landis, D.M.D., 1987, Morphology of striatal neurons containing VIP-like immunoreactivity, *J. Comp. Neurol.* 256:1-13.

Vincent, S.R., 1992, Histochemistry of endogenous enzymes, *in*: "Experimental Neuroanatomy, A Practical Approach," J.P. Bolam, ed., Oxford University Press, Oxford, pp. 153-171.

Vincent, S.R., and Johansson, O., 1983, Striatal neurons containing both somatostatin- and avian pancreatic polypeptide (APP)-like immunoreactivities and NADPH-diaphorase activity: A light and electron microscopic study, *J. Comp. Neurol.* 217:264-270.

Vincent, S.R., Staines, W.A., and Fibiger, H.C., 1983, Histochemical demonstration of separate populations of somatostatin and cholinergic neurons in the rat striatum, *Neurosci. Lett.* 35:111-144.

Wilson, C.J., Kita, H., and Kawaguchi, Y., 1989, GABAergic interneurones, rather than spiny cell axon collaterals, are responsible for the IPSP responses to afferent stimulation in neostriatal spiny neurons, *Soc. Neurosci. Abst.* 15:907.

MYELOARCHITECTONICS OF THE PRIMATE CAUDATE-PUTAMEN

Bruce Quinn and Ann M. Graybiel

Dept. of Brain and Cognitive Sciences
Massachusetts Institute of Technology
Cambridge, MA 02139

INTRODUCTION

In the early era of modern microscopic neuroanatomy, Gerlach, Flechsig, and Freud all developed exquisitely sensitive stains based on aqueous gold chloride which, due to their capriciousness, were completely succeeded by the methods of Weigert and others. We have returned to a variant of the gold chloride myelin stain, modified from the recent method of Schmued (1990), and show this technique reveals fiber-defined striosomes in the primate caudate-putamen.

HISTORICAL OVERVIEW: THE GOLD CHLORIDE MYELIN STAINS

From the 1870's through the turn of the century, fiber architectonics was an actively investigated field both in neuroanatomy and in neuropathology. While cortical cytoarchitectonic studies of that period are still referenced, such as the atlases of Brodmann, myeloarchitectonic works of the same era are largely forgotten, perhaps because optimal myelin stains have always been more complex and capricious than the straightforward Nissl stain (Beck, 1925; Kemper and Galaburda, 1984; Jones 1984).

Two classes of fiber stains predominate in modern neuropathology: pigment-based stains, such as the Luxol Fast Blue and Weigert stains, and silver-based impregnations, such as the Bielschowsky, Bodian, and Palmgren stains. However, remarkably simple gold chloride solutions can also provide detailed, exquisitely delicate fiber impregnations. The first reference to a simple, aqueous gold chloride fiber stain appears to be that of Gerlach (1870/1872), who briefly notes the recent discovery of this method in his chapter in Stricker's landmark encyclopedia of human and comparative histology. The gold chloride method was also used by Flechsig (1876), who devoted much of his long career to the study of myeloarchitectonics, both through serial developmental studies and in the adult central nervous system. Tantalizingly, Flechsig suggests that the sensitive gold chloride myelin stain would be a major advance for neuroanatomic technique, and that studies in his laboratory were underway to perfect the method; he states that many new studies would now be possible. However, Flechsig seems not to have used the gold chloride method in later publications.

The Basal Ganglia IV, Edited by
G. Percheron *et al.*, Plenum Press, New York, 1994

Sigmund Freud's period in a neuroanatomical laboratory research included both studies on invertebrates and papers on the human nervous system, such as the course of the acoustic pathway (Triarhou and del Cerro, 1985; Shepard, 1991). In 1884 he published three works on a simple gold chloride stain which he described as a modification of a unreliable procedure, given by Flechsig (1876) but later abandoned by him (*"er scheint auf dieser Methode verzichtet zu haben; auch ist mir keine Untersuchung eines Anderen bekannt, welche mit Hülfe derselben ausgeführt worden wäre"*; "Flechsig seems to have abandoned this method, and I know of no investigation by others which has utilized it"; Freud, 1884a) Neither Freud nor Flechsig cite Gerlach's mention of a gold chloride fiber stain (Gerlach, 1870/1872; 1872), which is puzzling since Stricker's handbook was well known and Freud had even worked as a summer student under Stricker (Bernfeld, 1949).

After months of laborious efforts to modify Flechsig's protocol, Freud published the method in a short preliminary paper (Freud, 1884a); a similar, short English version simultaneously appeared in *Brain* (Freud, 1884b). A much more detailed paper appeared later that year in the *Archiv für Anatomie und Physiologie, Anatomische Abteilung* (Freud, 1884c). The protocol Freud arrived at can be described briefly. Tissues were hardened in potassium dichromate or in Erlicki's fluid (2.5% potassium dichromate, 0.5% copper sulfate, in water). For staining, sections are transferred to 1% gold chloride in 25-50% alcohol for several hours, then transferred through sodium hydroxide to 10% potassium iodide, where they develop to a dark rose to magenta hue. Freud's improvement over the methods of Gerlach and Flechsig is the use of alcoholic gold chloride, and the potassium iodide development. The result was that "the fibers are made to show in a pink, deep purple, blue, or even black colour, and are brought distinctly into view, while the grey substance, vessels and neuroglia, lost in the slightly tinged background, are not intruded upon the attention of the observer" (Freud, 1884b). It should be noted that this protocol, based on simple aqueous or alcoholic gold chloride, is very different from the ammoniacal silver stains of Bielschowsky and Fontana, which sometimes used gold as a toning or blackening agent of a silver precipitate.

Freud was delighted in his discovery, and described his excitement and the enthusiasm of his colleagues in his letters to his fiance (Freud, 1960). However, Freud's method disappeared into obscurity, perhaps with changes in fixation methods--it may be dependent on potassium dichromate hardening--or perhaps through intrinsic capriciousness. Upson reviewed the field in 1888 and his opinion was stingingly critical; he states flatly that Freud's method was so unreliable as to be useless. Ironically, 1884 was also the publication year of Weigert's pre-eminently successful myelin stain based on a preparatory mordant bath and hematoxylin, which quickly entered universal use (Weigert, 1884, 1893; Barker, 1899; Spielmeyer, 1914). The intense interest aroused in this era by a usable myelin stain is witnessed in Weigert's acerbic comment that even if he devoted himself fully to replicating all the variants published on his method, he would not have time to test them all (Krücke, 1961).

The interval history of the gold chloride method, from 1888 to the present, can be summarized briefly. The gold chloride central nervous system method appears to have existed independently of a longer-lived gold chloride technique for peripheral nerve-muscle junctions as developed by Cohnheim and Ranvier in the 1870's (for review, Weddell and Zander, 1950); unlike the CNS methods, the peripheral nerve-muscle gold chloride method of Cole is still cited in contemporary textbooks and articles (see Cole, 1964; Silverberg et al., 1989). Citation to the Gerlach/Flechsig/Freud/Upson techniques, so similar to ours, seems to have disappeared near the turn of the century, being absent in compendia of the 1920's and 1930's. For example, Pollack (1905) says that the gold chloride methods of Gerlach and Flechsig have been replaced by the pigment method of Weigert, but that the older methods should not be forgotten entirely (*"so sollen doch die wesentlichsten alten Methoden nicht ganz enttront sein".*)

Given this background, we were intrigued by the gold chloride myelin stain recently reported independently by Schmued (1990). In our hands, we found that his protocol, in which sections are treated with dilute gold chloride in phosphate buffered saline, was improved and made more reliable by the addition of trace hydrogen peroxide (Quinn and Graybiel, 1991a). We found that this modified peroxide-gold myelin impregnation delineates striosomes clearly in select regions of the squirrel monkey caudate-putamen (Quinn and Graybiel, 1991b), as we discuss below.

METHODS AND RESULTS

Experimental Materials

Brain tissue from three male squirrel monkeys was used in this study; the monkeys also served as controls for other experiments in our laboratory. Animals were profoundly anesthetized and perfused with two liters of freshly prepared phosphate buffered 4% paraformaldehyde, postfixed overnight and cryoprotected in 20% glycerol (after Rosene et al., 1986). A few sections were also used from monkey brains prepared similarly but cryoprotected by 20-30% sucrose, with equivalent staining results. Frozen brains were sectioned at 30-40 microns on a sliding microtome and stored in 100 mM PBS, 0.1% thimerosol, at 4° C for up to several weeks before use. To examine possible species differences, sections were also obtained from several male Sprague-Dawley rats (150-200 g) which were processed by a similar protocol, and from routine postmortem autopsy brain (postmortem interval, 16-24 hours) fixed by immersion for two to four weeks.

Gold chloride myelin staining was conducted as described by Schmued (1990), or with the hydrogen peroxide modification we developed (Quinn and Graybiel, 1991a). In the Schmued procedure, sections are incubated for 3-4 hours in 0.2% gold chloride in 20 Mm neutral sodium phosphate buffer (Ph 7.5 in our work) and 0.9% NaCl. We pursued numerous variants of this procedure, systematically adjusting concentration of phosphate buffer, pH, salt concentration, and testing a wide variety of acids, bases, oxidants, and reducing agents for their ability to catalyze the reaction. Our optimized protocol utilizes trace hydrogen peroxide (Quinn and Graybiel, 1991a). Sections are washed for 30 minutes in distilled water and in 20 mM phosphate buffer, pH 7.2-7.4, and 0.6% NaCl. The gold solution is the same buffer-saline with 0.2% gold chloride and .012% hydrogen peroxide. The solution should be pale yellow and may grow more pale over the first 10 minutes. Myelinated pathways should begin to turn gray in 10 to 30 minutes. Total staining time varied between 30 and 90 minutes. Final color varied from dark magenta to cobalt blue. Sections could be examined under a dissecting microscope by transfer in puddles of the same dilute phosphate buffered saline.

As discussed above, the history of the gold chloride myelin stain, as used by Gerlach, Flechsig, Freud, and Upson, is hardly a testament to its reliability. Rarely, in individual vials, the reactants failed to yield myelin staining; this was usually overcome by returning the sections to dilute PBS and beginning again. We found the small modifications we made to the Schmued protocol, involving saline concentration and trace peroxide, made the technique quite reliable for daily use in our laboratory. After staining, sections were rinsed 10 minutes in normal 0.9% saline, rinsed in 5% sodium thiosulfate for 5 minutes, and washed in saline before standard mounting and resin coverslipping.

Selected sections of squirrel monkey brain, adjacent to myelin-stained sections, were immunostained by a standard procedure. Sections were incubated overnight in polyclonal rabbit antiserum to methionine-enkephalin (ENK; gift of Dr. R. Elde, University of Minnesota) and processed by the peroxidase-antiperoxidase method using diaminobenzidine as the chromogen. This procedure defines striosomes by the heterogeneous pattern of methionine-enkephalin staining.

Results

Using the Schmued (1990) protocol exactly as originally published, we obtained acceptable impregnation of the *heavily* myelinated pathways, such as the corpus callosum and the pencils of Wilson in the caudate-putamen. However, we found that under the reaction conditions described in our peroxide-enhanced protocol, the reaction was dramatically enhanced, proceeding faster and with much better reliability. We nearly always successfully stained not only heavily myelinated fibers, but a dense pattern of tangential and horizontal neocortical fibers and a fine, complexly organized feltwork of thin myelinated fibers in the basal ganglia, thalamus, and midbrain. In this report, we focus on results obtained in the caudate-putamen and comparison of staining patterns for myelinated fibers and a neuropeptide, enkephalin (ENK); ENK provides a recognized immunohistochemical standard for the delineation of striosome-matrix compartments.

When sections were stained by our peroxide-enhanced protocol, the extreme capsule, external capsule, internal capsule and pencils of Wilson were strongly stained, appearing dark purple to cobalt blue. Individual fibers were not readily appreciated, but fiber bundles or fascicles of differing orientation were conspicuous. There was a prominent global dorsoventral gradient in the staining of a fine latticework of myelinated fibers in the caudate-putamen (Figure 1). Against this background, zones or annuli formed by more dense fibers were clearly visible and formed oval or irregular elongated shapes. These structures were *most prominent in the dorsolateral putamen and the dorsal cap of the caudate nucleus*. They also displayed variation along the rostrocaudal axis as well, and were usually absent or very subtle in the anterior portion of the caudate-putamen, rostral to the anterior commissure. The dorsolateral myelin-defined figures became most prominent at the level of the globus pallidus. In contrast, ENK-defined striosomes were normally quite prominent in anterior part of the caudate-putamen, and sometimes may diminish in contrast in the caudal portions of the caudate-putamen. At all rostrocaudal levels of the putamen, there was a prominent dorsolateral "rim" defined by pale myelin staining running just medial to the external capsule. This rim was always complemented by a contrasting dorsolateral band of ENK-immunostaining. Wherever there was global overlap between clearly-defined myelin annuli and ENK-defined striosomes, the myelin annuli corresponded closely to striosomes (Figure 1; Quinn and Graybiel, 1991b).

No heterogeneous myelin figures could be identified in sections of rat caudoputamen processed under identical conditions. In preliminary studies in the human, fainter but easily detectable annuli of myelin fibers were seen, most prominent in the putamen and from the mid-length through caudal parts of the caudate-putamen. Further studies will need to characterize the relationship of these human myelin figures to striosomes as characterized by histochemical or immunohistochemical markers. The human myelin figures may represent a way to demonstrate human striosomes in routine autopsy material even after long-term immersion fixation, using only common reagents.

DISCUSSION

As described in our historical introduction, near the turn of the century, the analysis of myeloarchitectonics was considered of importance comparable to the analysis of cortical cytoarchitectonics; myeloarchitectonic maps defining regions of the human cerebrum develop in parallel with cytoarchitonic divisions such as the maps of Brodmann and von Economo.

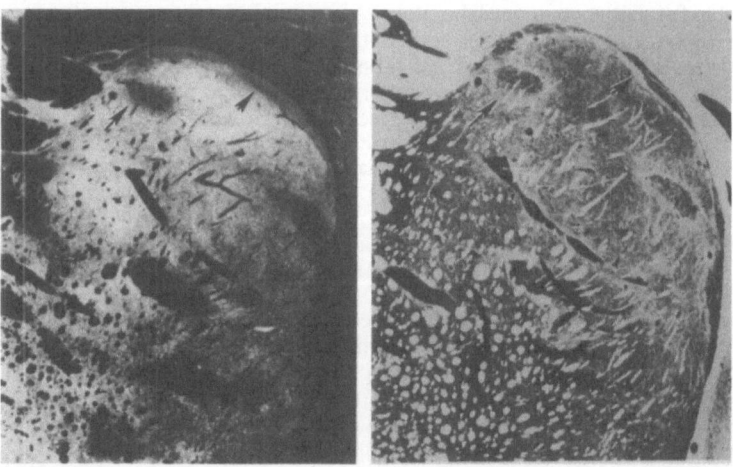

Figure 1. Adjacent sections demonstrate striosomes in the dorsolateral squirrel monkey putamen at a level slightly caudal of the anterior commissure, using direct-projection or reverse contrast photography. On the left, enkephalin-immunohistochemistry defines a number of striosomes as lightly-stained (dark) oval figures. On the right, several complementary strisomes are visualized as annuli of enhanced (white) staining and a less-stained central region (arrow). Note that both stains delineate a striosomal staining dorsolateral rim of the putamen, along the external capsule (arrowhead). In other, more rostral sections, enkephalin-defined striosomes remain prominent but myelin-defined striosomes are absent or nearly occult.

In contrast to the ease of producing adequate neuronal histologic preparations by Nissl techniques with one-bath stains compatible with a variety of embedding techniques, such as cresyl violet, thionin, and alcyan blue, the history of myelin-staining techniques reflects a series of relatively difficult methods. This technologic barrier, the variability of myelin stains, and their incompatibility with some sectioning methods such as paraffin embedding, may have contributed to the eventual preeminence of Nissl stains for neuroarchitectonic studies (for discussion see Beck, 1925). For many decades the major myelin staining methods were variations of the multi-step mordant-hematoxylin approach of Weigert (1884, 1893). In 1953, Klüver and Barrera introduced an alternative dye, Luxol Fast Blue, as a progressive one-bath myelin stain. This procedure, adaptable for both frozen and paraffin-embedded sections, rapidly became popular and was even heralded as one of the most significant advances in neurohistological technology (Margolis and Pickett, 1956). However, it has been noted that many so-called myelin stains *fail* to visualize *most* finely myelinated cortical fibers, particularly in paraffin-embedded material, and this is especially true for Luxol Fast Blue (Braitenberg, 1962; Novotny and Novotny, 1977; Braak, 1980; personal observation). Therefore, the readily executed gold chloride protocol described in this report represents not just a historical procedure but, in our peroxide-catalyzed protocol, comprises a distinctive addition to the present repertoire of research neuroanatomical techniques. Use of peroxide in heavy metal neural stains is very rare (Loots et al., 1977).

To our knowledge, no modern description of the striosome-related myelin figures in the primate caudate-putamen has been made. Similar myelin figures can occasionally be identified in the illustrations of some studies of the human basal ganglia where very sensitive classical myelin stains were used in the hands of expert technicians (one example is found in Vogt and Vogt (1942), their Figure 13). However, they remain occult in nearly all other publications such as the standard myelin-stain human brain atlas of Riley (1960; see also Brockhaus, 1942), where the striatal myelin-based rings are not discerned.

Nearly all modern specific neural markers in the caudate-putamen have been shown to have heterogeneous distributions (for review see Graybiel, 1991), reflecting enzymatic markers, neurotransmitters, neuropeptides, receptors, and efferent and afferent pathways of the neostriatum revealed by modern anterograde and retrograde tracer techniques. Moreover, combined physiologic and histochemical studies reveal the striosome-matrix system as well, as shown by the selective induction of protooncogene markers such as *jun* and *fos* under pharmacologic manipulation of the dopamine system (Graybiel et al., 1990).

Because these neurochemical systems are occult using the routine hematoxylin-eosin, glial, and silver stains in conventional neuropathology, this view of the caudate-putamen as a functionally heterogeneous matrix of interacting systems has only slowly begun to influence investigations of basal ganglia disease in man. There is recent evidence that these systems may play a role in human basal ganglia disease, since striatal neuronal islands seem to be spared in certain cases of Huntington's disease (Vonsattel et al., 1992), whereas induction of striosomal gliosis and neuronal loss (Hedreen and Folstein, 1992) is seen in other cases. There are very rare basal ganglia degenerations which seem to cause patchy or islandic disease (Gibb et al., 1992). In the MPTP-treated primate model of Parkinson's disease, the loss of nigrostriatal dopaminergic terminals may follow a striosome-matrix division as well, because MPTP can induce degeneration of dopaminergic terminals in the matrix compartment, sparing the patch compartment at least early in the course of toxicity (Moratalla et al., 1992). No human studies have yet addressed the possibility of similar differential vulnerability of the two compartments in idiopathic Parkinson's disease. The myelin fiber inhomogeneities of status marmoratus may on occasional be reminiscent of a striosome/matrix mosaic (Vogt and Vogt, 1942; and references therein to C. Vogt's earlier works), even with formation of a dorsolateral rim, but this is not always the case. Global gradients of vulnerability in human disease, such as the distinctive mediolateral gradient in Huntington's disease and the contrasting dorsolateral/lateromedial gradient in multisystem atrophy (Daniel, 1992) are likely to be reflected in natural gradients or regional specializations of striatal subsystems, such as the dorsoventral gradient of fine myelinated fibers and the dorsolateral zone of myelin-defined striosomes we have presented here. The importance of analysis of basal ganglia fiber system in conventional neuropathology has been emphasized occasionally (Oppenheimer, 1984), and the fiber-pattern defined striosomes we defined with the peroxide-gold technique are a robust histochemical marker which may facilitate the analysis of such gradients and the striosome-matrix system of the neostriatum in these diseases.

Acknowledgments

Dr. Quinn was supported by a fellowship from the American Parkinson Foundation (1989/1990). Current address: Division of Neuropathology, UCLA School of Medicine, Los Angeles, California 90024.

REFERENCES

Barker, L.F., 1899, " The Nervous System and its Constituent Neurones," Appleton, New York.

Beck, E., 1925, Zur exaktheit der myeloarchitektonischen Felderung des Cortex cerebri, *J . f. Psychol. Neurol.* 31:281-288.

Bernfeld, S., 1949, Freud's scientific beginnings, *American Imago* 6:162-196.

Braak, H., 1980, "Architectonics of the Human Telencephalic Cortex," Springer Verlag, Berlin.

Braitenberg, V., 1962, A note on myeloarchitecture, *J. Comp. Neurol.* 118:141-151.

Brockhaus, H., 1942, Zur feineren anatomie des Septum und des Striatum, *J. Psychol. Neurol.* 51:1-56.

Cole, W.V., 1964, A gold chloride method for motor end plates, *Stain Techn.* 21:23-25.

Daniel, S.E., 1992, The neuropathology and neurochemistry of multiple system atrophy, *in*: "Autonomic Failure," R. Bannister and C.J. Mathias, eds., Oxford, New York, pp. 564-585.

Flechsig, P., 1876, "Die Leitungsbahnen im Gehirn und Rückenmark des Menschen," Verlag Wilhelm Engelmann, Leipzig.

Freud, E.L., 1960, "The Letters of Sigmund Freud," Basic Books, New York.

Freud, S., 1884a, Eine neue Methode zum Studium des Faserverlaufes im Centralnervensystem, *Zentralblmed. Wiss* . 22:161-163.

Freud, S., 1884b, A new histological method for the study of nerve-tracts in the brain and spinal cord, *Brain* 7:86-88.

Freud, S., 1884c, Eine neue Methode zum Studium des Faserverlaufes im Centralnervensystem, *Archiv. f. Anatu. Physiol. Anat. Abt* .5-6:453-460.

Gerlach, J., 1870/1872, The spinal cord, *in*: "Manual of Human and Comparative Histology, vol. II (trans., H. Power)," S., Stricker, ed., The New Sydenham Society, London, (Original German edition, 1870), pp. 327-366.

Gerlach, J., 1872, Über die struktur der grauen Substanz des menschlichen Grosshirns, *Zentralbl. med. Wiss.* 10:273-275.

Gibb, W.R., Kilford, L., and Marsden, C.D., 1992, Severe generalized dystonia associated with a mosaic pattern of striatal gliosis, *Movement Disorders* 7:217-223.

Graybiel, A.M., Moratalla, R., and Robertson, H.A., 1990, Amphetamine and cocaine induce drug-specific activation of the c-fos gene in striosome-matrix compartments and limbic subdivisions of the striatum, *Proc. Natl. Acad. Sci.* 87:6912-6916.

Graybiel, A.M., 1991, Neurotransmitters and neuromodulators in the basal ganglia, *Trends Neurosciences* 13:244-254.

Hedreen, J.C., and Folstein, S.E., 1992, Early vulnerability of striosomes in Huntington's disease striatum, *Soc. Neurosci. Abst.* 18:168.

Jones, E.G., 1984, History of cortical cytology, *in*: "Cerebral Cortex, vol. 1: Cellular Components of the Cerebral Cortex," A. Peters and E.G. Jones, Plenum Press, New York, pp. 1-32.

Kemper, T.L., and Galaburda, A.M., 1984, Principles of cytoarchitectonics, *in*: "Cerebral Cortex, vol. 1: Cellular Components of the Cerebral Cortex," A. Peters and E.G.Jones, Plenum Press, New York; pp. 35-57.

Krücke, W., 1961, Carl Weigert, *in*: "50 Jahre Neuropathologie in Deutschland: 1885-1935", W. Scholz, ed., Georg Thieme Verlag, Stuttgart, pp. 5-19.

Loots, J.M., Loots, G.P., and Joubert, W.S., 1977, A silver impregnation method for nervous tissue suitable for routine use with mounted sections, *Stain Techn.* 52:85-87.

Moratalla, R., Quinn, B., Delaney, L., Langston, J.W., and Graybiel, A.M., 1992, Differential vulnerability of primate striosome-matrix and caudate-putamen dopamine systems to the neurotoxic effects of MPTP, *Proc. Natl. Acad. Sci.* 89:3859-3863.

Margolis, G., and Pickett, J.P., 1956, New applications of the luxol fast blue myelin stain, *Labor. Invest.* 56:459-474.

Novotny, G.E.K., and Novotny, E., 1977, Triple staining of normal and degenerating nervous tissue, *Stain Techn.* 52:97-99.

Oppenheimer, D.R., 1984, Diseases of the basal ganglia, cerebellum, and motor neurons, *in*: "Greenfield's Neuropathology", J.H. Adams, ed., John Wiley, New York, pp. 699-747.

Pollack, B., 1905, "Die Farbetechnik für das Nervensystems", Verlag S. Karger, Berlin.

Quinn, B., and Graybiel, A.M., 1991a, Characterization of an enhanced one-bath gold chloride impregnation selective for neural fibers in human brain, *J . Neuropath. Exp . Neurol.* 50:329.

Quinn, B., and Graybiel, A.M., 1991b, Fiber architectonics of the striosome-matrix organization of the primate caudate-putamen, *Soc. Neurosci. Abst.* 17:452.

Riley, H.A., 1960, "An Atlas of the Basal Ganglia, Brainstem, and Spinal Cord, based on Myelin-stained Material," Hafer, New York.

Rosene, D.L., Roy, N.J., and Davis, B.J., 1986, A cryoprotection method that facilitate cutting frozen sections of whole monkey brains for histological and histochemical processing without freezing artifact, *J. Histochem. Cytochem.* 34:1301-1315.

Schmued, L.C., 1990, A rapid, sensitive histochemical stain for myelin in frozen brain sections, *J Histochem. Cytochem.* 38:717-720.

Shepard, G.M., 1991, "Foundations of the Neuron Doctrine," Oxford, New York.

Silverberg, K.R., Ogilvy, C.S., and Borges, L.F., 1989, A modified gold chloride technique for optimal impregnation of nerves within corneal whole mounts and dura of the albino rat, *J. Histochem. Cytochem.* 37:269-271.

Spielmeyer, W., 1914, "Technik der mikroskopischen Untersuchung des Nervensystems", Julius Springer, Berlin.

Triarhou, L.C., and del Cerro, M., 1985, Freud's contributions to neuroanatomy, *Arch. Neurol.* 42:282-287.

Upson, H.S., 1888, On gold as a staining agent for nerve tissues, *J. Nerv. Ment. Dis.* 13:685-689.

Vogt, C., and Vogt, O., 1942, Morphologische Gestaltungen unter normalen und pathologischen Bedingungen: ein hirnanatomischer Beitrag zu ihrer Kenntnis, *J. f. Psychol. Neurol.* 50:11-310.

Vonsattel, J.P.G., Myers, R.H., Bird, E.D., Ge, P., and Richardson Jr, E.P., 1992, Maladie de Huntington: sept cas avec îlots néostriataux relativement préservés, *Revue Neurologique* 148:107-116.

Weddell, G., and Zander, E., 1950, A critical evaluation of methods used to demonstrate tissue neural elements, illustrated by references to the cornea, *J. Anat.* 88:168-195.

Weigert, C., 1884, Ausfuhrliche beschriebung der in No. 2 dieser Zeitschrift erwähnten neuen Farbungsmethode für das Centralnervensystem, *Fortschritte der Medizin* 2:190-194.

Weigert, C., 1893, Zur Markscheidenfarbung, *Deutsche Med Wochenschrift* 42:1184-1186.

THREE-DIMENSIONAL MORPHOLOGY OF STRIATAL NEURONS IN RELATION TO COMPARTMENTAL ORGANIZATION OF THE STRIATUM

Jérôme Yelnik, Chantal François and Gérard Percheron

Laboratoire de Neuromorphologie Informationnelle
et de Neurologie Expérimentale du Mouvement
INSERM U106, Hôpital de la Salpêtrière
75651 Paris Cedex 13 France

INTRODUCTION

The existence of a compartmentation of the striatum suggests that this neuronal set consists of different subdivisions, namely the striosomes and matrix (Graybiel and Ragsdale, 1978), which could process separately different types of information. Characteristically, this occurs in tight neuronal set-spaces (see Percheron et al. in this volume) which can constitute either closed or open nuclei according to whether the dendrites of their neurons cross or do not cross their cytoarchitectonic boundaries (Mannen, 1960). In the case of striatal compartmentation, the spatial organization of the dendrites of striatal neurons is crucial since it determines whether striosomes and matrix are informationally closed or open compartments.

In the rat, different combinations of methods were used in order to reveal on the same section the arrangement of the dendrites and the immunohistochemical contours of striosomes: intracellular biocytin and calbindin (Kawaguchi et al., 1989), intracellular HRP and enkephalin (Penny et al., 1988), intracellular lucifer yellow and enkephalin (Arts and Groenewegen, 1992). In cats and ferrets, Bolam et al. (1988) used Golgi impregnation and acetylcholinesterase or substance P revelation. The main conclusion of these studies was that the dendrites of individual neurons remain within their respective compartment i.e. that each compartment would be informationally closed. However, communication between the two compartments was said to be still possible. Penny et al. (1988) observed that the dendrites of large pallidallike aspiny neurons cross through compartmental boundaries and thus that they might serve to provide limited integration between compartments. Later, Kawaguchi (1992) noted that the axons of large aspiny cells arborize mostly in the matrix. On the other hand, Bolam et al. (1988) described two populations of spiny neurons. One, whose dendritic arborizations respect compartmental boundaries, would underlie segregation of information flow in striosomes and the extrastriosomal matrix. The other, whose dendritic arborizations are not influenced by compartmental boundaries, would permit communication between striosomes and matrix. Therefore, two concepts should be opposed. In the first one, spiny dendrites respect compartmental boundaries and the two compartments are closed nuclei which process information separately. Communication between the two compartments would be provided by large aspiny neurons. In the second one, spiny neurons can either respect compartmental boundaries or not, in such a way that the two compartments could function as either an open or a closed nucleus depending on the type of spiny neuron activated.

We have intended a series of studies in order to address this problem in the striatum of primates. The first step was to determine the most appropriate classification for striatal neurons of monkeys and human, which has been done in Yelnik et al. (1991a). A comparison with classifications previously elaborated (e.g. DiFiglia et al., 1976; Braak and Braak, 1982; Chang et al., 1982; Graveland et al., 1985) was also made (Yelnik et al., 1991b). These taxonomic studies confirmed previous results of our group (Yelnik et al., 1984, 1987) in showing that all morphological parameters have not the same significance. Indeed, topological (Percheron, 1982) and metrical (Yelnik et al., 1981) parameters appear to be stable features from which a taxonomic classification can be elaborated whereas geometrical parameters (Yelnik et al., 1983) are mainly influenced by environmental conditions. For this reason, classification and geometrical properties of striatal neurons are treated separately.

CLASSIFICATION OF STRIATAL NEURONS

Our classification was based on quantitative data and statistical arguments. Dendritic arborizations were reconstructed from serial sections and analyzed in three dimensions by using a computer-aided methodology (Yelnik et al., 1981, 1983). The striatum of monkey and human was found to be formed by four different neuronal species.

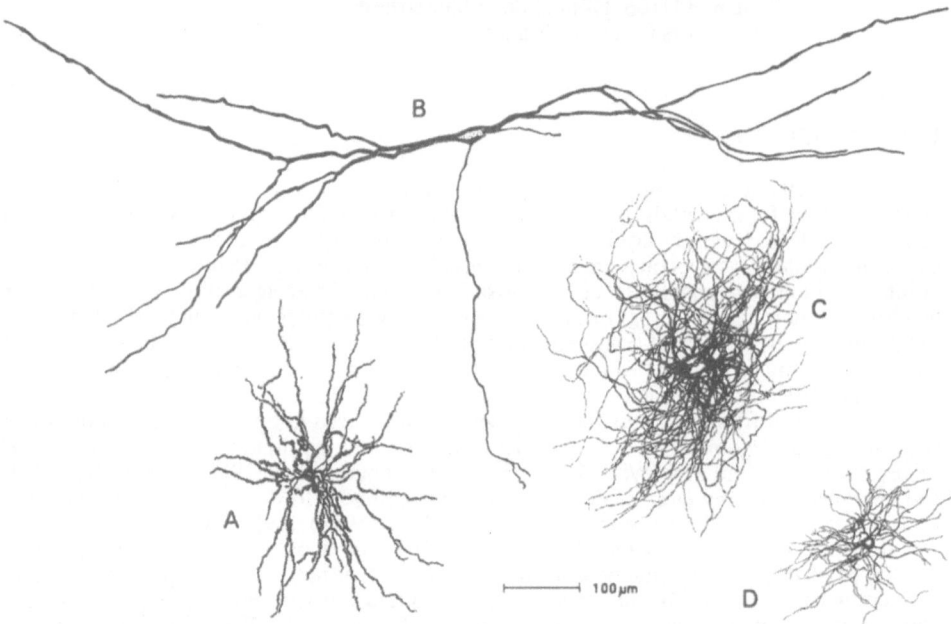

Figure 1. The four neuronal species of the striatum of monkeys (*Papio papio*) shown at the same magnification: (A) Spiny neuron, (B) Leptodendritic neuron, (C) Spidery neuron, (D) Microneuron. Camera lucida drawings from Golgi preparations. Note the hudge dimensions of the dendrites of the leptodendritic neuron.

Spiny Neurons

Spiny neurons (Fig. 1A) constitute 96% of striatal neurons. They are characterized by a high density of dendritic spines which cover the entire dendritic arborization except for the 31 most proximal micra. In contrast to what is commonly assessed, the localization of the first spine is not related to any topological order. Spiny neurons represent an extremely homogeneous neuronal population for which differences can be detected between different animal species. These differences bear on topological parameters but less markedly on metrical parameters. The dendritic formula (S-F) was 6-30 in the rat, 5-35 in the monkey and 6-42 in the human. The longest dendrite (Lm) was 220 μm in the rat for 275 μm in primates. The mean length of dendritic segments (Ln) was 67 μm in the rat, 93 μm in the monkey and 75 μm in human.

Leptodendritic Neurons

Leptodendritic neurons (Fig. 1B) account for 2% of striatal neurons. They are topologically and metrically indistinguishable from the large neurons of the globus pallidus and substantia nigra pars reticulata. They have very low topological parameters, a dendritic formula (S-F) of 4-13, and conversely very high metrical parameters: a mean length of dendritic segments (Ln) of 196 μm, a longest dendrite (Lm) of 900 μm. The geometry of striatal leptodendritic neurons is highly variable in comparison with that of pallidal neurons (Yelnik et al., 1984). While the latter have repeatedly the shape of a flat disc, the former can be either spherical, ellipsoidal, cylindrical or flattened. From that point of view, they resemble more closely the leptodendritic neurons of the substantia nigra (Yelnik et al., 1987) for which we demonstrated an influence of the environment, namely the geometry of the afferent striatopallidonigral bundle (François et al., 1987).

Spidery Neurons

Spidery neurons (Fig. 1C) represent only 1% of all striatal neurons. Nevertheless, they are the cholinergic neurons of the primate striatum and provide for a large part its cholinergic innervation. They have a voluminous and globular cell body which corresponds to the large cells of cytoarchitectonic preparations. They have a large number of dendritic stems (S=12) which ramify into a large number of varicose processes (F=129) which curve back toward the soma. There is often no axon visible but a large number of axonlike processes which could form a local circuitry. The total dendritic length (L=23,390 μm) is very high. Spidery neurons are specific to the striatum of primates since, to our knowledge, similar neurons have never been described elsewhere in the brain. Moreover, other cholinergic neurons, e.g. those of the basal nucleus of Meynert or of the medullary laminae of the lenticular nucleus, exhibit the entirely different and characteristic morphology of leptodendritic neurons.

Microneurons

Microneurons (Fig. 1D) are local circuit neurons which represent 1% of striatal neurons. They have a small cell body and thin dendritic stems (S=6). Each stem gives rise to very thin processes which bear swellings and closely resemble axonal branches. The mean number of tips is quite high (F=64) while the total dendritic length (L=4,680 μm) is low. Microneurons resemble local circuit neurons which can be found in several other cerebral regions (François et al., 1979).

Comparison with other classifications

An unexpected conclusion of this study was that several important differences existed between primate and non primate species. As these differences were not considered before, this led to a very confuse situation in terms of taxonomy. Comparison with other classifications was thus done separately for primate and non primate species. Differences with other classifications in primate bear predominantly on the existence of two types of spiny neurons referred to as spiny type I and II in the classification of DiFiglia et al. (1976). This subdivision was later used by Braak and Braak (1982) but the corresponding images are questionable. Another difference with the present classification is the subdivision of aspiny neurons into three different types. The aspiny type II corresponds unambiguously to spidery neurons (see DiFiglia et al., 1976; DiFiglia, 1987). The aspiny types I and III are very similar and it is very likely that they both correspond to microneurons. The main difference with classifications in non primate species concerns the morphology of cholinergic neurons. While they are spidery neurons in primates, they are leptodendritic neurons in rats and cats (Bolam et al., 1984). This suggests that striatal cholinergic neurons of rats and cats are similar to other cholinergic neurons (e.g. nucleus basalis) while they become a specific neuronal species in primates.

GEOMETRICAL PROPERTIES OF STRIATAL NEURONS

Methodological considerations

It is difficult to obtain a definite proof that the dendrites of a given type of neurons respect the boundaries of striatal compartments. If boundaries are clearly visualized by using immunohistochemical methods, it is not possible to stain large samples of neurons and to reconstruct their arborizations from serial sections. Conversely, if large numbers of neurons that can be reconstructed from serial sections are available, as in Golgi preparations, it is not possible to apply immunohistochemical methods for revealing compartmental boundaries. For this study, we have given priority to the three-dimensional analysis of groups of neighboring dendritic arborizations. The question that we addressed was to determine whether striosome-like structures could be traced on the basis of purely dendroarchitectonic criteria.

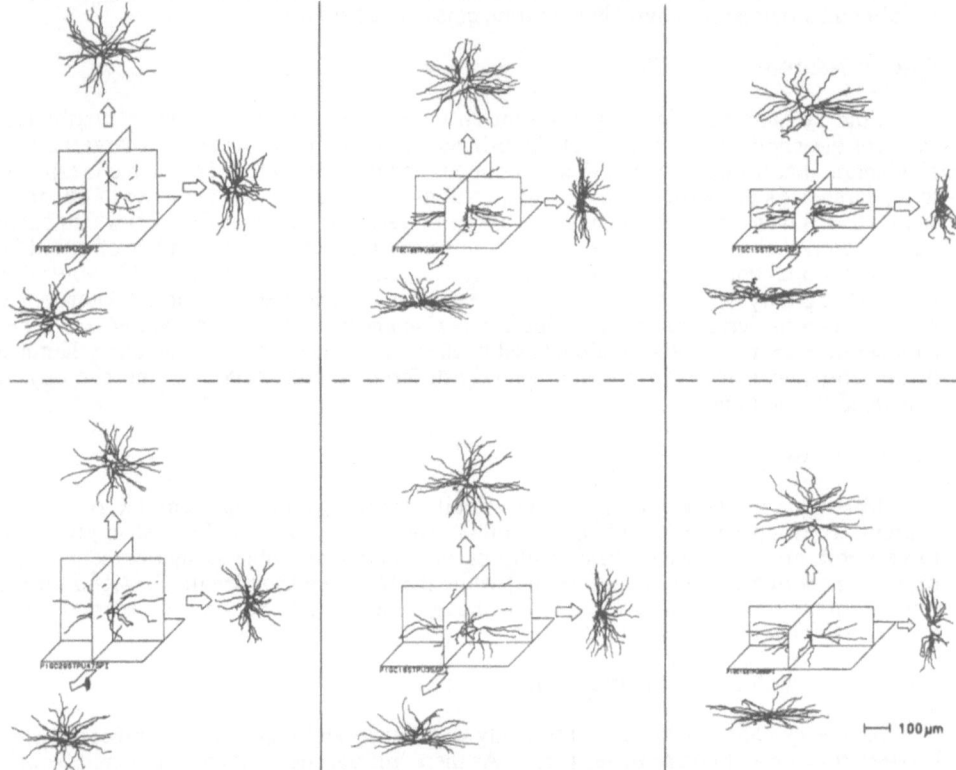

Figure 2. The three different shapes of spiny neurons: spherical (left column), half-spherical (middle column), disk-like (right column). In each case, two neurons are shown. Each neuron is represented in its principal system of axes, in perspective (central drawing) and in its three orthogonal projections. Note that projection in the principal plane (vertical arrows) is always circular while the two other projections (horizontal and oblique arrows) reveal the non sphericity of arborizations.

The geometry of dendritic arborizations of striatal neurons was studied by using the methodology described in Yelnik et al. (1983). Its specificity is that arborizations are reconstructed from serial sections and analyzed by using the statistical method called Principal Component Analysis. By this way, the geometry of arborizations which are larger than the thickness of histological sections, which have non-spherical shapes and which have complex orientations with reference to anatomical planes of section can be described without distortion (e.g. Yelnik et al., 1984; François et al., 1987).

Geometry of spiny neurons

Spiny striatal neurons could be subdivided into three different subgroups depending on the geometry of their dendritic arborization: spherical, half-spherical and disk-like (Fig. 2). These three types of arborizations were all circular in their principal plane but differed in the two other projection planes (Fig. 2). Conversely, some arborizations appeared to be half-spherical or disk-like shaped in a single section but proved to be spherical after reconstruction.

Spherical arborizations were by far the most numerous. They were found in both the caudate nucleus and putamen and in both associative and sensorimotor striatum. Half-spherical and disk-like arborizations were observed in three circumstances. (1) Dendrites of marginal neurons bent along the peripheral contours of the caudate nucleus and putamen that they did not cross. This is the sign that the striatum is a closed nucleus (Mannen, 1960). Dendritic arborizations of marginal spiny neurons were generally half-spherical. (2) Dendrites of striatal neurons did not invade the pencils of radial fibers which cross through the striatum. Therefore, neurons which lay close to a pencil had also a distorted arborization, sometimes disk-like shaped but more often half-spherical. (3) Half-spherical and disk-like arborizations were also observed within the striatum, at distance from both peripheral boundaries and pencils of radial fibers. In this case, they formed groups of neurons which were not disposed randomly but were aligned in forming, together, a kind of dendritic wall.

Such a group of neurons was analyzed in the anterior part of the putamen, close to the anterior commissure, i.e. in the associative striatum. Both spherical and non spherical neurons were selected in order to check the continuity of the dendritic wall. Each neuron was reconstructed separately from serial sections and its relative position within the group was precisely determined in three dimensions. The group, which finally comprised eighteen neurons reconstructed from ten 120 μm thick sections, was then examined through different orientations. It could also be cut into successive sections of different thickness through any desired direction (Fig. 3). By this mean, a dendritic border could be traced that no dendrite crossed through. This border delimited a kind of compartment which was about 500-600 μm large and 1.5 mm long. Spiny neurons of the group could then be separated into two different subgroups, the inside neurons and the outside neurons (Fig. 4). There was no significant differences between these two subgroups. Each group comprised neurons which were influenced by the dendritic border and neurons which were not. This geometrical influence was clearly dependant on neuron-to-border distance, i.e. an arborization was spherical if the cell body was distant from the border by more than half the size of the arborization (about 215 μm). Such striosome-like structures were not observed in the sensorimotor part of the striatum. In this region conversely, dendrites of Golgi-impregnated spiny neurons extend freely in all directions so that neighboring arborizations overlap each other in forming a continuous and homogeneous dendritic lattice. Groups of neurons were reconstructed in order to be sure that they were actually spherical, i.e. circular in their three principal projection planes.

Geometry of the other neuronal species

The geometry of the other striatal neuronal species was much more difficult to analyze than that of spiny neurons. A first reason is that they are far less numerous than spiny neurons (4% vs 96%). Also, they have much more complicated and/or variable morphology so that it is difficult to draw valuable statistical conclusions. Spidery neurons and microneurons have both been observed inside as well as outside striosome-like structures. They had generally spherical or ellipsoidal arborizations, which suggests that they might be little influenced by compartmental boundaries. Leptodendritic neurons had much more variable shapes and have not been observed within striosome-like structures. It seems likely, from the present material, that dendrites of these neurons tend to avoid striosome-like structures, which would be responsible of their geometrical variability.

The morphological characteristics of leptodendritic neurons, spidery neurons and microneurons greatly differ from those of spiny neurons. Computer simulations which took into account both the dendritic characteristics and the density of each neuronal species revealed that they also formed completely different dendritic lattices (see Yelnik et al., 1991a): the dendritic lattices formed by the three former neuronal species were about 50 times looser than the spiny lattice. In addition, the leptodendritic lattice was rather

500 µm

Figure 3. Computer analysis of a group of eighteen Golgi-impregnated neurons (14 spiny neurons and 4 leptodendritic neurons) of the striatum of a baboon. The figure shows the group as it appeared in two consecutive computer sections (thickness 100 µm). In each section a dendritic border was traced by following half-spherical and disk-like arborizations as a dendroarchitectonic criterion. Leptodendritic neurons are recognizable by their long and sparsely ramified dendrites (cell bodies are not represented).

homogeneous, while those formed by spidery neurons and microneurons were discontinuous. In other words, it is likely that portions of striatal tissue contain processes of spidery neurons and microneurons while adjacent portions do not. Therefore, a significant number of spiny dendrites might be unable to receive any innervation from microneurons and spidery neurons. These data suggest that some spiny neurons could receive very few or, possibly, no cholinergic innervation while others would receive a very dense one. This should be a point of future investigations.

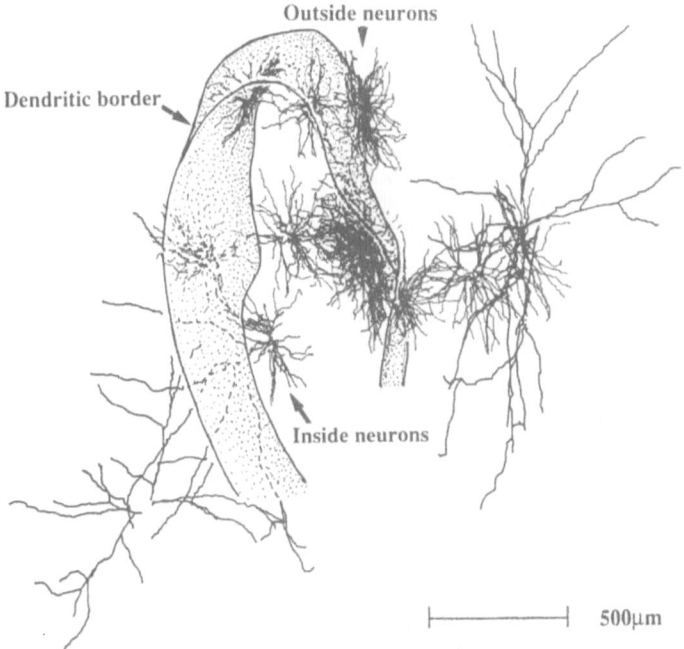

Figure 4. Computer representation of the group of eighteen neurons of figure 3 as it appeared after reconstruction from ten 120 μm thick successive Golgi sections. The three-dimensional aspect of the dendritic border, rendered with dots, was obtained by superposing the 10 successive contours. This border delimites a closed compartment whose dimensions fall within the size range of histochemically-defined striosomes.

CONCLUSIONS

We have been able to trace the border of a closed dendritic compartment which is very likely to correspond to a striosome in view of its dimensions. Since there is a compartmental organization which is built by the spatial arrangement of dendrites, this suggests that striosomes form a compartment which is closed in terms of information processing. Spiny and spidery neurons fall within the size range of the striosomal compartment. Leptodendritic neurons which have much longer dendrites are likely to remain outside. Bending of dendrites forming dendritic borders was not observed in the putamen, which suggests that this could be a pattern specific to the associative striatum.

REFERENCES

Arts, M.P.M., and Groenewegen, H.J., 1992, Relationships of the dendritic arborizations of ventral striatomesencephalic projection neurons with boundaries of striatal compartments. An in vitro intracellular labelling study in the rat, *Europ. J. Neurosci.* 4:574-588.

Bolam, J.P., Izzo, P.N., and Graybiel, A.M., 1988, Cellular substrate of the histochemically defined striosome/matrix system of the caudate nucleus: a combined Golgi and immunocytochemical study in cat and ferret, *Neuroscience* 24:853-875.

Bolam, J.P., Wainer, B.H., and Smith, A.D., 1984, Characterization of cholinergic neurons in the rat neostriatum. A combination of choline acetyltransferase immunocytochemistry, Golgi-impregnation and electron microscopy, *Neuroscience* 12:711-718.

Braak, H., and Braak, E., 1982, Neuronal types in the striatum of man, *Cell Tissue Res.* 227:319-342.

Chang, H.T., Wilson, C.J. and Kitai, S.T., 1982, A Golgi study of the rat neostriatal neurons: light microscopic analysis, *J. Comp. Neurol.* 208:107-126.

DiFiglia, M., 1987, Synaptic organization of cholinergic neurons in the monkey neostriatum, *J. Comp. Neurol.* 255:245-258.

DiFiglia, M., Pasik, P., and Pasik, T., 1976, A Golgi study of neuronal types in the neostriatum of monkeys, *Brain Res.* 114:245-256.

François, C., Percheron, G., Yelnik, J., and Heyner, S., 1979, Demonstration of the existence of local circuit neurons in the Golgi-stained primate substantia nigra, *Brain Res.* 172:160-164.

François, C., Yelnik, J., and Percheron, G., 1987, Golgi study of the primate substantia nigra II. Spatial organization of dendritic arborizations in relation to the cytoarchitectonic boundaries and to the striatonigral bundle, *J. Comp. Neurol.* 265:473-493.

Graveland, G.A., Williams, R.S., and DiFiglia, M., 1985, A Golgi study of the human neostriatum: neurons and afferent fibers, *J. Comp. Neurol.* 34:317-333.

Graybiel, A.M., and Ragsdale, C.W. Jr., 1978, Histochemically distinct compartments in the striatum of human, monkey and cat demonstrated by acetylcholinesterase staining, *Proc. Natl. Acad. Sci. U.S.A.* 75:5723-5726.

Kawaguchi, Y., 1992, Large aspiny cells in the matrix of the rat neostriatum in vitro: physiological identification, relation to the compartments and excitatory postsynaptic currents, *J. Neurophysiol.* 67:1669-1682.

Kawaguchi, Y., Wilson, C.J., and Emson, P.C., 1989, Intracellular recording of identified neostriatal patch and matrix spiny cells in a slice preparation preserving cortical inputs, *J. Neurophysiol.* 62:1052-1068.

Mannen, H., 1960, «Noyau fermé» et «noyau ouvert», *Arch. Ital. Biol.* 98:333-350.

Penny, G.R., Wilson, C.J., and Kitai, S.T., 1988, Relationship of the axonal and dendritic geometry of spiny projection neurons to the compartmental organization of the neostriatum, *J. Comp. Neurol.* 269:275-289.

Percheron, G., 1982, Principles and methods of the graph-theoretical analysis of natural binary arborescences, *J. Theoret. Biol.* 99:509-552.

Percheron, G., François, C., Yelnik, J., and Talbi, B., 1993, The basal ganglia related system (or cortico-baso-thalamo-cortical system) of primates. Definition, description and informational analysis. Affinity for "patchy" patterns, *in*: "The Basal Ganglia IV - New Ideas and Data on Structure and Function", G. Percheron, J.S. McKenzie, and J. Féger, eds, Plenum Press, New York.

Yelnik, J., François, C., Percheron, G., and Heyner, S., 1987, Golgi study of the primate substantia nigra I. Quantitative morphology and typology of nigral neurons, *J. Comp. Neurol.* 265:455-472.

Yelnik, J., François, C., Percheron, G., and Tandé, D., 1991a, Morphological taxonomy of the neurons of the primate striatum, *J. Comp. Neurol.* 313:273-294.

Yelnik, J., François, C., Percheron, G., and Lemonnier, E., 1991b, Cholinergic neurons of the rat and primate striatum are morphologically different, *in*: "Chemical Signalling in the Basal Ganglia", Satellite Symposium of the 14th meeting of the European Neuroscience Association, Babraham, Cambridge, 7-8 septembre 1991, Abstr. p3.

Yelnik, J., Percheron, G., and François, C., 1984, A Golgi analysis of the primate globus pallidus. II. Quantitative morphology and spatial orientation of dendritic arborizations, *J. Comp. Neurol.* 227:200-213.

Yelnik, J., Percheron, G., François, C., and Burnod, Y., 1983, Principal component analysis: A suitable method for the three-dimensional study of the shape, dimensions and orientation of dendritic arborizations, *J. Neurosci. Methods* 9:115-125.

Yelnik, J., Percheron, G., Perbos, J., and François, C., 1981, A computer-aided method for the quantitative analysis of dendritic arborizations reconstructed from serial sections, *J. Neurosci. Methods* 4:347-364.

THE EXTERNAL PALLIDUM AND THE SUBTHALAMIC NUCLEUS SEND CONVERGENT SYNAPTIC INPUTS ONTO SINGLE NEURONES IN THE INTERNAL PALLIDAL SEGMENT IN MONKEY: ANATOMICAL ORGANIZATION AND FUNCTIONAL SIGNIFICANCE

Yoland Smith[1], Thomas Wichmann[2] and Mahlon R. DeLong[2]

[1]Centre de Recherche en Neurobiologie, Hôpital de l'Enfant-Jésus and Université Laval, Québec, CANADA; [2]Department of Neurology, Emory University, Atlanta, GA, USA

INTRODUCTION

The recent introduction of powerful neuroanatomical techniques combined with extracellular recordings of neurones in the basal ganglia of normal animals and experimental models of basal ganglia diseases, have led to the construction of models that specify the hodology, chemistry and relative activity of the interconnected structures in the basal ganglia-thalamocortical circuits (see Albin et al., 1989; Alexander and Crutcher, 1990; DeLong, 1990 for reviews). According to these models, the information flow is transmitted along a "direct" and an "indirect" routes to the output structures of the basal ganglia, i.e. the internal segment of the globus pallidus (GPi) and the substantia nigra pars reticulata (SNr). The "direct" pathway arises largely from the GABA/substance P spiny neurones in the striatum that send an inhibitory projection to GPi and SNr. On the other hand, the "indirect" pathway arises from the GABA/enkephalin spiny neurones in the striatum that send an inhibitory projection to the external pallidum (GPe). The latter gives rise to a GABAergic input to the subthalamic nucleus which in turn sends a glutamatergic excitatory projection to GPe, GPi and SNr. Then, the pallidal and nigral output cells send the information to the thalamocortical neurones in the ventrolateral nucleus of the thalamus and to the tegmental pedunculopontine nucleus. A series of recent anatomical findings suggest that the inhibitory output from the GPe reaches the output structures of the basal ganglia, not only via the intercalated subthalamic nucleus, but also directly (Kitai and Kita, 1987b; Staines, 1988; Smith et al., 1988; Hazrati et al., 1990; Kincaid et al., 1990; Bolam and Smith, 1992; Smith et al., 1992). Furthermore, ultrastructural findings obtained in rodents suggest that the terminals from the globus pallidus (homologue of GPe in primates) occupy a strategic position on the perikaryon and the proximal dendrites of neurones in the entopeduncular nucleus (homologue of GPi in primates) and the SNr (Smith and Bolam, 1990, 1991; Bolam and Smith, 1992). However, because they were thought to exert a minor influence over the activity in GPi and SNr, the GPe-GPi and GPe-SNr projections were not included in the scheme of the basal ganglia circuitry described above (Albin et al., 1989; Alexander and Crutcher, 1990; DeLong, 1990).

In the light of the data obtained in rodents and the preliminary observations made in monkey (Smith et al., 1988; Hazrati et al., 1990), we decided to analyse the pattern of synaptic innervation of GPi neurones by GPe terminals in the squirrel monkey. Furthermore, the pattern of synaptic organization of the GPe terminals onto GPi neurones was compared with that displayed by the terminals from the subthalamic nucleus which is known as the excitatory drive of GPi neurones (Kitai and Kita, 1987a; Robledo and Féger, 1990; Féger et al., 1991; Kita and Kitai, 1991; Hamada and DeLong, 1992b; Kita, 1992).

MATERIALS AND METHODS

Injection of Anterograde Tracers and Preparation of Tissue

Two rhesus monkeys and four squirrel monkeys were used in these experiments. The rhesus monkeys were anesthetized with isofluorane and prepared for extracellular recordings in the subthalamic nucleus. A few days after the surgery, the animals were seated in a primate chair and the area of the subthalamic nucleus was mapped by using a combined injection-recording device (Hamada and DeLong, 1992a). Once the area of the subthalamic nucleus was determined, a volume of 1.8 to 3.0 µl of biocytin (Sigma Chemicals, St Louis, MI, USA; 3% in TRIS buffer 0.05M, pH 7.6) was delivered at different locations to cover the entire extent of the nucleus. The squirrel monkeys were anesthetized with a mixture of ketamine hydrochloride (Ketaset, 70 mg/kg, i.m.) and xylazine (10 mg/kg, i.m.). Then, they were fixed in a stereotaxic frame and received either bilateral injections of biocytin in the subthalamic nucleus (one monkey) or unilateral injection of *Phaseolus vulgaris*-Leucoagglutinin (PHA-L; Vector Laboratories, Burlingame, CA, USA; 2.5% in 0.01M phosphate buffer pH 8.0; three monkeys) in GPe. The PHA-L was loaded in glass micropipettes and delivered according to the injection parameters described in previous communications (Gerfen and Sawchenko, 1984; Smith and Bolam, 1990, 1991).

Following the appropriate survival periods (forty-eight hours for biocytin and ten days for PHA-L), the animals were deeply anesthetized with an overdose of pentobarbital and perfuse-fixed with a mixture of 2% paraformaldehyde and 1% glutaraldehyde in phosphate buffer (0.1M, pH 7.4). After the perfusion, the brains were cut in 7 mm-thick blocks, dissected out from the skull, placed in cold phosphate-buffered saline (PBS, 0.01M, pH 7.4) and sectioned at 60 µm in the transverse plane with a vibrating microtome. The sections were then pre-treated with sodium borohydride before being processed to reveal the anterograde tracers.

Methods to Reveal Biocytin and PHA-L

The biocytin was revealed according to the avidin-biotin complex (ABC) procedure as described previously (Smith and Bolam 1991; Bolam and Smith, 1992). In the material prepared for light microscopy, 0.3% Triton-X-100 was added to the ABC solution whereas the sections processed for electron microscopy were freeze-thawed at -80°C in order to improve the penetration of the ABC through the tissue. The biocytin-labelled structures were located by using 3,3'diaminobenzidine tetrahydrochloride (DAB; Sigma Chemicals Cie, St Louis, MI, USA) as chromogen. Once the biocytin-labelled elements revealed, the sections prepared for light microscopy were mounted onto gelatin-coated slides whereas the sections prepared for electron microscopy were post-fixed in osmium tetroxide, dehydrated in a graded series of alcohol and embedded in resin on microscope slides.

The PHA-L was revealed by immunohistochemical techniques with a primary antiserum raised in goat against PHA-L (Vector Labs, Burlingame, CA, USA; 1:2000), a biotinylated rabbit anti-goat IgG antiserum (Vector Labs; 1:200) and the ABC (Vector Labs; 1:100). The procedures to improve the penetration of the antibodies and the ABC through the tissue processed for light or electron microscopy were similar to those described for the biocytin. The PHA-L-immunoreactive elements were localized with the DAB method. The sections prepared for light or electron microscopic observations were then processed in the same way as were the sections containing the biocytin-labelled structures.

Analysis of the Material and Post-embedding Immunocytochemistry for GABA

The sections containing the biocytin or PHA-L injection sites were examined in the light microscope to determine the extent of the different tracers in their respective targets. The sections prepared for electron microscopic observations that contained the GPi were examined in the light microscope for the presence of close apposition between labelled varicosities and GPi neurones. Regions containing such appositions were photographed, cut out from the slides, re-embedded and cut in serial ultrathin sections on an ultramicrotome (Reichert-Jung Ultracut E). The sections were collected onto pioloform-coated single slot nickel grids, stained with lead citrate and examined in the electron microscope (Phillips 410 or Hitachi H-7100) for the presence of synaptic contacts between labelled terminals and GPi neurones. Some sections containing the GPi were processed for post-embedding immunocytochemistry for GABA using a primary antiserum raised in rabbits. The GABA-containing elements were located by using a gold-conjugated secondary antibody raised in goat (BioCell, Cardiff). The synthesis and purification of the GABA antiserum (Hodgson et al., 1985; Somogyi and Hodgson, 1985; Somogyi et al., 1985) as well as the procedures for the GABA immunostaining (Bolam and Ingham, 1990; Ingham, 1992) have been described in detail elsewhere. The size of the labelled terminals was measured from electron micrographs that were scanned and analysed with a computerized image analysis system.

RESULTS AND DISCUSSION

Labelling in GPe and GPi after Biocytin Injection in the Subthalamic Nucleus

Light Microscopic Observations. After biocytin injections confined to the subthalamic nucleus (Fig. 1A), rich plexuses of labelled axons and varicosities were visualized in both segments of the ipsilateral globus pallidus. The labelled elements aggregated in the form of bands that lay along the medullary lamina. These observations confirm previous data showing that the subthalamopallidal fibres display a band-like pattern in monkey (see Parent, 1990 for a review). Examination of these bands at high magnification showed the existence of two types of labelled varicosities (Fig. 1C). Varicosities of the first category which appeared to be as numerous in GPe as in GPi were relatively small and often associated with the dendritic shafts of pallidal neurones. A second group included large-sized varicosities that were more abundant in GPi than in GPe. These labelled elements formed pericellular baskets around the perikaryon and the proximal dendrites of GPi neurones (Fig. 1C). Although both types of varicosities were found in different regions of the pallidum, it appeared that the second category of labelled structures occurred less frequently in the rostromedial part of GPi. In addition to the labelled axons and varicosities, the GPe also contained biocytin-positive perikarya. It is likely that these neurones have been retrogradely labelled with biocytin, since the GPe is known as a major source of afferents to the subthalamic nucleus (see Parent, 1990 for a review).

Electron Microscopic Observations. In the electron microscope, the biocytin-labelled structures in both segments of the globus pallidus included myelinated and non-myelinated axons as well as two types of axon terminals. The morphological features and the pattern of synaptic innervation of the two categories of biocytin-containing terminals were strikingly different. The "type 1" terminals contained round and pleomorphic vesicles as well as mitochondria (Figs 2, 3B). These terminals abounded in both segments of the globus pallidus. In GPi, the "type 1" boutons had a mean cross-sectional area of $0.41 \pm 0.04~\mu m^2$ (Mean ± S.D.; N= 60) and formed asymmetric synapses with large (44.4%; N= 20) and small (51.1%; N= 23) dendritic shafts as well as spine-like structures (4.5%; N=2) (Fig. 4). In GPe, the morphological features and the pattern of synaptic innervation displayed by the "type 1" terminals were similar to that described in GPi. This population of terminals correspond to the subthalamopallidal boutons identified in other species (Kita and Kitai, 1987; Moriizumi et al., 1987; Okoyama et al., 1987). Furthermore, post-embedding immunocytochemistry for GABA revealed that these terminals did not display GABA immunoreactivity (Fig. 3B). These findings are in keeping with electrophysiological and immunocytochemical data suggesting that the subthalamic neurones exert an excitatory effect on their targets by using glutamate as neurotransmitter (see Féger et al., 1991 for a review). Therefore, it is likely that the "type 1" boutons identified in both GPe and GPi after biocytin injection in the subthalamic nucleus arise from the anterogradely labelled subthalamopallidal axons.

The "type 2" biocytin-labelled boutons were much more abundant in GPi than in GPe. These terminals contained pleomorphic vesicles and numerous mitochondria (Fig. 3A). In GPi, these boutons had a mean cross-sectional area of $0.9 \pm 0.4~\mu m^2$ (Mean ± S.D.; N=51)

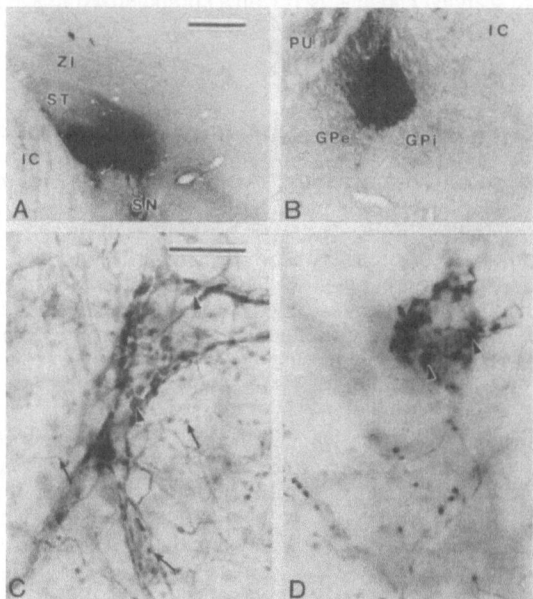

Figure 1. Light micrographs showing the location of a biocytin injection site in the subthalamic nucleus (A) and the resulting labelling in the internal pallidum (C). Note the presence of two types of labelled varicosities in the pallidum. A population of large-sized varicosities (arrowheads in C) is associated with the perikarya of GPi neurones whereas the small-sized varicosities (arrows in C) are often apposed to the surface of dendritic shafts. B shows the location of a PHA-L injection site in the dorsal part of the external pallidum. D illustrates a perikaryon of a GPi neurone tightly surrounded by anterogradely labelled varicosities (arrowheads) after the PHA-L injection site depicted in B. Note that the large-sized biocytin labelled terminals (arrowheads in C) and the PHA-L-immunoreactive terminals (arrowheads in D) display the same pattern of innervation. Scale markers: 500 μm in A (valid for B); 20 μm in C (valid for D). Abbreviations: GPe: external pallidum, GPi: internal pallidum, IC: internal capsule, PU: putamen, SN: substantia nigra, ST: subthalamic nucleus, ZI: zona incerta.

and formed symmetric synapses predominantly with the perikarya (40.7%; N=33) and proximal dendrites (56.7%; N= 46) of pallidal neurones (Fig. 4). Although they were less frequently visualized in GPe, the "type 2" boutons examined in this structure displayed the morphological features and a pattern of synaptic innervation similar to those examined in GPi (Fig. 3C). In many cases, the "type 1" and the "type 2" terminals were found to form synapses onto single GPi neurones (Fig. 2). In contrast to the "type 1" boutons, the post-embedding immunostaining for GABA revealed that the "type 2" terminals display GABA immunoreactivity (Figs 2,3A). On the basis of the neurotransmitter content and ultrastructural

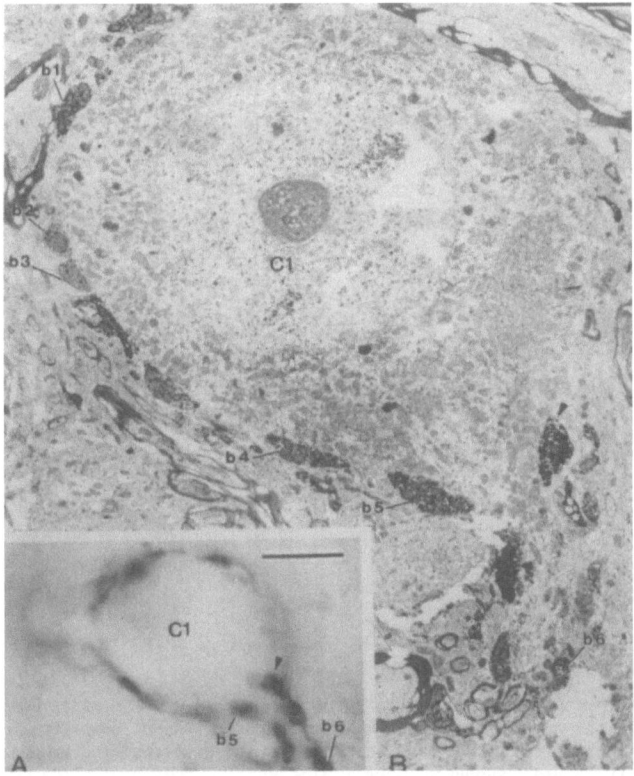

Figure 2. Correlated light (A) and electron (B) micrographs of a GPi neurone (C1) tightly surrounded by numerous biocytin-containing terminals after injection in the subthalamic nucleus. All the biocytin-labelled terminals around the perikaryon of this neurone form symmetric synapses and display the ultrastructural features of the type 2 boutons (b1-b5) whereas a single type 1 biocytin-containing bouton (b6) forms an asymmetric synapse with the proximal dendrite of the same neurone. The boutons b5 and b6 as well as the bouton indicated by an arrowhead can be seen in both the light and the electron micrographs. The ultrathin section (B) has been processed for the postembedding immunogold staining for GABA. The synaptic specialization and the density of gold particles associated with a type 2 (b1) and the type 1 (b6) biocytin-labelled terminals are shown at higher magnification in figure 3 A-B. Scale markers: 10 µm in A, 2 µm in B.

characteristics, we suggest that these "type 2" terminals do not arise from the anterogradely labelled subthalamopallidal axons. Although the exact source of these terminals still remains to be determined, it is reasonable to believe that the GPe is the source of these terminals. In fact, it is possible that the biocytin that has been delivered into the subthalamic nucleus was retrogradely transported along the axons of the pallidosubthalamic neurones and then anterogradely transported to GPi via axon collaterals. This idea is supported by the fact that

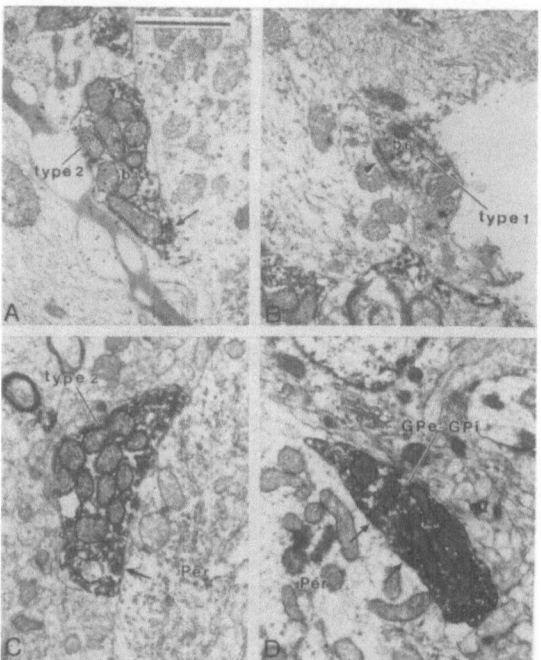

Figure 3. Electron micrographs showing ultrastructural features of the labelled terminals visualized in the pallidum after either biocytin injection in the subthalamic nucleus (A-C) or PHA-L injection in GPe (D). (A) illustrates a type 2 biocytin-containing terminal (b1) that forms a symmetric synapse (arrow) with the perikaryon of the GPi neurone shown in figure 2. This bouton is associated with a large number of gold particles (arrowhead) indicating that it is immunoreactive for GABA. (B) shows a type 1 biocytin-labelled terminal (b6) that forms an asymmetric synapse (arrowhead) with the dendritic shaft of the GPi neurone illustrated in figure 2. This bouton is devoid of gold particles indicating that it is non-immunoreactive for GABA. (C) depicts a type 2 biocytin-containing terminal that forms a symmetric synapse (arrow) with the perikaryon (per) of a GPe neurone. (D) shows a PHA-L-immunoreactive terminal forming a symmetric synapse (arrows) with the perikaryon (per) of a GPi neurone. Note that the type 2 biocytin-labelled terminals (A and C) and the PHA-L-immunoreactive bouton (D) display similar ultrastructural features. Scale marker: 1.0 μm in A (valid for B-D).

retrogradely labelled cells were found in GPe in the present material. Furthermore, the ultrastructural features and neurotransmitter content of the "type 2" biocytin-containing boutons visualized in our study were similar to those of the PHA-L-immunoreactive terminals examined in the rat entopeduncular nucleus (or GPi in primates) after injection in the globus pallidus (Kitai and Kita, 1987b; Bolam and Smith, 1992). The source of the "type 2" terminals in GPe might be the axon collaterals of the retrogradely labelled pallidosubthalamic neurones (Park et al., 1982; Falls et al., 1983; François et al., 1984). In order to resolve this issue, intracelllar injections of pallidosubthalamic neurones with biocytin or neurobiotin combined with the electron microscopic analysis of the labelled boutons must be done.

Labelling in GPi after PHA-L Injection in GPe

Light Microscopic Observations. After PHA-L injection in the external pallidum (Fig. 1B), rich plexuses of anterogradely labelled fibres and varicosities were visualized in the subthalamic nucleus, the substantia nigra, the reticular thalamic nucleus, the striatum and GPi. These findings confirm and extend previous anatomical studies suggesting that the subthalamic nucleus is the major but not the exclusive target of GPe neurones (see Parent, 1990). In GPi, the anterogradely labelled elements were visualized at the same rostrocaudal and dorsoventral levels as the PHA-L injection site. In general, the anterogradely labelled elements were distributed in GPi according to a mediolateral topography. However, the rostromedial third of GPi was almost completely devoid of anterogradely labelled elements and that even when the PHA-L injection site involved the medial part of the middle third of GPe. These findings suggest that the GPi neurones located in the rostromedial sector of the structure do not receive a prominent input from the post-commissural portion of GPe. It is possible that these neurones receive inputs from the pre-commissural GPe which was not injected in our study. A major feature of the GPe-GPi projection is the strict point to point topography, i.e the area of GPi that contained the anterogradely labelled elements after a particular injection in GPe was not much larger than the area occupied by the injection site itself (Fig. 1B).

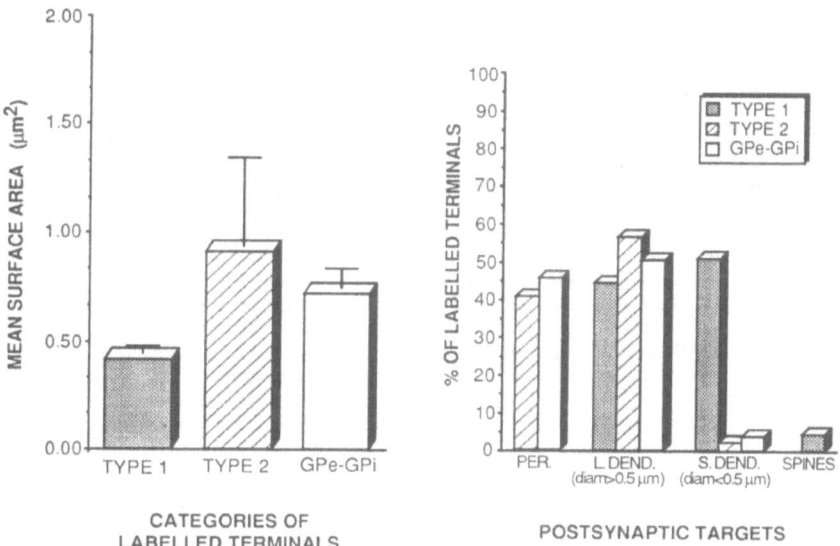

Figure 4. Left side: Histogram comparing the mean surface areas (Mean ± S.D.) of the different categories of labelled terminals visualized in GPi. Note that the type 2 biocytin-labelled boutons and the GPe-GPi terminals have a larger surface area than the type 1 biocytin-containing terminals. Right side: Histogram showing the relative distribution of the postsynaptic targets to the different categories of labelled terminals examined in GPi. Note that the biocytin-labelled type 2 boutons and the GPe-GPi terminals form synapses with the same target structures. However, the type 1 terminals form synapses with large and small dendritic shafts but not with perikarya.

In the light microscope, the anterogradely labelled varicosities observed in GPi were large and formed tight pericellular baskets around the perikarya and proximal dendrites of pallidal neurones (Fig. 1D). In cases in which the intervaricose segment of the axons were visualized, it appeared that single GPe axons gave rise to numerous varicosities surrounding the perikaryon of single GPi neurones.

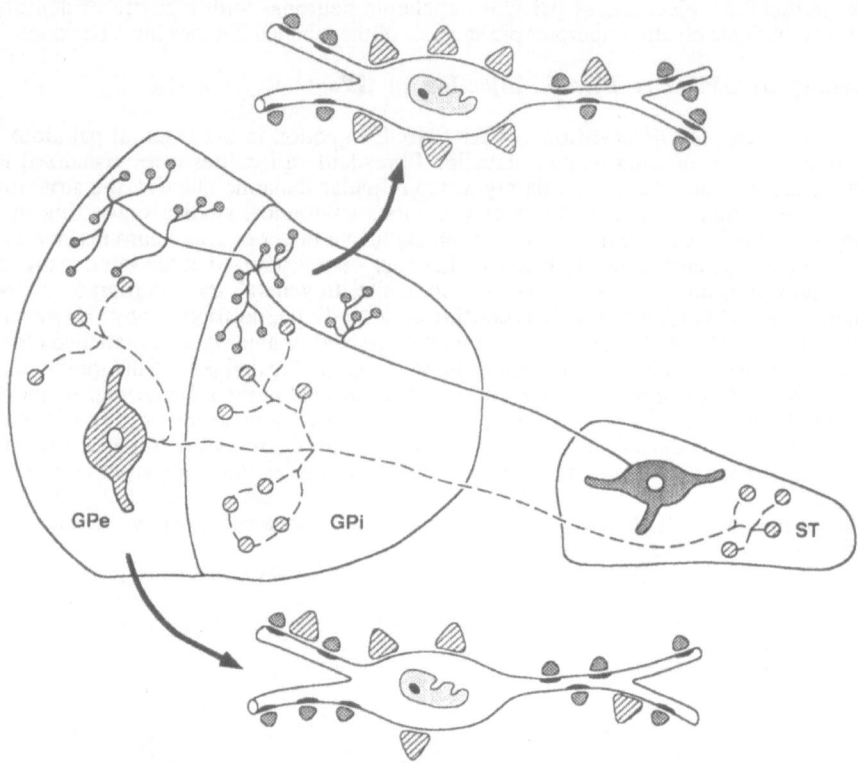

Figure 5. Schematic drawing to summarize the major findings of the present experiment. The neurones in the subthalamic nucleus project to both segments of the pallidum where they give rise to numerous medium-sized terminals (indicated dotted circles). The GPe projects to the GPi via collaterals of the pallidosubthalamic axons (indicated by a dashed line). The axons of the GPe projection neurones give rise to recurrent collaterals. In both pallidal segments, the GPe terminals are indicated by dashed circles. The subthalamic terminals form asymmetric synapses that are evenly distributed along the dendritic shafts of GPe and GPi neurones. In GPi, the large GABAergic GPe terminals form symmetric synapses with the perikarya and proximal dendrites In many cases, the GPe and the subthalamic terminals converge onto single GPi neurones. In GPe, the terminals of the recurrent axon collaterals form symmetric synapses with perikarya and dendritic shafts. Note that the density of GPe terminals around the perikaryon of GPi neurones is higher than that around GPe cells. Abbreviations: GPe: external pallidum; GPi: internal pallidum; ST: subthalamic nucleus.

Electron Microscopic Observations. Under the electron microscope, the PHA-L-immunoreactive varicosities appeared as terminal boutons that formed symmetric synapses predominantly with the perikarya (45.6%; N= 36) and proximal dendrites (50.6%; N= 40) of GPi neurones (Fig. 4). A few labelled terminals were found in contact with distal dendrites (3.8%; N= 3) but none formed synapses with spines or axonal structures. Morphologically, the GPe-GPi terminals were similar to the biocytin-labelled "type 2" boutons identified in GPi after injection in the subthalamic nucleus (compare Fig. 3A and C with Fig. 3D), i.e large-sized (mean cross-sectional area ± S.D. of 0.72 ± 0.09 μm^2; N= 32) containing pleomorphic vesicles and numerous mitochondria (Fig. 3D). The examination of two perikarya of GPi neurones through serial sections revealed that 85-90% of the boutons that formed synapses with these somata were either anterogradely labelled with PHA-L or displayed the morphological features of the GPe-GPi terminals. None of these perikarya were contacted by boutons that formed asymmetric synapses and displayed the morphological features of the type 1 terminals described above. This second series of experiments confirms our suggestion that the labelled type 2 boutons visualized in GPi after biocytin injection in the subthalamic nucleus were in fact GPe-GPi terminals that have been labelled by the retrograde transport of biocytin (King et al., 1989; Izzo, 1991) along the pallidosubthalamic axons.

FUNCTIONAL CONSIDERATIONS

The results of the present study confirm and extend previous data suggesting that the subthalamic nucleus is a major source of excitatory afferents to both segments of the globus pallidus in monkey. Moreover, our findings show that the GABAergic afferents from GPe are located in a rather powerful strategic position to influence the activity of GPi cells. Although the existence of a GPe-GPi connection has already been suggested in primates (Smith et al., 1988; Hazrati et al., 1990), this pathway was thought to exert at best a minor influence over GPi activity. However, the abundance of GPe terminals around the perikarya of GPi neurones suggest that the GPe exerts a more powerful direct inhibitory control on the basal ganglia output structures than previously thought. In fact, because they are located more proximally on GPi neurones than the subthalamic input, the GPe afferents may exert a shunting action which will reduce the excitatory influence generated more distally on the dendritic tree by the subthalamic neurones. In light of these anatomical data, it is necessary to revise the scheme of the basal ganglia connectivity outlined above, in that the inhibitory output from GPe reaches GPi not only via the intercalated subthalamic nucleus, but also directly (Fig. 6).

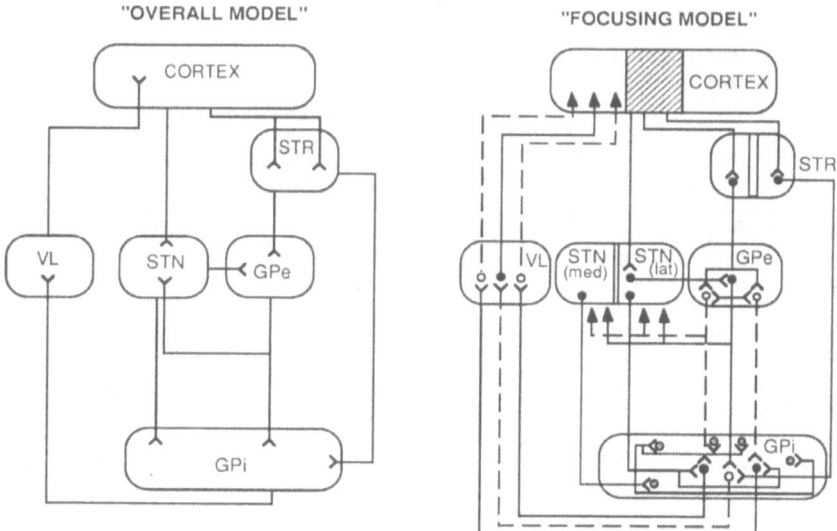

Figure 6. Proposed functional model suggesting that the GPe-GPi projection plays a crucial role in focusing the information facilitating a particular motor act to a restricted population of GPi neurones. On the left side, the model depicts the overall circuitry of the basal ganglia in primates. In the focusing model, the black and dotted circles with the full lines indicate populations of neurones and their axonal projections that become more active following a neuronal impulse from the cerebral cortex. On the other hand, the empty circles and the dashed lines indicate neuronal populations and their axonal projections that decrease their activity following a neuronal impulse from the cerebral cortex. The explanations of the focusing model are given in the section of the manuscript entitled "functional considerations".

Electrophysiological experiments have demonstrated that the information flow through the basal ganglia is a disinhibitory process by which neurones in the striatum inhibit tonically active neurones in the GPi/SNr, which in turn disinhibit neurones in the output stations of the basal ganglia; i.e. the superior colliculus, the thalamus and the pedunculopontine tegmental nucleus (see Chevalier and Deniau, 1990 for a review). However, it is important to remember that in addition to a conspicuous inhibitory influence, striatal stimulation results in the increased firing of a substantial proportion of GPi (Tremblay and Filion, 1989) and SNr (Chevalier and Deniau, 1990) neurones. Although various hypothesis have been suggested to explain this excitatory influence of the striatum on its target structures (see the discussion of Tremblay and Filion, 1989), we believe that the massive and powerful direct connection from GPe to GPi is the major source of these excitatory effects in the primate pallidum. It is possible that striatal stimulation inhibits tonically active neurones in GPe which would then disinhibit GPi neurones. However, the question that remains to be answered is the functional significance of this dual effect of striatal stimulation on pallidal neurones. In this regard, the electrophysiological findings obtained by Chevalier and Deniau (1990) in rodent SNr or Tremblay and Filion (1989) in monkey GPi demonstrate that when a pool of SNr or GPi neurones is silenced by striatal stimulation, the activity of the neighbouring cells increase in parallel. These observations suggest that the striatofugal inhibition is focused onto a small population of pallidal and nigral neurones with excitation of peripheral cells. Although there are still many issues that remain to be resolved about the microcircuitry of the GPe-GPi projection before being able to confirm its exact role in the functional circuitry of the basal ganglia, we suggest a model in which this projection plays a crucial role in the control of voluntary movements.

This model demands that the cortically initiated neuronal impulses related to the ongoing movements are transmitted along the striato-GPi and the subthalamo-GPe-GPi pathways, whereas the neuronal impulses controlling the antagonistic movements are transmitted along the GPe-GPi or subthalamo-GPi pathways. In this model, the information flowing along the so-called "direct" and "indirect" pathways must reach different sets of GPi neurones. For example, if a fast flexion of the arm must be done, a neuronal impulse from the arm area of the primary motor cortex related to the movement-facilitating activity reaches a population of striatal neurones projecting directly to GPi. Simultaneously, the movement-facilitating cortical information is sent to the dorsolateral part of the subthalamic nucleus where the sensorimotor neurones are located (DeLong et al., 1985). A selective pool of arm-related GPi neurones are then inhibited by both the direct striatal input and the GPe afferents which became activated by the excitatory drive from the arm-related neurones of the subthalamic nucleus. This decrease in activity of GPi neurones leads to the disinhibition of a population of movement-facilitating thalamic neurones which ultimately results in the activation of the prime mover muscles and the execution of the desired movement. In parallel, the neuronal impulses related to the antagonistic movement reach an adjacent but separate population of striatal neurones projecting to GPe (Flaherty and Graybiel, 1991, 1992). The activation of the inhibitory striato-GPe pathway leads to the disinhibition of a population of GPi neurones that closely surround those related to the facilitation of the ongoing movement. The activation of these GPi cells is further augmented by the excitatory input from neurones in the subthalamic nucleus related to the antagonistic movements. The disinhibited GPi neurones increase their inhibitory effect on a population of thalamocortical cells which results in the suppression of antagonistic movements. In addition, neurones located more medially in the subthalamic nucleus (Parent et al., 1989) send an excitatory input to GPi cells unrelated to the ongoing movement. This influence of the subthalamic nucleus on GPi increases the inhibitory effect of GPi neurones on thalamocortical cells, which suppresses unwanted movements from other body parts. With this model, the focusing for the intended movement is relatively narrow in comparison to the peripheral zone related to both the antagonistic and the unwanted movements. This suggestion is in keeping with electrophysiological data showing that a large proportion of GPi neurones increase their firing rate during fast voluntary movement (Mitchell et al., 1987). In Parkinson's disease, where the striato-GPe pathway becomes hyperactive and the striato-GPi pathway becomes hypoactive (see Gerfen, 1992 for a review), it is likely that the GPi neurones related to the ongoing movement are disinhibited like those in the peripheral zone. Therefore, the thalamocortical neurones become over-inhibited which ultimately results in the increase of stiffness and suppression of voluntary movements. On the other hand, in dyskinesia, the loss of the excitatory subthalamic influence on the peripheral GPi neurones and the central GPe neurones lead to an extended zone of disinhibition in the thalamus which results in decreased stiffness and appearance of involuntary movements. In addition to the spatial organization of the different afferents innervating GPi neurones, this model also relies on the temporal interplay between the inhibitory and excitatory influences reaching single GPi cells. The balance between these inputs is necessary for adequate "scaling" of movements (see Horak and Anderson, 1984).

In conclusion, the results of our study suggest that the GPe may play a crucial role in controlling the activity of GPi neurones in primates. Although the exact role of the GPe-GPi projection remains to be determined, the anatomical organization of this projection suggests that it is involved in focusing of information related to the intended motor act onto a restricted population of GPi neurones.

Acknowledgements

The authors would like to thank D. Lévesque and L.H. Rowland for technical assistance. Thanks are also due to Dr Peter Somogyi for the generous gift of his GABA antiserum. We also thank Dr Michel Filion for his helpful comments during the preparation of the manuscript. We acknowledge Mrs Isabelle Deaudelin and Louise Bertrand for the quantitative analysis and preparation of the schematic illustration. The department of pathology of the Énfant-Jésus hospital is acknowledged for the use of the Hitachi electron microscope. This research was supported by a NIH grant to M.R. Delong and a grant from the Medical Research Council of Canada (MRC) to Y. Smith. Y. Smith holds a scholarship from the MRC.

REFERENCES

Albin, R.L., Young, A.B., and Penney, J.B., 1989, The functional anatomy of basal ganglia disorders, *TINS* 12:366-375.

Alexander, G.E., and Crutcher, M.D., 1990, Functional architecture of basal ganglia circuits: neural substrates of parallel processing, *TINS* 13:266-271.

Bolam, J.P., and Ingham, C.A., 1990, Combined morphological and histochemical techniques for the study of neuronal microcircuits, *in*: "Handbook of Chemical Neuroanatomy, Analysis of Neuronal Microcircuits and Synaptic Interactions," A. Björklund, T Hökfelt, F.G. Wouterlood and A.N. van den Pol, eds., Elsevier, Amsterdam, pp. 125-198.

Bolam, J.P., and Smith, Y., 1992, The striatum and the globus pallidus send convergent synaptic inputs onto single cells in the entopeduncular nucleus of the rat: A double anterograde labelling study combined with postembedding immunocytochemistry for GABA, *J. Comp. Neurol.* 321:456-476.

Chevalier, G., and Deniau, J.M., 1990, Disinhibition as a basic process in the expression of striatal functions, *TINS* 13:277-280.

DeLong, M.R., 1990, Primate models of movement disorders of basal ganglia origin, *TINS* 13:281-285.

DeLong, M.R., Crutcher, M.D., and Georgopoulos, A.P., 1985, Primate globus pallidus and subthalamic nucleus: functional organization, *J. Neurophysiol.* 53:530-543.

Falls, W.M., Park, M.R., and Kitai, S.T., 1983, An intracellular HRP study of the rat globus pallidus. II. Fine structural characteristics and synaptic connections of medially located large GP neurons, *J. Comp. Neurol.* 220:229-245.

Féger, J., Robledo, P., and Renwart, N., 1991, The subthalamic nucleus: new data, new questions, *in*: "The Basal Ganglia III," G. Bernardi, M.B. Carpenter, G. DiChiara, M. Morelli and P. Stanzione, eds., Plenum Press, New York, pp. 99-108.

Flaherty, A.W., and Graybiel, A.M., 1991, Corticostriatal transformations in the primate somatosensory system. Projections from physiologically mapped body-part representations, *J. Neurophysiol.* 66:1249-1263.

Flaherty, A.W., and Graybiel, A.M., 1992, Multiple stages of sensorimotor processing in the primate basal ganglia, *Proceedings of the IVth International Basal Ganglia Society meeting*, p. 29.

François, C., Percheron, G., Yelnik, J., and Heyner, S., 1984, A Golgi analysis of the primate globus pallidus. I. Inconstant processes of large neurons, other neuronal types, and afferent axons, *J. Comp. Neurol.* 227:182-199.

Gerfen, C.R., 1992, The neostriatal mosaic: Multiple levels of compartmental organization in the basal ganglia, *Ann. Rev. Neurosci.* 15:285-320.

Gerfen C.R., and Sawchenko, P.E., 1984, An anterograde neuroanatomical tracing method that shows the detailed morphology of neurons, their axons and terminals: Immunohistochemical localization of an axonally transported plant lectin, *Phaseolus vulgaris* -leucoagglutinin (PHA-L), *Brain Res.* 290:219-238.

Hamada, I., and DeLong, M.R., 1992a, Excitotoxic lesions of the primate subthalamic nucleus result in transient dyskinesias of the contralateral limbs, *J. Neurophysiol.* 68:1850-1858.

Hamada, I., and DeLong, M.R., 1992b, Excitotoxic lesions of the primate subthalamic nucleus result in reduced pallidal neuronal activity during active holding, *J. Neurophysiol*. 68:1859-1866.

Hazrati, L-N., Parent, A., Mitchell, S., and Haber, S.N., 1990, Evidence for interconnections between the two segments of the globus pallidus in primates: a PHA-L anterograde tracing study, *Brain Res*. 533:171-175.

Hodgson, A.J., Penke, B., Erdei, A., Chubb, I.W., and Somogyi, P., 1985, Antisera to g-aminobutyric acid. I. Production and characterization using a new model system, *J. Histochem. Cytochem*. 33:229-239.

Horak, F.B., and Anderson, M.E., 1984, Influence of globus pallidus on arm movements in monkeys. I. Effects of kainic acid-induced lesions, *J. Neurophysiol*., 52:290-304.

Ingham, C.A., 1992, Immunocytochemistry II: Post-embedding staining, *in*: "Experimental Neuroanatomy: A Practical Approach," J.P. Bolam, ed., Oxford University Press, Oxford, pp.129-151.

Izzo, P.N., 1991, A note on the use of biocytin in anterograde tracing studies in the central nervous system: application at both light and electron microscopic level, *J. Neurosci. Methods* 36:155-166.

Kincaid, A.E., Newman, S.W., Young, A.B., and Penney, J.B., 1990, Evidence for a projection from the globus pallidus to the entopeduncular nucleus in the rat, *Neurosci. Lett*. 128:121-125.

King, M.A., Louis, P.M., Hunter, B.E., and Walker, D.W., 1989, Biocytin: a versatile neuroanatomical tract-tracing alternative, *Brain Res*. 497:361-367.

Kita, H., 1992, Responses of globus pallidus neurons to cortical stimulation: intracellular study in the rat, *Brain Res*. 589:84-90.

Kita, H., and Kitai, S.T., 1987, Efferent projections of the subthalamic nucleus in the rat: Light and electron microscopic analysis with the PHA-L method, *J. Comp. Neurol*. 260:435-452.

Kita, H., and Kitai, S.T., 1991, Intracellular study of rat globus pallidus neurons: membrane properties and responses to neostriatal, subthalamic and nigral stimulation, *Brain Res*. 564:296-305.

Kitai, S.T., and Kita, H., 1987a, Anatomy and physiology of the subthalamic nucleus: A driving force of the basal ganglia, *in*: "The Basal Ganglia II- Structure and Function: Current Concepts, Advances in Behavioral Biology," M.B. Carpenter and A. Jayaraman, eds., Plenum Press, New York, pp. 357-373.

Kitai, S.T., and Kita, H., 1987b, Dual striatonigral inhibitory actions, *Proc. International Conference Neural Mechanisms of Disorders of Movement*, p. 21.

Mitchell, S.J., Richardson, R.T., Baker, F.H., and DeLong, M.R., 1987, The primate globus pallidus: neuronal activity related to direction of movement, *Exp. Brain Res*. 68:491-505.

Moriizumi, T., Nakamura, Y., Okoyama, S., and Kitao, Y., 1987, Synaptic organization of the cat entopeduncular nucleus with special reference to the relationship between the afferents and entopedunculo-thalamic projection neurons: An electron microscope study by a combined degeneration and horseradish peroxidase tracing technique, *Neuroscience* 20:797-816.

Okoyama, S., Nakamura, Y., Moriizumi, T., and Kitao, Y., 1987, Electron microscopic analysis of the organization of the globus pallidus in the cat, *J. Comp. Neurol*. 265: 323-331.

Parent, A., 1990, Extrinsic connections of the basal ganglia, *TINS* 13:254-258.

Parent, A., Smith, Y., Filion, M., and Dumas, J., 1989, Distinct afferents to internal and external pallidal segments in the squirrel monkey, *Neurosci. Lett*. 96:140-144.

Park, M.R., Falls, W.M., and Kitai, S.T., 1982, An intracellular HRP study of the rat globus pallidus. I. Responses and light microscopic analysis, *J. Comp. Neurol*. 211: 284-294.

Robledo, P., and Féger, J., 1990, Excitatory influence of rat subthalamic nucleus to substantia nigra pars reticulata and pallidal complex: electrophysiological data, *Brain Res*. 518:47-54.

Smith, Y., and Bolam, J.P., 1990, The output neurones and the dopaminergic neurones of the substantia nigra receive a GABA-containing input from the globus pallidus in the rat, *J. Comp. Neurol*. 296:47-64.

Smith, Y., and Bolam, J.P., 1991, Convergence of synaptic inputs from the striatum and the globus pallidus onto identified nigrocollicular cells in the rat: a double anterograde labelling study, *Neuroscience* 44:45-73.

Smith, Y., Parent, A., and Dumas, J., 1988, Organization of efferent connections of the two pallidal segments in primate as revealed by PHA-L anterograde tracing method, *Soc. for Neurosci*. 14:719.

Smith, Y., Wichmann, T., and DeLong, M.R., 1992, Synaptic innervation of the globus pallidus by the subthalamic nucleus in monkey, *Proceedings of the IVth International Basal Ganglia Society meeting*, p. 73.

Somogyi, P., and Hodgson, A.J., 1985, Antisera to g-aminobutyric acid. III. Demonstration of GABA in Golgi-impregnated neurons and in conventional electron microscopic sections of cat striate cortex, *J. Histochem. Cytochem*. 33:249-257.

Somogyi, P., Hodgson, A.J., Chubb, I.W., Penke, B., and Erdei, A., 1985, Antisera to g-aminobutyric acid. II. Immunocytochemical application to the central nervous system, *J. Histochem. Cytochem*. 33:240-248.

Staines, W.A., 1988, PHA-L studies of the efferent connections of the rat globus pallidus, *Soc Neurosci. Abs*. 16:427.

Tremblay, L., and Filion, M., 1989, Responses of pallidal neurons to striatal stimulation in intact waking monkeys, *Brain Res*. 498:1-16.

FUNCTIONAL ARCHITECTURE OF THE RODENT SUBSTANTIA NIGRA PARS RETICULATA: EVIDENCE FOR SEGREGATED CHANNELS

Jean-Michel Deniau and Gilles Chevalier

Laboratoire des Communications et Régulations Cellulaires,
URA 1199, Université Pierre et Marie Curie, 4 Place Jussieu
75230 Paris (France)

While it is widely accepted that basal ganglia are involved in motor processes, the precise role of this system in motor functions is still unclear. Since the basal ganglia receive afferents from all the major functional sectors of the cortical mantle, it has been suggested that they constitute a suitable system to integrate motivation, cognition and perception for the elaboration of movement. A pertinent question however is to know exactly how the various cortical products are processed by the basal ganglia. Until recent past years, current views on the functional organization of basal ganglia were strongly influenced by their apparent funnel-like structure. Due to the reduction in nuclear volume along the cortico-striato-nigral and striato-pallidal circuits, the basal ganglia were regarded as an integrative device where the multiple cortical processings are compiled for the elaboration of a motor act. However, cumulative evidences in primate for a parallel arrangement of the transbasal ganglia pathways has led to the idea that basal ganglia maintain rather than mingle the functional heterogeneity of the cortical mantle (Alexander et al., 1986; Alexander and Crutcher, 1990). It is in fact well documented, in a variety of mammalian species, that the cortical functional mosaic is orderly mapped onto the striatal nuclear complex. While the sensorimotor cortices innervate predominantly the putamen (Kunzle, 1977; Flaherty and Graybiel, 1991), the associative frontal, parietal and temporal areas project mostly to the caudate (Goldman and Nauta, 1977; Yeterian and Pandya, 1991) and the allocortex to the accumbens (McGeorge and Faull, 1989; Berendse et al., 1992). Moreover, within each of these cortical representations there is further subdivision with each individual component projecting to a separate district of the main projection field (McGeorge and Faull, 1989; Berendse et al., 1992). In the sensorimotor division for example, the cortical head area projects to the ventral putamen and the limbs area, more dorsally. Besides the basic parallel arrangement, convergences have also been observed. As originally disclosed by Yeterian and Van Hoesen (1978) convergences concern mostly the cortical territories which are functionally related. The somatosensory and somatomotor areas for instance or the visual and cingulate areas which are known to be linked through direct cortico-cortical connections, project to a same striatal region. A summary of the way in which the functional cortical mosaic is mapped onto the rat striatal nuclear complex is given in figure 1.

Is such a segregation of cortical signals further maintained downstream the striatal processing stage? Despite a body of evidence that the striato-nigral and striato-pallidal pathways may continue the topologic arrangement of the cortico-striate projections, the way in which the striatal mosaic is further preserved in the basal ganglia outways is still poorly defined. In the present paper we will review our recent efforts to clarify this question in rodent.

The Basal Ganglia IV, Edited by
G. Percheron *et al.*, Plenum Press, New York, 1994

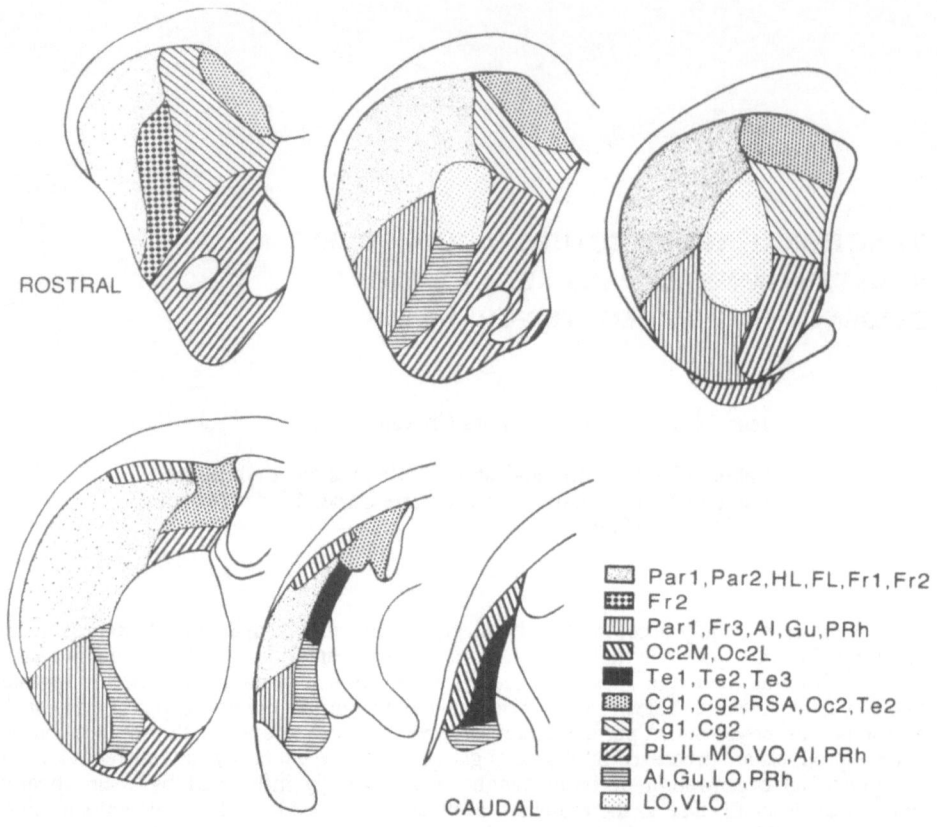

ROSTRAL

CAUDAL

▦	Par1,Par2,HL,FL,Fr1,Fr2
▦	Fr2
▥	Par1,Fr3,AI,Gu,PRh
▨	Oc2M,Oc2L
■	Te1,Te2,Te3
▦	Cg1,Cg2,RSA,Oc2,Te2
▧	Cg1,Cg2
▨	PL,IL,MO,VO,AI,PRh
▤	AI,Gu,LO,PRh
▦	LO,VLO

Figure 1. Representation of the cortical mosaic in striatum. The diagrams are equally spaced from rostral to caudal levels.
AI: agranular insular cortex; Cg1, Cg2: cingulate cortex area 1, area 2; FL: forelimb area; HL: hindlimb area; Fr1, Fr2, Fr3 : frontal areas 1, 2, 3; Gu : gustatory cortex; IL : infralimbic area; LO: lateral orbital area; MO: medial orbital area; Oc2: occipital cortical area 2; Oc2M, Oc2L: occipital area 2 medial, lateral parts; Par1, Par2 : parietal area 1, parietal area 2; PL : prelimbic area; PRh: perirhinal area; RSA : retrosplenial area; Te1, Te2, Te3 : temporal cortices, areas 1, 2, 3; VLO: ventrolateral orbital area; VO: ventral orbital area.

THE NIGRAL MOSAIC

In rodent, the substantia nigra pars reticulata and its small satellite the pars lateralis, are classically recognized as major relay stations for striatal outflow. Altogether, these nigral subdivisions are recipient zones for efferents of the caudate, putamen and nucleus accumbens. In turn via their ascending and descending projections, they open the striatum onto cortical motor, limbic and prefrontal areas as well as midbrain motor centers. Previous anatomical studies have provided evidence that striato-nigral projections are topologically ordered. As originally described by axonal transport of amino acids, parts of the striatum project to the substantia nigra under the form of longitudinal bands whose topographical organization is such that the mediolateral coordinates of striatum are roughly preserved in nigra and the dorsoventral striatal axis is reversed (Domesick, 1977; Nauta et al., 1978; Gerfen, 1985). Despite these descriptions the functional organization remained poorly defined. Since in these studies the tracers were generally applied irrespective of the detailed functional compartmentation of the striatum, the nigral representation of the striatal mosaic remained to be disclosed. So the nigral compartmentation was usually based on a crude dichotomy into a lateral sensorimotor division, innervated preferentially by the putamen and a medial one, devoted to prefronto-associative processes and innervated by the caudate (Ilinsky et al., 1985).

Injections of the lectin wheat germ agglutinin-peroxidase complex, restricted to defined divisions of striatum, clearly show that the striatal mosaic is mapped in nigra under the form of laminae. This lamellar architecture gives the nigra an onion like aspect. Within the lamellar nigral network each striatal district occupies a position conformable to the well known topographic rules in the striato-nigral connection. As documented in figure 2, while in the more ventral lamina are represented districts located dorsal in striatum (e.g. visual,

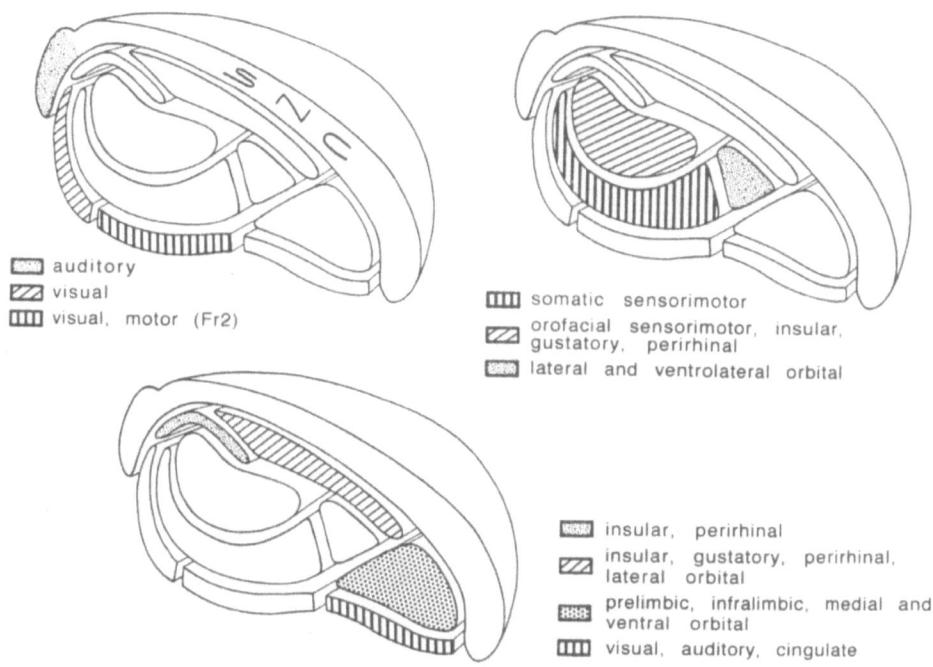

auditory
visual
visual, motor (Fr2)

somatic sensorimotor
orofacial sensorimotor, insular, gustatory, perirhinal
lateral and ventrolateral orbital

insular, perirhinal
insular, gustatory, perirhinal, lateral orbital
prelimbic, infralimbic, medial and ventral orbital
visual, auditory, cingulate

Figure 2. Representation of the cortical mosaic in the substantia nigra pars reticulata. SNC indicates the substantia nigra pars compacta.

oculomotor, cingulate), in the more dorsal nigral laminae are present districts ventral in striatum (e.g. insular, perirhinal, orbital, infralimbic). Interestingly, such a lamellar pattern of innervation has also been observed in the striato-pallidal projection. As reported in monkey by Smith and Parent (1986) and in rodent by Wilson and Phelan (1982), the striato-pallidal fibers terminate in the form of bands parallel to the medullary laminae. Hence lamination appears to be a characteristic feature of the striatal efferents and constitutes a suitable anatomical solution to maintain within basal ganglia output stations an accurate picture of the striatal functional mosaic.

THE LAMELLAR ORGANIZATION OF NIGRAL PROJECTION NEURONS

Does lamination in the striato-nigral innervation allow the striatum to segregate its influence on separate populations of nigral efferent neurons? To answer this question the regional distribution of the various nigral efferent cell groups that provide innervation of thalamus, superior colliculus and tegmentum has been examined using the wheat germ agglutinine peroxidase complex as a retrograde axonal tracer. Our results clearly demonstrate that, like striato-nigral terminals, the parent cells of nigral efferents obey a remarkable onion like organizational plan. They distribute along a series of curved laminae enveloping a longitudinal core located in the dorsolateral aspect of the substantia nigra pars reticulata (Deniau and Chevalier, 1992). As documented in figure 3, this lamination supports the topic organization of nigral efferents. Concerning the nigrocollicular cells, they occupy a more central position in the nigral onion as they innervate more lateral sites in the superior colliculus. Neurons in the more ventral lamina however are an exception to this rule. They innervate the whole collicular extent with a mediolateral topography. As for the nigrothalamic

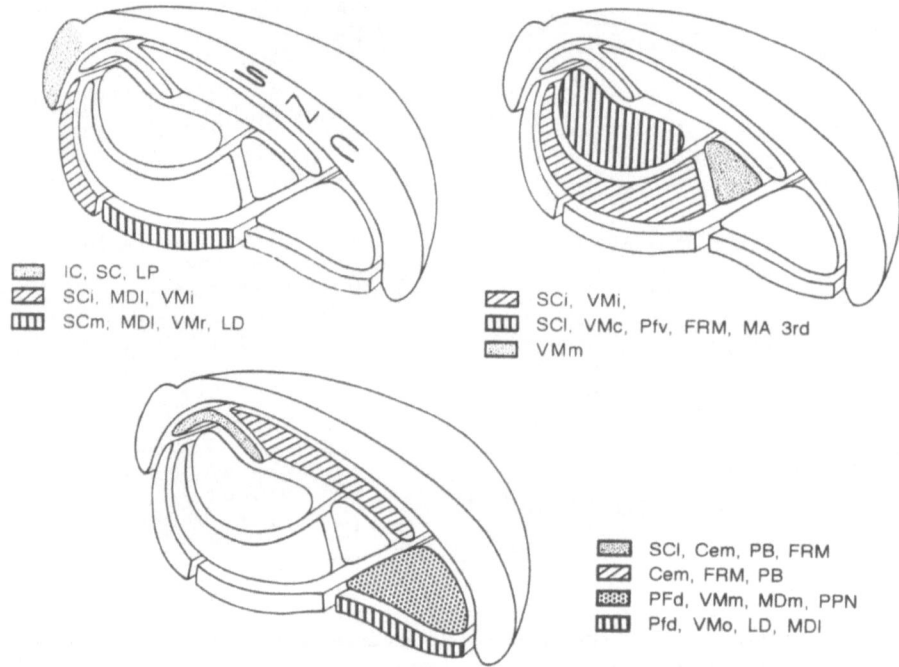

Figure 3. Lamellar distribution of the nigral efferent neurons which innervate the thalamus, tectum and tegmentum. As observed from retrograde axonal tracing, nigral efferent neurons are topographically ordered along a series of laminae enveloping a dorsolateral core. In this onion like structure, each lamella is composed of a specific set of neurons innervating conjointly particular areas in thalamus, tectum and/or tegmentum.

Cem: central medial thalamic nucleus; FRM: mesencephalic reticular formation; IC: inferior colliculus; LD: lateral dorsal thalamic nucleus; LP: lateral posterior thalamic nucleus; MA 3rd: medial accessory oculomotor nucleus; MDm, MDl: mediodorsal thalamic nucleus medial, lateral; PB: parabrachial nuclei; Pfd, Pfv: parafascicular thalamic nucleus dorsal, ventral; PPN: pedunculopontine nucleus; SC: superior colliculus; SCi, SCl, SCm: superior colliculus, intermediate, medial, lateral parts; SNC: substantia nigra pars compacta; VMc, VMi, VMm, VMo, VMr: ventral medial thalamic nucleus caudal, intermediate, medial, oral, rostral parts;

66

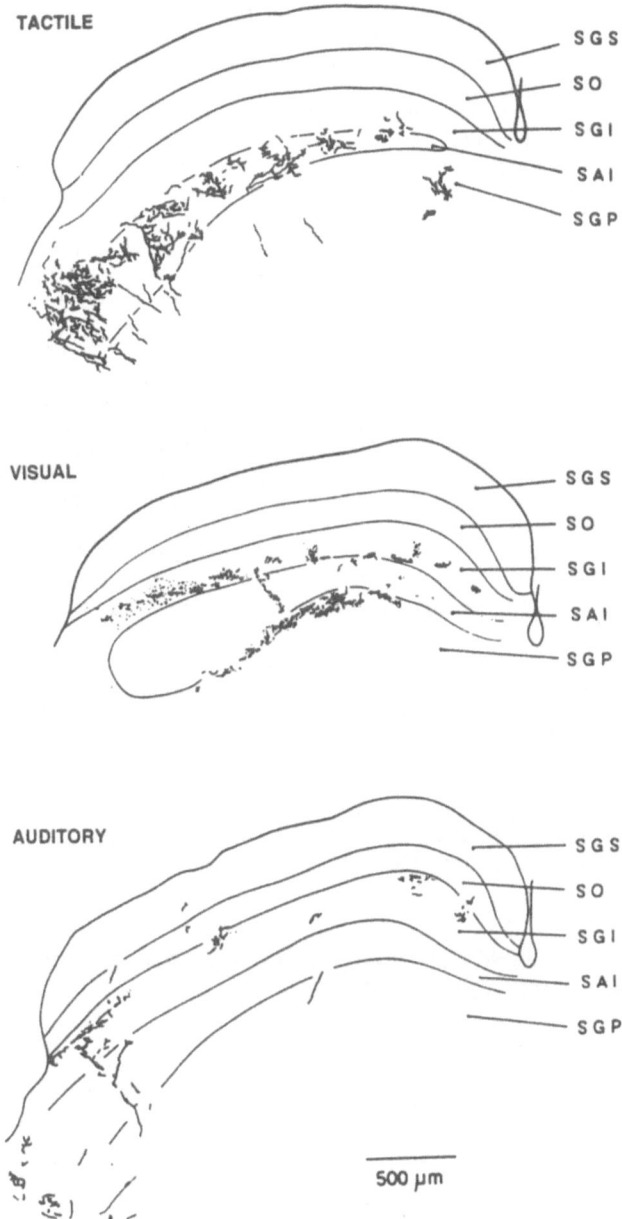

Figure 4. Differential innervation of the rat superior colliculus by the tactile, visual and auditory nigral divisions. SAI: stratum album intermedium; SGS : stratum griseum superficiale; SO : stratum opticum; SGI: stratum griseum intermedium; SGP : stratum griseum profundum.

neurons, while the cells innervating the lateral mediodorsal, the lateral dorsal and the rostral aspect of the ventral medial nucleus lie in the more ventral lamina, those innervating the paracentral, the rostral part of parafascicular and the caudo-central regions of the ventral medial nucleus are clustered in the central core. Finally the nigrotegmental neurons also segregate into laminae whose dorsoventral position varies with the region they innervate in reticular formation. Since the parent cells of the three main nigral efferent systems are largely mixed within the susbtantia nigra pars reticulata, we see this nucleus as an associative onion-like structure whose lamellae are composed of sets of neurons innervating conjointly particular loci of thalamus, tectum and/or tegmentum. This model contrasts with the previous conceptions of nigral partitioning which were based on a strict segregation of nigral efferent cell groups (Faull and Mehler, 1978; Beckstead and Frankfurter, 1982; François et al., 1984). Although the logic of such associations remain conjectural, a key for interpretation is provided by the observation that associated nigral populations act on thalamic, collicular and/or tegmental areas which are in other respects linked through direct connections. The nigro-collicular cells for instance connect neurons that in turn project on thalamic nuclei ventral medial, lateral mediodorsal and paracentral (Chevalier and Deniau, 1984; Bickford and Hall, 1989) which are just the target for their nigrothalamic partners. The same remark applies for the nigro-thalamic/nigro-tegmental coupling since it associates the pedunculo-pontine and the ventral medial thalamic nucleus which are in other respects mutually connected (Jimenez-Castellanos and Reinoso-Suarez, 1985). From a functional point of view, the substantia nigra pars reticulata is known to be the major vehicle of the striatal disinhibitory influence on thalamic, collicular and tegmental networks. We have documented in rat, that striatum exerts a gating influence on these networks by phasically interrupting the tonic inhibitory discharge of nigral cells (Chevalier and Deniau, 1990). Thus a logic for the associative arrangement of nigral efferents would be to ensure a coherent gating of the information flow in functionally associated cell groups of thalamus, colliculus and tegmentum.

The present data disclose that the parallel architecture of the basal ganglia is at least in rodent, based on a lamellar arrangement of the substantia nigra pars reticulata. Consequently it is thanks the nigral laminae that the outflow of each individual striatal district will be dispatched to particular subregions in thalamus, tectum and tegmentum. So for instance while the ventral nigral lamina would dispatch the visual and associated visuomotor, visuolimbic striatal processings to the superior colliculus and the thalamic nuclei paralamellar mediodorsal, lateral dorsal and ventral medial, by contrast the dorsomedial nigral compartment would distribute the outflow of the prefronto-associative striatal district to the thalamic nuclei parafascicular, mediodorsal, ventral medial (medial aspect) and the peribrachial area.

WHY DO BASAL GANGLIA SEGREGATE THEIR INFLUENCE ON EXECUTIVE MOTOR CENTERS?

The nigro-collicular pathway offers a good opportunity to address this question since it directly connects basal ganglia to an executive motor center. Functionally this pathway opens the various sensorimotor districts of the nigral onion onto premotor neurons of the intermediate collicular layers. To disclose how the nigral channels enter the collicular circuitry, each of the various sensorimotor sectors of the nigral onion has been carefully infiltrated with high resolution and low diffusion anterograde tracers (biocytin, WGA-HRP). The results show that neurons in each individual nigral lamina have their own collicular recipient zone (figure 4). As a rule, neurons in the tactile nigral channel innervate massively the stratum album intermedium in the form of a medio-lateral series of patches. By contrast, axon terminal from neurons of the visual channel exhibit a complementary pattern of innervation by lining in both its dorsal and ventral aspects the recipient area of the somatic channel. Finally the pars lateralis auditory channel ends more dorsal, in the interface between the stratum opticum and the stratum griseum intermedium and along the lateral edge of the intermediate strata. Considering such a dissociated pattern of nigral terminals, a reasonable prediction is that each nigral channel appear to be in a position to connect a particular subset of collicular output neurons (Redgrave et al., 1986). So the visual and auditory nigro-collicular channels would engage the ipsilateral descending pathways whose cells of origin lie mainly in the stratum griseum intermedium, whereas the somatic channel would connect the crossed colliculo-reticulo-spinal tract in the stratum album intermedium. Behavioural observations in rodent have stressed the possibility that these two main collicular efferent systems may support different classes of orienting behaviour: avoidance/escape behaviour for the ipsilateral pathway and approach for the crossed system (Sahibzada et al., 1986; Westby et al., 1990). Recalling the arousing function of the basal ganglia on collicular effectors, we see the nigro-collicular channels as specialized actuators for distinct classes of orienting behaviors.

CONCLUDING REMARKS

The present data obtained in rodent corroborate the previous proposals in primate that basal ganglia are organized into a multitude of parallel circuits. In rodent, this parallel arrangement is underlied by a remarkable lamellar architecture of the substantia nigra pars reticulata. Thanks to the lamellar nigral architecture, the functional partitioning that is imposed on the striatum by the cortical mantle is further preserved in the basal ganglia outflow. It is worth noting that such a parallel arrangement is not conflicting with the notion of convergence. As illustrated figures 1, 2, convergences in basal ganglia also obey the principle of parallel processing since they have their own channels. Apparently they reflect no more than the cortico-cortical associations. In this respect, it is predictible that from rodent to primate, convergent channels increase in number and variety.

Acknowledgements

This work was supported by Esprit Basic Reseach Action No. 3149 and the Human Frontier Program.

REFERENCES

Alexander, G. E., and Crutcher, M.D., 1990, Functional architecture of basal ganglia circuits: neural substrates of parallel processing, *Trends Neurosci.* 13:266-271.

Alexander, G.E., DeLong, M.R., and Strick, P.L., 1986, Parallel organization of functionally segregated circuits linking basal ganglia and cortex, *Ann. Rev. Neurosci.* 9:357-381.

Beckstead, R.M., and Frankfurter, A., 1982, The distribution and some morphological features of substantia nigra and ventral tegmental area in the rat, *Neuroscience* 7:2377-2388.

Berendse, H.W., Galis de Graaf, Y., and Groenewegen, H.J., 1992, Topographical organisation and relationship with ventral striatal compartments of prefrontal corticostriatal projections in the rat, *J. Comp. Neurol.* 316:314-347.

Bickford, M. E., and Hall, W. C., 1989, Collateral projections of predorsal bundle cells of the superior colliculus in the rat, *J. Comp. Neurol.* 283:86-106.

Chevalier, G., and Deniau, J. M., 1984, Spatio-temporal organization of a branched tecto-spinal/tecto diencephalic neuronal system, *Neuroscience* 12:427-439.

Chevalier, G., and Deniau J.M., 1990, Disinhibition as a basic process in the expression of striatal functions, *Trends Neurosci.* 13:277-280.

Deniau, J.M., and Chevalier, G., 1992, The lamellar organization of the rat substantia nigra pars reticulata: distribution of projection neurons, *Neuroscience* 46:361-377.

Domesick, V.B., 1977, The topographical organization of the striatonigral connection in the rat, *Anat. Rec.* 187:567.

Faull, R.L.M., and Mehler, W.R., 1978, The cells of origin of nigrotectal, nigrothalamic and nigrostriatal projection in the rat, *Neuroscience* 3:989-1002.

Flaherty, A.W., and Graybiel, A.M., 1991, Corticostriatal transformations in the primate somatosensory system. Projections from physiologically mapped body part representations, *J. Neurophysiol.* 66: 1249-1263.

François, C., Percheron, G., and Yelnik, J., 1984, Localization of nigrostriatal, nigrothalamic and nigrotectal neurons in ventricular coordinates in macaques, *Neuroscience* 13:61-76.

Gerfen, C.R., 1985, The neostriatal mosaic. I Compartmental organization of projections from the neostriatum to the substantia nigra in the rat, *J. Comp. Neurol.* 236:454-476.

Goldman, P.S., and Nauta, W.J.H. 1977, An intricately patterned prefrontocaudate projection in the rhesus monkey, *J. Comp. Neurol.* 171:369-386.

Ilinsky, I. A., Jouandet, M. L., and Goldman-Rakic P. S., 1985, Organization of the nigrothalamocortical system in the rhesus monkey, *J. Comp. Neurol.* 236:315-330.

Jimenez-Castellanos, J., and Reinoso-Suarez, F., 1985, Topographical organization of the afferent connections of the principal ventromedial thalamic nucleus in the cat, *J. Comp. Neurol.* 236:297-314.

Kunzle, H., 1977, Projections from primary somatosensory cortex to basal ganglia and thalamus in the monkey, *Exp. Brain Res.* 30:481-492.

McGeorge, A.J., and Faull, R.L.M., 1989, The organization of the projection from the cerebral cortex to the striatum in the rat, *Neuroscience* 29:503-537.

Nauta, W.J.H., Smith, G.P., Faull, R.L.M., and Domesick, V.B. 1978, Efferent connections and nigral afferents of the nucleus accumbens septi in the rat, *Neuroscience* 3:385-401.

Redgrave, P., Odenkule, A., and Dean, P., 1986, Tectal cells of origin of the predorsal bundle in rat: location and segregation from ipsilateral descending pathway, *Exp. Brain Res.* 63:279-293.

Smith, Y., and Parent, A., 1986, Differential connections of caudate nucleus and putamen in the squirrel monkey (Saimiri sciureus), *Neuroscience* 18:347-371.

Sahibzada, N., Dean, P., and Redgrave, P., 1986, Movements resembling orientation and avoidance elicited by electrical stimulation of the superior colliculus in rats, *J. Neurosci.* 6:723-733.

Westby, G.W.M., Keay, K.A., Redgrave, P., Dean, P., and Bannister M., 1990 Output pathways from the rat superior colliculus mediating approach and avoidance have different sensory properties, *Exp. Brain. Res.* 81:626-638.

Wilson, C. J., and Phelan, K. D., 1982, Dual topographic representation of neostriatum in the globus pallidus of rats, *Brain Res.* 243:354-359.

Yeterian, E.H., and Pandya, D.N., 1991, Prefrontostriatal connections in relation to cortical architectonic organization in rhesus monkeys, *J. Comp. Neurol.* 312:43-67.

Yeterian, E. H., and Van Hoesen, G. W., 1978, Cortico-striate projections in the rhesus monkey: the organization of certain cortico-caudate connections, *Brain Res.* 139:43-63.

INTEGRATIVE ASPECTS OF BASAL GANGLIA CIRCUITRY

Suzanne N. Haber, Eileen Lynd-Balta and Will P.J.M. Spooren

Department of Neurobiology and Anatomy
University of Rochester School of Medicine
Rochester, NY 14642

INTRODUCTION

The basal ganglia is a set of interconnected subcortical structures that influence cortical activity through corticobasal ganglia loops. Information is thought to flow from the cortex via the striatum to the pallidum and substantia nigra, from the pallidum and the substantia nigra to the thalamus, and finally back to the cortex. Based on some of the pathways and physiology of the basal ganglia sensorimotor circuit in the primate, several theories concerning the organization of the primate basal ganglia have been suggested. A current one proposes that the frontal cortex and the basal ganglia are arranged in parallel, functionally segregated circuits (Alexander et al., 1990). In this concept, each individual circuit has a specific, functionally distinct region of the frontal cortex as a nodal point, and projects to anatomically distinct sectors of the striatum, the pallidum, the substantia nigra, and the thalamus. From there, the thalamocortical pathway completes the loop to cortex. In addition to the sensorimotor circuit, others have been proposed that involve distinct regions of the basal ganglia associated with particular cortical regions including the dorsolateral prefrontal cortex, the orbitofrontal cortex, and the cingulate cortex.

In both Parkinson's disease and Huntington's chorea, a clear pathology is evident in basal ganglia structures, and patients have severe motor disabilities. In addition, cognitive and emotional problems are described as fundamental features of these diseases. This combination of symptoms occurs regardless of the region of degeneration. A great deal of emphasis has been placed on the motor pathways and physiology of basal ganglia structures, and as a result, we know a great deal about this part of the system in primates. However, even within this system several aspects have not been fully explored. For example, projections to the substantia nigra are assumed to be organized in a similar fashion as the projections to the internal segment of the globus pallidus. According to the organizational principles stated above, the substantia nigra should segregate pathways according to cortical function. There is strong anatomical support that different circuits converge in the substantia nigra. We hypothesize that the substantia nigra plays an important role in integrating information between different cortical circuits.

This chapter addresses the issue of the basal ganglia as a complex structure with both segregated and convergent pathways. Much of the territory of basal ganglia structures is devoted to information from cortical regions not directly associated with the motor system. In particular a large ventromedial region of the striatum processes information from the limbic lobe and the amygdala. The central region receives input from association cortex. Our experiments show that some basal ganglia connections maintain a general separation of different cortical circuits while in others there is a convergence of information from different circuits. Circuits through the striato/pallidal, and pallido/subthalamic or thalamic pathways remain relatively separate. However, through the substantia nigra and feedback

The Basal Ganglia IV, Edited by
G. Percheron *et al.*, Plenum Press, New York, 1994

loops via regions of the globus pallidus, an integration of different circuits takes place that is likely to play an important role in the basic function of the basal ganglia. We have focused on areas of potential interaction between the limbic and motor systems since these are the most anatomically separated systems.

Corticostriatal, striatopallidal, and pallido/subthalamo/thalamic pathways are topographically organized conferring potential separation of circuits

Corticostriatal pathways. Anatomical and physiological studies show that different cortical areas project onto the striatum in an organized and topographic manner (Kemp and Powell, 1970; Künzle, 1975; Künzle and Akert, 1977; Jones et al., 1977; Yeterian and Van Hoesen, 1978; Van Hoesen et al., 1981; Selemon and Goldman-Rakic, 1985; Yeterian and Pandya, 1991) and this topography is maintained through the pallidal complex and the thalamus. In broad terms, the striatum is divided into a dorsolateral region primarily receiving input from sensorimotor cortex, a central territory receiving information from association cortex, and a ventromedial part that receives input from the limbic lobe and subcortical structures associated with the limbic system such as the amygdala (Fig. 1). The dorsolateral striatum receives inputs from the supplementary motor cortex, arcuate premotor cortex, and primary motor cortex. Interdigitated with these projections are inputs from the primary somatosensory cortex. The anatomical and physiological features of this "sensorimotor circuit" in relation to body and limb movements have been characterized in some detail (DeLong, 1971; Künzle, 1975; Alexander and DeLong, 1985; Liles and Updyke, 1985; Kimura, 1990) demonstrating that the somatotopic organization of the sensorimotor cortex is maintained in the striatum.The nucleus accumbens is that part of the striatum that has long been associated with the limbic lobe, receiving inputs from the anterior cingulate cortex, the orbitofrontal cortex, the hippocampus, as well as the amygdala

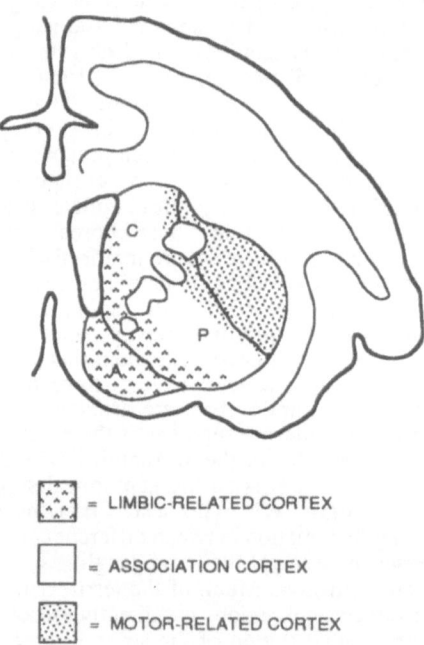

= LIMBIC-RELATED CORTEX

= ASSOCIATION CORTEX

= MOTOR-RELATED CORTEX

Figure 1. Cortical inputs to the striatum are topographically organized such that limbic, association, and motor territories can be identified. Afferent projections from association cortex overlap with inputs from motor-related cortex in the dorsolateral striatum, and with inputs from limbic-related structures in the ventromedial striatum.

(Groenewegen et al., 1980; Groenewegen et al., 1982; Kelley and Domesick, 1982; Kelley et al., 1982; Phillipson and Griffiths, 1985; McGeorge and Faull, 1989). In monkeys, these inputs are not restricted to the nucleus accumbens, but project to most of the ventromedial caudate nucleus, the rostral pole of the caudate nucleus and the rostroventral putamen (Van Hoesen et al., 1981; Hemphill et al., 1981; Russchen et al., 1984; Selemon and Goldman-Rakic, 1985; Yeterian and Pandya, 1991; Mizobuchi and Haber, 1991, 1992). Our work focuses on the ventral striatal system. The medial orbitofrontal cortex and the anterior part of anterior cingulate cortex project to the ventromedial part of the caudate nucleus, the dorsomedial nucleus accumbens, and the ventromedial edge of the rostral putamen. No terminal fields are seen in the dorsolateral part of the caudate nucleus. In the putamen, terminals are located in only a small, ventromedial edge of the nucleus adjacent to the nucleus accumbens. Although the dorsal part of the nucleus accumbens contains dense fiber terminals, there are very few fibers in the ventromedial part. Different injection sites reveal a mediolateral topography such that, for example, medial area 13 terminates medial to the fibers arising from injections placed more laterally in area 13.

Striatopallidal pathways. The striatopallidal projections are also topographically arranged such that striatopallidal fibers arising from different parts of the striatum terminate in distinct regions of the external, internal, and ventral segments of the pallidum (Fig. 2). Dorsolateral parts of the striatum terminate topographically in the lateral, dorsal, and central portions of the internal and external segments (Johnson and Rosvold, 1971; Carpenter, 1976; Parent et al., 1984). Here, as in the striatum, specific groups of pallidal neurons respond to the movement of a particular part of the body (DeLong, 1971). The ventral striatal fibers terminate in discrete ventral and medial pallidal regions including the subcommissural part of the globus pallidus, the rostral pole of the external segment and the rostral medial portion of the internal segment of the globus pallidus (Haber et al., 1990).

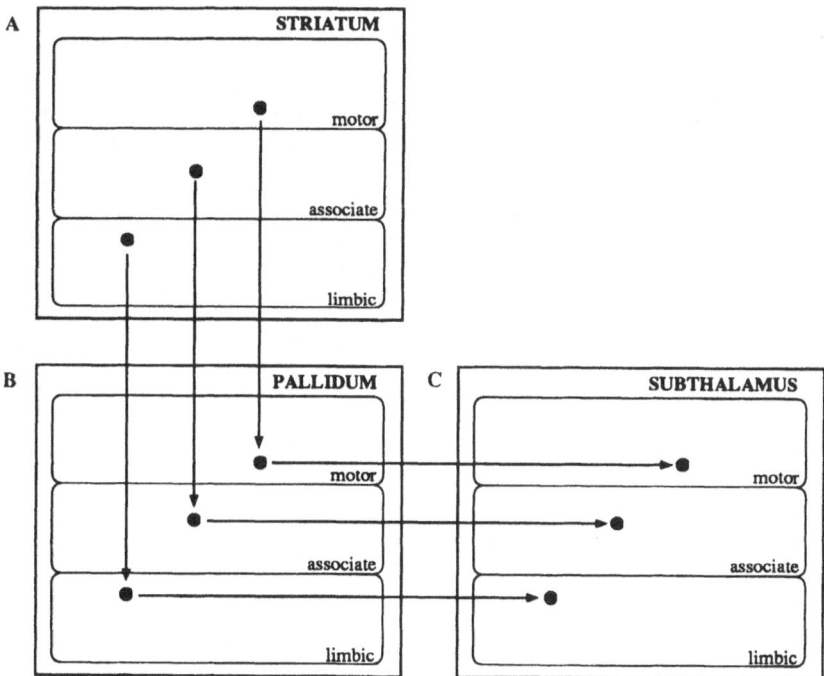

Figure 2. Schematic drawing of the segregated parallel pathways through the basal ganglia from striatum to pallidum (Figs. A,B) and from pallidum to the subthalamic nucleus (Figs. B,C). The striatopallidal and pallidosubthalamic projetions are topographically organized such that striatopallidal and pallidosubthalamic fibers terminate in distinct regions of the pallidum and subthalamic nucleus, respectively.

This entire region is referred to as the ventral pallidum in that it is the main target for the entire ventral striatum. The efferent pathways from the ventral striatum to the ventral pallidum are also organized in a rather strict medial-to-lateral and dorsal-to-ventral topographic manner.

Pallidosubthalamic and pallidothalamic pathways. Pallidosubthalamic and pallidothalamic projections are also topographically organized (Kuo and Carpenter, 1973; Kim et al., 1975; Carpenter, 1976; Parent and De Bellefeuille, 1982; Harnois and Filion, 1982; DeVito and Anderson, 1982). The sensorimotor part of the internal pallidal segment projects to the ventrolateral and ventral anterior nuclei of the thalamus. These thalamic regions send projections to the supplementary motor cortex (Strick, 1976; Jurgens, 1984; Wiesendanger and Wiesendanger, 1985). The external segment of the globus pallidus projects topographically to the subthalamic nucleus. The descending efferent projections from the ventral pallidum of primates terminate primarily in the subthalamic nucleus and the adjacent lateral hypothalamus, in the substantia nigra, and in the lateral habenular nucleus (Haber et al., 1993). Unlike the situation in rodents, few fibers are found in the thalamus. The projections to the STN and the lateral hypothalamus are dependent on the location of the injection site, with a clear mediolateral topographic organization (Haber et al., 1993). Thus, both within the ventral and dorsal pallidum as well as between them there is a separation of terminals in the STN (Fig. 2). These results together with the data from striatopallidal projections indicate that the separate lines of conduction from the various regions of the striatum, projecting topographically to the pallidum, maintain this organization through the pallidal projections to the STN and thalamus.

Striatonigral, pallidonigral, and nigrostriatal pathways are not topographically organized suggesting potential integration of circuits

Striatonigral Pathways. Striatal fibers in rats terminate in the substantia nigra with an inverse dorsal-ventral topography, such that the dorsal striatum projects to the ventral substantia nigra, and the ventral striatum projects to the dorsal substantia nigra (Nauta et al., 1978). In general, there is also a mediolateral topography in rats (Gerfen, 1985; Nauta et al., 1978). Various organizational schemes have been suggested in primates. Some studies demonstrate that the rostral striatum projects to the rostral substantia nigra, and the caudal striatum projects to the caudal substantia nigra, while others show that the caudate nucleus projects to the rostral substantia nigra and the putamen to the caudal substantia nigra (Szabo, 1962; Szabo, 1970; Selemon and Goldman-Rakic, 1990). We addressed the issue of striatonigral projections with respect to the corticostriatal pathways by placing tracers into different ventral striatal regions and into different regions of the dorsolateral striatum. There is little topography from projections arising from within these "functional" domains of the striatum. For example, in contrast to the ventral striatopallidal projection, fibers from specific regions of the ventral striatum are not confined to a specific region of the substantia nigra. Descending fibers from all experiments, regardless of tracer placement within the ventral striatum, result in dense terminal fields in the medial substantia nigra. These terminals extend dorsolaterally throughout the substantia nigra (Haber et al., 1990; Lynd-Balta et al., 1991; Lynd-Balta and Haber, 1992) (Fig. 3). Likewise, when we injected into the leg, or head region of the dorsolateral striatum, fibers from each area terminate in the same rostrocaudal, mediolateral region (Lynd-Balta et al., 1991; Lynd-Balta and Haber, 1992). Thus, unlike the terminals in the pallidum, striatal efferent fibers within each functional domain overlap extensively throughout the nigra.

Furthermore, the distribution of striatonigral terminals does not show an inverse dorsoventral topography, but is related to functional domains of the striatum. We tested this by placing injections into different dorsal or ventral regions. Terminal fibers from an injection into the medial caudate nucleus, dorsal to the nucleus accumbens were compared to fibers from a more ventral injection but into the motor part of the putamen. Fibers from the more ventral, but motor part of the putamen terminate ventral to those arising from the medial caudate nucleus injection. Thus, the location of terminal fields in the substantia nigra from a particular striatal locus can best be predicted by a knowledge of the cortical inputs to that striatal region, and not simply its dorsoventral, mediolateral, or rostrocaudal position in the striatum (Lynd-Balta et al., 1991; Lynd-Balta and Haber, 1992). A common

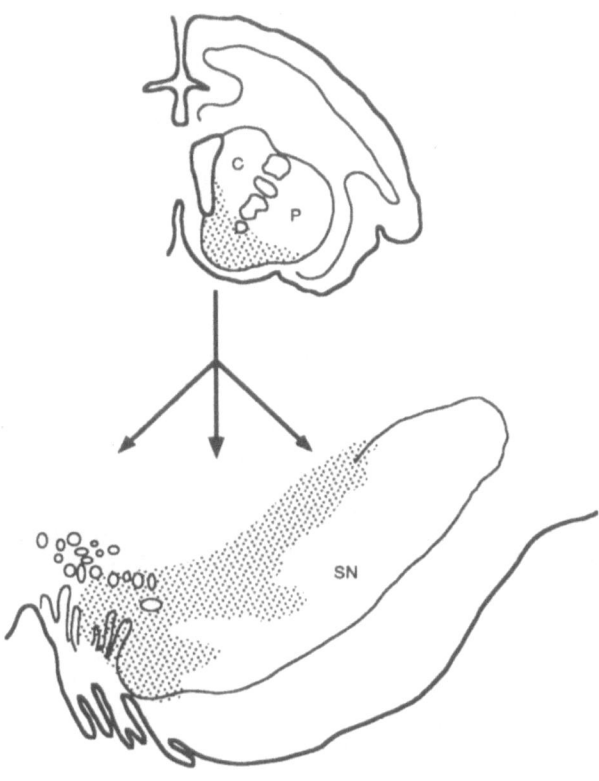

Figure 3. Ventral striatonigral projections are not topographically organized. The ventral striatum innervates the medial pars reticulata and a large mediolateral extent of the pars compacta in the dorsal substantia nigra.

feature of the ventral and dorsolateral striatonigral connections is that both areas project throughout the rostrocaudal extent of the substantia nigra.

Pallidonigral Projections. Projections from different regions of the ventral pallidum or dorsal pallidum to the substantia nigra also do not maintain a mediolateral topographic arrangement. Terminals from all injection sites in the ventral pallidum, including the most medial one, stretch laterally along most of the dorsal tier of the melanin-positive cells. There is also overlap between the pallidal projections from different regions of the dorsal pallidum to the substantia nigra (Spooren and Haber, in preparation). This supports the notion that within the motor circuit as well as within the limbic circuit, pallidal output to the STN and substantia nigra may have different roles. Pallidal projections through the STN maintain separate loops. Convergence of fibers in the substantia nigra suggests that the nigra may integrate information from related pallidal regions. However, as observed in the striatonigral projections, fibers projecting to the substantia nigra from the dorsal pallidum primarily remain separate from projections of the ventral pallidum. Thus, although within the same functional domain (limbic or motor) there is a large degree of convergence, the two systems remain separate in the substantia nigra.

An important distinction between the dorsal striatum and pallidum and the ventral striatum and pallidum is that projections from both the striatum and pallidum in the dorsal system terminate primarily in the pars reticulata. The dorsal system projects outside basal ganglia structures, to the thalamus. In contrast, the ventral system projects not only to the medial pars reticulata, but also extensively throughout the pars compacta. Therefore a major component of this pathway remains within basal ganglia circuitry. These terminals distribute over a wide mediolateral range of dopamine neurons and are in a position to not only influence the dopamine neurons projecting to the ventral striatum, but also those that project to the dorsolateral portions of the striatum. The idea that the ventral striatum is in a position to modulate the dorsal striatum via the dopaminergic neurons of the pars compacta was first proposed by Nauta et al. (Nauta et al., 1978). At that time the observation was made that in rats, the ventral striatum projects to the dorsal part of the substantia nigra in the area of the dopaminergic neurons.

Nigrostriatal pathways. Despite the extensive work on the nigrostriatal pathways in rats, there is little agreement on the organization of mesencephalic projections to the striatum in the primate (Carpenter and Peter, 1971; Szabo, 1980; Parent et al., 1983; Smith and Parent, 1986; Hedreen and DeLong, 1991). The nigrostriatal projections have been described as topographic, with the rostral substantia nigra projecting to the caudate nucleus, and the caudal substantia nigra projecting to the putamen (Szabo, 1980) or that the medial and lateral substantia nigra are connected to the rostral and caudal striatum, respectively (Hedreen and DeLong, 1991). Other studies show that nigral neuron projecting to the caudate nucleus and putamen overlap extensively, suggesting that nigrostriatal projections are not topographically organized (Parent et al., 1983; Smith and Parent, 1986). Recently, we compared injections of retrograde tracers into the sensori-

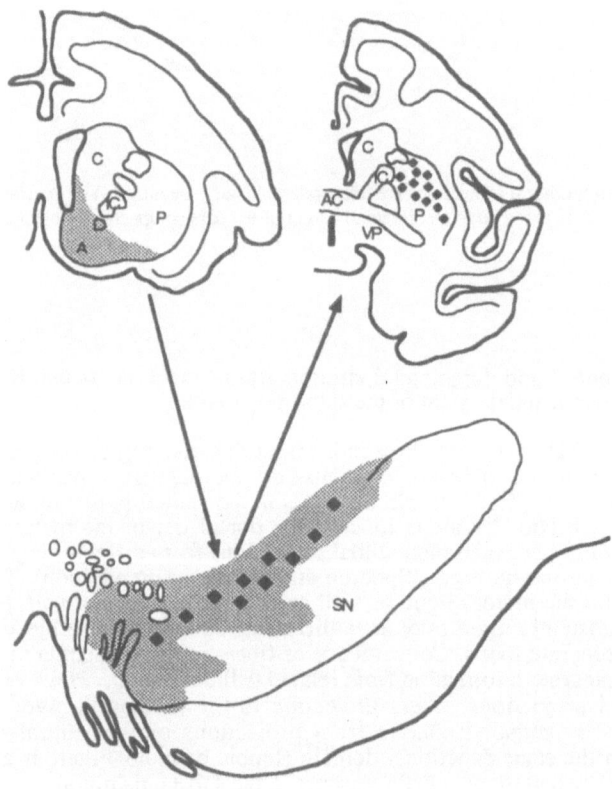

Figure 4. The ventral striatum may modulate the dorsal striatum by projecting to substantia nigra neurons that in turn project to the dorsal striatum. Ventral striatonigral projections terminate in the dorsal substantia nigra (shading), in the area of compacta neurons projecting to the dorsolateral striatum (diamonds).

motor-related striatum with injections into the ventral striatum. In general, the nigrostriatal pathway appears to be less topographically arranged in the primate than reported in the rodent. Our results show that the majority of the mesencephalic projections to the ventral striatum originate from the medial half including a central portion of the substantia nigra and are found in the entire rostrocaudal extent of the substantia nigra. They do not have a mediolateral topography. Rather, retrogradely labeled neurons are found in similar mediolateral positions following injections into more medial or lateral regions of the ventral striatum. Likewise tracer injections into three different regions of the putamen, (i.e. representing head, and leg areas), result in retrogradely labeled neurons throughout the entire rostrocaudal extent of the substantia nigra.

Unlike ventral striatal projecting neurons, those to the dorsolateral striatum arise from a wide mediolateral range (Fig. 4). A consistent labeling of many neurons in the medial portion of the substantia nigra after injections into motor striatal regions is surprising. The relationship between ventral striatal inputs to the dorsal region of the pars compacta, and the extent of the medial-lateral neurons that project to the dorsolateral striatum supports the notion that the ventral striatum may modulate the dorsal striatum via dopaminergic neurons (Fig. 4). There is direct experimental support for a modulation of the dorsal striatum by the ventral striatum in rats. Somogyi et al. showed synaptic contact at the electron-microscopic level of degenerating ventral striatal fibers onto retrogradely labeled nigral neurons that project to the dorsal striatum (Somogyi et al., 1981). Recently we placed an anterograde tracer into the ventral striatum and a retrograde tracer into the dorsolateral striatum in monkeys. Our results show that at the light microscopic level, fibers from the ventral striatum terminate in very close proximity to retrogradely labeled cells projecting to the dorsolateral striatum (Fig. 4). In a related experiment, we placed an anterograde tracer into the ventral pallidum and a retrograde tracer into the dorsolateral striatum. Again, we found that fibers from the ventral pallidum terminate in very close proximity to many retrogradely labeled cells 5. These two experiments strongly support the idea that regions of the substantia nigra are in an excellent position to modulate incoming information from the ventral basal ganglia and convey that information to more dorsal regions of the basal ganglia.

Integration between circuits can take place via feedback loops

An important issue is to what extent are the separate cortical basal ganglia pathways maintained in the so called feedback loops? There are at least three important, often disregarded feedback projections that are likely to provide important links between different cortical circuits.

STN Projections. The subthalamic nucleus projects primarily back to the pallidal complex, including the ventral pallidum, the external segment, and the internal segment. Injections of retrograde tracers in different parts of the ventral pallidum result in a consistent labeling of cells in the entire medial one half of the STN. This indicates that specific regions of the STN project to a large area of the ventral pallidum while discrete pallidal regions project to restricted regions of the STN. Separate circuits are maintained in one direction (pallidosubthalamic) while a more general modulatory influence exists in the other (subthalamopallidal) (Fig. 5B, 5C). Thus, specific parts of the subthalamic nucleus modulate not only the output of the internal segment of the pallidum, but also regions of the external segment and ventral pallidum from which it does not receive input from.

Pallidostriatal Projection. The globus pallidus, particularly the external segment, issues a dense projection to the striatum. After injections of anterograde tracers into various regions of the ventral pallidum and the external segment, we found that discrete regions of the pallidum project to a wide area of the striatum. Unlike the striatopallidal connections which are topographically organized and maintain strictly confined projections, the pallidostriatal projections innervate extensive areas of the striatum. These projections freely cross the boundaries of different functional striatal domains. For example, projections from a restricted region of the ventral striatum projects also to a restricted region of the ventral pallidum. However, the projections from that restricted pallidal region are widely distributed throughout the ventral striatum. In addition, it also sends fibers into more dorsal striatal regions. Thus, the ventral pallidum which receives inputs from specific ventral striatal regions, is in a position to modulate a large area of striatum (Fig. 5A, 5B).

Striatostriatal Internal Connections. We have also analyzed intrastriatal projections from various anterograde tracer injections placed in different parts of the ventral and dorsolateral striatum. Based on fiber and terminal arborizations there is an extensive intrastriatal network of connections. These connections bridge the various "functional domains" such that a relatively small injection into the ventral striatum sends fibers dorsalward reaching a large area of the dorsal striatum. The ventral striatum is in a position to communicate with the dorsal striatum through these connections. This provides the most direct way in which various cortical circuits can communicate.

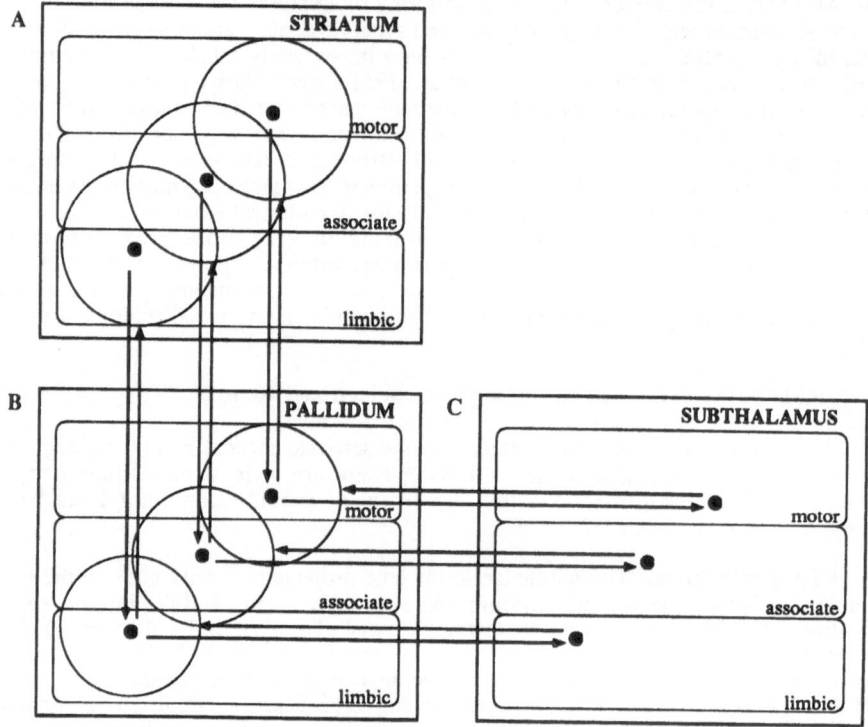

Figure 5. Schematic drawing of the bidirectional relationship of the striatum with the pallidum (A, B) and the pallidum with the subthalamus (B,C). The ascending projections of the subthalamus and the pallidum to the pallidum and striatum respectively are organized such that distinct subthalamic and pallidal regions (solid dots) innervate extensive and widespread areas of the pallidum and striatum (open circles) respectively. These ascending projections freely cross the boundaries of different functional striatal domains.

SUMMARY

Basal ganglia pathways are organized in at least two different ways. There is a general topographic organization of different cortical circuits through some channels and a convergence of circuits through other channels. The flow of information from the cortex to the pallidum, from the pallidum to the STN/thalamus, and from the thalamus back to cortex is separated into different channels that are related to specific cortical regions. In contrast, the flow of information to the substantia nigra integrates different circuits to modulate the general activity of the basal ganglia. Feedback loops from the STN to both pallidal segments, from the globus pallidus to the striatum, and from intrastriatal projections also provide important inter-circuit modulation. It is likely that through these integrative mechanisms that the symptomatology seen in basal ganglia disease is varied. If indeed all circuits were segregated, one would expect a clear clinical/pathological correlation. In fact this does not occur. Rather, patients exhibit a remarkable variety and severity of symptoms that do not always reflect the region of cell loss. An interesting and puzzling feature of Parkinson's disease is that before severe symptoms occur, a rather large number of dopaminergic neurons must degenerate. The estimated percentage of neuronal loss is approximately 80%. The fact that the patient appears to be relatively unaffected until a large percentage of neurons are lost may be due to the extensive integration of circuits. That is, modulation of the striatum, and therefore the output of the basal ganglia via the internal segment of the globus pallidus and the substantia nigra, pars reticulata can be maintained through the extensive feedback loops and through the few dopaminergic neurons remaining.

Acknowledgments

This work was supported by PHS Grants, NS22511 and MH 45573.

REFERENCES

Alexander, G.E., and DeLong, M.R., 1985, Microstimulation of the primate neostriatum. II. Somatotopic organization of striatal microexcitable zones and their relation to neuronal response properties, *J. Neurophysiol.* 53:1417-1430.

Alexander, G.E., Crutcher, M.D., and DeLong, M.R., 1990, Basal ganglia-thalamocortical circuits: Parallel substrates for motor, oculomotor, "prefrontal" and "limbic" functions, *Prog. Brain Res.* 85:119-146.

Carpenter, M.B., 1976, Anatomical organization of the corpus striatum and related nuclei, *in*: "The Basal Ganglia," M.D. Yahr, ed., Raven Press, New York, pp. 1-36.

Carpenter, M.B., and Peter, P., 1971, Nigrostriatal and nigrothalamic fibers in the rhesus monkey, *J. Comp. Neurol.* 144:93-116.

DeLong, M.R., 1971, Activity of pallidal neurons during movement, *J. Neurophysiol.* 34:414-427.

DeVito, J.L., and Anderson, M.E., 1982, An autoradiographic study of efferent connections of the globus pallidus in *Macaca mulatta*, *Exp. Brain Res.* 46:107-117.

Gerfen, C.R., 1985, The neostriatal mosaic. I. Compartmental organization of projections from the striatum to the substantia nigra in the rat, *J. Comp. Neurol.* 236:454-476.

Groenewegen, H.J., Becker, N.E.H.M., and Lohman, A.H.M., 1980, Subcortical afferents of the nucleus accumbens septi in the cat, studied with retrograde axonal transport of horseradish peroxidase and bisbenzimid, *Neuroscience* 5:1903-1916.

Groenewegen, H.J., Room, P., Witter, M.P., and Lohman, A.H.M., 1982, Cortical afferents of the nucleus accumbens in the cat, studied with anterograde and retrograde transport techniques, *Neuroscience* 7:977-996.

Haber, S.N., Lynd, E., Klein, C., and Groenewegen, H.J., 1990, Topographic organization of the ventral striatal efferent projections in the rhesus monkey: An anterograde tracing study, *J. Comp. Neurol.* 293:282-298.

Haber, S.N., Lynd, E., and Mitchell, S.J., 1993, The organization of the descending ventral pallidal projections in the monkey, *J. Comp. Neurol.* 329:111-128.

Harnois, C., and Filion, M., 1982, Pallidofugal projections to thalamus and midbrain: A quantitative antidromic activation study in monkeys and cats, *Exp. Brain Res.* 47:277-285.

Hedreen, J.C., and DeLong, M.R., 1991, Organization of striatopallidal, striatonigral, and nigrostriatal projections in the macaque, *J. Comp. Neurol.* 304:569-595.

Hemphill, M., Holm, G., Crutcher, M., DeLong, M., and Hedreen, J., 1981, Afferent connections of the nucleus accumbens in the monkey, *in*: "The Neurobiology of the Nucleus Accumbens," R.B. Chronister and J.F. DeFrance, eds., Maine: Haer Institute, Brunswick, pp. 75-81.

Johnson, T.N., and Rosvold, H.E., 1971, Topographic projections on the globus pallidus and the substantia nigra of selectively placed lesions in the precommissural caudate nucleus and putamen in the monkey, *Exp. Neurol.* 33:584-596.

Jones, E.G., Coulter, J.D., Burton, H., and Porter, R., 1977, Cells of origin and terminal distribution of corticostriatal fibers arising in the sensory-motor cortex of monkeys, *J. Comp. Neurol.* 173:53-80.

Jurgens, U., 1984, The efferent and afferent connections of the supplementary motor area, *Brain Res.* 300:63-81.

Kelley, A.E., and Domesick, V.B., 1982, The distribution of the projection from the hippocampal formation to the nucleus accumbens in the rat: An anterograde and retrograde-horseradish peroxidase study, *Neuroscience* 7:2321-2336.

Kelley, A.E., Domesick, V.B., and Nauta, W.J.H., 1982, The amygdalostriatal projection in the rat. An anatomical study by anterograde and retrograde tracing methods, *Neuroscience* 7:615-630.

Kemp, J.M., and Powell, T.P.S., 1970, The cortico-striate projection in the monkey, *Brain* 93:525-546.

Kim, R., Nakano, K., Jayaraman, A., and Carpenter, M.B., 1975, Projections of the globus pallidus and adjacent structures: An autoradiographic study in the monkey, *J. Comp. Neurol.* 169:263-290.

Kimura, M., 1990, Behaviorally contingent property of movement related activity of the primate putamen, *J. Neurophysiol.* 63:1277-1296.

Kuo, J., and Carpenter, M.B., 1973. Organization of pallidothalamic projections in the rhesus monkey, *J. Comp. Neurol.* 151:201-236.

Künzle, H., 1975, Bilateral projections from precentral motor cortex to the putamen and other parts of the basal ganglia: an autoradiographic study in *Macaca fascicularis.*, *Brain Res.* 88:195-209.

Künzle, H., and Akert, K., 1977, Efferent connections of cortical, area 8 (frontal eye field) in *Macaca fascicularis:* a reinvestigation using the autoradiographic technique, *J. Comp. Neurol.* 173:147-164.

Liles, S.L., and Updyke, B.V., 1985, Projection of the digit and wrist area of precentral gyrus to the putamen: relation between topography and physiological properties of neurons in the putamen, *Brain Res.* 339:245-255.

Lynd-Balta, E., and Haber, S.N., 1992, Organization of dopaminergic striatal projections in relation to peptidergic innervation of the substantia nigra in primate, *Soc. for Neurosci. Abst.* 18:306.

Lynd-Balta, E.L., Mitchell, S.J., and Haber, S.N., 1991, Organization of nigrostriatal and striatonigral projections in the primate, *Abst. Soc. Nsc.* 17:1300.

McGeorge, A.J., and Faull, R.L.M., 1989, The organization of the projection from the cerebral cortex to the striatum in the rat, *Neuroscience* 29:503-537.

Mizobuchi, M.M., and Haber, S.N., 1991, Topographical organization of the medial orbitofronto-striatal projection and its relationship to striosomes and enkephalin patches in the monkey, *Abst. Soc. Nsc.* 17:470.

Mizobuchi, M.M., and Haber, S.N., 1992, The cingulostriatal projection in the monkey: topography and relationship to chemical organization, *Soc. for Neurosci. Abst.* 18:306.

Nauta, W.J.H., Smith, G.P., Faull, R.L.M., and Domesick, V.B., 1978, Efferent connections and nigral afferents of the nucleus accumbens septi in the rat, *Neuroscience* 3:385-401.

Parent, A., and De Bellefeuille, L., 1982, Organization of efferent projections from the internal segment of the globus pallidus in the primate as revealed by fluorescence retrograde labeling method, *Brain Res.* 245:201-213.

Parent, A., Bouchard, C., and Smith, Y., 1984, The striatopallidal and striatonigral projections: two distinct fiber systems in primate, *Brain Res.* 303:385-390.

Parent, A., Mackey, A., and De Bellefeuille, L., 1983, The subcortical afferents to caudate nucleus and putamen in primate: a fluorescence retrograde double labeling study, *Neuroscience* 10:1137-1150.

Phillipson, O.T., and Griffiths, A.C., 1985, The topographic order of inputs to nucleus accumbens in the rat, *Neuroscience* 16:275-296.

Russchen, F.T., Bakst, I., Amaral, D., and Price, J.L., 1984, Amygdalostriatal projections in the monkey (*Macaca fascicularis*). An anterograde tracing study, *Brain Res.* 329:241-257.

Selemon, L.D., and Goldman-Rakic, P.S., 1985, Longitudinal topography and interdigitation of corticostriatal projections in the Rhesus monkey, *J.Neurosci.* 5:776-794.

Selemon, L.D., and Goldman-Rakic, P.S., 1990, Topographic intermingling of striatonigral and striatopallidal neurons in the rhesus monkey, *J. Comp. Neurol.* 297:359-376.

Smith, Y., and Parent, A., 1986, Differential connections of caudate nucleus and putamen in the squirrel monkey (*saimiri sciureus*), *Neuroscience* 18:347-371.

Somogyi, P., Bolam, J.P., Totterdell, S., and Smith, A.D., 1981, Monosynaptic input from the nucleus accumbens-ventral striatum region to retrogradely labelled nigrostriatal neurones, *Brain Res.* 217:245-263.

Strick, P.L., 1976, Anatomical analysis of ventrolateral thalamic input to primate motor cortex, *J. Neurophysiol.* 39:1020-1031.

Szabo, J., 1962, Topical distribution of the strital efferents in the monkey, *Exp. Neurol.* 5:21-36.

Szabo, J., 1970, Projections from the body of the caudate nucleus in the rhesus monkey, *Exp. Neurol.* 27:1-15.

Szabo, J., 1980, Organization of the ascending striatal afferents in monkeys, *J. Comp. Neurol.* 189:307-321.

Van Hoesen, G.W., Yeterian, E.H., and Lavizzo-Mourney, R., 1981, Widespread corticostriate projections from temporal cortex of the rhesus monkey, *J. Comp. Neurol.* 199:205-219.

Wiesendanger, R., and Wiesendanger, M., 1985, The thalamic connections with medial area 6 (supplementary motor cortex) in the monkey (*Macaca fascicularis*), *Exp. Brain Res..* 59:91-104.

Yeterian, E.H., and Pandya, D.N., 1991, Prefrontostriatal connections in relation to cortical architectonic organization in Rhesus monkeys, *J. Comp. Neurol.* 312:43-67.

Yeterian, E.H., and Van Hoesen, G.W., 1978, Cortico-striate projections in the rhesus monkey: the organization of certain cortico-caudate connections, *Brain Res.* 139:43-63.

ORGANIZATION OF THE PROJECTIONS FROM THE VENTRAL STRIATO-PALLIDAL SYSTEM TO VENTRAL MESENCEPHALIC DOPAMINERGIC NEURONS IN THE RAT

Henk J. Groenewegen, Henk W. Berendse and Floris G. Wouterlood

Graduate School of Neurosciences Amsterdam
Research Institute Neurosciences Vrije Universiteit
Faculty of Medicine
Department of Anatomy and Embryology
1081 BT Amsterdam
The Netherlands

INTRODUCTION

The connections between the basal ganglia, i.e. the striatum, the pallidum, the substantia nigra, and the subthalamic nucleus, as well as the connections between these structures and their input and output areas, e.g. the cortex and the thalamus, show distinct topographies. It has been proposed that this organization forms the structural basis for a number of parallel, functionally segregated basal ganglia-thalamocortical circuits (Alexander et al., 1986, 1990; Groenewegen et al., 1990). Circuits that involve the dorsolateral sector of the caudate-putamen complex, receiving predominantly inputs from the (pre)motor and somatosensory cortical areas, are associated with sensorimotor processes. Loops that involve more ventral and medial regions of the caudate nucleus and putamen, receiving inputs from the prefrontal, parietal and temporal association cortices, are involved in complex behavioral and cognitive processes. Finally, circuits involving the ventral striatum (constituted by the most ventral and medial parts of the caudate-putamen, the nucleus accumbens, and the striatal parts of the olfactory tubercle) which receives inputs from limbic cortical areas, is implicated in affective, emotional, and visceral functions. Since the ascending dopaminergic system reaches the entire striatum, this might explain, at least in part, that dopamine is implicated in a wide variety of behavioral and cognitive functions (Robbins [1992] and references therein).

The organization of the connections between the striatum and the dopaminergic neurons in the ventral mesencephalon, i.e. the ventral tegmental area (VTA/A10), the substantia nigra pars compacta (A9), and the retrorubral cell group (A8), does not comply in a simple way with the above-outlined parallel organization. First, the ascending dopaminergic projections to the striatum are rather diffusely organized (Björklund and Lindvall, 1984; Fallon and Loughlin, 1987). Second, there is a difference in the distribution of the fibers from the dorsal striatum and those from the ventral striatum to the ventral mesencephalon. A shown by Nauta and Domesick (1979) and Gerfen (1985), dorsal striatal areas project predominantly to ventral parts of the substantia nigra, in particular its pars reticulata, whereas progressively more ventrally located striatal areas project to successively more dorsal parts of the substantia nigra. It can thus be presumed that the dorsal striatal areas have much less access to the dopaminergic neurons than the ventral striatal areas: whereas dorsal striatal fibers may reach the (distal) dendrites of a ventral tier of neurons in the pars compacta of the substantia nigra (A9; cf., also Gerfen, 1985), ventral striatal fibers have terminations among the perikarya of the dorsally located dopaminergic cell groups (including A10, A9, A8; Nauta et al., 1978). Also the ventral pallidum, main recipient of ventral striatal fibers, appears to project to the VTA and the substantia nigra pars compacta (Haber et al., 1985; Zahm, 1989; Groenewegen et al., 1993). The possible consequence of this arrangement is that the ventral striatopallidal system has a strong influence on the ascending dopaminergic system which, for the major part, is directed at the dorsal striatum.

In this paper we will first briefly review and illustrate the main features of the organization of the projections to the ventral mesencephalic dopaminergic cell groups stemming from ventral striatopallidal regions. Subsequently, we will provide some new data from which we conclude that the ventral striatum may indeed influence dorsal striatal areas by means of the dopaminergic system.

EXPERIMENTAL PROCEDURES

Retrograde tracing experiments

Injections of the retrograde tracer choleratoxin, subunit B (CTb, 1% in Tris buffer) were placed in the ventral mesencephalon of 10 rats using an air-pressure system (glass micropipettes; injection volumes 5-25 nl). The animals were allowed to survive for 6-11 days and were subsequently perfused with 150 ml saline and 500 ml of a fixative containing 4% paraformaldehyde and 15% picric acid in phosphate buffer. Frozen sections were cut at 35 µm in a frontal plane, and incubated in a sequence of goat anti-CTb, donkey anti-goat whole serum, and goat peroxidase-antiperoxidase. They were then treated with Ni-enhanced diaminobenzidine (Ni-DAB, 12,5 mg in 25 ml 0,1M phosphate buffer with 1 ml 15% ammonium-Ni-sulfate). A number of sections were counterstained with cresyl violet prior to coverslipping. For details of the surgical-, injection-, and incubation procedures, we refer to Berendse et al. (1992).

Anterograde tracing experiments

For the analysis of the projections from the basal forebrain to the ventral mesencephalon more than 50 experiments with injections of anterograde tracers in the ventral striatum or the ventral pallidum (VP) were available.

Phaseolus vulgaris-leucoagglutinin. Injections of the tracer *Phaseolus vulgaris*-leucoagglutinin (PHA-L, 2.5% in Tris buffer) were placed by applying a 5 µA positive pulsed current to the PHA-L solution in a glass micropipette for 15-30 min. After 7-14 days survival, the rats were perfused with the same fixative as used for the CTb-experiments (see above). In the series of experiments in which it was planned to double-immunostain for dopamine, the fixative consisted of 4% glutaraldehyde and 0.2% sodium bisulphite (Na$_2$S$_2$O$_5$; reductor) in 0.1 M cacodylate buffer. Following postfixation, frontal sections were cut at 35 µm on a freezing microtome. Sections were incubated in a sequence of rabbit anti-PHA-L, swine anti-rabbit whole serum, and rabbit peroxidase-antiperoxidase. After the incubations, the sections were treated with DAB or Ni-DAB (see above). A number of incubated sections were counterstained with cresyl violet. For further details of the surgical-, injection-, and incubation procedures, see Groenewegen et al. (1993).

In several cases, sections were used for electron microscopy (incubations according to the protocol of Wouterlood and Groenewegen, 1985).

Biotinylated dextran-amine. In an additional series of experiments, injections of the tracer biotinylated dextran-amine (BDA, 5%; Veenman et al., 1992) were placed iontophoretically (6.5 µA positive pulsed current during 10 min) in the ventral striatum. After 7-14 days survival the animals were perfused with the same fixative as used for CTb-injected animals (see above). BDA was visualized in frozen sections (35 µm) by incubation with an avidin-biotin-horseradish peroxidase complex (ABC-kit), followed by a DAB- or a Ni-DAB-staining (see above).

Combined anterograde and retrograde tracing experiments

In order to identify presumptive contacts between ventral striatal fibers and mesencephalic (presumptively dopaminergic) neurons projecting to the dorsal striatum, injections of PHA-L in the ventral striatum were combined with injections of the retrograde tracer Fluorogold (FG, 2% in acetate buffer) in different parts of the dorsal striatum. FG was chosen as a tracer for its ability to label dendrites over considerable lengths. One microliter injections of FG were placed using a 10 μl Hamilton syringe. Following a 7-10 day survival period, the animals were perfused with the same fixative as used for CTb-tracing (see above). The general procedures for double-immunohistochemical staining have previously been described in detail (Groenewegen and Wouterlood, 1990). In brief, frozen sections of 35 μm were incubated in a cocktail of primary antibodies consisting of goat anti-PHA-L and rabbit anti-FG (kindly provided by Dr. H.T. Chang, Memphis, TN). Thereafter, the sections were incubated in the appropriate secondary antibodies, namely donkey anti-goat IgG and swine anti-rabbit IgG. Subsequently, they were incubated with a goat peroxidase-antiperoxidase complex, followed by treatment with Ni-DAB (see above; blue-black reaction product in PHA-L-containing fibers). Lastly, the sections were incubated with rabbit peroxidase-antiperoxidase complex, followed by a reaction with DAB, resulting in a brown reaction product (FG-labeling).

Anterograde tracing experiments in combination with tyrosine hydroxylase- or dopamine-immunocytochemistry

In virtually all experiments in which anterograde tracers had been injected, sections through the ventral mesencephalon were treated according to a double-immunohistochemical protocol to show the tracer in combination with dopaminergic neurons. In brief, sections were incubated in a cocktail of primary antibodies consisting of goat anti-PHA-L and either rabbit anti-dopamine or mouse anti-tyrosine hydroxylase (TH). Subsequently, the sections were incubated in a mixture of appropriate secondary antibodies: donkey anti-goat IgG and swine anti-rabbit IgG or goat anti-mouse IgG. Thereafter, the sections were incubated with goat peroxidase-antiperoxidase complex, followed by a reaction with Ni-DAB (see above; blue-black precipitate). Lastly, the sections incubated in rabbit peroxidase-antiperoxidase complex (dopamine-immunostaining) or rat peroxidase-antiperoxidase (TH-immunostaining) and then they were treated with DAB (brown reaction product).

Electron microscopy

Immuno-electron microscopic double-label experiments were carried out to study the relationships of fibers from the ventral striatum, anterogradely labeled with BDA, and TH-positive neurons in the substantia nigra. To stain the BDA, we used the protocol described by Wouterlood and Jorritsma-Byham (1993). The sections were first cryoprotected (for details see Wouterlood et al., 1993) and subjected to overnight incubation with the ABC solution. Next, the sections were treated with DAB and then subjected to silver-gold enhancement of the DAB reaction product (for details, see Jorritsma-Byham et al., 1993). Subsequently, the sections were incubated with mouse anti-TH antibodies, goat anti-mouse IgG and mouse-peroxidase-antiperoxidase. Finally, the sections were treated with DAB (see above), and postfixed (1 hr) in 1% OsO4 in 100 mM cacodylate buffer, block-stained (1 hr) with 2% aqueous uranyl acetate, and further processed for electron microscopy. Ultrathin sections were contrasted with lead citrate and investigated in a Philips EM 301 electron microscope

RESULTS AND COMMENTS

Topographical organization of ventral striatopallidal projections to the ventral mesencephalon

Distribution of retrogradely labeled ventral striatal and ventral pallidal neurons projecting to the ventral mesencephalon. Injections of CTb in different parts of the substantia nigra/VTA complex result in distinct patterns of retrogradely labeled

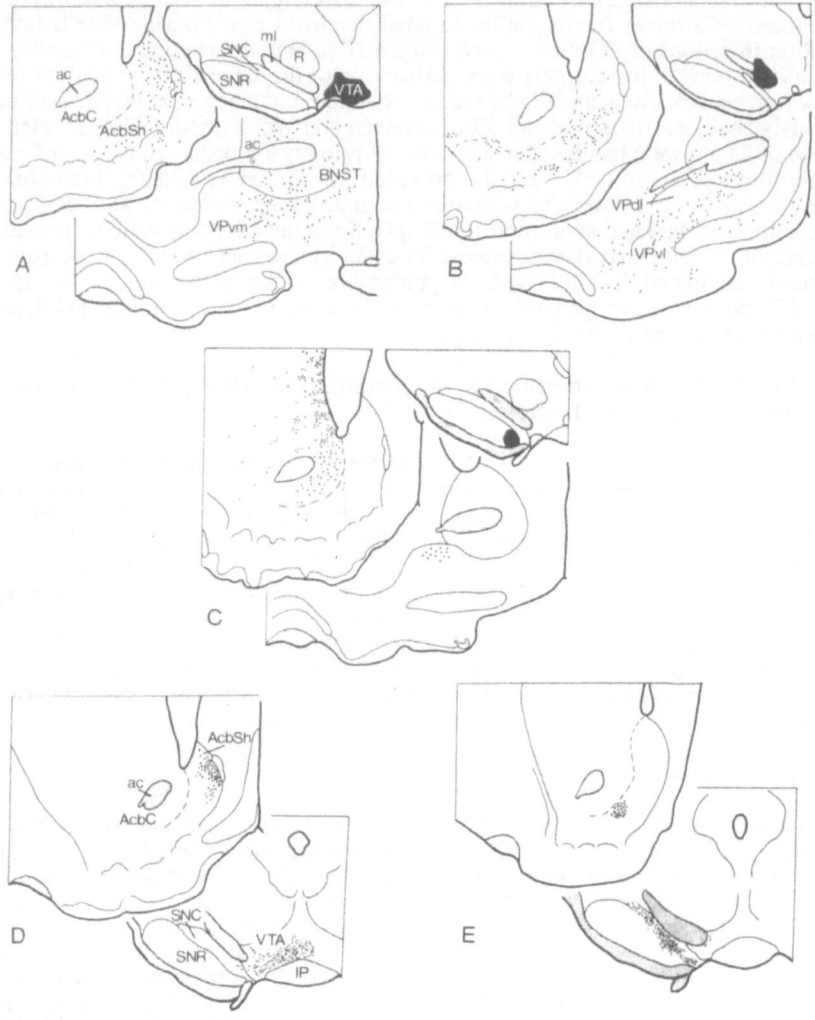

Figure 1. Chartings of the distribution of retrogradely labeled neurons in the basal forebrain following injections in the ventral mesencephalon (A-C), and anterogradely labeled fibers in the ventral mesencephalon following injections in the Acb (D/E). A-C, injections of Ctb in the medial part of the VTA (A), the lateral part of the VTA (B), and the medial part of the pars reticulata of the substantia nigra (C). D, injection of BDA in the caudal, dorsomedial part of the shell of the Acb. E, injection of PHA-L in the ventral part of the shell of the Acb. *Abbreviations:* ac, anterior commissure; Acb, nucleus accumbens; AcbC, core of the Acb; AcbSh, shell of the Acb; BNST, bed nucleus of the stria terminalis; ml, medial lemniscus; R, nucleus ruber; SNC, pars compacta of the substantia nigra; SNR, pars reticulata, VP, ventral pallidum; VPdl, dorsolateral part of VP; VPvm, ventromedial part of VP; VTA, ventral tegmental area.

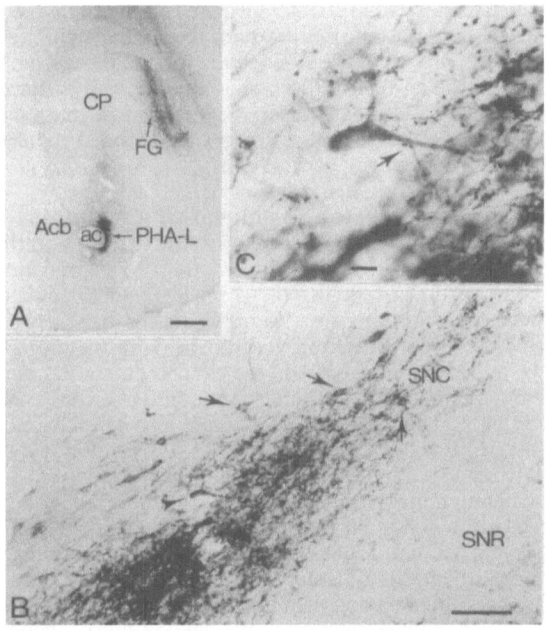

Figure 2. Photomicrographs showing the light microscopical relationships of anterogradely, PHA-L-labeled fibers and terminals from the ventral striatum with FG-positive neurons retrogradely labeled following an injection in the dorsal striatum. A, injection sites of PHA-L (DAB-Ni precipitate) in the core of the Acb, just lateral to the anterior commissure and FG (DAB reaction product) in the dorsolateral part of the caudate-putamen complex. B, low power micrograph of a double-immunostained section through the substantia nigra. Note the dense plexus of darkly stained (DAB-Ni) PHA-L labeled fibers and terminals in the ventral part of pars compacta and the adjacent dorsal part of the pars reticulata of the substantia nigra. Dorsal to and amidst the PHA-L labeled plexus lie more lightly appearing FG-labeled neurons (arrows). C, high power micrograph of a retrogradely labeled neuron (indicated with arrowhead in B). One of the main dendrites shows close appositions with PHA-L-labeled varicosities (arrow). *Abbreviations:* ac, anterior commissure; Acb, nucleus accumbens; CP, caudate-putamen; SNC, pars compacta of the substantia nigra; SNR, pars reticulata of the substantia nigra. Bars represent 1 mm (A), 10 μm (B), and 100 μm (C).

cells in both the ventral striatum and the VP. Following an injection in the medial part of the VTA, labeled neurons in the ventral striatum are predominantly present in the medial part of the shell of the nucleus accumbens (Acb), whereas labeled neurons in the VP are found in its ventromedial part (VPvm; Figure 1A; Berendse et al., 1992; Groenewegen et al., 1993). An injection of CTb in the lateral part of the VTA and in the adjacent medial part of the substantia nigra pars compacta results in labeled neurons more ventrally and laterally in the shell of the Acb as well as in sporadic clusters of cells in the core of the Acb and the ventrolateral part of the caudate-putamen complex. In this case, retrogradely labeled cells in the VP are located in its ventromedial and ventrolateral parts, leaving its dorsolateral part free of labeling (Figure 1B). A more lateral injection in the ventral mesencephalon, primarily involving the pars compacta of the substantia nigra just lateral to the medial terminal nucleus, results in a pattern of labeling in both the ventral striatum and the VP similar to that of the previous case (Berendse et al., 1992; Groenewegen et al., 1993). Injections of CTb in the adjacent area of the substantia nigra pars reticulata, which contains the ventrally extending dendrites of dopaminergic neurons in the medial part of the pars compacta, result in labeling of neurons in the core of the Acb and in the dorsally adjacent medial part of the caudate-putamen complex. Labeled neurons in the VP are now located in its dorsolateral part (VPdl; Figure 1C), whereas also the medial part of the dorsal pallidum contains labeling (Berendse et al., 1992; Groenewegen et al., 1993).

Distribution in the substantia nigra/VTA complex of anterogradely labeled fibers from the ventral striatum and the ventral pallidum. In agreement with previous studies (Nauta et al., 1978; Heimer et al., 1991), injections of PHA-L or BDA in the ventral striatum result in labeling of fibers in the VTA and the dorsal part of the substantia nigra. Since our injections are generally smaller than those of Nauta et al. (1978), a certain topography in these descending ventral striatal projections can be discerned. As shown in Figure 1D, an injection of BDA in the caudal, dorsomedial part of the shell of the Acb results in labeling of fibers and terminals predominantly in the medial part of the VTA and to a lesser extent in the lateral part. This labeling also extends into more caudal parts of the VTA. By contrast, following an injection of PHA-L into more ventral and lateral parts of the shell of the Acb (Figure 1E) most of the labeled fibers and terminals are found in the pars compacta of the substantia nigra. A much smaller amount of labeling is present in the lateral part of the VTA and in the dorsal part of the pars reticulata of the substantia nigra (Figure 1E). Caudal to the level shown in Figure 1E, labeling extends into the retrorubral cell group (A8). Relatively large injections of anterograde tracers in the core of the Acb result in labeling of both the pars reticulata and the pars compacta of the substantia nigra; labeling in the pars reticulata is mostly concentrated in its medial one-third. According to Berendse et al. (1992), the projections to the two parts of the substantia nigra are probably derived from different clusters of neurons which inhabit different ventral striatal compartments (cf., also Gerfen, 1985; Gerfen et al., 1987).

Injections of anterograde tracers in the VP result in labeling of fibers and terminals in both parts of the substantia nigra, the VTA, and the retrorubral cell group (not illustrated; Haber et al., 1985; Grove, 1988; Groenewegen et al., 1993). Injections in the medial part of the VP leads to labeling more medially, whereas more lateral injections in the VP result in labeling located more laterally in the VTA/substantia nigra complex (see also results of retrograde tracing experiments above; Groenewegen et al., 1993).

Main organizational features of the ventral striatopallidal connections with the ventral mesencephalic dopaminergic cell groups. The organization of the ventral striatopallidal pathways to the dopaminergic cell groups is schematically shown in Figure 5A. There appears to exist a rough mediolateral topographical organization in these projections and, furthermore, interconnected ventral striatal and ventral pallidal regions project in an overlapping fashion to distinct regions in the substantia nigra/VTA complex. Thus, the medial part of the shell of the Acb projects predominantly to the VTA (cf., Berendse et al., 1992). Projections from the VP to the VTA stem predominantly from the ventromedial part (VPvm) which receives afferents mainly from the medial shell of the Acb (Zahm, 1989; Zahm and Heimer, 1990; Heimer et al., 1991; Groenewegen et al., 1993). In turn, VTA dopaminergic neurons reciprocate the projections from the medial shell of the Acb but, in addition, project to more lateral parts of the ventral striatum, including the lateral shell and the core of the Acb and ventromedial parts of the caudate-putamen complex. The ventral

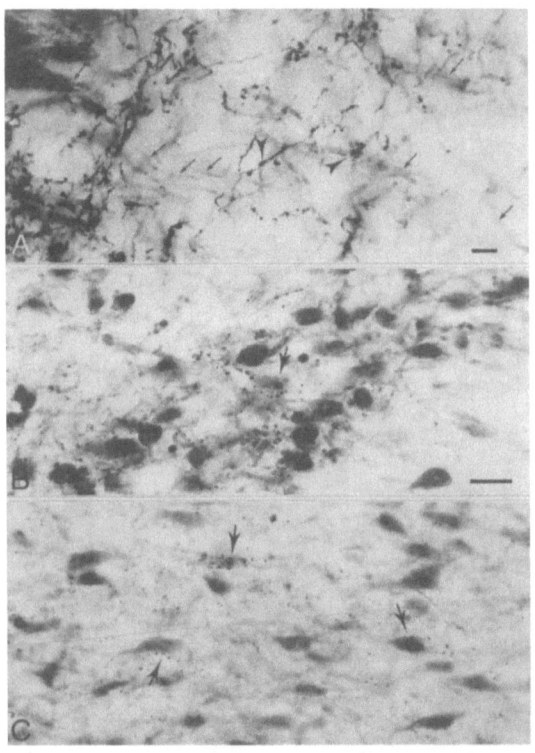

Figure 3. Photomicrographs of sections through the ventral mesencephalon, double-immunostained for dopamine or TH (DAB reaction product) and anterogradely PHA-L-labeled fibers (DAB-Ni reaction product). A, high power micrograph of the dorsomedial part of the pars reticulata showing the dorsoventrally oriented TH-positive dendrites (arrows) and the predominantly mediolaterally running PHA-L labeled, varicose fibers from the Acb. Examples of close appositions between dendrites and boutons of an afferent fibers are indicated with arrowheads. B, close appositions between a dopaminergic cell body in A9 (arrow) and PHA-L-filled boutons following an injection in the VP. C, various close appositions between TH-positive neurons in the VTA and PHA-L-filled afferents from the VP (arrows). Bars represent 10 μm (A) and 25 μm (B), holds also for C.

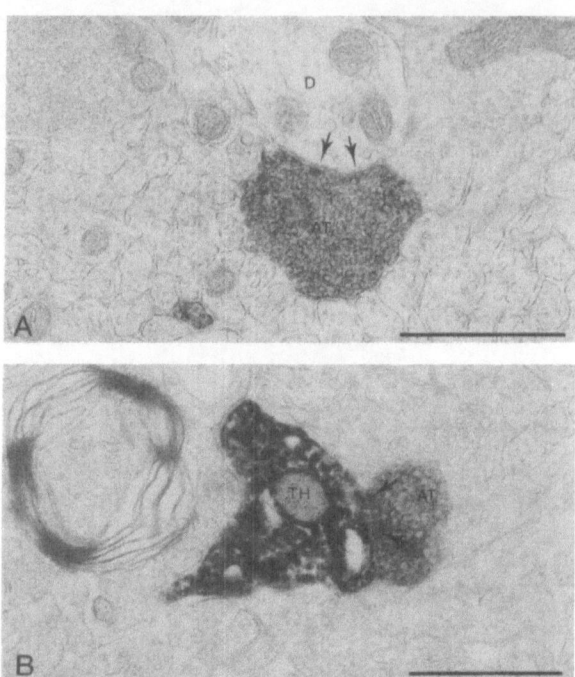

Figure 4. Electron microscopy. A, Labeled axon terminal (AT) in the substantia nigra following injection of BDA in the Acb. This terminal which is typical for the population of ventral striatonigral axon terminals, is packed with synaptic vesicles, and forms a symmetrical synaptic contact (arrows) with a small-caliber dendrite (D). Bar = 1.0 μm. B, BDA tracing (injection in the Acb) combined with TH-immunocytochemistry in the substantia nigra pars compacta. The BDA-labeled axon terminal (AT, DAB reaction product superposed with grains produced by silver-gold enhancement) forms a synaptic contact (arrows) with a TH-immunoreactive (DAB precipitate) dendrite (TH). Bar = 1.0 μm.

and lateral parts of the shell of the nucleus accumbens in turn project to lateral parts of the VTA and, in addition, reach both the dorsal and ventral tiers of the A9 cell group and the A8 retrorubral cell group (Heimer et al., 1991; Berendse et al., 1992). The ventral and ventrolateral parts of the VP project to a common area of dopaminergic neurons (cf., also Groenewegen et al., 1993). In addition, clusters of neurons in the core of the Acb and in the ventrolateral parts of the caudate-putamen complex have a pattern of projections to these dopaminergic cell groups very similar to that of the neurons in the lateral shell of the Acb (Berendse et al., 1992). It may thus be concluded that the ventral striatum, in particular the shell of the Acb, and the ventral pallidum, especially its ventromedial and ventrolateral parts, have ample access to the ventral mesencephalic dopaminergic neurons.

Relationships of ventral striatopallidal afferents with dopaminergic nigrostriatal neurons: light and electron microscopy

Experiments in which PHA-L was injected in the ventral striatum and FG in the dorsal striatum result in most cases in an extensive overlap in the substantia nigra of anterogradely PHA-L labeled fibers and terminals and retrogradely FG-labeled neurons. Figure 2 shows an experiment in which the FG-injection is located in the dorsolateral part of the caudate-putamen complex and the PHA-L injection in the lateral part of the core of the Acb. In this case, at a mid-rostrocaudal level of the substantia nigra, retrogradely labeled neurons are present throughout the pars compacta of the substantia nigra (A9), whereas the main terminal field of fibers from the ventral striatum is located ventrally in A9 (Figure 2B). Occasionally, close appositions between varicosities of ventral striatal fibers and cell bodies or dendrites of neurons retrogradely labeled from the dorsal striatum can be observed (Figure 2C).

In sections double-immunostained for TH (or dopamine) and PHA-L (transported from the Acb or the VP), close appositions between TH-positive elements and PHA-L labeled varicosities are present (Figure 3). Our results suggest that a difference exists in the organization of the ventral striatal and the ventral pallidal afferents at the cellular level. Although no quantitative analysis was done, we observed that ventral striatal fibers more frequently apposed dendrites than perikarya of TH-immunoreactive neurons. Figure 3A illustrates a few close appositions of fibers from the Acb with dendrites in the dorsomedial part of the pars reticulata; similarly, close appositions between labeled ventral striatal fibers and TH-positive dendrites and, to a much lesser degree perikarya were observed in the cell body regions of the substantia nigra/VTA complex. By contrast, fibers from the VP more frequently establish close appositions with the perikarya of TH-positive neurons, both in the pars compacta of the substantia nigra and in the VTA (Figure 3B, C). Occasionally, multiple appositions of a single beaded fiber with one cell body and/or its main dendrites were found (Figure 3B, C).

In the electron microscope, we have sofar studied only the afferents from the ventral striatum. Ventral striatal axon terminals in the substantia nigra, whether labeled with PHA-L or BDA, have a similar morphology: they are medium-sized, packed with synaptic vesicles, and contain only a few mitochondria. The mode of termination is remarkably uniform, i.e., with the shafts of large or small-caliber dendrites, and symmetrical synaptic contacts (Figure 4A). Usually, the target dendrites receive many contacts from unlabeled terminals as well. In double-immunostained preparations (transported BDA and endogenous TH) we found proof of the existence of synaptic contacts between terminals of fibers from the Acb and the dendrites of local, TH-containing neurons (Figure 4B).

CONCLUSIONS

On the basis of the data presented above and in line with previous studies (e.g., Nauta et al., 1978; Nauta and Domesick, 1979) it can be concluded that the ventral striatopallidal system has a rather specific relationship with the dopaminergic neurons in the ventral mesencephalon. Whereas dorsal striatal fibers have a relationship only with the dopaminergic cells of the substantia nigra, which predominantly concerns the dendrites of the ventral tier of neurons of the pars compacta (A9; Gerfen, 1985), the ventral striatal and VP fibers reach the dopaminergic neurons in A10, A9 as well as in A8, and have appositions not only with the dendrites of these neurons but also with their perikarya. The results of the combined retrograde/anterograde tracing experiments confirm that neurons in the substantia

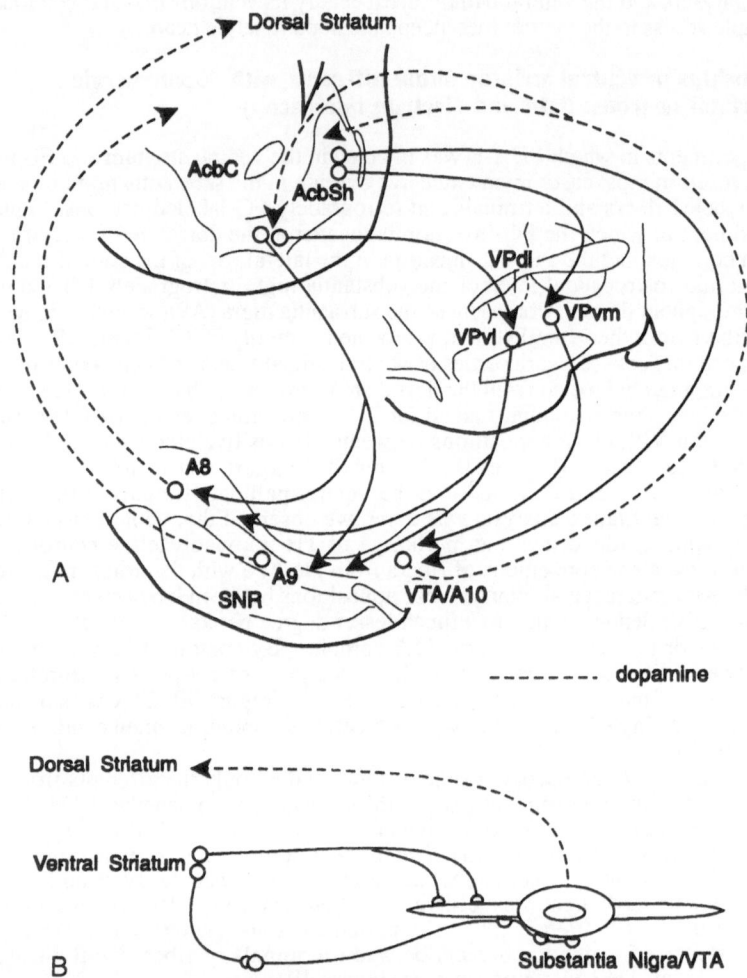

Figure 5. Schematic representation of the main features of the ventral striatopallidal projections to the dopaminergic neurons that issue ascending projections to both the dorsal and the ventral striatum (A). In B the presumptive differential relationships of the ventral striatal and ventral pallidal afferents with the dopaminergic neurons are outlined at the cellular level. *Abbreviations:* see Figure 1.

nigra/VTA complex that project to all parts of the dorsal striatum may be contacted by ventral striatal efferents. We also provided evidence that (at least several of) the close appositions between ventral striatal afferents and dopaminergic dendrites seen in the light microscope, represent true synaptic contacts (cf. also Wouterlood et al., 1992).

The present results thus confirm and extend previous assumptions that the ventral striatum, and also the VP, can communicate with the dorsal striatum via pathways through the ventral mesencephalon (Nauta et al., 1978; Somogyi et al., 1981). This implies that the ascending dopaminergic system, that modulates the information transfer through the various functionally different basal ganglia-thalamocortical circuits (see Introduction), is under the influence of the ventral striatopallidal system. Since limbic structures constitute the main inputs to the ventral striatum, distributing in particular to the shell region of the Acb (for reviews: Groenewegen et al., 1991; Zahm and Brog, 1992), the pathways through the ventral mesencephalon presumably form (part of) the neuronal substrate for limbic influences on motor and cognitive functions mediated by the dorsal striatal system.

Our present light microscopical results indicate that at the cellular level the ventral striatum and the ventral pallidum have different relationships with the dopaminergic neurons in the ventral mesencephalon. Whereas fibers from the ventral striatum terminate predominantly on (distal) dendrites, fibers from the ventral pallidum have the tendency to form more frequently contacts with cell bodies as well as proximal dendrites of dopaminergic neurons (Figure 5B). In a number of cases, ventral pallidal fibers displayed multiple appositions with single perikarya or dendrites of dopaminergic neurons. It is of interest to note that this arrangement of ventral striatal and ventral pallidal afferents with dopaminergic neurons is very similar to the arrangement that dorsal striatal and dorsal pallidal fibers have with neurons of the substantia nigra pars reticulata projecting to the superior colliculus (Smith and Bolam, 1991). Since most neurons in the ventral striatum are GABAergic, the present ultrastructural findings of symmetrical contacts between ventral striatal afferents and TH-positive dendrites strongly suggest that the ventral striatum exerts an inhibitory influence on the dopaminergic neurons in the ventral mesencephalon. Also the output neurons of the ventral pallidum are presumably GABAergic and inhibitory. It must be pointed out that the exact electrophysiological effect of activation of the ventral striatopallidal system on the dopaminergic neurons is unknown. The fact that ventral pallidal neurons have a high rate of spontaneous activity (Yim and Mogenson, 1983), in conjunction with the present anatomical observations, suggests that the ventral pallidum exerts a spontaneous strong (inhibitory) influence on, at least a population of, dopaminergic neurons. This implies that the possibility exists that activity of inhibitory ventral striatal neurons projecting to the ventral pallidum leads to a disinhibition of the dopaminergic neurons. At the same time, the direct projections of the ventral striatum to the dendrites of dopaminergic cells would lead to a presumably less powerful inhibition. In view of the potential significance of the influence of the ventral striatopallidal system on cognitive and behavioral functions mediated by the dorsal striatum, our present ignorance of the precise influence of the ventral striatopallidal system on the dopaminergic neurons calls for further electrophysiological and behavioral studies.

Acknowledgements

We wish to thank Drs. A.H.M. Lohman and M.P. Witter for critically reading the manuscript, and Mrs. M.P.M. Arts, Mrs. Y. Galis-de Graaf, Mr. P.H. Goede, Mrs. A Pattiselanno, Mr. D. de Jong, Mrs. J. Hage and Mrs. J. Kos for technical assistance. This work was supported in part by NWO-Program Grant #900-050-093.

REFERENCES

Alexander, G.E, Crutcher, M.D., and DeLong, M.R., 1990, Basal ganglia-thalamocortical circuits: parallel substrates for motor, oculomotor, 'prefrontal' and 'limbic' functions, *in:* "The Prefrontal Cortex: its Structure, Function and Pathology," H.B.M. Uylings, C.G. VanEden, J.P.C. De Bruin, M.A. Corner and M.G.P. Feenstra, eds., Elsevier, Amsterdam, *Prog. Brain Res.* 85:119-146.

Alexander, G.E., DeLong, M.R., and Strick, P.L., 1986, Parallel organization of functionally segregated circuits linking basal ganglia and cortex, *Ann. Rev. Neurosci.* 9:357-381.

Berendse, H.W., Groenewegen, H.J., and Lohman, A.H.M., 1992, Compartmental distribution of ventral striatal neurons projecting to the ventral mesencephalon in the rat, *J. Neurosci.* 12:2079-2103.

Björklund, A., and Lindvall, O., 1984, Dopamine-containing systems in the CNS, *in*: "Handbook of Chemical Neuroanatomy," Vol. 2, "Classical Neurotransmitters in the CNS," A. Björklund and T. Hökfelt, eds., Elsevier, Amsterdam, pp. 55-122.

Fallon, J.H., and Loughlin, S.E., 1987, Monoamine innervation of cerebral cortex and a theory of the role of monoamines in cerebral cortex and basal ganglia, *in*: "Cerebral Cortex, Vol. 6" E.G. Jones and A. Peters, eds., Plenum Press, New York, pp. 41-127

Gerfen, C.R., 1985, The neostriatal mosaic. I. Compartmental organization of projections from the striatum to the substantia nigra in the rat, *J. Comp. Neurol.* 236:454-476.

Gerfen, C.R., Herkenham, M., and Thibault, J., 1987, The neostriatal mosaic: II. Patch- and matrix-directed mesostriatal dopaminergic and non-dopaminergic systems, *J. Neurosci.* 7:3915-3934.

Groenewegen, H.J., and Wouterlood, F.G., 1990, Light and electron microscopic tracing of neuronal connections with Phaseolus vulgaris-leucoagglutinin (PHA-L), and combinations with other neuroanatomical techniques, *in*: "Handbook of Chemical Neuroanatomy, Vol. 8: Neuronal Microcircuits," A. Björklund, T. Hökfelt, F.G. Wouterlood and A. VandenPol, eds., Elsevier, Amsterdam, pp. 47-124.

Groenewegen, H.J, Berendse, H.W., and Haber, S.N., 1993, Organization of the output of the ventral striatopallidal system in the rat. Ventral pallidal efferents, *Neuroscience* in press.

Groenewegen, H.J., Berendse, H.W., Meredith, G.E., Haber, S.N., Voorn, P., Wolters, J.G., and Lohman, A.H.M., 1991, Functional anatomy of the ventral, limbic system-innervated striatum, *in*: "The Mesolimbic Dopamine System: From Motivation to Action", P. Willner, and J. Scheel-Krüger, eds., Wiley, Chichester, pp. 19-59.

Groenewegen, H.J., Berendse, H.W., Wolters, J.G., and Lohman, A.H.M., 1990, The anatomical relationship of the prefrontal cortex with the striatopallidal system, the thalamus and the amygdala: Evidence for a parallel organization, *in*: "The Prefrontal Cortex: Its Structure, Function, and Pathology," H.B.M. Uylings, C.G. Van Eden, J.P.C. De Bruin, M.A. Corner and M.G.P. Feenstra, eds., Elsevier, Amsterdam, *Prog. Brain Res.* 85: 95-118.

Grove, E.A., 1988, Efferent connections of the substantia innominata in the rat, *J. Comp. Neurol.* 277:347-364.

Haber, S.N., Groenewegen, H.J., Grove, E.A., and Nauta, W.J.H., 1985, Efferent connections of the ventral pallidum: Evidence of a dual striato pallidofugal pathway, *J. Comp. Neurol.* 235:322-335.

Heimer, L., Zahm, D.S., Churchill, L., Kalivas, P.W., and Wohltman, C., 1991, Specificity in the projection patterns of accumbal core and shell in the rat, *Neuroscience* 41:89-125.

Jorritsma-Byham, B., Witter, M.P., and Wouterlood, F.G., 1993, Combined anterograde tracing with biotinylated dextran-amine, retrograde tracing with Fast blue and intracellular filling of neurons with Lucifer yellow. An electron microscopic method, *J. Neurosci. Meth.* submitted.

Nauta, W.J.H., and Domesick, V.B., 1979, The anatomy of the extrapyramidal system, *in*: "Dopaminergic Ergot Derivates and Motor Function," K. Fuxe and D.B. Calne, eds., Pergamon Press, Oxford, pp. 3-22.

Nauta, W.J.H., Smith, G.P., Faull, R.L.M., and Domesick, V.B., 1978. Efferent connections and nigral afferents of the nucleus accumbens septi in the rat, *Neuroscience* 3:385-401.

Robbins, T.W., 1992, Introduction: milestones in dopamine research, *Semin. Neurosci.* 4:93-97.

Smith, Y., and Bolam, J.P., 1991, Convergence of synaptic inputs from the striatum and the globus pallidus onto identified nigrocollicular cells in the rat: a double anterograde labelling study, *Neuroscience* 44:45-73.

Somogyi, P., Bolam, J.P., Totterdell, S., and Smith, A.D., 1981, Monosynaptic input from the nucleus accumbens-ventral striatum region to retrogradely labelled nigrostriatal neurones, *Brain Res.* 217:245-263.

Veenman, C.L., Reiner, A., and Honig, M.G., 1992, Biotinylated dextran amine as an anterograde tracer for single- and double labeling studies, *J. Neurosci. Meth.* 41:239-254.

Wouterlood, F.G., and Groenewegen, H.J., 1985, Neuroanatomical tracing by use of *Phaseolus vulgaris*-leucoagglutinin (PHA-L): Electron microscopy of PHA-L filled neuronal somata, dendrites, axons and axon terminals, *Brain Res.* 326:188-191.

Wouterlood, F.G., and Jorritsma-Byham, B., 1993, The anterograde tracer biotinylated dextran-amine: Comparison with the tracer PHA-L in preparations for electron microscopy, *J. Neurosci. Meth.* in press.

Wouterlood, F.G., Goede, P., Arts, M.P.M., and Groenewegen, H.J., 1992, Simultaneous characterization of efferent and afferent connectivity, neuroactive substances, and morphology of neurons, *J. Histochem. Cytochem.* 40:457-465.

Wouterlood, F.G., Pattiselanno, A., Jorritsma-Byham, B., Arts, M.P.M., and Meredith, G.E., 1993, Connectional, immunocytochemical and ultrastructural characterization of neurons injected intracellularly in fixed brain tissue, *in*: "Morphological Investigations of Single neurons *in vitro*, IBRO Handbook Series: Methods in the Neurosciences," G.E. Meredith, and G.W. Arbuthnott, eds., Wiley, Chichester, in press.

Yim, C.Y., and Mogenson, G.J., 1983, Response of ventral pallidal neurons to amygdala stimulation and its modulation by dopamine projections to nucleus accumbens, *J. Neurophysiol.* 50:148-161.

Zahm, D.S., 1989, The ventral striatopallidal parts of the basal ganglia in the rat - II. Compartmentation of ventral pallidal efferents, *Neuroscience* 30:33-50.

Zahm, D.S., and Brog, J.S., 1992, On the significance of subterritories in the "accumbens" part of the rat ventral striatum. Commentary, *Neuroscience* 50:751-761.

Zahm, D.S., and Heimer, L., 1990, Two transpallidal pathways originating in nucleus accumbens, *J. Comp. Neurol.* 302:437-446.

TRANSITION AREAS OF THE STRIATOPALLIDAL SYSTEM WITH THE EXTENDED AMYGDALA IN THE RAT AND PRIMATE: OBSERVATIONS FROM HISTOCHEMISTRY AND EXPERIMENTS WITH MONO- AND TRANSSYNAPTIC TRACER

George F. Alheid[1], Carlos Beltramino[2,6], Alex Braun[3],
Richard R. Miselis[4], Chantal François[5] and Jose de Olmos[6]

[1]Depts of Psychiatric Medicine and [2]Otolaryngology
  University of Virginia Health Sciences Center
  Charlottesville, Virginia 22908
[3]Department of Pathology
  VA Medical Center
  Northport, New York 11768
[4]Dept. Animal Biology
  School of Veterinary Medicine
  University of Pennsylvania
  Philadelphia, Pennsylvania, 19104
[5]Laboratoire de Neuromorphologie informationnelle et
  de Neurologie expérimentale du mouvement
  Pav. INSERM Claude Bernard, Hôpital de la Salpêtrière
  Paris Cedex 13 France
[6]Instituto de Investigacion Medica
  Mercedes y Martin Ferrerya, Cordoba, Argentina

INTRODUCTION

In several papers we have extensively reviewed the concepts of the ventral striato-pallidal system and the extended amygdala (de Olmos et al., 1985; Heimer et al., 1985; Alheid and Heimer, 1988; Alheid et al., 1990; Heimer and Alheid, 1991; Heimer et al., 1991a, 1993), and these topics are only briefly recapitulated here. In some of these (e.g. Alheid and Heimer, 1988; Heimer and Alheid, 1991; Heimer et al., 1993) we have pointed out areas where these two structures are difficult to distinguish; these are the problem areas that we wish to confront in this chapter. In some instances, it is clear that the extended amygdala occupies portions of the forebrain normally considered part of the basal ganglia and more speculatively, we believe that some unusual features in other areas of the basal ganglia might reflect some ectopic elements of the extended amygdala.

THE VENTRAL STRIATOPALLIDAL SYSTEM
AS AN ANATOMICAL THEORY

In recent years it has become generally accepted that the definition of the basal ganglia can be stretched to include the nucleus accumbens and olfactory tubercle, as well as the

The Basal Ganglia IV, Edited by
G. Percheron *et al.*, Plenum Press, New York, 1994

associated ventral pallidum (Heimer and Wilson, 1975; Heimer, 1978). This latter structure is the ventral extension of the pallidum that is found below the anterior commissure at the caudal end of the nucleus accumbens. This flattens out rostrally to form a layer of pallidal neurons generally dividing the nucleus accumbens from the olfactory tubercle (e.g. figures 2, 3 in Heimer and Alheid, 1991). Conceptually, this reunification of what had been considered unrelated parts of the basal forebrain has presented a type of anatomical theory that has allowed the analysis of this part of the basal forebrain in terms of better known cortico-striato-pallidal circuits.

THE EXTENDED AMYGDALA AS A SECOND ANATOMICAL THEORY

The *extended amygdala* is a novel term suggested (Alheid and Heimer, 1988) to describe the complex interrelated columns of neurons that extend from the centromedial amygdala to the rostro-medial portions of the subcortical telencephalon (de Olmos, 1972; Schwaber et al., 1982; de Olmos et al., 1985; Grove, 1988a, b). This term is admittedly difficult but was chosen to reflect the prediction that the functions of this structure when appreciated in its entirety should closely resemble many of those that have been traditionally associated with the amygdala. These include psychoendocrine, psychomotor, autonomic, and emotional behavior that have generally sustained a prominent place for the amygdala in conceptualizations of the "limbic system". We have recently argued (Heimer and Alheid, 1991; Heimer et al., 1993) that the limbic system is a term that may have little relevance to the emerging picture of basal forebrain anatomy. From an anatomical point of view, the loose inclusion of various structures with the limbic system runs contrary to the specific and systematic relations that characterize the basal ganglia, extended amygdala, and hypothalamus. With progress in understanding these systematic relations it would seem to be more reasonable to use these insights as a guide to functional-anatomical experiments, rather than an ill-defined association of well-defined forebrain nuclei that are sometimes included on behavioral grounds with little anatomical justification, and sometimes included on anatomical grounds with little functional support.

An unfortunately confusing aspect of using the term extended amygdala is that, as we have defined this structure, it excludes the nuclei of the basolateral complex of the amygdala. A survey of the structure, connections, and histochemistry of the basolateral complex suggests that it may be practically considered as a cortical structure (de Olmos et al., 1985; Carlsen and Heimer, 1988; Alheid and Heimer, 1988; Carlsen, 1989; McDonald, 1992). Analogous to the basolateral complex, McDonald (1992) recently suggested that additional areas such as the cortical nuclei, amygdalo-piriform transition area, and the amygdalo-hippocampal transition area, could also be conveniently grouped as specialized cortical zones, based on the morphology and arrangement of their constituent cell populations, and their associative interconnections. Based on this view, these latter nuclei might also be excluded from the extended amygdala which would then be limited to the central and medial nuclei, as well as their "symmetrical" counterparts in the rostral forebrain.

The extended amygdala consists of two major divisions, one including the central amygdala, and therefore termed the "central division" and another including the medial amygdala and termed the "medial division" (de Olmos et al., 1985; see also Grove, 1988a, b). The extension of these two prototypical structures are found in neurons accompanying the stria terminalis in its supracapsular course en route to the bed nucleus of the stria terminalis, and in ventrally coursing neurons that traverse the area underneath the globus pallidus (and behind the caudal surface of the ventral pallidum). In the rat the sublenticular gray columns join the central and medial amygdaloid nuclei with the lateral and medial divisions of the bed nucleus of the stria terminalis respectively (figure 1A). These latter two compartments of the bed nucleus are included within the particular division of the extended amygdala with which they are immediately adjacent. As with the "theory" of the ventral striato-pallidal system, the theory of the extended amygdala can be used to predict the anatomical relations of caudomedial accumbens, the bed nucleus of the stria terminalis, and the subpallidal cell columns based on the known relations of the central and medial nuclei of the amygdala. Several of the relations which seem to distinguish the extended amygdala from the adjoining basal ganglia are: the projections to the brainstem from the central division of extended amygdala; projections to the medial hypothalamus from the medial division of the extended amygdala; and the large network of interconnecting axons within the extended amygdala (figure 1B, and de Olmos et al., 1985; Grove, 1988a, b). This network is in stark contrast to the generally columnar although potentially convergent organization found in the striatum (e.g. Percheron et al., 1984; Yelnik et al., 1984; Alexander et al., 1986; François et al., 1987; Alexander et al., 1990; Zemanick et al., 1991).

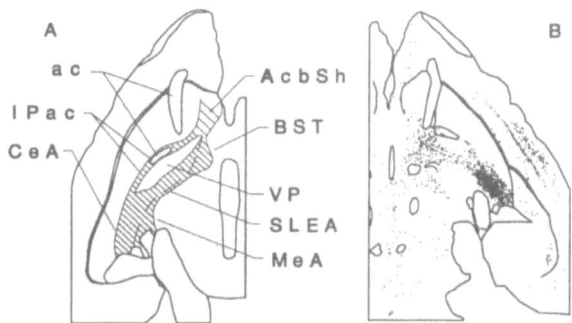

Figure 1. 1A; schematic diagram of extended amygdala of the rat (hatched areas). This is a composite drawing made from several horizontal sections where the outline of the extended amygdala was estimated from Nissl and adjacent immunostained sections for angiotensin II, and projected onto a single horizontal section. 1B; retrograde neuronal labeling in the rat extended amygdala after a Fast Blue injection in the central nucleus of the amygdala. This diagram is a composite from two adjacent horizontal sections. The long interconnections of the extended amygdala in that retrogradely labeled neurons are apparent throughout the extent of the central division of this structure, including the caudomedial accumbens, bed nucleus of the stria terminalis, sublenticular extended amygdala and along the interstitial nucleus of the posterior limb of the anterior commissure. AcbSh-accumbens shell; BST-bed nucleus of the stria terminalis; CeA-central amygdaloid nucleus; IPac-interstitial nucleus of the posterior limb of the anterior commissure; MeA-medial amygdaloid nucleus; SLEA-sublenticular extended amygdala, VP-ventral pallidum.

While one might predict relative homogeneity along the rostro-caudal axis of extended amygdala, this is complicated by the variety of internal subnuclei that are found within the extended amygdala. Aside from the gross separation into medial and central divisions, many smaller subdivisions within these broadly defined areas are apparent. For example, the central nucleus of the amygdala is readily divided into three or more zones based on cytoarchitecture and histochemistry (e.g. McDonald, 1982; de Olmos et al., 1985; Cassell et al., 1986; Price et al., 1987) including prominent lateral and medial portions. The lateral portion may be further parcellated into a central lateral, and a lateral capsular part. The medial nucleus of the amygdala is similarly divisible into anterior and posterior portions, with dorsal and ventral subnuclei in the posterior portion. All of these subnuclei appear to have their counterpart in the rostral forebrain, particularly in the bed nucleus of the stria terminalis (e.g. McDonald, 1983; de Olmos et al., 1985; Price et al., 1987; Moga et al., 1989). However, it is not the case that corresponding columns for all of these subnuclei may be identified within the interconnecting gray matter that courses both dorsal and ventral to the internal capsule.

A SURVEY OF TRANSITION TERRITORIES BETWEEN THE VENTRAL STRIATOPALLIDAL SYSTEM AND THE EXTENDED AMYGDALA

Our inability to certify with precision the exact borderline between the extended amygdala and striatopallidal system has led us to use the term "transition area" where the

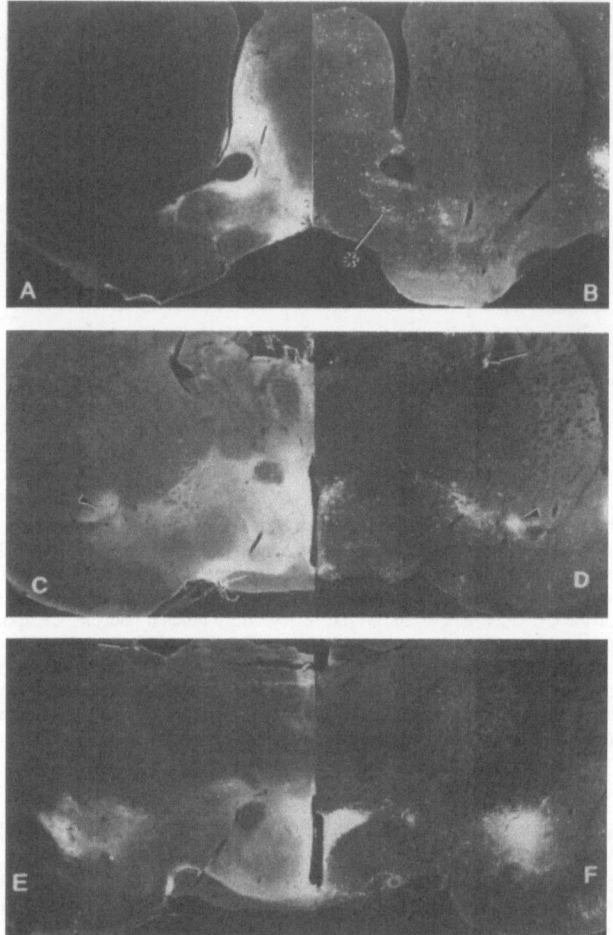

Figure 2. Angiotensin II immunoreactivity in rat extended amygdala, and retrograde labeling of rat extended amgydala after transsynaptic transport of pseudorabies virus from the stomach. Figures 2A, 2C, and 2E are coronal sections immunostained for angiotensin II, while 2B, 2D, and 2F are comparable sections from a rat in which pseudorabies virus was injected in the stomach and esophagus. Immunodetection was used to demonstrate the presence of the virus in forebrain neurons. The asterisk in 2B indicates retrograde labeling of cells in a ventral area that appears to represent a small tongue of the extended amygdala that invades the caudal parts of the ventral pallidum (see text). Arrowheads in 2C, D indicate the retrograde labeling along the interstitial nucleus of the posterior limb of the anterior commissure and the corresponding zone of angiotensin II immunoreactivity. The arrows in 2D and 2E indicate pockets of neurons along the course of the stria terminalis that are retrogradely labeled by the virus or within a nest of angiotensin II immunoreactive terminals. It should be noted that the retrograde viral labeling is confined to the central division of the extended amygdala, while angiotensin II immunoreactivity is found in both the central and medial divisions (see text).

morphological, connectional, and neurohistochemical data are ambivalent. In a sense, one could be more restrictive and not use this term unless individual neurons actually possessed characteristics of both adjacent territories. At the present time insufficient data exist to always distinguish between these two alternatives.

Caudomedial Accumbens, the Interstitial Nucleus of the Posterior Limb of the Anterior Commissure, and the Amygdalostriatal Transition Area of the Temporal Lobe

In the rat, a strong association with the extended amygdala exists for the column of cells that begins in the caudal portion of the medial "shell" of the nucleus accumbens and which is contiguous with the bed nucleus of the stria terminalis at its lateral edge, but also with an ill-defined aggregation of cells found above and below the posterior limb of the anterior commissure. These amygdala related neurons traverse the caudal surface of the accumbens and follow the posterior limb of the anterior commissure in an unbroken column back into the central nucleus of the amygdala, and include at least the rostral portion of the area adjacent to the central nucleus of the amygdala that has been termed the amygdalostriatal transition area.

Very little, if any, angiotensin II immunoreactivity is found within the rat striatum or pallidum (Lind et al., 1985; Lind and Ganten, 1990), but dense reactivity characterizes the centromedial portions of the amygdala. Comparatively dense immunoreactivity is also found in the accumbens shell, and in the interstitial nucleus of the posterior limb of the anterior commissure (IPac) of de Olmos (1972), (figure 2; and Alheid and Heimer, 1988; Heimer and Alheid, 1991). This appears continuous with the rostral portions of the amygdalo-striatal transition area and the caudally contiguous central amygdaloid nucleus. This observation is reinforced by the occurrence of retrograde labeling in these zones when retrograde tracers are placed within the general vicinity of the parabrachial nucleus (Jackson and Crossman, 1981; Alheid and Heimer, 1988 figure 10C; Grove, 1988b; Steininger et al., 1992; Semba and Fibiger, 1992) or in the central nucleus of the amygdala (figure 1B, and Ottersen, 1980; Grove, 1988b).

While retrograde labeling from the tegmentum may well involve areas in and around the pedunculopontine nucleus that may be associated with the basal ganglia, transsynaptic labeling of the extended amygdala, including the caudomedial accumbens, and interstitial

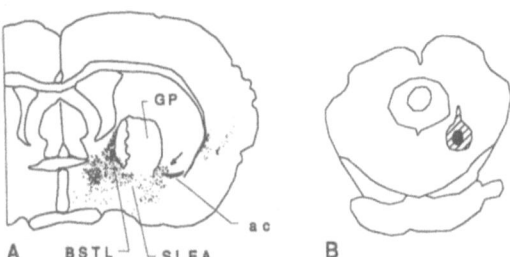

A   BSTL   SLEA   B

Figure 3. Retrograde labeling in the extended amygdala of the rat after Fast Blue injections in the mesopontine reticular formation (3B) and parabrachial nucleus (not shown). 3A is plotted from a single coronal section and demonstrates the continuity of projection neurons extending from the lateral division of bed nucleus of the stria terminalis (BSTL) into the sublenticular extended amygdala (SLEA) just beneath the globus pallidus (GP). The arrow indicates some of the numerous cells labeled in the territory adjacent to the posterior limb of the anterior commissure that could variously be considered an anterior division of the amygdalo-striatal transition area, or a dorsal extension of the interstitial nucleus of the posterior limb of the anterior commissure.

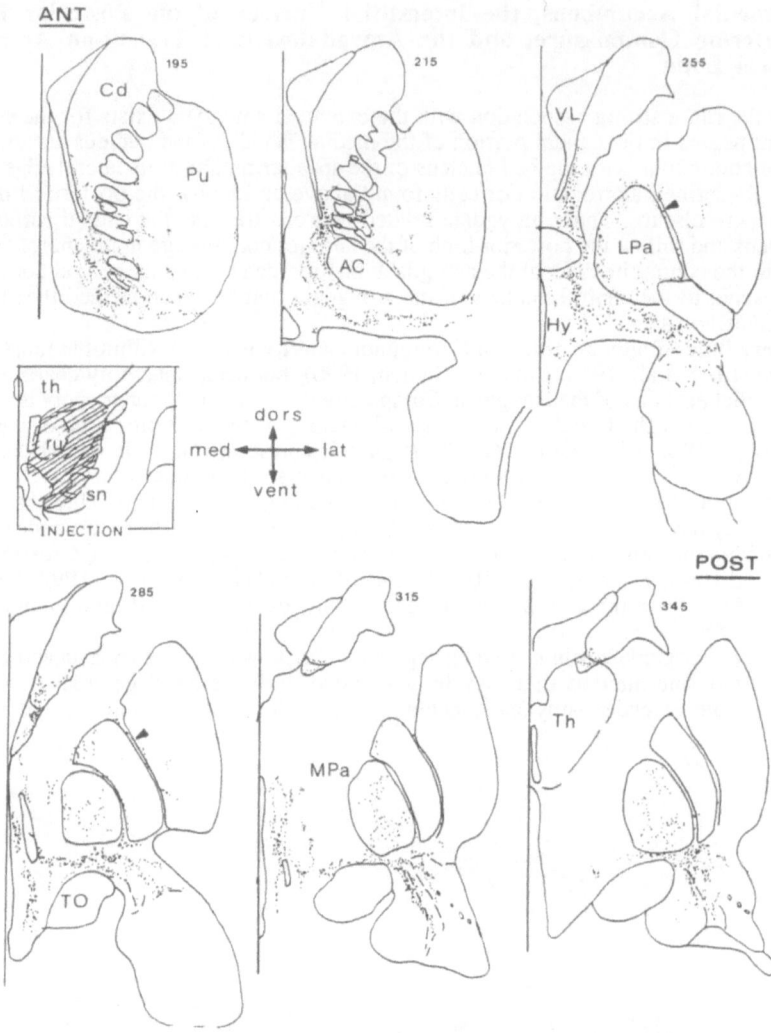

Figure 4. Retrograde labeling in the basal forebrain of the macaque after horseradish peroxidase injections in the mesopontine reticular formation. The inset shows the extent of the halo surrounding the injection site. Only minor involvent of the substantia nigra is suggested by the paucity of labeling in the caudate and putamen. Densely labeled are cells throughout the extended amygdala including a continuous column of labeled neurons accompanying the stria terminalis (the supracapsular portion of the extended amygdala). Also noteworthy are the neurons found in the external medullary lamina of the globus pallidus (arrowheads) that are suggestive of cells that could lie along the course of errant stria terminalis fibers (see text).

nucleus of the posterior limb of the anterior commissure can be obtained after injections of herpes type viruses in the visceral organs innervated by the autonomic nervous system (figure 2BDF; and Blessing et al., 1991; Jansen et al., 1992). It is unlikely that this labeling is related to so-called extrapyramidal midbrain areas.

It should be pointed out that the retrograde labeling found in the caudal part of the accumbens shell does not appear to be completely coextensive with the entire area that is relatively rich in angiotensin immunoreactivity. The retrogradely labeled cells projecting to the amygdala, brainstem, or transsynaptically to the autonomic nervous system, rapidly diminish in density as one moves rostrally in the accumbens shell. The same might be said for the retrograde labeling in the interstitial nucleus of the posterior limb of the anterior commissure, where enriched angiotensin II immunoreactivity is found across a larger area than the cells projecting to the dorsal brainstem (figure 2C, D). This could be taken to indicate the transitional nature of these structures, or alternatively, that more than one layer of extended amygdalar neurons occur in these areas, with only one of these giving rise to caudal brainstem projections. In the rat, neurons within zones corresponding to these angiotensin II immunoreactive areas apparently project densely to the retrorubral area, and nearby mesopontine reticular formation (Berendse et al., 1992), targets that also receive extended amygdala inputs in both the rat and the primate (figure 3; Krettek and Price, 1978; Hopkins and Holstege, 1978; Price and Amaral, 1981; Holstege et al., 1985 (bed nucleus of stria terminalis)). It should be noted that projections caudal to substantia nigra are generally not observed for neurons in dorsal striatum (e.g. figures 3, 4; although see Haber et al., 1990).

The Supracapsular Portion of the Extended Amygdala

This designation identifies neurons that accompany the stria terminalis along its dorsal course to and from the temporal lobe. In the rat this consists of small pockets of cells that may be identified with the medial or central division of the extended amygdala depending on their histochemistry and projections to the medial hypothalamus or parabrachial nucleus respectively (de Olmos et al., 1985). While these form only an archipelago in the rat, histochemical observations in the human and monkey brain (Strenge et al., 1977; Martin et al., 1991), and preliminary observations from retrograde labeling in the monkey (figure 4), indicate that in the primate these neurons comprise a continuous column of cells adjacent to the stria terminalis. This larger gray mass of the primate therefore would seem to be more amenable to functional assessment using available physiological and pharmacological tools.

In addition to retrograde labeling from the hypothalamus or brainstem, the supracapsular portion of the bed nucleus of the stria terminalis in the rat is readily identified with the extended amygdala on the basis of immunohistochemistry for angiotensin II (figure 2E), or by retrograde transsynaptic labeling (figure 2D) from the peripheral autonomic nervous system (i.e. for the cells related to the central division of extended amygdala).

The Marginal Zone of the Striatum

This territory identified in the rat by Shu and colleagues as a unique striatal zone at the rostral surface of the globus pallidus, is distinguished by an unusual pattern of substance P immunoreactivity (Shu et al., 1988), especially dense staining for zinc as demonstrated with the Timm's stain for metals (figure 5; Shu et al., 1990; Heimer and Alheid, 1991), and poor staining with acetylcholinesterase histochemistry (Shu et al., 1988). The marginal zone appears to send projections to the vicinity of magnocellular cholinergic cells at the caudal end the basal ganglia and extended amygdala since the unusual immunohistochemical features of of the dorsal pallidum and apparently it also projects near the caudal end of the substantia nigra (Shu et al., 1988). This area is worth considering in the context of transitions between this zone are similar and contiguous with that found at the caudal face of the nucleus accumbens (ventrally, figure 5A) and in the bed nucleus of the stria terminalis (medially figure 5B). This relationship is best appreciated in sagittal or horizontal sections for both the rat and the primate (figure 5; and Heimer and Alheid, 1991).

One possible anatomical basis for the relationship of this zone with the extended amygdala is the little appreciated fact that aside from the more compact bundles of fibers that interconnect the rostral and caudal portions of the extended amygdala via the dorsal running stria terminalis and the ventral coursing association pathways, many errant fibers take a more direct route to (and possibly from?) the forebrain via the external and internal medullary laminae of the globus pallidus (e.g. Price and Amaral, 1981). Thus some of the neurons within the marginal zone of the striatum could represent path neurons along these amygdala association fibers in a manner similar to the supracapsular cell islands accompanying the stria terminalis. Accordingly, after central amygdaloid nucleus injections of anterograde tracers, scattered axonal branches and apparent terminals can be found in the external medullary lamina of the pallidum, apparently including the marginal zone of the striatum (Alheid, Beltramino, unpublished observations). Several recent reports have also argued for central amygdala or bed nucleus of the stria terminalis projections to the medial part of the dorsal pallidum in the rat (Arluison et al., 1990) and cat (Shinonaga et al., 1992), and this projection is also evident in our material; it should be kept in mind, however, that these may be targeting corticopetal or brainstem projecting cells located with this precinct.

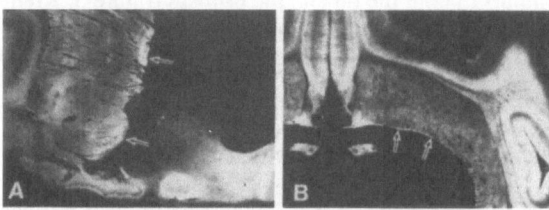

Figure 5. Timm's staining in the marginal zone of the rat (5A) and primate striatum 5B).
Both the rat and the monkey brains in this figure were stained with the sodium selenite version of the Timm's stain for Zinc (Danscher, 1982). 5A is a sagittal section of the rat brain where the dense Zinc labeling is evident at the caudal end of the nucleus accumbens and at the rostral face of the globus pallidus (arrows). 5B is a horizontal section through the squirrel monkey brain at the level of the bed nucleus of the stria terminalis. The dense staining of the bed nucleus is apparent, but also a contiguous band of dense (bright) Zinc labeling that extends lateral along the anterior commissure and in front of the globus pallidus (arrows).

Borders of the Ventral Pallidum with the Extended Amgydala in the Rat

A projection from the ventral pallidum to the midbrain is frequently cited as evidence for the similarity between the dorsal and ventral pallidal pathways (Haber et al., 1985; Swanson et al., 1984; Mogenson et al., 1985). In the dorsal pallidum most thalamic and brainstem projecting neurons are found in the internal pallidal segment, suggesting that in the ventral pallidum, elements of both the internal and external pallidal segments might be found together. This is the position that we adopted in our earlier exploration of the thalamic output of the ventral pallidum. The dense projection that originates from the area of the ventral pallidum and projects to the mediodorsal thalamus (Young et al., 1984) was seen as evidence for the dual nature of the ventral pallidum and its essential parallelism with the dual divisions of the dorsal pallidal complex. Unfortunately, in the rat we were generally unable to find a significant amount of retrogradely labeled neurons confined to the ventral pallidum when retrograde tracers were injected in the vicinity of the pedunculopontine tegmental nucleus. A result that is consistent with most reports using brainstem injections of retrograde tracers (e.g Jackson and Crossman, 1981; Swanson et al., 1984; see also Semba and Fibiger, 1992; Steininger et al., 1992). Cells surround the ventral pallidum rostrally and caudally, appearing to avoid the ventral pallidum when it is simultaneously marked with immunohistochemistry (figure 10c in Alheid and Heimer, 1988). In the rat a few neurons projecting to the brainstem can be identified ventrally, that the casual observer might relate to the ventral pallidum, but these occur in an histochemically mixed area suggesting intermingled elements of ventral pallidum and extended amygdala. In sagittal sections this appears to result from a small tongue of the extended amygdala that interrupts the caudal surface of the ventral pallidum and which may form a loose bridge between the extended amygdala and related neurons at the caudal surface of the nucleus accumbens (Heimer et al., 1993 figure 6b). In coronal sections of the rat brain this isthmus often appears as a small window generally positive for various immunohistochemical stains typical of the extended amygdala (e.g. angiotensin II, figure 2A; and Lind et al., 1985; or norepinephrine, Swanson and Hartman, 1974 figure 15; Grove, 1988a figure 2a). It is this complex topography of the ventral pallidum and adjacent extended amygdala that mitigates against successful use of anterograde tracers alone (e.g. see discussion in Haber et al., 1993) in order to define the exact targets of the ventral pallidum. Even when these are combined with complementary experiments using retrograde labeling, the complexity of the area can be misleading.

Faced with this interpretation of the data, from brainstem injections of retrograde tracers, and our general working hypothesis of parallels between the dorsal and ventral striatum, one might ask; where is the counterpart of the brainstem projecting internal pallidal segment for the ventral striatopallidal system? The answer to this seems to be, that for the most part it can be found within the internal pallidal segment. In the monkey it has been recently shown (Haber et al., 1990) that the nucleus accumbens sends a projection to the ventral pallidum but also to the anteromedial part of the medial pallidal segment (see also, Alheid et al., 1990 figure 38). In the rat this problem is more difficult, but the accumbens terminates densely in the lateral hypothalamus just medial to the entopeduncular nucleus (Heimer et al., 1991b). It has been noted that this zone resembles the adjacent entopeduncular nucleus immunohistochemically (Haber and Nauta, 1983; Heimer et al., 1991) except that the neurons are not encapsulated by the myelinated bundles of the peduncle. Our current supposition then, is that the medial pallidal segment for the rat ventral striatopallidal system is found within this extension of the entopeduncular nucleus into the lateral hypothalamus. As an alternative, the ventral equivalent of the internal pallidal segment could be split between the substance P rich caudomedial (and ventral) parts of ventral pallidum and the ectopic entopeduncular cells in the lateral hypothalamus.

Since the internal pallidal segment includes pallidothalamic projecting neurons, this leaves us with an embarassing excess of pallidothalamic pathways. The projection from ventral forebrain to mediodorsal thalamus is very prominent in the rat and appears to originate within pallidal areas (Young et al., 1985; Price and Slotnik, 1983; Mogenson et al., 1987; Hallenger et al., 1987; Groenewegen, 1990). A similar dense projection to dorsal thalamus is also observed in the monkey (Russchen et al., 1987; Hreib et al., 1988; Parent et al., 1988; Haber et al., 1993) but it is clear that these cells are also found outside the territory of the ventral pallidum. In fact, it has been argued that for the primate, mediodorsal projecting neurons may be preferentially located outside the ventral pallidum (Hreib et al., 1988; Haber et al., 1993). While this may be an instance of a pronounced species difference, it could imply that the significance of the mediodorsal projection and its specific relation to the remainder of forebrain needs to be reassessed in both species. Even within the ventral pallidum of the rat a preferential medial localization of mediodorsal projecting neurons is evident (Young et al., 1985; Mogenson et al., 1987; Groenewegen, 1990), and this may argue for a particular targeting of these cells by the shell of the accumbens (Zahm and Heimer, 1990; Zahm and Brog, 1992). It also should be recalled that the dorsal thalamic projecting neurons in the rat also include considerable numbers within the extension of the pallidum into the olfactory tubercle, and that the special significance of olfactory stimuli for such species may be reflected by the prominence of this pathway.

THE VENTRAL STRIATOPALLIDAL SYSTEM AND EXTENDED AMYGDALA AS TWO COMPETING THEORIES FOR "LIMBIC-MOTOR" INTEGRATION

One problem that confronts functional studies of the basal forebrain is the over generalization of the ventral striatopallidal system to encompass portions of the caudal sub-pallidal areas that are occupied by the extended amygdala. Intense interest is justifiably focussed on the functions of the nucleus accumbens as a subcortical target of the mesolimbic dopaminergic projection which, in turn, has been identified as a probable target of anti-dopaminergic neuroleptic drugs. The accumbens is also considered as an important component in reinforcing electrical self-stimulation, or the effective target for drugs of abuse. For example, the role of the nucleus accumbens as a substrate for the increased motor activation occurring in response to a variety of drugs is well documented (e.g. Mogenson et al., 1980; Austin and Kalivas, 1990; Pulvirenti et al., 1991). Often overlooked is the fact that the ventral tegmental dopamine projections also include the extended amygdala and especially its central division. This confound is amplified if one takes into account our discrimination that the extended amygdala invades the caudal and medial portions of the nucleus accumbens.

The unusual neurochemistry, psychopharmacology, and anatomy that characterize portions of the nucleus accumbens and distinguish it from dorsal striatum are often cited as evidence for its special role in the basal ganglia with respect to processing emotional and motivated behaviors. While this is not unreasonable if one views the accumbens as part of a "limbic" striatum, it is striking how much of the exceptional anatomical characteristics of the caudomedial and "shell" portions of the accumbens might be considered as part of the normal constituents of the extended amygdala. This includes in the rat, projections to the remainder of the extended amygdala, and to rostral and caudal lateral hypothalamic areas not associated with the entopeduncular nucleus (Heimer et al., 1991), projections to the retrorubral area, central gray, and mesopontine reticular formation (Heimer et al., 1991; Haber et al., 1990; Berendse et al., 1992). These apparent relations of the extended amygdala, not only with the autonomic nervous system, but also with somatomotor targets in brainstem has induced Holstege (1990) to include this part of the forebrain within his third or "limbic" motor system.

It is useful to note that some projections do not appear to provide a ready distinction between the ventral striatopallidal system and the extended amygdala. Most notable is the descending projections to the dopaminergic cells in the midbrain (e.g. Hopkins and Holstege, 1978; Holstege et al., 1985; Gonzales and Chesselet, 1990; Wallace et al., 1992). While it may well turn out that efferents from the extended amygdala particularly target the same neurons that give rise to the dopaminergic amygdalopetal projections, it may also be the case that nigrostriatal neurons are also the recipient of extended amygdala afferents. This projection from areas that comprise the extended amygdala could be substantial, from our perspective that this would include the caudomedial accumbens, bed nucleus of the stria terminalis, the central nucleus of the amygdala, and the interconnecting subpallidal and supracapsular cell columns. At least one attraction of this type of input would be that the dopaminergic cells would provide one avenue for the information processed in the extended amygdala to reach broad portions of the caudate and putamen, a useful corridor for motivational influences on motor integration, but a possibility that except for the nucleus accumbens, seems to await direct anatomical confirmation at the ultrastructural level.

The identification of the extended amygdala as a macrostructure in basal forebrain presents opportunities for physiological and behavioral studies to analyze functions from what seemed to be widely disparate areas of the forebrain, in terms of a more unified processing system with predictable anatomical relations. The caveat is in order, that the numerous internal subdivisions of the extended amygdala present a uniquely challenging structure for analysis. With respect to the basal ganglia, the ventral striatopallidal system would seem to offer similar opportunities, although our anatomical insights would suggest that care should be taken to avoid the transition zones most closely aligned with the extended amygdala if a relatively uncontaminated basal ganglia response is sought. One might alternatively see the identification of transition zones as an additional opportunity where behavioral and physiological responses might also demonstrate a transition between functions characteristic of the basal ganglia and those associated with the extended amygdala.

Acknowledgements

This work was supported by USPHS grant NS17743 (GFA) and GM27739 (RRM), and by Consejo Nacional de Investigaciones Cientificas y Tecnicas of Argentina (CAB and J. de O).

REFERENCES

Alexander, G.E., DeLong, M.R., and Strick, P.L., 1986, Parallel organization of functionally segregated circuits linking basal ganglia and cortex, *Ann. Rev. neurosci.* 9:357-381.

Alexander, G.E., Crutcher, M.D., and Delong, M.R., 1990, Basal Ganglia-thalamocortical circuits-parallel substrates from motor, oculomotor, prefrontal, and limbic functions, *Prog. Brain Res.* 85:119-146.

Alheid, G.F., and Heimer, L., 1988, New perspectives in basal forebrain organization of special relevance for neuropsychiatric disorders; the striatopallidal, amygdaloid, and corticopetal components of substantia innominata, *Neurosci.* 27:1-39.

Alheid, G.F., Heimer, L., and Switzer, R.C., 1990, The basal ganglia, *in*: "The Human Nervous System," G. Paxinos, ed., Academic Press, San Diego, pp. 483-582.

Arluison, M., Vankova, M., Cesselin, F., and Leviel, V., 1990, Origin of some enkephalin-containing afferents to the ventro-medial region of the globus pallidus in the rat, *Brain Res. Bull.* 25:25-34.

Austin, M.C., and Kalivas, P.W., 1990, Enkephalinergic and GABAergic modulation of motor activity in the ventral pallidum, *J. Pharm. Exper. Ther.* 252:1370-1377.

Berendse, H.W., Groenewegen, H.J., and Lohman, A.H.M., 1992, Compartmental distribution of ventral striatal neurons projection to the mesencephalon in the rat, *J. Neurosci.* 12:2079-2103.

Blessing, W.W., Li, Y.-W, and Wesselingh, S.L., 1991, Transneuronal transport of herpes simplex virus from the cervical vagus to brain neurons with axonal inputs to central vagal sensory nuclei in the rat, *Neurosci.* 42:261-274.

Carlsen, J., 1989, New perspectives on the functional anatomical organization of the basolateral amygdala, *Acta Neurol. Scand.* suppl. 122, 79:5-27.

Carlsen, J., and Heimer, L., 1988, The basolateral amygdaloid complex as a cortical-like structure, *Brain Res.* 441:377-380.

Cassell, M.D., Gray, T.S., and Kiss, J.Z., 1986, Neuronal architecture in the rat central nucleus of the amygdala: A cytological, hodological, and immunocytochemical study, *J. Comp. Neurol.* 246:478-499.

Danscher, G., 1982, Exogenous selenium in the brain; a histochemical technique for light and electron microscopical localization of catalytic selenium bonds, *Histochem.* 76:281-293.

de Olmos, J.S., 1972, The amygdaloid projection field in the rat as studied with the cupric-silver method, *in*: "The Neurobiology of the Amygdala," B.E. Elefteriou, ed., Plenum, New York, pp. 145-204.

de Olmos, J.S., Alheid, G.F., and Beltramino, C.A., 1985, Amygdala, *in*: "The Rat Nervous System," G. Paxinos, ed., Academic Press, New York, pp. 223-334.

François, C., Yelnik, J., and Percheron, G., 1987, Golgi study of the primate substantia nigra. II. Spatial organization of dendritic arborizations in relation to the cytoarchitectonic boundaries and to the striatonigral bundle, *J. Comp. Neurol.* 265:473-493.

Gonzales, C., and Chesselet, M.F., 1990, Amygdalonigral pathway: An anterograde study in the rat with phaseolus vulgaris leucoagglutinin (PHA-L), *J. Comp. Neurol.* 297:182-200.

Groenewegen, H.J., 1990, Organization of the afferent connections of the mediodorsal thalamic nucleus in the rat, related to the mediodorsal-prefrontal topography, *Neurosci.* 24:379-431.

Grove, E.A., 1988a, Neural associations of the substantia inominata in the rat; afferent connections, *J. Comp. Neurol.* 277:315-346.

Grove, E.A., 1988b, Efferent connections of the substantia innominata in the rat, *J. Comp. Neurol.* 277:347-364.

Haber, S.N., Groenewegen, H.J., Grove, E.A., and Nauta, W.J.H., 1985, Efferent connections of the ventral pallidum; evidence of a dual striatopallidofugal pathway, *J. Comp. Neurol.* 235:322-335.

Haber, S.N., Lynd, E., Klein, C., and Groenewegen, H.J., 1990, Topographic organization of the ventral striatal efferent projections in the rhesus monkey: an anterograde tracing study, *J. Comp. Neurol.* 293:282-298.

Haber, S.N., Lynd-Balta, E., and Mitchell, S.J., 1993, The organization of the descending ventral pallidal projections in the monkey, *J. Comp. Neurol.* 329:111-128.

Haber, S.N., and Nauta, W.J.H., 1983, Ramifications of the globus pallidus in the rat as indicated by patterns of immunohistochemistry, *Neurosci.* 9:245-260.

Hallanger, A.E., Levey, A.I., Henry, J.L., Rye, D.B., and Winer, B.H., 1987, The origins of cholinergic and other subcortical afferents to the thalamus in the rat, *J. Comp. Neurol.* 262:105-124.

Heimer, L., 1978, The olfactory cortex and the ventral striatum, *in*: "Limbic Mechanisms," K.E. Livingston and O. Hornykiewicz, eds., Plenum, New York, pp. 95-187.

Heimer, L., and Alheid, G.F., 1991, Piecing together the puzzle of basal forebrain anatomy, *in*: "The Basal Forebrain: Anatomy to Function", T.C. Napier, P.W. Kalivas and I. Hanin, eds., Plenum Press, New York, pp. 1-44.

Heimer, L., Alheid, G.F., and Zaborszky, L., 1985, Basal Ganglia, *in*: "The Rat Nervous System," Volume 1, Forebrain and Midbrain, G. Paxinos, ed., Academic Press, Sydney, pp. 37-86.

Heimer, L., Alheid, G.F., and Zahm, D.S., 1993, Basal forebrain organization: an anatomical framework for motor aspects of drive and motivation, *in*: "Limbic motor circuits and neuropsychiatry," P. Kalivas, ed., CRC reviews, in press.

Heimer, L., de Olmos, J.S., Alheid, G.F., and Záborszky, L., 1991a, Perestroika in basal forebrain; opening the borders between neurology and psychiatry, *in*: "Role of the forebrain in sensation and behaviour," G. Holstege, ed., Progr. Brain Res. 87:109-165.

Heimer, L., and Wilson, R.D., 1975, The subcortical projections of allocortex; similarities in the neural associations of the hippocampus, the piriform cortex and the neocortex, *in*: "Golgi Centennial

Symposium Proceedings," M. Santini, ed., Raven Press, New York, pp. 177-193.

Heimer, L., Zahm, D.S., Churchill, L., Kalivas, P.W., and Wohltmann, C., 1991b, Specificity in the projection patterns of accumbal core and shell in the rat, *Neurosci.* 41:89-125.

Holstege, G., 1990, Descending motor pathways and the spinal motor system: Limbic and non-limbic components, *in*: "Role of the forebrain in sensation and behavior," G. Holstege, ed., Progr. Brain Res. 87:307-421.

Holstege, G., Meiners, L., and Tan, K., 1985, Projections of the bed nucleus of the stria terminalis to the mesencephalon, pons, and medula oblongata in the cat, *Exp. Brain Res.* 58:379-391.

Hopkins, D.A., and Holstege, G., 1978, Amygdaloid projections to the mesencephalon, pons, and medulla oblongata in the cat, *Exp. Brain Res.* 32:529-547.

Hreib, K.K., Rosene, D.L., and Moss, M.B., 1988, Basal forebrain efferents to the medial dorsal thalamic nucleus in the rhesus monkey, *J. Comp. Neurol.* 277:365-390.

Jackson, A., and Crossman, A.R., 1981, Basal ganglia and other afferent projections to the peribrachial region in the rat; a study using retrograde and anterograde transport of horseradish peroxidase, *Neurosci.* 6:1537-1549.

Jansen, A.S.P., Terhorst, G.J., Mettenleiter, T.C., and Loewy, A.D., 1992, CNS cell groups projecting to the submandibular parasympathetic preganglionic neurons in the rat-- A retrograde transneuronal viral cell body labeling study, *Brain Res.* 572:253-260.

Krettek, J.E., and Price, J.L., 1978, Amygdaloid projections to subcortical structures within the basal forebrain and brainstem in the rat and cat, *J. Comp. Neurol.* 178:225-254.

Lind, R.W., Ganten, D., 1990, Angiotensin, *in*: "Handbook of Chemical Neuroanatomy," V 9, Neuropeptides in the CNS part II., A. Björklund, T. Hökfelt, and M.J. Kuhar, eds., Elsevier, Amsterdam, pp. 165-286.

Lind, R.W., Swanson, L.W., and Ganten, D., 1985, Organization of angiotensin II immunoreactive cells and fibers in the rat central nervous system, *Neuroendocrinology* 40:2-24.

Martin, Lee J., Powers, R.E, Dellovade, T.L., and Price, D.L., 1991, The Bed Nucleus-Amygdala Continuum in Human and Monkey, *J. Comp. Neurol.* 309:445-485.

McDonald, A.J., 1982, Cytoarchitecture of the central amygdaloid nucleus of the rat, *J. Comp. Neurol.,* 208:401-418.

McDonald, A.J., 1983, Neurons of the bed nucleus of the stria terminalis: a Golgi study in the rat, *Brain Res. Bull.* 10:111-120.

McDonald, A.J., 1992, Cell types and Intrinsic Connections the Amygdala, *in*: "The Amygdala: Neurobiological Aspects of Emotion, Memory, and Mental Dysfunction," J. Aggleton, ed., Wiley-Liss, New York, pp. 67-96.

Moga, M.M., Saper, C.B., and Gray, T.S., 1989, Bed nucleus of the stria terminalis: cytoarchitecture, immunohistochemistry, and projection to the parabrachial nucleus in the rat, *J. Comp. Neurol.* 283:315-332.

Mogenson, G.J., Ciriello, J., Garland, J., and Wu, M., 1987, Ventral pallidum projections to mediodorsal nucleus of the thalamus: an anatomical and electrophysiological investigation in the rat, *Brain Res.* 404:221-230.

Mogenson, G.J., Jones, D.L., and Yim, C.Y., 1980, from motivation to action: functional interface between the limbic system and the motor system, *Prog. Neurobiol.* 14:69-97.

Mogenson, G.J., Swanson, L.W., and Wu, M., 1985, Evidence that projections from substantia innominata to zona incerta and mesencephalic locomotor region contribute to locomotor activity, *Brain Res.* 334:65-76.

Ottersen, O.P., 1980, Afferent connections to the amygdaloid complex of the rat and cat. II. Afferents from the hypothalamus and the basal telencephalon, *J. Comp. Neurol.* 194: 267-289.

Parent, A., Paré, D., Smith, Y., and Steriade, M., 1988, Basal forebrain cholinergic and noncholinergic projections to the thalamus and brainstem in cats and monkeys, *J. Comp. Neurol.* 277:281-391.

Percheron, G., Yelnik, J., and François, C., 1984, Golgi analysis of the primate globus pallidus. III. Spatial organization of the striato-pallidal complex, *J. Comp. Neurol.* 227:214-227.

Price, J.L., and Amaral, D.G., 1981, An autoradiographic study of the projections of the central nucleus of the monkey amygdala, *J. Neurosci.* 11:1242-1259.

Price, J.L., Russchen, F.T., and Amaral, D.G., 1987, The limbic region. II. The amygdaloid complex, *in*: "Handbook of Chemical Neuroanatomy: Integrated Systems of the CNS," Vol.5, Part 1. A. Björklund, T. Hökfelt, and L.W. Swanson, eds., Elsevier Science Publishers, Amsterdam, pp. 279-386.

Price, J.L., and Slotnick, B.M., 1983, Dual olfactory representation in the rat thalamus; an anatomical and electrophysiological study, *J. Comp. Neurol.* 215:63-77.

Pulvirenti, L., Swerdlow, N.R., and Koob, G.F., 1991, Nucleus accumbens NMDA antagonist decreases locomotor-activity produced by cocaine, heroin or accumbens dopamine, but not caffeine, *Pharmacol. Biochem. Beh.* 40:841-845.

Russchen, F. T., Amaral, D. G., and Price, J. L., 1987, The afferent input to the magnocellular division of the mediodorsal thalamic nucleus in the monkey, *Macaca fascicularis, J. Comp. Neurol.* 256:175-210.

Schwaber, J.S., Kapp, B.S., Higgins, G.A., and Rapp, P.R., 1982, Amygdaloid and basal forebrain direct connections with the nucleus of the solitary tract and the dorsal motor nucleus of the vagus, *J. Neurosci.* 2:424-1438.

Semba, K., and Fibiger, H.C., 1992, Afferent connections of the laterodorsal and the pedunculopontine tegmental nuclei in the rat: A retro- and anterograde transport and immunohistochemical study, *J. Comp. Neurol.* 323:387-410.

Shinonaga, Y., Takada, M., and Mizuno, N., 1992, Direct projections from the central amygdaloid nucleus to the globus pallidus and substantia nigra in the cat, *Neurosci.* 51:691-703.

Shu, S.Y., McGinty, J.F., and Peterson, G.M., 1990, High density of zinc- containing and dynorphin B- and substance P-immunoreactive terminals in the marginal division of the rat striatum, *Brain Res. Bull.* 24:201-205.

Shu, S.Y., Penny, G.R., and Peterson, G.M., 1988, The "marginal division"; a new subdivision in the neostriatum of the rat, *J. Chem. Neuroanat.* 1:147-163.

Steininger, T.L., Rye, D.B., and Wainer, B.H., 1992, Afferent projections to the cholinergic pedunculopontine tegmental nucleus and adjacent midbrain extrapyramidal area in the albino rat. I. Retrograde tracing studies, *J. Comp. Neurol.* 321:515-543.

Strenge, H., Braak, E., and Braak, H., 1977, Über den Nucleus Striae Terminalis im Gehirn des Erwachsenen Menschen, Z. Mikrosk, *Anat. Forsch.* (Leipzig) 91:105-118.

Swanson, L.W., and Hartman, B.K., 1974, The central adrenergic system. An immunofluroescence study of the location of cell bodies and their efferent connections in the rat utilizing dopamine-ß-hydroxylase as a marker, *J. Comp. Neurol.* 163:467-506.

Swanson, L.W., Mogenson, G.J., Gerfen, C.R., and Robinson, P.,1984, Evidence for a projection from the lateral preoptic area and substantia innominata to the "mesencephalic locomotor region" in the rat, *Brain Res.* 295:161-178.

Wallace, D.M., Magnuson, D.J., and Gray, T.S., 1992, Organization of amygdaloid projections to brain-stem dopaminergic, noradrenergic, and adrenergic cell groups in the rat, *Brain Res. Bull.* 28:447-545.

Yelnik, J., Percheron, and, G., François, C., 1984, Golgi analysis of the primate globus pallidus. II. Quantitative morphology and spatial orientation of dendritic arborizations, *J. Comp. Neurol.* 227:200-213.

Young, W.S. III, Alheid, G.F., and Heimer, L., 1984, The ventral pallidal projection to the mediodorsal thalamus; a study with fluorescent retrograde tracers and immunohistofluorescence, *J. Neurosci.* 4:1626-1638.

Zahm, D.S., and Brog, J.S., 1992, Commentary: On the significance of the core-shell boundary in the rat nucleus accumbens, *Neurosci.* 50:751-767.

Zahm, D.S., and Heimer, L., 1990, Two transpallidal pathways originating in nucleus accumbens, *J. Comp. Neurol.* 302:437-446.

Zemanick, M.C., Strick, P.L., and Dix, R.D., 1991, Direction of transneuronal transport of Herpes-simplex virus-1 in the primate motor system is strain-dependent, *Proc. Nat. Acad. Sci.* (USA) 88:8048-8051.

SYNAPTIC ORGANIZATION OF THE AMYGDALAR INPUT TO THE NUCLEUS ACCUMBENS IN THE RAT

Luke R. Johnson, Rebecca L.M. Aylward and Susan Totterdell

Department of Pharmacology
Mansfield Road
Oxford, OX1 3QT
England

The amygdala has long been implicated in the recognition of emotionally relevant stimuli (for review see Everitt & Robins, 1992) and there is now considerable evidence to support the concept of the nucleus accumbens (NA) as a site of limbic - motor interface (Mogenson et al., 1980). Clinically the amygdala has been implicated in such disorders as schizophrenia (Gray et al., 1991) and depression (Drevets et al.,1992). The caudal basolateral amygdaloid nucleus (BLa) sends a very heavy projection, via the stria terminalis, to the bed nucleus of the stria terminalis, the medial edge of the dorsal striatum and NA shell while the rostral BLa sends ascending axons on a ventral path to the lateral NA and prefrontal cortex (De Olmos and Ingram, 1972; Groenewegen et al., 1980; McDonald, 1991).

Ragsdale and Graybiel (1988) found in the cat that the distributions of anterior dorsal BLa afferents in the dorsal striatum and NA are preferentially to acetylcholinesterase-poor striosomes. The caudomedial NA is innervated by the medial and ventral amygdala, probably the parvicellular region. However, a PHA-L study in rat (Kita and Kitai, 1990) has shown a projection to the matrix compartment of the rostro-ventral caudate-putamen. In the rat amygdalar-accumbens neurons have been shown to be large, spiny neurons sometimes described as "pyramidal" (Carlsen, 1988).

Electrical stimulation of amygdalar projection neurons has been shown to activate medium sized spiny neurons in the NA (De France et al., 1980; Callaway et al., 1991). Yim and Mogenson (1986) found that stimulation of the ventral tegmental area modulated this excitatory response, presumably via its dopaminergic input. This mesolimbic projection has been shown to mediate conditioned reinforcement (Everitt and Robins, 1992). Systemic amphetamine supports the acquisition of place preference, while BLa lesion abolishes it. These data would suggest a direct anatomical interaction in the accumbens between the mesolimbic dopamine and the amygdala input. In addition to the intense dopamine staining described in the NA (Voorn et al., 1986), immunocytochemistry has revealed, amongst others, the presence of somatostatin (Johansson et al.,1984) and GABA (Meredith and Wouterlood, 1991). Both are present in interneuronal populations, but GABA is also thought to be the transmitter of the medium size projection neurons. GABA is found to decrease amphetamine induced locomotion (Scheel-Krüger, 1986) and stimulation of NA somatostatin receptors has been reported to increase locomotion (Bell and Reisine, 1993).

Using a variety of tracing techniques combined with immunocytochemistry we examined the excitatory amygdalar input to the NA and its possible modulation either by local interneurons or by inhibitory dopaminergic fibres from the ventral tegmental area.

The Basal Ganglia IV, Edited by
G. Percheron *et al.*, Plenum Press, New York, 1994

Amygdalar Efferents to the NA

Injections of the anterograde tracer biocytin into the caudal BLa confirmed that fibres passed dorsally and caudally into the stria terminalis, before turning rostrally and running down into the NA, where they distribute over both core and shell regions, but mainly the latter. It was also possible to follow some fibres along a ventral route where they finally reached the fundus striati and then the most lateral aspects of the NA. These fibres were sparse, much finer than those in the stria terminalis, and varicose along their entire projection. They may be a part of the extended amygdala as proposed by Alheid and Heimer (1988).

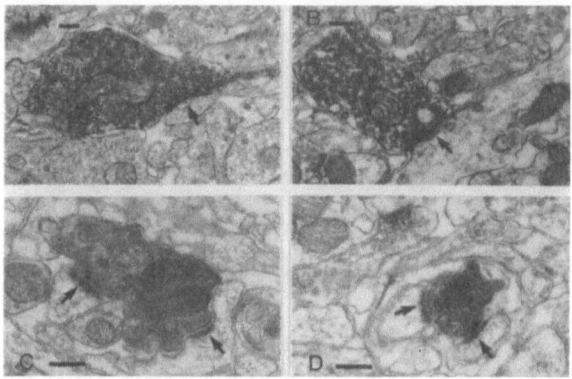

Figure 1. Anterogradely labelled boutons in the nucleus accumbens from the basolateral nucleus of the amygdala. Biocytin labelled boutons in the shell in asymmetrical synaptic contact with a spine head (A) and a dendrite (B). Degenerating boutons in the shell (C) and core (D) in asymmetrical synaptic contact with two spines simultaneously. Scale bar = 0.2μm

In addition to the anterograde transport of biocytin, we also used the anterograde transport of horseradish peroxidase (HRP) and degeneration following electrolytic lesions of the BLa to identify amygdalar terminals in the NA. All methods appeared to label similar populations of terminals (Figure 1). Of the 112 boutons examined, although absolute numbers were higher in the shell, there was homogeneity of targets between the core and shell with 87% synaptic on to spines and 13% on to dendrites in the shell, and 82% on to spines and 18% on to dendrites in the core. The proportion of boutons where two spines were simultaneously contacted was also found to be the same for core and shell. These proportions, similar to those found by Kita and Kitai (1990) using PHA-L, were consistent for all three labelling methods.

Interaction between Amygdalar Input and GABA, Somatostatin and TH

Antibodies to GABA, somatostatin 28 (1-12) (SOM) and tyrosine hydroxylase (TH) were used to examine the relationship between the excitatory input from the amygdala and inhibitory neurotransmitters. These results are summarized diagrammatically in Figure 3. In the light microscope GABA immunoreactivity was seen only in neurons with the morphology of interneurons (Meredith and Wouterlood, 1991), despite the fact that the medium-sized spiny neurons are also GABAergic. Their distribution in both core and shell seemed homogeneous. Immunoreactive varicosities, which we have assumed probably originate from this interneuronal population, were seen in the whole of the NA, sometimes apparently clustering around cell bodies. Despite poor penetration and sub-optimal ultrastructure, in the electron microscope GABA-immunoreactive boutons were found in symmetrical synaptic contact with somata (16% core, 7% shell), dendrites, large (56% core, 63% shell) and small (20% core and shell) and some of which are spiny, and occasionally spines (7% core, 9% shell). An example was found of a GABA-immunoreactive bouton in symmetrical synaptic contact with a spine which was also post-synaptic to a bouton degenerating following a lesion of the amygdala (Figure 2A).

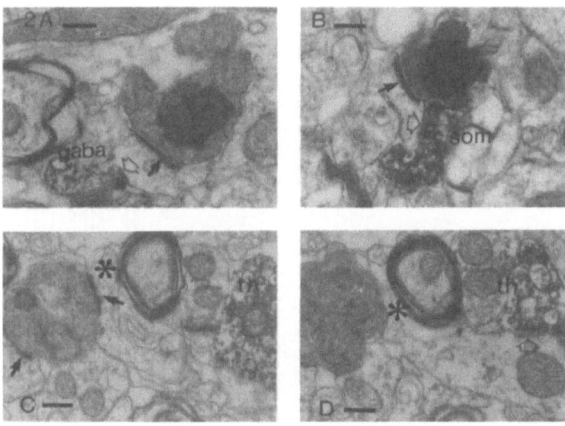

Figure 2. Boutons degenerating after a lesion of the basolateral amygdala in asymmetrical synaptic contact (arrows) with profiles in rat NA that are also postsynaptic to transmitter immunolabelled boutons forming symmetrical synapses (open arrows). A. Degenerating amygdala input to a spine head and adjacent is a GABA immunopositive bouton making a symmetrical synapse. B. Amygdalar input to a small dendrite or spine which is also postsynaptic to symmetrical somatostatin input. C. A degenerating bouton in asymmetrical synaptic contact with two targets (arrows), one of which is to spine head (asterisk) which is shown in serial section (D) to be postsynaptic to a bouton immunopositive for tyrosine hydroxylase. Scale bar = 0.2μm

Somatostatin-immunoreactive neurons and fibres were also found in both the core and shell of the NA, although neurons were most common in the lateral aspects of the nucleus. They were usually bipolar, with long, smooth dendrites, branching infrequently. Immunoreactive fibres were somewhat hetereogeneously distributed, occurring in dense patches for example, at the medial edge of the shell, below the septal hook. In the electron microscope SOM-immunoreactive boutons make symmetrical synaptic contacts mainly with small diameter dendrites (58%), and spines (27%), but occasionally with somata (1%) and larger dendrites (14%). The postsynaptic target distribution was the same for core and shell. Somatostatin immunoreactive input was found onto a spine that was also postsynaptic to a degenerating bouton of amygdalar origin (Figure 2B).

An antibody to tyrosine hydroxylase was used as a marker of the dopaminergic input to the NA from the ventral tegmental area. The distribution of fibres was dense throughout the

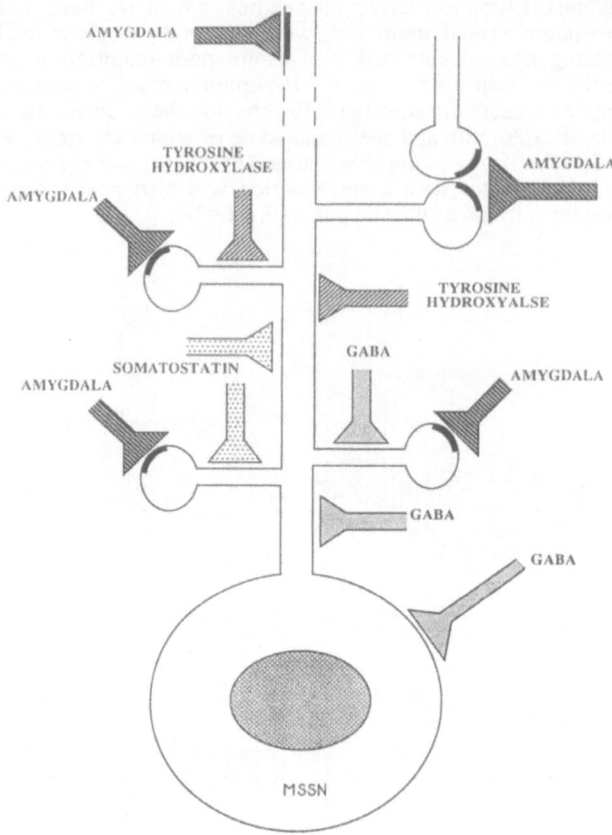

Figure 3. Summary of the input to the medium size spiny neuron of the nucleus accumbens in relation to the amygdala input.

nucleus but, as described by Voorn et al. (1986) for dopamine, particularly dense just below the septal hook of the shell. In the electron microscope TH-immunoreactive fibres were seen to make en-passant symmetrical synaptic contacts with dendrites and spines, including spine necks. Notably these synaptic specializations are relatively small, often visible in only two or three ultrathin sections, and involving quite short appositions of membrane (Sesack and Pickel, 1992).

After lesions of the amygdala, degenerating boutons were found quite frequently in the shell of the NA and in this tissue examples were found of boutons immunoreactive for TH in symmetrical synaptic contact with profiles which were also postsynaptic to degenerating boutons (Figures 2C & D). As the chances of finding examples of convergence where the two boutons are in synaptic contact with the same profile are inevitably limited by technical aspects of the methods, we also processed tissue where the postsynaptic neuron had been identified by the *in vitro* intracellular filling method (Buhl et al., 1989) using biotinylated Lucifer Yellow (Molecular Probes) as the fluorescent label, and the ABC method for EM localization. In total we have found 10 examples of convergence in 4 different rats. This confirms and extends our earlier observations (Aylward and Totterdell, 1991).

Functional Significance

We have shown that amygdala input to the NA distributes over the whole nucleus. However, the projection is topographic and from the caudal BLa the majority of fibres seem to terminate in the shell. In the shell we have shown that convergence occurs between this excitatory input and the presumably inhibitory dopaminergic input from the VTA. The location of the TH-immunoreactive boutons on the same spines or spine necks as the amygdalar input, is similar to that found for the inputs to the NA from the hippocampus (Totterdell and Smith, 1989), the prefrontal cortex (Sesack and Pickel, 1992) and the cortical input to the dorsal striatum (Freund et al., 1984), where it is thought to suggest a modulatory function for dopamine. The targets are medium size spiny neurons, which presumably give rise to the projections from the NA. Our observations agree with those of Zahm (1992) that, in the shell, more of the DA input is to dendritic shafts than is seen in the core. This is relevant as a study by Meredith et al., (1992) has shown that projection neurons of the shell are less spiny and have fewer branches than those in the core. However, despite these differences, the dopamine input to spiny neurons in the shell still occupies a strategic position in relation to the incoming information from the amygdala. From a physiological viewpoint it is interesting that the caudal BLa and the ventral subiculum both terminate in the NA shell, which is devoid of calcium binding proteins (Voorn et al., 1989).

The unusual role of the amygdala as a limbic input to the *dorsal* striatum remains to be elucidated. Firstly, does amygdala input converge with A9 dopamine as a striosomal projection target would predict, or do the striatal postsynaptic neurons receive input from A8 or A10, more like their counterparts in the ventral striatum? Secondly, does the amygdala information then return to the *ventral* striatofugal pathway, where it gains access to the prelimbic cortex as well as feeding indirectly to the basal ganglia via the substantia nigra (SN) pars compacta, or does it join the dorsal striatal output, where it gains direct access to the basal ganglia output nuclei, feeding into motor system or the premotor nuclei of the brainstem by way of the SN pars reticulata?

In addition to the convergence in the NA of TH immunoreactive boutons and excitatory input from the amygdala, we have also shown that, in the shell, GABA and SOM, presumably both as part of the intrinsic system of the NA, make synaptic contact with neurons which receive an amygdalar input. Although much less dense, the SOM input, like TH, is mainly to small diameter dendrites and spines. As in the dorsal striatum, the GABA input is more commonly found on the large dendrites and somata, where it is likely to exercise a more general influence on the excitation level of the cell.

The role of the NA as a site of limbic-motor interaction highlights the importance of establishing the ultrastructural interactions between it's different elements. If one function of DA in the NA is to filter incoming sensory information (Van Kammen and Kelley, 1991) from limbic areas such as the subiculum, prefrontal cortex or amygdala, a disturbance in either the incoming information or the filter might be implicated in schizophrenia (Gray et al., 1991). We have thus addressed one question posed by Mogenson in the Symposium on the Nucleus Accumbens held in 1980, in that we have shown that limbic inputs to the NA do converge onto neurons that are also postsynaptic to dopamine. However, it is still not yet known whether single NA neurons are postsynaptic to inputs from more than one limbic area.

REFERENCES

Alheid, G.F., and Heimer, L., 1988, New perspectives in basal forebrain organization of special relevance for neuropsychiatric disorders: the striatopallidal, amygdaloid, and corticopetal components of the substantia innominata, *Neuroscience* 27:1-39.
Aylward, R.L.M., and Totterdell, S., 1991, The nucleus accumbens as a site of interaction between limbic inputs and mesolimbic dopamine, *Proc. 3rd IBRO World Congress of Neuroscience* P49.13.
Bell, G.I., and Reisine, T., 1993, Molecular biology of somatostatin receptors, *Trends Neurosci.* 16:34-38.
Buhl, E.H., Schwerdtfeger, W.K., Germroth, P., and Singer, W., 1989, Combining retrograde tracing, intracellular injection, anterograde degeneration and electron microscopy to reveal synaptic links, *J. Neurosci. Meth.* 29: 241-250.
Callaway, C.W., Hakan, R.L., and Henriksen, S.J., 1991, Distribution of amygdala input to the nucleus accumbens septi - an electrophysiological investigation, *J. Neural Trans.-Gen. Section.* 83: 215-225.
Carlsen, J., 1988, Immunocytochemical localization of glutamate decarboxylase in the rat basolateral amygdaloid nucleus, with special reference to GABAergic innervation of amygdalostriatal projection neurons, *J. Comp. Neurol.* 273: 513-526.
De France, J.F., Marchand, J.E., Stanley, J.C., Sikes, R.W., and Chronister, R.B., 1980, Convergence of excitatory amygdaloid and hippocampal input in the nucleus accumbens septi, *Brain Res.* 185:183-186.

de Olmos, J.C. and Ingram, W.R., 1972, The projection field of the stria terminalis in the rat brain. An experimental study, *J. Comp. Neurol.* 146:303-334.

Drevets, W.C., Videen, T.O., Price, J.L., Preskorn, S.H., Carmichael, T., and Raichle, M.E., 1992, A functional anatomical study of unipolar depression, *J. Neurosci.* 12:3628-3461.

Everitt, B.J., and Robbins, T.W., 1992, Amygdala-ventral striatal interactions and reward related processes, *in*: "The Amygdala: Neurobiological Aspects of Emotion, Memory, and Mental Dysfunction," J.P. Aggleton, ed., Wiley-Liss, Chichester, pp. 401-429.

Freund, T.F., Powell, J.F., and Smith, A.D., 1984, Tyrosine hydroxylase-immunoreactive boutons in synaptic contact with identified striatonigral neurons, with particular reference to dendritic spines, *Neuroscience* 13:1189-1215.

Gray, J.A., Feldon, J., Rawlins, J.N.P., Hemsley, D.R., and Smith, A.D., 1991, The neuropsychology of schizophrenia, *Behav. Brain Sci.* 14:1-84.

Groenewegen, H.J., Room, P., Witter, M.P., and Lohman, A.H.M., 1980, Subcortical afferents of the nucleus accumbens septi in the cat, studied with retrograde axonal transport of horseradish peroxidase and bisbenzimid, *Neuroscience* 5:1903-1916.

Johansson, O., Hökfelt, T., and Elde, R.P.,1984, Immunohistochemical distribution of somatostatin-like immunoreactivity in the central nervous system of the adult rat, *Neuroscience* 13:265-339.

Kita, H., and Kitai, S.T.,1990, Amygdaloid projections to the frontal cortex and the striatum in the rat, *J. Comp. Neurol.* 298:40-49.

McDonald, A.J., 1991, Topographical organization of amygdaloid projections to the caudatoputamen, nucleus accumbens, and related striatal-like areas of the rat brain, *Neuroscience* 44:15-33.

Meredith, G.E., Agolia, R., Arts, M.P.M., Groenewegen, H.J., and Zahm, D.S., 1992, Morphological differences between projection neurons of the core and shell in the nucleus accumbens of the rat, *Neuroscience* 50:149-162.

Meredith, G.E., and Wouterlood, F.G., 1991, Synaptic organization of nucleus accumbens (ventral striatum), *in*: "The Basal Ganglia III," G. Bernardi, M.B. Carpenter, G. DiChiara, M. Morelli and P Stanzione eds., Plenum Press, New York, pp. 167-176.

Mogenson, G.J., Jones, D.L., and Yim, C.Y., 1980, From motivation to action: functional interface between the limbic system and the motor system, *Progr. Neurobiol.* 14: 69-97.

Ragsdale, C.W., and Graybiel, A.M., 1988, Fibers from the basolateral nucleus of the amygdala selectively innervate striosomes in the caudate nucleus of the cat, *J. Comp. Neurol.* 269:506-522.

Scheel-Krüger, J., 1986, Dopamine-GABA interactions: evidence that GABA transmits, modulates and mediates dopaminergic functions in basal ganglia and limbic system, *Acta Neurol. Scand.* 73:suppl.107:9-54.

Sesack, S.R., and Pickel, V. M., 1992, Prefrontal cortical efferents in the rat synapse on unlabeled neuronal targets of catecholamine terminals in the nucleus accumbens septi and on dopamine neurons in the ventral tegmental area, *J. Comp. Neurol.* 320:145-160.

Totterdell, S., and Smith, A.D., 1989, Convergence of hippocampal and dopaminergic input onto identified neurons in the nucleus accumbens of the rat, *J. Chem. Neuroanat.* 2:285-298.

Vankammen, D.P., and Kelley, M., 1991, Dopamine and norepinephrine activity in schizophrenia - an integrative perspective, *Schizoph. Res.* 4:173-191.

Voorn, P., Gerfen, C.R., and Groenewegen, H.J., 1989, Compartmental organization of the ventral striatum of the rat- immunohistochemical distribution of enkephalin, substance P, dopamine and calcium-binding protein, *J. Comp. Neurol.* 289:189-201.

Voorn, P., Jorritsma-Byham, B., Van Dijk, C., and Buijs, R.M., 1986, The dopaminergic innervation of the ventral striatum in the rat: a light- and electron-microscopical study with antibodies against dopamine, *J. Comp. Neurol.* 251:84-99.

Yim, C.Y., and Mogenson, G.J., 1986, Mesolimbic dopamine projection modulates amygdala-evoked EPSP in nucleus accumbens neurons: an in vivo study, *Brain Res.* 369:347-352.

Zahm, D.S., 1992, An electron microscopic morphometric comparison of tyrosine hydroxylase immunoreactive innervation in the neostriatum and the nucleus accumbens core and shell, *Brain Res.* 575:341-346.

TOPOGRAPHIC ARRANGEMENT OF COLLATERAL PROJECTIONS FROM THE BASOLATERAL AMYGDALOID NUCLEUS TO BOTH THE PREFRONTAL CORTEX AND NUCLEUS ACCUMBENS

Yasuhide Shinonaga, Masahiko Takada, Yoshiaki Ikai and Noboru Mizuno

Department of Morphological Brain Science
Faculty of Medicine, Kyoto University
Kyoto 606-01, Japan

INTRODUCTION

The basolateral amygdaloid nucleus (ABL) sends projection fibers mainly to the prefrontal cortex (PFC) and striatum (Nauta, 1961; Krettek and Price, 1977, 1978a,b; Divac et al., 1978; Macchi et al., 1978; Porrino et al., 1981). The ABL-derived amygdalocortical projection fibers terminate in both the medial and lateral aspects of the prefrontal cortex; densely in the prelimbic (PL) and dorsal agranular insular (AId) cortices, and sparsely in the infralimbic and ventral agranular insular cortices (Krettek and Price, 1977; Groenewegen et al., 1990). On the other hand, the amygdalostriatal projection fibers terminate predominantly in the ventral striatum including the nucleus accumbens (NA), and to a lesser extent in the dorsal striatum (Krettek and Price, 1978a; Royce, 1978; Groenewegen et al., 1980; Veening et al., 1980; Kelly et al., 1982; Parent et al., 1983; Russchen et al., 1985). Anatomical studies have indicated in the rat that both the amygdalocortical and amygdalostriatal projections which arise from the ABL are topographically organized (Krettek and Price, 1977, 1978a; Russchen and Price, 1984; Phillipson and Griffiths, 1985; Groenewegen et al., 1990; McDonald, 1987, 1991a,b); the more rostral portion of the ABL projects to the lateral aspect of the PFC, particularly to the PL, and the lateral part of the NA (NAl), while the more caudal portion of the ABL projects to the medial aspect of the PFC, especially to the AId, and the medial part of the NA (NAm). In the present study, we investigated whether or not single neurons in the rat ABL project to both the PFC and NA by way of axon collaterals.

METHODS

Under general anesthesia with sodium pentobarbital (40 mg/kg b.wt., i.p.), adult male Wistar rats (250-450 g b.wt.) received combined injection of two kinds of fluorescent retrograde tracers, True Blue (TB; Illing) and Diamidino Yellow (DY; Illing). These tracers were stereotaxically infused by pressure through a fine needle attached to a 1-µl Hamilton microsyringe. A volume of 0.1-0.15 µl of a 5% aqueous suspension of TB was injected into the AId or PL on one side of the brain, while a volume of 0.15-0.2 µl of a 3% aqueous suspension of DY was injected into the NAm or NAl on the same side of the brain. Four different types of combined injections of TB and DY were made in the present study: (1) TB injection into the AId combined with DY injection into the NAm; (2) TB injection into the AId combined with DY injection into the NAl; (3) TB injection into the

PL combined with DY injection into the NAm; (4) TB injection into the PL combined with DY injection into the NAl. Each tracer was slowly deposited over 10 min, and the injection needle was kept in place for an additional 10 min.

After a survival of 3-4 days, the rats were reanesthetized deeply, and perfused transcardially with 300 ml of 10% formalin dissolved in 0.05 M phosphate buffer (pH 7.3). The brains were removed immediately, and saturated with 20% sucrose in the same buffer at 4°C. Serial brain sections were cut transversely at 40 μm thickness on a freezing microtome. The sections through the injection sites and amygdala were observed with a Zeiss epifluorescence microscope. An ultraviolet filter providing excitation light of about 360 nm wavelength was used to examine blue-emitting TB-positive cells and yellow-emitting DY-positive ones.

Based on the cytoarchitectonic classification of Krettek and Price (1978b), the ABL was divided into two parts; the anterior one (ABLa) that comprises medium-sized neurons, and the posterior one (ABLp) that comprises small neurons. The PFC was also demarcated according to Krettek and Price (1977).

RESULTS

TB Injection into the AId Combined with DY Injection into the NAm

After TB injection into the AId, many neuronal cell bodies were retrogradely labeled in the ABL on the side ipsilateral to the injection. These TB-positive cells were seen predominantly in the rostral 2/3, and to a lesser extent in the caudal 1/3 of the ABL; they were more frequently distributed in the ventral 1/2 of the ABL than in its dorsal counterpart, except at the rostralmost level where TB-positive cells were diffusely distributed over the nucleus. A greater number of ABL cells were retrogradely labeled with DY injected ipsilaterally into the NAm. These DY-positive cells were found mainly in the caudal 2/3 of the ABL; almost no DY labeling occurred in the rostralmost part of the ABL. A small number of cells were double-labeled with both TB and DY; they constituted only less than 5% of the total TB-positive cells (Table 1). Most of the double-labeled cells were small in size, and were located in the ABLp, particularly in its ventral part. In addition, some double-labeled cells were observed in the lateral and basomedial amygdaloid nuclei.

TB Injection into the AId Combined with DY Injection into the NAl

TB-positive cells were observed densely in the rostral 2/3, and sparsely in the caudal 1/3 of the ABL. Similarly, DY-positive cells were found predominantly in the rostral 2/3 of the ABL; no DY-positive cells were detected in the caudalmost part of the ABL. In the ABL, cells double-labeled with both TB and DY amounted to more than 30% of the total

Table 1. Double-labeled ABL neurons after TB injection into the AId or PL combined with DY injection into the NAm or NAl.

Injection sites	TB + DY (n)	Total TB (n)	TB + DY / total TB (%)
AId (TB) + NAm (DY)	56	1267	4.4
AId (TB) + NAl (DY)	296	907	32.6
PL (TB) + NAm (DY)	518	1062	48.8
PL (TB) + NAl (DY)	242	652	37.1

Counts of ABL neurons double-labeled with both TB and DY (TB+DY) and single-labeled with TB only were obtained from the series of every third section through the ABL.

TB-positive cells (Table 1). The vast majority of the double-labeled cells were medium-sized (Fig. 1A), and were distributed in the ABLa, especially in its ventral part (Fig. 2A). The intermediodorsal thalamic nucleus also contained many TB-positive and DY-positive cells. However, all of them were single-labeled with either tracer.

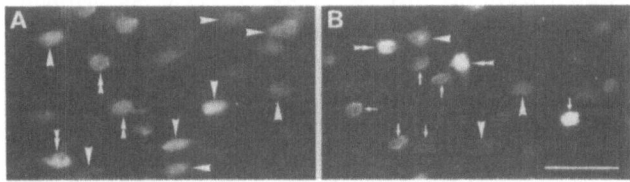

Figure 1. Photomicrographs of retrogradely labeled neurons in the ABL. A: TB injection into the AId combined with DY injection into the NAl. B: TB injection into the PL combined with DY injection into the NAl. Double arrowheads indicate neurons double-labeled with both TB and DY; single arrowheads point to neurons single-labeled with TB only; arrows indicate neurons single-labeled with DY only. Scale bar = 50 μm.

TB Injection into the PL Combined with DY Injection into the NAm

DY labeling was more marked in the caudal 2/3 of the ABL. TB-positive cells were also distributed densely in the caudal 2/3, and sparsely in the rostral 1/3 of the ABL. No distinct topography in the dorsoventral axis was evident in the distribution of TB-positive cells. As many as 50% of these TB-positive cells were further labeled with DY injected into the NAm (Table 1). The double-labeled cells were seen frequently in the ABLp, and occasionally in the ABLa. They were more numerously found ventrally throughout the entire rostrocaudal extent of the ABL; many of them were aggregated in the medial half of

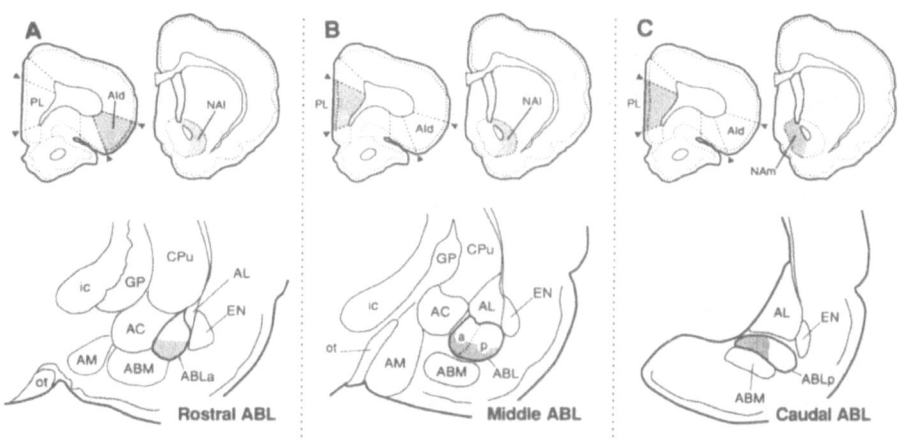

Figure 2. Schematic diagram showing the topographic arrangement of collateral projections from the ABL to both the PFC and NA. Upper row: Sites of TB injection in the AId or PL, and DY injection in the NAm or NAl (stippled areas). Lower row: Distribution of ABL neurons double-labeled with both TB and DY (stippled area). A: TB injection into the AId combined with DY injection into the NAl. B: TB injection into the PL combined with DY injection into the NAl. C: TB injection into the PL combined with DY injection into the NAm. a, ABLa; ABM, basomedial amygdaloid nucleus; AC, central amygdaloid nucleus; AL, lateral amygdaloid nucleus; AM, medial amygdaloid nucleus; CPu, caudate-putamen; EN, endopiriform nucleus; GP, globus pallidus; ic, internal capsule; ot, optic tract; p, ABLp.

the ABLp at the caudalmost level of the ABL (Fig. 2C), and were small in size. Additionally, a number of cells in the paraventricular thalamic nucleus were single-labeled with either TB or DY.

TB Injection into the PL Combined with DY Injection into the NAl

TB labeling was marked in the caudal 2/3 of the ABL, while DY labeling was prominent in the rostral 2/3 of the ABL. A certain overlap of the distribution areas of TB-positive and DY-positive cells was seen in the middle 1/3 of the ABL. In the ABL, cells double-labeled with both tracers amounted to almost 40% of the total TB-positive cell population (Table 1). Both medium-sized and small cells were double-labeled (Fig. 1B). These double-labeled cells were found throughout the entire rostrocaudal extent of the ABL with the highest density in its middle 1/3. They were distributed evenly in both the ABLa and ABLp, in particular abundance in their ventral parts (Fig. 2B).

DISCUSSION

It has recently been shown in the rat that single ABL neurons send their axons simultaneously to both the PFC and NA (McDonald, 1991a). The present results have further indicated that such collateral projections are topographically organized. The axons bifurcating to both the AId and NAl arise preferentially from the rostral ABL (from the ABLa), those bifurcating to both the PL and NAl from the middle ABL (from both the ABLa and ABLp), and those bifurcating to both the PL and NAm from the caudal ABL (from the ABLp). At least 30-50% of ABL neurons projecting to the PFC (to the AId or PL) issue axon collaterals to the NA (to the NAl or NAm). On the other hand, only less than 5% of ABL neurons projecting to the AId issue axon collaterals to the NAm. The single neurons giving rise to collaterals projecting to both the PFC and NA are distributed mainly in the ventral part of the ABL.

The differential localization of ABL neurons projecting to both the PFC and NA by way of axon collaterals may reflect the functional segregation of the ABL. Considerable amounts of anatomical and physiological data have indicated that each of the AId and PL subserves a specific autonomic function; the AId is involved in the visceral sensory function on the basis of its input from the parabrachial nuclei and its output to the nucleus of the solitary tract (Saper, 1982a,b; van der Kooy et al., 1982, 1984; Neafsey et al., 1986; Zhang and Sasamoto, 1990; for review, see Neafsey, 1990), while the PL in the visceral motor function on the basis of anatomical and physiological data (Terreberry and Neafsey, 1983, 1987; Hurley-Gius and Neafsey, 1986; Neafsey et al., 1986; for review, see Neafsey, 1990). The fact that the NAl and NAm receive cortical afferent fibers originating respectively from the AId and PL (Beckstead, 1979; Berendse et al., 1992) favors the notion that the visceral sensory and motor functions are segregated in the NA as well. Thus, it is likely that the rostral ABL which contains single neurons projecting to both the AId and NAl is implicated in the visceral sensory processings, whereas the caudal ABL which contains single neurons projecting to both the PL and NAm plays important roles in the visceral motor controls. In addition, the present results suggest that the middle ABL which contains single neurons projecting to both the PL and NAl may simultaneously control both the motor and sensory aspects of the visceral functions through branched axons.

Groenewegen et al. (1990) have recently introduced two circuits concerning the basal ganglia-thalamocortical link. One is the "prelimbic circuit" that represents the PL-NAm-ventral pallidum-mediodorsal thalamic nucleus-PL loop, and the other is the "dorsal agranular insular circuit" that represents the AId-NAl-ventral pallidum-mediodorsal thalamic nucleus-AId loop. The identified three modes of collateral projections from the ABL to both the PFC and NA may exert modulatory influences upon these two circuits which probably underlie the limbic and/or autonomic functions. The collateral projections from the rostral ABL to both the AId and NAl may modulate the "dorsal agranular insular circuit", whereas those from the caudal ABL to both the PL and NAm may modulate the "prelimbic circuit". The collateral projections from the middle ABL to both the PL and NAl may modulate both circuits.

Acknowledgements

We are grateful for the photographic help of Mr. Akira Uesugi. This work has been supported in part by Grants-in-Aid for Special Research on Priority Areas 04246106 and 04255204, and Scientific Research (B) 02454113 from the Ministry of Education, Science and Culture of Japan.

REFERENCES

Beckstead, R.M., 1979, An autoradiograpphic examination of corticocortical and subcortical projections of the mediodorsal-projection (prefrontal) cortex in the rat, *J. Comp. Neurol.* 184:43-62.

Berendse, H.W., Galis-de Graaf, Y., and Groenewegen, H.J., 1992, Topographical organization and relationship with ventral striatal compartments of prefrontal corticostriatal projections in the rat, *J. Comp. Neurol.* 316:314-347.

Divac, I., Kosmal, A., Björklund, A., and Lindvall, O., 1978, Subcortical projections to the prefrontal cortex in the rat as revealed by the horseradish peroxidase technique, *Neuroscience* 3:785-796.

Groenewegen, H.J., Becker, N.E.H.M., and Lohman, A.H.M., 1980, Subcortical afferents of the nucleus accumbens septi in the cat, studied with retrograde axonal transport of horseradish peroxidase and bisbenzimid, *Neuroscience* 5:1903-1916.

Groenewegen, H.J., Berendse, H.W., Wolters, J.G., and Lohman, A.H.M., 1990, The anatomical relationship of the prefrontal cortex with the striatopallidal system, the thalamus and the amygdala: evidence for a parallel organization, *in*: "The Prefrontal Cortex: its Structure, Function and Pathology, Progress in Brain Research, vol. 85," H.B.M. Uylings, C.G. van Eden, J.P.C. de Bruin, M.A. Corner and M.G.P. Feenstra, eds., Elsevier, Amsterdam, pp. 95-118.

Hurley-Gius, K.M., and Neafsey, E.J., 1986, The medial frontal cortex and gastric motility: microstimulation results and their possible significance for the overall pattern of organization of rat frontal and parietal cortex, *Brain Res.* 365:241-248.

Kelly, A.E., Domestick, V.B., and Nauta, W.J.H., 1982, The amygdalostriatal projection in the rat. An anatomical study by anterograde and retrograde tracing methods, *Neuroscience* 7:615-630.

Krettek, J.E., and Price, J.L., 1977, Projections from the amygdaloid complex to the cerebral cortex and thalamus in the rat and cat, *J. Comp. Neurol.* 172:687-722.

Krettek, J.E., and Price, J.L., 1978a, Amygdaloid projections to the subcortical structures within the basal forebrain and brainstem in the rat and cat, *J. Comp. Neurol.* 178:225-254.

Krettek, J.E., and Price, J.L., 1978b, A description of the amygdaloid complex in the rat and cat with observations on intra-amygdaloid axonal connections, *J. Comp. Neurol.* 178:255-280.

Macchi, G., Bentivoglio, M., Rossini, P., and Tempesta, E., 1978, The basolateral amygdaloid projections to the neocortex in the cat, *Neurosci. Lett.* 9:347-351.

McDonald, A.J., 1987, Organization of amygdaloid projections to the mediodorsal thalamus and prefrontal cortex: a fluorescence retrograde transport study in the rat, *J. Comp. Neurol.* 262:46-58.

McDonald, A.J., 1991a, Organization of amygdaloid projections to the prefrontal cortex and associated striatum in the rat, *Neuroscience* 44:1-14.

McDonald, A.J., 1991b, Topographical organization of amygdaloid projections to the caudatoputamen, nucleus accumbens, and related striatal-like areas of the rat brain, *Neuroscience* 44:15-33.

Nauta, W.J.H., 1961, Fibre degeneration following lesions of the amygdaloid complex in the monkey, *J. Anat.* 95:515-531.

Neafsey, E.J., 1990, Prefrontal cortical control of the autonomic nervous system: anatomical and physiological observations, *in*: "The Prefrontal Cortex: its Structure, Function and Pathology, Progress in Brain Research, vol. 85," H.B.M. Uylings, C.G. van Eden, J.P.C. de Bruin, M.A. Corner, and M.G.P. Feenstra, eds., Elsevier, Amsterdam, pp. 147-166.

Neafsey, E.J., Hurley-Gius, K.M., and Arvanitis, D., 1986, The topographical organization of neurons in the rat medial frontal, insular and olfactory cortex projecting to the solitary nucleus, olfactory bulb, periaqueductal gray and superior colliculus, *Brain Res.* 377:261-270.

Parent, A., Mackey, A., and De Bellefeuille, L., 1983, The subcortical afferents to caudate nucleus and putamen in the primate: a fluorescence retrograde double labeling study, *Neuroscience* 10:1137-1150.

Phillipson, O.T., and Griffiths, A.C., 1985, The topographic order of inputs to the nucleus accumbens in the rat, *Neuroscience* 5:275-296.

Porrino, L.J., Crane, A.M., and Goldman-Rakic, P.S., 1981, Direct and indirect pathways from the amygdala to the frontal lobe in rhesus monkey, *J. Comp. Neurol.* 198:121-136.

Royce, G.J., 1978, Cells of origin of subcortical afferents to the caudate nucleus: a horseradish peroxidase study in the cat, *Brain Res.* 153:465-475.

Russchen, F.T., Bakst, I., Amaral, D.G., and Price, J.L., 1985, The amygdalostriatal projections in the monkey. An anterograde tracing study, *Brain Res.* 329:241-257.

Russchen, F.T., and Price, J.L., 1984, Amygdalostriatal projections in the rat. Topographical organization and fiber morphology shown using the lectin PHA-L as an anterograde tracer, *Neurosci. Lett.* 47:15-22.

Saper, C.B., 1982a, Convergence of autonomic and limbic connections in the insular cortex of the rat, *J. Comp. Neurol.* 210:163-173.

Saper, C.B., 1982b, Reciprocal parabrachial-cortical connections in the rat, *Brain Res.* 242:33-40.

Terreberry, R.R., and Neafsey, E.J., 1983, Rat medial frontal cortex: a visceral motor region with a direct projection to the solitary nucleus, *Brain Res.* 278:245-249.

Terreberry, R.R., and Neafsey, E.J., 1987, The rat medial frontal cortex projects directly to autonomic regions of the brainstem, *Brain Res. Bull.* 19:639-649.

van der Kooy, D., Koda, L.Y., McGinty, J.F., Gerfen, C.R., and Bloom, F.E., 1984, The organization of projections from the cortex, amygdala, and hypothalamus to the nucleus of the solitary tract in rat, *J. Comp. Neurol.* 224:1-24.

van der Kooy, D., McGinty, J.F., Koda, L.Y., Gerfen, C.R., and Bloom, F.E., 1982, Visceral cortex: a direct connection from prefrontal cortex to the solitary nucleus in rat, *Neurosci. Lett.* 33:123-127.

Veening, J.G., Cornelissen, F.M., and Lieven, P.A.J.M., 1980, The topical organization of the afferents to the caudoputamen of the rat. A horseradish peroxidase study, *Neuroscience* 5:1253-1268.

Zhang, G., and Sasamoto, K., 1990, Projections of two separate cortical areas for rhythmical jaw movements in the rat, *Brain Res. Bull.* 24:221-230.

PALLIDAL AFFERENTS TO THE NEURONS IN THE ANTERIOR THALAMIC RETICULAR NUCLEUS PROJECTING TO THE CENTROMEDIAN NUCLEUS

Tetsuro Kayahara, Yukihiko Yasui and Katsuma Nakano

Department of Anatomy, Faculty of Medicine, Mie University
Tsu, Mie 514, Japan

INTRODUCTION

The thalamic reticular nucleus is a thin layer of neurons interposed between the internal capsule laterally and the external medullary lamina medially, and covers the rostral and lateral aspects of the thalamus. This nucleus receives inputs from collaterals of both corticothalamic and thalamocortical axons (Jones, 1985). Cells of this nucleus in turn send their axons back into the thalamus, and innervate both relay cells and local GABAergic interneurons (for review see Steriade and Deschênes, 1984)

A discrete topographical organization of the thalamocortical system to and from neurons of the thalamic reticular nucleus has been reported (Jones, 1975; Ohara and Lieberman, 1985). Collaterals associated with any particular system terminate in a defined sector of the thalamic reticular nucleus. The thalamic retuicular nucleus may be subdivided into a number of sectors, visual, auditory, and somatosensory reticular nuclei (Ohara and Lieberman, 1985). The rostral sector of the reticular nucleus is associated with various motor and limbic centers (Jones, 1975). The thalamic reticular nucleus acts as a pacemaker of thalamic rhythmic activities, and seems to be involved in the behavioral state (Steriade and Deschênes, 1984).

Recently, increasing attention has been directed to the reticular nucleus. In particular, the visual reticular nucleus has been studied in detail electron microscopically (Ohara and Lieberman, 1985). The rostral sector of the reticular nucleus receives projections from the lateral pallidal segment (Haber et al., 1985; Cornwall et al., 1990), and projects to the centromedian nucleus (Steriade et al., 1984). However, this sector has not been studied in detail in terms of synaptic organization. We have attempted to define direct synaptic contacts between the axon terminals of pallidal origin and the centromedian nucleus projecting neurons in the thalamic reticular nucleus on the basis of electron microscopic observation combined with lesioning and retrograde neuronal labeling techniques.

MATERIALS AND METHODS

Fourteen adult cats (weighing 2.2 - 3.0 kg) were used in this study. All cats were anesthetized with ketamin (10 mg/kg, i.m.) followed by pentobarbital (15 mg/kg, i.p.). In order to degenerate neurons in the globus pallidus (GP) region, a single injection (0.02 - 0.1µl) of 2.5% kainic acid dissolved in phosphate buffer was made stereotaxically into the GP region through a glass micropipette attached to a pressure delivery apparatus. After 24 hours, 3 - 5% horseradish peroxidase (HRP) conjugated to wheat germ agglutinin (WGA-HRP) (Toyobo) dissolved in tris-buffer was injected stereotaxically into the centromedian nucleus by pressure (volume, 0.02 - 0.01μl) or iontophoretic injections (5μA, 60 min) on the side ipsilateral to the kainic acid injection.

Following a survival time of 3 to 4 days of the kainic acid injection, the animals were deeply reanesthetized and perfused transcardially using a peristaltic pump with 500 ml of 0.9% saline in 0.1 M sodium phosphate buffer (PB) followed by 300 ml of a mixture of 1 - 2% paraformaldehyde and 1.25 - 2.5% glutaraldehyde in 0.1 M PB. Brains were immediately removed from the skull, cut into three pieces, and postfixed for 1 - 2 hours in the same cold fixative at 4°C. The pieces were sectioned serially at 120 μm in the frontal plane with a vibratome. These sections were incubated for visualization of HRP with tetramethyl benzidine (TMB), and reaction product was stabilized with cobalt chloride-diaminobenzidine-tetrahydrochloride (Co-DAB). Electron microscopic (EM) specimens were punched out from 120-μm thick vibratome sections after the thalamic reticular nucleus, containing retrogradely HRP-labeled neurons, were identified under a low-power dissecting microscope. The specimens were postfixed with buffered 1.0% osmium tetraoxide solution for 1 hour, stained en block with 2% uranyl acetate, then embedded in Epon 812 after dehydration with graded ethanol and propylene oxide. EM blocks were thin sectioned on a Reichert Ultracut ultramicrotome, mounted onto mesh grids, stained with lead citrate and finally viewed and photographed with a Hitachi H-800 electron microscope. The lesion sites in the GP, WGA-HRP injection sites in the centromedian nucleus and the vibratome sections remaining after removal of the EM samples were observed light microscopically following staining with Cresyl violet for verification of the location.

RESULTS

A great number of labeled neurons were found in the anterior portion of the thalamic reticular nucleus following WGA-HRP injection in the centromedian nucleus. These neurons were variable in size (13 x 18 μm to 24 x 46 μm), and multipolar or fusiform in shape, giving off primary, secondary or sometimes tertiary dendrites containing HRP reaction product. In electron microscopic feature, they were rich in highly organized rough endoplasmic reticulum, Golgi apparatus, and other cell organella. Kainic acid-degenerating boutons of pallidal origin were concentrated in the anterior part of the reticular nucleus ipsilaterally. These degenerating boutons made synaptic contacts with somata and dendrites of labeled and unlabeled neurons in the anterior part of the reticular nucleus. Electron

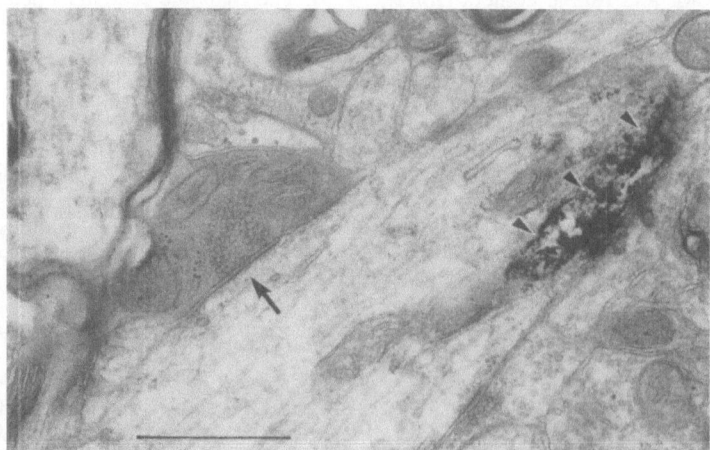

Figure 1. Electron micrograph showing a degenerating bouton of pallidal origin (arrow) making an asymmetrical synaptic contact with the secondary dendrite containing HRP reaction product (arrowheads) in the anterior part of thalamic reticular nucleus following a kainic acid injection in the globus pallidus and an HRP injection in the centromedian nucleus ipsilaterally. Bar: 1 μm.

microscopic profiles of degenerating boutons exhibited a moderate or marked electron dense degeneration. The darkly degenerating boutons contained closely packed, round or heteromorphic synaptic vesicles and a few swollen or shrunken mitochondria. Although the postsynaptic apposition was not obvious, these boutons made mainly asymmetrical synaptic contact with various levels of the dendrites and somata of HRP-labeled neurons in the reticular nucleus (Fig. 1). They distributed on the somata (10%), the proximal primary dendrites (18%) and on the distal dendrites (72%). Some 42% of synapses on the distal dendrites attached to the secondary dendrites (larger than about 2 μm in diameter) and 30% to the dendrites more distal than the secondary dendrites. There were also small degenerating boutons (about 1.0 μm in diameter) making asymmetrical synaptic contact on the more distal dendrites (less than 2 μm in diameter) unlabeled with HRP. It is not definite whether these small degenerating boutons make synaptic contacts with the centromedian nucleus projection neurons because of the difficulty of the HRP labeling in the more distal dendritic branches. Degenerating axo-spinosus synapses were not detected. Anterogradely HRP-labeled axon terminals were seen contacting varicose parts of dendrites of retrogradely labeled neurons in the thalamic reticular nucleus. These boutons contained loosely packed, round synaptic vesicles, and made asymmetrical synaptic contacts.

DISCUSSION

The thalamic reticular nucleus is the interface between thalamocortical and corticothalamic systems and modulates the activity of thalamic relay neurons and interneurons to influence the transfer of information from thalamus to cerebral cortex. The reticular nucleus is a collection of GABA-containing neurons located at a strategic position in thalamocortical systems (Houser et al., 1980; de Biasi et al., 1986). Coexistence of glutamic acid decarboxylase (GAD) and somatostatin-like immunoreactivity was also demonstrated in the nucleus (Oertel et al., 1983). The reticular nucleus is densely innervated by axon collaterals of both thalamocortical relay neurons and cortical neurons projecting to the thalamus. Terminals of these axon collaterals contain glutamate as excitatory neurotransmitter (see review by McCormick, 1992). The reticular nucleus is also densely innervated by cholinergic fibers arising from both the pedunculopontine tegmental nucleus in the brainstem and the nucleus basalis in the forebrain (Levey et al., 1987; Hallanger et al., 1987; Steriade et al., 1987), by noradrenergic fibers from the locus coeruleus (Asanuma, 1992), and GABAergic fibers from the recurrent collaterals of the reticular cell itself. Cells in the reticular nucleus are interconnected via inhibitory synapses (for reviews see McCormick, 1992). The reticular nucleus is moderately innervated by serotoninergic fibers which are suggested to arise from the raphe nuclei (Lavoie and Parent, 1991).

Ohara (1988) identified three types of terminals in the reticular nucleus; a small terminal with densely packed spherical vesicles (D-terminal), a large terminal with loosely packed spherical vesicles (L-terminal) and a F-terminal with flattened vesicles. D-terminal originates from the cerebral cortex, and contacts with dendrites of all sizes and dendritic spines, but only rarely contacts with neuronal somata. L-terminal originates in the dorsal thalamus and contacts predominantly with dendritic shafts. F-terminal is likely to be a recurrent collateral of reticular axon and makes synaptic contacts with the somata of reticular neurons (Ohara and Lieberman, 1985). In the rat reticular nucleus, all synaptic appositions with somatic profiles exhibit symmetric membrane specializations (Ohara and Lieberman, 1985), whereas the choline acetyltransferase (CAT) positive terminals contact dendritic profiles, not somata (Hallanger and Wainer, 1988). CAT positive synaptic profiles contain multiple mitochondria and densely packed, round vesicles, and make asymmetric synaptic contact on dendrites of all but the finest diameter (Hallanger and Wainer, 1988).

In the present study, the monosynaptic connection has been firstly demonstrated between pallido-reticular and reticulo-centromedian projection cells electron microscopically. In addition, small degenerating synapses were observed making synaptic contacts on the more distal unlabeled dendritic branches of reticular neurons. It is not clear whether these neurons are projection neurons to the centromedian nucleus or not, because of the difficulty of labeling on the distal levels of dendritic branches. In our cases the synaptic profiles of pallidal afferents in the reticular nucleus look like those of CAT-positive synapses. In the monkey, cholinergic neurons were found in the caudoventral part of the lateral pallidal segment and a small rostral apical region near the internal capsule (Armonda and Carpenter,

1991). In the non-primate, CAT positive neurons were scattered throughout the globus pallidus (unpublished data). Cholinergic projection from the globus pallidus to the reticular nucleus might be expected, although CAT immunostaining has not been done in our cases. However, GABAergic terminals of the pallidoreticular afferents can not be denied, especially in the axosomatic synapses, as GABAergic terminals with round or pleomorphic vesicles have been indicated. Moreover, these terminals made some asymmetric synaptic contacts (de Biasi et al., 1986).

According to Hazrati and Parent (1991) the reticular nucleus projection from the lateral pallidal segment is not cholinergic. They postulated that this projection is GABAergic. An extrinsic GABAergic input to the reticular nucleus was traced from the caudal basal nucleus (Asanuma and Porter, 1990). In our findings, the postsynaptic thickening of the pallidal afferent synapses was not obvious, and it was difficult to differentiate from the symmetrical synapse. There is considerable variation in the thickness of the postsynaptic densities associated with CAT-immunoreactive fibers. According to de Lima et al. (1985), the appearance of postsynaptic cholinergic terminals in the reticular nucleus was variable, with symmetric configurations being most frequent. Heterosynaptic modulation of local circuit processing was suggested. The muscarinic effects of acetylcholine (Ach) can be both excitatory and inhibitory depending on the nature of the postsynaptic target cell. A dual action of Ach in the thalamus was suggested, and it was implied that Ach affects interneurons and relay cells differently (de Lima et al., 1985). The functional role of cholinergic transmission within the reticular nucleus revealed conflicting results.

Ach is involved in higher neuronal processing such as attention, learning, memory and in certain behavioral state control (sleep and waking). Ach may modulate information processed by the thalamus. Kolasiewicz et al. (1992) suggest that activation of the cholinergic and muscarinergic receptors in the rostral reticular nucleus modulates the motor function of rats. According to Cornwall et al. (1990), the medial part of the rostral reticular nucleus seems to act in the gustatory autonomic and visceral information processing, while the lateral part seems to be involved in the visual attention (higher regulation of eye movements). The lateral part of the rostral reticular nucleus receives non-dopaminergic afferents from the substantia nigra and retrorubral area. The medial part receives dopaminergic afferents from the ventral tegmental area, interfascicular nucleus and pretectal area.

In the present study, pallido-reticulo-centromedian nucleus connections were indicated in the rostral sector of the thalamic reticular nucleus. The rostral reticular nucleus projection from the medial part of the GP was demonstrated in the rat by using PHA-L method (Cornwall et al., 1990). On the other hand, the rostrolateral part of the reticular nucleus projects centromedian-parafascicular nuclei (Steriade et al., 1984). The rostral reticular nucleus receives cortical afferents from the frontal cortex, and thalamic afferents from the motor thalamic nuclei. In our previous autoradiographic studies on the thalamocortical projections, massive terminations originating from the motor thalamic nuclei were observed in the rostral reticular nucleus as illustrated only in figures (Nakano et al., 1992). Although their functional significance is not clear, the pallido-reticulo-centromedian connections seem to modulate information related to motor behavior.

Acknowledgements

This study has been supported in part by Grant-in-Aid for Scientific Research from the Ministry of Education, Science and Culture of Japan (03670022).

REFERENCES

Armonda, R.A., and Carpenter, M.B., 1991, Distribution of cholinergic pallidal neurons in the squirrel monkey (Saimiri sciureus) based upon choline acetyltransferase, *J. Hirnforsch.* 32:357-367.

Asanuma, C., 1992, Noradrenergic innervation of the thalamic reticular nucleus: A light and electron microscopic immuno-histochemical study in rats, *J. Comp. Neurol.* 319:299-311.

Asanuma, C., and Porter, L.L., 1990, Light and electron microscopic evidence for a GABAergic projection from the caudal basal forebrain to the thalamic reticular nucleus in rats, *J. Comp. Neurol.* 302:159-172.

Cornwall, J., Cooper, J.D., and Phillipson, O.T., 1990, Projections to the rostral reticular thalamic nucleus in the rat, *Exp. Brain Res.* 80:157-171.

de Biasi, S., Frassoni, C., and Spreafico, R., 1986, GABA immuno-reactivity in the thalamic reticular nucleus of the rat. A light and electron microscopical study, *Brain Res.* 399:143-147.

de Lima, A.D., Montero, V.M., and Singer, W., 1985, The cholinergic innervation of the visual thalamus: an EM immunocytochemical study, *Exp. Brain Res.* 59:206-212.

Haber, S.N., Groenewegen, H.J., Grove, E.A., and Nauta, W.J.H., 1985, Efferent connections of the ventral pallidum: Evidence of a dua striatopallidofugal pathway, *J. Comp. Neurol.* 235:322-335.

Hallanger, A.E., and Wainer, B.H., 1988, Ultrastructure of ChAT- immunoreactive synaptic terminals in the thalamic reticular nucleus of the rat, *J. Comp. Neurol.* 278:486-497.

Hallanger, A.E., Levey, A.I., Lee, H.J., Rye, D.B., and Wainer, B.H., 1987, The origins of cholinergic and other subcortical afferents to the thalamus in the rat, *J. Comp. Neurol.* 262:105-124.

Hazrati, L.-N., and Parent A., 1991, Projection from the external pallidum to the reticular thalamic nucleus in the squirrel monkey, *Brain Res.* 550:142-146.

Houser, C.R., Vaughn, J., Barber, R.P., and Roberts, E., 1980, GABA neurons are the major cell type of the nucleus reticularis thalami, *Brain Res.* 200:341-354.

Jones, E.G., 1975, Some aspects of the organization of the thalamic reticular complex, *J. Comp. Neurol.* 162:285-308.

Jones, E.G., 1985, "The Thalamus," Plenum Press, New York.

Kolasiewicz, W., Sauss, C., Block, F., and Sontag, K.-H., 1992, Behavioural effects after cholinergic stimulation of the reticular thalamic nucleus in rats, *J. Neural. Transm.* 87:163-173.

Lavoie, B., and Parent, A., 1991, Serotoninergic innervation of the thalamus in the primate: an immunohistochemical study, *J. Comp. Neurol.* 312:1-18.

Levey, A.I., Hallanger, A.E., and Wainer, B.H., 1987, Choline acetyltransferase immunoreactivity in the rat thalamus, *J. Comp. Neurol.* 257:317-332.

McCormick, D.A., 1992, Neurotransmitter actions in the thalamus and cerebral cortex and their role in neuromodulation of thalamocortical activity, *Prog. Neurobiol.* 39:337-388.

Nakano, K., Tokushige, A., Kohno, M., Hasegawa, Y., Kayahara, T., and Sasaki, K., 1992, An autoradiographic study of cortical projections from motor thalamic nuclei in the macaque monkey, *Neurosci. Res.* 13:119-137.

Oertel, W.H., Graybiel, A.M., Mugnaini, E., Elde, R.P., Schmechel, D.E., and Kopin, I.J., 1983, Coexistence of glutamic acid decarboxylase and somatostatin-like immunoreactivity in the neurons of the feline nucleus reticularis thalami, *J. Neurosci.* 3:1322-1332.

Ohara, P.T., 1988, Synaptic organization of the thalamic reticular nucleus, *J. Electron Microscop. Techn.* 10:283-292.

Ohara, P.T., and Lieberman, A.R., 1985, The thalamic reticular nucleus of the adult rat: experimental anatomical studies, *J. Neurocytol.* 14:365-411.

Steriade, M., and Deschênes, M., 1984, The thalamus as a neuronal oscillator, *Brain Res. Rev.* 8:1-63.

Steriade, M., Parent, A., and Hada, J., 1984, Thalamic projections of nucleus reticularis thalami of cat: A study using retrograde transport of horseradish peroxidase and fluorescent tracers, *J. Comp. Neurol.* 229:531-547.

Steriade, M., Parent, A., Paré, D., and Smith, Y., 1987, Cholinergic and non-cholinergic neurons of cat basal forebrain project to reticular and mediodorsal thalamic nuclei, *Brain Res.* 408:372-376.

THE RETICULAR THALAMIC NUCLEUS AND THE OUTPUT NUCLEI OF THE BASAL GANGLIA: A NEUROANATOMICAL VIEW BASED ON HODOLOGICAL STUDIES

Silvano de las Heras[1], José Antonio Gandia[1,2]
and José Manuel Giménez-Amaya[1]

[1]Departamento de Morfología
 Facultad de Medicina
 Universidad Autónoma de Madrid, Spain
[2] Departamento de Traumatología y Cirugía Ortopédica
 Hospital Ramón y Cajal
 Madrid, Spain

In the past, the so-called "extrapyramidal disorders" have been placed among the motor disturbances that are not articulated with the "pyramidal" corticospinal organization of the motor function (Wilson, 1912, 1914). More recently, it was demonstrated that the "extrapyramidal circuit" is organized, at least in part, starting and ending upon the cerebral cortex including the primary motor cortex (Alexander et al., 1986; Hoover and Strick, 1993). The basal ganglia would be at the base of this major subcortical loop through the thalamic relays of their output nuclei.

In the last decade, the idea that the basal ganglia are a complex set of subcortical structures that not only integrate motor functions, but also other sorts of neural information has become more and more evident. The striatum, for example, is a clear instance of this fact. This subcortical structure is the main recipient of the basal ganglia receiving a great variety of connections from places that are, by no means, considered "motor" in the anatomical and clinical meaning of the word. Thus, the striatum presents afferents from almost every parcel of the neocortex and from other telencephalic structure linked to the limbic system: the amygdala. The hippocampus, another important telencephalic structure associated with limbic processes projects to the ventral striatum in the rat (Kelley and Domesick, 1982) and it might also reach more dorsal striatal territories through its relationships with the cingulate gyrus (Graybiel, 1990). The striatum has also input from a great variety of thalamic nuclei, very prominently from the rostral and caudal intralaminar system. It finally receives fibers from the dopaminergic zones of the mesencephalon, and from the serotoninergic area in the mesencephalic raphe nuclei (Nauta and Feirtag, 1986).

It is clear, therefore, that the basal ganglia might be seen as a group of subcortical structures involved in the organization of complex behaviors and, perhaps, as a place where brain mechanisms of movement, thought and motivation intersect (Nauta and Feirtag, 1986). This contribution is aimed to suggest that the output connectivity of the basal ganglia to the reticular thalamic nucleus, might be another important route to integrate a great constellation of neural information before to complete the corticosubcortical loops through the basal ganglia, and another way to interact with premotor targets of the basal ganglia in the brainstem (Gandia et al., 1993). First, we shall briefly review the thalamic connectivity of the basal ganglia, dealing especially with the reticular thalamic nucleus. Then, we shall suggest, in a speculative way, some functional implications of these projections.

THALAMIC CONNECTIVITY OF THE BASAL GANGLIA

As mentioned above, a very substantial part of the afferentation of the striatum is coming from the thalamus. In fact, the rostral and caudal intralaminar nuclei represent the most prominent groups projecting upon the striatal tissue (see references in Alheid et al., 1990; see also Fénelon et al., 1991 and François et al., 1991, for the primate). There are, however, other thalamic nuclei that also send projections to the striatum. Thus, other thalamic nuclei belonging to the medial, lateral and ventral nuclear groups have neurons that project to the striatum (Beckstead, 1984; Heimer et al., 1985; Jayaraman, 1985; Parent, 1986; Alheid et al., 1990; Berendse and Groenewegen, 1990; Nakano et al., 1990). In some instances, the striatal innervation of these nuclei is made by means of collaterals of cortical projecting-cells in those thalamic nuclei (Royce, 1983). This thalamic afferentation of the striatum might interact very actively with the terminals from other sources of striatal afferents, prominently with cortical and nigral striatal fibers.

The internal circuit of the basal ganglia is organized in such a way that the striatal projecting-cells may reach very specific areas in the output nuclei of the basal ganglia that, in turn, can innervate concrete targets outside the basal ganglia (Gandia and Giménez-Amaya, 1991). In fact, a modular organization has been proposed as the basic design of the mammalian striatum that could integrate a varied sources of neural information, and channeled it to very particular targets either in the thalamus or in the brainstem through the output nuclei of the basal ganglia, or may participate in several internal loops within the basal ganglia themselves (Malach and Graybiel, 1986; Giménez-Amaya and Graybiel, 1990; Hontanilla and Giménez-Amaya, 1990; Giménez-Amaya and Graybiel, 1991; Graybiel et al., 1991).

The output nuclei of the basal ganglia have been classically described as the internal segment of the globus pallidus (or its homologue in rodents and carnivores, the entopeduncular nucleus) and the pars reticulata of the substantia nigra. These two nuclei project extensively upon the thalamus. The primate brain has been the more studied in this regard. Thus, as summarized by Alheid et al. (1990), the internal segment of the globus pallidus projects principally to the ventral anterior (small-celled part), ventral lateral (anterior part), mediodorsal, centromedian and lateral habenula thalamic nuclei. Collaterals to the reticular formation in the brainstem in addition to this thalamic connectivity are often present. The pars reticulata of the substantia nigra projects mainly to the ventral anterior (magnocellular part), ventromedial (principal portion) and mediodorsal (paralaminar, medial magnocellular and lateral small-celled parts) thalamic nuclei. There also exists collaterals of these nigral fibers reaching the superior colliculus and the reticular formation in the brainstem.

Moreover, one of us has proposed to include the external segment of the globus pallidus (globus pallidus in rodents and carnivores) as another output nucleus of the basal ganglia (Giménez-Amaya, 1991a, 1991b, 1992). The reasons for this inclusion rest not only in the thalamic connectivity described from cells located in this nucleus in the rat, that comprises the mediodorsal, ventromedial, paraventricular and lateral habenula thalamic nuclei (Hattori and Sugimoto, 1983; Price and Slotnick, 1983; Sugimoto and Hattori, 1984; Haber et al., 1985; Mogenson et al., 1987; Russchen et al., 1987; Groenewegen, 1988; Parent et al., 1988), but mainly for its projections to the reticular thalamic nucleus in the rat, cat and monkey (Carter and Fibiger, 1978; Nauta, 1979; Haber et al., 1985; Parent et al., 1988; Cornwall et al., 1990; Hazrati and Parent, 1991; Gandia et al., 1992, 1993).

RETICULAR CONNECTIVITY OF THE BASAL GANGLIA

The reticular thalamic nucleus is a sheet of gray matter that covers the thalamus all through its rostral pole and in its ventral and lateral portions (Fig. 1). The reticulation of this nucleus is clear because it is pierced by fibers entering or leaving the internal capsule to establish reciprocal connections between the thalamus and the neocortex (Nauta and Feirtag, 1986). As described by Scheibel and Scheibel (1966), the cellular contingent of the reticular thalamic nucleus is made by neurons ranged in size from 20-40 μm. Dendrites of these neurons are arranged parallel to the long axis of the nucleus and therefore, orthogonally to the thalamocortical and corticothalamic fibers crossing the nucleus. Their surfaces are generally covered by long and filamentous spines enlarging the area where the axodendritic contacts occur. The axons can be followed caudally where they innervate areas in the thalamus and a more limited group of them (20%) extends on the mesencephalic tegmentum. These axons also give numerous collaterals within the reticular thalamic nucleus itself. The neuropil of this nucleus is made up by collaterals of the thalamocortical and corticothalamic fibers which cross through the nucleus, and also by the first axonal portion of its cells (Ramón y Cajal, 1904; Scheibel and Scheibel, 1966).

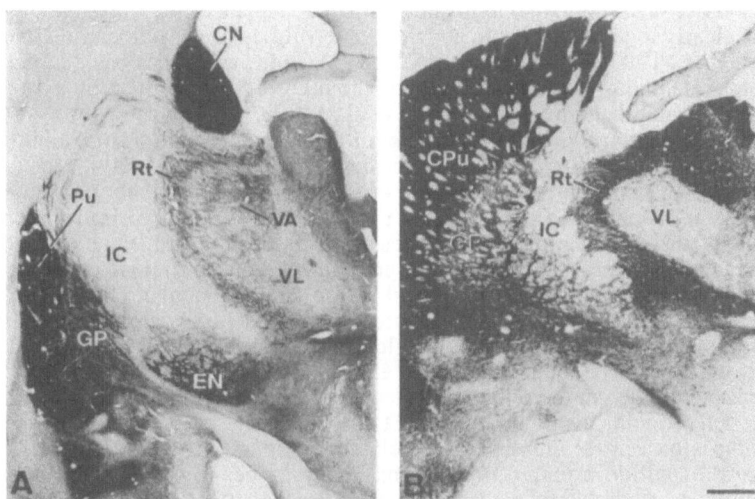

Figure 1. Coronal sections of the cat (A) and the rat (B) processed for acetylcholinesterase histochemistry illustrating the reticular thalamic nucleus and some other nuclei of the thalamus and the basal ganglia. Abbreviations: CN: caudate nucleus; CPu: caudate putamen complex; EN: entopeduncular nucleus; GP: globus pallidus; IC: internal capsule; Pu: putamen; Rt: reticular thalamic nucleus; VA: ventral anterior thalamic nucleus; VL (cat): ventral lateral thalamic complex; VL (rat): ventral lateral thalamic nucleus. Scale bar: 1.19 mm (A) and 0.43 mm (B).

The presence of neurotransmitters and neuropeptides in the reticular thalamic nucleus has begun to be known from the seminal study of Houser et al. (1980). They observed the existence of immunoreactivity for glutamic acid decarboxylase spread in all axes of the reticular thalamic nucleus. Moreover, they identified "punctate" structures attributed to axon terminals next to the neuronal body of its neurons. Interestingly, it has been observed these terminals with almost an absence of immunoreactivity in the cells in other thalamic nuclei. The coexistence of the neuropeptide, somatostatin, in the reticular neurons has been demonstrated in the cat and in the monkey as well (Oertel et al., 1983). In conclusion, the reticular cells use gamma-aminobutyric acid as neurotransmitter which exert an inhibitory influence upon the thalamic targets of the reticular thalamic nucleus (Yen et al., 1985).

The reticular thalamic nucleus receives projections from a great variety of corticothalamic and thalamocortical pathways (Jones, 1975, 1985; Yen et al., 1985; Cornwall and Phillipson, 1988). In sharp contrast, it does project only to the thalamus and other structures at the level of the brainstem. In fact, it appears to reciprocate the thalamoreticular connectivity with a topographical ordering of its efferent fibers (Jones, 1975; Steriade et al., 1984; Crabtree and Killackey, 1989; Velayos et al., 1989). Remarkably, two brainstem structures seem to receive substantial connections from the reticular thalamic nucleus: the superior colliculus and the reticular formation in the mesencephalon (Tortelly and Reinoso-Suárez, 1980; Parent and Steriade, 1984; Giménez-Amaya et al., 1987).

The reticular thalamic nucleus is related to the basal ganglia through the projections of the output nuclei of the basal ganglia (Giménez-Amaya, 1991a, 1991b, 1992). Several groups have described in the rat the existence of clear pallidoreticular projections, although quite restricted in the rostrocaudal and dorsoventral coordinates (Carter and Fibiger, 1978; Haber et al., 1985; Cornwall et al., 1990; Gandia et al., 1992, 1993). Nauta (1979) first illustrated those projections in the cat, and Hazrati and Parent (1991) have shown a more extensive reticular distribution of the pallidal cells innervating this thalamic nucleus in the primate. Moreover, at least for the reticular projections from the globus pallidus in the rat, they are organized with a direct dorsoventral topographical order (Haber et al., 1985; Gandia et al., 1992, 1993). Also in the rat, the pars reticulata of the substantia nigra appears to project clearly to the rostral pole and ventral territories of the reticular thalamic nucleus (Cornwall et al., 1990; Gandia et al., 1992, 1993). In the cat, it has been demonstrated by both anatomical and physiological methods the nigroreticular projection upon the rostral pole of the reticular thalamic nucleus (Paré et al., 1990).

Regarding the intranuclear distribution of reticular afferents from nuclei included in the basal ganglia, it appears to exist interspecies differences. In the rat, for instance, a high percentage of the pallidoreticular fibers cross the internal capsule and enter into the reticular thalamic nucleus caudally to the portions of the globus pallidus explored. They preferentially tend to innervate central zones of the reticular thalamic nucleus following the intranuclear mediolateral coordinates. In contrast, the nigroreticular projections tend to innervate preferentially the more peripheral regions of the reticular thalamic nucleus, both medially and laterally. It appears, therefore, that in rodents a spatial segregation of the reticular projections coming from the globus pallidus and the pars reticulata of the substantia nigra occur (Gandia et al., 1992, 1993).

Moreover, in the cat the reticular projections from the globus pallidus tend to extend more in the reticular thalamic nucleus than do in the rat, with also a preferential distribution upon central areas of this thalamic nucleus (Nauta, 1979). In the monkey, however, the pallidoreticular projections seem to occupy the entire rostrocaudal, dorsoventral and mediolateral extent of the reticular thalamic nucleus (Hazrati and Parent, 1991).

FUNCTIONAL IMPLICATIONS AND SPECULATIONS

The anatomical and physiological data aforementioned, allow us to speculate upon the functional importance of the reticular connections of the basal ganglia. Thus, the connections from the external segment of the globus pallidus (globus pallidus in nonprimates) with the reticular thalamic nucleus are well established in mammals (see also the studies of Hässler, 1949, in the man) and, most probably, they appear to convey neural information already processed in the striatum. The mammalian striatum, on the other hand, presents a clear heterogeneous structure, with two most-visible compartments, according to its distribution of acetylcholinesterase and other substances: the striosomal/patch compartment immersed within a larger matrix compartment (Pert et al., 1976; Graybiel and Ragsdale, 1978; Gerfen, 1992).

The matrix compartment appears to be the main origin of the striatopallidal pathway, although the striosomal/patch compartment could have a small representation as well (Giménez-Amaya and Graybiel, 1990). At any rate, the striatal information going to the external segment of the globus pallidus is mainly directed to the massive key modulatory side-pathway, the subthalamic loop. The fact that part of this information also leaves the internal circuits of the basal ganglia, may indicate a tune regulatory action in the final thalamic flow upon the cortical mantle after the response of the basal ganglia. In addition, it seems conceivable that a massive pathway from the pallidum to the reticular thalamic nucleus is not necessary to truly affect the cortical realm after the activation of these subcortical structures. Besides, the pars reticulata of the substantia nigra appears also to project to the reticular thalamic nucleus. The spatial segregation of the pallidoreticular and the nigroreticular projections seems to indicate a different action of the basal ganglia upon the reticular thalamic nucleus.

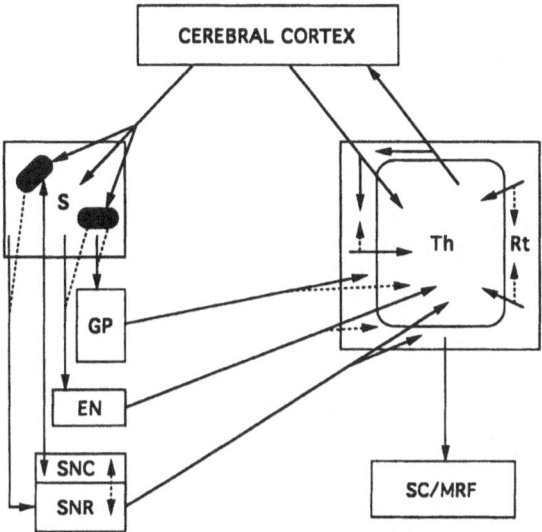

Figure 2. Highly schematic diagram illustrating in nonprimates species the hodological relationships dealt with in this contribution. The cerebral cortex sends projections upon the two compartments of the striatum (jet-black for striosomal/patch compartment) and to the thalamus including the reticular thalamic nucleus. The striatal projections reach the output nuclei of the basal ganglia in which the innervation upon the thalamus and the reticular thalamic nucleus is organized. The thalamocortical projections complete this major corticosubcortical loop. Dashed lines show additional putative pathways. Abbreviations: EN: entopeduncular nucleus; GP: globus pallidus; MRF: mesencephalic reticular formation; Rt: reticular thalamic nucleus; S: striatum; SC: superior colliculus; SNC: substantia nigra, pars compacta; SNR: substantia nigra, pars reticulata; Th: thalamus.

In summary, these distinct reticular systems of projection from the basal ganglia may organize a different set of neural influences upon the thalamic and brainstem targets of the reticular thalamic nucleus, and also upon the thalamocortical and corticothalamic routes related to the reticular sectors that receive innervation from the globus pallidus/globus pallidus and the pars reticulata of the substantia nigra (Fig. 2). The hypothetical modular organization of the striatal projection-neurons could reach particular zones in the output nuclei of the basal ganglia and, therefore, activate very concrete thalamocortical modules related with the basal ganglia function as a whole. It is interesting to note, that the effect of these presumably gabaergic projections from the globus pallidus/globus pallidus and the pars reticulata of the substantia nigra upon the gabaergic cells of the reticular thalamic nucleus might have the global action of inhibiting or releasing different targets of those reticular cells depending on the corticostriatal excitation and the modulation of the nigrostriatal and thalamostriatal systems (Gandia et al., 1993).

Acknowledgements

The authors express their sincere appreciation to Professor J.L. Velayos and A. Rosell for reading the manuscript. This work was supported by DGICYT PB88-0170 and FIS 93/0337.

REFERENCES

Alexander, G.E., DeLong, M.R., and Strick, P.L., 1986, Parallel organization of functionally segregated circuits linking basal ganglia and cortex, *Ann. Rev. Neurosci.* 9:357-381.

Alheid, G.F., Heimer, L., and Switzer, R.C. III, 1990, Basal ganglia, *in*: "The Human Nervous System", G. Paxinos, ed., Academic Press, San Diego, pp. 183-582.

Beckstead, R.M., 1984, The thalamostriatal projection in the cat, *J. Comp. Neurol.* 223:313-346.

Berendse, H.W., and Groenewegen, H.J., 1990, Organization of the thalamostriatal projections in the rat, with special emphasis on the ventral striatum, *J. Comp. Neurol.* 299:187-228.

Carter, D.A., and Fibiger, H.C., 1978, The projections of the entopeduncular nucleus and globus pallidus in rat as demonstrated by autoradiography and horseradish peroxidase histochemistry, *J. Comp. Neurol.* 177:113-124.

Cornwall, J., and Phillipson, O.T., 1988, Mediodorsal and reticular thalamic nuclei receive collateral axons from prefrontal cortex and laterodorsal tegmental nucleus in the rat, *Neurosci. Lett.* 88:121-126.

Cornwall, J., Cooper, J.D., and Phillipson, O.T., 1990, Projections to the rostral reticular thalamic nucleus in the rat, *Exp. Brain Res.* 80:157-171.

Crabtree, J.W., and Killackey, H.P., 1989, The topographic organization and axis of projection within the visual sector of the rabbit's thalamic reticular nucleus, *Eur. J. Neurosci.* 1:94-109.

Fénelon, G., François, C., Percheron, G., and Yelnik, J., 1991, Topographic distribution of the neurons of the central complex (centre médian-parafascicular complex) and of other thalamic neurons projecting to the striatum in macaques, *Neuroscience* 45:495-510.

François, C., Percheron, G., Parent, A., Sadikot, A.F., Fénelon, G., and Yelnik, J., 1991, Topography of the projection from the central complex of the thalamus to the sensorimotor striatal territory in monkeys, *J. Comp. Neurol.* 305:17-34.

Gandia, J.A., and Giménez-Amaya, J.M., 1991, A neuroanatomical analysis of the rostral striatopallidal pathway in the rat, *J. Hirnforsch.* 32:79-88.

Gandia, J.A., García, M., and Giménez-Amaya, J.M., 1992, Afferent projections to the reticular thalamic nucleus from the globus pallidus and the substantia nigra pars reticulata in the rat, *4th Meeting of the International Basal Ganglia Society* , *Abstr.* p.32.

Gandia, J.A., de las Heras, S., García, M., and Giménez-Amaya, J.M., 1993, Afferent projections to the reticular thalamic nucleus from the globus pallidus and the substantia nigra in the rat, *Brain Res. Bull.* in press.

Gerfen, C.R., 1992, The neostriatal mosaic: multiple levels of compartmental organization, *Trends Neurosci.* 15:133-139.

Giménez-Amaya, J.M., 1991a, The association cortex and the basal ganglia: a neuroanatomical view upon their relationship based on hodological studies, *J. Hirnforsch.* 32:501-510.

Giménez-Amaya, J.M., 1991b, Una visión neuroanatómica de los ganglios basales con algunas implicaciones en su fisiopatología, *Rev. Med. Univ. Navarra* 36:131-136.

Giménez-Amaya, J.M., 1992, Some observations upon the distribution of cytochrome oxidase activity in the globus pallidus of the rat, *Acta Anat. (Basel)* 143:246-252.

Giménez-Amaya, J.M., and Graybiel, A.M., 1990, Compartmental origins of the striatopallidal projection in the primate, *Neuroscience* 34:111-126.

Giménez-Amaya, J.M., and Graybiel, A.M., 1991, Modular organization of projection neurons in the matrix compartment of the primate striatum, *J. Neurosci.* 11:779-791.

Giménez-Amaya, J.M., Tortelly, A., and Reinoso-Suárez, F., 1987, Topographical organization of the connections from the diencephalon to the superior colliculus in the cat, *Soc. Neurosci. Abstr.* 17:1435.

Graybiel, A.M., 1990, Neurotransmitters and neuromodulators in the basal ganglia, *Trends Neurosci.* 13:244-254.

Graybiel, A.M., and Ragsdale, C.W. Jr., 1978, Histochemically distinct compartments in the striatum of human, monkey and cat demonstrated by acetylthiocholinesterase staining, *Proc. Natl. Acad. Sci. USA* 75:5723-5726.

Graybiel, A.M., Flaherty, A.W., and Giménez-Amaya, J.M., 1991, Striosomes and matrisomes, *in*: "Basal Ganglia III", G. Bernardi, M.B. Carpenter, G. Di Chiara, M. Morelli, and P. Stanzione, eds., Plenum Press, New York. pp. 3-12

Groenewegen, H.J., 1988, Organization of the afferent connections of the mediodorsal thalamic nucleus in the rat, related to the mediodorsal-prefrontal topography, *Neuroscience* 24:379-431.

Haber, S.N., Groenewegen, H.J., Grove, E.A., and Nauta, W.J.H., 1985, Efferent connections of the ventral pallidum: evidence of a dual striatopallidofugal pathway, *J. Comp. Neurol.* 235:322-335.

Hässler, R., 1949, Uber die Rinden- und Stammhirnanteile des menschlichen Thalamus. *Psychiat. Neurol. Med. Psychol.* 1:181-187.

Hattori, T., and Sugimoto, T., 1983, Direct projections of the globus pallidus to the medial thalamus in the rat, *Soc. Neurosci. Abstr.* 9:1230.

Hazrati, L.-N., and Parent, A., 1991, Projection from the external pallidum to the reticular thalamic nucleus in the squirrel monkey, *Brain Res.* 550:142-146.

Heimer, L., Alheid, G.F., and Zaborszky, L., 1985, Basal ganglia, *in*: "The Rat Nervous System", G. Paxinos, ed., Academic Press Australia, North Ryde, pp. 37-86.

Hontanilla, B., and Giménez-Amaya, J.M., 1990, La proyección talamoestriatal y su relación con las proyecciones eferentes del estriado: un estudio anatómico en el gato, XIV Congreso Nacional de la Sociedad Anatómica Española, p.24.

Hoover, J.E., and Strick, P.L., 1993, Multiple output channels in the basal ganglia, *Science* 259:819-821.

Houser, C.R., Vaughn, J.E., Barber, R.P., and Roberts, E., 1980, GABA neurons are the major cell type of the nucleus reticularis thalami, *Brain Res.* 200:341-354.

Jayaraman, A., 1985, Organization of thalamic projections in the nucleus accumbens and the caudate nucleus in cats and its relation with hippocampal and other subcortical afferents, *J. Comp. Neurol.* 231:396-420.

Jones, E.G., 1975, Some aspects of the organization of the thalamic reticular complex, *J. Comp. Neurol.* 162:285-308.

Jones, E. G., 1985, "The Thalamus," Plenum Press, New York.

Kelley, A.E., and Domesick, V.B., 1982, The distribution of the projection from the hippocampal formation to the nucleus accumbens in the rat: an anterograde- and retrograde horseradish peroxidase study, *Neuroscience* 7:2321-2335.

Malach, R., and Graybiel, A.M., 1986, Mosaic architecture of the somatic sensory-recipient sector of the cat's striatum, *J. Neurosci.* 6:3436-3458.

Mogenson, G.J., Ciriello, J., Garland, J., and Wu, M., 1987, Ventral pallidum projections to mediodorsal nucleus of the thalamus: an anatomical and electrophysiological investigation in the rat, *Brain Res.* 404:221-230.

Nakano, K., Hasegawa, Y., Tokushige, A., Nakagawa, S., Kayahara, T., and Mizuno, N., 1990, Topographical projections from the thalamus, subthalamic nucleus and pedunculopontine tegmental nucleus to the striatum in the Japanese monkey, Macaca fuscata, *Brain Res.* 537:54-68.

Nauta, H.J.W., 1979, Projections of the pallidal complex: an autoradiographic study in the cat, *Neuroscience* 4:1853-1873.

Nauta, W. J. H., and Feirtag, M., 1986, "Fundamental Neuroanatomy," W. H. Freeman, New York.

Oertel, W.H., Graybiel, A.M., Mugnaini, E., Elde, R.P., Schmechel, D.E., and Kopin, I.J., 1983, Coexistence of glutamic acid decarboxylase- and somatostatin-like immunoreactivity in neurons of the feline nucleus reticularis thalami, *J. Neurosci.* 3:1322-1332.

Paré, D., Hazrati, L.-N., Parent, A., and Steriade, M., 1990, Substantia nigra pars reticulata projects to the reticular thalamic nucleus of the cat: a morphological and electrophysiological study, *Brain Res.* 535:139-146.

Parent, A., 1986, "Comparative Neurobiology of the Basal Ganglia," John Wiley, New York.

Parent, A., and Steriade, M., 1984, Midbrain tegmental projections of nucleus reticularis thalami of cat and monkey: a retrograde transport and antidromic invasion study, *J. Comp. Neurol.* 229:548-558.

Parent, A., Paré, D., Smith, Y., and Steriade, M., 1988, Basal forebrain cholinergic and noncholinergic projections to the thalamus and brainstem in cats and monkeys, *J. Comp. Neurol.* 277:281-301.

Pert, C.B., Kuhar, M.J., and Snyder, S.H., 1976, Opiate receptor: autoradiographic localization in rat brain, *Proc. Natl. Acad. Sci. USA* 73:3729-3733.

Price, J.L., and Slotnick, B.M., 1983, Dual olfactory representation in the rat thalamus: an anatomical and electrophysiological study, *J. Comp. Neurol.* 215:63-77.

Ramón y Cajal, S., 1904, "Textura del Sistema Nervioso del Hombre y de los Vertebrados", N. Moya, Madrid.

Royce, G.J., 1983, Single thalamic neurons which project to both the rostral cortex and caudate nucleus studied with the fluorescent double labeling method, *Exp. Neurol.* 79:773-784.

Russchen, F.T., Amaral, D.G., and Price, J.L., 1987, The afferent input to the magnocellular division of the mediodorsal thalamic nucleus in the monkey, Macaca fascicularis, *J. Comp. Neurol.* 256:175-210.

Scheibel, M.E., and Scheibel, A.B., 1966, The organization of the nucleus reticularis thalami: a Golgi study, *Brain Res.* 1:43-62.

Steriade, M., Parent, A., and Hada, J., 1984, Thalamic projections of nucleus reticularis thalami of cat: a study using retrograde transport of horseradish peroxidase and fluorescent tracers, *J. Comp. Neurol.* 229:531-547.

Sugimoto, T., and Hattori, T., 1984, Direct projections from the globus pallidus to the paraventricular nucleus of the thalamus in the rat, *Brain Res.* 323:188-192.

Tortelly, A., and Reinoso-Suárez, F., 1980, Projections to the superior colliculus from the dorsal hypothalamic area and other prosencephalic structures derived from the embryonic subthalamic longitudinal band of the diencephalon, *Neurosci. Lett.* 18:257-260.

Velayos, J.L., Jiménez-Castellanos, J. Jr., and Reinoso-Suárez, F., 1989, Topographical organization of the projections from the reticular thalamic nucleus to the intralaminar and medial thalamic nuclei in the cat, *J. Comp. Neurol.* 279:457-469.

Wilson, S.A.K., 1912, Progressive lenticular degeneration: a familiar nervous disease associated with cirrhosis of the liver, *Brain* 34:295-509.

Wilson, S.A.K., 1914, An experimental research into the anatomy and physiology of the corpus striatum, *Brain* 36:427-492.

Yen, C.T., Conley, M., Hendry, S.H.C., and Jones, E.G., 1985, The morphology of physiologically identified GABAergic neurons in the somatic sensory part of the thalamic reticular nucleus in the cat, *J. Neurosci.* 5:2254-2268.

DEVELOPMENT, DEGENERATION AND TRANSPLANTS

IN THE BASAL GANGLIA

DEVELOPMENTAL PLASTICITY OF DOPAMINE SYSTEMS: PRE- AND POST-SYNAPTIC COMPONENTS OF THE DOPAMINE SYSTEM ARE MODIFIED BY NEONATAL LESIONS

Jeffrey N. Joyce[1] and Bethany Neal-Beliveau[2]

[1]Departments of Psychiatry and Pharmacology
  University of Pennsylvania School of Medicine
  Philadelphia, PA 19104-6141
[2]Department of Psychology
  Indiana University

INTRODUCTION

Over the past 15 years, a model for examining the plasticity of DA systems, involving the administration of the neurotoxin 6-hydroxydopamine (6-OHDA) to damage the mesostriatal DA system, has been well characterized. This model benefits from the considerable information available regarding the organization of the pre- and post-synaptic components of the DA system (Fallon and Moore, 1978; Boyson et al., 1986; Gerfen et al., 1987; Joyce, 1991a), their development (Murrin et al., 1985; Murrin and Zeng, 1989; Rao et al., 1991), and their regulation following lesions of the mature DA system (Marshall et al., 1989; Joyce, 1991a,b). In adult rats, extensive lesions of DA neurons produce profound behavioral impairments, such as aphagia, adipsia, akinesia, and sensory neglect, which mimic changes observed in Parkinson's disease (Marshall, 1979; Ungerstedt, 1971; Zigmond and Stricker, 1973). In contrast, it has been known for many years that extensive neonatal 6-OHDA lesions do not induce a parkinsonian-like syndrome (Breese et al., 1984,1987; Bruno et al., 1985, 1986; Criswell et al, 1992), and produce distinctly different alterations of DA receptor function as identified by behavioral pharmacologic probes. Adult-lesioned rats were found to be hyporesponsive to d-amphetamine, but exhibited a potentiated response to the DA precursor L-DOPA or direct dopamine agonists. The potentiated effect of L-DOPA suggests that D1 and/or D2 receptors are supersensitive in adult-lesioned rats. Adult-lesioned rats were subsequently found to exhibit behavioral supersensitivity to the selective D2 receptor agonist quinpirole, with no change in their response to the selective D1 receptor agonist SKF38393 (Breese et al., 1987; Duncan et al., 1987). Thus, adult-lesioned rats exhibit a functional supersensitivity to D2 receptor stimulation. Furthermore, adult 6-OHDA lesioned rats show elevations in D2 receptor number and reductions in D1 receptor number (Marshall et al, 1989; Joyce, 1991a,b). This increase in D2 receptor density in the lateral striatum is partly responsible for the behavioral recovery after DA depletion (Creese et al 1977; Marshall, 1984).

Unlike adult-lesioned rats, neonatally-lesioned rats respond differently to challenges with DA agonists and antagonists. Neonatally-lesioned rats are not supersensitive to D2 receptor agonists. Rather, they exhibit an increased sensitivity to D1 receptor agonists (Breese et al., 1984, 1987; Criswell et al, 1992). Furthermore, the neonatally-lesioned rats exhibit self-biting and self-mutilatory behaviors following treatment with L-DOPA and D1 receptor agonists, behaviors not observed in either intact or adult-lesioned animals (Breese et al., 1984, 1987; Criswell et al, 1992). There is an increasing response to the D1 receptor agonists with repeated treatment ("priming phenomenon"), whereas the response to the D2 receptor agonist quinpirole does not change (Criswell et al., 1989). Neonatally-lesioned rats are less sensitive than are intact rats to the behavioral effects of DA receptor antagonists such as haloperidol, fluphenazine and SCH23390 (Breese et al., 1985; Bruno et al., 1985; Weihmuller and Bruno, 1989). Thus, neonatally-lesioned rats exhibit a functional supersensitivity to D1 receptor agonists and subsensitivity to D1 antagonists. These results suggest that the neural mechanisms underlying the sparing of function in neonatally lesioned rats as adults are disimilar from those following such lesions made in the adult. Moreover, direct evidence for receptor plasticity following neonatal 6-OHDA lesions is inconsistent (Breese et al, 1987). The mechanisms underlying the differential neurochemical and behavioral effects of neonatal as compared to adult 6-OHDA lesions have, until recently, remained unclear. It has been our hypothesis that the effects of neonatal 6-OHDA lesions can only be understood within the context of the ontogeny of the DA system; specifically, that lesions made at the earliest postnatal stage alter the subsequent development of both the pre- and postsynaptic components of the DA system because of the critical role DA plays in this development (Rao et al, 1991; Neal and Joyce, 1991, 1992; Neal-Beliveau and Joyce, 1993). Since the ontogeny of the DA system obeys an explicit compartmental organization of the striatum, the effects of lesions to this system are necessarily related to this organization.

Ontogenetic Development of Dopamine Receptors

The first evidence for the compartmental organization of the striatum came from the observations that DA histofluoresence had an islandic appearance in the striatum of the developing rat (Olson et al., 1972). It has since been established that these "dopamine islands" coincide with the acetylcholinesterase-poor zones circumscribing the patches in the adult rat (Herkenham and Pert, 1981; Graybiel, 1984). Furthermore, the development of the striatum, including the histogenesis and migration of striatal neurons (van der Kooy and Fishell, 1987), DA input (Specht et al., 1981a,b; van der Kooy, 1984; Voorn et al., 1988) and cortical innervation (Donoghue and Herkenham, 1986), also follows this compartmental pattern. There is also a patch/matrix development of DA receptors that develops in concert with this. We have used the technique of quantitative autoradiography with the iodinated ligands [^{125}I]SCH 23982 and [^{125}I]IBZM to characterize the independent development of the two receptors both quantitatively and with anatomical precision (Rao et al, 1991). In addition, the development of the two subtypes have been examined for their relationship to independent markers of striatal organization, DA and acetylcholine terminals. D1 receptors appear initially as dense patches in the ventral and lateral caudate-putamen (CPu). Patches of dense binding are visible in other regions of the CPu (but not the nucleus accumbens) by P3. The D1 receptor patches are aligned with the patchy distribution of [^{3}H]mazindol binding to DA terminals at this early age suggesting that the binding of [^{125}I]SCH 23982 during the first week postnatally is predominantly within the patch compartment. Murrin and Zeng (1989) have independently reported that patches of dense D1 receptors (labeled with [^{3}H]SCH 23390) observed during early postnatal development are aligned with highly delineated patches of DA fluoresence and μ-opiate receptors. Since the patches of DA fluoresence and μ-opiate receptors are markers for the patch compartment in early development (Murrin and Ferrer, 1984; van der Kooy, 1984), this further corroborates the hypothesis that D1 receptors are initially expressed in higher amounts by neurons of the patch compartment than of the matrix compartment. The patchy appearance of D1 receptor binding becomes obscured between P7 and P10, and completely disappears by P16. This is due to a large increase in the density of sites in the matrix with no reduction of sites in the patch compartment (Fig. 1). The change in the topography of [^{3}H]mazindol binding to striatal DA terminals from a patchy to homogeneous pattern (P7 - P10) appeared to coincide with or just precede the similar change in the patterning of D1 receptors. D2 receptors become expressed in the SNpc at the time (P7) when the pattern of [^{3}H]mazindol labeling of striatal DA fibers changes from patchy to homogeneous and matrix-directed neurons of the midbrain region are undergoing significant changes in levels of TH mRNA (Solberg et al., 1992). This expression of D2 autoreceptors in the SNpc DA neurons (Mengod et al., 1989; LeMoine and Bloch, 1991) is consistent with the evidence that D2 autoreceptor effects can occur as early as postnatal day 11 (McDougall and Bardo, 1991). Thus, there is a coincident expression of high-affinity DA uptake sites on the matrix-directed DA fibers, D2 autoreceptors in the SN, and D1 receptors in the matrix compartment between P7 and P10 that ushers in the second stage of development of the DA system.

Our results show that D2 receptors develop independently of the D1 receptor (Rao et al., 1991). During the first week postnatally, the topographical development of D2 receptors occurs as along the lateral-to-medial axis. D2 receptors are initially visible in the dorsolateral CPu at P3 to P5 and during development the density of D2 receptors remains highest in the lateral CPu. Importantly, while there is evidence for microzones of higher D2 receptor density in the dorsolateral CPu at the earliest ages, they are always complementary to the D1 receptor-enriched patches. The topography of [3H]HC-3 binding, an index of the density of cholinergic terminals, develops as a lateral-to-medial gradient and is not patchy like that of D1 receptors or [3H]mazindol binding at similar early postnatal ages (Rao et al., 1991). The maturation of the patterning of [3H]HC-3 binding, and generation and maturation of the large cholinergic interneurons (Phelps et al., 1989) of the striatum, correlates well with

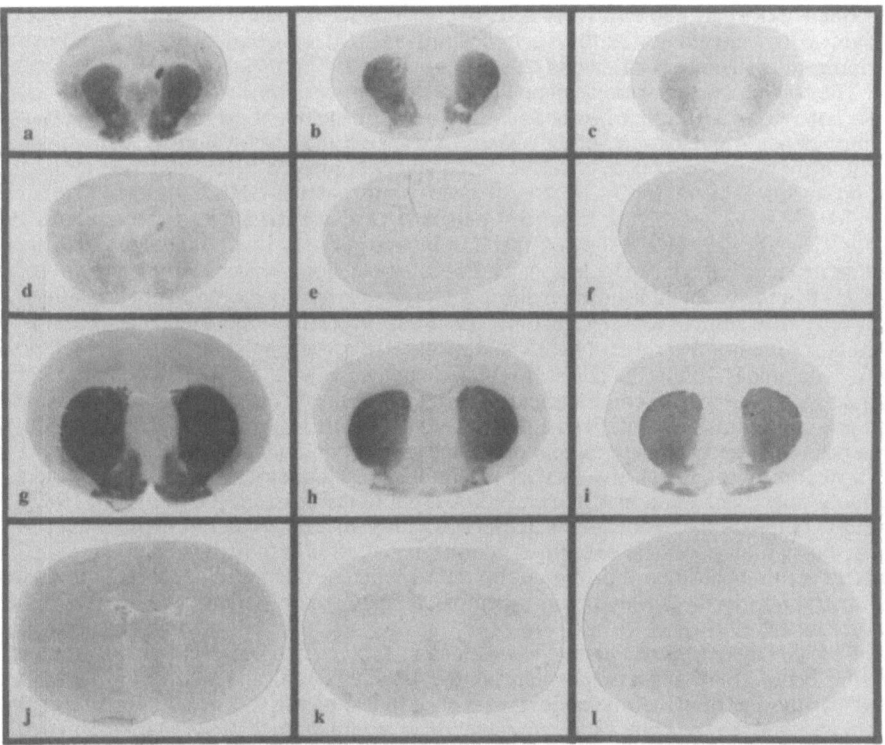

Figure 1. Photomicrographs of autoradiographs depicting [125I]SCH23982 binding to the D1 receptors (a,d,g,j), [125I]IBZM D2 receptors (b,e,h,k) and [3H]mazindol to DA uptake sites (c,f,i,l) at posnatal day 7 (a,b,c) and postnatal day 30 (g,h,i). Total (a,b,c,g,h,i) and nonspecific binding (d,e,f,j,k,l) are shown for each radioligand.

the developmental pattern of D2 receptors. This suggests that the development of D2 receptors may initially be related to the maturation of cholinergic terminals. However, other cellular components of the matrix compartment of the striatum that are closely associated with these cholinergic terminals also play a role in the postnatal development.

The results of our work and other studies indicate that the developmental pattern of DA innervation <u>does not</u> correspond to the pattern of D2 receptor development (Murrin, 1982; Murrin et al., 1985; Rao et al., 1991) and <u>does</u> for D1 receptor expression (Caboche et al., 1991; Rao et al., 1991). We have proposed that the development of D1 receptors in the patch and matrix compartments occurs independently and in relation to maturation of the DA fibers within each compartment. In contrast, the initial development of the D2 receptors appears unrelated to the maturation of the DA afferents, complementary to the patch- and matrix-related D1 receptor systems, and in concert with the maturation of the cholinergic neuropil of the matrix compartment.

Effects of Neonatal 6-OHDA-Induced Dopamine Loss on Adult Dopamine Receptor Expression: Neurochemical and Behavioral Studies

The mechanisms that underlie the differential behavioral effects of neonatal and adult 6-OHDA lesions remain unclear. We have proposed that the neurochemical and functional consequences of neonatal 6-OHDA lesions are dependent upon the timing of the lesions. This hypothesis has been explored in several stages: (1) receptor adaptations as a consequence of early and relatively selective lesions to the patch-directed DA system; (2) behavioral consequences of the early lesion; and (3) consequences for the postnatal development of the matrix-oriented DA system.

The present studies were designed to directly compare the pharmacological specificity of the behavioral differences in neonatally and adult-lesioned rats and the alterations in receptor density and pattern of DA loss. Lesions to the DA system in early postnatal development have different behavioral consequences compared to lesions made in the adult. The prototypical behavioral effect of the early postnatal 6-OHDA lesion is self-biting behavior (SBB). In a series of important papers Breese and associates (Breese et al., 1984, 1987; Criswell et al., 1992) showed that administration of L-DOPA induces SBB in adult rats that received extensive DA lesions on P5. The behavioral pharmacologic profile of SBB indicates that the neonatal lesions results in a more selective D1 receptor supersensitivty than observed with adult 6-OHDA lesions. In those investigations the authors employed techniques for administering 6-OHDA that resulted in a widespread and near total depletion of DA when measured in the adult. It remains unclear to what extent the loss of DA to the separate patch and matrix compartments might play in the differences between neonate and adult lesioned animals. We utilized a technique that results in a relatively selective reduction in the patch-directed DA system, that of the intrastriatal administration of 6-OHDA at P0 or P1 (Neal and Joyce, 1991). If 6-OHDA is administered intrastriatally prior to P3, there is a relatively selective depletion of the early DA input to the striatum (Gerfen et al, 1987). We wished to compare the neonatally lesioned rats to rats that received DA depleting lesions as adults, but which could be examined for the behavioral responses to DA agonists. We utilized the intraventricular injection of 6-OHDA which produces bilateral reductions in the DA innervation to the striatum and receptor changes similar to those with intranigral 6-OHDA (Joyce, 1991a).

Neonatal rats received bilateral intrastriatal injections of 6-OHDA (4 µg per striatum) or VEHICLE on day of birth or postnatal day 1 (P0/P1) (Neal and Joyce, 1991) and were tested 90 days later for their behavioral response to DA agonists. The adult rats (60 days of age) were given intracerebral ventricular injections of 6-OHDA (150 µg/10 ml each ventricle, Joyce, 1991a) to mimic the pattern of DA loss in the neonates and tested 90 days later for their behavioral response to DA agonists. Additional animals were chronically administered reserpine to produce a loss of striatal DA without damaging DA terminals (Joyce, 1991b). Animals were administered the mixed D1/D2 agonist apomorphine (0.5 mg/kg), the D1 agonist SKF38393 (10 mg/kg) or the D2 agonist quinpirole (1.0 mg/kg) in a counter balanced order. They were placed into a large observation chamber and scored for the appearance of a number of behaviors (see Neal and Joyce, 1991). Cohorts were killed and processed for receptor autoradiography as described in detail elsewhere (Joyce, 1991a,b; Neal and Joyce, 1991, 1992). DA uptake sites were labeled with [3H]mazindol, D1 receptors with [3H]SCH23390, and D2 receptors with [3H]spiroperidol.

Effects of Early Postnatal Damage to the Patch-Directed DA system on DA Receptor Expression. Based on our observations of the development of the DA systems in the normal rat and the behavioral consequences of the intrastriatal 6-OHDA lesions at P0/P1 we reasoned that early neonatal lesions should primarily affect DA D1 receptors expressed in the patch compartment. We examined the long-term effects of early DA loss to the patch compartment on DA receptor expression (Neal and Joyce, 1992). The rats were killed at 90 days of age and the brains processed for quantitative autoradiography. There was a patchy loss of TH-like immunoreactivity and of [3H]mazindol binding as determined by autoradiography, particularly evident in the dorso-medial CPu (Table 1). There was also a reduction in the number of μ-opioid receptor patches (labeled with [3H]naloxone), a marker for the striatal patch compartment, and a similar patchy loss of D1 binding sites. The number of D2 sites was not altered (Table 1). In the adult rat, greater than 80% losses of DA are required before changes in D2 receptor density can be observed (Joyce, 1991a). The introcerebroventricular injections of 6-OHDA produce greater than 90% losses of DA uptake sites in the medial CPu but a less substantial loss in the lateral CPu (Table 1). This results in a marked loss of D1 receptors in the dorsal CPu and an increase of D2 receptors in the lateral CPu (Joyce, 1991b). Chronic reserpine (1.0 mg/kg, every second day) results in a greater than 90% loss of DA and a marked increase in D2 receptor number (Joyce, 1991b). In contrast to 6-OHDA lesions of the mesostriatal DA system, reserpine produces an increase in D1 receptor number. To test if larger depletions of DA at P0/P1 would induce a change in D2 receptor density we examined animals at P90 that had been given differing doses of 6-OHDA bilaterally instrastriatally at P0/P1 (Neal-Beliveau and Joyce, 1993). Even with near total losses of [3H]mazindol binding to DA uptake sites there was not an increase in D2 receptor number. However, the degree of D1 loss was enhanced with higher doses of 6-OHDA.

TABLE 1. Numbers of dopamine uptake sites, D1 and D2 receptors in regions of striatum in unlesioned control cases, animals receiving neonatal 6-OHDA lesions or animals receiving adult 6-OHDA lesions.

BINDING SITES	STRIATAL	REGION		
	Dorsomedial	Dorsolateral	Ventromedial	Ventrolateral
DA Uptake Sites				
Unlesioned Controls	2579 ± 127	2710 ± 145	2232 ± 94	2292 ± 119
Neonatal 6-OHDA	1620 ± 131[1]	2041 ± 140[1]	1528 ± 145[1]	1965 ± 114
Adult 6-OHDA	185 ± 101[2]	635 ± 111[2]	125 ± 101[2]	685 ± 122[2]
D1 Receptors				
Unlesioned Controls	2181 ± 75	2185 ± 63	2222 ± 79	2299 ± 74
Neonatal 6-OHDA	1897 ± 66[1]	1990 ± 63[1]	2046 ± 65[1]	2162 ± 75
Adult 6-OHDA	1461 ± 49[2]	1486 ± 58[2]	2299 ± 54[2]	2295 ± 95
D2 Receptors				
Unlesioned Controls	407 ± 9	697 ± 12	401 ± 7	617 ± 14
Neonatal 6-OHDA	401 ± 9	677 ± 12	391 ± 6	600 ± 22
Adult 6-OHDA	399 ± 11	899 ± 22[2]	411 ± 8	802 ± 23[2]

[1]Significant at $p < 0.05$ as compared to unlesioned controls.
[2]Significant at $p < 0.05$ as compared to unlesioned controls and to neonatal 6-OHDA group.

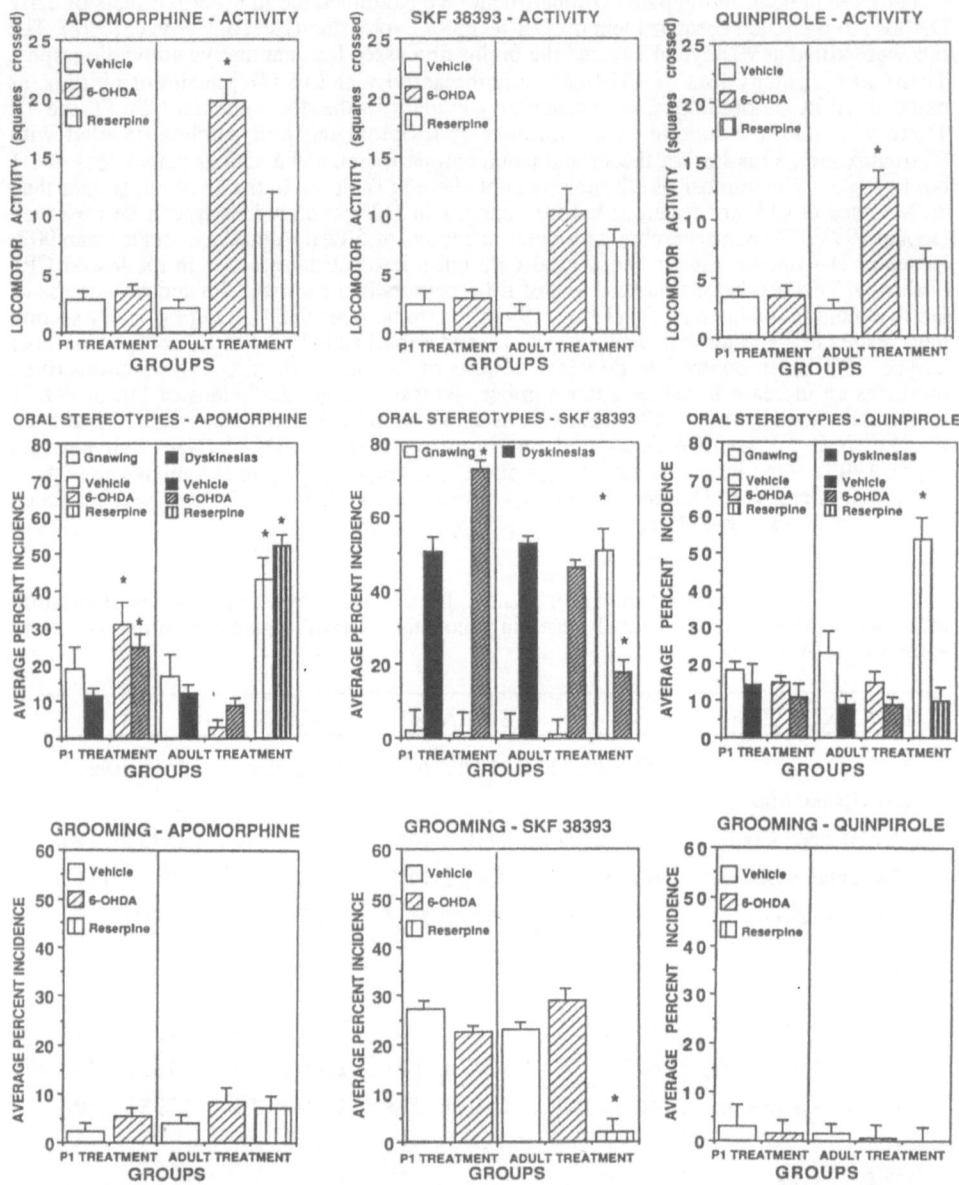

Figure 2. Frequency of stereotypic behaviors or amount of locomotor activity in the vehicle control group (Vehicle), neonatal 6-OHDA group (P1 treatment), adult 6-OHDA group (Adult treatment) and adult reserpine treatment group. All animals were treated with the D1/D2 agonists apomorphine, the D1 agonist SKF38393 or the D2 agonist quinpirole. Significant differences for treatment groups from vehicle group indicated by asterik.

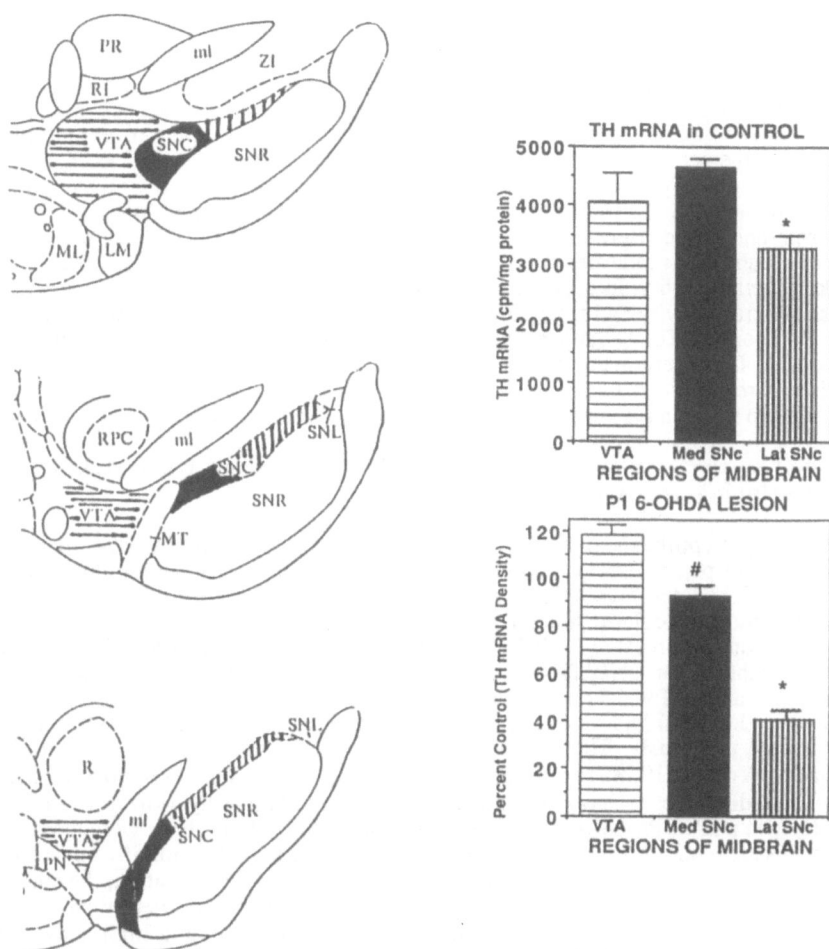

Figure 3. Shown on the left are the regions of the midbrain analyzed for levels of TH mRNA as labeled by a [^{35}S]oligonucleotide. Densitometric analysis of the amount of labeling is shown on the right. In controls the lateral SNpc shows lower amounts of labeling than the medial SNpc or the VTA. In the adults receiving neonatal 6-OHDA lesions the lateral SNpc shows significantly less labeling (*) than the medial SNpc or VTA.

Effects of Early Postnatal Damage to the Patch-Directed DA system on DA Receptor-Mediated Behaviors. The control vehicle animals exhibited increased locomotor activity, sniffing stereotypies and mild increases in oral stereotypies to apomorphine or quinpirole but not to SKF38393. Animals with intrastriatal 6-OHDA lesions made on P0 or P1 show SBB to L-DOPA and to the selective D1 agonist SKF38393. However, the early neonatal intrastriatal lesions also resulted in a subsensitivity to the cataleptic effects of the D1 antagonist SCH23390, with no altered response to the D2 antagonist haloperidol (Neal and Joyce, 1991). In response to apomorphine the neonatal lesioned rats showed increased amounts of oral stereotypies, both gnawing and oral dyskinesias (Fig. 4). There were not alterations in locomotor activity or grooming responses. No gnawing was observed following treatment with SKF38393 but other oral stereotypies were elevated. Thus, the selective D1 receptor agonist SKF38393 (10 mg/kg) but not the D2 agonist quinpirole produced self-biting and oral dyskinesias (Fig. 2). In the adult 6-OHDA rats there was a reduction in gnawing stereotypies to apomorphine as compared to the vehicle-controls or reserpine treated rats. In contrast, the locomotor activity response to apomorphine or quinpirole was elevated in the adult 6-OHDA group. The reserpine treated animals showed elevated oral stereotypies to apomorphine, SKF38393 and quinpirole and reduced grooming responses to SKF38393 (Fig. 4). Thus, the behavioral responses to the drugs indicated that the neonatally lesioned rats showed a relatively selective D1-like supersensitivity. In contrast, the adult 6-OHDA group showed a relatively selective D2-like supersensitivity and the reserpine group showed enhanced oral stereotypies to both D1 or D2 agonists. The D2 supersensitivity in the adult 6-OHDA group is likely related to the elevated D2 receptor number. The DAergic supersensitivity in the reserpine group, likewise, is related to the elevation in D1 and D2 receptors. In contrast, the neonatal 6-OHDA group shows a D1 supersensitivity in the face of a small reduction in D1 receptors.

Effects of Early Postnatal Damage to the Patch-Directed DA system on the later-developing DA system

Another hypothesis derived from our model is that the damage to the early (predominantly patch-directed) maturing DA system could result in an alteration in the maturation of the later-developing DA innervation. We have tested the possibility that the alteration of the patch-related striatonigral pathway could lead to changes in the matrix-related DA innervation (Frohna et al., 1992). We examined levels of TH mRNA, D2 mRNA and D2 receptor number in the SN/VTA of 6 neonatally lesioned rats (P1 6-OHDA) and 5 Vehicle counterparts at age P90. The oligonucleotide probes used for the D2 receptor mRNA (Mengod et al., 1989) and rat TH cDNA (Grima et al., 1985; Han et al., 1987) have been characterized previously. The labeling procedures were a modification of the method of Gerfen and Young (1988) for 3'-end labeled with [^{35}S]-dATP. 160-180 ml of labeled oligonucleotide probe, at a concentration of 1×10^6 cpm/40 ml of hybridization buffer, were added to each pair of ProbeOn Plus slides. Pairs of ProbeOn Plus slides were then clamped together, tissue sections facing each other, suspended in a sealed, humidified chamber, and allowed to hybridize for 18-24 hours at 37ºC. Following the hybridization, the slides were processed through routine post-hybridization protocols. The slides were then apposed to ^{3}H-Hyperfilm (Amersham) for densotometric analysis by including high activity [^{3}H]-containing plastic standards (American Radiolabeled Chemicals, St. Louis, MO). The high activity [^{3}H]-containing plastic standards were standardized against [^{35}S]containing tissue standards. Following film development the films were digitized with a computer-based image analysis system (BRAIN) and the mean optical density of each brain region of interest converted to a value of cpm/mg protein.

Thionin staining of the region revealed that the neonatal 6-OHDA animals had substantially fewer neurons in the SNpc, particularly the ventral tier, and in the SNpr. In situ hybridization histochemistry revealed that the SNpc had few neurons expressing TH mRNA over them but the VTA and remaining neurons of the medial SNpc showed increased grain density (See Fig. 3). The autoradiographs of the D2 receptors ([^{125}I]epidepride binding) showed a thinner rim of binding in the SNpc and lower binding in the VTA than in the controls. There was also reduced densities of grains in sections labeled for the D2 receptor mRNA (Fig. 4). This suggests that the removal of the patch-directed DA afferents in early development leads to enhanced synthesis of TH and lower synthesis of D2 autoreceptors in the remaining neurons of the SN/VTA. Previous studies have shown that the projections of neurons from the medial SNpc and the lateral VTA are more highly collateralized than those in the SNpl (Fallon and Loughlin, 1982; Loughlin and Fallon, 1984). The neurons situated in the medial SNpc and VTA normally express higher levels of TH mRNA than other subpopulations of neurons in the SNpc (Weiss-Wunder and Chesselet, 1991), suggesting they respond to an increased demand for the enzyme based on the need for transport of more TH molecules to a larger terminal field. It is possible that these neurons are hypercollateralizing following the PO/P1 lesions to fill in "missing" synapses in the striatum. Then the TH mRNA is higher as a consequence of the increased demand. It is also known that TH mRNA in neurons of the VTA and medial SNpc stabilized at a later time-point than that for neurons in the SNpc (Solberg et al., 1992). Perhaps, loss of the early DA system provides stabilization of TH mRNA at elevated levels.

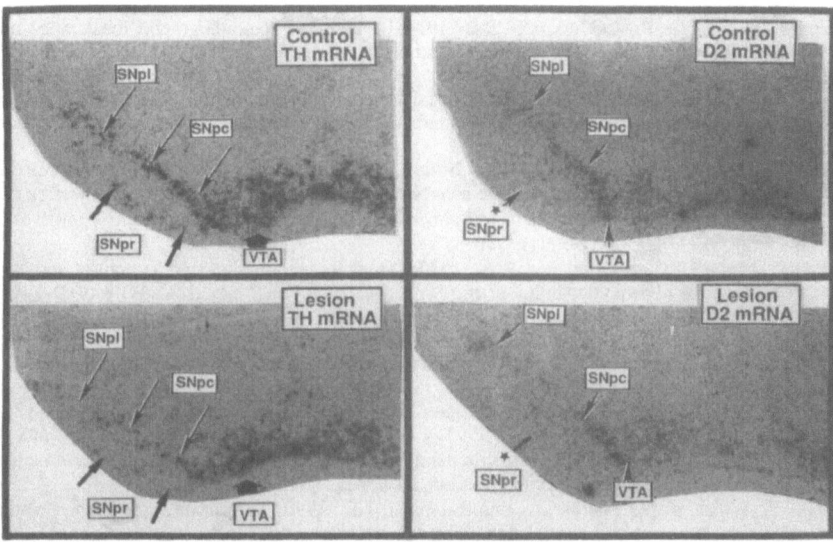

Figure 4. Photomicrographs of autoradiographic images of [^{35}S]labeled oligonucleotides for labeling of TH mRNA and the D2 receptor mRNA in the midbrain region of an adult rat receiving 6-OHDA (bottom) or vehicle (top) as a neonate.

Our results indicate that the effects of the neonatal 6-OHDA lesions on expression of TH mRNA are very different from those in the adult rat. Following adult 6-OHDA lesions, there is a significant correlation between the degree of DA depletion and the loss of TH activity within the striatum (Pasinetti et al., 1992). TH activity and the amount of TH protein were found to be significantly decreased 3 weeks post-lesion. However, the levels had returned to normal by 90 days post-lesion (Pasinetti et al., 1992). This suggests that there may be an increased activation of the synthetic enzyme TH. The normalization of TH coincides with the recovery of striatal DA levels, but contrasts to a decrease in TH mRNA concentration per neuron by 90 days post-lesion (Pasinetti et al., 1992). Thus, it has been assumed that the remaining DA neurons are more efficient in translating TH mRNA for TH synthesis, and ultimately the final product DA (Pasinetti et al., 1989, 1992).

Model. The DA innervation to the striatum arising from the medial SNpc and VTA may be releasing higher amounts of DA at basal levels and in response to l-dopa administration. This may be due to a number of factors. The neonatal 6-OHDA lesions result in a failure of the neurons of the patch compartment to express D1 receptors. These neurons provide the major striatonigral inhibitory feedback system and have reduced Substance P activity with the neonatal 6-OHDA lesions (Sivam et al, 1987; Sivam, 1989). This would presumably lead to reduced inhibitory feedback on dopamine neurons of the SN-VTA complex. The midbrain neurons remaining after the neonatal 6-OHDA lesions have higher levels of TH mRNA, reduced expression of high-affinity DA uptake sites and lower expression of D2 autoreceptors. Thus, they are likely to have increased extracellular DA levels (Castaneda et al., 1990) This altered baseline in DA levels would provide for tonically higher activity at D1-receptor mediated systems. In contrast, D2 receptor-mediated events are unlikely to be supersensitive. We hypothesize that the neonatal lesions lead to an overreactivity of D1 receptor-mediated efferent systems that innervate pallidum and SNpr but a subsensitivity to the D1 receptor mediated inhibitory feedback system. This would allow for a greater contribution of D1 receptor-mediated systems to the induction of oral stereotypies.

REFERENCES

Boyson, S.J., McGonigle, P., and Molinoff, P.B., 1986, Quantitative autoradiographic localization of the D1 and D2 subtypes of dopamine receptors in rat brain, *J. Neurosci.* 6:31177-3188.

Breese, G.R., Baumeister, A.A., McCown, T.J., Emerick, S.G., Frye, G.D., Crotty, K., and Mueller, R.A., 1984, Behavioral differences between neonatal and adult 6-hydroxy-dopamine-treated rats to dopamine agonists: Relevance to neurological symptoms in clinical syndromes with reduced brain dopamine, *J. Pharmacol. Expt. Ther.* 231:343-354.

Breese, G.R., Duncan, G.R., Napier, T.C., Bondy, S.C., Iorio, L.C., and Mueller, R.A., 1987, 6-Hydroxydopamine treatments enhance behavioral responses to intracerebral microinjection of D_1 and D_2-dopamine agonists into nucleus accumbens and striatum without changing dopamine antagonist binding, *J. Pharmacol. Exp. Ther.* 240:167-176.

Breese, G.R., Napier, T.C., and Mueller, R.A., 1985, Dopamine agonist-induced locomotor activity in rats treated with 6-hydroxydopamine at differing ages: functional supersensitivity of D-1 dopamine receptors in neonatally-lesioned rats, *J. Pharmacol. Exp. Ther.* 234:447-455.

Bruno, J.P., Stricker, E.M., and Zigmond, M.J., 1985, Rats given dopamine-depleting brain lesions as neonates are subsensitive to dopaminergic antagonists as adults, *Behav. Neurosci.* 99:771-775.

Bruno, J.P., Zigmond, M.J., and Stricker, E.M., 1986, Rats given dopamine-depleting brain lesions as neonates do not respond to acute homeostatic imbalances as adults, *Behav. Neurosci.* 100:125-128.

Caboche, J., Rogard, M., and Besson, M.-J., 1991, Comparative development of D_1-dopamine and μ opiate receptors in normal and in 6-hydroxydopamine-lesioned neonatal rat striatum: dopaminergic fibers regulated μ but not D_1 receptor distribution, *Devl. Brain Res.* 58:111-122.

Castaneda, E., Whishaw, I.Q., Lerner, L., and Robinson, T.E., 1990, Dopamine depletion in neonatal rats: effects on behavior and striatal dopamine release assessed by intracerebral microdialysis during adulthood, *Brain Res.* 508:30-39.

Creese, I., Burt, D.R., and Snyder, S.H., 1977, Dopamine receptor binding enhancement accompanies lesion-induced behavioral supersensitivity, *Science* 197:596-598.

Criswell, H.E., Mueller, R.A., and Breese, G.R., 1992, Pharmacologic evaluation of SCH-39166, A-69024, NO-0756 and SCH-23390 in neonatal-6-OHDA-lesioned rats: further evidence that self-mutilatory behavior induced by L-DOPA is related to D_1 dopamine receptors, *Neuropsychopharm.* 7:95-103.

Criswell, H.E., Mueller, R.A., and Breese, G.R., 1989, Priming of D_1-dopamine responses: Long-lasting behavioral supersensitivity to a D_1-dopamine agonist following repeated administration to neonatal 6-OHDA-lesionsed rats, *J. Neurosci.* 9:125-133.

Donoghue, J.P., and Herkenham, M., 1986, Neostriatal projections from individual cortical fields conform to histochemically distinct striatal compartments in the rat, *Brain Res.* 365:397-403.

Duncan, G.E., Criswell, H.E., McCown, T.J., Paul, I.A., Mueller, R.A., and Breese, G.R., 1987, Behavioral and neurochemical responses to haloperidol and SCH-23390 in rats treated neonatally or as adults with 6-hydroxydopamine, *J. Pharmacol. Exp. Ther.* 243:1027-1033.

Fallon, J.H., and Loughlin, S.E., 1982, Monoamine innervation of the forebrain: Collateralization, *Brain Res. Bull.* 9:925-307.

Fallon, J.H., and Moore, R.Y., 1978, Catecholamine innervation of the basal forebrain. IV. Topography of the dopamine projection to the basal forebrain and neostriatum, *J. Comp. Neurol.* 180:545-580.

Frohna, P.A., Rioux, L., Neal, B.S., and Joyce, J.N., 1992, Damage to the striatal dopamine systems in early development selectively spares the matrix-directed dopamine system, *Soc. Neurosci. Abstr.* 18:622.

Gerfen, C.R., Baimbridge, K.G., and Thibault, J., 1987, The neostriatal mosaic: III. Biochemical and developmental dissociation of patch-matrix mesostriatal systems, *J. Neurosci.* 7:3935-3944.

Gerfen, C.R., and W.S. Young, 1988, Distribution of striatonigral and striatopallidal peptidergic neurons in both patch and matrix compartments: an in situ hybridization histochemistry and fluorescent retrograde tracing study, *Brain Res.* 460:161-167.

Graybiel, A.M., 1984, Correspondence between the dopamine islands and striosomes of the mammalian striatum, *Neurosci.* 13 :1157-1187.

Grima, B., Lamarou, A., Blanot, F., Biguet, F.N., and Mallet, J., 1985, Complete coding sequence of rat tyrosine hydroxylase mRNA, *Proc. Natl. Acad. Sci. U.S.A.* 82:617-621.

Han, V.K.M., Snouweart, J., Towle, A.C., Lund, P.K., and Lauder, J.M., 1987, Cellular localization of tyrosine hydroxylase mRNA and its regulation in the rat adrenal medulla and brain by in situ hybridization with an oligodeoxyribonucleotide probe, *J. Neurosci. Res.* 17: 11-18.

Herkenham, M., and Pert, C.B., 1981, Mosaic distribution of opiate receptors, parafascicular projections and acetylcholinesterase in rat striatum, *Nature (Lond.)* 291:415-417.

Joyce, J.N., 1991a, Differential response of striatal dopamine receptor subtypes to the loss of dopamine following 6-hydroxydopamine or reserpine treatment, *Exp. Neurol.* 113:261-276.

Joyce, J.N., 1991b, Relationship of striatal dopamine receptor subtypes and cholinergic muscarinic receptors to their uptake sites and regulation following 6-hydroxydopamine lesions of the mesostriatal dopamine system, *Exp. Neurol.* 113:277-290.

Le Moine, C. and Bloch, B, 1991, Rat striatal and mesencephalic neurons contain the long isoform of the D2 dopamine receptor mRNA, *Mol. Brain Res.* 10:283-289.

Loughlin, S.E., and J.H. Fallon, 1984, Substantia nigra and ventral tegmental area projections to cortex: topography and collateralization, *J. Neurosci.* 11:425-435.

Marshall, J.F., 1984, Behavioral consequences of neuronal plasticity following injury to nigrostriatal dopaminergic neurons, *in*: "Aging and Recovery of Function in the Central Nervous System," S.W. Scheff, ed., Plenum Press, New York, pp. 101-127.

Marshall, J.F., 1979, Somatosensory inattention after dopamine depleting 6-OHDA injections: spontaneous recovery and pharmacological control, *Brain Res.* 177:311-324.

Marshall, J.F., Navarrete, R., and Joyce, J.N., 1989, Decreased striatal D1 binding density following mesotelencephalic 6-hydroxydopamine injections: an autoradiographic analysis, *Brain Res.* 493:247-257.

McDougall, S.A., and Bardo, M.T., 1991, Ontogenetic changes in dopaminergic pre and postsynaptic elements in rat brain: effects of quinpirole and sulpiride, *Neuropharmacol.* 30:531-534.

Mengod, G., Martinez-Mir, M I., Vilaro, M T., Palacios, J M, 1989, Localization of the mRNA for the dopamine D2 receptor in the rat brain by in situ hybridization histochemistry, *Proc. Natl. Acad . Sci . U.S.A.* 86:8560-8564.

Murrin, L.C., 1982, In vivo studies of dopamine receptor ontogeny, *Life Sci.* 31:971-980.

Murrin, L.C., and Ferrer, J.R., 1984, Ontogeny of the rat striatum: correspondence of dopamine terminals, opiate receptors and acetylcholinesterase, *Neurosci. Lett.* 47: 155-160.

Murrin, L.C., Gibbens, D.L., and Ferrer, J.R., 1985, Ontogeny of dopamine, serotonin and spirodecanone receptors in rat forebrain - an autoradiographic study, *Brain Res.* 23:91-109.

Murrin, L.C., and Zeng, W., 1989, Dopamine D1 receptor development in the rat striatum: early localization in striosomes, *Brain Res.* 480:170-177.

Neal, B.S., and Joyce, J.N., 1991, Dopamine D1 receptor behavioral responsivity following selective lesions of the striatal patch compartment during development, *Devl. Brain Res.* 60:105-113.

Neal, B.S., and Joyce, J.N., 1992, Neonatal 6-hydroxydopamine lesions differentially affect striatal D1 and D2 receptors, *Synapse* 11:35-46.

Neal-Beliveau, B.S., and Joyce, J.N., 1993, D1 and D2 dopamine receptors do not up-regulate in response to neonatal intrastriatal 6-hydroxydopamine lesions, *Neurosci. Lett.*, In Press.

Olson, L., Seiger, A., and Fuxe, K., 1972, Heterogeneity of striatal and limbic dopamine innervation: highly fluorescent islands in developing and adult rats, *Brain Res.* 44:283-288.

Pasinetti, G.M., Lerner, S.P., Johnson, S.A., Morgan, D.G., Telford, N.A., and Finch, C.E., 1989, Chronic lesions differentially decrease tyrosine hydroxylase messenger RNA in dopaminergic neurons of the substantia nigra, *Mol. Brain Res.* 5:203-209.

Pasinetti, G.M., Osterburg, H.H., Kelly, A.B., Kohama, S., Morgan, D.G., Reinhard Jrs, J.F., Stellwagen, R.H., and Finch, C.E., 1992, Slow changes of tyrosine hydroxylase gene expression in dopaminergic brain neurons after neurotoxin lesioning: a model for neuron aging, *Mol. Brain Res.* 13:63-73.

Phelps, P.E., Brady, D.R., and Vaughn, J.E., 1989, The generation and differentiation of cholinergic neurons in rat caudate-putamen, *Devl. Brain Res.* 46:47-60.

Rao, P.A., Molinoff, P.B., and Joyce, J.N., 1991, Ontogeny of dopamine D1 and D2 receptor subtypes in rat basal ganglia: a quantitative autoradiographic study, *Devl. Brain Res.* 60:167-177.

Sivam, S.P., 1989, D1 dopamine receptor-mediated substance P depletion in the striatonigral neurons of rats subjected to neonatal dopaminergic denervation: implications for self-injurious behavior, *Brain Res.* 500:119-130.

Sivam, S.P., Breese, G.R., Krause, J.E., Napier, T.C., Mueller, R.A., and Hong, J.-S., 1987, Neonatal and adult 6-hydroxydopamine-induced lesions differentially alter tachykinin and enkephaline gene expression, *J. Neurochem.* 49:1623-1633.

Solberg, Y., Pollack, Y, and Silverman,W.F., 1992, Differential expression of tyrosine hydroxylase mRNA in the developing rat mesencephalon, *Cell. Mol. Neurobiol.* 12:569-580.

Specht, L.A., Pickel, V.M., Joh, T.H., and Reis, D.J., 1981a, Light microscopic immuno-cytochemical localization of tyrosine hydroxylase in prenatal rat brain. I. Early ontogeny, *J. Comp. Neurol.* 199:233-253.

Specht, L.A., Pickel, V.M., Joh, T.H., and Reis, D.J., 1981b, Light microscopic immunocytochemical localization of tyrosine hydroxylase in prenatal rat brain. II. Late ontogeny, *J. Comp. Neurol.* 199:255-276.

Ungerstedt, U., 1971, Adipsia and aphagia after 6-hydroxydopamine-induced degeneration of the nigro-striatal dopamine system, *Acta Physiol. Scand. Suppl.* 367:95-122.

Van der Kooy, D., 1984, Developmental relationships between opiate receptors and dopamine in the formation of caudate/putamen patches, *Devl. Brain Res.* 14:300-303.

Van der Kooy, D., and Fishell, G., 1987, Neuronal birthdate underlies the development of striatal compartments, *Brain* 401:155-161.

Voorn, P., Kalsbeek, A., Jorritsma-Byham, B., and Groenewegen, H.J., 1988, The pre-and postnatal development of the dopaminergic cell groups in the ventral mesencephalon and the dopaminergic innervation of the striatum of the rat, *J. Neurosci.* 25:857-887.

Weihmuller, F.B., and Bruno, J.P., 1989, Age-dependent plasticity in the dopaminergic control of sensorimotor development, *Behav. Brain Res.* 35:95-109.

Weiss-Wunder, L.T., and Chesselet, M.-F., 1991, Subpopulations of mesencephalic dopaminergic neurons express different levels of tyrosine hydroxylae messenger RNA, *J. Comp. Neurol.* 303:478-488.

Zigmond, M.J., and Stricker, E.M., 1973, Recovery of feeding and drinking by rats after intraventricular 6-hydroxydopamine or lateral hypothalamic lesions, *Science* 182:717-720.

MECHANISMS OF NEURONAL DEGENERATION IN HUNTINGTON'S DISEASE

Robert J. Ferrante[1,2] M. Flint Beal[3,4] and Neil W. Kowall[1,2]

Geriatric Research Education Clinical Center, Bedford VA Medical Center[1], Bedford, Massachusetts; Departments of Neurology and Pathology, Boston University School of Medicine[2]; Neurology Service, Massachusetts General Hospital[3]; and Harvard Medical School[4], Boston, Massachusetts

INTRODUCTION

Two complementary research strategies have been undertaken to define the mechanisms of neuronal degeneration in Huntington's disease (HD) (Kowall et al.,1987; Ferrante, 1991). Because HD is an autosomal dominant disease caused by a specific mutation localized to the telomeric region of the short arm of chromosome four, a great deal of effort has been directed to finding the gene locus responsible for the disease (Gusella et al., 1986). Although the gene has recently been identified, the specific biochemical abnormality that causes HD and the mechanism of neuronal degeneration has not yet been determined (Huntington's Disease Collaborative Research Group, 1993). This strategy has made it possible to diagnose presymptomatic individuals and has contributed to mapping studies of the human genome. A second complementary strategy that we and others have used has been to define the specific changes that occur in HD brain and explore animal models that recreate these pathological alterations (Kowall et al., 1987; Beal et al., 1989a and b, 1991a and b; Albin et al., 1990; DiFiglia, 1990; Ferrante, 1991; Ferrante et al., 1992, 1993;). Human postmortem studies suffer from being largely descriptive rather than mechanistic but, by using markers that define populations of neurons, we have been able to answer important preliminary questions during the past few years: Is their differential neuronal vulnerability in HD and, if so, which neurons are especially sensitive to degeneration and which neurons are resistant? What characteristics define these neurons and how do they relate to mechanisms of selective vulnerability? Why is neuronal degeneration delayed in onset and expressed so selectively? What are the earliest changes found in HD brain? Can these changes be reproduced in experimental animal models *in vivo*? And, finally, can we ameliorate the neurodegenerative process in these models? We have made substantial progress towards answering these questions and have reached some general conclusions: 1) Specific subsets of neurons resist degeneration in HD (Ferrante, 1991); 2) Proliferative changes affect damaged neurons early in the course of HD (Ferrante et al., 1991); 3) Patterns of selective vulnerability can be reproduced in experimental animals with NMDA-type glutamate excitotoxins (Beal et al., 1991a and b; Ferrante et al., 1993); 4) Similar patterns of differential vulnerability can be produced with specific electron transport chain inhibitors that compromise oxidative phosphorylation (Beal et al., 1991a; Ferrante et al., 1992; Storey et al., 1992; Brouillet et al., 1993). Because of our detailed understanding about the patterns of neuronal degeneration in HD, we are now able to develop new animal models that may more closely reproduce the patterns of neuronal degeneration found in HD.

The Basal Ganglia IV, Edited by
G. Percheron *et al.*, Plenum Press, New York, 1994

The background and evidence supporting each of these four major conclusions and the specific observations we and others have made are discussed below.

ANATOMY OF THE NORMAL HUMAN STRIATUM

The Patch-Matrix Organization of Human Striatum

The generally accepted contemporary view of striatal organization designates two compartments based on histochemical and hodological criteria; the patch, or striosome, and the surrounding matrix. This dichotomic subdivision is mainly based on observations in the rat and cat (Graybiel and Ragsdale, 1983; Gerfen et al., 1985). In our studies of human striatum we have been impressed by the complexity of substance P and met-enkephalin staining patterns, both considered to be classical markers for striatal patches (Ferrante et al., 1986). We have stained serial sections of normal human striatum at the level of the nucleus accumbens for acetylcholinesterase (AChE), NADPH diaphorase, tyrosine hydroxylase, calbindin D28K, enkephalin, substance P, cholecystokinin, transforming growth factor alpha, and microtubule associated protein-2 (MAP-2). Substance P, enkephalin, transforming growth factor alpha, and cholecystokinin all show a similar core of low immunoreactivity surrounded by a rim of intense staining. Dorsally these patches are smaller and are often cut tangentially resulting in a small immunoreactive rim without a core. A superimposed gradient of matrix staining is maximal ventrally where it merges with the immunoreactive rim of the patches from which it could not be distinguished. Calbindin staining is uniform in the matrix without a dorsoventral gradient. Patches of low calbindin staining correspond exactly with the unstained cores of substance P, enkephalin, transforming growth factor alpha, and chole-cystokinin. As previously reported, the areas of low AChE, NADPH diaphorase, MAP-2 and tyrosine hydroxylase correspond. These patches, however are larger than the calbindin patches and correspond exactly to the core plus the immunoreactive rim of substance P, enkephalin, transforming growth factor alpha, and cholecystokinin. The core region of patches did not intensely stain with any method. The striatal organization in the human is more complex than previously reported such that a dyadic patch-matrix schema does not adequately define the histochemical heterogeneity. Faull and colleagues have identified a third neurochemical compartment using autoradiographic techniques for neurotensin receptors (Faull et al., 1989). The outer border of striatal patches is marked by thin rings (annuli) of high neurotensin receptor density which corresponds to an AChE-negative border zone between the AChE-poor patch and the AChE-rich matrix regions. This overlapping zone may have important functional significance in compartmental integration and regulation of activity. Further studies of the annulus compartment may resolve the incongruities of exact compartmental localization of specific neurochemical markers within the striatum.

Neuronal Classes within the Striatum

Neuronal populations of the striatum can be characterized with respect to their cytoarchitectonic features and the neurochemical compounds contained within them. One basis for classification is the relative pressence or absence of dendritic spines. Striatal neurons are referred to as spiny and aspiny neurons, as identified in Golgi impregnations (Graveland et al., 1985b). Biochemical characteristics further distinguish these neuronal subtypes. Enzyme and immunohistochemical methods demonstrate that striatal neurons contain various neurotransmitter substances, neuropeptides, and related enzymes which are specific to particular neuronal subsets. Aspiny striatal neurons contain somatostatin-neuropeptide Y-NADPH diaphorase (Ferrante et al., 1985, 1987a), acetylcholinesterase (Parent et al., 1984; Ferrante et al, 1987c), choline acetyltransferase (Ferrante et al., 1987d), vasoactive intestinal polypeptide (Theriault and Landis, 1987), and cholecystokinin (Takagi et al., 1984); while gama-aminobutyric acid (Ribak et al., 1979), substance P (Izzo et al., 1987), enkephalins (Izzo et al., 1987), dynorphin (Besson et al., 1990), transforming growth factor alpha (Fallon, 1987) and calbindin D28k (Gerfen et al., 1985; Ferrante et al., 1991) are among those substances contained within spiny neurons.

We have recently performed *in situ* hybridzation histochemical studies of substance P mRNA in postmortem human striatum using a riboprobe labeled with ^{35}S and confirmed the localization of substance P in medium sized striatal neurons as initially reported by Chesselet (Chesselet and Affolter, 1987).

Pathological Anatomy of the Striatum in Huntington's Disease

Huntington's disease (HD) is an autosomal dominant disorder that leaves a very characteristic pathological signature on the affected brain. The neuropathology has identified certain clues to help explain the pathogenesis of the disorder. The most striking neuro-pathological changes are found within the neostriatum in which there is gross atrophy of the caudate nucleus and putamen accompanied by marked neuronal loss and astrogliosis. The extent of striatal pathology, neuronal loss, and gliosis provide a basis for grading the severity of HD pathology (Vonsattel et al., 1985). There is a dorso-ventral progression of neuronal death with the dorsal striatum effected earliest. General cell stains disclose a regional pattern of striatal degeneration with relative sparing of the ventromedial striatum and the nucleus accumbens. Radioimmunoassay analysis of HD brain tissue demonstrates a marked disparity between the loss and preservation of neurochemical substances within the basal ganglia (Beal and Martin, 1986), suggesting that the destructive process is not equally expressed in all striatal neurons and that there is a selective pattern of neuronal vulnerability. Consistent with this, an extensive enzyme- and immuno-histochemical examination of the cell types in HD striatum has shown that the medium-sized spiny striatal neurons are disproportionately affected early and most severely, while large and medium-sized aspiny neurons are relatively spared (Dawbarn et al., 1985; Ferrante et al., 1985, 1987a, c, and e, 1991a, Ferrante, 1991). Markers for spiny neurons are depleted in a gradient similar to that seen with general cell stains (Ferrante et al., 1986, 1991; Ferrante, 1991). Substance P, which is in both spiny and aspiny neurons in a 70:30 % ratio respectively (Bolam et al., 1983) is depleted in HD. Some substance P neurons, however, are still present throughout HD striatum. The majority of these preserved neurons have been characterized as aspiny, using light and electron microscopic methods (Ferrante et al., 1987e).

Early morphologic alterations of spiny striatal neurons have been described using Golgi and immunocytochemical methods (Graveland et al., 1985a; Ferrante et al., 1991; Ferrante 1991). Degeneration and regeneration both occur. There are no qualitative differences between those aspiny neurons impregnated in control and HD striatum. Recent findings suggest that proliferative changes precede degenerative changes in HD (Ferrante et al., 1991). Proliferative changes, found primarily in moderate grades (grade 2) of HD, include prominent recurving of distal dentritic segments, short-segment branching along the length of dendrites, and increased numbers and size of dendritic spines. Degenerative alterations consist of truncated dendritic arbors, focal dendritic swelling, and marked spine loss which are almost entirely found in severe cases (grades 3 and 4) of HD. We have suggested that these newly formed dendritic arbors and increased numbers of dendritic spines found early on in the disorder may form functional connections which could facilitate neuronal excitabilty and exacerbate cell death (Ferrante et al., 1991).

Calbindin-positive neurons in HD have similar dysmorphic alterations to those observed in Golgi preparations (Ferrante et al., 1991). The extent to which enkephalin and substance P subsets develop proliferative changes is not known. It is also not clear if similar dendritic changes affect NADPH diaphorase aspiny neurons which are spared in HD. Transforming growth factor alpha, which colocalizes with enkephalin in the striatum, more clearly labels discrete dendritic processes than either enkephalin or substance P. In HD, transforming growth factor alpha neurons are depleted in a dorsoventral gradient, similar to that seen with enkephalin and substance P antisera. The staining intensity of individual neurons, however, is somewhat increased over controls. Proliferative dendritic abnormalities and dysmorphic dendritic recurving affects transforming growth factor alpha but not NADPH diaphorase neurons. Our observations show that dendritic proliferation involves enkephalin neurons that contain a growth factor, transforming growth factor alpha, possibly not present in substance P neurons. If transforming growth factor alpha plays a causal role in the development of dendritic proliferation, spiny substance P neurons should not show evidence of dendritic proliferation in HD. Studies of neural cell adhesion molecule (NCAM), an axonal growth cone marker, suggest that axonal plasticity is also increased in HD (Nihei and Kowall, 1992).

In addition to the neuronal susceptibilty, there is a differential pattern of patch-matrix involvement in HD striatum. Although the patch-matrix compartments persist throughout the caudate and putamen, the pattern is altered in that the total area of matrix is significantly reduced, while the size and total area of patch is unchanged as compared to the normal striatum (Ferrante et al., 1986, 1987c; Kowall et al., 1987). These findings have been confirmed using calbindin immunocytochemistry (Seto-Oshima et al., 1988) and in GABA receptor studies (Faull et al., 1993).

Pathological Anatomy of other Subcortical Structures in Huntington's Disease

A cardinal feature of HD is chorea, a disorder characterized by irregularly timed, abrupt and randomly distributed involuntary movements. Both the neuropathologic findings in HD (Reiner et al., 1988; Albin et al., 1990) and experimental primate models (Crossman, 1987; DeLong, 1990) suggest chorea may result from a loss of striatal projections to the external pallidum. The release of the external pallidum from inhibitory striatal input results in excessive inhibition on the subthalamic nucleus which in turn causes decreased activation of the internal segment of the pallidum and reduced inhibition of the thalamus, leading to increased cortical excitation and chorea. Consistent with this hypothesis, bicuculline blockade of the GABAergic input to the external pallidum causes chorea in primates (Crossman et al., 1988).

It has been suggested that there is a differential vulnerability of enkephalin and substance P striatal spiny projection neurons in HD based on terminal staining within the globus pallidus (Reiner et al., 1988). Under normal conditions, substance P immunoreactivity is preferentially located in the internal segment of globus pallidus (GPi), while metenkephalin (ME) immunoreactivity is primarily within the external segment of the globus pallidus (GPe). Both neuropeptides are distributed within the substantia nigra (SN). Reiner and coworkers have reported that the density of ME immunoreactivity (ir) in the GPe and substance P-ir in the SN reticulata (SNr) is decreased early on in HD, while the density of substance P-ir in the GPi and SN compacta (SNc) is decreased late in the disorder. We have been unable to confirm these results (Ferrante et al., 1990; Ferrante, 1991). The GP from 17 HD patients of all grades (G0-G4) and 14 controls were immunocytochemically stained, using immunosera against substance P, ME, and transforming growth factor alpha (TGF). There was a gradient loss of substance P-, ME-, and TGF-ir in the GP of HD patients that corresponded to the severity of the disorder, with the most marked reductions occurring in the most severe grades. Within each grade of HD, however, no difference could be identified between the density of immunoreactive staining of substance P in the GPi and the density of ME and TGF in the GPe. These findings were supported by densitometric image analysis in which the density of the staining in each HD grade was measured as a percentage of the normal controls (Ferrante, 1991). These percentages were nearly equal in each grade. Immunocytochemical and neurochemical studies provide evidence that there is an early and comparable loss of substance P and ME in HD striatum (Marshall et al., 1983; Ferrante et al., 1986). It is possible that treatment of patients with dopamine receptor antagonists, such as haloperidol, affects the patterns of postmortem immunostaining.

We have studied the distribution of DARP-32, a phosphoprotein localized in dopaminoceptive neurons, in postmortem human brain and find that it intensely labels neurons that project to both segments of the pallidum equally, suggesting that it may be localized in both enkephalinergic and substance P positive neurons.

Extrastriatal Involvement in Huntington's Disease

There does not appear to be a consensus of opinion with regard to the extrastriatal pathologic changes within the SN in HD. Early histopathologic reports are divided with regard to nerve cell loss (Forno and Jose, 1973; Ferrante, 1991). The SNr is reported to show fibrillary gliosis with no changes within the SNc in the majority of cases. A recent analysis of the SN in 4 HD patients, demonstrated a significant reduction in size and number of both pigmented and non-pigmented neurons (Oyanagi et al., 1989). Neurochemical analyses of nigral homogenates from HD patients have demonstrated marked reductions in concentrations of substance P, enkephalin and GABA (Beal and Martin, 1986). These findings have been attributed to a loss of the striatal projection neurons that contain these substances. It has been suggested, however, that substance P immunoreactivity in SNr is lost early in the disease as compared to that in the SNc (Reiner et al., 1988).

We have examined the SN of 42 HD patients with moderate, severe, and very severe grades (Ferrante et al., 1989; Richardson, 1990). Sections were all at the level of the red nucleus and oculomotor nucleus. Counts of pigmented and nonpigmented neurons were made within the SN and respective areas of SNc and SNr were computed using an image analysis system. The total area of SN was significantly reduced in HD. The SNr, however, had a greater area loss than the SNc. Nonpigmented neurons associated with the SNr were significantly reduced in number, while pigmented neurons of the SNc were significantly increased in number. The increase in pigmented neurons most likely reflects cell sparing along with moderate neuropil loss of striatonigral terminal projections to the SNc. The significant reductions of nonpigmented neurons and neuropil in the SNr in HD may reflect transneuronal degeneration and the loss of matrix afferents, respectively. The matrix compartment of the striatum, which is severely affected in the disorder, projects primarily to the SNr (Gerfen, 1985). When deprived of their afferent input, target neurons may die or atrophy. This is consistent with striatal excitotoxic lesions in rodents in which there is neuronal degeneration in the SNr (Saji and Reis, 1987). The administration of muscimol, a specific GABA agonist, prevented degeneration of these neurons in this experiment, suggesting that neuronal death may be due to a loss if inhibitory GABAergic input with subsequent excessive excitation of SNr neurons.

In normal SN, there is a regional variation in the distribution of substance P and ME, such that the most intense immunoreactivity of both neuropeptides is found within the SNr and the pars alpha and beta of the SNc (Ferrante, 1991). The nigral patterns of substance P and ME overlap, with the intensity of substance P much greater than ME. We have shown that both substance P and ME immunoreactivities in the SN are severely and equally affected in HD (Ferrante, 1991). The loss of activity correlates with the severity of the disease. These findings have been confirmed, using neurochemical analysis for substance P, in graded cases of HD (Beal et al., 1988b).

MECHANISMS OF NEURONAL DEGENERATION

Excitotoxin Lesions in Animals as a Model of Huntington's Disease

The McGeers (1976) and Coyle and Schwarcz (1976) first showed that injections of the excitatory glutamate-type neurotoxin, kainic acid, produced axon sparing lesions of the striatum that resembled HD. This model was refined by using n-methyl-d-aspartate (NMDA) type excitotoxins, including the endogenous toxin quinolinic acid which, unlike kainic acid, spare NADPH diaphorase neurons (Beal et al., 1986, 1989b). Chronic lesions, where months have passed to allow resorption of the necrotic injection site, even more closely reproduce the patterns of selective neuronal sparing in the rat striatum and cerebral cortex found in HD (Beal et al., 1991b). Excitotoxin lesions in the monkey provide an experimental primate model which more closely resembles the neuropathologic, neurochemical and clinical features of Huntington's disease (Ferrante et al., 1993). These lesions were characterized by a central zone of intense astrogliosis and marked neuronal depletion. Immunocytochemical and enzyme histochemical markers for both large and medium-sized aspiny- and spiny-striatal neurons clearly demonstrated a selective pattern of neuronal vulnerability and resistance to the excitotoxic effects of quinolinic acid within lesioned striata. There was a disproportionate involvement of the matrix compartment similar to that seen in HD. These changes were accompanied by beha-vioral changes suggestive of a hyperkinetic movement disorder. Ultrastructural analysis confirmed axon sparing lesions with neuronal loss and astrogliosis. Nonphosphorylated neurofilament immunoreactivity was reduced in cell bodies but increased in axons, consistent with NMDA-mediated activation of the calcium-dependent phosphatase, calcineurin, which dephosphorylates neurofilament protein (Nihei and Kowall, 1992). These animal models demonstrate a characteristic profile consistent with the features of HD and strongly suggest that an excitotoxic process plays a role in the pathogenesis of this disorder.

Oxidative Phosphorylation and Selective Neuronal Vulnerability in Huntington's Disease

An abnormality affecting the NMDA receptor or an endogenous excitotoxin could be responsible for HD but candidate toxins, such as quinolinic acid, are not increased in this disease. A novel hypothesis explaining the pattern of degeneration in HD has recently evolved based on recent animal studies and observations in HD. It suggests that impaired cellular energy may be involved in the degenerative process (Albin and Greenamyre, 1992; Beal, 1992). The initial relevant observations were made by Olney who showed that partial membrane depolarization can produce NMDA receptor mediated excitotoxicity by removing the voltage-dependent magnesium block of the NMDA linked calcium channel (Olney, 1978). The open calcium channel permits normal amounts of endogenous glutamate to produce NMDA receptor-mediated neurotoxicity. Recent evidence suggests that other mechanisms producing partial membrane depolarization also produce NMDA-type excitotoxin lesions (Novelli et al., 1988; Zeevalk and Nicklas, 1991). Storey et al. showed that 1-methyl-4-phenylpyridinium (MPP+) can produce excitotoxic lesions in the striatum (Storey et al., 1992). MPP+ toxicity is mediated by inhibition and free-radical mediated damage of the electron transport chain Complex I (ubiquinone oxidoreductase) (Mariani et al., 1986; Singer and Ramsay, 1990; Ramsay et al., 1991). Impaired electron transport chain function produces energy failure that results in membrane depolarization, removal of magnesium from the NMDA-linked calcium channel, and subsequent excitotoxic injury. Twenty years ago Mettler showed that sodium azide, a Complex IV (cytochrome oxidase) inhibitor, produced striatal damage and a hyperkinetic movement disorder in primates (Mettler, 1972). Animal studies show that striatal injections of the mitochondrial toxins produce differential neuronal toxicity identical to that produced by NMDA receptor agonists (Beal et al., 1991a; Ferrante et al., 1992; Storey et al., 1992; Brouillet et al., 1993). The systemic administration of the succinate dehydrogenase (Complex II) inhibitor 3-nitropropionic acid (3NP) causes striatal degeneration in animals and humans. Recent studies, which demonstrate the effects of intrastriatal, subacute systemic, and chronic systemic administration of 3NP in rats, show selective striatal involvement (Ferrante et al., 1992; Brouillet et al., 1993). Intrastriatal and subacute systemic 3NP lesions were similar to those we found with striatal lesions produced by non-NMDA agonists such as kainate and AMPA. There was a comparable loss of markers for both spiny striatal neurons (GABA, substance P, calbindin) and medium-sized aspiny striatal neurons (somatostatin, neuropeptide Y, NADPH-diaphorase). Histopathologic examination of the lesions showed dose-dependent bilateral striatal lesions with neuronal loss and gliosis. These lesions were age-dependent. Four to twelve month old rats were much more vulnerable to the effects of 3NP than were one month old animals. Prior decortication, which removes cortico-striatal glutamatergic input, ameliorated these lesions. In contrast, chronic (1 month) low-dose systemic administration produced lesions confined to the striatum in which there was marked gliosis and loss of Nissl-stained neurons with selective sparing of NADPH-diaphorase and large Nissl-stained neurons, consistent with an NMDA excitotoxic process. Growth-related proliferative changes in dendrites of spiny striatal neurons, similar to those found in HD, were identified in Golgi preparations. These included recurved dendritic arbors with increased dendritic spine density. The chronic administration of 3NP produced striatal lesions replicating many of the characteristic histologic and neurochemical features of HD. These results are consistent with *in vitro* studies suggesting that mild metabolic compromise can selectively activate NMDA receptors (Beal et al., 1991a).

Decreased activity of Complex II/III of the electron transport chain has been found in the caudate nucleus but not in other brain areas in HD (Mann et al., 1990). Cytochrome oxidase (Complex IV) abnormalities have also been reported in the caudate nucleus (Brennan et al., 1985). This loss of enzymatic activity in damaged brain areas may be a consequence of neuronal loss rather than a cause. Studies of platelets from HD patients suggest that Complex I activity may be selectively decreased in HD patients although it is normal in at risk family members (Parker et al., 1990). Complex I is composed of over 30 subunits, the majority of which are encoded by nuclear DNA (Hatefi, 1985). It is possible that the mutation in HD affects a nuclear encoded component of Complex I or the complex process of protein translocation and import into mitochondria. Other electron transport chain complexes, including Complex II, Complex III (ubiquinol cytochrome c reductase) and

Complex IV (cytochrome oxidase) are normal in blood platelets (Parker et al., 1990) We initially examined patterns of Complex IV activity in human striatum because animal studies showed that it was concentrated in dendritic spines which are metabolically active and contain a large number of mitochondria (Ferrante et al., 1988). We hypothesized that it might demarcate early sites of damage. Our observations showed that depletion of Complex IV activity parallels the pattern of neuronal loss in HD striatum.

Other indirect lines of evidence suggest a possible role for mitochondria in the pathogenesis of HD. The disease onset is earlier and the clinical course is more severe if the gene is inherited from the father, suggesting that a maternally transmitted factor such as mitochondria might modify the course of the disease (Myers et al., 1988). Tellez-Nagel et al. performed ultrastructural studies on brain biopsies of four HD patients and found evidence of mitochondrial abnormalities and increased lipofuscin, a pigment that accumulates as a consequence of free radical mediated membrane damage (Tellez-Nagel et al., 1973). The delayed onset of HD may be due to the contribution of age-related mutations of mitochondrial DNA that cause decreased mitochondrial function, as shown by Trounce and coworkers in skeletal muscle (Trounce et al., 1989). Even though the neuron is postmitotic, the mitochondria in neurons continue to proliferate (Linnane et al., 1989). It is not known if specific neurons are disproportionately sensitive to electron transport chain inhibition. Tissue specific expression patterns of mitochondrial proteins have been defined but the regional activity of specific electron transport proteins in the brain have not been thoroughly studied (Wallace, 1991). Recently, focal putaminal degeneration has been reported in a family with Leber's disease due to a point mutation in NADH-dehydrogenase (Larsson et al., 1991). Encephalopathy also occurs in patients with MELAS, another disease characterized by a deficiency in Complex I (Wallace, 1991). In addition, Koroshetz and coworkers have found increased anaerobic metabolism in the cerebral cortex of patients with HD using MR spectroscopy, further suggesting that an abnormality of oxidative phosphorylation occurs in this disease (Jenkins, 1992).

Since antioxidative agents do not protect against quinolinic acid-induced striatal damage (Beal et al., 1988a), is energy failure a significant causal factor in the pathogenesis of Huntington's disease? The quinolinic acid model of Huntington's disease strongly suggests that excitotoxicity may play an important role in the pathogenesis of the disorder. Quinolinic acid acts directly at the NMDA receptor, causing a cascade of events leading to cell death. Excitotoxic mechanisms may be the final common pathway in neuronal death, not only in Huntington's disease but other neurologic disorders. We hypothesize that impaired electron transport chain activity is the trigger that initiates excitotoxic damage in HD. Metabolic stress results in membrane depolarization which secondarily results in a slow excitotoxic process. In the hypothetical cascade of events leading towards cell death, the impairment of energy metabolism is antecedent to excitotoxic damage. NMDA antagonists, such as MK-801, provide complete protection against either toxins that impair mitochondiral energy production or NMDA agonists. Antioxidant agents, on the other hand, cannot prevent damage caused by directly acting NMDA-type excitotoxins such as quinolinic acid.

Free Radicals and Reactive Oxygen Species in the Pathogenesis of Huntington's Disease

NADPH diaphorase neurons were the first subset of neurons shown to be selectively spared in HD striatum (Ferrante et al., 1985). Recently, NADPH diaphorase enzyme activity has been identified as an isoform of NO synthase (Dawson et al., 1991a and b; Hope et al., 1991). The NO radical produced by the synthase reacts with the superoxide anion radical to form the highly reactive and toxic hydroxyl radical (Bast et al., 1991). NO has been shown to mediate glutamate neurotoxicity in vitro (Dawson et al., 1991b). If energy failure is a primary pathogenetic event in HD, there may be accelerated production of reactive free radical species exacerbated by the relative sparing of NO synthase neurons that may further contribute to free radical damage (Halliwell and Gutteridge, 1990). It has been recently shown that the NO radical specifically inhibits Complex I, creating a vicious cycle of electron transport chain inhibition (Stadler et al., 1991). Free radicals and reactive oxygen species damage membrane lipids leading to the formation of lipofuscin (Harman, 1989). Friede showed that lipofuscin accumulates in neurons with greater levels of oxidative activity (Friede, 1962). It has been demonstrated that the Complex I inhibitor MPP+ causes increased neuronal lipofuscin formation in experimental animals (Hadjiconstantinou et al., 1987; Mann et al., 1990). If free radical damage is increased in HD, there may be increased lipofuscin formation in neurons sensitive to degeneration. The distribution of NADPH diaphorase neurons may match the distribution of neuronal degeneration in HD. In the striatum, spared NADPH diaphorase neurons are mainly found in the matrix, the striatal compartment most severely affected in HD.

Evidence suggests that under certain circumstances NO toxicity may be limited in the cell of synthesis (Stamler et al., 1992). A non-toxic early intermediate of NO, N-hydroxy-arginine may be produced within the neuron and that successive oxidation outside the cell, in a milieu containing Fe-heme, may facilitate NO liberation and other highly toxic species. The substrate availability in the extracellular environment can alter the toxic impact. Varying forms of NO have different responses. Specific thiol groups of NO, S-nitrosothiols, can modulate NMDA glutamate receptors and ameliorate receptor-mediated excitotoxicity (Stamler et al., 1992). It appears as though the form in which NO is presented reflects its toxicity. Alternatively, aspiny striatal neurons, including somatostatin/NADPH-diaphorase neurons, may lack or have far fewer NMDA receptors than spiny striatal neurons. This may confer resistance to this subset of striatal neurons and has been supported by quinolinic acid lesion studies.

Recent studies suggest that free radical toxicity is involved in familial amyotrophic lateral sclerosis (Rosen et al., 1993). The genetic defect is on chromosome 21q and is associated with a mutation in Cu/Zn-binding superoxide dismutase (SOD). SOD functions to reduce super oxide and free radical levels, preventing the subsequent formation of highly toxic radicals of NO. Alterations in other SOD proteins encoded by the human genome, such as manganese-dependent SOD and extracellular SOD, may be involved in selective neurodegenerative disorders, particularly HD.

Cascades of Cellular Activation: From Cell Damage to Cell Death

Irrespective of the primary pathogenetic mechanisms operative in HD, it is likely that an array of responses result that are eventually overwhelmed, leading to cell death. Membrane receptor stimulation activates second messengers that postranslationally modify proteins, including transcription regulating factors that control immediate early gene expression (Morgan and Curran, 1991). NMDA receptor activation leads to phosphorylation of response-element binding proteins that regulate the transcription of the basic-zipper family of DNA binding proteins (Cole et al., 1989). These transiently expressed immediate early genes bind to upstream promoter-enhancer regions that control the synthesis of neurotrophins, cytoskeletal proteins and neuropeptide transmitters such as proenkephalin (Morgan and Curran, 1991). If abnormal proteins are gen-erated by a pathological stimulus, such as ischemia or heat stress, a heat shock or stress response is initiated (Lowe and Mayer, 1990). Some inducible heat shock proteins that are expressed regulate protein folding and transport into mitochondria (Lindquist and Craig, 1988). The heat shock protein ubiquitin is covalently bound to denatured proteins marked for ATP-dependent non-lysosomal proteolysis. Ubiquitin immunoreactivity is increased in subsets of neurons in HD cortex, striatum and the subthalamic nucleus, suggesting that a neuronal stress response occurs in HD (Evans and Kowall, 1991; Harrington et al., 1991).

We sought to further examine the neuronal stress response by determining whether the 72 kD heat shock protein (HSP 72) was also induced in the striatum and neocortex in HD. In a preliminary study, we examined samples of striatum and neocortex from 8 patients with HD (1 grade I, 1 grade II, 2 grade III and 4 grade IV) and 5 age matched controls using a monoclonal antibody raised against purified human HSP 72 (Stress Gen). In normal striatum and neocortex, there was no evidence of neuronal HSP 72 immuno-reactivity. In all HD samples, HSP 72 immunoreactive neurons were present in the striatum and neocortex. There was no clear correlation between the severity of striatal atrophy and the amount of HSP 72 staining. In HD striatum, immunoreactive medium sized neurons were most prominent in the body of the caudate nucleus and the putamen. In the cerebral cortex, pyramidal and nonpyramidal neurons in all layers were labeled. These observations provide further evidence that a stress response affects specific neuronal subsets in the striatum and neocortex in HD.

CONCLUSION

Whether or not a better understanding of specific neuronal vulnerability and resistance within HD striatum might in any way lead to discovering the pathogenesis of this disorder is still not known. There is no direct evidence that the neuropathological and neurochemical alterations which have been identified play a role in the etiology of HD. The studies from human postmortem brain tissue, however, have been instrumental in establishing an excitotoxic model of the disease and emphasize the value of such studies. Continued examination of the postmortem and experimental alterations may provide clues to the nature of the underlying degenerative process.

Why nerve cells die in certain diseases of the brain is unclear and remains a great challenge to neuroscientists investigating the causes of brain disorders. There is a growing body of evidence that suggests that cellular energy impairment may play an important role in degenerative diseases of the nervous system. Inhibitors of mitochondrial function in animals and in accidental human ingestion produce changes in nerve cells which make them susceptible to excitotoxic injury. The cellular changes observed share a number of common features found in human nervous system disorders. Impairment of mitochondrial function in experimental animals closely resembles the pattern of nerve cell death and chemical changes observed in HD patients. These observations imply that the primary genetic defect in HD may cause mitochondrial abnormalities with subsequent changes in energy metabolism that lead to a destruction of nerve cells within the brain. In addition, if this hypothesis is true, it may help to explain the delayed onset of the disease which occurs in mid-life. A gradual loss of energy metabolism occurs with increasing age. A partial defect in mitochondrial function at birth may clinically go unrecognized. Coupled with the normal loss in energy, however, the disorder may present itself during mid-life or later when mitochondrial function has sufficiently decreased to cause disease expression.

These findings have important clinical significance which may have wide ranging implications in how we treat patients with chronic destructive brain diseases. If subsequent findings demonstrate that mitochondrial alterations in energy metabolism are present in HD patients, there are a number of therapeutic approaches which may ameliorate nerve cell death. The administration of compounds which bridge bioenergic defects or block excitatory mechanisms may be affective. Secondly, there are rapid and reliable methods to screen for mitochondrial diseases. These procedures are relatively inexpensive and would involve a simple blood test. If a defect were identified, then treatment could proceed early on and retard or possibly prevent the disease.

REFERENCES

Albin, R.L., Reiner, A., Anderson, K.D., Penney, J.B., and Young, A.B., 1990, Striatal and nigral neuron subpopulations in rigid Huntington's disease: Implications for the functional anatomy of chorea and rigidity-akinesia, *Ann. Neurol.* 27:357-365.

Albin, R., and Greenamyre, J.T., 1992, Alternative excitotoxic hypotheses, *Neurology* 42:733-738.

Bast, A., Haenen, G., and Doelman, C., 1991, Oxidants and antioxidants: State of the art, *Am. J. Med.* 91(3C):2S-13S.

Beal, M.F., Kowall, N.W., Ellison, D.W., Mazurek, M.F., Swartz, K.J., and Martin, J.B., 1986, Replication of the neurochemical characteristics of Huntington's disease with quinolinic acid, *Nature* 321:168-171.

Beal, M.F., and Martin, J.B., 1986, Neuropeptides in neurological disease, *Ann. Neurol.* 20:547-565.

Beal, M.F., Kowall, N. W., Swartz, K. J., Ferrante, R. J., and Martin, J. B., 1988a, Systemic approaches to modifying quinolinic acid striatal lesions in rats, *J. Neurosci.* 10:3901-3908.

Beal, M.F., Ellison, D.W., Mazurek, M.F., Swartz, K.S., Malloy, J.R., Bird, E.D., and Martin, J.B.,1988b, A detailed examination of substance P in pathologically graded cases of Huntington's disease, *J. Neurol. Sci.* 84:51-61.

Beal, M.F., Kowall, N.W., Ferrante, R.J., and Cipolloni, P.B., 1989a, Quinolinic acid striatal lesions in primates as a model of Huntington's disease, *Ann. Neurol.* 26:137.

Beal, M.F., Kowall, N.W., Swartz, K.J., Ferrante, R.J., and Martin, J.B., 1989b, Differential sparing of somatostatin-neuropeptide Y and cholinergic neurons following striatal excitotoxin lesions, *Synapse* 3:38-47.

Beal, M.F., Swartz, K.J., Hyman, B.T., Storey, E., Finn, S.F., and Koroshetz, W., 1991a, Amino-oxyacetic acid results in excitotoxin lesions by a novel indirect mechanism, *J. Neurochem.* 57:1068-1073.

Beal, M.F., Ferrante, R.J., Swartz, K.J., and Kowall, N.W., 1991b, Chronic quinolinic acid lesions in rats closely resemble Huntington's disease, *J. Neurosci.* 11:1649-1659.

Beal, M.F., 1992, Does impairment of energy metabolism result in excitotoxic neuronal death in neurodegenerative disease?, *Ann. Neurol.* 31:119-130.

Besson, M.J., Graybiel, A.M., and Quinn, B., 1990, Co-expression of neuropeptides in the cat's striatum: an immunohistochemical study of substance P, dynorphin and enkephalin, *Neurosci.* 39:33-58.

Bolam, J.P., Somogyi, P., Takagi, H., Fodor, I., and Smith, A.D., 1983, Localization of substance P-like immunoreactivity in neurons and nerve terminals in the neostriatum of the rat: a correlated light and electron microscopic study, *Neurocytol.* 12:325-344.

Brennan, W.A.J., Bird, E.D., and Aprille, J.R.,1985, Regional mitochondrial respiratory activity in Huntington's disease brain, *J. Neurochem.* 44:1948-1950.

Brouillet, E., Jenkins, B.G., Hyman, B.T., Ferrante, R.J., Kowall, N.W., Srivastava, R., Roy, D.S., Rosen, B.R., and Beal, M.F., 1993, Age-dependent vulnerability of the striatum to the mitochondrial toxin 3-nitropropionic acid, *J. Neurochem.* 60:356-359.

Chesselet, M.F., and Affolter, H.U.,1987, Preprotachykinin messenger RNA detected by insitu hybridization in striatal neurons of the human brain, *Brain Res.* 410:83-88.

Cole, A.J., Saffen, D.W., Baraban, J.M., and Worley, P.F., 1989, Rapid increase of an immediate early gene messenger RNA in hippocampal neurons by synaptic NMDA receptor activation, *Nature* 340:474-476.

Coyle, J.T., and Schwarcz, R., 1976, Lesion of striatal neurons with kainic acid provides a model for Huntington's chorea, *Nature* 263:244-246.

Crossman, A.R., 1987, Primate models of dyskinesia: The experimental approach to the study of basal ganglia-related involuntary movement disorders, *Neurosci.* 21:1-40.

Crossman, A.R., Mitchell, I.J., Sambrook, M.A., and Jackson, A.,1988, Chorea and myoclonus in the monkey induced by gamma-aminobutyric acid antagonism in the lentiform complex, *Brain* 111:1211-1233.

Dawbarn, D., DeQuidt, M.E., and Emson, P.C., 1985, Survival of basal ganglia neuropeptide Y-somatostatin neurones in Huntington's disease, *Brain Res.* 340:251-261.

Dawson, T.M., Bredt, D.S., Fotuhi, M., Hwang, P., and Snyder, S.H.,1991a, Nitric oxide synthase and neuronal NADPH diaphorase are identical in brain and peripheral tissue, *Proc. Natl. Acad. Sci. USA* 88:7797-7801.

Dawson, V. L., Dawson, T.M., London, E.D. Bredt, D.S., and Snyder, S.H., 1991b, Nitric oxide mediates glutamate neurotoxicity in primary cortical cultures, *Proc. Natl. Acad. Sci. USA* 88:6368-6371.

DeLong, M.R., 1990, Primate models of movement disorders of basal ganglia origin, *Trends Neurosci.* 13:281-285.

DiFiglia, M., 1990, Excitotoxic injury of the neostriatum: a model for Huntington's disease, *Trends neurosci.*13:286-289.

Evans, A.J., and Kowall, N.W., 1991, Ubiquitin immunoreactivity is increased in Huntington's disease cerebral cortex, *Soc. Neurosci. Abstr.* 17:1450.

Fallon, J.H., 1987, Growth factors in the basal ganglia, *in:* "Neurotoxins and Their Pharmocologic Implications," P. Jenner, ed., Raven Press, New York, pp. 247-260.

Faull, R.L.M., Dragunow, M., and Villiger, J.W.,1989, The distribution of neurotensin receptors and acetylcholinesterase in the human caudate nucleus: evidence for the existence of a third neurochemical compartment, *Brain Res.* 488:381-385.

Faull, R.L.M., Waldvogel, H.J., Nicholson, L.F.B., and Synek, B.J.L., 1993, The distribution of GABA-bezodiazepine receptors in the basal ganglia in Huntington's disease and the quinolinic acid lesioned rat, *Prog. Brain Res.* in press.

Ferrante, R.J., Kowall, N.W., Beal, M.F., Richardson, E.P. Jr, Bird, E.D., and Martin, J.B., 1985, Selective sparing of a class of striatal neurons in Huntington's disease, *Science.* 230:561-563.

Ferrante, R.J., Kowall, N.W., Beal, M.F., Richardson, E.P. Jr, Bird, E.D., and Martin, J.B., 1986, Topography of enkephalin, substance P, and acetylcholinesterase staining in Huntington's disease striatum, *Neurosci. Lett.* 71: 283-288.

Ferrante, R.J., Kowall, N.W., Beal, M.F., Martin, J.B., Bird, E.D., and Richardson, E.P. Jr, 1987a, Morphologic and histochemical characteristics of a spared subset of striatal neurons in Huntington's disease, *J. Neuropathol. Exp. Neurol.* 46:12-27.

Ferrante, R.J., and Kowall, N.W., 1987b, Tyrosine hydroxylase-like immunoreactivity is distributed in the matrix compartment of normal human and Huntington's disease striatum, *Brain Res.* 416:141-146.

Ferrante, R.J., Beal, M.F., Kowall, N.W., Richardson, E.P. Jr, and Martin, J.B., 1987c, Sparing of acetylcholinesterase-containing striatal neurons in Huntington's disease, *Brain Res.* 411:162-166.

Ferrante, R.J., Kowall, N.W., Hersh, L.B., Bruce, G., and Richardson, E.P. Jr, 1987d, Colocalization of cholineacetyltransferase- and acetyl-cholinesterase-containing neurons in Huntington's disease, *Soc. Neurosci. Abstr.* 13:1030.

Ferrante, R.J., Kowall, N.W., Martin, J.B., and Richardson, E.P. Jr, 1987e, Substance P-containing striatal neurons in Huntington's disease, *J. Neuropathol. Exp. Neurol.* 46:375.

Ferrante, R.J., Kowall, N.W., and Richardson, E.P. Jr,1988, Patch-matrix distribution of cholecystokinin and cytochrome oxidase activity in normal and Huntington's disease striatum, *Soc. Neursci. Abstr.* 14:1046.

Ferrante, R.J., Kowall, N.W., and Richardson, E.P. Jr, 1989, Neuronal and neuropil loss in the substantia nigra in Huntington's disease, *J. Neuropathol. Exp. Neurol.* 48:380.

Ferrante, R.J., Kowall, N.W., Harrington, K., and Richardson, E.P. Jr, 1990, Terminal striatal substance P and met enkephalin projections in the globus pallidus are equally affected in Huntington's disease, *J. Neurosci.* 16:1120.

Ferrante, R.J., Kowall, N.W., and Richardson, E.P. Jr, 1991, Proliferative and degenerative changes in striatal spiny neurons in Huntington's disease: A combined study using the section Golgi method and calbindin D28k immunocytochemistry, *J. Neurosci.* 11:3877-3887.

Ferrante, R.J., 1991, Huntingtons disease: morphometric and immuno- cytochemical alterations, *in:* "New Issues in Neuroscience, Basal Ganglia and Movement Disorders.," A. Bignami, ed., Thieme, New York, pp. 191-201.

Ferrante, R.J., Kowall, N.W., Brouillet, E., and Beal, M.F., 1992, Impaired mitochondrial metabolism reflects the striatal pathology in Huntington's disease, *Soc. Neurosci. Abstr.* 18:167.

Ferrante, R.J., Kowall, N.W., Cipolloni, P.B., Storey, E., and Beal, M.F., 1993, Excitotoxin lesions in primates as a model for Huntington's disease: Histopathologic and neurochemical characterization, *Exper. Neurol.* 119:46-71.

Forno, L.S., and Jose, C., 1973, Huntington's chorea: a pathological study, *in:* "Huntington's Chorea," A. Barbeau, T. N. Chase and G.W. Paulson, eds.,1872-1972, Raven Press, New York, pp. 453-470.

Friede, R.L., 1962, The relation of the formation of lipofuscin to the distribution of oxidative enzymes in the human brain, *Acta Neuropathol.* 2:113-125.

Gerfen, C.R., 1985,The neostriatal mosaic: I. Compartmental organizatioon of projections from the striatum to the substantianigra in the rat, *J. Comp. Neurol.* 236:454-476.

Gerfen, C.R., Baimbridge, K.G., and Miller, J.J., 1985, The neostriatal mosaic: compartmental distribution of calcium-binding protein and parvalbumin in the basal ganglia of the rat and monkey, *Proc. Natl. Acad. Sci. USA* 82:8780-8784.

Graveland, G.A., Williams, R.S., and DiFiglia, M.A., 1985a, Evidence for degenerative and regenerative changes in neostriatal spiny neurons in Huntington's disease, *Science* 227:770-773.

Graveland, G.A., Williams, R.S., and DiFiglia, M.A., 1985b, A Golgi study of the human neostriatum: Neurons and afferent fibers, *J. Comp. Neurol.* 234:317-333.

Graybiel, A.M., and Ragsdale, C.W., 1983, Biochemical anatomy of the striatum, *in:* "Chemical Neuroanatomy," P.C. Emson, ed., Raven Press, New York, pp. 427-504.

Gusella, J.F., Gilliam, T.C., Tanzi, R.E., MacDonald, M.E., and Cheng, S. V., 1986, Molecular genetics of Huntington's disease, *Cold Spring Harb. Symp. Quant. Biol.* 1:359-364.

Hadjiconstantinou, M., Tjioe, S., Alho, H., Miller, C., and Neff, N.H., 1987, MPTP accelerates the accumulation of lipofuscin in mouse adrenal gland, *Neurosci. Lett.* 83:1-6.

Halliwell, B., and Gutteridge, J., 1990, Role of free radicals and catalytic metal ions in human disease: An overview, *Meth. Enzymol.* 186:1-85.

Harman, D., 1989, Lipofuscin and ceroid formation: the cellular recycling system, *Adv. Exp. Med. Biol.* 266:3-15.

Harrington, K.M., Ferrante, R.J., and Kowall, N.W., 1991, Evidence for neuronal degeneration in the subthalamic nucleus in Huntington's disease, *Soc. Neurosci. Abs.* 17:1449.

Hatefi, Y., 1985, The mitochondrial electron transport and oxidative phosphorylation system, *Ann. Rev. Biochem.* 54:1015-1069.

Hope, B.T., Michael, G.J., Kniggen K.M., and Vincent, S.R.,1991, Neuronal NADPH diaphorase is a nitric oxide synthase, *Proc. Natl. Acad. Sci. USA* 88:2811-2814.

Huntington's Disease Collaborative Research Group, 1993, A novel gene containing a trinucleotide repeat that is expanded and unstable on Huntington's disease chromosomes, *Cell* in press.

Izzo, P.N., Graybiel, A.M., and Bolam, J.P., 1987, Characterization of substance P- and [met]enkephalin-immunoreactive neurons in the caudate nucleus of the cat and ferret by a single section Golgi procedure, *Neurosci.* 20:577-587.

Jenkins, B.G., Koroshetz, W.J., Beal, M.F., and Rosen, B.R., 1992, Localized proton-NMR spectroscopy in patients with Huntington's disease (HD) demonstrates abnormal lactate levels in occipital cortex: Evidence for compromised metabolism in HD, *Neurology* in press.

Kowall, N.W., Ferrante, R.J., and Martin, J.B.,1987, Patterns of cell loss in Huntington's disease, *TINS* 10:24-29.

Larsson, N.-G., Andersen, O., Horne, E., Oldfors, A., and Wahlstrom, J., 1991, Leber's hereditary optic neuropathy and Complex I deficiency in muscle, *Ann. Neurol.* 30:701-708.

Lindquist, S., and Craig, E.A., 1988, The heat shock proteins, *Ann. Rev. Genetics* 22:631-677.

Linnane, A.W., Marzuki, S., Ozawa, T., and Tanaka, M., 1989, Mitochondrial DNA mutations as an important contributor to ageing and degenerative diseases, *Lancet* i:642-645.

Lowe, J., and Mayer, R.J., 1990, Ubiquitin, cell stress and diseases of the nervous system, *Neuropath. Appl. Neurobiol.* 16:281-291.

Mann, V.M., Cooper, J.M., Javoy-Agid, F., Agid, Y., Jenner, P., and Schapira, A.H.V., 1990, Mitochondrial function and parental sex effect in Huntington's disease, *Lancet* 336:749.

Mariani, A.P., Neff, N.H., and Hadjiconstantinou, M., 1986, MPTP treatment decreases dopamine and increases lipofuscin in mouse retina, *Neurosci. Lett.* 72:221-226.

Marshall, P.E., Landis, D.M.D., and Zlaneritis, E.L., 1983, Immunocytochemical studies of substance P and leu-enkephalin in Huntington's disease, *Brain Res.* 289:11-26.

McGeer, E.G., and McGeer, P.L., 1976, Duplication of biochemical changes of Huntington's disease by intrastriatal injections of glutamic and kainic acids, *Nature* 263:517-519.

Mettler, F.A., 1972, Choreoathetosis and striopallidal necrosis due to sodium azide, *Exp. Neurol.* 34:291-308.

Morgan, J.I., and Curran, T., 1991, Stimulus-transcription coupling in the nervous system: involvement of the inducible proto-oncogenes *fos* and *jun*, *Ann. Rev. Neurosci.* 14:421-452.

Meyers, R.H., Vonsattel, J.P., Stevens, T.J., Cupples, L.A., Richardson, E.P., Martin, J.B., and Bird, E.D., 1988, Clinical and neuropathologic assessment of severity in Huntington's disease, *Neurology* 38:341-347.

Nihei, K., and Kowall, N.W.,1992, Neurofilament and neural cell adhesion molecule immunocytochemistry of Huntington's disease striatum, *Ann. Neurol.* 31:59-63.

Novelli, A., Reilly, J.A., Lysko, P.G., and Henneberry, R.C., 1988, Glutamate becomes neurotoxic via the N-methyl-D-aspartate receptor when intracellular energy levels are reduced, *Brain Res.* 451:205-212.

Olney, J.W. (1978) Neurotoxicity of excitatory amino acids, *in*: "Kainic Acid as a Tool in Neurobiology," J.W., Olney and R.L. McGeer, eds., Raven Press, New York, pp. 95-122.

Oyanagi, K., Takeda, S., Takahashi, H., Ohama, E., and Ikuta, F., 1989, A quantitative investigation of the substantia nigra in Huntington's disease, *Ann. Neurol.* 26:13-19.

Parent, A., Csonka, C., and Etienne, P., 1984, The occurrence of large acetylcholinesterase-containing neurons in human neostriatum as disclosed in normal and Alzheimer's diseased brains, *Brain Res.* 291:154-158.

Parker, W.J.R., Boyson, S.J., Luder, A.S., and Parks, J.K., 1990, Evidence for a defect in NADH: ubiquinone oxidoreductase (complex I) in Huntington's disease, *Neurology* 40:1231-1234.

Ramsay, R.R., Krueger, M.J., Youngster, S.K., and Singer, T.P.,1991, Interaction of 1-methyl-4-phenylpyridium ion (MPP=) and its analogs with the rotenone/piericidin binging site of NADH dehydrogenase, *J. Neurochem.* 56:1184-1190.

Reiner, A., Albin, R.L., Anderson, K.D., D'Amato, C.J., Penney, J.B., and Young, A.B., 1988, Differential loss of striatal projection neurons in Huntington's disease, *Proc. Natl. Acad. Sci. USA* 85:5733-5737.

Ribak, C.E., Vaughn, J.E., and Roberts, E., 1979, The GABA neurons and their axon terminals in the rat corpus striatum as demonstrated by GAD immunocytochemistry, *J. Comp. Neurol.* 187:261-284.

Richardson, E.P. Jr.,1990, Third Dorothy S. Russell memorial lecture. Huntington's disease: some recent neuropathological studies, *Neuropathol. Appl. Neurobiol. 16:* 451-460.

Rosen, D.R., Siddique, T., Patterson, D., Figlewicz, D.A., Sapp, P., Hentati, A., Donaldson, D., Goto, J., O'Regan, J.P., Deng, H.-X., Rahmani, Z., Krizus, A., McKenna-Yasek, D., Cayabyab, A., Gaston, S.M., Berger, R., Tanzi, R.E., Halperin, J.J., Herzfelt, B., Van den Bergh, R., Hung, W.-Y., Bird, T., Deng, G., Mulder, D.W., Smyth, C., Laing, N.G., Soriana, E., Pericak-Vance, M.A., Haines, J., Rouleau, G.A., Gusella, J.A., Horvitz, H.R., and Brown, R.H., 1993, Mutations in Cu/Zn superoxide dismutase gene are associated with familial amyotrophic lateral sclerosis, *Nature* 362:59-62.

Saji, M., and Reis, D.J.,1987, Delayed transneuronal death of substantia nigra neurons prevented by gamma-aminobutyric acid agonist, *Science* 235:66-69.

Seto-Ohshima, A., Emson, P.C., Lawson, E., Mountjoy, C.Q., and Carrasco, L.H., 1988, Loss of matrix calcium-binding protein-containing neurons in Huntington's disease, *Lancet* 1252-1255.

Singer, T.P., and Ramsay, R.R., 1990, Mechanism of the neurotoxicity of MPTP. An update, *Febs. Lett.* 274:1-8.

Stadler, J., Curran, R.D., Ochoa, J.B., Harbrecht, B.G., Hoffman, R.A., Simmons, R.L., and Billiar, T.S., 1991, Effect of endogenous nitric oxide on mitochondrial respiration of rat hepatocytes in vitro and in vivo, *Arch. Surg.* 126:186-191.

Stamler, J.S., Singel, D.J., and Loscalzo, J., 1992, Biochemistry of nitric oxide and its redox-activated forms, *Science* 258:1898-1902.

Storey, E., Hyman, B.T., Jenkins, B., Brouillet, E., Miller, J.M., Rosen, B.R., and Beal, M.F., 1992, MPP+ produces excitotoxic lesions in rat striatum due to impairment of oxidative metabolism, *J. Neurochem.* 58:1975-1978.

Takagi, H., Mizuta, H., Matsuda, T., Inagaki, S., Tateishi, K., and Hamaoka, T., 1984, The occurrance of cholecystokinin-like immunoreactive neurons in the rat neostriatum: light and electron microscopic analysis, *Brain Res.* 309:346-349.

Tellez-Nagel, I., Johnson, A.B. and Terry, R.D., 1973, Ultrastructural and histochemical study of cerebral biopsies in Huntington's Chorea, *Adv. Neurol.* 1:387-398.

Theriault, E., and Landis, D.M.D. 1987, Morphology of neurons containing VIP-like immunoreactivitry, *J. Comp. Neurol.* 256:1-13.

Trounce, I., Byrne, E., and Marzuki, S., 1989, Decline in skeletal muscle mitochondrial respiratory chain function: possible factor in ageing, *Lancet* i:637-639.

Vonsattel, J.-P., Meyers, R.H., Stevens, T.J., Ferrante, R.J., Bird, E.D., and Richardson, J. Edward P., 1985, Neuropathological classification of Huntington's disease. *J. Neuropathol. Exp. Neurol.* 44:559-577.

Wallace, D.C., 1991, Mitochondrial genes and neuromuscular disease, *Res. Publ. Assoc. Res. Nerv. Ment. Dis.* 69:101-120.

Zeevalk, G.D., and Nicklas, W.J.,1991, Mechanisms underlying initiation of excitotoxicity associated with metabolic inhibition, *J. Pharm. Exp. Ther.* 257:870-878.

NITRIC OXIDE: A NOVEL INTERCELLULAR MESSENGER IN THE STRIATUM

Piers C. Emson, Rosalinda Guevara Guzman, Rosa Señaris, Jiro Kishimoto, Weiming Xu, Liu Lizhi, Paula Norris and Keith M. Kendrick

MRC - Molecular Neuroscience Group and Department of Neurobiology
AFRC, Babraham Institute
Babraham
Cambridge
CB2 4AT, U.K.

INTRODUCTION

The free radical gas nitric oxide (NO) has recently been proposed as a messenger, or novel type of neurotransmitter in the brain (Bredt et al., 1991; Snyder and Bredt, 1992). The initial evidence for a role for NO in the central nervous system came when Garthwaite et al (1988) showed that cerebellar neurones would synthesize NO in response to the excitatory neurotransmitter glutamate, an observation confirmed by Bredt and Snyder (1989) amongst others. Subsequent work by Bredt and Snyder (1990) characterized the enzyme responsible for nitric oxide production, nitric oxide synthase (NOS) from rat brain and they cloned and isolated NOS (Bredt et al., 1991). The cloning and isolation of brain NOS revealed that the enzyme was structurally related only to cytochrome P450-oxido-reductase (CP-450 OR) which like NOS has an electron-transferring/accepting carboxy-terminal sequence (Bredt et al., 1991). The ability of NOS to reduce the dye nitro-blue tetrazolium accounting for a significant amount of brain diaphorase activity and diaphorase staining reflects NOS containing neurones (Figure 1) (Hope et al., 1991). We have confirmed these observations using *in situ* hybridization with antisense probes for NOS and NOS specific antibodies which visualize the same population of neurones in the brain and striatum (Figure 1).

A feature of NOS/diaphorase containing neurones in the striatum is that they are relatively resistant to ischaemia (Thomas and Pearse, 1964) and that they survive the pathology of Huntington's disease (Dawbarn et al., 1985; Ferrante et al., 1985; Kowall et al., 1987). These observations suggested to us that it would be worthwhile to consider whether overactivity of NOS containing neurones (NO is a toxic gas) might be implicated in human Huntington's disease and what role NO may have in striatal signalling.

NOS and Huntington's Disease

The hypothesis to be considered was whether NOS neurones might be overactive in Huntington's disease producing neurotoxic NO, and whether human brain NOS might be a candidate gene for the gene involved in Huntington's Disease. In order to consider this possibility we used RNA-PCR to clone a fragment of rat brain NOS and used this fragment to obtain a human brain NOS cDNA from a human cerebellar cDNA library. This cDNA was used to screen a panel of mouse-human cell hybrids (Kishimoto et al., 1992) to determine the chromosomal localization of the human brain NOS gene (the Huntington's Disease gene is on chromosome 4). This study revealed that human brain NOS was on chromosome 12 and subsequent fluorescent *in situ* hybridization further localized the gene to the 12q 24.2-12q 24.31 region of chromosome 12. This location indicated that brain NOS gene is not directly implicated in Huntington's disease, but it is still a possibility that overactivity of NOS may contribute to cell loss in Huntington's Disease. It is also of considerable interest to understand how these neurones protect themselves against the toxic effects of "NO" they produce. It may be that it is this metabolic resistance which underlies these cells survival in Huntington's Disease, although alternative explanations involving a relative lack of excitotoxic glutamate receptors on NOS cells (Beal et al., 1986, 1988) are also a possibility.

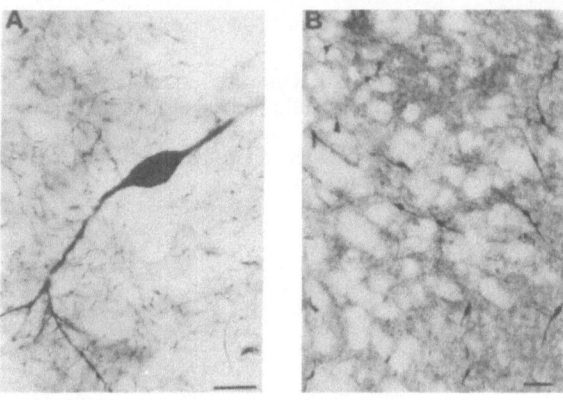

Figure 1. (A) Demonstration of NADPH-diaphorase staining in a medium sized striatal neurones. (B) Nitric oxide synthase immunoreactivity in striatal interneurones. Note the resemblance between the NADPH-diaphorase positive neurone (A) and those visualized with the anti-NOS antibody (B). Scale bar A=20μm and B=100μm.

Nitric Oxide in the Striatum. Diaphorase staining (Figure 1) identifies a population of 1-2% of medium to large aspiny interneurones which also contain the two neuropeptides, somatostatin and neuropeptide Y (Vincent and Johansson, 1983; Vincent et al., 1983; Ferrante et al., 1985; Hope et al., 1991). These cells do not however, contain a classical transmitter candidate such as GABA or acetylcholine and it is suggested that nitric oxide may fulfil this "transmitter" role in these cells. In this sense NO is a very different transmitter from candidates such as GABA and acetylcholine as it is a freely diffusible gas which diffuses down a concentration gradient to influence haem containing molecules such as guanylate cyclase and the cytochromes, and to nitrosylate a number of proteins (Snyder and Bredt, 1992). NO would, however, be an ideal candidate to integrate neuronal activity and transmitter release within striatal patch or matrix compartments (Wilson, 1990). Release of NO probably occurs following elevation of intracellular Ca^{++} as brain NOS is Ca^{++}/calmodulin activated enzyme. The location of striatal NOS cells is also suited to a role of integrating striatal activity as they are known to receive prominent cortical (excitatory) (Takagi et al., 1983) and pallidal (inhibitory) inputs (Staines and Hincke, 1991). The surrounding cells, the inhibitory GABA-ergic spiny output neurones are also extremely rich in guanylate cyclase (a major target for NO) to produce cGMP (Ariano and Matus, 1981; Ariano et al., 1982; Ariano, 1983, 1984; Matsuoka et al., 1992).

It was therefore of considerable interest to consider in detail what role local NO release has on neuronal properties and transmitter release in the striatum in addition to its established effects as EDRF (Knowles and Moncada, 1992). The effects of NO on foetal striatal

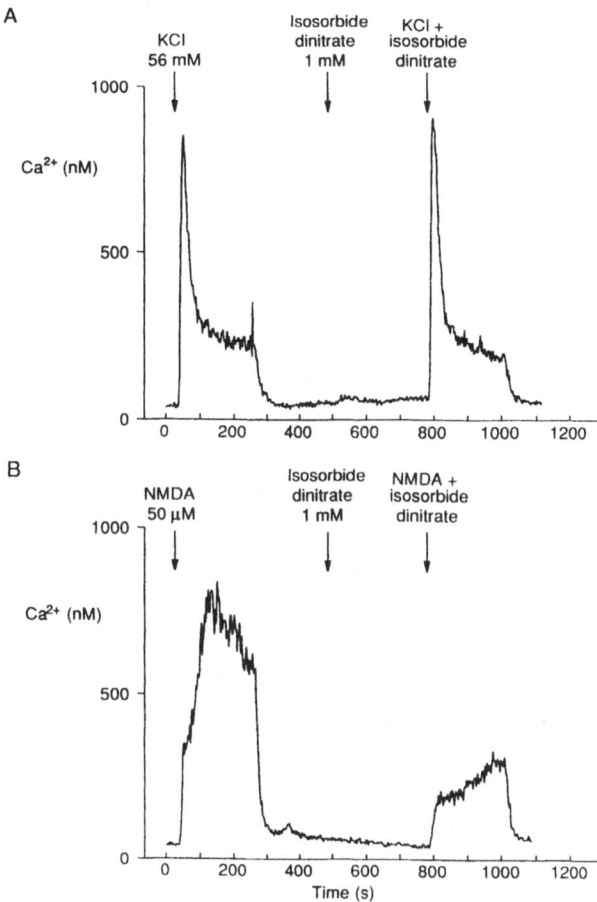

Figure 2 Video imaging of a Fura-2 loaded cultured striatal neurones. (A) The response of the neurone to the application of a depolarizing pulse of KCl is not influenced by the presence of the nitric oxide releaser isosorbide dinitrate. The response to KCl indicates that NO does not influence the opening of voltage sensitive Ca^{++} channels in the neurone. (B) The response of the same striatal neurone imaged in (A) to application of NMDA. The application of NMDA produces a rapid rise in intracellular Ca^{++}. The response is however substantially reduced when NMDA is applied together with isosorbide. Cells were washed 200 seconds after the application of each drug or agonist.

neurones was considered *in vitro*. Cultured embryonic rat striatal neurones were loaded with the calcium indicator dye fura-2 and changes in intracellular Ca^{++} were monitored by quantitative ratio imaging. Application of N-methyl-D aspartate, kainic acid or potassium chloride increased intracellular Ca^{++}. Application of these agonists together with the nitric oxide releasers isosorbide and s-nitroso-acetylpencillamine (SNAP) showed that the effects of NMDA application (but not of kainate or potassium chloride) were blocked or substantially reduced in a dose dependent manner by application of NO releasers (Figure 2). This effect was reversible after washing the cells. These experiments confirm and extend the observation of Manzoni et al. (1992) who showed that an additional NO releaser 3-morpholino-syndnonimine (SIN-I) blocked the NMDA channel apparently by direct chemical modification of the NMDA channel protein. These data suggest that one effect of local NO release in response to excitatory glutamatergic activity from cortical inputs to the striatum

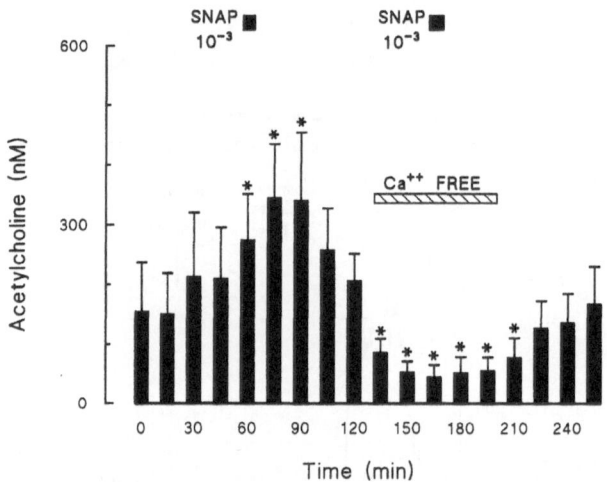

Figure 3. Effects of the nitric oxide (NO) releaser S-nitroso-penicillamine (SNAP) on acetylcholine release from the rat striatum *in vivo*. Note the effect of SNAP is abolished in Ringer lacking Ca^{++}.

may be to reduce intracellular Ca^{++} entry by blocking the NMDA receptors on local striatal neurones. This would mean that NOS cells would tend to provide a local inhibitory feedback mechanism to reduce the effects of cortical inputs onto GABA-ergic output neurones. In this case brief local NO release would be neuroprotective (reducing the excitotoxic effects of glutamate), however prolonged exposure of cells to NO was toxic.

These data indicate that local NO release might be expected to influence striatal transmitter release either by effects on guanylate cyclase (NO elevates cGMP), by blockade of NMDA receptors, or by direct effects on other proteins involved in transmitter release or uptake. To consider these possibilities in detail we have used *in vivo* microdialysis together with sensitive electrochemical or fluorescence detection for monoamines (ACh and dopamine) and amino acid (GABA, glutamate) neurotransmitters. Results with nitric oxide releasers such as SNAP, isosorbide and s-nitrosoglutathione (SNOG) were complicated as many of these NO releasers contained compounds such as penicillamine, and glutathione which had direct effects on striatal transmitter release separate from any effects of "NO" itself. In general NO releasers produced Ca^{++} dependent release of acetylcholine and GABA (Figure 3) DA release was also increased by high concentrations of NO releasers (10^{-3}-10^{-2}M), however DA and GABA release were also produced by penicillamine and glutathione. In consequence we considered the effects of NO itself by bubbling NO gas through the Ringer immediately prior to microdialysis.

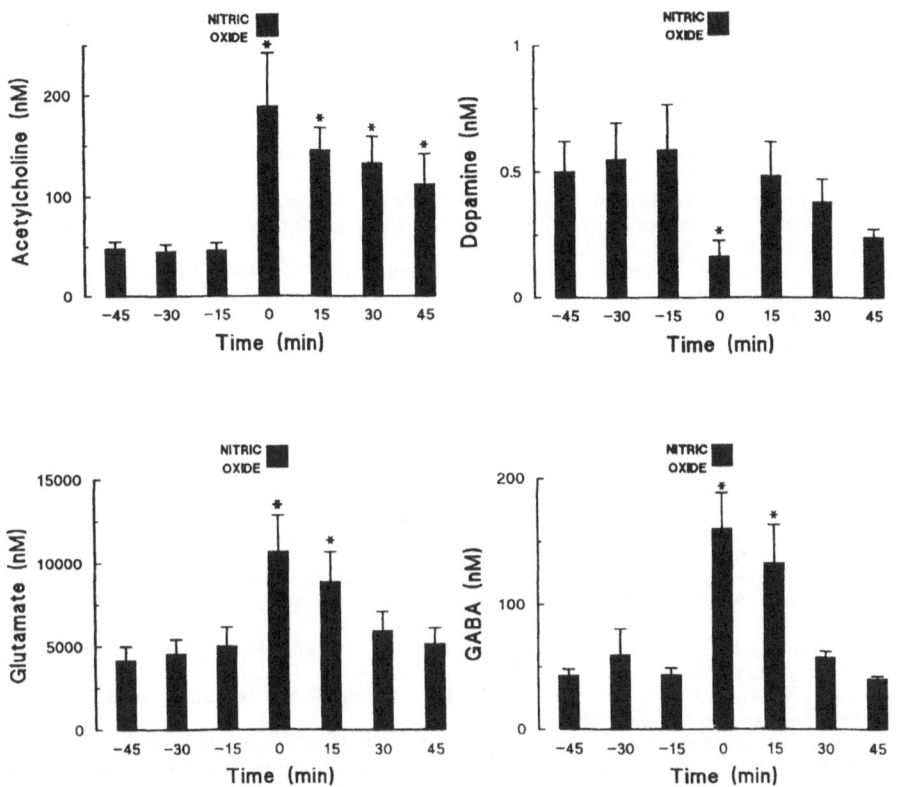

Figure 4. Effects of nitric oxide (NO) gas infused into the striatum of the rat *in vivo*. Note that NO releases acetylcholine, glutamate and GABA but strongly inhibits dopamine release.

Surprisingly, in view of the expected short half life of "NO" in solution, delivery of NO gassed Ringer through a gas tight syringe produced substantial effects on striatal transmitter release (the Ringer was buffered to pH 7.0). Direct application of NO itself stimulated ACh and GABA release (Figure 4) as observed with NO releasers and this effect was Ca^{++} dependent. Dopamine release was strongly inhibited by NO (Figure 4) indicating these effects of NO on striatal neurotransmitter release were not a general non-specific release of all transmitter candidates. Release of 5-HT and glutamate was also increased but in a non-Ca^{++} dependent fashion.

Interpretation of this data is not straight forward as we are so far uncertain of the local concentrations of NO in the striatum *in vivo*, however it is clear that local release of transmitters can be substantially modified by locally released NO. The ability of NO to reduce striatal DA release and enhance ACh release may indicate that NOS containing striatal neurones may provide important local mechanisms by which the dopamine: acetylcholine balance in striatal function may be maintained (Stoof et al 1992). Certainly the ability of released NO to diffuse locally to influence cortical inputs and striatal transmitter release provides a means by which activity in striatal compartments may be integrated both in space and time.

REFERENCES

Ariano, M.A., and Matus, A.I., 1981, Ultrastructural localization of cyclic GMP and cyclic AMP in rat striatum, *J. Cell Biol.* 91:287-292.

Ariano, M.A., Lewicki, J.A., Branswein, H.J., and Murad, F., 1982, Immunohistochemical localization of guanylate cyclase within neurons of rat brain, *Proc. Natl. Acad. Sci. USA* 79:1316-1320.

Ariano, M.A., 1983, Distribution of components of the guanosine 3',5'-phosphate system in rat caudate-putamen, *Neurosci.* 10:707-723.

Ariano, M.A., 1984, Rat striatal cyclic nucleotide-reactive cells and acetylcholinesterase reactive interneurons are separate populations, *Brain Res.* 296:160-163.

Beal, M.F., Kowall, N.W., Ellison, D.W., Mazurek, M.F., Swartz, K.J., and Martin, J.B., 1986, Replication of the neurochemical characteristics of Huntington's disease by quinolinic acid, *Nature* 321:168-171.

Beal, M.F., Kowall, N.W., Swartz, K.J., Ferrante, R.J., and Martin, J.B., 1988, Systemic approaches to modifying quinolinic acid striatal lesions in rats, *J. Neurosci.* 8:3901-3908.

Bredt, D.S., and Snyder, S.H., 1989, Nitric oxide mediates glutamate-linked enhancement of cGMP levels in the cerebellum, *Proc. Natl. Acad. Sci. USA* 86:9030-9033.

Bredt, D.S., and Snyder, S.H., 1990, Isolation of nitric oxide synthase, a calmodulin-requiring enzyme, *Proc. Natl. Acad. Sci. USA* 87:682-685.

Bredt, D.S., Hwang, P.M., Glatt, C.E., Lowenstein, C.L., Reed, R.R., and Snyder, S.H., 1991, Cloned and expressed nitric oxide synthase structurally resembles cytochrome P-450 reductase, *Nature* 351:714-719.

Dawbarn, D., DeQuit, M.E., and Emson, P.C., 1985, Survival of basal ganglia neuropeptide Y-somatostatin neurones in Huntington's disease, *Brain Res.* 340:251-261.

Ferrante, R.J., Kowall, N.W., Beal, M.F., Richardson, Jr. E.P., Bird, E.D., and Martin, J.B., 1985, Selective sparing of a class of striatal neurons in Huntington's disease, *Science* 230:561-563.

Garthwaite, J., Charles, S.L., and Chess-Williams, R., 1988, Endothelium-derived relaxing factor release on activation of NMDA receptors suggests role as intercellular messenger in the brain, *Nature* 336:385-388.

Hope, B.T., Michael, G.J., Knigge, K.M., and Vincent, S.R., 1991, Neuronal NADPH-diaphorase is a nitric oxide synthase, *Proc. Natl. Acad. Sci. USA* 88:2811-2814.

Kishimoto, J., Spurr, N., Emson, P.C., and Xu, W., 1992, Localization of brain nitric oxide synthase to human chromosome 12, *Genomics* 14:802-804.

Knowles, R.G., and Moncada, S., 1992, Nitric oxide as a signal in blood vessels, *TINS* 17:399-402.

Kowall, N.W., Ferrante, R.J., and Martin, J.B., 1987, Patterns of cell loss in Huntington's disease, *Trends Neurosci.* 10:24-29.

Manzoni, O., Prezeau, L., Marin, P., Deshager, S., Bockaert, J., and Fagni, L., 1992, Nitric oxide-induced blockade of NMDA receptors, *Neuron* 8:653-662.

Matsuoka, I., Giuili, G., Poyard, M., Stengel, D., Parma, J., Guellaen, G., and Hanoune, J., 1992, Localization of adenylyl and guanylyl cyclase in rat brain by *in situ* hybridization: comparison with calmodulin mRNA distribution, *J. Neurosci.* 12:3350-3360.

Nakajima, K., Harada, K., Ebina, Y., Yoshimura, T., Ito, H., Ban, T., and Shingai, R., 1993, Relationship between resting cytosolic Ca^{2+} and responses induced by N-methyl-D-aspartate in hippocampal neurons, *Brain Res.* 603:321-323.

Staines, W.A., and Hincke, M.T.C., 1991, Substantial alterations in neurochemical and metabolic indices in select basal ganglia neurons follow lesions of globus pallidus neurons in rats, *Soc. Neurosci. Abst.* 17 (1):456.

Stoof, J.C., Drukarch, B., De Boer, P., Westerink, B.H.C., and Groenewegen, H.J., 1992, Regulation of the activity of striatal cholinergic neurons by dopamine, *Neurosci.* 47:755-770.

Snyder, S.H., and Bredt, D.S., 1991, Nitric oxide as a neuronal messenger, *Trends Pharmacol. Sciences* 12:125-128.

Snyder, S.H., and Bredt, D.S., 1992, Biological roles of nitric oxide, *Scientific American* May :28-35.

Takagi, H., Somogyi, P., Somogyi, J., and Smith, A.D., 1983, Fine structural studies on a type of somatostatin-immunoreactive neuron and its synaptic connections in the rat neostriatum: a correlated light and electron microscopic study, *J. Comp. Neurol.* 214:1-16.

Thomas, E., and Pearse, A.G.E., 1964, The solitary active cells, Histochemical demonstration of damage-resistant nerve cells with a TPN-diaphorase reaction, *Acta Neuropathol.* 3:238-249.

Vincent, S.R., and Johansson, O., 1983, Striatal neurons containing both somatostatin- and avian pancreatic polypeptide (APP)-like immunoreactivities and NADPH-diaphorase activity: a light and electron microscopic study, *J. Comp. Neurol.* 217:264-270.

Vincent, S.R., Johansson, O., Hökfelt, T., Skirboll, L., Elde, R.P., Terenius, L., Kimmel, J., and Goldstein, M., 1983, NADPH-diaphorase: a selective histochemical marker for striatal neurons containing both somatostatin- and avian pancreatic polypeptide (APP)-like immunoreactivities, *J. Comp. Neurol.* 217:252-263.

Wilson, C.J., 1990, Basal ganglia, *in*: "The Synaptic Organisation of the Brain," G.M. Shepherd, ed., Oxford University Press, Oxford, pp. 279-317.

SOME CONSEQUENCES OF LOCAL BLOCKADE OF NITRIC-OXIDE SYNTHASE IN THE RAT NEOSTRIATUM

Gordon W. Arbuthnott[1], Paul A.T. Kelly[2] and Ann K. Wright[1]

[1]Preclinical Veterinary Sciences
[2]Clinical Neurosciences
  University of Edinburgh
  Edinburgh
  Scotland, U.K.

INTRODUCTION

In addition to the spiny cells of the neostriatum, which compose more than 95% of the population of neurones, there are a number of interneurones whose chemical identity is known but whose roles in striatal neurophysiology are still controversial.

The acetylcholine containing interneurones (Bolam et al., 1984) have long been implicated in the generation of Parkinsonian symptoms. As long ago as 1962, Barbeau suggested the idea that the balance between dopamine and acetylcholine was vital for normal movement and that the disturbance of this balance resulted in tremor (if dopamine prevailed) and immobility (if acetylcholine prevailed) (Barbeau, 1962). Our understanding of the mechanism of interaction between these two transmitters is still incomplete (Consolo et al., 1992) but we now have the anatomical basis (or lack of it, see Lehmann and Langer, 1983) with which to derive possible cellular interactions that could underlie the 'balance'.

The GABA/parvalbumin containing interneurones (Bolam et al., 1983; Cowan et al., 1990) are smaller in cell body diameter but are very large in their potential influence on spiny cells with which they make extensive connections (Bolam et al., 1983; Cowan et al., 1990; Kita et al., 1990). They seem to have a role in a feedforward inhibition of the spiny output cells (Kita et al., 1990) which could have powerful control of the action of cortical input upon striatal efferent neurones.

The smallest of the neostriatal interneurones are an important group which can be stained by antibodies to NPY and to Somatostatin and also with the histochemical procedure for NADPH-diaphorase (Takagi et al., 1983; Vincent and Johansson, 1983; Vincent et al., 1983; Chesselet and Graybiel, 1986; Kowall et al., 1987; Desjardins and Parent, 1992). Their function in the striatal circuitry is obscure. The discovery that NADPH-diaphorase and NO-synthase are the same enzyme (Dawson et al., 1991; Hope et al. 1991) suggests that these cells are a source of this enigmatic neuronal messenger in the striatum. We already had evidence that there were long term changes in the efficacy of the cortico-striatal pathway (unpublished observations; but see Calabresi et al. (1992a, b and c); Garcia-Munoz et al. (1992) and this volume). The recent suggestions that NO might be important in the long term changes in synaptic plasticity in the hippocampus (Schuman and Madison, 1991; O'Dell et al., 1991; Mizutani et al., 1993) prompted us to try to see the effect of inhibition of NO synthesis on diaphorase staining in the neostriatum. We used the nitro substituted arginine L-G-Nitroarginine methyl ester (L-NAME) which is a potent inhibitor of NO production in peripheral tissues (Dwyer et al. 1991).

The Basal Ganglia IV, Edited by
G. Percheron *et al.*, Plenum Press, New York, 1994

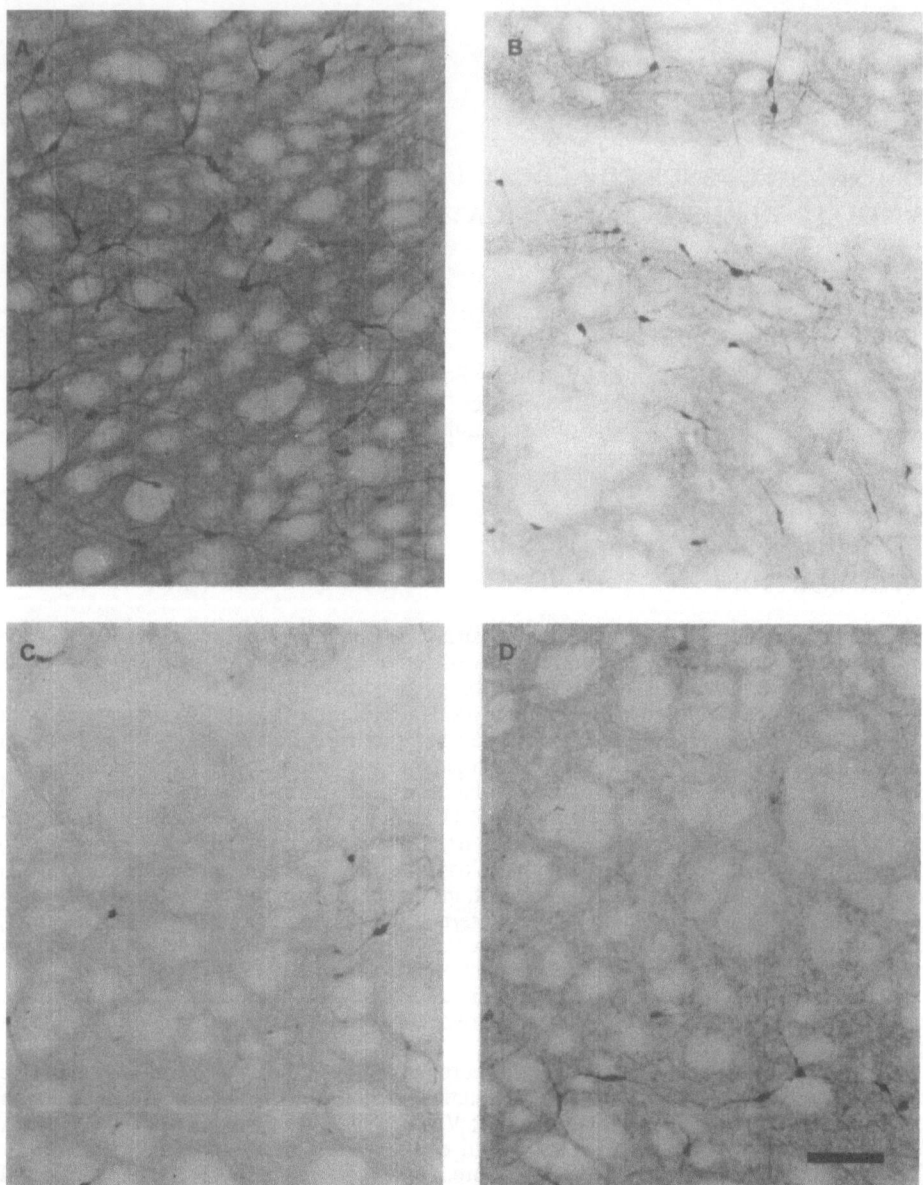

Figure 1. Photomicrographs of sections of striatum from rats treated with the diaphorase method. The striatal interneurones are visible in all the photographs and in A, which comes from a control animal the densely stained fibre plexus is also obvious. The background neuropile is much less darkly stained in the sections taken at 0.5 and 1 hour after the injection of L-NAME (B and C). The appearance of individual interneurones in B suggests that they too have been damaged by the injection. They recover quickly (C and D) while the neuropile staining does not recover completely until 24 hours after the injection as seen in D. The scale bar is relevant to all the photomicrographs, and represents 100μ.

METHODS AND RESULTS

We first demonstrated that there was no interference of the synthesis inhibitor L-NAME on the histochemical reaction which we used to display the diaphorase activity in tissue sections (Hope and Vincent, 1989). Including L-NAME at any stage in the incubation procedure even up to 125mM did not affect the staining intensity during the reaction in fixed tissue sections .

Then we studied the reaction of diaphorase containing neurones *in vivo* to direct injection of L-NAME into the striatum. Adult Han Wistar rats (300-400g weight) were anaesthetised with halothane (1% in inspired air) and mounted in a stereotaxic frame. The skull was exposed and through a small burr hole positioned over the head of the striatum a Hamilton syringe needle was introduced into the nucleus (coordinates A.P. + 0.7, L. 2.0, with respect to Bregma; 76.0mm ventral to the cortical surface). 1μl of physiological saline containing 20μmole L-NAME was then injected over 5 minutes and the animals perfused through the heart with 4% Paraformaldehyde, 0.05% Gluteraldehyde at chosen times after the injection. Histochemical processing of 50μ sections of the striatum of animals treated in this way revealed that diaphorase staining was reduced from 30 min after the injections until about 4 hours later but subsequently recovered completely. Figure 1 shows the effect of the injections in striatal tissue close to the injected region at various times after completion of the L-NAME injection.

The results suggested that in spite of there being no sign of an interaction with the diaphorase reaction there was still an effect of the inhibition of NO *in vivo*. At first we though that it was a direct effect on the cells in that their capability of synthesising NO was vital to their normal staining properties. Since it seems likely that these cells survive in Huntington's Chorea (Aronin et al., 1983; Marshall and Landis, 1985), when all the spiny cells surrounding them are dying, the change in the cell staining was surprising. Could it be that the NO cells are so dependent on NO-synthase that their survival depends on the activity of this potentially toxic enzyme reaction?

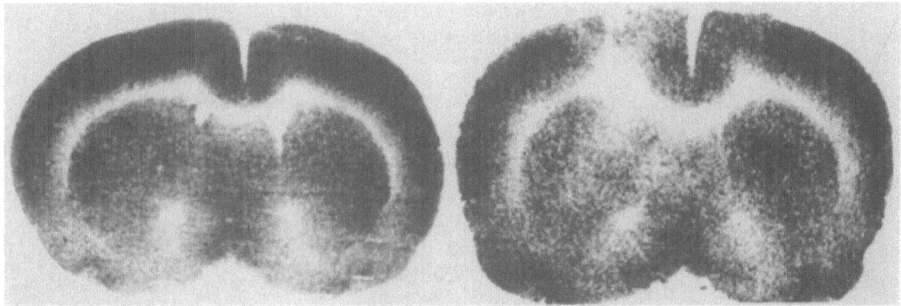

Figure 2. Autoradiograms from control sections of rat brain at the level of the striatum from ACSF-(left) and L-NAME-injected animals (right). In each case the injected striatum appears on the left of the photograph. In general terms, grain density of the images is proportional to blood flow. Note relative homogeneity of grain density in ACSF-injected animals and increased heterogeneity evident in the L-NAME injected striatum. The scale bar represents 1mm.

In the periphery the physiological function NO is clear. It is the endothelium derived relaxing factor (EDRF) of peripheral blood vessels (Palmer et al., 1987). It seemed prudent to us that we study the action of our direct injections of L-NAME upon the flow through cerebral vessels, especially in view of the sensitivity of the striatum to anoxic damage in middle cerebral artery occlusion models of cerebral ischaemia. There was direct evidence that

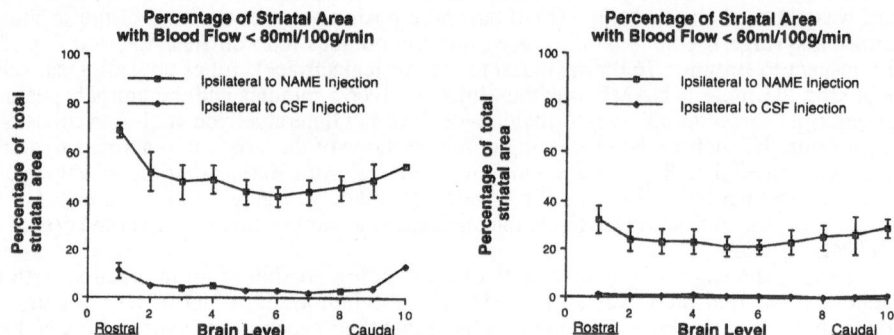

Figure 3. Quantitative planimetry of striatal autoradiographic images showing percentage of total striatal tissue with blood flow less than 80ml/100g/min (left) and less than 60ml/100g/min (right) in L-NAME- and ACSF-injected animals.

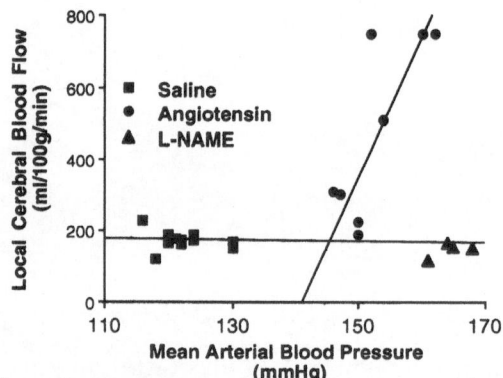

Figure 4. The relationship between local blood flow and mean arterial blood pressure (MABP) in conscious rats treated with saline, angiotensin-II, or L-NAME. Note that in angiotensin-treated rats, blood flow is markedly increased when blood pressures exceed approximately 150 mmHg (the upper limit of autoregulation). In contrast, there is no evidence of a breakdown in autoregulation in L-NAME-treated rats, even at much higher levels of MABP.

anaesthesia compromised the responsiveness of cerebral vessels (Edvinsson and McCulloch, 1981) and so these experiments were designed to be performed in conscious rats. Under general anaesthesia a guide canula with a guard wire was implanted 10 days prior to the experiments. On the day of the experiment the animals were restrained and a fine canula introduced into the guide tube so that its tip was within the striatum in 10 rats. In 5 animals 2μl of artificial cerebrospinal fluid (ACSF) was injected while in the other 5 L-NAME was injected in the same volume and at a dose of 20μmole. One hour after the intra-cerebral injection the animals received a 45 sec infusion of iodo-[^{14}C]-antipyrine i.v. and were immediately painlessly killed and their brains removed. The brains were frozen and then mounted on a cryostat chuck. 20μ sections were cut and mounted on coverslips and closely opposed to sheet film in the dark for 7 days. The resulting autoradiographs were developed and analysed in a Quantimet image analyser so that the iodoantipyrene concentrations could be estimated and the blood flow through the striatum and overlying cortex thus could be calculated (Sakurada et al., 1978). Figure 2 shows two autoradiographs from the series, one control after the injection of 2μl ACSF and one after L-NAME. Striatal blood flow was measured from three consecutive coronal sections taken at 10 preset levels through the head of the striatum at 200μ apart. Because striatal blood flow is inherantly inhomogeneous, and because we had no reason to suppose that the diffusion of the L-NAME would be reproducibly regular, the total area of the striatum was measured from each of these sections and the proportion of this area perfused at less than 50% or 75% of the mean value was derived. Both mean arterial blood pressure and blood gas tensions were monitored and remained unchanged after the intrastriatal injections. Figure 3 illustrates the mean effect on blood flow at the 10 levels through the striatum. The action of L-NAME applied peripherally is well known. Its action on blood vessels leads to a huge increase in systemic blood pressure (Rees et al., 1991). Such an action would itself be expected to compromise cerebral blood flow since the pressure rises beyond the normal threshold for autoregulation. In a series of rats threated with i.v. L-NAME (30 mg/Kg in 1 ml) the mean peripheral blood pressure was raised to 164±1mmHg but the cerebral blood flow remained in the normal range in spite of this (Fig. 4). A similar rise in systemic blood pressure (to 157±1mmHg) caused by the infusion of angiotensin-11 certainly led to the blood flow in the brain breaking out of the autoregulatory control and rising focally in parietal and occipital cortex, for example. The suggestion has been made (Thomas et al., 1993) that L-NAME increases the upper limit of autoregulation to higher values of systemic blood pressure.

DISCUSSION

The discovery of the ability of diaphorase staining cells to synthesise NO in brain had many consequences. Our excited attempts to change enzyme activity in the hope either to develop a method of permanently disabling the diaphorase cells in the striatum, or to uncover their physiological actions, seems to have met a premature end in the discovery of the profound actions of the inhibitor on cerebral blood flow.

On the other hand this, perhaps prosaic, action may be of vital importance. There are several aspects of particular interest. Just as in the periphery where the vasoconstrictor tone derived from sympathetic activity had held centre stage for too long, so we have to face the possibility that the blood flow in the brain has both vasoconstrictor and vasodilator tone. The rapid and profound vasoconstriction which follows L-NAME application begs the question of the source of the vasoconstrictor activity. Is the cerebral circulation, like that in the periphery, in constant balance between sympathetic tone and NO production? At least deep in the brain (for example in the striatum, where we have looked) it seems unlikely, since sympathetic nerve fibres do not to penetrate beyond the pia and ependyma. What other activity can be responsible for the vasoconstrictor tone? Does the change in the autoregulation set point hold a clue? All this seems a long way from how the interneurones in the striatum are connected with the function of the basal ganglia.

One suggestion for a function for NO is as the long sought 'presynaptic modifier' in long term potentiation (LTP). This model of memory formation has received rather a lot of exposure recently and at least a portion of the recent interest stems from increasing evidence that much of the time the increased efficacy of synapses in the hippocampus might be due to presynaptic modification of release parameters (Malinow and Tsein, 1990; Bekkers and Stevens, 1990; Larkman et al., 1991). A test of the two favourite mechanisms by which a change in postsynaptic response might generate a presynaptic change came down in favor of NO as the signal which, produced in postsynaptic cells might influence presynaptic terminals in the vicinity to increase transmitter release. The fact that there is no evidence that the particular postsynaptic cells involved (the hippocampal pyramidal cells) produce NO (Vincent and Kimura, 1992) has not inhibited a flood of experiments on this idea (Izumi et al., 1992; Mizutani et al., 1993). The recent suggestion that the action might be particularly clear on near threshold stimulation parameters and the difficulty with replicating any of the findings in many labs suggests that we do not yet understand how this action is produced.

On the other hand when we studied local blood flow using the autoradiographic iodoantipyrine technique at one hour after the injection there were obvious reductions in local blood flow in the region of the injection. It is just possible that the actions observed after NOS inhibition in all the experiments so far are a reflection of this compromised blood flow. Of course many of the definitive experiments have been done *in vitro*, where an action on blood vessels is unlikely to cause any major change. The effect *in vivo* is difficult to obtain and although there may be an effect on learning the problems of the experimental design are very real. How does the huge increase in blood pressure affect the animals ability to perform the behavioural task? Certainly the changes of perfusion/metabolism in the brain after these large doses of NO synthesis inhibitor make the interpretation of all of these experiments fraught with difficulties.

Our results also suggest caution in the application of the neuroprotective action of NOS inhibitors to clinical practice. The large increase in peripheral blood pressure is not a good side effect for a drug suggested to be useful for giving to patients with stroke. Of course the increase in the threshold for the loss of autoregulation is some encouragement as is the discovery of an inhibitor of NOS which does not raise arterial blood pressure, at least in the mouse (Moore et al., 1993). On the other hand it is hard to see how a drastic reduction of cerebral blood flow could be helpful in cases of cerebrovascular accident in brain either. If our results of locally applied L-NAME generalise to other areas of brain it may be that the neuroprotection will be an *in vitro* phenomenon only; it is certain that the agents will need to be extensively tested before they become of routine clinical use.

How then are our results relevant to the action of the striatal interneurones? At first sight the technical difficulties have defeated our investigation. The enzyme inhibitor does inhibit the formation of citrulline from arginine but not the colour reaction. The application to the striatum leads to extensive depletion of the diaphorase staining. This does not last more than about 4 hours and the recovery is, as far as we can see, complete by 24 hours.

At the time of maximum decrease in the diaphorase staining there is a marked reduction in the blood flow in the region which may itself be responsible for the change in staining which we see. Obviously there are a great many experiments still to be done, but at first glance these early results suggest that the output from the diaphorase cells may well control local blood flow rather than the transfer of information between compartments in the striatum by other (more conventional) means. We must look elsewhere for the substrate for long term changes in synaptic efficacy in the striatum, as well as in the hippocampus.

The important cortical input to these cells and their powerful pericellular baskets of input from the globus pallidus may have an important functional relevance. The excitatory cortical input could signal to the blood vessels nearby that more blood flow is required ('because' a cortical input in the neighbourhood is creating a requirement for energy supply). On the other hand the inhibitory baskets may indicate that blood can be redistributed ('because' the signal has already been received downstream in the globus pallidus). Wild speculation it is true, but at least a line for future investigation of these fascinating interneurones.

REFERENCES

Aronin, N., Cooper, P.E., Lorenz, L.J., Bird, E.D., Sagar, S.M., Leeman, S.E., and Martin, J.B., 1983, Somatostatin is increased in the basal ganglia in Huntington disease, *Ann. Neurol.* 13:519-526.

Barbeau, A., 1962. The pathogenesis of Parkinson's disease: a new hypothesis, *Can. Med. Assoc. J.* 87:802-807.

Bekkers, J.M., and Stevens, C.F., 1990, Presynaptic mechanisms for long-term potentiation in the hippocampus, *Nature* 346:724-729.

Bolam, J.P., Clarke, D.J., Smith, A.D., and Somogyi, P., 1983, A type of aspiny neuron in the rat neostriatum accumulates ^{3}H-gamma-amino-butyric acid: combination of Golgi-staining, autoradiography and electron microscopy, *J . Comp. Neurol* 213:121-134.

Bolam, J.P., Ingham, C.A., and Smith, A.D., 1984, The section-Golgi-impregnation procedure 3. Combination of Golgi-impregnation with enzyme histochemistry and electron microscopy to characterize acetylcholinesterase-containing neurons in the rat neostriatum, *Neurosci* 12: 687-709.

Calabresi, P., Maj, R, Mercuri, N.B., and Bernardi, G., 1992a, Co-activation of D_1 and D_2 dopamine receptors is required for long-term synaptic depression in the striatum, *Neurosci. Lett.* 142:95-99.

Calabresi, P., Maj, R., Pisani, A., Mercuri, N.B. and Bernardi, G., 1992b, Long-term synaptic depression in the striatum: Physiological and pharmacological, *J. Neurosci.* 12:4224-4233.

Calabresi, P., Pisani, A., Mercuri, N.B., and Bernardi, G., 1992c, Long-term potentiation in the striatum is unmasked by removing the voltage-dependent magnesium block of NMDA receptor channels, *Eu. J. Neurosci.* 4:929-935.

Chesselet, M.-F., and Graybiel, A.M., 1986. Striatal neurons expressing somatostatin-like immunoreactivity: evidence for a peptidergic interneuronal system in the cat, *Neurosci.* 17:547-571.

Consolo, S., Girotti, P., Russi, G., and Di Chiara, G., 1992, Endogenous dopamine facilitates striatal in vivo acetylcholine release by acting on D_1 receptors localized in the striatum, *J. Neurochem.* 59:1555-1557.

Cowan, R.L., Wilson, C.J., Emson, P.C., and Heizmann, C.W., 1990, Parvalbumin-containing GABAergic interneurons in the rat neostriatum, *J. Comp. Neurol.* 302:197-205.

Dawson, T.M., Bredt, D.S., Fotuhi, M., Hwang, P.M., and Snyder, S.H.,1991, Nitric oxide synthase and neuronal NADPH diaphorase are identical in brain and peripheral tissues, *Proc. Natl. Acad. Sci. USA* 88:7797-7801.

Desjardins, C., and Parent, A, 1992, Distribution of somatostatin immuno-reactivity in the forebrain of the squirrel monkey: Basal ganglia and amygdala, *Neurosci.* 47:115-133.

Dwyer, M.A., Bredt, D.S., and Snyder, S.L., 1991, Nitric oxide synthase: irreversible inhibition by L-NG-nitro arginine in brain in vitro and in vivo, *Biochem, Biophys. Res. Commun.* 176:1136-1141.

Edvinsson, L., and McCulloch, J., 1981, Effects of pentobarbital on contractile responses of feline cerebral arteries, *J. Cereb. Blood Flow Metab.* 1:437-440.

Garcia-Munoz, M., Young, S.J., and Groves, P.M., 1992, Presynaptic long-term changes in excitability of the corticostriatal pathway, *Neuroreport.* 3:357-360.

Hope, B.T., Michael, G.J., Knigge, K.M., and Vincent, S.R., 1991, Neuronal NADPH-diaphorase is a nitric oxide synthase, *Proc. Natl. Acad. Sci. USA* 88:2811-2814.

Hope, B.T., and Vincent, S.R., 1989, Histochemical characterization of neuronal NADPH-diaphorase, *J. Histoche. Cytochem.* 37:653-661.

Izumi, Y., Clifford, D.B., and Zorumski, C.F., 1992, Inhibition of long-term potentiation by NMDA-mediated nitric oxide release, *Science* 257:1273-1276.

Kita, H., Kosaka, T., and Heizmann, C.W., 1990, Parvalbumin-immunoreactive neurons in the rat neostriatum: a light and electron microscopic study, *Brain Res.* 536:1-15.

Kowall, N.W., Ferrante, R.J., Beal, M.F., Richardson, E.P., Jr., Sofroniew, M.V., Cuello, A.C., and Martin, J.B., 1987, Neuropeptide Y, somatostatin, and reduced nicotinamide adenine dinucleotide phosphate diaphorase in the human striatum: a combined immuno-cytochemical and enzyme histochemical study, *Neurosci.* 20:817-828.

Larkman, A., Stratford, K., and Jack, J., 1991, Quantal analysis of excitatory synaptic action and depression in hippocampal slices, *Nature* 350:344-347.

Lehmann, J., and Langer, S.Z., 1983, The striatal cholinergic interneuron: synaptic target of dopaminergic terminals, *Neurosci.* 10:1105-1120.

Malinow, R., and Tsein, R.W., 1990, Presynaptic enhancement shown by whole-cell recordings of long-term potentiation in hippocampal slices, *Nature* 346:177-180.

Marshall, P.E., and Landis, D.M.D., 1985, Huntington's disease is accompanied by changes in the distribution of somatostatin-containing neuronal processes, *Brain Res.* 329:71-82.

Mizutani, A., Saito, H., and Abe, K., 1993, Involvement of nitric oxide in long-term potentiation in the dentate gyrus in vivo, *Brain Res.* 605:309-311.

Moore, P.K., Babbedge, R.C., Wallace, P., Gaffen, Z.A., and Hart, S.L., 1993, 7-nitro indazole, an inhibitor of nitric oxide synthase, exhibits anti-nociceptive activity in the mouse without increasing blood pressure, *Br. J. Pharmacol.* 108:296-297.

O'Dell, T.J., Hawkins, R.D., Kandel, E.R., and Arancio, O., 1991, Tests of the roles of two diffusible substances in long-term potentiation: Evidence for nitric oxide as a possible early retrograde messenger, *Proc. Natl. Acad. Sci. USA* 88:11285-11289.

Palmer, R.M.J., Ferrige, A.G., and Moncada, S., 1987, Nitric oxide release accounts for the biological activity of endothelium-derived relaxing factor, *Nature* 327:524-526.

Rees, D.D., Palmer, R.M.J., and Moncada, S., 1991, Role of endothelium-derived nitric oxide in the regulation of blood pressure, *Proc. Natl. Acad. Sci. USA* 86:3375-3378.

Sakurada, O., Kennedy, C., Jehle, J., Brown, J.D., Carbin, G.L., and Sokoloff, L., 1978, Measurement of local cerebral blood flow with iodo-[^{14}C]-antipyrine, *Am. J. Physiol.* 234:487-499.

Schuman, E.M., and Madison, D.V., 1991, A requirement for the inter-cellular messenger nitric oxide in long-term potentiation, *Science* 254:1503-1506.

Takagi, H., Somogyi, P., Somogyi, J., and Smith, A.D., 1983, Fine structural studies on a type of somatostatin-immuno-reactive neuron and its synaptic connections in the rat neostriatum: a correlated light and electron microscopic study, *J. Comp. Neurol.* 214:1-16.

Thomas, C.L., Kelly, P.A.T., Ritchie, I.M., Sharkey, J., and Arbuthnott, G.W., 1993, Cerebrovascular autoregulation to hypertension induced by inhibition of nitric oxide synthesis, *Br. J. Pharmacol.* 108:129P.

Vincent, S.R. and Johansson, O., 1983. Striatal neurons containing both somatostatin and avian pancreatic polypeptide (APP)-like immuno-reactivities and NADPH diaphorase activity: A light and electron microscopic study, *J. Comp. Neurol.* 217:264-270.

Vincent, S.R., Johansson, O., Hokfelt, T., Skirboll, L., Elde, R.P., Terenius, L., Kimmel, J., and Goldstein, M., 1983, NADPH-diaphorase: A selective histochemical marker for striatal neurons containing both somatostatin and avian pancreatic polypeptide (APP)-like immunoreactivities, *J. Comp. Neurol.* 217:252-263.

Vincent, S.R., and Kimura, H., 1992, Histochemical mapping of nitric oxide synthase in the rat brain, *Neurosci.* 46:755-784.

TRANSSYNAPTIC MICROGLIAL REACTIONS IN THE QUINOLINIC ACID MODEL OF HUNTINGTON'S DISEASE

Rudolf Töpper[1,2], Jochen Gehrmann[1], Michael Schwarz[2], Frank Block[2], Johannes Noth[2] and Georg W. Kreutzberg[1]

[1]Dept. of Neuromorphology, Max-Planck-Institute of Psychiatry, 8033 Martinsried, Germany
[2]Dept. of Neurology, University of Aachen, 5100 Aachen, Germany

INTRODUCTION

Recent neuropathological studies have indicated that the degenerative process in patients with early Huntington's disease is restricted to a subpopulation of caudate neurons projecting to the internal pallidal segments (Albin et al., 1990). In more advanced cases of Huntington's disease degenerative changes are found in other parts of the basal ganglia as well as in cortical and thalamic areas (for review Richardson, 1990). The question what causes the progression of the neuropathology has not yet been answered. Does the hitherto unknown pathogenetic mechanism affect these extrastriatal areas at a later timepoint or do transneuronal degenerative processes play a role here?

Since patients with Huntington's disease are characterized by distinct neurophysiological abnormalities (Noth et al., 1984; Töpper et al., 1993) we recently did neurophysiological recordings in an established animal model of Huntington's disease, namely rats with intrastriatal quinolinic acid injections (DiFiglia, 1990). Although neuronal cell death is restricted to the striatum in these rats (Sanberg et al., 1989), amplitudes of somatosensory evoked potentials (SEP) were diminished (Schwarz et al., 1992). Since there are no direct anatomical connections between the striatum and the pathways which transmit somatosensory information, the modulatory effect of the basal ganglia on SEPs must therefore be transmitted over an indirect pathway. Our SEP recordings indicated that the modulation must take place either in the thalamus or in the cortex. A good candidate for gating basal ganglia input onto the somatosensory system is the reticular thalamic nucleus which receives collaterals of pallidothalamic fibres and which has GABAergic projections to the somatosensory relay nuclei of the thalamus.

Our hypothesis is in line with a recent model trying to explain the pathophysiology of the choreatic movement disorder of Huntington's disease (Albin et al., 1989). Chorea was attributed to a decreased activity of inhibitory pallidothalamic and nigrothalamic projections, which leads to a disinhibition of thalamic neurons in the ventroanterior and ventrolateral nuclei.

In the present study we were interested whether these postulated changes in activity in basal ganglia circuits are associated with morphologically detectable changes. Activated microglia served as the "activity marker". Activation of microglia is charaterized by morphological as well as immunological changes. In recent years a number of studies have established the important role of activated microglia in different experimental neuropathologies such as cerebral ischemia and immune-mediated lesions (for review Streit et al., 1988; Gehrmann et al. 1992a, b). It is known that injection of a neurotoxin leads to a local microglial reaction in the injected area (Coffey et al., 1990; Marty et al., 1991). The present study was undertaken to look for the presence of activated microglia in striatal projection areas of quinolinic acid injected rats.

METHODS

Male Wistar rats (10 weeks old, weighing 220-250 g) were anesthetized with chloral hydrate 4% (1 ml/100 g body weight) and mounted in a small animal stereotaxic frame. They received right intrastriatal injections of 240 nmol quinolinic acid in 1 µl phosphate buffered saline. Quinolinic acid was brought into solution in 0.1 M PBS to which NaOH was added to dissolve the toxin.

Quinolinic acid was injected at the following coordinates: AP +1.2, L +2.5, V -6.3, according to Watson and Paxinos. After survival times of 1, 3, 5, 10, 16, and 21 days three animals for each time point were killed by decapitation under deep ether anaesthesia, the brain quickly removed and immediately snap-frozen in high pressure CO_2. Cryostat sections (20 µm) were collected on gelatin-coated glass slides and stored at -80° C until immunocytochemical investigation.

In addition to three rats injected with quinolinic acid one control was used for each timepoint. The control animal received the same volume of solvent into the striatum.

Sections were made from the following planes: At the level of the striatum at the lesion site (+1.2), through the striatum caudal to the lesion (+0.2), through the globus pallidus and anterior thalamus (-1.5), through the medial thalamus and the entopeduncular nucleus (-2.8), and at the level of substantia nigra and the superior colliculus (-5.8).

Immunocytochemistry

The microglial response was studied using mouse monoclonal antibodies MUC 101 and MUC 102 (dilution 1:1600 and 1:1000, respectively), which are markers of microglial activation (Gehrmann and Kreutzberg, 1991). Cryostat sections which were fixed in formalin and acetone were incubated with the primary antibody at 4° C overnight. Antibody detection was carried out using biotinylated horse anti mouse antibody (Serotec) and the Vectastain ABC-Elite kit (Vector Labs) with 3,3'-diaminobenzidine as peroxidase substrate. Appropriate controls were made by replacing the primary antibody with PBS.

RESULTS

Sham Operated Controls

Resting microglia were stained weakly with MUC 101 and more intensly with MUC 102 throughout the rat brain. The injection of solvent resulted in a localized activation of microglial/macrophage cells along the needle tract and at the injection site.

Injection of Quinolinic Acid

From day 1 onwards strongly MUC 101- and MUC 102-immunoreactive cells were present throughout the entire ipsilateral striatum. On the contralateral side no increased staining on microglia was observed at any timepoint.

Remote microglial activation was found only in areas ipsilateral to the quinolinic acid injection. It was not observed in control animals. In the globus pallidus MUC 101- and 102-immunoreactive cells were detected as early as day 1. In the entopeduncular nucleus and in the substantia nigra pars reticulata microglial activation started at day 3. Whereas in the globus pallidus and the entopeduncular nucleus the microglial response was present until day 21, the last time point studied, activated microglia were no longer present in the substantia nigra at that time. In contrast to the striatum, where immunoreactive cells had a round, ameboid-like morphology suggesting a peripheral origin, immunoreactive cells in areas remote from the lesion had stout, hypertrophic processes, which are morphological features of activated microglia.

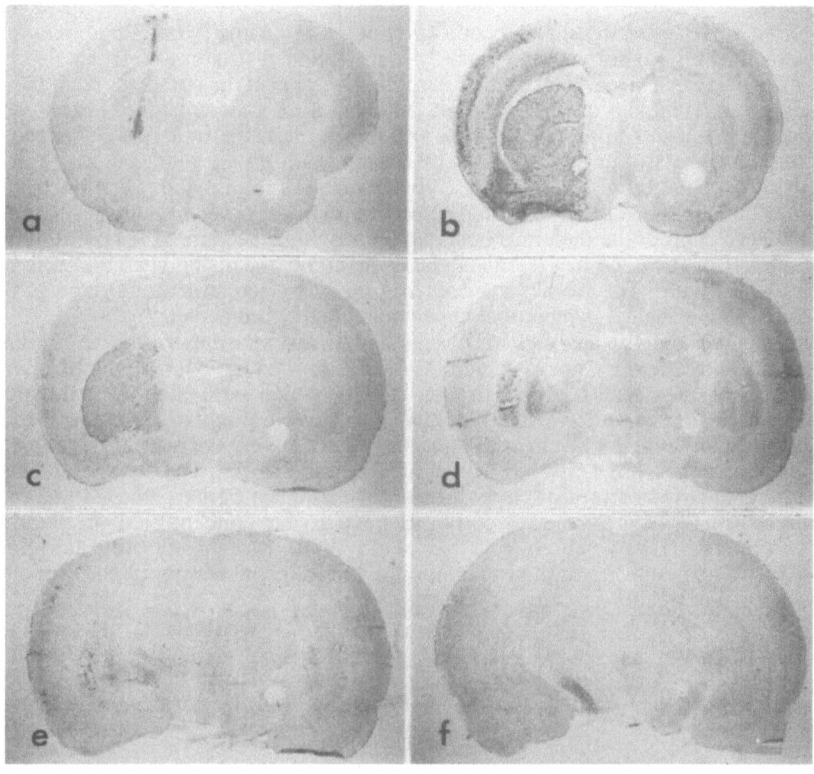

Figure 1. Overviews 5 days after injection of solvent (a) or quinolinic acid (b-f). MUC-101 immunoreactivity. After solvent injection the microglial reaction is localized around the needle tract (a). After quinolinic acid injection MUC-101 positive cells are found in the striatum itself (b-d), the overlying cortex (b), the globus pallidus (c,d), the entopeduncular nucleus (e), the ventroanterior and ventromedial nucleus of the thalamus (e), and the substamtia nigra (f). The hippocampus and the superior colliculus remained free of activated microglia. Magnification x16.8.

A microglial reaction was seen in the thalamus starting at day 5. Immunoreactive ramified hypertrophic microglial cells were mainly found in the anterior part of the thalamus. Within the thalamus, the ventromedial nucleus (VM) as well as the ventrolateral (VL) and the ventroanterior (VA) nucleus were affected in all rats. Immunoreactive microglia was also detected in the anterior part of the reticular thalamic nucleus (RT) in all rats. In addition, in some rats immunoreactive cells were found in the lateral and medial ventroposterior nucleus (VPL, VPM) and in the mediodorsal nucleus (MD).

Microglial activation could also be detected in the cortex overlying the quinolinic acid lesion. The symmetrical distribution of microglia around the needle tract makes it likely that diffusion of the toxin along the needle tract into the surrounding cortex is the cause of the cortical microglial activation. The laminar pattern of microglial activation in the cortex with a cluster in laminae 1 and 2 is paralleled by the differential laminar distribution of NMDA receptors in the rat cortex (Maragos et al., 1988) which are activated by the NMDA agonist quinolinic acid.

There was, however, no microglia activation in the hippocampus which is very sensitive to glutamatergic damage suggesting that diffusion of the toxin played a role only for the microglial activation in close vicinity to the injection site.

The superior colliculus which receives afferents from the substantia nigra remained free of activated microglia.

DISCUSSION

This study demonstrates that injection of quinolinic acid into the rat striatum is followed by a pronounced microglial reaction not only in the lesioned striatum itself but also in areas remote from the lesion site. Whereas the microglial reaction in the striatum and in the globus pallidus is observed by day 1, reactive microglia in the substantia nigra and the entopeduncular nucleus appears by day 3 and in the thalamus by day 5. This sequential activation along the striato-nigro/pallido-thalamic circuit projections suggests a neuronal mechanism for the microglial activation. Whereas the microglial activation in striatal first order projection areas can be explained as a reaction to the degeneration of striatal terminals, microglial activation in the thalamus occurs too early to be accounted for by transsynaptic degeneration. It can be hypothesized that hyperexcitation of the disinhibited thalamic neurons is a possible stimulus for the observed microglial activation, since microglia have been shown to become activated by neuronal hyperexcitation (Gehrmann et al., 1993).

Recent morphometric analyses of substantia nigra and prefrontal cortex in Huntington's disease brains have shown a neuronal cell loss which correlated with the degree of cell loss in the caudate (summarized in Richardson, 1990), which was taken as indication that transneuronal degenerative processes may play a role in Huntington's disease. The present finding, that in an animal model of Huntington's disease very early after degeneration of striatal projection neurons activated microglia are found in first and second order striatal projection areas supports this hypothesis. Neuropathological studies in early or presymptomatic Huntington's disease brains have so far concentrated on pathological changes in the caudate nucleus (e.g. Albin et al., 1990). Since activated microglia are a sensitive marker for subtle neuronal pathology it would be interesting to perform a similar analysis of the microglial response in these brains.

Acknowledgments

We appreciate the technical assistance of Dietmute Büringer, Irmtraud Milojevic, and Karin Brückner.

REFERENCES

Albin, R.L., Young, A.B., and Penney, J.P., 1989, The functional anatomy of basal ganglia disorders, *Trends Neurosci.* 12:366-375.

Albin, R.L., Young, A.B., Penney, J.P., Handelin, B., Balfour, R., Anderson, K.D., Markel, D.S., Tourtelotte, W.W., and Reiner, A., 1990, Abnormalities of striatal projection neurons and N-Methyl-D-Aspartate receptors in presymptomatic Huntington's disease, *New Engl. J. Med.* 322:1293-1298.

Coffey, P.J., Perry, V.H., and Rawlins, J.N.P., 1990, An investigation into early stages of the inflammatory response following ibotenic acid-induced neuronal degeneration, *Neuroscience* 35:121-132.

DiFiglia, M., 1990, Excitotoxic injury of the neostriatum: a model for Huntington's disease, *Trends Neurosci.* 13:286-289.

Gehrmann, J., and Kreutzberg, G.W., 1991, Characterisation of two new monoclonal antibodies directed against rat microglia, *J. Comp. Neurol.* 313:409-430.

Gehrmann, J., Gold, R., Linington, C., Lannes-Vieira, J., Wekerle, H., and Kreutzberg, G.W., 1992a, Spinal cord microglia in experimental allergic neuritis: evidence for fast and remote activation, *Lab. Investig.* 67:100-113.

Gehrmann, J., Banati, R., and Kreutzberg, G.W., 1992b, The reactive microglia: involvement in diseases of the nervous system with special reference to the aquired immune deficiency syndrome (AIDS), *in:* "HIV infection of the central nervous system," S. Weis and H. Hippius, eds., Hogrefe and Huber, Seattle, Toronto, Göttingen, Bern.

Gehrmann, J., Mies, G., Bonnekoh, P., Banati, R., Iijima, T., Kreutzberg, G.W., and Hossmann, K.-A., 1993, Microglial reaction in the rat cerebral cortex induced by cortical spreading depression, *Brain Pathol.* 3:11-18.

Maragos, W.F., Penney, J.B., and Young, A.B., 1988, Anatomic correlation of NMDA and ^{3}H-TCP-labeled receptors in rat brain, *J. Neurosci.* 8:493-501.

Marty, S., Dusart, I., and Peschanski, M., 1991, Glial changes following an excitotoxic lesion in the CNS-I. microglia/macrophages, *Neuroscience* 45:529-539.

Noth, J., Engel, L., Friedemann, H.H., and Lange, H.W., 1984, Evoked potentials in patients with Huntington's disease, *Electroenceph. clin. Neurophysiol.* 59:134-141.

Richardson, E.P., 1990, Third Dorothy S. Russell Memorial lecture: Huntington's disease: some recent neuropathological studies, *Neuropathol. and Appl. Neurobiol.* 16:451-460.

Sanberg, P.R., Calderon, S.F., Giardano, M., Tew, J.M., and Norman, A.B., 1989, The quinolinic acid model of Huntington's disease: locomotor abnormalities, *Exp. Neurol.* 105:45-53.

Schwarz, M., Block, F., Töpper, R., Sontag, K-H, and Noth, J., 1992, Abnormalities of somatosensory evoked potentials in the quinolinic acid model of Huntington's disease: evidence that basal ganglia modulate sensory cortical input, *Ann. Neurol.* 32:358-364.

Streit, W., Graeber, M.B., and Kreutzberg, G.W., 1988. Functional plasticity of microglia: A review, *Glia* 1:301-307.

Töpper, R., Schwarz, M., Podoll, K., Dömges, F., and Noth, J., 1993, Absence of frontal somatosensory evoked potentials in Huntington's disease, *Brain* 116:87-101.

NEURONAL NECROSIS AFTER FOREBRAIN ISCHEMIA AFFECTS

GABAERGIC NEURONS IN THE SUBSTANTIA NIGRA AND STRIATUM

Frank Block[1], Maria Sieklucka[2], Christine Heim[3] and Karl-Heinz Sontag[4]

[1]Dept. Neurology, Univ Essen, Hufelandstr. 55, D-4300 Essen, FRG
[2]Dept. Pharmacology, Lublin, Poland
[3]Dept. Psychiatry, D-3400 Goettingen, FRG
[4]Max-Planck-Institute for Experimental Medicine, D-3400 Goettingen, FRG

INTRODUCTION

Transient forebrain ischemia is known to produce selective neuronal cell death in the CA1 sector of the hippocampus in rats (Pulsinelli et al., 1982; Schmidt-Kastner and Freund, 1991). As the cell death within the CA1 sector of the hippocampus takes more than one day to develop it has been described as delayed neuronal cell death (Kirino, 1982). Neuronal damage also occurs in the striatum following global ischemia (Pulsinelli et al., 1982; Smith et al., 1984). Furthermore, neuronal cell death or even infarction have been observed in the substantia nigra pars reticulata (Smith et al., 1984; Araki et al., 1990).

Excitatory amino acids which are excessively released during ischemia (Benveniste et al., 1984) appear to be one pathogenetic factor for neuronal damage following cerebral ischemia. Using the the excitotoxic index Globus et al. (1991) described an imbalance between excitatory and inhibitory amino acids which might account for neuronal vulnerability to ischemia. This finding and the fact that GABAergic substances provide protection against neuronal damage (Johansen and Diemer, 1991; Shuaib et al., 1992) suggest that the inhibitory transmitter GABA is also involved in the pathogenesis of post-ischemic neuronal damage.

In the present study neuronal damage and GABA content were evaluated in substantia nigra, striatum and hippocampus at 4 different intervals after transient global ischemia to further characterize the pathogenetic role of GABA in neuronal damage following forebrain ischemia

METHODS

Cerebral ischemia was induced by four-vessel-occlusion (4VO) in male Wistar rats (aged 15 weeks) anesthetized with pentobarbital (60 mg/kg i.p.). Both vertebral arteries were occluded by electrocauterization.The animals were allowed to recover for 24 hours with free access to water but not food. On the next day, the carotid arteries were exposed under halothane/nitrous oxide anesthesia and were occluded for 20 min using microvascular clamps. Subsequently, both clamps were removed and both arteries were inspected for immediate reperfusion. During occlusion, the head temperature (monitored by a thermoprobe located in the temporal muscle) was kept at 37°C with a heating lamp. For control, in sham operated animals both vertebral arteries were cauterized in pentobarbital anesthesia and both common carotid arteries were exposed but not clamped in halothane anesthesia the following day.

For histology, animals were sacrificed under deep anaesthesia by perfusion-fixation through the heart 1, 3, 5 and 9 days after operation. Following a rinse with Ringer-solution 200-300 ml neutral-buffered 4% formaldehyde was infused. After the brains had been removed from the skull and stored in the same fixative, they were embedded in paraffin and 5 μm frontal sections at the level of the dorsal hippocampus, striatum and substantia nigra were prepared according to a stereotaxic atlas (Paxinos and Watson, 1982). Sections were stained with cresyl-violet for conventional histology.

For determination of regional GABA content rats were decapitated 1, 3, 5 and 9 days after operation. The heads were immediately placed in liquid nitrogen for 6 sec. The brains were removed and the different structures dissected on an ice plate. GABA content was measured according to the method of Lowe et al. (1958) and modified by Sutton and Simmonds (1974) via a fluometric assay which depends on the formation of a fluorescent product from the reaction between GABA and ninhydrin at alkaline pH and in the presence of glutamate. GABA content is expressed in μmol/g of fresh tissue.

RESULTS

One day after ischemia neuronal damage was observed in striatum where mostly medium-seized neurons were affected (Fig. 1A). Neuronal damage was aslo present in the subsatantia nigra pars reticulata at the first day following ischemia but not in the substantia nigra pars compacta (Fig. 1B). The damage in striatum and substantia nigra pars reticulata did not increase any further at the following days. In the hippocampus, some pyknotic cells were found in the CA1 sector one day after ischemia, from the third day on after 4VO a marked neuronal necrosis in the CA1 sector was present which did not progress any further (Fig. 1C).

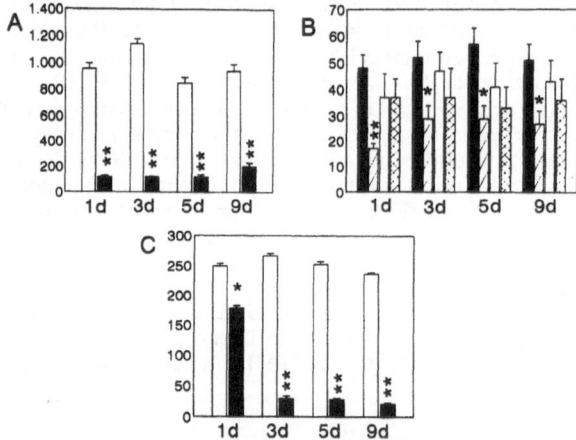

Figure 1. Cell counts of striatum (A) and CA1 sector of hippocampus (C) of controls (black bars) and ischemic animals (white bars) and of substantia nigra (B) of controls (black bars for pars reticulata, white bars for pars compacta) and ischemic animals (hatched bars for pars reticulata, crossed bars for pars compacta) 1, 3, 5 and 9 days after operation. Significances: * p < 0.01; ** p < 0.001 versus control, Student's t test.

The GABA content was reduced in the substantia nigra from the first day on after 4VO whereas in the hippocampus and striatum no significant changes cold be observed at the first day (Fig. 2). From the third day on after 4VO a redcued GABA content was measured in the striatum. In contrast, no changes in GABA content were found in the hippocampus throughout 9 days following 4VO.

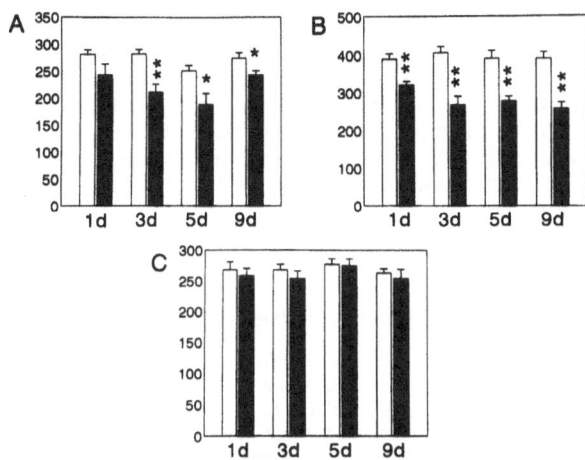

Figure 2. GABA content (ug GABA/g tissue) within striatum (A), substantia nigra (B) and hippocampus (C) of controls (black bars) and ischemic animals (white bars) 1, 3, 5 and 9 days after operation. Significances: * p < 0.05; ** p < 0.01 versus control, Student's t test.

DISCUSSION

The present results demonstrate that neuronal damage following cerebral ischemia in substantia nigra pars reticulata and striatum is accompanied by a decrease in GABA content. The ischemic cell injury in striatum affects mostly medium-seized, projection neurons whereas the large cell types which are interneurones are spared (Francis and Pulsinelli, 1982). In substantia nigra pars reticulata the output neurones to thalamic nuclei degenerate. Both, striatal and nigral projection neurons are GABAergic (Albin et al., 1989). Thus, the observed decrease in GABA content appears to reflect the cell death of these GABAergic neurons. In contrast, the delayed neuronal death in the CA1 sector of hippocampus is not accompanied by a decrease in GABA content. Using immunohistochemistry for GAD (Nitsch et al., 1989) and using 3H-GABA binding (Johansen et al., 1991) it could be demonstrated that GABAergic interneurons which provide the GABAergic input into the CA1 sector are preserved following global ischemia. The fact that the GABA content in the hippcampus is not altered by ischemia seems to reflect the survival of the GABAergic neurons.

Comparison of the pattern of vulnerability of hippocampus, substantia nigra and striatum reveals some similarities: (a) all nuclei recieve a excitatory, glutamatergic input (Collingridge and Lester, 1989) and (b) the vulnerable neurons are projection neurons (Walaas, 1983; Albin et al., 1989). Besides these similarities a striking difference exists: in substantia nigra and striatum the cell death occurs within 24 hrs and affects GABAergic neurons whereas in hippocampus the damage develops over a period of 72 hrs and the GABAergic neurons are spared. The preservation of these inhibitory GABAergic neurons may counteract the excitation induced by ischemia for a certain period and by this it may be one factor contributing to the delay in hippocampal neuronal damage. This notion is supported by the finding that enhancement of GABAergic transmission reduces neuronal damage in the CA1 sector of hippocampus following ischemia (Johansen and Diemer, 1991).

Transient occlusion of both common carotid arteries in normotensive rats (BCCA) is characterized by transient metabolic changes like increase in extracellular lactate and decrease in local PO2 (Block et al., 1993). Following this oligemic situation no consistent neuronal damage can be observed but longlasting functional changes like deficits in spatial learning (Jaspers et al., 1990) and an increase in inositol phosphate metabolism at the hippocampal level (Sieklucka et al., 1991). Up to 3 months following BCCA an increased GABA content was measured substantia nigra, hippocampus and frontal cortex (Sieklucka et al., 1992). This increase in GABA content is accompanied by a decreased susceptibility to bicuculline-induced seizures and a delayed appearance of seizure activity after hippocampal kindling

(Bortolotto et al., 1991). In contrast to the present data which demonstrate a decrease in GABA content and a neuronal necrosis in response to forebrain ischemia this moderate reduction in cerebral blood flow results in an increase in GABA content without neuronal damage. On the basis of the findings in these two models of reduced cerebral blood flow it appears that GABAergic transmission and/or neurons are very susceptible to reduced cerebral blood flow and that the observed long-term changes following a reduction in cerebral blood flow are at least partly due to these changes in GABAergic transmission.

REFERENCES

Albin, R.L., Young, A.B., and Penney, J.B., 1989, The functional anatomy of basal ganglia disorders, *TINS* 12:366-375.

Araki, T., Inoue, T., Kato, H., Kogure, K., and Murakami, M., 1990, Neuronal damage and calcium accumulation following transient cerebral ischemia in the rat, *Mol. Chem. Neuropathol.* 12:203-213.

Benveniste, H., Drejer, J., Schousboe, A., and Diemer, N,H., 1984, Elevation of the extracellular concentrations of glutamate and aspartate in rat hippocampus during transient cerebral ischemia monitored by intracerebral microdialysis, *J. Neurochem.* 43:1369-1374.

Block, F., Sieklucka, M., Schmidt-Kastner, R., Heim, C., and Sontag, K.-H., 1993, Metabolic changes during and after transient clamping of carotid arteries in normotensive rats, *Brain Res. Bull.* in press.

Bortolotto, Z., Heim, C., Sieklucka, M., Block, F., Sontag, K.-H., and Cavalheiro, E.A., 1991, Effects of bilateral clamping of carotid arteries on hippocampal kindling in rats, *Physiol. Behav.* 49:667-672.

Collingridge, G.L., and Lester, R.A., 1989, Excitatory amino acid receptors in the vertebrate central nervous system, *Pharmacol. Rev.* 40:143-210.

Francis, A., and Pulsinelli, W.A., 1982, The response of GABAergic and cholinergic neurons to transient cerebral ischemia, *Brain Res.* 243:271-278.

Globus, M.Y., Ginsberg, M.D., and Busto, R., 1991, Excitotoxic index - a biochemical marker of selective vulnerability, *Neurosci. Lett.* 127:39-42.

Jaspers, R.M.A., Block, F., Heim, C., and Sontag K-H, 1990, Spatial learning is affected by transient occlusion of common carotid arteries (2VO):comparison of behavioural and histopathological changes after '2VO' and 'four-vessel-occlusion' in rats, *Neurosci. Lett.* 117:149-153.

Johansen, F.F., and Diemer, N.H., 1991, Enhancement of GABA neurotransmission after cerebral ischemia in the rat reduces loss of hippocampal CA1 pyramidal cells, *Acta Neurol. Scand.* 84:1-6.

Johansen, F.F., Christensen, T., Jensen, M.S., Valente, E., Jensen, C.V., Nathan, T., Lambert, J.D., and Diemer, N.H., 1991, Inhibition in postischemic rat hippocampus: GABA receptors, GABA release, and inbitory postsynaptic potentials, *Exp. Brain Res.* 84:529-537.

Kirino, T., 1982, Delayed neuronal death in the gerbil hippocampus following ischemia, *Brain Res.* 239:57-69.

Nitsch, C., Goping, G., and Klatzo, I., 1989, Preservation of GABAergic perikarya and boutons after transient ischemia in the gerbil hippocampal CA1 field, *Brain Res.* 495:243-452.

Lowe, J.P., Robinson, E., and Eyerman, G.S., 1958, The fluorimetric measurements of glutamic decarboxylase and its distribution in brain, *J. Neurochem.* 3:8-18.

Paxinos, G., and Watson, C., 1982, "The Rat Brain in Stereotaxic Coordinates," Academic Press, Sydney.

Pulsinelli, W.A., Brierley, J.B., and Plum, F., 1982, Temporal profile of neuronal damage in a model of transient forebrain ischemia, *Ann. Neurol.* 11:491-498.

Schmidt-Kastner, R., and Freund, T.F., 1991, Selective vulnerability of the hippocampus in brain ischemia, *Neurosci.* 40:599-636.

Shuaib, A., Ijaz, S., Hasan, S., and Kalra, J., 1992, Gamma-vinyl GABA prevents hippocampal and substantia nigra reticulata damage in repetitive transient forebrain ischemia, *Brain Res.* 590:13-17.

Sieklucka, M., Heim, C., Block, F., and Sontag, K.-H., 1992, Transient reduction of cerebral blood flow leads to longlasting increase in GABA content in vulnerable structures and decreased susceptibility to bicuculline induced seizures, *J. Neural Transm. [GenSect]* 88:87-94.

Sieklucka, M., Heim, C., Sontag, K-.H., and Osborne, N.N., 1991, Transient occlusion of carotid arteries increases formation of inositol phosphate: Evidence for a specific effect on alpha-1-receptors, *Neurochem. Int.* 18:175-189.

Smith, M.-L., Auer, R.N., and Siesjî, B.K., 1984, The density and distribution of ischemic brain injury in the rat following 2-10 min of forebrain ischemia, *Acta Neuropathol.* 76:253-264.

Sutton, J., and Simmonds, M., 1974, Effects of acute and chronic pentobarbitone on the gamma-aminobutyric acid system in rat brain, *Biochem. Pharmacol.* 23:1801-1808.

Walaas, I., 1983, The hippocampus, *in*: "Chemical Neuroanatomy", P.C. Emson, ed., Raven Press, New York, pp. 337-358.

INTERNAL COMPOSITION OF STRIATAL GRAFTS:
LIGHT AND ELECTRON MICROSCOPY

Deborah J. Clarke[1], Klas Wictorin[2], Stephen B. Dunnett[3] and
J. Paul Bolam[4]

[1]Dept. Human Anatomy & [4]MRC Anat. Neuropharmacol. Unit,
  Dept. Pharmacol., Oxford; [2]Dept. Med. Cell Res., Lund, Sweden &
[3]Dept. Exp. Psychol., Cambridge

INTRODUCTION

Grafts of fetal rat striatal primordium implanted into a host rat striatum, previously lesioned using the excitotoxins kainic, ibotenic or quisqualic acids, develop features reminiscent of adult neostriatum with respect to cell types, transmitters and receptors (Isacson et al., 1986; Clarke et al., 1988; DiFiglia et al.,1988; Graybiel et al., 1989; Helm et al., 1990; Liu et al., 1990; Zhou, 1990). Evidence is also accruing for the formation of afferent and efferent graft-host connections (Pritzel et al., 1986; Clarke et al., 1988; Wictorin et al., 1989, 1990a and b, 1991; Xu et al., 1990). A striking feature of the grafts is, however, the clearly heterogeneous distribution of many of the markers, with areas or "patches" containing high concentrations of eg. AChE or dopaminergic afferents separated by "non-patch" regions with low concentrations of these markers. Simple explanations for this heterogeneity such as the intermingling of neurons of both graft and host origin or the presence of large areas of non-neuronal tissue, have been ruled out using tritiated thymidine pre-labelling of either the donor tissue or recipient animals (Liu et al., 1990) and cross-species grafting followed by identification of the graft tissue with a species-specific neuronal marker (Wictorin et al., 1991). Furthermore, it seems unlikely that this "patch/non-patch" arrangement could be explained by the well-known organization of the mature striatum of striosomes and matrix (Graybiel, 1990), since in the normal rat striatum this patterning is not as easily identifiable using the marker AChE or staining for dopaminergic afferents as in the striatal grafts and, moreover, markers of both the adult striosome and matrix compartments have been located together within the grafts' "patch" regions (Graybiel et al., 1989).

Two other theories have been proposed to explain these heterogeneities within striatal grafts. First, it has been suggested that the "patch/non-patch" arrangement could be due to a retention of immature features normally only associated with fetal tissue. Thus, Isacson et al. (1987) demonstrated correspondence between areas of high AChE staining and areas with dense met-enkephalin immunoreactivity or dense dopamine receptor density; a staining pattern reminiscent of the immature striatum during the early post-natal period. Further support for this theory comes from Golgi-impregnation studies (McAllister et al., 1985) and experiments in which electrophysiological recordings have been made (Rutherford et al., 1987). The second theory puts forward the possibility that the implants actually consist of a mixture of different tissue types, due to the inclusion in the graft tissue of cells that are other than striatal precursor cells, found within the ganglionic eminences. This idea was first reported by Isacson et al. (1987) and Walker et al. (1987) and verified by DiFiglia et al. (1988) who all found pallidal-like structures within their striatal grafts. Graybiel et al. (1989)

used somatostatin- and calbindin-immunostaining and, within AChE-poor (i.e. "non-patch") zones, identified neurons normally found within globus pallidus, basolateral amygdala and ventrolateral cortex. In our own studies we have used DARPP-32 as a marker of striatal tissue, since this marker is evenly distributed within the normal striatum (Ouimet et al., 1984) and have found "patches" rich in DARPP-32 immunoreactivity surrounded by "non-patch" areas which do not contain any of the densely DARPP-32-positive cells typically found in neostriatum (Wictorin et al., 1989).

In the present light and electron microscopical study we have used tyrosine hydroxylase (TH) or DARPP-32 immunocytochemistry to identify the "patches" in combination with two anatomical markers to further characterize the cellular composition of the grafts. Thus, glutamate decarboxylase (GAD) immunocytochemistry was used to examine the "patch" and "non-patch" zones since the staining patterns of striatum, pallidum and cortex are highly distinctive with respect to this transmitter. Golgi-impregnation was also used were used to assess the morphology of the neuronal types present within the two graft regions.

METHODS

Fourteen Sprague-Dawley (ALAB, Stockholm, Sweden) and 12 female Wistar adult rats (Olac, Bicester, UK) received unilateral ibotenic acid lesions (14μg/rat) of the right striatum. Two-four weeks later suspensions of E14-15 fetal rat striatal primordium were grafted into the lesioned striatum. All surgery was performed under Equithesin anaesthesia (0.3ml/100g). Lesioning and graft preparation techniques have been described in detail elsewhere (Wictorin et al., 1989). Following 6-8 months survival, the rats were perfused through the ascending aorta with a buffered fixative containing 4% paraformaldehyde and 0.05% glutaraldehyde in 0.1M phosphate buffer. The brains were sectioned at 70μm on a vibrating microtome and processed to reveal tyrosine hydroxylase (TH; Eugene-Tech., 1:1500, raised in rabbit; 12 rats) or DARPP-32 (gift of Dr. P. Greengard; 1:20,000, raised in mouse; 14 rats) immunoreactivity using a standard avidin-biotin-peroxidase protocol and diaminobenzidine (DAB) as the chromagen in the visualisation reaction, as described previously (Hsu et al., 1981; Wictorin et al., 1989). After the DAB visualisation step, some of the TH-immunoreactive sections were further processed for GAD immunocytochemistry (gift of Dr. W. Oertel.1:1000, raised in sheep) using benzidine dihydrochloride (BDHC) as the chromagen in the visualisation step (Levey et al., 1986). Alternatively, in another series of sections from 5 specimens, the GAD-immunoreactivity was first visualised using DAB followed sequentially by visualisation of the DARPP-32-immunoreactivity using BDHC. All sections were then treated with 1% osmium tetroxide in 0.1M phosphate buffer and selected sections which had received a single immunostain using either TH or DARPP-32 immunocytochemistry were processed for Golgi-impregnation. These sections were immersed in a 3.5% potassium dichromate solution for a minimum of 4 hours and then assembled between glass microscope slides and placed in beakers of 1.5% silver nitrate, according to the single section Golgi method of Izzo et al. (1987). During the silver nitrate impregnation stage the sections were viewed at intervals under the light microscope to observe the progress of the Golgi-impregnation. The sections were removed from the silver nitrate when sufficient neurons were Golgi-impregnated. These sections were then illuminated using a fibre-optic light source and subsequently gold-toned (Bolam and Ingham, 1990). Finally, all material was routinely dehydrated, with the inclusion of 1% uranyl acetate at the 70% ethanol stage, and embedded in epoxy resin (Durcupan ACM, Fluka) for light and electron microscopy.

The material was examined in the light microscope and neurons of interest drawn using a camera lucida and photographed. The TH and DARPP-32 immunocytochemistry served as good markers for the "patches" observed within the striatal grafts. Thus, Golgi-impregnated neurons could be categorized as being "patch" or "non-patch". Neurons outside TH or DARPP-ir patches could be further classified by examination of their ultrastructural morphology as being pallidal-like or cortical-like. The patterns of GAD-immunoreactivity were also examined inside and outside the TH or DARPP-immunoreactive (ir) patches and compared with that seen in control striatum, globus pallidus and a variety of cortical regions. These examinations were carried out at both the light and electron microscope levels.

RESULTS

Surviving grafts were found in all animals and they all showed good TH or DARPP-immunoreactivity which highlighted the "patch" and "non-patch" regions of the grafts (Figure 1A, B). In double-labelled sections the 2 chromogens, DAB and BDHC, were readily distinguishable from each other both at the light and electron microscopic levels; DAB giving its characteristic brown colour whereas BDHC appeared blue-black and crystalline.

GAD Immunoreactivity

GAD-immunoreactivity was present both inside and outside the regions defined as patches, although the patterns of staining appeared markedly different. Inside the patches, the GAD immunoreactivity displayed a diffuse distribution of fine fibre staining punctuated by many medium-sized GAD-ir neuronal perikarya (Figure 1C). Outside the patches however, the fibres were also found ensheathing large dendrites (Figure 1E) or forming pericellular nets around both unstained and GAD-ir medium-and large-sized neurons (Figure 1D). The GAD-ir perikarya were generally larger than those seen inside patches and obvious indentations in the nuclear membrane could be seen at the EM level (Figure 2B).

At least 3 populations of GAD-ir neurons could be detected at the electron microscopic level within the striatal grafts. Of these, 2 were medium-sized and were found predominantly within the patch areas. One class had a smooth nuclear envelope and was reminiscent of the medium-sized spiny projection neuron of the striatum (Figure 2A), whereas the other showed deep nuclear indentations and appeared similar to the population of striatal GABA-ergic interneurons (Figure 2B). The other type was found exclusively outside the TH-ir or DARPP-ir patches and comprised larger neurons with indented nuclei, some of which were surrounded by multiple synaptic inputs many of them from GAD-ir boutons (Figure 3A). These features are typical of cortical and pallidal tissue. Occasional axo-axonic contacts between GAD-ir and unlabelled terminals were observed (Figure 3C), another feature consistent with the presence of cortical tissue. Ensheathment of presumed large dendrites with GAD-positive boutons was confirmed at the electron microscopical level, with numerous GAD-ir boutons seen to form symmetric synapses with dendritic shafts which were often interspersed with synaptic boutons forming asymmetrical specializations. Some of the latter displayed subjunctional bodies, a feature that often occurs in globus pallidus (Figure 3B).

Morphological Cell Types/Golgi-impregnation

Golgi-impregnation tissue that was sufficiently good to warrant further study, was achieved in five animals. From these animals a total of 46 Golgi-impregnated neurons were drawn, of which 31 were found within TH-ir patches and 15 were outside patches. Seven of the "non-patch" neurons had features typical of cortical pyramidal neurons (Figure 4A-C), whereas 8 appeared similar to neurons of pallidal origin (Figure 4H-K). Of the 31 "patch" neurons (Figure 4D-G), 5 were both DARPP-ir and Golgi-impregnated, thus confirming their striatal origin.

Two of the neurons that had a pallidal-like morphology at the LM level, as revealed by Golgi-impregnation, were examined at the EM level. Their ultrastructural characteristics were typical of pallidal neurons, having an ovoid soma with deep nuclear indentations and the primary and secondary dendrites ensheathed by synaptic boutons. Both symmetric and asymmetric synaptic specializations were present.

DISCUSSION

The presence of patches within the grafts was readily apparent using either TH or DARPP-32 immunoreactivity as markers. The patches occupied approximately one-third to one-half of the cross-sectional area of the implants, similar to previous reports (see review by Wictorin, 1992) and suggest a heterogeneous composition of the tissue. However, in contrast, the GAD-immunoreactivity appeared uniform in distribution; only the staining patterns observed at high magnification varied between the patch and non-patch areas.

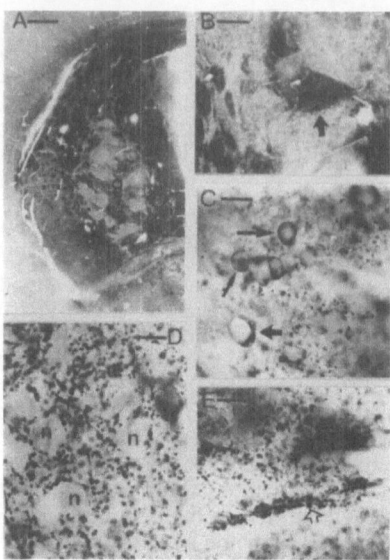

Figure 1. Light micrographs of striatal grafts showing the patchy appearance revealed with TH immunoreactivity and the GAD immunoreactivity, both inside and outside those patches. A. Size and position of the striatal graft (g) stained with TH-immunoreactivity. B. Higher magnification of the TH-ir patches (one patch indicated by an arrow). C. Graft GAD-ir neurons. In the patches, one type is the medium spiny neuron (double arrow), whilst the other type is an aspiny interneuron (arrow). Outside the patches, a larger neuron is present which resembles those of cortical or pallidal origin (short arrow). D. Neurons (n) in non-patch regions surrounded by pericellular nets of GAD-ir boutons. E. Large dendrite (arrow) in non-patch region ensheathed by GAD-ir boutons. Scale bars: A:750µm. B:200µm. C-E:15µm.

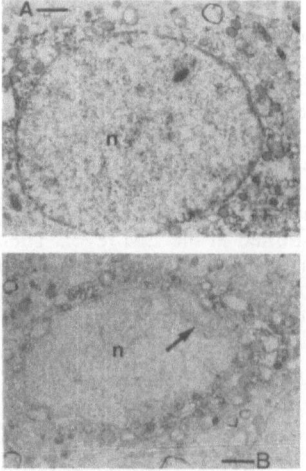

Figure 2. Electron micrographs of the 2 types of GAD-ir neuron found inside patches. A. Example of a weakly stained neuron of medium-sized densely spiny type with sparse cytoplasm and smooth nuclear envelope. n=nucleus. B. Example of a GAD-ir aspiny interneuron with moderate amount of cytoplasm and deep indentations (arrow) in nuclear membrane. n=nucleus. Scale bars: A & B:1.5µm.

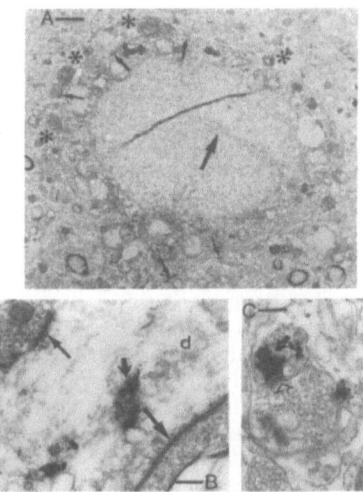

Figure 3. Electron micrographs of GAD-ir structures in non-patch regions. A. GAD-ir neuron ensheathed by synaptic input (curved arrows) and GAD-ir boutons (asterisks). B. GAD-ir dendrite containing crystals of BDHC reaction product (thick arrow) receiving synaptic inputs (arrows) associated with dense subjunctional bodies. C. GAD-ir bouton in probable axo-axonic contact (arrow) with an unlabelled bouton. Scale bars: A:1.5μm. B & C:0.25μm.

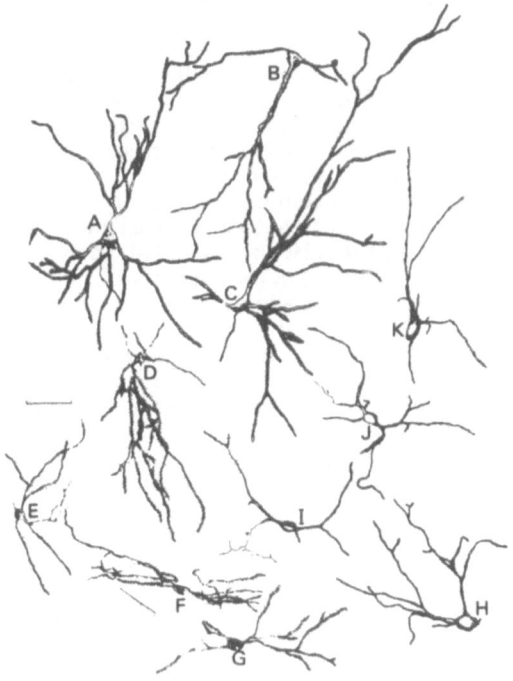

Figure 4. Camera lucida drawings of Golgi-impregnated neurons in patch and non-patch areas.
A-C. Examples of neurons resembling cortical pyramidal neurons in non-patch areas.
D-G. Examples of patch neurons, mainly of medium size spiny type.
H-K. Examples of non-patch neurons resembling neurons of pallidal origin. Scale bar: 50μm.

Roberts and DiFiglia (1990) reported similar findings with respect to their GAD-ir neurons whereas those immunoreactive for enkephalin were organized into dense clusters. Within the patches, 2 classes of GAD-ir neuron were distinguishable surrounded by a plexus of GAD-ir fibres. These 2 types of GAD-ir neuron have been described in normal striatum; one represents the major striatal projection neuron, the medium-sized densely spiny neuron whilst the other is a GABAergic interneuron (Bolam et al., 1983, 1985; Oertel and Mugnaini, 1984; Kita and Kitai, 1988). The GAD-ir fibres within the patch regions of the grafts may represent axon collaterals of these 2 neurons or, alternatively, some may represent an input from either the host striatum or the non-patch regions of the graft.

The GAD-ir staining pattern in the non-patch regions was reminiscent of that seen in both globus pallidus and areas of neocortex. Globus pallidus is one of the brain areas richest in the transmitter GABA with the vast majority of its neurons being GABAergic (Smith et al., 1987) and its major input from neostriatum also containing that transmitter (Graybiel, 1990). Pallidal neurons receive abundant synaptic inputs to their somata and primary dendrites, many of which are themselves GABAergic (Ingham et al., 1988), a feature illustrated in this chapter. Another feature typical of pallidal tissue, but also found to a small degree in striatum, is the presence of large dendrites ensheathed in synaptic inputs (Fox et al., 1974; Totterdell et al., 1984), a feature frequently observed within the non-patch areas. The proximity of these pallidal-like areas to the patch (striatal-like) zones may suggest the presence of mini 'striato-pallido-striatal' systems within the grafts with the GAD-ir neurons of the patches providing a GABAergic input to the pallidal-like non-patch regions and reciprocal inputs contributing to the dense GAD-ir fibre network of the patches. This theory is supported by the observation that striatal grafts are very rich in internal projections (Wictorin et al., 1989; Xu et al., 1992).

The appearance of cell bodies ensheathed in GAD-ir boutons is more commonly associated with cortical areas (Freund et al., 1983) where GAD-ir boutons of GABAergic interneurons contact the perikarya of pyramidal cells. However, a similar pattern is also seen in globus pallidus, as described above. Another ultrastructural feature associated with GABAergic cortical interneurons is the presence of axo-axonic contacts formed between their axon terminals and other unlabelled axonal elements. These features were, again, seen in the non-patch regions of the grafts and are illustrated in this chapter. Since these cortical neurons do not normally contribute to striatal circuitry, it is perhaps unlikely that such neurons are responsible for a significant proportion of the GAD-ir innervation of the patches. These results thus demonstrate the presence of tissue of both cortical and pallidal origin, as assessed from their patterns of GAD-immunoreactivity at both the LM and EM levels, within the non-patch areas of the striatal grafts.

Further characterization of these cellular constituents was revealed using Golgi-impregnation. Previous studies using this technique in striatal grafts (McAllister et al., 1985; DiFiglia et al., 1988; Clarke et al., 1988; Helm et al., 1990) have reported the presence of neurons which closely resemble striatal neurons. These include the medium-sized densely spiny neuron and several interneuron populations, all of which have been described within normal striatum (Chang et al., 1982). Similar neuron types were found in this study within the patch regions. In the non-patch regions, however, neurons which were not typical of striatal tissue were observed, most notably those resembling cortical pyramidal neurons (Ramon y Cajal, 1911). The other illustrated group of neurons found within the non-patch regions resembled those found in globus pallidus (Fox et al., 1974; Totterdell et al., 1984; Ingham et al., 1988), although neurons of similar morphology do occur within neostriatum itself and represent a second class of striatonigral projection neuron (Bolam et al., 1981).

Many of the studies using Golgi-impregnation of striatal grafts have suggested that some of the neurons that display morphologies inconsistent with those of normal striatum represent immature striatal neurons. Our results provide evidence that this is not necessarily the case as several non-patch neurons displayed features of typical mature cortical pyramidal neurons. It is thus more logical to assume that these heterogeneities in striatal grafts are not concerned with the developmental state of the tissue, nor with the striosome/matrix compartmentalization, but reflect 'contamination' of the implanted striatum with areas known to develop alongside the striatal anlage within, or in close proximity to, the ganglionic eminences. These areas include lateral zones of cortex, amygdala and globus pallidus (see Wictorin, 1992 for more discussion). This mixture of striatal and non-striatal tissue within 'striatal' grafts could explain the discrepancies often seen in the literature where seemingly conflicting results concerning the anatomical composition of the grafts are reported. If only the patch regions are taken into consideration, the grafts do closely resemble normal striatum in both neuronal composition and synaptic connectivity.

CONCLUSIONS

Intrastriatal striatal grafts are of great interest both as a model system for the study of striatal development and specificity of synaptic interactions as well as a potential therapy for Huntington's disease. The present findings stress the importance of identifying the 2 distinct graft compartments when analyzing the anatomical characteristics of the implants. Furthermore, the probable mixture of striatal and non-striatal tissue within the graft underlines the importance of trying to develop alternative procedures for tissue preparation through which purer striatal grafts could be obtained.

REFERENCES

Bolam, J.P., Clarke, D.J., Smith, A.D., and Somogyi, P., 1983, A type of aspiny neuron in the rat neostriatum accumulates [3H]-γ-aminobutyric acid: combination of Golgi-staining, autoradiography and electron microscopy, *J. Comp. Neurol.* 213:121-134.

Bolam J.P., and Ingham, C.A., 1990, Combined morphological and histochemical techniques for the study of neuronal microcircuits, *in:* "Handbook of Chemical Neuroanatomy, Volume 8, Neuronal microcircuits- combined morphological, immunocytochemical and electrophysiological techniques for the study of synaptic interactions between identified CNS neurons," A. Van den Pol and F. Wouterlood, eds., Elsevier, Amsterdam, pp. 125-198.

Bolam, J.P., Powell, J.F., Wu, J-Y. and Smith, A.D., 1985, Glutamate decarboxylase immunoreactive structures in the rat neostriatum: A correlated light and electron microscopic study including a combination of golgi impregnation with immunocytochemistry, *J. Comp. Neurol.* 237:1-20.

Bolam J.P., Somogyi, P., Totterdell, S., and Smith, A.D., 1981, A second type of striatonigral neuron: a comparison between retrogradely labelled and golgi-stained neurons at the light and electron microscope levels, *Neuroscience* 6:2141-2157.

Chang, H.T., Wilson, C.J., and Kitai, S.T., 1982, A Golgi study of rat neostriatal neurons: light microscopic analysis, *J. Comp. Neurol.* 208:107-126.

Clarke, D.J., Dunnett, S.B., Isacson, O., Sirinathsinghji, D.J.S., and Bjorklund, A., 1988, Striatal grafts in rats with unilateral neostriatal lesions - 1. Ultrastructural evidence of afferent synaptic inputs from the host nigrostriatal pathway, *Neuroscience* 24:791-801.

DiFiglia, M., Schiff, L. and Deckel, A.W., 1988, Neuronal organization of fetal striatal grafts in kainate- and sham-lesioned rat caudate: light and electron microscopic observations, *J. Neurosci.* 8:1112-1130.

Freund, T.F., Martin, K.A.C., Smith, A.D., and Somogyi, P., 1983, Glutamate decarboxylase-immunoreactive terminals of Golgi-impregnated axo-axonic and of presumed basket cells in synaptic contact with pyramidal neurons of the cat's visual cortex, *J. Comp. Neurol.* 221:263-279.

Fox, C.A., Andrade, A.N., Lu Qui, I.F., and Rafols, J.A., 1974, The primate globus pallidus: a Golgi and electron microscopic study, *J. Hirnforsch*, 15:75-93.

Graybiel, A.M., Liu, F.C., and Dunnett, S.B., 1989, Intrastriatal grafts derived from fetal striatal primordia. I. phenotopy and modular organization, *J. Neurosci.* 9:3250-3271.

Graybiel, A.M., 1990, Neurotransmitters and neuromodulators in the basal ganglia, *Trends Neurosci.* 13:244-253.

Helm, G.A., Palmer, P.E., and Bennett, J.P., 1990, Fetal neostriatal transplants in the rat: a light and electron microscopic Golgi study, *Neuroscience* 37:735-756.

Hsu, S.M., Raine, L., and Fanger, H., 1981, The use of avidin-biotin peroxidase complex (ABC) in immunoperoxidase techniques: a comparison between ABC and unlabelled antibody (PAP) procedures, *J. Histochem. Cytochem.*, 29:577-580.

Ingham, C.A., Bolam, J.P., and Smith, A.D., 1988, GABA-immunoreactive synaptic boutons in the rat basal forebrain: comparison of neurons that project to the neocortex with pallidosubthalamic neurons, *J. Comp. Neurol.* 273:263-282.

Isacson, O., Dawbarn, D., Brundin, P., Gage, F.H., Emson, P.C., and Bjorklund, A., 1987, Neural grafting in a rat model of Huntington's disease: striosomal-like organization of striatal grafts as revealed by immunocytochemistry and receptor autoradiography, *Neuroscience* 22:481-497.

Isacson, O., Dunnett, S.B., and Bjorklund, A., 1986, Behavioural recovery in an animal model of Huntington's disease, *Proc. Natl. Acad. Sci. USA*, 83:2728-2732.

Izzo, P.N., Graybiel, A.M., and Bolam, J.P., 1987, Characterization of substance P-and [met]enkephalin-immmunoreactive neurons in the caudate nucleus of cat and ferret by a single section Golgi procedure, *Neuroscience* 20:577-587.

Kita, H., and Kitai, S.T., 1988, Glutamate decarboxylase immunoreactive neurons in rat neostriatum: their morphological types and populations, *Brain Res.* 447:346-352.

Levey, A.I., Bolam, J.P., Rye, D.B., Hallanger, A., Mesulam, M-M., and Wainer, B.H., 1986, A light and electron microscopic procedure for sequential double antigen localization using diaminobenzidine and benzidine dihydrochloride, *J. Histochem. Cytochem.* 34:1449-1457.

Liu, F-C., Graybiel, A.M., Dunnett, S.B., and Baughman, R.W., 1990, Intrastriatal grafts derived from fetal striatal primordia: 2. Reconstitution of cholinergic and dopaminergic systems, *J. Comp. Neurol.* 295:1-14.

McAllister, J.P., Walker, P.D., Zemanick, M.C., Weber, A.B., Kaplan, L.I., and Reynolds, M.A., 1985, Morphology of embryonic neostriatal cell suspensions transplanted into adult neostriata, *Dev. Brain Res.* 23:282-286.

Oertel, W.H., and Mugnaini, E., 1984, Immunocytochemical studies of GABAergic neurons in rat basal ganglia and their relation to other neuronal systems, *Neurosci. Lett.* 47:233-238.

Ouimet, C.C., Miller, P.E., Hemmings, Jr., H.C., Walaas, S.I., and Greengard, P., 1984, DARPP-32, a dopamine-and adenosine 3':5'-monophosphate-regulated phosphoprotein enriched in dopamine-innervated brain regions. III. Immunocytochemical localization in rat brain, *J. Neurosci.* 4:111-124.

Pritzel, M., Isacson, O., Brundin, P., Wiklund, L., and Bjorklund, A., 1986, Afferent and efferent connections of striatal grafts implanted into the ibotenic acid lesioned neostriatum in adult rats, *Exp. Brain Res.* 65:112-126.

Ramon-y-Cajal, S., 1911, "Histologie du Système Nerveux de l'homme et des Vertébrés", Maloine, Paris.

Roberts, R.C., and DiFiglia, M., 1990, Long term survival of GABA-, enkephalin-, NADPH-diaphorase-and calbindin-d28k-containing neurons in fetal striatal grafts, *Brain Res.* 532:151-159.

Rutherford, A., Garcia-Munoz, M., Dunnett, S.B., and Arbuthnott, G.W., 1987, Electrophysiological demonstration of host cortical inputs into striatal grafts, *Neurosci. Lett.* 83:276-281.

Smith, Y., Parent, A., Seguela, P., and Descarries, L., 1987, Distribution of GABA-immunoreactive neurons in the basal ganglia of the squirrel monkey, *J. Comp. Neurol.* 259:50-64.

Totterdell, S., Bolam, J.P., and Smith, A.D., 1984, Characterization of pallidonigral neurons in the rat by a combination of Golgi-impregnation and retrograde transport of horseradish peroxidases: their monosynaptic imput from the neostriatum, *J. Neurocytol.* 13:593-616.

Walker, P.D., Chovanes, G.I., and McAllister II, J.P., 1987, Identification of acetylcholine-reactive neurons and neuropil in neostriatal transplants, *J. Comp. Neurol.* 259:1-12.

Wictorin, K., 1992, Anatomy and connectivity of intrastriatal striatal transplants, *Prog. in Neurobiol.* 38:611-639

Wictorin, K., Simerly, R.B., Isacson, O., Swanson, L.W., and Bjorklund, A., 1989a, Connectivity of striatal grafts implanted into the ibotenic acid lesioned striatum. III. Efferent projecting graft neurons and their relation to host afferents within the grafts, *Neuroscience* 30:313-330.

Wictorin, K., Clarke, D.J., Bolam, J.P., and Bjorklund, A., 1990a, Fetal striatal neurons grafted into the ibotenate lesioned striatum: efferent projections and synaptic contacts in the host globus pallidus, *Neuroscience* 37:301-315.

Wictorin, K., Brundin, P., Gustavii, B., Lindvall, O., and Bjorklund, A., 1990b, Reformation of long axon pathways in adult rat CNS by human forebrain neuroblasts, *Nature* 347:556-558.

Wictorin, K., Lagenaur, C.F., Lund, R.D., and Bjorklund, A., 1991, Efferent projections to the host brain from intrastriatal striatal mouse-to-rat grafts: timecourse and tissue-type specificity as revealed by a mouse specific marker, *Eur. J. Neurosci.* 3:86-101.

Xu, Z.C., Wilson, C.J., and Emson, P.C., 1990, Restoration of thalamostriatal projections in rat neostriatal grafts: an electron microscopic analysis, *J. Comp. Neurol.* 303:2-14.

Xu, Z.C., Wilson, C.J., and Emson, P.C., 1992, Morphology of intracellularly stained spiny neurons in rat striatal grafts, *Neuroscience* 48:95-110.

Zhou, F.C., 1990, Connectivities of the striatal grafts and laminin guiding, *in*: "Neural Transplantation. From Molecular Basis to Clinical Application," *Prog. Brain Res.* 82:441-458.

FURTHER INVESTIGATIONS ON THE MECHANISMS INVOLVED IN INTRASTRIATAL MESENCEPHALIC GRAFTS IN THE RAT, WITH SPECIAL REFERENCE TO DOPAMINE-NEUROPEPTIDE Y INTERACTIONS

Annie Daszuta, Hakima Moukhles, Jacqueline Vuillet and André Nieoullon

Laboratoire de Neurobiologie Cellulaire et Fonctionnelle du CNRS
31 Chemin J. Aiguier, Marseille, France

A number of transplantation studies on an animal model for Parkinson's disease bearing a 6-hydroxydopamine (6-OHDA) lesion of the nigrostriatal dopaminergic pathway were carried out since the late 70's (Björklund and Stenevi, 1979, Perlow et al., 1979). Implanting foetal dopamine (DA) neurons into the striatum of these rats was found to alleviate some Parkinson's disease like symptoms in the DA deficient recipients. In rats with unilateral lesion, motor impairments such as drug-induced rotational asymmetry were found to be completely abolished after DA grafts, while in more complex conditioned behavior tests, rats remained severely impaired (Björklund et al., 1987). Despite these limitations, human foetal ventral mesencephalic tissue has been implanted into the brain of Parkinsonian patients with variable success. Although there is some ambiguity in the statement that brain grafts can have functional effects, because it contains spontaneously active neurons and releases neurotransmitter, or because it may trigger some adaptive behavioral responses in the host animals, the cellular mechanisms whereby ventral mesencephalic transplants act need to be exactly determined for research on neural transplantation to be able to progress.

GENERAL FEATURES OF DA GRAFTS

Regulation of the metabolism and release of DA by grafted cells

Fetal dopaminergic neurons in mesencephalic cell suspensions have been found to synthesize and release DA when grafted into 6-OHDA-denervated striata in adult rats (Zetterström et al., 1986; Strecker et al., 1987; Forni et al., 1989). Although grafts of this kind usually restore only low DA levels in the target, the turnover of the amine as expressed by the tissular DOPAC/DA ratio and the DA release rate was restored to the control levels in the transplanted rats. These data indicate that the functional recovery observed after these grafts may be facilitated by dopaminergic hyperactivity which can persist for up to 2 years after transplantation (Nishino et al., 1990). In these grafts, autoregulation of DA cell activity has been found to occur involving D1 and D2 receptor subtypes as in normal rats. Mixed D1-D2 receptor agonist (apomorphine) or antagonist (haloperidol) can partially decrease or increase, respectively, the release or metabolism of DA in the transplants (Herman et al., 1985; Strecker et al., 1987). In connection with these data, we have recently described differential effects of selective D1 and D2 receptor agents on the release of DA as monitored using in vivo voltammetry (Daszuta et al., 1992). While D2 receptors involved in the direct autoregulation process appeared to show some subsensitivity to D2 antagonist, sulpiride, injected into the grafted striata, the normalization of DA release in response to D1 antagonist, SCH 23390, treatment suggested that postsynaptic D1 receptors in the host cell populations might control the dopaminergic activity as much as in the control striata : data demonstrating that grafted neurons responded to electrical stimulation of the prefrontal cortex and the striatum (Fisher et al., 1991) are, in fact, consistent with the hypothesis that DA cells may receive functional inputs from host neurons (Doucet et al., 1989). Preliminary results from our laboratory point to similar conclusions, since the pharmacological stimulation or blockade of various sub-types of excitatory amino-acid receptors may induce similar changes in DA release in transplanted as in control striata, which means that presumably glutamatergic cortical afferents from the host might be able to modulate the DA activity of the grafts. All these data are consistent with the idea that the dopaminergic cells become much more fully integrated into the host circuitry than what was previously suspected to be the case, due to the ectopic position of the graft.

Normalization of DA receptor sensitivity

In grafted animals, both presynaptic and postsynaptic events are known to have been restored. At the cellular level, long-term survival intrastriatal DA grafts have been found to reverse the effects of DA denervation on various neuronal populations of the host striatum. For instance, intraventricular solid transplants containing DA cells have been reported to partially reverse the lesion-induced increase in striatal glutamate decarboxylase (Segovia et al., 1989). DA grafts have also been found to exert functional effects on efferent striatal populations of the host containing enkephaline or substance P (Manier et al., 1991; Mendez et al., 1991). Lastly, the inhibitory influence normally exerted by the nigrostriatal DA input onto striatal interneurons containing either acetylcholine (Herman et al., 1988; Jackish et al., 1991), or neuropeptide Y (Moukhles et al., 1992b) were found to be restored after intraparenchymatal DA grafts.

These data are consistent with the normalization of striatal D1 and D2 dopaminergic receptor densities and D2 gene expression in grafted animals (Dawson et al., 1991; Gagnon et al., 1991; Chritin et al., 1992). Although the changes observed in the D1 receptor density after lesions and grafts have not yet been fully explained, dopaminergic transplants also normalized the amphetamine and apomorphine-induced Fos expression which is presumed to be primarily mediated by D1 receptors in striata with 6-OHDA lesions (Cenci et al., 1992). Moreover, in the globus pallidus, one of the primary targets of the striatal output neurons, the fos expression was found to be normalized by the graft (Cenci et al., 1992). As the density of D1 receptors also returned to the prelesion values in the substantia nigra (Dawson et al., 1991), it seems likely that DA cells grafted into the striatum may have exerted functional effects over a larger area than that innervated by their corresponding terminals.

Graft-induced recoveries were also evaluated by measuring motor behavioral responses to D1 and D2 DA receptor agonists in transplanted rats. These results are still a matter of controversy however, since DA grafts either reduced D1 but not D2 receptor-mediated rotation (Herman et al., 1991; Robertson et al., 1991), decreased the supersensitivity of the D2 receptors in terms of motor behavior and to a lesser extent that of the D1 receptors (Rioux et al., 1991), or normalized both (Becker and Ariano, 1991). DA grafts may in fact differentially affect the D1 and D2 DA receptors, and the existence of differences in the experimental procedures or in the time-course of the normalization of DA receptor sensitivity between the D1 and D2 subtypes might explain the differences between the results. Although these data suggest that it may be difficult to relate behavioral improvements to appropriate local physiological changes induced by the grafts, it is important to determine the various factors or processes involved in their modes of action.

MECHANISMS OF ACTION OF DA GRAFTS

The restoration of synaptic transmission was found to be an essential criterion for characterizing the functional state of DA grafts. By comparing various types of catecholaminergic grafts, such as those containing dopaminergic cells from the ventral mesencephalon (VM) with those containing noradrenergic cells originating from the central (locus coeruleus) or peripheral nervous system (sympathetic ganglia or adrenal medulla), correlations between the functional recovery and the number of surviving cells expressing TH immunoreactivity have been found to exist in all these grafts (Nishino et al., 1991). The highest level of recovery was obtained, however, with VM grafts which re-established synaptic connections with the host. The fact that adrenal medulla transplants were still able to partly attenuate rotational behavior raised the question as to the mechanisms whereby which endocrine cells may affect the working of the host brain.

Restoration of synaptic dopaminergic transmission

Upon studying the functional effects of fetal mesencephalon transplanted into 6-OHDA-denervated striatum of adult rats, it was rapidly demonstrated that the substantial effects of the grafts on spontaneous motor and sensorimotor behavior may depend on the reinnervation of the host by the grafted DA neurons. In intrastriatal grafts consisting of mesencephalic cell suspensions, the extent of the axonal growth and the occurrence of TH-positive synapses in the host striatum have been correlated with the reversal of the lesion-induced rotational asymmetry after amphetamine administration (Clarke et al., 1988, Nishino et al., 1990). The restoration of neuronal interactions at the ultrastructural level is, therefore, among the cellular mechanisms involved in this example of functional recovery.

Although data have been collected describing the ultrastructural features of graft to host relationships, the targets of newly formed DA contacts have rarely been identified from the point of view of their neurotransmitter. Data from our laboratory have suggested that interneurons containing NPY may occupy a strategic position in the striatal network (Kerkerian et al., 1991): in particular, they constitute targets for dopaminergic and glutamatergic inputs, and for intrinsic GABAergic and cholinergic neurons. As previously mentioned, we have investigated the possible restoration of cell interactions between DA fibers arising from the graft and striatal cells containing NPY, at the ultrastructural level. Single and double immunolabelling was performed to detect TH and NPY containing neurons at various times after transplantation and in various areas of the grafted striatum.

At all the post-graft times tested, TH-positive fiber outgrowth was particularly obvious in the immediate vicinity of the graft and the extent of reinnervation of the striatum by graft-derived DA fibers decreased over long distances in the host tissue (Fig. 1A). Within

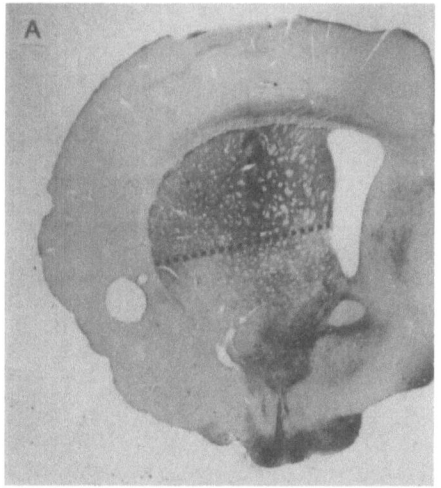
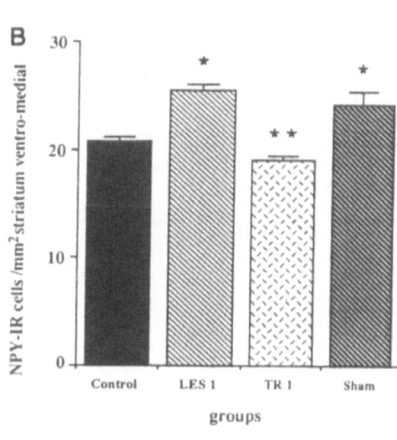

Figure 1. A : Example of tyrosine hydroxylase (TH) immunoreactive neurons grafted into the dorsal part of the striatum. Dashed line delineates the two parts of the structure densely (dorsal) or poorly (ventral) reinnervated by the graft.
B : Histograms showing changes in the densities of NPY immunoreactive cells in the striatal area poorly reinnervated by the ventral mesencephalic graft (TR1) 1 month post transplantation, as compared with normal (control), 6-OHDA lesioned (Les 1) and rats transplanted with fetal spinal cord (Sham). p<0.01 as compared with control (*) or with lesion group (**).

one month of the grafting, TH-immunoreactive neurons showed most of the normal intrinsic morphological features characteristic of adult rat neurons in the substantia nigra. By this time, they had already established direct contacts with various striatal neuronal populations, such as those containing NPY (Fig. 2B). The relative frequency of TH-NPY relationships in the area most densely reinnervated by the graft was similar to that determined in the intact striatum and decreased thereafter, suggesting that a further elongation of the TH axons may result in a wider distribution of the TH-NPY associations over the host striatum (Fig. 2A). Meanwhile, some signs of continuing maturation were observed, such as a decrease in the proportion of axosomatic and a corresponding increase in the axodendritic TH-NPY relationships. In the zones most distal from the graft however, the reinnervation was always far from complete and the few TH immunoreactive fibres projected to unlabelled dendritic elements, but were never observed contacting the NPY neurons examined in these areas. The synaptic relationships between grafted DA cells and striatal NPY interneurons are therefore, mainly re-established within the area proximal to the graft, while in the more distal territories,

A

	Frequency of relationships between TH varicosities and NPY cells		
	soma	dendrite	total
Normal	(4/22) 18%	(6/38) 16%	17%
Graft 1 month	(10/31) 32%	(3/42) 7%	18%
Graft 3 months	(4/57) 7%	(5/44) 11%	9%

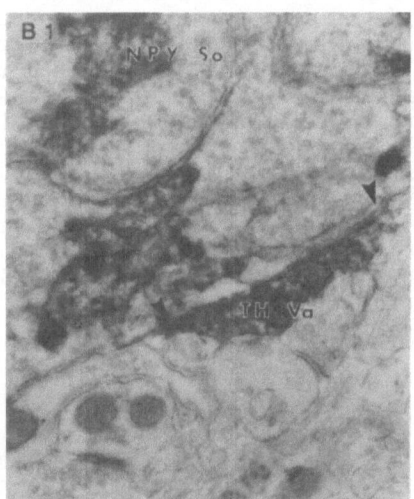

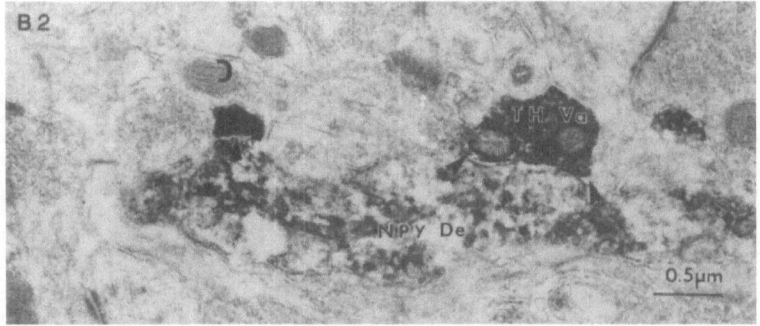

Figure 2. A : Frequency of relationships between graft-derived TH varicosities and host striatal NPY cells in the striatal area densely reinnervated by the graft. Sections have always been collected at the surface of the blocks from 2 animals of each group.
B : DAB-labelled TH varicosities (TH Va) forming a presumed synaptic contact of the symmetrical type with BDHC-labelled NPY soma (NPY So) in **B1**, and a close apposition with NPY dendrite (NPY De) in **B2**.

DA-NPY interactions appear to depend on the restoration of multiple cellular interactions involving, for instance, neurons containing acetylcholine or GABA, which are scattered throughout the whole striatum, or may result from other neuronal processes.

The next question was to determine the degree of functional recovery of DA/NPY cellular relationships in the transplanted rats. Previous results obtained at our laboratory have in fact demonstrated that neurons containing NPY may be under the inhibitory control of the nigrostriatal DA pathway (Kerkerian et al., 1991). Intranigral injection of 6-OHDA resulted in a marked increase of about 30% in the number of NPY immunopositive cells, as compared with the control values. In our study, a complete reversal of the increase in the number of NPY immunopositive neurons occurring after DA denervation was detected in DA grafted animals (Fig. 1B). This "normalization" was observed as early as one month post-grafting, in spite of the partial reinnervation of the striatum as shown by using TH immuno-cytochemistry or by measuring the DA levels in similar groups of transplanted rats. The effects on the host NPY immunoreactivity were found to be specific to DA grafts, since no reversal was observed in sham-spinal-cord-transplanted rats. Above all, similar degrees of normalization were recorded in the proximal area densely reinnervated by the graft and in the ventro-medial distant zone, and no further functional effects worthnoting were detected three months after transplantation. These data demonstrated that grafted DA neurons are able to induce an early and widespread normalization of the DA-NPY interactions in the striatum of transplanted rats, which appeared to be unrelated to the density of the DA reinnervation of the host. They, therefore, raised the question as to what mechanisms, besides the restoration of synaptic transmission, may be involved in these functional effects of DA grafts.

Paracrine mode of action of DA released by grafted cells

The various mechanisms whereby DA transplants can act have to be taken into account to explain the normalization of host NPY immunoreactivity induced by grafts, and the apparent lack of relationship between the outspread of DA reinnervation and the extent of the host responsiveness. In our study, one possible action pathway might pass through the blood vessel network, since at the light microscopic level we have frequently observed TH-IR fibres running in close apposition to striatal capillaries. On the other hand, close associations between NPY-IR neurons and blood vessels seem to be a feature of the cellular interactions involving NPY cells. This may therefore constitute a favourable situation for DA/NPY interactions via the circulatory system. In view of these data, both the restoration of synaptic transmission and paracrine DA action may be simultaneously involved in the mechanisms whereby DA transplants can act.

The possibility that a neuro-humoral action and/or a diffuse release of DA by the grafted cells may occur has been suggested since the first reports describing the functional effects of adrenal medulla grafts. This has by now become one of the standard explanations for the mode of action of DA grafts, since an increasing number of data have shown that functional effects occur over areas extending well beyond that reached by TH fibres. In graft-reinnervated striata with 6-OHDA lesions, the clearance of DA by neuronal uptake may be less efficient than in normal striata, facilitating the diffusion of DA via extracellular fluids into large areas and the consecutive activation and normalization of DA receptors. The relative importance of synaptic versus paracrine DA transmission for the occurrence of functional effects as the result of intrastriatal grafting has also been re-evaluated using cell lines such as PC12 cells encapsulated in polymers (Winn et al., 1991) or genetically engineered cells producing catecholamines to be transplanted to the denervated striatum (Horellou et al., 1990). These cells expressed TH immunoreactivity, secreted low or high levels of DOPA and/or DA and the intrastriatal grafts partially reversed the apomorphine-induced motor asymmetry. Although the long-term state and effect of grafts of genetically modified cells are still under investigation, these studies again suggest that DA diffusion may account for transplant-induced recovery in experimental parkinsonism.

Trophic effects

The mechanisms of the recovery induced by DA grafts may also involved trophic influences by molecules secreted by various categories of cells in the graft and in the host populations. Trophic factors do in fact contribute to the developmental processes as well as to the maintenance of adult brain after injury. It has long been established that denervation of the structure prior to grafting may facilitate the neuronal survival rate or the fiber outgrowth from grafted cells, presumably due to the central role of astrocytes (Nieto-Sampiedro et al., 1984). In the case of intrastriatal DA grafts, the fact that 6-OHDA-induced denervation of the striatum did not affect the survival rate of grafted catecholamine neurons but increased the fiber outgrowth (Doucet et al., 1990), suggests that both membranous and diffusible factors may be involved in these reactions. Conversely, in young MPTP-treated mice, in which nigrostriatal pathway had not been completely eliminated, the recovery of endogenous dopaminergic fibers was enhanced by adrenal medulla grafts and induced the restoration of motor function (Bohn et al., 1987). Trophic molecules may help grafts to survive and to develop better and systemic treatment of DA transplanted rats with gangliosides (Nishino et al., 1991) or co-grafting DA cells with striatal (Yurek et al., 1990) or peripheral nerve tissue (Van Horne et al., 1991) have been used to promote graft effects. Besides the expected specific effects of embryonic striatal glia on the development of mesencephalic DA cells (Prochiantz et al., 1979), the mechanism whereby adult sciatic nerve enhances rotational reduction when co-grafted with VM tissue in the striatum remains to be determined. In fact, molecules from the immune (interleukines) and blood (PDGF) systems, and others (BDNF, FGF, IGF, etc...) are all candidate agents which may either directly, or undirectly through glia, affect DA neurons. Interestingly, Loughlin and coll. have reported at the present meeting, that the TGF α secreted by striatal astrocytes, may contribute to the efficiency of DA transplants, since intrastriatal infusion of TGFα partly decreased apomorphine-induced rotation in 6-OHDA lesioned rats. This is probably the most interesting aspect of these recent studies on the growth factors, with the aim of stimulating regeneration of the dying nigrostriatal neurons. Moreover, searching for the molecules which are involved in the processes underlying CNS development or aging, such as the selective neuronal death occurring in response to specific environmental signals, might help us to reach a better understanding of the etiology of Parkinson's disease.

In conclusion, DA grafts on animal models for Parkinson's disease can help to restore motor capacities via various mechanisms, probably acting in synergy. These transplants might predictably be improved by implanting DA cells at their original location, the substantia nigra, in order to structurally repair the neuronal circuitry. This would be facilitated if trophic factors were discovered which specifically promote the survival and development of DAergic neurons in vivo. It will also be necessary to work on several different models of striatal DA denervation. For instance, we are now studying the functional effects of DA transplants after intrastriatal 6-OHDA injection, in order to simulate a more progressive degeneration of the host DA neurons. Under these conditions, we have demonstrated that grafted rats can recover not only from spontaneous motor disturbances, but can also regain more integrative abilities such as those involved in a conditioned reaction time motor task (Moukhles et al., 1992a). These data can again, be interpreted as resulting from a functional synergy between endogenous and graft-derived DA fibers.

Although the processes underlying Parkinson's disease have not yet been elucidated, there exists considerable evidence that transplants of fetal DA neurons can partly act as a substitute for the missing dopaminergic innervation. This should encourage further efforts towards a fuller documentation of the subject at animal research level, with a view to developing clinical applications. The multiple possibilities offered by gene therapy should also be kept in mind, however, pending a more thorough understanding of the factors that lead to the loss of DA cells.

REFERENCES

Becker, J.B., and Ariano, M.A., 1991, Behavioral effects of fetal substantia nigra tissue grafted into the dopamine-denervated striatum: responses to selective D1 and D2 dopamine receptor agonists, *Rest. Neurol. Neurosci.* 3:187-195

Björklund, A., Lindvall, O., Isacson, O., Brundin, P., Wictorin, K., Strecker, R.E., Clarke, D.J., and Dunnett, S. B., 1987, Mechanisms of action of intracerebral neural implants: studies on nigral and striatal grafts to the lesioned striatum, *Trends Neurosci.* 10:509-516.

Björklund, A., and Stenevi, U., 1979, Reconstruction of the nigrostriatal dopamine pathway by intracerebral nigral transplants, *Brain Res.* 177:555-560.

Bohn, M.C., Cupit, L., Marciano, F., and Gash, D.M., 1987, Adrenal medulla grafts enhance recovery of striatal dopaminergic fibers, *Science* 237:913-916.

Cenci, M.A., Kalen, P., Mandel, R.J., Wictorin, K., and Björklund, A., 1992, Dopaminergic transplants normalize amphetamine- and apomorphine-induced Fos expression in the 6-hydroxydopamine-lesioned striatum, *Neuroscience* 46:934-957.

Chritin, M., Savasta, M., Mennicken, F., Bal, A., Abrous, D.N., Le Moal, M., Feuerstein, C., and Herman, J.P., 1992, Intrastriatal dopamine-rich implants reverse the increase of dopamine D2 receptor mRNA levels caused by lesion of the nigrostriatal pathway: a quantitative in situ hybridization study, *Eur. J. Neurosci.* 4:663-672.

Clarke, D. J., Brundin, P., Strecker, R. E., Nilsson, O. G., Björklund, A., and Lindvall, O., 1988, Human fetal dopamine neurons grafted in a rat model of Parkinson's disease: Ultrastructural evidence for synapse formation using tyrosine hydroxylase immunocytochemistry, *Exp. Brain Res.* 73:115-126.

Daszuta, A., Moukhles, H., Forni, C., Dusticier, N., and Nieoullon A., 1992, Regulation of dopamine release as monitored by in vivo voltammetry in intrastriatal grafts of fetal mesencephalon neurons, *Rest. Neurol. Neurosci.* 4:15P32.

Dawson, T. M., Dawson, V. L., Gage, F. H., Fisher, L. J., Hunt, M. A., and Wamsley, J.K., 1991, Functional recovery of supersensitive dopamine receptors after intrastriatal grafts of fetal substantia nigra, *Exp. Neurol.* 111:282-292.

Doucet, G., Brundin, P., Descarries, L., and Björklund, A., 1990, Effect of prior dopamine denervation on survival and fiber outgrowth from intrastriatal fetal mesencephalic grafts, *Eur. J. Neurosci.* 2:279-290.

Doucet, G., Murata, Y., Brundin, P., Bosler, O., Mons, N., Geffard, M., Ouimet, C.C., and Björklund, A., 1989, Host afferents into intrastriatal transplants of fetal ventral mesencephalon, *Exp. Neurol.* 106:1-19.

Fisher, L.J., Young, S.J., Tepper, J.M., Groves, P.M., and Gage, F.H., 1991, Electrophysiological characteristics of cells within mesencephalon suspension grafts, *Neuroscience* 40:109-122.

Forni, C., Brundin, P., Strecker, R.E., El Ganouni, S., Björklund, A., and Nieoullon A., 1989, Time-course of recovery of dopamine neuron activity during reinnervation of denervated striatum by fetal mesencephalic grafts as assessed by in vivo voltametry, *Exp. Brain Res.* 76:75-87.

Gagnon, C., Bédard, P.J., Rioux, L., Gaudin, D., Martinoli, M.G., Pelletier, G., and Di Paolo, T., 1991, Regional changes of striatal dopamine receptors following denervation by 6-hydroxydopamine and fetal mesencephalic grafts in the rat, *Brain Res.* 558:251-263.

Herman, J.P., Abrous, D.N., and Le Moal, M., 1991, Anatomical and behavioral comparison of unilateral dopamine-rich grafts implanted into the striatum of neonatal and adult rats, *Neuroscience* 40:465-475.

Herman, J.P., Choulli, K., and Le Moal, M., 1985, Activation of striatal dopaminergic grafts by haloperidol, *Brain Res. Bull.* 15:543-546.

Herman, J. P., Lupp, A., Abrous, N., Le Moal, M., Hertting, G., and Jackisch, R., 1988, Intrastriatal dopaminergic grafts restore inhibitory control over striatal cholinergic neurons, *Exp. Brain Res.* 73:236-248.

Horellou, P., Brundin, P., Kalen, P., Mallet, J., and Björklund, A., 1990, In vivo release of DOPA and dopamine from genetically engineered cells grafted to the denervated rat striatum, *Neuron* 5:393-402.

Jackisch, R., Duschek, M., Neufang, B., Rensing, H., Hertting, G., and Herman, J.P., 1991, Long term survival of intrastriatal dopaminergic grafts: Modulation of acetylcholine release by graft-derived dopamine, *J. Neurochem.* 57:267-276.

Kerkerian-Le Goff, L., Salin, P., Vuillet, J., and Nieoullon, A., 1991, NeuropeptideY neurons in the striatal network. Functional adaptive responses to impairment of striatal inputs, *in*: "The Basal Ganglia III," G. Bernardi, M.B. Carpenter, G. Di Chiara, M. Morelli and P. Stanzione, eds., Plenum Press, New York, pp. 49-61.

Manier, M., Abrous, D.N., Feuerstein, C., Le Moal, M., and Herman, J. P., 1991, Increase of striatal methionin enkephalin content following lesion of the nigrostriatal dopaminergic pathway in adult rats and reversal following the implantation of embryonic dopaminergic neurons: a quantitative immunohistochemical analysis, *Neuroscience* 42:427-439.

Mendez, I., Elisevich, K., and Flumerfelt, B., 1991, Dopaminergic innervation of substance P-containing striatal neurons by fetal nigral grafts : an ultrastructural double-labeling immunocytochemical study, *J. Comp. Neurol.* 308:66-78.

Moukhles, H., Amalric, M., Nieoullon, A., and Daszuta, A., 1992a, Partial recovery of sensorimotor deficits in a conditioned task induced by mesencephalic cells grafted to locally dopamine deafferented striatum in the rat, *Rest. Neurol. and Neurosci.* 15P30.

Moukhles, H., Nieoulllon, A., and Daszuta, A., 1992b, Early and widespread normalization of dopamine-neuropeptide Y interaction in the rat striatum after transplantation of fetal mesencephalon cells, *Neuroscience* 4:781-792.

Nieto-Sampiedro, M., Whittemore, S.R., Needels, D.L., Larson, J., and Cotman, C., 1984, The survival of brain transplants is enhanced by extracts from injured brain, *Proc. Natl., Acad. Sci.* 81:6250-6254.

Nishino, H., Hashitani, T., and Kumazaki, M., 1991, Grafting of catecholaminergic cells in the mammalian brain and reconstruction of disturbed function: basic problems to be solved, *J. Comp. Biochem. Physiol.* 98:211-220.

Nishino, H., Hashitani, T., Kumazaki, M., Sato H, Furuyama F., Isobe Y., Watari N., Kanai M., and Shiosaka S., 1990, Long-term survival grafted cells, dopamine synthesis/release, synaptic connections, and functional recovery after transplantation of fetal nigral cells in rats with unilateral 6-OHDA lesions in the nigrostriatal dopamine pathway, *Brain Res.* 534:83-93.

Perlow, M.J., Freed, W.J., Hoffer, B.J., Seiger, A., Olson, L., and Wyatt, R.J., 1979, Brain grafts reduce motor abnormalities produced by destruction of nigrostriatal dopamine system, *Science* 204:643-646.

Prochiantz, A., Diporzio, U., Kato, A., Berger, B., and Glowinski J., 1979, In vitro maturation of mesencephalic dopamine neurons from mouse embryos is enhanced in presence of their striatal target cells, *Proc. Natl. Acad. Sci.* 76:5387-5391.

Rioux, L., Gaudin, C., Bui, L.K., Gregoire, L., Di Paolo, T., and Bédard, P.J., 1991, Correlation of functional recovery after 6-hydroxydopamine lesion with survival of grafted fetal neurons and release of dopamine in the striatum of the rat, *Neuroscience* 40:123-131.

Robertson, G.S., Fine, A., and Robertson, H.A., 1991, Dopaminergic grafts in the striatum reduce D1 but not D2 receptor-mediated rotation in 6-OHDA-lesioned rats, *Brain Res.* 539:304-311.

Segovia, J., Meloni, R., and Gale, K., 1989, Effect of dopaminergic denervation and transplant-derived reinnervation on a marker of striatal GABAergic function, *Brain Res.* 493:185-189.

Strecker, R. E., Sharp, T., Brundin, P., Zetterström ,T., Ungerstedt, U., and Björklund, A., 1987, Autoregulation of dopamine release and metabolism by intrastriatal nigral grafts as revealed by intracerebral dialysis, *Neuroscience* 22:169-178.

Van Horne, C.G., Strömberg, I., Young, D., Olson, L., and Hoffer, B., 1991, Functional enhancement of intrastriatal dopamine-containing grafts by the co-transplantation of sciatic nerve tissue in 6-hydroxydopamine-lesioned rats, *Exp. Neurol.* 113:143-154.

Winn, S.R., Tresco, P.A., Zielinski, B., Greene, L.A., Jaeger, C.B., and Aebischer, P., 1991, Behavioral recovery following intrastriatal implantation of microencapsulated PC12 cells, *Exp. Neurol.* 113:322-329.

Yurek, D.M., Collier, T.J., and Sladek, J.R., 1990, Embryonic mesencephalic and striatal co-grafts : development of grafted dopamine neurons and functional recovery, *Exp. Neurol.* 109:191-199.

Zetterström, T., Brundin, P., Gage, F.H., Sharp, T., Isacson, O., Dunnett, S.B., Ungerstedt, U., and Björklund, A., 1986, In vivo measurement of spontaneous release and metabolism of dopamine from intrastriatal nigral grafts using intracerebral dialysis, *Brain Res.* 362:344-349.

EFFICACY OF INTRASTRIATAL TRANSPLANTS:
ROLE OF TROPHIC FACTORS

Sandra E. Loughlin, Tiffany P. Lee and James H. Fallon

Anatomy and Neurobiology Department
University of California, Irvine
Irvine, CA 92717

INTRODUCTION

The dopaminergic (DA) substantia nigra-ventral tegmental area (SN-VTA) projection to the caudate-putamen (striatum) is critical in the control of motor behavior. Clinically, Parkinson's disease reflects a massive degeneration of the nigrostriatal system (Bernheimer et al., 1973). The replacement of the system by intrastriatal transplants of fetal midbrain suspensions has resulted in improvements in some patients (Sladek and Shoulson, 1988; Lindvall et al., 1989; Freed et al., 1990). In experimental animals, it has been shown that fetal midbrain cells survive transplantation, extend processes and synthesize DA (Perlow et al, 1979; Bjorklund et al, 1980, 1983a, 1983b). Such transplants ameliorate certain deficits associated with the loss of DA input to the striatum (Bjorklund et al., 1987; Yurek and Sladek, 1990). Similar transplants reverse motor abnormalities produced by unilateral denervation (Perlow et al., 1979; Bjorklund et al., 1980, 1983a, 1983b) including apomorphine or amphetamine -induced rotation behavior. Transplants of adrenal medulla have also been shown to ameliorate some behavioral deficits (Allen et al., 1989; Hansen et al., 1989; Lieberman et al., 1989).

A number of studies, however, suggest that the simple replacement of dopamine may not fully predict efficacy of transplants. Following adrenal medulla transplants, behavioral improvement may occur in the absence of surviving chromaffin cells (Bohn et al., 1987). Transplants may promote survival and axonal regeneration of damaged host cells (Kromer et al., 1981; Bregman and Reier, 1986), or may normalize dopamine receptor populations in denervated regions (Freed and Cannon-Spoor, 1988). In MPTP-treated primate, transplants of fetal cerebellum or spinal cord, which do not synthesize dopamine, induce behavioral improvement which has been attributed to sprouting of intrinsic dopaminergic fibers (Bankiewicz et al., 1991). While it has been suggested that the mechanisms by which transplants produce these effects include trophic influences (Kromer et al., 1981; Bregman and Reier, 1986; Yurek and Sladek, 1990), the neurochemical mediators of such effects remain unknown.

It is important to determine the contributions of trophic influences to the efficacy of intrastriatal transplants. The effects of neurotrophic factors in normalizing function following loss of the nigrostriatal system might, thus, be to 1) increase survival of transplanted cells, 2) induce regeneration, sprouting, reinstatement of phenotypic expression or increased efficacy of dopaminergic cells remaining in the host brain, and/or 3) induce compensatory changes in other neurochemical systems. To achieve the clinical goal of identifying neurotrophic factors which might ameliorate symptoms of Parkinson's disease, it would be useful to pursue studies aimed simply at identification of factors which have behavioral efficacy. In order to understand the mechanism of action of such factors following intrastriatal transplants, however, it is first necessary to identify the neurotrophic factors in the nigrostriatal system and to characterize the effects of defined trophic factors in nigrostriatal function. It is then possible to design direct tests of the efficacy of physiologically relevant neurotrophic factors in the normalization of nigrostriatal function. While such studies serve the clinical goal of developing strategies to treat Parkinson's disease, they also offer basic data relevant to understanding the plasticity of a chemically defined system in the adult brain.

We will review here the evidence in support of the hypothesis that some of the efficacious effects of intrastriatal transplants might be transduced by the synthesis and release of neurotrophic factors, either in the transplant itself or in the host brain as a result of placement of the transplant. While the effects of these neurotrophic factors might be to induce changes in other neurochemical systems or receptor populations, we will concentrate on the possibility that neurotrophic factors might cause regeneration of remaining substantia nigra cells. We will review the evidence consistent with a role of two families of growth factors, the fibroblast growth factors and those which bind to the epidermal growth factor receptor, including transforming growth factor alpha.

NEUROTROPHIC FACTORS IN THE NIGROSTRIATAL SYSTEM

Recent studies have revealed the presence of a large number of characterized neurotrophic factors in the central nervous system (see Loughlin and Fallon, 1993 and Fallon and Loughlin, 1993 for review). While neurotrophic effects might be considered to encompass a broad continuum of changes, ranging from modification of neurochemical balance to actual neurogenesis, a useful definition has been proposed by Hefti et al, (1993). According to this definition, a neurotrophic factor is an endogenous, soluble protein regulating survival, growth, morphological plasticity, or synthesis of proteins for differentiated functions of neurons. Neurotrophic effects of transplants might, thus, be considered to be neurotrophic factor-induced changes in the morphological or neurochemical status of the host brain.

The ways in which neurotrophic factors affect target neurons are complex (see Fallon and Loughlin, 1993, for review). These might include effects transduced by the binding of neurotrophic factors produced in the substantia nigra cells themselves, in nearby neurons or glial cells, or in target tissues. While little is known of the changes which occur in neurotrophic factor synthesis and secretion following transplants, the localization of certain neurotrophic factors in the adult nigrostriatal system has been studied. In order to hypothesize which neurotrophic factors mediate the effects of transplants, it is, perhaps, more important to determine which growth factor receptors are localized to the nigrostriatal system. Unfortunately, little is known of the precise cellular localization of neurotrophic receptors in the central nervous system.

In the adult substantia nigra, dopaminergic cells have been shown to synthesize characterized growth factors. Brain-derived neurotrophic factor and neurotrophin-3 mRNAs (Gall et al., 1992) have been detected in substantia nigra cells, as have basic and acidic fibroblast growth factor mRNAs and protein (Bean et al., 1991). While the localization of basic and acidic fibroblast growth factors remains controversial (Fallon et al., 1992), brain-derived neurotrophic factor mRNA has been found in the striatum (Hofer, et al., 1990). Both mRNA and the precursor for transforming growth factor alpha (TGFa), which binds to the epidermal growth factor (EGF) receptor, have been found in striatal astrocytes (Fallon et al., 1990, Seroogy et al., 1993). The presence of these neurotrophic factors in the nigrostriatal system is consistent with a role in the support of dopaminergic cells.

Survival of dopaminergic cells in ventral mesencephalic cultures is enhanced by plating at higher density, especially in the presence of glial cells, and by the addition of media conditioned on mesencephalic or striatal glia (Rousselet et al., 1988, O'Malley et al., 1991). This suggests that cultured substantia nigra dopamine cells respond to neurotrophic factors derived from these regions. Recent studies of substantia nigra cells *in vitro* have allowed the examination of their responsiveness to characterized neurotrophic factors. Insulin and insulin-like growth factor have been shown to stimulate dopamine uptake and cell proliferation in ventral mesencephalic cultures (Knusel et al., 1990). Basic fibroblast growth factor and EGF also increase dopamine uptake and dopamine cell survival, though this effect may be dependent on glial proliferation, suggesting that these are not direct effects (Knusel et al., 1990; Casper et al., 1991). Brain-derived neurotrophic factor also increases the survival of dopaminergic cells (Hyman et al., 1991). In addition, this growth factor protects dopaminergic neurons against the effects of selective neurotoxins (Spina et al., 1992). TGFa has also been shown to support dopaminergic cells (Alexi and Hefti, 1993). While data from such culture studies must be interpreted with caution, they suggest that dopaminergic cells respond to a number of neurotrophic factors, including fibroblast growth factors and those which bind to the EGF receptor.

SOURCE OF NEUROTROPHIC FACTORS FOLLOWING INTRASTRIATAL TRANSPLANTS

Following placement of transplants, several changes in the neurochemical environment of the striatum occur. Obviously, any neurochemicals synthesized and secreted by the transplanted cells themselves can interact with nearby host tissue. It has also been shown that transplants alter the blood-brain barrier such that peripherally circulating compounds can affect the central nervous system (Sanberg et al., 1988). Thus, circulating neurotrophic factors might contribute to transplant effects. Intrastriatal transplants also are associated with increases in the population of microglia and macrophages in the surrounding tissue (Lu et al., 1991). Such cell types have been shown to express TGFa and other neurotrophic factors (Rappolee et al., 1988). In addition, the placement of intrastriatal transplants has been shown the increase the population of astrocytes in the host brain (Whitaker-Azmitia et al., 1987). In particular, we have shown that a large increase in the population of astrocytes expressing TGFa is associated with transplants of a number of fetal and adult tissues, including fetal ventral mesencephalon and adrenal medulla (Loughlin et al., in preparation and see below).

ADMINISTRATION OF NEUROTROPHIC FACTORS IN *IN VIVO* NEUROTOXIC LESION MODELS

In order to determine whether any of the efficacious effects of transplants might be transduced by neurotrophic factors, a few studies have examined the effects of administration of neurotrophic factors in neurotoxic lesion models. It has been reported that addition of nerve growth factor (NGF) to transplants of adrenal medulla decreases rotational behavior and increases sprouting of tyrosine hydroxylase immunoreactive fibers (Pezzolli et al., 1988). NGF alone, however, did not improve behavior. Peripheral injections of GM-1 ganglioside have recently been shown to accelerate recovery in MPTP treated primates (Schneider et al., 1992). Peripheral injections of substance P shortly after 6-OHDA lesions decreased behavioral deficits in animals which have subtotal, but not complete, depletion of striatal dopamine (Mattioli et al., 1992). Intraventricular infusions of EGF, which binds to the same receptor as TGFa in other tissues (Korc et al., 1991), have been shown to increase striatal dopamine content in MPTP lesioned mice (Hadjiconstantinou et al., 1991). EGF also increased the number of surviving tyrosine hydroxylase immunoreactive cells in the substantia nigra and fibers in the striatum when infused after knife cuts of the nigrostriatal pathway (Pezzoli et al., 1991). Untreated animals showed 80% loss of dopamine neurons. EGF treated animals also showed fewer rotations in response to amphetamine than untreated animals.

Transforming growth factor alpha

Several lines of experiments in our laboratory have suggested that one nigrostriatal growth factor, TGFa, may play an important role in the efficacious effects of transplants and may act by promoting regeneration of dopaminergic cells (Loughlin et al., 1989; Fallon et al., 1990). While TGFa is similar to EGF in many of its actions, TGFa is more potent than EGF in a number of systems (Ebner and Derynck, 1991). Furthermore, TGFa has recently been shown to support a population of dorsal root ganglion neurons which

EGF does not support (Chalazonitis et al., 1992). EGF and TGFa may act via a common receptor and/or may act on separate receptor populations in brain.

TGFa is a 50 amino acid peptide, originally isolated from media conditioned by transformed mouse cells (De Larco and Todaro, 1978; Twardzik et al., 1982) and human tumor cells (Todaro et al., 1980). TGFa has also been identified in embryonic tissue (Lee et al., 1985a) and, therefore, may act as a growth factor on normal cells (Proper et al., 1980, Tam, 1985; Twardzik, 1985). The TGFa precursor (TGFaP) has been isolated and characterized (Derynck et al., 1984; Twardzik and Ranchalis, 1987; Burgess, 1989; Puolakkainen and Twardzik, 1993). The 159 amino acid precursor (Gentry et al., 1987) contains a transmembrane sequence and cytoplasmic domain which is highly conserved between species (Derynck et al., 1984; Derynck, 1987; Bringman et al., 1987). The active TGFa peptide is proteolytically cleaved, released (Bringman et al., 1987), and exerts many of its effects by binding to the EGF-receptor (Massague, 1983; Korc et al., 1991).

TGFa m-RNA is present in developing (Junier et al., 1991) and adult rat brain (Lee et al., 1985b; Lazar and Blum, 1992) and is heterogeneously distributed (Loughlin et al., 1989; Kudlow et al,. 1989), exhibiting higher levels in basal ganglia and ventral mesencephalon. In situ hybridization autoradiography has also demonstrated the presence of TGFa m-RNA in several areas of mouse and rat brain, including striatum (Wilcox and Derynck, 1988; Seroogy et al., 1993). TGFa immunoreactivity (Code et al., 1987; Kudlow et al., 1989) and TGFa precursor-like immunoreactivity (TGFaP-IR) is present in discrete brain regions in adult rat brain. TGFaP-IR is localized to a subpopulation of astrocytes and may also be present in neurons (Loughlin et al., 1989; Fallon et al., 1990; Junier et al., 1991). We have previously shown that five to ten percent of astrocytes in a number of brain regions, including striatum and SN-VTA, exhibit TGFaP-IR (Fallon, et al 1990). These results, and those of others, demonstrating the presence of TGFa in brain are consistent with a neurotrophic role of TGFa, for example, in maintenance and/ or plasticity in the adult nigrostriatal system.

TGFa has been shown to have trophic effects on CNS cells, including SN-VTA cells. Treatment of neuronal cultures with TGFa results in an increase in survival and process outgrowth (Chalazonitis et al., 1992). TGFa and EGF, which binds to the same receptor (Korc et al., 1991), cause an increase in dopamine uptake in ventral mesencephalic cultures (Knusel et al., 1990; Alexi et al., 1991). Since EGF does not have this effect in cultures treated with cytosine arabinoside, it has been suggested that it acts indirectly through glial production of another growth factor. However, in some cell lines, EGF has also been shown to stimulate expression of the dopamine synthetic enzyme, tyrosine hydroxylase (TH) (O'Malley and Wagner, 1988) and dopamine receptors (Missale, et al, 1991). In rat brain, EGF decreases some of the deleterious effects of dopaminergic toxins (Hadjiconstantinou et al., 1991). TGFa has been shown to cause both gliogenesis and neurogenesis from striatal embryonic precursor cells in vitro (Reynolds et al, 1990). Both neurons and glia have also been generated from adult striatal tissue under the influence of EGF (Reynolds and Weiss, 1992).

TGFa and intrastriatal transplants

Studies on the endogenous expression of growth factors are complementary to those which examine the effects of exogenous application of growth factors. Lesions of the nigro-striatal projection resulted in a minor increase in TGFaP-IR astrocytes, while transplants of adrenal medulla suspension caused a greater, though still modest, increase in TGFaP-IR. Transplants of fetal cortex suspension, which have been shown to cause motor deficits (Lu et al., 1990), produced a major increase in TGFaP-IR, but greatly disrupted the normal striatal architecture. Following intrastriatal transplantation of fetal ventral mesencephalon suspension into the striatum, a marked increase in TGFaP-IR was observed in the region of the transplant. Double-labeling experiments showed that virtually all TGFaP-IR was co-localized with glial fibrillary acidic protein-IR, suggesting that astrocytes were the source of TGFa in ventral mesencephalic transplants. Preliminary in situ hybridization studies suggested that TGFa mRNA was also upregulated by transplants, in parallel with TGFaP-IR. This suggests that the TGFaP is present in striatal glia and levels are greatly increased following transplants.

TGFa infusions

Since increases in TGFa are associated with efficacious transplants, we propose that TGFa may play a role in the ameliorative effects of transplants. To test this hypothesis, lesioned animals were given cell-free infusions of TGFa. Rats received unilateral 6-OHDA lesions of the dopaminergic projection to the striatum and apomorphine (0.25 mg/kg ip) -induced rotation behavior was quantified. One group of animals then received intrastriatal infusions of TGFa (0.05 µg/µl) in artificial cerebrospinal fluid (aCSF) via an Alzet minipump (2002). 200 µl was infused into the striatum on the lesion side over a two week period. A control group received infusions of aCSF alone. Rotation behavior was quantified one and two weeks after placement of the pump and animals were sacrificed. Brains were processed for localization of TGFaP-IR and TH-IR.

aCSF infusions produced a modest increase in endogenous TGFaP-IR. Infusions of TGFa caused a greater increase in TGFaP-IR and TGFa mRNA, indicating that the adult brain is capable of increasing the population of TGFa synthesizing astrocytes in the absence of fetal cell transplants. The effects of direct intrastriatal infusions on 6-OHDA induced motor deficits were also examined. Apomorphine induced rotation behavior was unchanged in animals which received aCSF infusions ($p > 0.4$). Infusions of TGFa, however, caused a significant, 40% decrease in rotation ($p < 0.005$).

The mechanism by which TGFa infusions improve motor behavior in 6-OHDA lesioned animals remains unknown. Behavioral recovery may reflect regeneration of dopaminergic afferents, increased efficacy of remaining dopaminergic terminals, normalization of dopamine receptor populations or compensatory changes in other neurochemical systems. The output of the caudate-putamen is modulated by a number of neurotransmitter systems, including monoaminergic, cholinergic, glutamatergic and peptide systems (Fallon and Loughlin, 1987) which exhibit marked plasticity. While rotation behavior has only limited value in the examination of nigrostriatal function, the results suggest that TGFa might be efficacious in normalizing behavior. In our TGFa infusion studies, immunocytochemical localization of TH suggested that dopaminergic fibers might have regenerated into the 6-OHDA lesioned striatum. In brains which received TGFa infusions, a population of TH-IR fibers was observed which was localized to the region surrounding infusions. Whether these fibers represent regenerating substantia nigra afferents or sprouting of other TH-IR fibers remains unknown. However, since endogenous TGFaP-IR is increased by efficacious transplants, and TGFa infusions reduce lesion-induced motor deficits, it is possible that the efficacious effects of transplants are mediated directly by TGFa. The role of other neurotrophic factors remains to be determined.

Acknowledgements

The support of the Parkinson's Disease Foundation is gratefully acknowledged. This research was also supported by NS 26761 to SEL and NS 15321 to JHF. Active TGFa was generously provided by Dr. Daniel Twardzik at Bristol-Myers Squibb.

REFERENCES

Alexi, T., and Hefti, F., 1993, Trophic actions of transforming growth factor alpha on mesencephalic dopaminergic neurons developing in culture, *Neuroscience* in press.

Alexi, T., Denton, T.L., and Hefti, F., 1991, Effects of TGF-alpha and TGF-beta on ventral mesencephalic dopaminergic cultures, *Neurosci. Abst.* 17.

Allen, G.S., Burns, S., Tulipan, N.B., and Parker, R.A., 1989, Adrenal medullary transplantation to the caudate nucleus in Parkinson's Disease, *Arch Neurol.* 46:487-491.

Bankiewicz, K.S., Plunkett, R.J., Jacobowitz, D.M., Kopin, I.J., and Oldfield, E.H., 1991, Fetal nondopaminergic neural implants in parkinsonian primates, *J. Neurosurg.* 74:97-104.

Bean, A.J., Elde, R., Cao, Y., Oellig, C., Tamminga, C., Goldstein, M., Pettersson, R.F., and Hokfelt, T., 1991, Expression of acidic and basic fibroblast growth factors in the substantia nigra of rat, monkey, and human, *Proc. Natl Acad. Sci.* 88:10237-10241.

Bernheimer, H., Birkmayer, W., Hornykiewicz, O., Jellinger, K., and Seitelberger, K., 1973, Brain dopamine and the syndrome of Parkinson and Huntington: Clinical, morphological, and neurochemical correlations, *J. Neurol. Sci.* 20:415-455.

Bjorklund, A., Schmidt, R.H., and Stenevi, U., 1980, Functional re-innervation of the neostriatum in the adult rat by use of intraparenchymal grafting of dissociated cell suspensions from the substantia nigra, *Cell Tiss. Res.* 212:39-45.

Bjorklund, A., Stenevi, U., Schmidt, R.H., Dunnett, S.B., and Gage, F.H., 1983a, Intracerebral grafting of neuronal cell suspensions. I. Introduction and general methods of preparation, *Acta Physiol. Scand., Suppl.* 522:1-7.

Bjorklund, A., Stenevi, U., Schmidt, R.H., Dunnett, S.B., and Gage, F.H., 1983b, Intracerebral grafting of neuronal cell suspensions. II. Survival and growth of nigral cell suspensions implanted in different brain sites, *Acta Physiol. Scand., Suppl.* 522:9-16.

Bjorklund, A., Lindvall, O., Isacson, O., Brundin, P., Wictorin, K., Strecker, R.E., Clarke, D.J., and Dunnett, S.B., 1987, Mechanisms of action of intracerebral neural implants: studies of nigral and striatal grafts to the lesioned striatum, *Trends Neurosci.* 10:509-516.

Bohn, M.C., Cupit, L., Marciano, F., and Gash, D.M., 1987, Adrenal medulla grafts enhance recovery of striatal dopaminergic fibers, *Science* 237:913-916.

Bregman, B.S., and Reier, P.J., 1986, Neural tissue transplants rescue axotomized rubrospinal cells from retrograde death, *J. Comp. Neurol.* 244:86-95.

Bringman, T.S., Lindquist, P.B., and Derynck, R., 1987, Different transforming growth factor-alpha species are derived from a glycosylated and palmitoylated transmembrane precursor, *Cell* 48:429-440.

Burgess, A.W., 1989, Epidermal growth factor and transforming growth factor alpha, *British Medical Bulletin* 45(2):401-424.

Casper, D., Mytilineou, C., and Blum, M., 1991, EGF enhances the survival of dopamine neurons in rat embryonic mesencephalon primary cell culture, *J. Neurosci. Res.* 30:372-381.

Chalazonitis, A., Kessler, J.A., Twardzik, D.R., and Morrison, R.S., 1992, Transforming growth factor alpha, but not epidermal growth factor, promotes the survival of sensory neurons in vitro, *J. Neurosci.* 12(2): 583-594.

Code, R.A., Seroogy, K.B., and Fallon, J.H., 1987, Some transforming growth factor-alpha connections and their colocalization with enkephalin in the rat central nervous system, *Brain Res.* 421:401-405.

De Larco, J.E., and Todaro, G.J., 1978, Growth factors from murine sarcoma virus-transformed cells, *Proc. Nat. Acad. Sci.* 75:4001-4005.

Derynck, R., Roberts, A.B., Winkler, M.E., Chen, E.Y., and Goeddel, D.V.,1984, Human transforming growth factor alpha: Precursor structure and expression in E. coli, *Cell* 38:287-297.

Derynck, R., 1987, Structure of transforming growth factors -a and -ß and their precursors, *in* : "Oncogenes, Genes, and Growth Factors," G. Guroff, ed., John Wiley and Sons, New York, pp. 133-163.

Ebner, R., and Derynck, R., 1991, Epidermal growth factor and transforming growth factor-alpha: Differential intracellular routing and processing of ligand-receptor complexes, *Cell Regulation* 2:599-612.

Fallon, J.H., Di Salvo, J., Loughlin, S.E., Gimenez-Gallago, G., Seroogy, K.B., Bradshaw, R.A., Morrison, R.S., Ciofi, P., and Thomas, K.A., 1992, Localization of acidic fibroblast growth factor within the mouse brain using biochemical and immunocytochemical techniques, *Growth Factors* 6:139-157.

Fallon, J.H., Loughlin, S.E., Annis, C.M., Gentry, L., and Twardzik, D., 1990, Localization of cells containing transforming growth factor alpha precursor immunoreactivity in the basal ganglia of the adult rat brain, *Growth Factors* 2:241-250.

Fallon, J.H., and Loughlin, S.E., 1987, Monoamine innervation of cerebral cortex and a theory of the role of monoamines in cerebral cortex and basal ganglia, *in*: "Cerebral Cortex, Vol. 6," E.G. Jones and D. Peters, eds., Plenum Press, New York pp. 41-127.

Fallon, J.H., and Loughlin, S.E., 1993, Functional implications of the anatomical localization of neurotrophic factors, *in*: "Neurotrophic Factors," S.E. Loughlin and J.H. Fallon, eds., Academic Press, San Diego pp. 1-24.

Freed, C.R., Breeze, R.E., Rosenberg, N.L., Schneck, S.A., Wells, T.H., Barrett, J.N., Grafton, S.T., Mazziotta, J.C., Eidelberg, D., and Rottenberg, D.A., 1990, Therapeutic effects of human fetal dopamine cells transplanted in a patient with Parkinson's disease, *Progress Brain Res.* 82:715-721.

Freed, W.J., and Cannon-Spoor, H.E., 1988, Cortical lesions increase reinnervation of the dorsal striatum by substantia nigra grafts, *Brain Res.* 446:133-143.

Gall, C.M., Gold, S.J., Isackson, P.J., and Seroogy, K.B., 1992, Brain-derived neurotrophic factor and neurotrophin-3 mRNAs are expressed in ventral midbrain regions containing dopaminergic neurons, *Mol. Cell. Neurosci.* 3:56-63.

Gentry, L.E., Twardzik, D.R., Lim, G.J., Ranchalis, J.E., and Lee, D.C., 1987, Expression and characterization of transforming growth factor precursor protein in transfected mammalian cells, *Mol. and Cell. Biol.* 7(5):1585-1591.

Hadjiconstantinou, M., Fitkin, J.G., Dalia, A., and Neff, N.H., 1991, Epidermal growth factor enhances striatal dopaminergic parameters in the 1-methyl-4-phenyl-1,2,3,6-tetrahydropyridine-treated mouse, *J. Neurochem.* 57(2):479-482.

Hansen, J.T., Bing, G., Notter, M.F.D., and Kordower, J.H., 1989, Adrenal chromaffin cells as transplants in animal models of Parkinson's Disease, *J. Electron Microscopy Technique* 12:308-315.

Hefti, F., Denton, T.L., Knusel, B., and Lapchak, P.A., 1993, Neurotrophic factors: What are they and what are they doing?, *in*: "Neurotrophic Factors," S.E. Loughlin and J.H. Fallon, eds., Academic Press, San Diego, pp. 25-50.

Hofer, M., Pagliusi, S.R., Hohn, A., Leibrock, J., and Barde, Y.-A., 1990, Regional distribution of brain-derived neurotrophic factor mRNA in the adult mouse brain, *EMBO J* 9:2459-2464.

Hyman, C., Hofer, M., Barde, Y.-A., Juhasz, M., Yancopoulos, G.D., Squinto, S. P., Lindsay, R.M., 1991, BDNF is a neurotrophic factor for dopaminergic neurons of the substantia nigra, *Nature* 350:230-232.

Junier, M.-P., Ma, Y.J., Costa, M.E., Hoffman, G., Hill, D.F., Ojeda, S.R., 1991, Transforming growth factor alpha contributes to the mechanism by which hypothalamic injury induces precocious puberty, *Proc. Natl. Acad. Sci. USA* 88:9743-9747.

Knusel, B., Michel, P.P., Schwaber, J.S., Hefti, F., 1990, Selective and nonselective stimulation of central cholinergic and dopaminergic development in vitro by nerve growth factor, basic fibroblast growth factor, epidermal growth factor, insulin and the insulin-like growth factors I and II, *J. Neurosci.* 10(2):558-570.

Korc, M., Chandrasekar, B., and Shah, G., 1991, Differential binding and biological activities of epidermal growth factor and transforming growth factor alpha in a human pancreatic cancer cell line, *Cancer Res* 51, 6243-6249.

Kromer, L.F., Bjorklund, A., and Stenevi, U., 1981, Regeneration of the septohippocampal pathways in adult rats is promoted by utilizing embryonic hippocampal implants as bridges, *Brain Res.* 210:173-200.

Kudlow, J.E., Leung, A.W.C., Kobrin, M.S., Paterson, A.J., and Asa, S.L., 1989, Transforming growth factor-a in the mammalian brain, *J. Biol. Chem.* 264:3880-3883.

Lazar, L.M., and Blum, M., 1992, Regional distribution and developmental expression of epidermal growth factor and transforming growth factor - alpha mRNA in mouse brain by a quantitative nuclease protection assay, *J. Neurosci.* 12(5):1688-1697.

Lee, D.C., Rochford, R., Todaro, G.J., and Villarreal, L.P., 1985a, Developmental expression of rat transforming growth factor-alpha mRNA, *Mol. and Cell. Biol.* Dec.:3644-3646.

Lee, D.C., Rose, T.M., Webb, N.R., and Todaro, G.J., 1985b, Cloning and sequence analysis of a cDNA for rat transforming growth factor-a, *Nature* 313:489-491.

Lieberman, A., Ransohoff, J., Berczeller, P., Brous, P., Eng, K., Goldstein, M., Kaufman, B., Koslow, M., and Chin, L., 1989, Adrenal medullary transplants as a treatment for advanced Parkinson's Disease, *Acta Neurol. Scand.* 126:189-196.

Lindvall, O., Rehncrona, S., Brundin, P., Gustavii, B., Astedt, B., Widner, H., Lindholm, T., Bjorklund, A., Leenders, K., Rothwell, J., Frackowiak, R., Marsden, C.D., Johnels, B., Steg, G., Freedman, R., Hoffer, B., Seiger, A., Bygdeman, M., Stromberg, I., and Olson, L., 1989, Human fetal dopamine neurons grafted into the striatum in two patients with severe Parkinson's disease: A detailed account of methodology and a 6-month follow-up, *Arch. Neurol.* 46:615-631.

Loughlin, S.E., Baratta, J., Gentry, L., Twardzik, D., Annis, C.M., and Fallon, L.H., Transforming growth factor alpha precursor immunoreactive astrocytes: Modulation by injury and transplants of fetal suspensions, in preparation.

Loughlin, S.E., Annis, C.M., Twardzik, D.R., Lee, D.C., Gentry, L., and Fallon, J.H., 1989, Growth factors in opioid rich brain regions: Distribution and response to intrastriatal transplants, *Adv. Biosci.* 75:403-406.

Loughlin, S.E., and Fallon, J.H., 1993, "Neurotrophic Factors," Academic Press, San Diego.

Lu, S.Y., Shipley, M.T., Norman, A.B., and Sanberg, P.R., 1991, Striatal, ventral mesencephalic and cortical transplants into the intact rat striatum: A neuroanatomical study, *Exp. Neurol.* 113:109-130.

Lu, S.Y., Giordano, M., Norman, A.B., Shipley, M.T., and Sanberg, P.R., 1990, Behavioral effects of fetal tissue transplants into the intact striatum, *Pharm. Biochem. Behav.* 27:1-14.

Mattioli, R., Schwarting, R.K.W., and Huston, J.P., 1992, Recovery from unilateral 6-hydroxydopamine lesion of substantia nigra promoted by the neurotachykinin substance P_{1-11}, *Neuroscience* 48(3):595-605.

Massague, J., 1983, Epidermal growth factor-like transforming growth factor receptors in human plasma membranes on A432 cells, *J. Biol. Chem.* 258:13614-13620.

Missale, C., Castelletti, L., Boroni, F., Memo, M., and Spano, P., 1991, Epidermal growth factor induces the functional expression of dopamine receptors in the GH3 cell line, *Endocrinology* 128(1):13-20.

O'Malley, E.K., Black, I.B., and Dreyfus, C.F., 1991, Local support cells promote survival of substantia nigra dopaminergic neurons in culture, *Exp. Neurol.* 112:40-48.

O'Malley, K.L., and Wagner, C., 1988, Growth factor effects on tyrosine hydroxylase messenger RNA in rat pheochromocytoma cell lines, *in*: "Progress in Catecholamine Research," A. Dahlstrom, R. Belmaker and M. Sandler, eds., Alan R. Liss, New York, pp. 35-40.

Perlow, M.J., Freed, W.J., Hoffer, B.J., Seiger, A., Olson, L., and Wyatt, R.J., 1979, Brain grafts reduce motor abnormalities produced by destruction of nigrostriatal dopamine system, *Science* 204:643-646.

Pezzoli, G., Fahn, S., Dwork, A., Truong, D.D., de Yebenes, J. G., Jackson-Lewis, V., Herbert, J., and Cadet, J.L., 1988, Non-chromaffin tissue plus nerve growth factor reduces experimental parkinsonism in aged rats, *Brain Res.* 459:398-403.

Pezzoli, G., Zecchinelli, A., Ricciardi, S., Burke, R.E., Fahn, S., Scarlato, G., and Carenzi, A., 1991, Intraventricular infusion of epidermal growth factor restores dopaminergic pathway in hemiparkinsonian rats, *Movement Disorders* 6(4):281-287.

Proper, J.A., Bjornson, C.L., and Moses, H.L., 1980, Mouse embryos contain polypeptide growth factors capable of inducing a reversible neoplastic phenotype in non-transformed cells in culture, *J. Cell Physiol.* 110:169-174.

Puolakkainen, P., and Twardzik, D.R., 1993, Transforming growth factors alpha and beta, *in*: "Neurotrophic Factors," S.E. Loughlin and J.H. Fallon, eds., Academic Press, San Diego, pp. 359-390.

Rappolee, D.A., Mark, D., Banda, M.J., and Werb, Z., 1988, Wound macrophages express TGF alpha and other growth factors in vivo: Analysis by mRNA phenotyping, *Science* 241:708-712.

Reynolds, B.A., Tetzlaff, W., and Weiss, S., 1990, EGF- and TGF alpha - responsive striatal embryonic progenitor cells produce both neurons and astrocytes, *Abstr. Soc. Neurosci.* 474(2).

Reynolds, B.A., and Weiss, S., 1992.,Generation of neurons and astrocytes from isolated cells of the adult mammalian central nervous system, *Science* 255:1707-1709.

Rousselet, A., Fetler, L., Chamak, B., and Prochiantz, A., 1988, Rat mesencephalic neurons in culture exhibit different morphological traits in the presence of media conditioned on mesencephalic or striatal astroglia, *Developmental Biology.* 129:495-504.

Sanberg, P.R., Nash, D.R., Calderon, S.F., Giordano, M., Shipley, M.T., and Norman, A.B., 1988, Neural transplants disrupt the blood-brain barrier and allow peripherally acting drugs to exert a centrally mediated behavioral effect, *Exp. Neurol.* 102:149-152.

Schneider, J.S., Pope, A., Simpson, K., Taggart, J., Smith, M.G., and DiStefano, L., 1992, Recovery from experimental Parkinsonism in primates with G_{M1} ganglioside treatment, *Science* 256:843-846.

Seroogy, K.B., Lundgren, K.H., Lee, D.C., Guthrie, K.M., and Gall, C.M., 1993, Cellular localization of ttransforming growth factor alpha mRNA in rat forebrain, *J. Neurochem.* in press.

Sladek, J.R., Jr., and Shoulson, I., 1988, Neural transplantation: A call for patience rather than patients, *Science* 240:1386-1388.

Spina, M.B., Squinto, S.P., Miller, J., Lindsay, R.M., and Hyman, C., 1992, Brain-derived neurotrophic factor protects dopamine neurons against 6-hydroxydopamine and N-methyl-4-phenylpyridinium ion toxicity: Involvement of the glutathione system, *J. Neurochem.* 59:99-106.

Tam, J.P., 1985, Physiological effects of transforming growth factor in the newborn mouse, *Science* 229:673-675.

Todaro, G.J., Fryling, C., and De Larco, J.E., 1980, Transforming growth factors produced by certain human tumor cells: polypeptides that interact with epidermal growth factor receptors, *Proc. Natl. Acad. Sci. (USA)* 77:5258-5262.

Twardzik, D.R., 1985. Differential expression of transforming growth factor-a during prenatal development of the mouse, *Cancer Res.* 45:5413-5416.

Twardzik, D.R., Todaro, G.J., Reynolds, F.H., Jr., and Stephenson, J.R., 1982, Abelson MuLV induced transformation involves production of a polypeptide growth factor, *Science* 216:894-897.

Twardzik, D.R., and Ranchalis, J.E., 1987, Growth modulating peptides that utilize the receptor for epidermal growth factor, *Bone Matrix* 421-437.

Whitaker-Azmitia, P.M., Ramirez, A., Noreika, L., Gannon, P.J., and Azmitia, E.C., 1987, Onset and duration of astrocytic response to cells transplanted into the adult mammalian brain, *in*: "Cell Tissue Transplantation Into the Adult Brain," E.C. Azmitia and A. Bjorklund, eds., New York Academy Press, New York., pp. 10-23.

Wilcox, J.N., and Derynck, R., 1988, Localization of cells synthesizing transforming growth factor-alpha mRNA in the mouse brain, *J. Neurosci.* 8(6):1901-1904.

Yurek, D.M., and Sladek, J.R., Jr., 1990, Dopamine cell replacement: Parkinson's Disease, *Annu. Rev. Neurosci.* 13:415-440.

NEUROPHYSIOLOGY OF THE BASAL GANGLIA

A DISTRIBUTED NETWORK OF CONTEXT-DEPENDENT FUNCTIONAL UNITS IN THE RAT NEOSTRIATUM

Lucy L. Brown[1], Samuel M. Feldman[2], Ivan Divac[3], Peter J. Hand[4] and Theodore I. Lidsky[5]

[1]Albert Einstein College of Medicine, Bronx, NY 10461 USA
[2]New York University, New York, NY 10003 USA
[3]University of Copenhagen, Panum Institute, Copenhagen 2200, Denmark
[4]University of Pennsylvania School of Veterinary Medicine, Philadelphia, PA 19104 USA
[5]New York State Institute for Basic Research, Staten Island, NY 10314 USA

INTRODUCTION

Since the early descriptions of the uneven distribution of dopamine terminals (Olson et al., 1972), opiate receptors (Pert et al., 1976) and acetylcholinesterase (Graybiel and Ragsdale, 1978), a significant chemoarchitectonic organization of the striatum has been well documented, with an ever-growing number of markers found in either one of two compartments: patch (striosome), and matrix (reviews in Graybiel, 1990; Gerfen, 1992). In contrast to the advances in knowledge about chemoarchitecture, functional microanalysis of the striatum is lagging. There is no functional parcellation that can be compared to the chemoarchitectonics. However, a number of studies in which different techniques were employed do suggest a certain degree of functional heterogeneity (for an early review, see Divac, 1968). For example, lesions induce locus-specific effects in all species studied (see reviews in Öberg and Divac, 1979 and Divac and Öberg, 1992). In addition, the consequences of electrical stimulation (Alexander and DeLong, 1985), or movement during cell recording (Crutcher and DeLong, 1984; Carelli and West, 1991) also depend upon the sampled site. These techniques, however, have some drawbacks: they are invasive, may affect unspecified neighboring tissue, and often demand *a priori* decisions about the site to be manipulated. Getting around some of these problems can be a difficult task. For example, elaborate controls were needed to eliminate the possibility that behavioral effects of striatal lesions were not an artifact resulting from accidental damage of passing fibers (Divac et al., 1978; Divac and Diemer, 1980). More recent techniques, based on regional changes of metabolism or blood flow, avoid the drawbacks of the classical techniques, and have opened new ways to study functional heterogeneity of the striatum. We have used one of these techniques, ^{14}C deoxyglucose autoradiography, to look for functionally defined zones ("units", "modules") in the striatum. The problem has been to locate *functionally* defined units, compartments or other descriptive zones in striatum that may resemble, parallel or coincide with the chemoarchitectural compartments.

Corticostriate anterograde tracing studies have provided important clues to a functional unit of the striatum, the cortex being a major source of striatal afferents. Somatosensory and motor cortex projections to striatum are generally somatotopically organized, but they extend anteroposteriorly and mediolaterally for several millimeters, well beyond what one would expect from a non-distributed, single linear representation (Künzle, 1975; Künzle, 1977; Flaherty and Graybiel, 1991a; Flaherty and Graybiel, 1991b). In addition, the projection is distributed into discrete zones (Künzle, 1977; Flaherty and Graybiel, 1991a). The somatosensory projections are also anisotropic: the somatotopic arrangement varies from one anteroposterior level to another (Malach and Graybiel, 1986). Furthermore, the discrete zones of both motor and sensory cortex projections into the striatum vary in concentration of labelled terminals, and appear to be localized predominantly in the matrix compartment (Malach and Graybiel, 1986; Flaherty and Graybiel, 1991a; Flaherty and Graybiel, 1991b). Thus a functional unit of the cortex (a primary motor or sensory cortex representation of one body region) distributes its influence anteroposteriorly into multiple discrete striatal zones.

EXPERIMENTAL APPROACH

With the goal of defining a functional unit in striatum that can be related to global behavioral variables and striatal chemoarchitecture, we have concentrated on the somatosensory-motor system and the ^{14}C 2-deoxyglucose (DG) technique (Sokoloff et al., 1977). The somatosensory-motor system is well understood both anatomically and functionally, second only to the visual system. In the rat and in other species, electrophysiological studies have localized hindlimb, trunk, forelimb and vibrissae regions in cortex (Hall and Lindholm, 1974; Chapin and Lin, 1984). The somatotopic organization permits a simple and reliable correlation of structure and function. Furthermore, anatomic-functional units in the cortical somatosensory area that represent vibrissae in rats are easily recognized, and enable a search for corresponding "units" in striatum. Finally, the sensori-

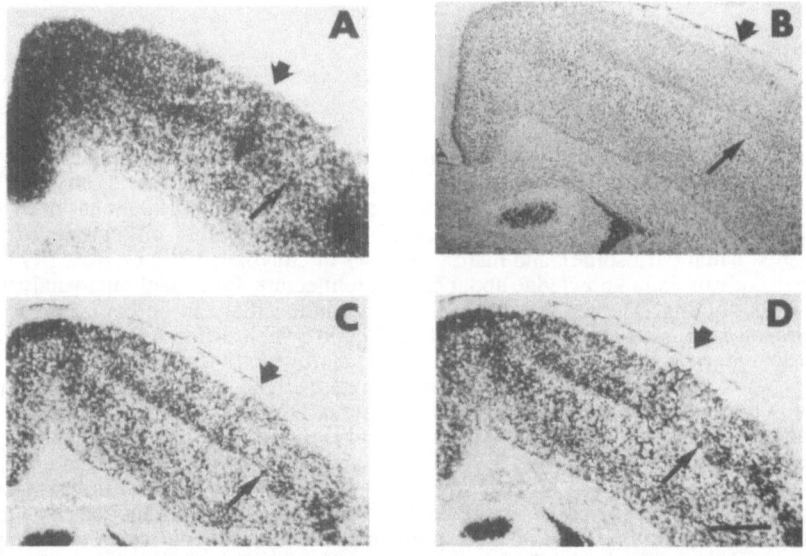

Figure 1. Representation of increased cortical metabolic activity associated with tactile stimulation of forelimb. **A.** Autoradiogram digitized for glucose utilization rate. The region of activation is indicated by the top arrow. **B.** Adjacent thionin-stained section. Arrows indicate the same regions as in A. (The digitized sections are not at the same magnification.) **C.** Digitized image of stained section in B to permit alignment with autoradiogram in A. Arrows as in A and B. **D.** The region of activation outlined on the digitized stained section seen in C. A computer-assisted algorithm was used to determine the edges and outline the region of increased glucose utilization. Activation is in granular cortex and includes layers II-IV, and parts of layers Va and Vb. It is greatest in III and IV and upper Va. Arrows as in A-C. From Feldman and Brown, 1992. Bar = 1mm in A,C,D.

motor cortex has rich and elaborate bilateral projections to the striatum, unlike the primary visual cortex in primates (Saint-Cyr et al. 1990).

The use of the DG technique for striatal studies has several advantages. The technique estimates the level of neural activity at the same level of anatomical resolution as striatal chemoarchitecture. The technique allows work with fully conscious animals, permits measurements throughout the entire striatum and minimizes artifactual changes of striatal activity. To make measurements throughout the striatum is especially important because single cortical regions appear to project over an extensive anteroposterior region. It has been shown that regional variations of activity in the striatum can be detected (Divac and Diemer, 1980; Collins and Divac, 1984), even at a level of reduction that reveals "patchiness" (Divac, 1983; Brown et al., 1987). Such patchy areas of relatively high glucose utilization rates in striatum probably reflect predominantly activity in afferent systems.

Regional variations in glucose utilization observed with the DG technique appear to reflect, predominantly, axon terminal activity (Mata et al., 1980; Kadekaro et al., 1985; Kadekaro et al., 1987). Thus the activity in afferents to a nucleus, and the terminals of its intrinsic axon collaterals, are the principal source of metabolic demand reflected by the technique. The basal ganglia supply an instructive example. Electrophysiological studies find that in awake, resting animals, striatal cells are relatively inactive compared to globus pallidus cells. Yet, striatal glucose utilization rates are high, and globus pallidus rates are low. This is because the pallidal glucose utilization rate reflects the low firing rate of the striatal cells; even the spontaneous activity of pallidal cells does not bring glucose utilization rates up much over those of white matter. By contrast, although striatal cells have low rates of spontaneous activity, the glucose utilization rate is relatively high compared to the rest of

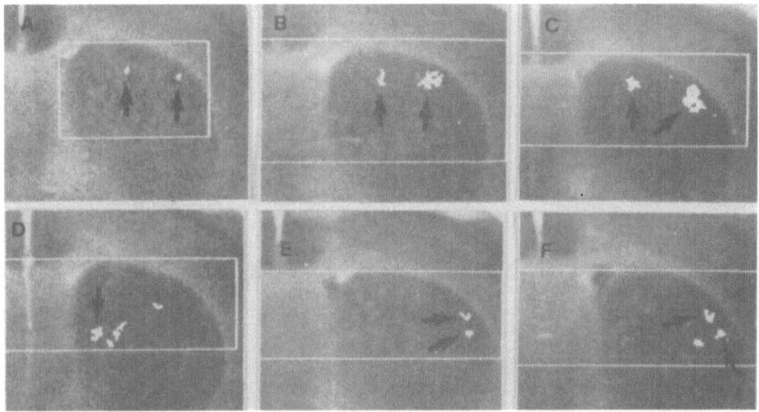

Figure 2. Digitized autoradiograms of coronal sections through right striatum in rat. The midline is to the left. The white areas (arrows) mask what we have called "features", discrete zones of 30 or more contiguous pixels ($\geq$ 10,000 μm^2) with optic densities greater than a pre-set criterion. The size of each feature was measured, and its position was defined according to a Cartesian coordinate system. A. Features in an animal stimulated with a nylon bristle on the left hindlimb footpad. The arrows indicate the two largest features. B. Features in an animal stimulated on the left hindlimb above the foot. The two largest features are in the same striatal regions as in the animal shown in A.. C. Features in another animal stimulated on the hindlimb above the foot. Localization of the two features is similar to that in animals shown in A and B. All of the five animals in the group showed features in these regions. D. A control animal. Controls showed patterns that were different from stimulated animals, and more within-group variability. E. Features in an animal stimulated on the left forelimb, on the wrist. Note that the two features form a dyad (arrows) in a striatal region different from the hindlimb-stimulated animals, in A-C. F. Features in another animal stimulated on the left forelimb, just above the wrist. Two features are in similar positions to the animal in E (arrows). The features also form a triad. All five animals in the forelimb group showed features in these same regions.

the brain; it is comparable to layer IV in cortex. This is apparently because striatal cells receive continuous input from cortex and elsewhere. The contribution of axons to regional metabolism appears to be limited. Changes in glucose utilization in the corpus callosum are rare, except with electrical stimulation of the cortex (e.g. Sharp et al., 1988). Thus the effects of physiologic stimuli on striatal glucose utilization reflect predominantly the activity of axon terminal activity from cortex, thalamus, substantia nigra, and intrinsic axon collaterals.

RESULTS

Somatotopic Representation in Cortex and Striatum

In a recent series of studies using the DG technique, we investigated somatotopy in cortex and striatum. We described cortical and striatal functional units using a nylon bristle brushing stimulus (2.5g) to the hindlimb, trunk, or forelimb of partially restrained, awake rats (Brown, 1992; Feldman and Brown, 1992). Cortical activation, observed in granular regions, formed apparent columns through layers III, IV, and parts of V (Fig. 1). The "columns" extended 200-800 μm mediolaterally, and 200-1000 μm anteroposteriorly (Feldman and Brown, 1992). The anteroposterior and mediolateral localization of DG activation related to body region agreed well with electrophysiological studies (Hall and Lindholm, 1974; Chapin and Lin, 1984).

Unlike cortex, striatal activation in these animals could not be detected by inspection with unaided eyes, because patchy activation is also present in control animals (Brown, et al., 1987). However, striatal activation could be objectively defined with computer-assisted techniques. Digitized autoradiograms of coronal sections through striatum were processed to detect the most dense areas (areas of highest metabolic rate), and the resulting images revealed multiple discrete zones of activity related to the stimuli (Fig. 2). We have called

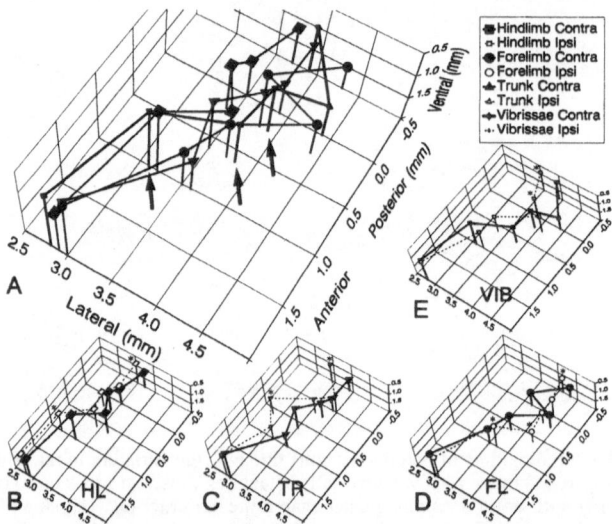

Figure 3. Three-dimensional graphical representations of mean positions of the largest features in striatum following hindlimb, trunk, forelimb, or vibrissae stimulation. The shifts in relative position of body part representation suggest the existence of a striatal map that permits different somatotopic juxtapositions: a combinational map. **A.** Contralateral to stimulation. Features for each body region reliably shift their position relative to each other at different anteroposterior levels (P<.01, group x anteroposterior interaction effect). The three arrows show where vibrissae locations are juxtaposed to either hindlimb, forelimb, or trunk positions. **B.-E.** Feature centroid position contralateral (solid lines) and ipsilateral (dotted lines) to stimulation in each group. At some anteroposterior levels the contra and ipsi features overlap, while at others they are offset (p<.05, matched t). Connecting lines are used only to emphasize the shifting positions and do not necessarily indicate continuous activation along the path of the line. * p<.05, matched t. HL=hindlimb. TR=trunk. FL=forelimb. VIB=vibrissae.

such discrete zones "features" (Brown, 1992; Brown and Feldman, 1993), for which the defining criterion is thirty or more contiguous pixels with optical densities greater than a preset level above detection threshold. The advantage of our computer-assisted detection technique is that the expectation of the investigator is not imposed on a pre-selected "region of interest" approach. Furthermore, the size and shape of functional units (contiguous pixels in the range of the highest densities) is revealed (Fig. 2), and both size and position of the centroid of each feature is measured. The technique is basically a form of "thresholding" (Russ, 1990) that we described in greater detail elsewhere (Brown and Feldman, 1993). It is important to note that features are also seen in unstimulated controls, which is to be expected in awake animals. In controls, however, the features are not localized as in stimulated animals, do not show within-group clustering of localized points around a single striatal region, and are smaller in size (Fig. 2; Brown, 1992; Brown and Feldman, 1993).

Features associated with somatosensory stimuli were found bilaterally throughout the middle 2/3 of striatum, anteroposteriorly, for each of the three body regions stimulated (Brown, 1992). Several features were localized in regions different from controls at most of the anteroposterior levels analyzed (Figs. 2 and 3). When features were ranked in each coronal section, from largest to smallest, we found that the first, second, third and, in one case, sixth largest were differently localized, compared to control. Furthermore, the relative arrangement of features representing different body parts varied at different anteroposterior levels (Fig. 3). This anisotropy of functional zones was consistent with the anisotropic arrangement of corticostriatal projections seen by Malach and Graybiel (1986) in the cat. These observations suggest that the somatosensory representation in striatum forms a combinational map (Brown, 1992) that allows input from each body region to combine with input from other body region representations at different anteroposterior levels. For example, the forelimb representation, elongated in the anteroposterior axis, is juxtaposed to the hindlimb region at one anteroposterior level, and to trunk at another. Eleven of the 15 possible pairs of feature centroids for six body regions studied fell within 300 μm at at least one anteroposterior level, reliably maintaining separation at others. (300 μm was chosen as an approximation of the diameter of the dendritic field formed by a medium spiny cell.) When data from a more recent study that used vibrissae stimulation (Brown and Lidsky, 1993) are added to the map in Figure 3, it can be seen that this additional body region is also arranged to allow juxtaposition to each of the other body regions, at different anteroposterior levels, which adds confirmation that the striatal somatosensory representation forms a combinational map. In addition, ipsilateral and contralateral representations of the same body part were not symmetrical in the striatum at several anteroposterior levels (Fig. 3; Brown, 1992). The offset of the two locations for the forelimb was as great as 450 μm, for hindlimb 450 μm, for trunk 600 μm, and for vibrissae 650 μm.

The Effect of Context on the Distribution and Pattern of Activation

In another series of studies, we varied the context in which stimuli were applied (Brown and Lidsky, 1993). Using subcutaneous electrodes, the vibrissae pad was stimulated electrically at 2/sec, to induce a twitch and somatosensory activation of two to three posterior vibrissae. One group had received eight days of prior training, during which a 30 minute period of stimulation was followed by presentation of a sucrose solution for drinking. Another group received the same treatment every day, but without the sucrose solution. A third group received no stimulation prior to the 2DG experimental session and was naive to the stimulation (data shown in Fig. 3). There was also a surgical control group. The vibrissae twitch stimulation produced a modest (10 %) increase in glucose utilization rate in the spinal nucleus of V and primary somatosensory cortex. In striatum, the positions of features related to the stimulation were seen in the dorsolateral quadrant at several anteroposterior levels (Fig. 4). All stimulation groups showed similarities as well as differences. At bregma, they exhibited one feature, or a cluster of three to four features in two regions, one ventrolateral and one dorsolateral (Fig. 5). The clusters were typically triads (Fig. 5B) or dyads (Fig. 5C). However, the sucrose group differed from the no-sucrose control group with the largest features located ventrally, while in the naive group the largest features were more dorsally located. The distributed features appear to have functional meaning related to context: Subsets of the larger distributed network appear to be more or less activated under different conditions.

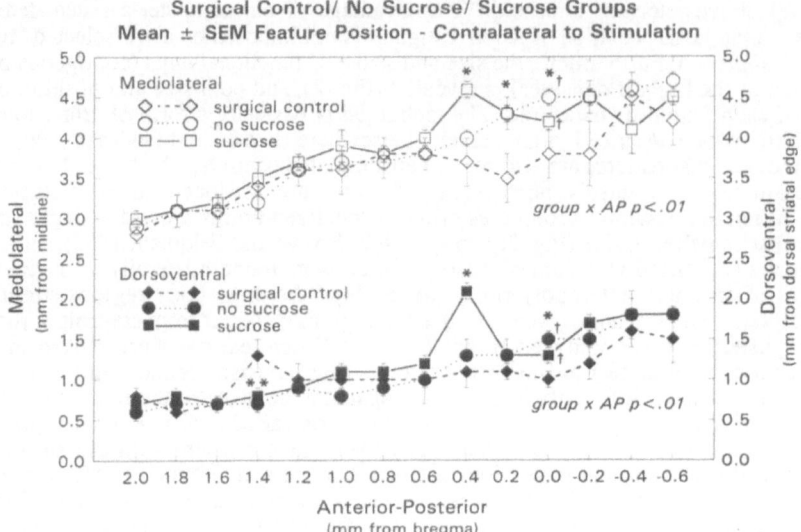

Figure 4. Mean mediolateral and dorsoventral position of the centroids of the largest striatal features identified in coronal sections through fourteen anteroposterior levels, contralateral to subcutaneous vibrissae pad stimulation. One group was a surgical control; one received sucrose at the end of 30 min training sessions; a third group was stimulated but did not receive sucrose at the end of the 30 min training sessions. The no-sucrose and sucrose stimulation groups are both different from surgical controls, and from each other (treatment group x anteroposterior level, p<.01). The anteroposterior regions where feature locations were different among groups are +1.4, +0.4, +0.2 and 0.0. * p<.05 compared to control, or, at +0.4, sucrose is compared to control and no sucrose at +0.4; † p<.05, ipsi compared to contra, matched t.

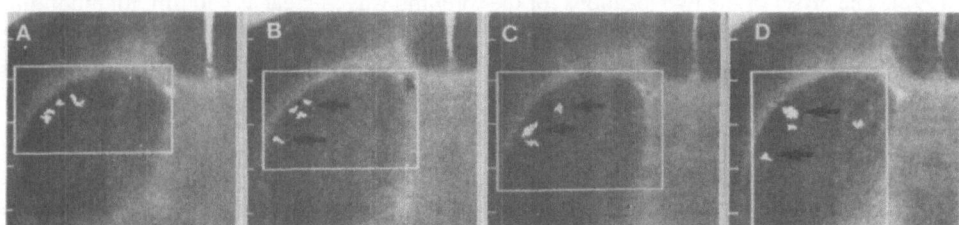

Figure 5. Features in left striatum of rat seen in coronal sections at bregma (AP 0.0) in individual animals stimulated via subcutaneous electrodes that caused vibrissae to move. Digitized images of glucose utilization autoradiograms are shown with features masked in white. Dorsal and ventral features (arrows) were seen in all stimulated animals. **A.** Surgical control. Four features are seen dorsolaterally. **B.** Stimulated, no-sucrose animal. A triad of features is seen dorsally, and one ventrolaterally. This pattern was repeated for each animal in the group. The mean largest feature for the group was the ventrolateral one. **C.** Stimulated, sucrose animal. Ventrolaterally, two features form a dyad, the dorsal one being the largest, a pattern that was present in every animal in the group. Note that dorsolaterally there is a feature in the region where there is a triad in the no-sucrose animal in *B*. **D.** Stimulated, naive to stimulation. The three major features are distributed dorsolaterally and ventrolaterally. The largest one in the naive animals tends to be in the dorsolateral region. A medially located feature was seen in some animals.

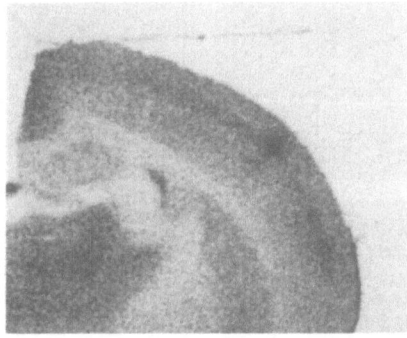

Figure 6. Autoradiogram of a coronal section through rat vibrissae somatosensory cortex following stimulation of vibrissa C3. The "column" of high activity corresponds to a cortical "barrel" (Kossut, et al., 1988).

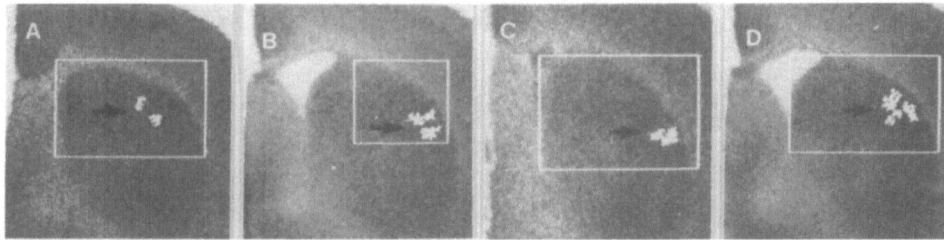

Figure 7. Digitized autoradiograms of coronal sections through right rat striatum, contralateral to mechanical brushing of vibrissa C3. The largest features are masked in white. The location and pattern of dyads and triads of features was similar in three out of four animals studies. **A.** Control animal at bregma. All vibrissae were trimmed, none brushed. **B.** An animal with Vibrissa C3 brushed. The dyad of features in the lateral striatum seen here was also seen in other animals, but not in the control group. **C.** Another animal with vibrissa C3 stimulated provides another example of a dyad in the lateral striatum at the level of bregma. **D.** A triad of features. Such triads were observed in three out of the four stimulated animals.

The Functional Unit Associated with Stimulation of a Single Vibrissa

In an attempt to identify a striatal somatosensory functional unit at a finer level of analysis, we have examined the effects of stimulation of a single vibrissa on patterns of activation in the cortex and striatum (Divac et al., 1992). In four rats, a single vibrissa (C3) was isolated by clipping (acutely) all other vibrissae. The vibrissa was then mechanically stimulated at 3-5 Hz with a rostrocaudal stroking motion (See Kossut et al., 1988). Four control animals also had all their vibrissae clipped acutely, but did not receive stimulation. The columnar or spindle-like cortical activation typically associated with vibrissa stimulation was observed (Fig. 6; Kossut et al., 1988).

In the striatum, features in the stimulated group were consistently localized in the dorsolateral quadrant over several anteroposterior levels, from 200 μm anterior to bregma to 200 μm posterior (Fig. 7). Controls did not show this pattern. In addition, feature size in stimulated animals was greater than in unstimulated controls (Fig. 7). A second active region, one mm rostral to bregma, was seen in all stimulated animals. These locations are similar to those found with electrical stimulation of the vibrissae pad in naive rats in the context study described above (Brown and Lidsky, 1993). The features consistently formed dyads and triads (Fig. 7B and 7C), as in the context study (Fig. 5B and 5C). Thus in spite of substantial procedural differences, the same patterns of striatal activation were seen in both studies.

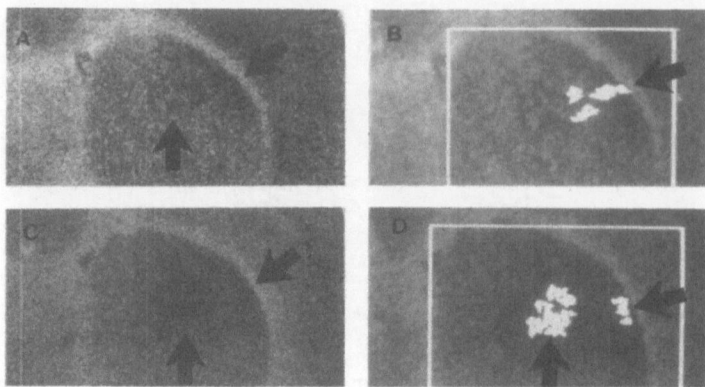

Figure 8. Autoradiograms of coronal sections through the right striatum of two rats that received electrical stimulation of either vibrissae somatosensory cortex (*A, B*), or vibrissae motor cortex (*C, D*). The two conditions produced patterns of activation with similarities and differences. **A.** Autoradiogram from a rat with activation of the entire vibrissae somatosensory cortex. The darkest regions form a lateral strip under the external capsule (arrow), where physiological stimulation of vibrissae also produces activation. Other, less activated regions are seen medially (arrow). **B.** Digitized image of *A* shows the largest features masked in white (arrows). **C.** Autoradiogram from a rat stimulated in motor cortex. Visible vibrissae movements were induced by the stimulation. A dark lateral strip under the external capsule, similar to that seen in *A*, above, may be the major input field for vibrissae *sensory* cortex. A large medial region is also activated (arrow), which corresponds to a less activated region in *A*, above. This may be the major input field for vibrissae *motor* cortex. **D.** Digitized image of *C* shows the largest features. (Adapted from Brown and Sharp, 1993).

Electrical Stimulation of Cortex and the Relationship of Somatosensory to Motor Activity in Striatum

Striatal activation produced by electrical stimulation of the cortex was consistent with the general somatotopy for vibrissae, hindlimb, and forelimb in striatum found with physiological stimuli (Brown and Sharp, 1993). The distribution of features in striatum resulting from cortical activation also confirmed the anteroposterior extent of the functional activity associated with physiologically applied stimuli (Brown and Sharp, 1993). Furthermore, electrical stimulation of the vibrissae motor cortex in one group, and stimulation of the vibrissae sensory cortex in another, revealed very similar fields of activation in striatum. The fields were much larger than those shown for physiologic stimulation, but so too were the cortical fields of activation. What is important is that the map in striatum includes not only a somatotopic representation of the body, but also provides for sensory and motor interaction in overlapping and closely juxtaposed regions in striatum (Fig. 8). Another important point is that the sensory and motor representations are also different in their anteroposterior course through striatum. For example, while the somatosensory activation pattern for forelimb shifts its position from anterior to posterior, the motor activation pattern for forelimb does not. It stays separate from the hindlimb activation region. Only the somatosensory features move to be juxtaposed to other somatosensory body region representations, or to their own motor representations.

DISCUSSION

The studies described here indicate that the striatal counterpart of a cortical functional unit consists of a set of distributed discrete zones (features) that can combine with, and presumably influence, other zones that represent sensory, motor, and perhaps novel or "significant" input to striatum. The combinations appear to take place over all three planes: mediolateral, dorsoventral, and anteroposterior. These functional discrete zones are similar in size and distribution to the anatomical corticostriate projection patterns described in other species (Jones et al., 1977, Malach and Graybiel, 1986; Flaherty and Graybiel, 1991a); they could coincide with patches, or could form subcompartments in the matrix, or could be distributed in both compartments. Importantly, the features can vary in size or distribution under different conditions; they are modifiable, or context-dependent, as part of their normal function. This context dependency may depend upon events in the somatosensory cortex, other cortical regions, thalamus and/or striatum.

Uncontrolled Movements

When experiments are done in awake animals, the effect of movements must be taken into account. In the somatotopy experiments (Brown, 1992), in which animals were lightly restrained, electromyographic activity was measured from the limb being stimulated. Movements occurred sporadically but were not consistent over the 45 minutes of the experiment, nor were they time-locked to the stimuli. In the context experiments (Brown and Lidsky, 1993), the animals were free to move in an experimental chamber. No differences in overall movement among the groups were observed. The animals sat quietly; it is unlikely that each of the animals in one of the context groups, and not in the others, performed the same movement continuously for the 30 min of the experiment, which would have been necessary to obtain the consistent feature patterns across animals. In addition, studies of electrical stimulation of the motor cortex have identified the striatal regions associated with movement of the vibrissae, forelimb and hindlimb (Brown and Sharp, 1993). None of the localization effects we describe here for the functional unit associated with somatosensory stimulation is the same as those associated with movement.

The Functional Unit

The smallest functional unit in the striatum we have observed is associated with stimulation of a single vibrissa in naive animals. One of us (Divac) suggested that this functional unit may be a striatal "module" or "barreloid" (after Van der Loos, 1976). At this time we can define the vibrissa "barreloid" as restricted to the dorsolateral striatum, consisting of two to three discrete zones at any one level in an extensive anteroposterior representation. The size is not beyond the range of a single striatal cell's dendritic and/or axon collateral field in rats (Kawaguchi et al., 1990). However, as a module it is not of uniform size and strictly repeatable in all its size and shape characteristics, especially under different conditions. Under different conditions, the relative activity within a dyad or triad may shift. We do not yet know to what extent the "barreloid" occupies the patch or matrix compartment, but it is possible that one discrete zone in a triad coincides with the patch and another with the matrix, just as different layers of the cortex appear to coincide principally with one or the other compartment (Gerfen, 1989).

The exact relationship of the functional zones we have described to corticostriate fields seen with anterograde tracing studies is difficult to determine without further studies. There is one study of corticostriate projections in rodents for which the cortical injection was into a vibrissa barrel (Welker et al., 1988). A single barrel projection covered a dorsal striatal region that was extensive anteroposteriorly. The region was very similar to that seen with electrical stimulation of the vibrissae cortex (Brown and Sharp, 1993). However, few discrete zones were seen anteriorly; rather, a continuous innervation of a field was described. This may be because all cortical layers and cells that project to striatum were labelled, but functionally activated cortex "labels" a subset of cortical cells that project to striatum and thus form discrete zones. Anterograde tracing studies in rat that label one to two layers have described the appearance of discrete zones in striatum (Gerfen, 1989). It would not be surprising if the anatomic projection field from all layers of a cortical body representation were wider than any one functional field because the anatomic field might include a number of projections that are not all physiologically active at once.

Stimulation of two vibrissae would be an interesting test of the resolution of the functional unit and combinational organization of striatum. Our current results indicate that the field of vibrissae representation extends both dorsolaterally and ventrolaterally, the most anterior vibrissae being represented most dorsally. Within this field, features seen after stimulation of several vibrissae, or of vibrissa C3 alone, suggest that representation of neighboring vibrissa is overlapping in striatum (Fig. 8). In mice, neighboring vibrissae projections also appear to overlap (Welker et al., 1988), and in cats, cortical projections from neighboring body regions overlap in striatum (Malach and Graybiel, 1986). To what extent the most anterior and posterior vibrissae overlap or are separate at any point remains to be determined.

The finding that one or another subset of activated features is increased in size as a function of context provides further insight into the nature of the functional units and the striatal combinatorial map. The features that were detected during application of weak subcutaneous shocks in the region of the vibrissae represent the major barrage of afferent terminal activation in these striatal regions. The size of each feature is dependent upon the amount of terminal activity (number of active terminals, rate of discharge) in that discrete zone which, in turn, is dependent upon the source of the activity. Thus, under different contextual conditions, the same stimulus causes different features to increase or to decrease in size, because not all of the input to each of them is the same. The striatum receives input from the entire cortex, including motor, sensory, association and limbic, in addition to thalamus and substantia nigra. The effects of context can be seen as terminal afferent activity resulting from any of the foregoing, either via relatively direct pathways, or via indirect activation. Corticostriate systems activated by different motivational or emotional states, for example, might be convergently affecting feature size and location in one context, and not another. Similarly, pathways activated only after modification by training would also affect convergent interaction of terminal afferents differentially, according to contextual conditions. It appears likely that some contextual aspects of striatal activation may be mediated by afferents other than those from the cortex because the striatal target regions innervated by the limbic cortex and somatosensory motor areas are spatially segregated (e.g. Divac and Diemer, 1980) and because most intrinsic axon collaterals of the striatum are too short to connect medial and lateral striatal regions.

Considerations for the Combinational Map

The combinational anatomic organization of striatum could provide the substrate for the singular electrophysiological responses observed in this structure during behavior. Although the reactions of many striatal cells can be defined as either "sensory" or "motor" in the traditional sense, the activities of the majority do not fit neatly into either category. In the "motor" category, many cells respond to whole body movements (Carelli and West, 1991) or to movement of several body regions (Crutcher and DeLong, 1984), making them candidates for cells that combine inputs. Striatal cell activity also appears to combine the properties of both sensory and motor functions (Manetto and Lidsky, 1986; Lidsky and Manetto, 1987). For example, some cells discharge only during sensory-triggered movements, rather than during movements *per se* (Manetto and Lidsky, 1986). Moreover, such cells are stimulus-specific; movements of similar topography, initiated by dissimilar stimuli, are not associated with cellular discharge (Lidsky and Manetto, 1987). Such cells may be "combinationally" specific. Also, the magnitude of responses of striatal cells to somatosensory stimuli can be altered by the behavioral context in which the stimuli are presented, such as the movement in which the animal is engaged (Lidsky and Manetto, 1989). A combinational organization of inputs from sensory and motor cortex could produce such variations in response. Finally, the output from a combinational map in striatum would cause alteration of the discharge rate of cells in the globus pallidus to occur under many different conditions, rather than to a single condition (Mink and Thatch, 1991).

That somatosensory-motor representation in striatum might form a combinational map is not surprising in view of the fact that one of the functions attributed to the basal ganglia is regulation of complex movements. This would require a controlling system capable of storing either movements or, more likely, positions, as integrated units (see Smyth, 1984; Rosenbaum et al., 1992). Representation of a sequence of movements or positions would necessitate the possibility of complex interactions among cells that represent all of the different body regions involved in any of the wide range of skeletomuscular behaviors in the mammalian repertoire.

The Combinational Somatosensory Map and Disease

If the cortical commands for complex movements are monitored by the proprioceptive and tactile feedback from the movement in progress, appropriate positions might be defined by the afferent convergence in a specific region of the combinatorial map. In this way, the map would effectively regulate the movement by continuously monitoring its progress via a system specialized to register the coincidence of feedback from several parts of the body that occur uniquely with specific body positions or movements. The kinds of motor activity associated with normal basal ganglia function involve movements that include extensive sensory feedback. For example, animals with restricted striopallidal lesions show deficits in drinking only under the circumstance that requires a skilled positioning of the mouth and head to get water (Labuszewski et al., 1981). In the human, a typical example of disrupted behavior in parkinsonian patients is buttoning one's shirt. The case of the hemiparkinsonian patient may be instructive. In four hours of taped interviews with one of us (Brown), a hemiparkinsonian patient described how she "loses" her affected leg. She does not "lose" it while standing still, being perfectly aware of her leg at rest. Instead, she "loses" the leg when she goes upstairs: she does not predict correctly where her leg is as she goes to take the next step. One could say that the deficit appears in a specific context, the context being movement on stairs, which involves the coordination of several different joint angles of both sides of the body and all limbs, usually without visual feedback. Such behaviors require continuous somatosensory guideposts during movement, unlike the kinds of ballistic movements that are affected by cerebellar damage. It is therefore interesting that our data suggest that the primary source of combinatorial interaction may be somatosensory, versus motor, representation in striatum. It suggests that the feedback during movement provides guideposts to reinforce the appropriate sequential program to emerge. If the feedback is novel, or is significant in some other way, it may combine with other elements to allow a different action. Deficits in movement initiation may be the result of the system being uninformed about aspects of the environment from which it normally receives input.

In conclusion, the arrangement of distributed modules that can combine with other modules, and be modified, is an appropriate substrate for the organization of behavior that requires flexibility, both in the performance of a specific movement, and in adapting to the environment.

Acknowledgements

This work was supported in part by USPHS grant NS21356. We thank Diane Smith and Kim Gallo for expert technical assistance.

REFERENCES

Alexander, G.E. and DeLong, M.R., 1985, Microstimulation of the primate neostriatum. II. Somatotopic organization of striatal microexcitable zones and their relation to neuronal response properties, *J. Neurophysiol.* 53:1417-1443.

Brown, L.L., Wolfson, L.I., and Feldman, S.M., 1987, Functional neuroanatomic mapping of the rat striatum: regional differences in glucose utilization in normal controls and after treatment with apomorphine, *Brain Res.* 411:65-71.

Brown, L.L., 1992, Somatotopic organization in rat striatum: evidence for a combinational map, *Proc. Natl. Acad. Sci. USA* 89:7403-7407.

Brown, L.L. and Feldman, S.M., 1993, The organization of somatosensory activity in dorsolateral striatum of rat, *in:* "Chemical Signalling in the Basal Ganglia," G. Arbuthnott and P.C. Emson, eds., in press.

Brown, L.L., and Lidsky, T.I., 1993, Context-dependent activity in rat neostriatum, submitted.

Brown, L.L., and Sharp, F.R., 1993, Somatotopic organization of rat striatum mapped with ^{14}C deoxyglucose during movements elicited by cortical stimulation, submitted.

Carelli, R.M., and West, M.O., 1991, Representation of the body by single neurons in the dorsolateral striatum of the awake, unrestrained rat, *J. Comp. Neurol.* 309:231-249.

Chapin, J.K., and Lin, C.-S., 1984, Mapping the body representation in the SI cortex of anesthetized and awake rats, *J. Comp. Neurol.* 229:199-213.

Collins, R.C., and Divac, I., 1984, Neostriatal participation in prosencephalic systems: evidence from deoxyglucose autoradiography, *Adv. Neurol.* 40:117-122.

Crutcher, M.D., and DeLong, M.R., 1984, Single cell studies of the primate putamen. I. Functional organization, *Exp. Brain Res.* 53:233-243.

Divac, I., 1968, Effects of prefrontal and caudate lesions on delayed response in cats, *Acta Biol. Exp.* (Warszawa), 28:149-167.

Divac, I., Markowitsch, H.J., and Pritzel, M., 1978, Behavioral and anatomical consequences of small intrastriatal injections of kainic acid in the rat, *Brain Res.* 151:523-532.

Divac, I., and Diemer, N.H., 1980, Prefrontal system in the rat visualized by means of labelled deoxyglucose - further evidence for functional heterogeneity of the neostriatum, *J. Comp. Neurol.* 190:1-13.

Divac, I., 1983, Two levels of functional heterogeneity of the neostriatum, *Neurosci.* 10:1151-1155.

Divac, I., and Öberg, R.G.E., 1992, Subcortical mechanisms in cognition, *in:* "Neuropsychological Disorders Associated with Subcortical Lesions," G. Vallar, S.F. Capra, and C.-W. Wallesch, eds., Oxford University Press, Oxford, pp 42-60.

Divac, I., Brown, L.L., and Hand, P.J., 1992, The search for the neostriatal "barreloid", *Abstracts, Fourth Triennial Meeting of the International Basal Ganglia Society,* p. 25.

Feldman, S.M., and Brown, L.L., 1992, Columns in rat somatosensory cortex seen with deoxyglucose autoradiography, *Soc. Neurosci. Abstr.* 18:1387.

Flaherty, A.W., and Graybiel, A.M., 1991a, Corticostriatal transformations in the primate somatosensory system. Projections from physiologically mapped body-part representations, *J. Neurophysiol.* 66:1249-1263.

Flaherty, A.W., and Graybiel, A.M., 1991b, A second input system for body representations in the primate striatal matrix, *Soc. Neurosci. Abstr.* 15:285.

Gerfen, C.R., 1989, The neostriatal mosaic: striatal patch-matrix organization is related to cortical lamination, *Science* 246:385-388.

Gerfen, C.R., 1992, The neostriatal mosaic: multiple levels of compartmental organization in the basal ganglia, *Annu. Rev. Neurosci.* 15:285-320.

Graybiel, A.M., and Ragsdale, C.W. Jr., 1978, Histochemically distinct compartments in the striatum of human, monkey, and cat demonstrated by acetylcholinesterase staining, *Proc. Natl. Acad. Sci. USA* 75:5723-5726.

Graybiel, A.M., 1990, Neurotransmitters and neuromodulators in the basal ganglia, *TINS* 13:244-254.

Hall, R.D., and Lindholm, E.P., 1974, Organization of motor and somatosensory neocortex in the albino rat, *Brain Res.* 66:23-38.

Jones, E.G., Coulter, J.D., Burton, H., and Porter, R., 1977, Cells of origin and terminal distribution of corticostriatal fibers arising in the sensory-motor cortex of monkeys, *J. Comp. Neurol.* 173:53-80.

Kadekaro, M., Crane, A.M., and Sokoloff, L., 1985, Differential effects of electrical stimulation of sciatic nerve on metabolic activity in spinal cord and dorsal root ganglion in the rat, *Proc. Natl. Acad. Sci. USA* 82:6010-6013.

Kadekaro, M., Vance, W.H., Terrell, M.L., Gary, H. Jr., Eisenberg, H.M., and Sokoloff, L., 1987, Effects of antidromic stimulation of the ventral root on glucose utilization in the ventral horn of the spinal cord in the rat, *Proc. Natl. Acad. Sci. USA*, 84:5492-5495.

Kawaguchi, Y., C.J. Wilson, and Emson, P.C., 1990, Projection subtypes of rat neostriatal matrix cells revealed by intracellular injection of biocytin, *J. Neurosci.* 10:3421-3438.

Kossut, M., Hand, P.J., Greenberg, J., and Hand, C.L., 1988, Single vibrissal cortical column in SI cortex of rat and its alterations in neonatal and adult vibrissa-deafferented animals: a quantitative 2DG study, *J. Neurophysiol.* 60:829-852.

Künzle, H., 1975, Bilateral projections from precentral motor cortex to the putamen and other parts of the basal ganglia. An autoradiographic study in Macaca fascicularis, *Brain Res.* 88:195-209.

Künzle, H., 1977, Projections from the primary somatosensory cortex to basal ganglia and thalamus in the monkey, *Exp. Brain Res.* 30:481-492.

Labuszewski, T., Lockwood, R., McManus, F.E., Edelstein, L.R., and Lidsky, T.I., 1981, Role of postural deficits in oro-ingestive problems caused by globus pallidus lesions, *Exp. Neurol.* 74:93-110.

Lidsky, T.I., and Manetto, C., 1987, Context dependent activity in the striatum of behaving cats, *in:* "BasalGanglia and Behavior: Sensory Aspects of Motor Functioning," J.S. Schneider and T.I. Lidsky, eds., Hans Huber, Bern, pp 123-133.

Lidsky, T.I., and Manetto, C., 1989, The effects of movements on caudate sensory responses, *Neurosci. Letters* 96:295-299.

Malach, R., and Graybiel, A.M., 1986, Mosaic architecture of the somatic sensory-recipient sector of the cat's striatum, *J. Neurosci.* 6:3436-3458.

Manetto, C., and Lidsky, T.I., 1986, Caudate neuronal activity in cats during head turning: selectivity for sensory-triggered movements, *Brain Res. Bull.* 16:425-428.

Mata, M., Fink, D.J., Gainer, H., Smith, C.B., Davidsen, L., Savaki, H., Schwartz, W.J., and Sokoloff, L., 1980, Activity-dependent energy metabolism in rat posterior pituitary reflects sodium pump activity, *J. Neurochem.* 34:213-215.

Mink, J.W., and Thach, W.T., 1991, Basal ganglia motor control. I. Nonexclusive relation of pallidal discharge to five movement modes, *J. Neurophysiol.*, 65:273-300.

Öberg, R.G.E., and Divac, I., 1979, Cognitive functions of the neostriatum, *in:* "The Neostriatum," I. Divac and R.G.E. Öberg, eds., Pergamon Press, Oxford, pp 291-313.

Olson, L., Seiger, A., and Fuxe, K., 1972, Heterogeneity of striatal and limbic dopamine innervation: Highly fluorescent islands in developing and adult rats, *Brain Res.* 44:283-288.

Pert, C.B., Kuhar, M.J., and Snyder, S. H., 1976, Opiate receptor: autoradiographic localization in rat brain, *Proc. Nat. Acad. Sci. USA* 73:3729-3733.

Rosenbaum, D.A., Engelbrecht, S.E., Bushe, M., and Loukopoulos, L.D.,1993, Knowledge model for selecting and producing reaching movements, *in*: "Control of arm and hand posture and movement. Journal of Motor Behavior," T. Flash, ed., Heldref Publication, Washington, in press.

Russ, J.C., 1990, "Computer-assisted Microscopy. The Measurement and Analysis of Images," Plenum Press, New York.

Saint-Cyr, J.A., Ungerleider, L.G., and Desimone, R., 1990, Organization of visual cortical inputs to the striatum and subsequent outputs to the pallido-nigral complex in the monkey, *J. Comp. Neurol.* 298:129-156.

Sharp, J.W., Gonzalez, M.F., Morton, M.T., Simon, R.P., and Sharp, F.R., 1988, Decreases of cortical and thalamic glucose metabolism produced by parietal cortex stimulation in the rat, *Brain Res.* 438:357-362.

Sokoloff, L., Reivich, M., Kennedy, C., DesRosiers, M.H., Patlak, C.S., Pettigrew, K.D., Sakurada, O., and Shinohara, M., 1977, The ^{14}C deoxyglucose method for the measurement of local cerebral glucose utilization: theory, procedure, and normal values in the conscious and anesthetized albino rat, *J. Neurochem.* 28:897-916.

Smyth, M.M., 1984, Memory for movements, *in*: "The psychology of human movements," M.M. Smyth and A.M. Wing, eds., Academic Press, London, pp 83-117.

Van der Loos, H., 1976, Barreloids in mouse somatosensory thalamus, *Neurosci. Lett.* 2:1-6.

Welker, E., Hoogland, P.V., and Van der Loos, H., 1988, Organization of feedback and feedforward projections of the barrel cortex: a PHA-L study in the mouse, *Exp. Brain Res.* 73:411-435.

BASIC ELECTROPHYSIOLOGY AND POSSIBLE NEW THERAPEUTIC APPROACHES TO MOVEMENT DISORDERS

Alessandro Stefani, Paolo Calabresi, Nicola B. Mercuri, Francesca Stratta, Antonio Pisani, Antonello Bonci and Giorgio Bernardi

Clinica Neurologica, Università di Roma Tor Vergata, Rome, Italy

INTRODUCTION

What the basal ganglia do, is it the on-going question? New models have reevaluated the input/output ratio of single structures as inserted in parallel, functional systems (Alexander and Crutcher, 1990). These models have reinforced the assumption that the basal ganglia are a key station for the execution of organized movements (DeLong, 1990; Goldman-Rakic and Selemon, 1990). At the molecular level, new families of receptors are explored. The cloning of glutamate metabotropic receptors is heading the surprising multiplicity of the neurobiology of excitatory transmission (Gasic, 1992). The definition of new subclasses of dopamine receptors is an invitation to reconsider the pharmacology of the amine (Surmeier et al., 1992). Radical changes, however, in the therapy of movement disorders have barely taken place, being the introduction of levo-dopa still a "cornerstone" of the therapy of the parkinsonian patient (Hornykiewicz, 1966). Whatever are the fundamental functions of the basal ganglia, a striking dichotomy risks to develop between basic research acquisition and the daily urgency of patient's quality of life. In presenting our recent findings, we aim to highlight those aspects of mesencephalic, neostriatal and pallidal physiology whose clinical impact could be relevant.

MATERIAL AND METHODS

Adult Wistar rats (2-4 months) are used for all experiments. Brains are quickly removed and 200-350 μm thick coronal slices, containing either the mesencephalon or the neostriatum, are prepared from tissue blocks with the use of a vibratome. For neostriatal recordings, the slices usually include the corpus callosum and neocortex. A single slice is transferred to the recording chamber and submerged in a continuously flowing Krebs solution (36°C, 2-2.5 ml/min.) gassed with a 95% O_2, 5% CO_2 mixture. Intracellular recording electrodes are filled with 2M KCl or 2M K-acetate (30-60 Mohms); extracellular recording electrodes are filled with 2M NaCl (5-10 Mohms). Bipolar electrodes are used to evoke synaptic potentials of the corticostriatal pathway. These electrodes are located either in cortical areas adjacent to the recording electrode or in the white matter between the cortex and the striatum. To study synaptic plasticity in the striatum, we used a conditioning tetanus composed by three trains of stimuli of 3 second duration, 100 Hz frequency, at 20 second interval.

To perform whole-cell recordings from isolated neurons, 500 μm thick external globus pallidus (GPe) microslices are incubated in oxygenated media containing pronase E (1.5 mg/ml; incubation time=20-25 min); then, two microslices are mechanically triturated and the supernatant finally splashed into a Petri dish mounted on the stage of an inverted microscope; a small sample of healthy cells - somata and eventually 2-4 brief processes - has become available; medium to large neurons, of fusiform or poligonal shape, are chosen for recordings, which are performed at room temperature with fire-polished pipettes (Corning 7052, 2-7 Mohms).

RESULTS

Pharmacology of DA-autoreceptors

A great deal of informations has recently emerged regarding the membrane physiology of dopamine(DA)-containing cells of the rat ventral mesencephalon. The DAergic neurons of substantia nigra pars compacta (SNc) and ventral tegmental area (VTA) provide DA to most of the brain. These cells have receptors for their own transmitters on the somata, dendrites and terminals which control the synthesis and regulate the extracellular content of DA in both their terminal fields and in the ventral mesencephalon itself (Robertson, 1992). In vitro intracellular recordings from these structures have provided a rather detailed analysis of dopamine pharmacology (Lacey et al., 1987; Mercuri et al., 1992b). A G-protein mediated outward potassium current promoted by DA through activation of D2 receptors was, in fact, characterized (Fig. 1). In this regard, we also analysed the effects of drugs affecting either the release or the re-uptake of DA (Mercuri et al., 1991, 1992b). The indirect agonist amphetamine was responsible of an increased release of DA from cell dendrites into the extracellular space, thereby leading to the a membrane hyperpolarization and a consequent inhibition of the spontaneous repetitive firing (Fig. 1A). The amphetamine response persisted in low-calcium media and was antagonized by sulpiride, suggesting the effect was dependent upon the activation of D2(-D3) receptors. DA uptake blockers, like amineptine, nomifensine and cocaine, instead of simply mimicking the DA response, did potentiate cellular responses to exogenously applied DA (Mercuri et al., 1991) (Fig.1B, see conclusions).

Mammalian mesencephalon receive an abundant impingement from cortical and subthalamic areas releasing excitatory amino acids (EAA). It is also well known that the electrical stimulation of mesencephalic slices produces excitatory post-synaptic potentials on DA cells (Mereu et al., 1991). EAA-related neurotoxicity was implicated in neurodegenerative disorders (Choi, 1988; Sonsalla et al., 1989). Noticeably, antagonists of N-methyl-D-aspartate (NMDA) protected SN from MPP+-induced toxicity (Turski et al., 1991). We did analyse the full spectrum of ionotropic and metabotropic agonists on SNc/VTA slices with single electrode voltage-clamp techniques (Fig. 2) (Mercuri et al., 1992a; Mercuri et al., 1993).

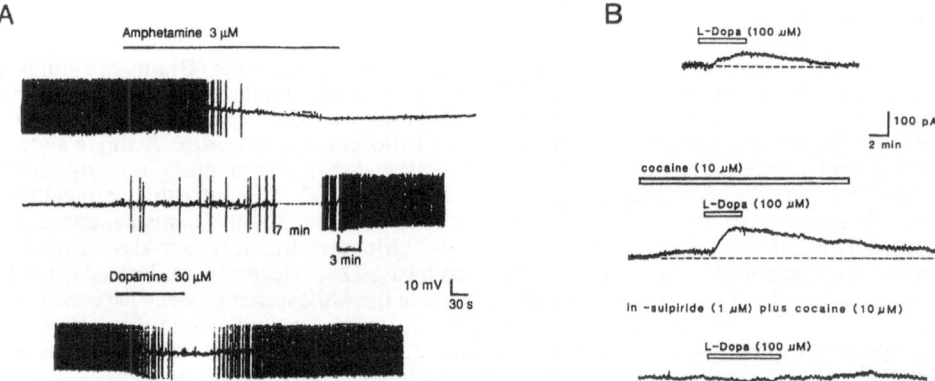

Figure 1. Current and voltage-clamp recordings of dopaminergic cells. A: amphetamine and dopamine caused inhibition of the firing rate and membrane hyperpolarization. B: the outward current promoted by L-Dopa was potentiated by the pre-incubation in cocaine. The response is blocked by the D2(D3)-antagonist sulpiride.

NMDA depolarized the membrane potential and increased the firing rate. When the neurons were clamped at resting potential (-50/-60 mV), NMDA produced an inward current (Fig. 2). This inward current did not desensitize for prolonged applications and was dose-dependent. The apparent membrane input conductance was decreased, because of the fall of the slope conductance over a range of potentials from about -100 to - 45 mV (Fig. 2). APV selectively reduced the NMDA response. Interestingly, the peak amplitude of the NMDA response was already maximal near rest (Mercuri et al., 1993).

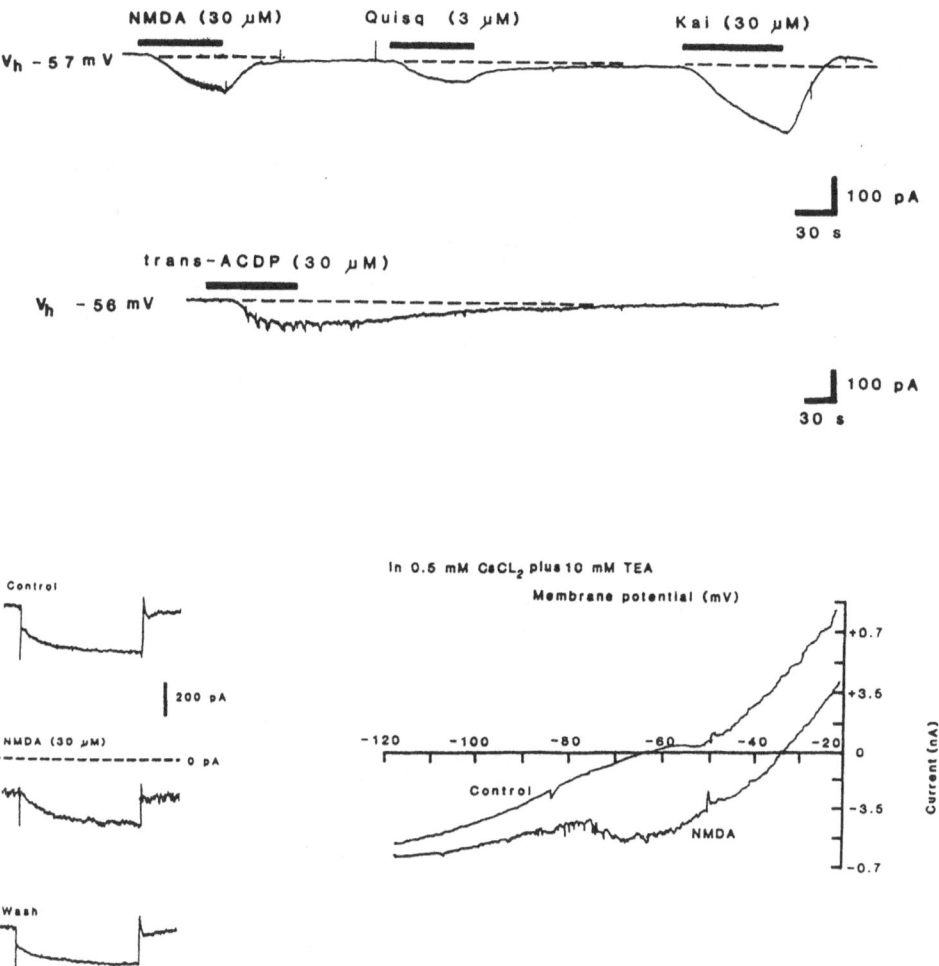

Figure. 2. Upper part: inward responses induced in a DAergic cells by EAA receptor agonists.Lower part: conductance change and inward current produced by NMDA 30 μM; note the decrease of the input conductance. To the right: current/voltage curves obtained before and under NMDA 30 μM. NMDA evoked a region of negative slope conductance. Its maximal effect peaked between -70 and -50 mV (from Mercuri et al., 1992a and from Mercuri et al., 1993).

Synaptic transmission in the striatum

The pharmacology of the synaptic transmission in the mammalian neostriatum has been one of the main goals of our laboratories (Calabresi et al., 1990). Glutamate and GABA mediated synaptic potentials, activated either by intrastriatal or cortico-callosal stimulation, are modulated by dopaminergic agonists, muscarinic agents, excitatory amino acids coupled to metabotropic receptors and GABA-b agonists. Recently, we have approached striatal synaptic plasticity and focused it as a possible target of DA modulation.

Use-dependent alterations in synaptic efficacy are the core of many theories regarding the storage of informations in the brain (Collingridge and Singer, 1990). Long-term potentiation (LTP) in hippocampus has been postulated to be associated to memory. Long-term depression (LTD) in cerebellum could represent the cellular substrate of some forms of motor learning (Ito, 1989). Tetanic stimulation (see methods) of the corticostriatal pathway did produce long-lasting depression in all cells tested (>2 hours) (Fig. 3) (Calabresi et al., 1992c). In some neurons, a short-term post-tetanic potentiation was also observed after the conditioning tetanus but it was consistently followed by significant depression of the potentials. As Figure 3 shows, LTD was also measured by extracellular recordings of field potentials.

LTD was: i) not associated with changes of intrinsic membrane properties (membrane potential, apparent membrane input resistance, current-induced firing pattern) (Calabresi et al., 1992c), ii) not modified by incubation of the slice in bicuculline (Calabresi et al., 1992c), iii) reversed in LTP when magnesium was omitted from the bath (Calabresi et al., 1992b).

The initial approach to LTD pharmacology moved from the consideration that DA modulates the processing of cortical input in the striatum, through at least two classes of well characterized receptors, D1 and D2. SCH 23390, a D1 receptor antagonist, prevented the generation of LTD in both extracellularly recorded field potentials and intracellularly recorded

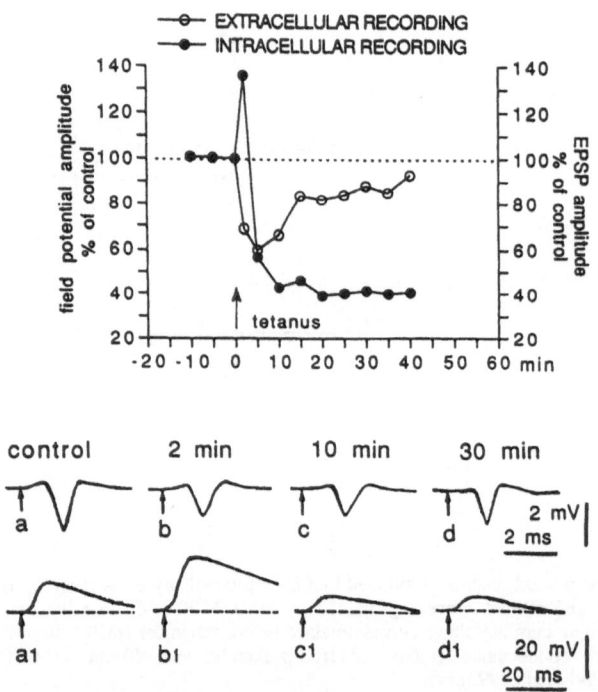

Figure. 3. Effect of tetanic stimulation of corticostriatal fibers on the amplitude of field potentials (open squares) and EPSPs (filled circles); each point represents several experiments. At the bottom, field potential (a/d), cortically driven EPSPs (b/e) and DC current-induced repetitive firing (c/f) are shown, before (pre) and after (post) the LTD induction (from Calabresi et al., 1992c).

EPSPs. Interestingly, also the perfusion of the slice with sulpiride, a rather selective D2 antagonist, blocked LTD.

These evidences show that coactivation of both classes of DA receptors is required for striatal LTD. They also suggest a role for endogenous DA in this form of synaptic plasticity. We therefore tested such a possibility on 6-OHDA lesioned animals (Fig. 4) (Calabresi et al., 1992a; Calabresi et al., 1992c). In DA-depleted slices, tetanic stimulation failed to induce LTD, but LTD could be restored by the coadministration of D1 and D2 agonists (Fig. 4) (Calabresi et al., 1992a; Calabresi et al., 1992c).

At present, we are investigating the possible involvement of different second messenger systems in the generation of striatal LTD. The lithium-induced block of striatal LTD suggests that alterations of phosphoinositide (PI) turnover may affect this form of synaptic plasticity (Calabresi et al., 1993).

Voltage-dependent currents in globus pallidus

External globus pallidus is currently indicated as a crucial area in the "filtering" of neostriatal message throughout the basal ganglia (Alexander and Crutcher, 1990; DeLong,

DA·LESIONED RATS

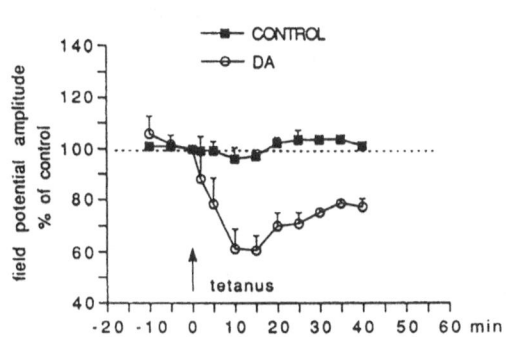

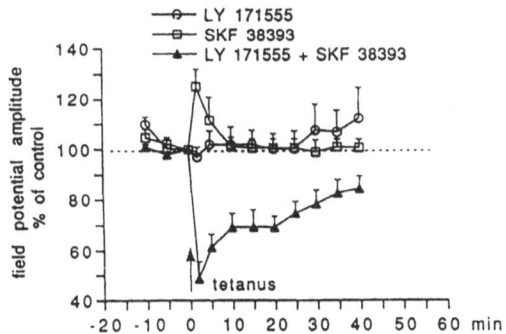

Figure. 4. Graphic representation of field potential reduction vs. time in slices of 6-OHDA-lesioned animals. Upper part: exogenously applied dopamine (30 µM) restored LTD. Lower part: pretreatment of DA-depleted slices either with SKF 38393 (10 µM) or with LY 171553 (3 µM) alone failed to induce striatal LTD; only the coadministration of D1 and D2 dopamine receptor agonists reproduced LTD (From Calabresi et al., 1992a).

1990; Goldman-Rakic and Selemon, 1990). Surprisingly, many recent interpretations are based on a profound histochemical and anatomical knowledge (Percheron et al., 1984), but they are scarsely provided by physiological and pharmacological informations at the cellular level. Our whole-cell patch-clamp recordings of rat GPe were directed towards the study of voltage-gated conductances and their modulation by opioids.

A fast A-current was initially isolated in all the patched neurons according to its voltage-dependency (Fig. 5) (Stefani et al., 1992). Activation and inactivation curves configured its role in the voltage trajectory leading to spike discharge; half-inactivation voltage was around 60 mV, suggesting an active contribution to the resting membrane potential (Kita and Kitai, 1991). Fastly-inactivating A-current may interfere in the fine control of regular, rhytmical firing behaviour (Stefani et al., 1992). The whole pattern of outward potassium currents activated by depolarization was however dominated by a slow, sustained conductance, whose activation threshold (around -20 mV) and TEA-sensitivity suggested its identity as a delayed rectifier (IDr). Asides from fast A and IDr-like currents, we analysed a third component, which shares kinetics and voltage-dependency with a similar conductance already described in striatum (Surmeier et al., 1991). This "slow-A-current" manifested a peculiar sensitivity to both TEA and 4-aminopyridine (4-AP), but in the low micromolar range; moreover, its half-range; moreover, its half-inactivation voltage sit around -90, what makes its actual contribution to membrane physiology negligible unless deep hyperpolarizations occur, and recovered from inactivation with a time-constant one-order magnitude slower than fast-A-current (about 500 ms vs. 40) (Fig. 5C). Interestingly, this "slow-A" was present in 45% of neurons; the functional implications are currently examined. We could postulate the presence of slow-A as a sort of additional mechanism in order to dampen "unrequested overexcitability". On the other hand, this current can be suppressed if a quick robust repetitive firing is taking place.

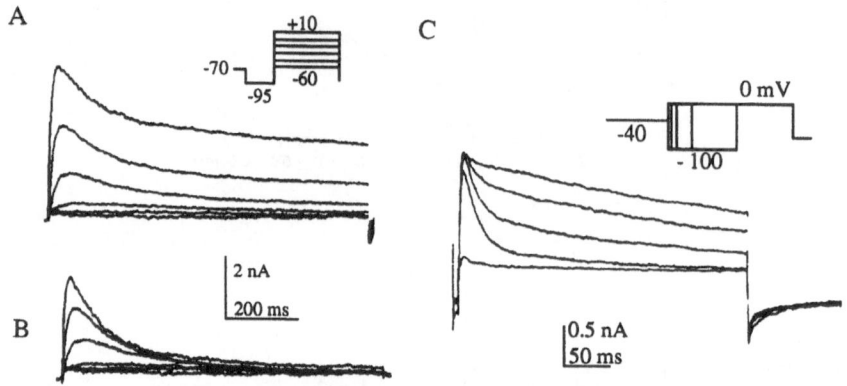

Figure. 5. Outward potassium currents activated by depolarization in GP neurons. A-B: the recordings are taken in TEA(10 mM)-added media. A: inactivating as well as sustained conductances are evoked by depolarizing steps. B: a transient, fastly-inactivating A-like current is isolated when the currents evoked by the same depolarizing steps, but relatively positive potentials (-40 mV, not shown) are subtracted from A. C: recovery from inactivation protocols unmask fastly as well as slowly recovering components. (from Stefani et al., 1992).

Pallidal cells possess sustained not-inactivating, calcium-dependent, outward currents. Preliminary results have shown that enkephalins decreased these conductances. Furthermore, we have investigated wheter enkephalins directly affected inward calcium currents. Barium was chosen as the main charge carrier through calcium channels in order to minimize run-down; ramp-activated currents (1mV/1ms) were preferentially studied (Stefani et al., 1993).

Enkephalin, at concentrations ranging from 10 to 1000 nM, caused a reversible and dose-dependent decrease of calcium currents (Fig. 6). The identification of the channel preferentially modulated, as well as the definition of the receptors involved by the use of selective antagonists, is under current investigation.

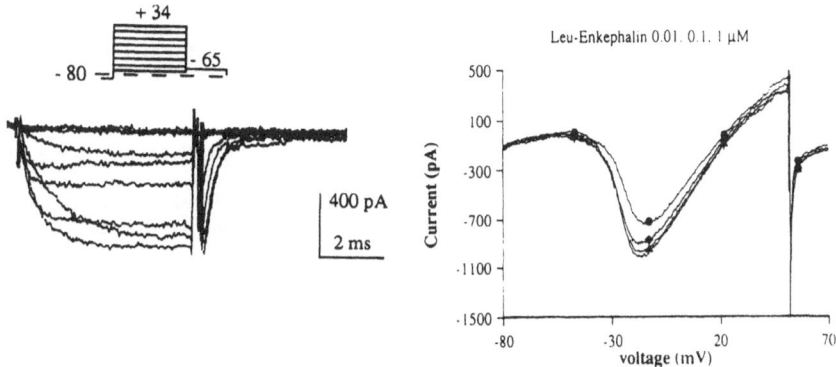

Figure 6. Inward barium currents in a GPe neuron, activated either by step pulses or ramps (1mV/1 ms). To the right: Leu-enkephalin applications (.01, .1 and 1 µM) reduced the ramp-evoked inward conductances.

CONCLUSIONS

We have briefly presented some aspects of the pharmacology of the basal ganglia, which seemed to be worth analysing in light of their potential for predictable new therapeutic approaches. The data obtained from mesencephalic DAergic neurons provided interesting insights concerning possible implications in the therapy of Parkinson disease and in the treatment of drug abuse. In the case of amineptine, a tricyclic antidepressant, we proposed its coadjuvant use in the early therapy of Parkinson's disease. Moreover, being similar to amineptine the cocaine effects, we suggested the eventual introduction of analogue molecules in the treatment of abuse disorders. On the other hand, the SN/VTA neurons showed peculiar sensitivity to EAAs, and in particular to NMDA receptor activation; the overstimulation of these could very well cause and/or sustain the cellular damage, because of facilitation of excessive loading of calcium into the cell. Moreover, spontaneous slow calcium waves (normally generated by these neurons) might further facilitate calcium entry during NMDA-induced depolarization. These evidences provide an electrophysiological explanation of the high sensitivity of DAergic mesencephalic neurons to the neurodegenerative processes causing parkinsonism and suggest a therapeutic approach to cell death by using NMDA antagonists.

In neostriatum, we showed a long-term modification of synaptic efficacy as regulated by endogenous dopamine. The clinical implications seem attractive. Instead of being considered only as a phasic modulator, DA could exert a role in long-term regulation of striatal excitatory synaptic transmission. Striatal DA release could contribute to the acquisition of new motor performances. Another functional implication of LTD in neostriatum regards the synergistic effects of D1 and D2 agonists, which probably raise the expectancy of newly-synthesized molecules.

Finally, we have introduced some recent acquisitions on GP membrane physiology. We emphasized the importance of the relative contribution of specific outward conductances to the whole pattern of currents activated by depolarization. The state of modulation of the A-channels would contribute to set the cellular behaviour in terms of firing pattern and, consequently, neurotransmitter release. Enkephalin modulation of calcium currents, so extensively studied in different structures, may provide a possible new perspective in the understanding of some aspects of hyperkinetic disorders.

In conclusion, we hope that our basic electrophysiological approaches to the pharmacology of the basal ganglia may help to develop new therapeutic strategies in the treatment of basal ganglia diseases.

Acknowledgements

We are greatly thankful to Giuseppe Gattoni and Massimo Tolu for their technical help.

REFERENCES

Alexander, G.E., and Crutcher, M.D., 1990, Functional architecture of basal ganglia circuits; neural substrate of parallel processing, *Trends Neurosci.* 13:266-271.

Calabresi , P., Mercuri, N.B., and Bernardi, G., 1990, Synaptic and intrinsic control of membrane excitability of neostriatal neurons. II: an in vitro study, *J. Neurophysiol.* 63:663-675.

Calabresi, P., Maj, R., Mercuri, N.B., and Bernardi, G, 1992a, Coactivation of D1 and D2 dopamine receptors is required for long-term depression in neostriatum, *Neurosci. Lett.* 142:95-99.

Calabresi, P., Pisani, A., Mercuri, N.B., and Bernardi, G., 1992b, Long-term potentiation in the striatum is unmasked by removing the voltage-dependent block of NMDA receptors channels, *Eur. J. Neurosci.* 4:929-935.

Calabresi, P., Pisani, A., Mercuri, N.B., and Bernardi, G., 1992c, Long-term synaptic depression in the striatum: physiological and pharmacological characterization, *J. Neuroscience* 92:4224- 4333.

Calabresi, P., Pisani, A., Mercuri, N.B., and Bernardi, G., 1993, Lithium-treatment blocks LTD in the striatum, *Neuron* in press.

Choi, D.W., 1988, Glutamate neurotoxicity and disease of the nervous system, *Neuron* 7:357-367.

Collingridge, G.L., and Singer, W., 1990, Excitatory amino acid receptors and synaptic plasticity, *Trends Pharmacol. Sci.* 11:290-296.

DeLong, M.R., 1990, Primate models of movement disorders of basal ganglia origin, *Trends. Neurosci* 13:281-285.

Gasic, G.P., 1992, Molecular neurobiology of glutamate receptors, *Ann. Rev. Physiol.* 54:507- 536.

Goldman-Rakic, P.S., and Selemon, L.D., 1990, New frontiers in basal ganglia research, *Trends Neurosci.* 13:241-244.

Hornykiewicz, O., 1966, Dopamine and brain function, *Pharmacol. Rev.* 18:925-964.

Ito, M., 1989, Long-term depression, *Ann. Rev. Neurosci.* 12:85-102.

Kita, H., and Kitai, S.T., 1991, Intracellular study of rat globus pallidus neurons: membrane properties and responses to neostriatal, subthalamic and nigral stimulation, *Brain Res.* 564:296-305.

Lacey, M.G., Mercuri, N.B., and North, R.A., 1987, Dopamine acts on D2 receptors to increase potassium conductance in neurones of the rat substantia nigra zona compacta, *J. Physiol.* 392:397-416.

Mercuri, N.B., Stratta, F., Calabresi, P., and Bernardi, G., 1991, Electrophysiological effects of amineptine on neurones of the rat substantia nigra pars compacta: evidence for an inhibition of the dopamine uptake system, *Br. J. Pharmacol.*, 104:700-704.

Mercuri, N.B., Stratta, F., Calabresi, P., and Bernardi, G., 1992a, Electrophysiological evidences for the presence of ionotropic and metabotropic excitatory amino acid receptors on dopaminergic neurons of the rat mesencephalon: an in vitro study, *Funct. Neurol.* 7:231-234.

Mercuri, N.B., Calabresi, P., and Bernardi, G, 1992b, Electrophysiological actions of dopamine and dopaminergic drugs on neurones of the substantia nigra pars compacta and ventral tegmental area, *Life Sci.* 51:711-718.

Mercuri, N.B., Stratta, F., Calabresi, P., and Bernardi, G., 1993, A voltage-clamp analysis of NMDA-induced responses on dopaminergic neurons of the rat substantia nigra zona compacta and ventral tegmental area, *Brain Res.* 593:51-56.

Mereu, G., Costa, E., Armstrong, D.M., and Vicini, S., 1991, Glutamate receptor subtypes and excitatory synaptic currents of dopamine neurons in midbrain slices, *J. Neurosci.* 11:1350-1356.

Percheron, G., Yelnik, J., and Francois, C., 1984, A Golgi analysis of the primate globus pallidus. III: spatial organization of the striatopallidal complex, *J. Comp. Neurol.* 227:214-227.

Robertson, H.A., 1992, Dopamine receptor interactions: some implications for the treatment of Parkinson's disease, *Trends Neurosci.* 15:201-206.

Sonsalla, P.K., Riordan, D.E., and Heikkila, R.E., 1989, Role for excitatory amino acids in methamphetamine-induced nigrostriatal dopaminergic neurotoxicity, *Science* 243:398-400.

Stefani, A., Calabresi, P., Mercuri, N.B., and Bernardi, G., 1992, A-current in rat globus pallidus: a whole-cell voltage-clamp study on acutely dissociated neurons, *Neurosci. Lett.* 144:4-8.

Stefani, A., Surmeier, D.J., and Bernardi, G., 1993, The μ-agonist DAGO decreases HVA calcium currents in acutely dissociated neostriatal neurons, *Eur. J. Neuroscience* submitted.

Surmeier, D.J., Stefani, A., Foehring, R., and Kitai, S.T., 1991, Developmental expression of a slowly-inactivating current in rat neostriatal neurons, *Neurosci. Lett.* 122:41-46.

Surmeier, D.J., Eberwine, J., Wilson, C.J., Cao, Y., Stefani, A., and Kitai, S.T., 1992, Dopamine receptor subtypes colocalize in rat striatonigral neurons, *Proc. Nat. Acad. Sci.USA* 89:10178-10182.

Turski, L., Bressler, K., Retting, K.J., Loschmann, P.A., and Wachtel, H., 1991, Protection of substantia nigra from MPP+ neurotoxicity by N-methyl-d-aspartate antagonists, *Nature* 349:415-418.

RESPONSES OF PALLIDAL NEURONS TO MICROELECTRO-PHORETICALLY APPLIED GLUCOSE AND NEUROCHEMICALS

László Lénárd, Béla Faludi, Zoltán Karádi, András Czurkó, Imre Vida
and Csaba Niedetzky

Neurophysiology Research Group of the Hungarian Academy
of Sciences at Institute of Physiology
Pécs University Medical School
H-7643 Pécs, Hungary

INTRODUCTION

The globus pallidus (GP) is one of the major centers of the central nervous system controlling movements and postures. In addition to these motor functions, the GP is basically involved in the regulation of feeding, body weight, metabolic processes and sensorimotor integration. Electrical stimulation of the GP in rats elicits feeding related consummatory motor responses (Szabó et al., 1977). On the other hand, bilateral electrolytic destructions or neurotoxic lesions of the GP with 6-OHDA or kainic acid cause aphagia, adipsia, sex-dependent metabolic consequences, rapid weight loss and death of animals (Lénárd et al., 1975, 1981; Lénárd, 1977; Hahn et al. 1979; Sándor et al., 1992). Those rats that recover to spontaneous feeding and drinking exhibit deficits to different regulatory challenges (Marshall et al., 1974; Neill and Linn, 1975). Despite the large amount of data concerning the central regulatory functions of the GP, little is known about the specific neurochemical attributes of local cells in this structure.

Metabolic deficits and the severity of feeding disturbances observed after bilateral GP lesions are similar to those seen after bilateral damages of the lateral hypothalamus (LH) (Kent and Grossmann, 1973; Marshall et al., 1974; Lénárd et al., 1988, 1991). It has been suggested that these complex symptoms caused by lesions of the LH and GP might be due to the destruction of the ascending nigrostriatal and mesolimbic dopaminergic fibers (Marshall et al., 1974; Lénárd, 1977; Lénárd et al., 1981, 1988).

In the LH specific glucose-sensitive (GS) neurons monitoring feeding related endogenous chemosensory signals were described in both rats and monkeys (Oomura et al., 1969; Aou et al., 1984; Nishino et al., 1988; Karádi et al., 1988, 1990, 1992b). Firing rate of the GS neurons decreases in response to systemic or local, electrophoretic administration of glucose, and these neurons respond to various feeding-associated endogenous substances, including metabolites and hormones (Oomura, 1980).

The main goal of the present experiments was to study whether the GP of rats and monkeys contain similar GS neurons. By means of the multibarrel microelectrophoretic method, glucose and various neurotransmitters were applied to pallidal cells and their activity was recorded simultaneously.

METHODS

Eleven anesthetized male CFY rats (290-350 g) and 3 awake monkeys (one male and two females, Macaca mulatta, weighing 4.5 - 8 kg) were used. Experimental arrangements and details of recording and data analysis have been described previously (Karádi et al., 1990, 1992b). Animals were cared for in accordance with the NIH Guidelines. Rats were operated stereotaxically under urethane anesthesia (1.2 g/kg) to drill a small, 3 mm-diam hole in the skull for inserting multibarreled microelectrodes. Recordings were made from the area of the globus pallidus corresponding to the stereotaxic coordinates of De Groot (1959): A:5.8-7.2; L:2.5-3.5; V:3-(-)1. Monkeys were operated on under pentobarbital anesthesia (40 mg/kg) and aseptic conditions to stereotaxically fix a plastic plate to their skull. The plate restrained the monkey in the stereotaxic apparatus during the daily experimental sessions and thus, allowed insertion of microelectrodes through a small hole (3 mm in diam) in the skull and the incised dura mater.

Extracellular recording of pallidal single neuron activity and microelectrophoretic application of chemicals were accomplished with nine-barreled glass micropipettes in which the central barrel contained a carbon fiber (7 um diam). Drugs were applied electrophoretically through the micropipettes surrounding the recording electrode. Application current was produced by a constant current device (NeuroPhore BH-2). Each barrel was filled with one of the following solutions: 0.5 M monosodium L-glutamate; 0.2 M N-methyl-D-aspartate (NMDA); 0.5 M GABA; 0.5 M D-glucose (in 0.15 M NaCl); 0.5 M noradrenaline hydrochloride (NA; in 2% ascorbic acid solution) and 0.5 M dopamine hydrochloride (DA; in 2% ascorbic acid solution). Extracellular action potentials were passed into a preamplifier, a high gain amplifier with low and high cut filters, and to a window discriminator to form standard pulses. Neuronal spikes and formed pulses were continuously monitored on oscilloscopes (HAMEG HM-203). Pulses were fed into an IBM AT computer to construct frequency histograms. All raw data, the original action potentials, formed pulses and marker signals were stored on magnetic tapes and floppy disks for off-line analyses. Coordinates for electrode placements were chosen according to the atlas of Snider and Lee (1961): A:11-14; L:6-9; V:3-7. Cells that showed nonspecific current effects were excluded from the analysis. Neurons were considered to be responsive to glucose and neurochemicals if their firing rates were significantly different from the baseline level within 5-15 s of current onset, and if the activity changes to drugs by different current intensities proved to be dose-dependent and replicable. Recording sites were reconstructed by means of X-ray photographs and histological identifications of electrolytic microlesions made by elgiloy electrodes.

Table 1. Activity changes of rat pallidal neurons to electrophoretically applied glucose and neurochemicals

	Glucose	GABA	Glutamate	NMDA	DA	NA
↑	0 (0)	0 (0)	29 (36)	4 (6)	2 (4)	2 (4)
↓	9 (11)	38 (58)	6 (7)	20 (36)	19 (38)	20 (36)
↑↓	0 (0)	0 (0)	0 (0)	10 (18)	0 (0)	0 (0)
∅	71 (89)	28 (42)	46 (57)	22 (40)	29 (58)	33 (60)
Total	80 (100)	66 (100)	81 (100)	56 (100)	50 (100)	55 (100)

(Numbers in parentheses are percentages. Abbreviations: GABA, gamma-amino-butyric-acid; NMDA, N-methyl-D-aspartate; DA, dopamine; NA, noradrenaline. ↑, excitation; ↓, inhibition; ↑↓, excitation followed by inhibition; ∅, no effect.)

RESULTS

Activity of 88 neurons was recorded from the GP of rats. In monkeys 67 pallidal neurons were tested. Neurons exhibited distinct firing characteristics to different neurochemicals and glucose. In rats, 9 (11%) pallidal cells of 80 showed definite inhibitory responses to the electrophoretically applied glucose. These GS neurons exhibited activity decrease or inhibition in a dose-dependent manner (Table 1, Fig. 1). The remaining 71 cells did not show activity changes to glucose and were classified as glucose-insensitive (GIS) neurons. Application of GABA (66 cells) resulted in sharp inhibition with a short latency in 38 cells (58%) (Table 1, Fig. 2). No activity increase was observed when GABA was used. The facilitatory effect of glutamate was seen only in 29 pallidal neurons (36%), whereas in 6 cases (8%) a definite activity decrease was demonstrated (Table 1). Almost 2/3 of the pallidal cells tested, exhibited activity changes to NMDA administration (Fig. 2). The result was surprising, because the predominant response was inhibition (20 of 56 neurons examined, 36%), or a biphasic response occurred: facilitation was followed by inhibition (10 neurons, 18%). Only 4 cells (6%) exhibited activity increase to NMDA. When DA or NA was applied the predominant response was activity decrease (Figs. 1 and 2). Facilitation to cate-cholamines was observed only in several cases (Table 1). GS neurons exhibited a definite responsiveness to GABA. As far as the different types of the NMDA effects are concerned, no difference was found in the responsiveness of GS or GIS neurons. Similar observations were made when responses to catecholamines were considered.

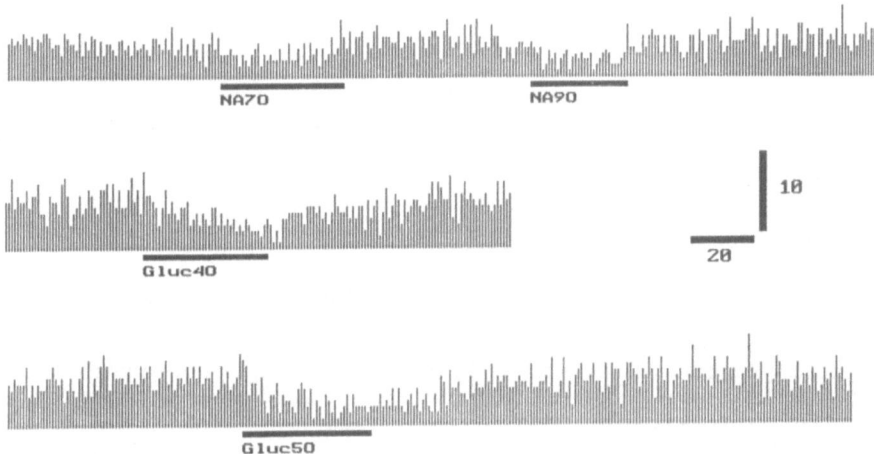

Figure 1. Activity changes of a glucose-sensitive (GS) neuron recorded from the ventromedial part of the rat globus pallidus (GP). Dose-dependent activity decrease to electrophoretically applied noradrenaline (NA) and glucose (Gluc). Horizontal lines, duration of drug applications. Numbers, electrophoretic current intensities in nA. Calibrations: abscissa, time (s); ordinate, impulses/s.

In monkeys, glucose was administered to 58 pallidal neurons. Six cells (10%) showed a definite dose-dependent activity decrease or inhibition to glucose. GABA was used in 52 cells, and 22 (42%) exhibited inhibitory responses. Both GS and GIS neurons responded to GABA. Predominant response to glutamate was facilitation, though inhibition occurred in 5 neurons (8%). NMDA was applied only in 10 cells. One neuron responded with facilitation, 3 cells with inhibition, and biphasic effects were recorded from 2 pallidal neurons. Catecholamine sensitive neurons exhibited activity decrease to DA (15/48 cells, 31%) or NA (16/53, 30%), though in several cases activity increase was found (3 and 2 cells; 6% and 4%, respectively). In 3 GS neurons DA or NA evoked inhibitory responses.

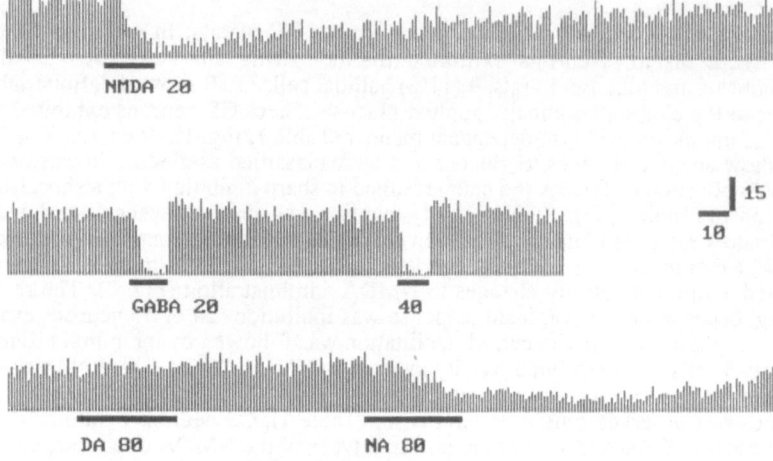

Figure 2. Activity changes of a glucose-insensitive (GIS) neuron recorded from the rat GP. Activity decrease to N-methyl-D-aspartate (NMDA) and noradrenaline (NA) and inhibition to GABA. No response to dopamine (DA). Abbreviations and symbols are identical to those in Figure 1.

Reconstruction of the recording sites based on X-ray pictures and histological sections showed that GS cells were located in the ventromedial part of the GP of rats. In monkey experiments, GS neurons were found exclusively at the ventromedial apex of the internal segment of the GP.

DISCUSSION

By means of the microelectrophoretic method, the neurochemical sensitivity of pallidal cells was studied in both rats and monkeys. In order to control the micropipette tip proximity to neurons, usually excitatory or inhibitory amino acids are administered. Since our pilot experiments (Faludi et al., 1992) showed that sensitivity of pallidal neurons to glutamate was somewhat different from that seen in other brain regions, the effects of glutamate, NMDA and GABA were studied simultaneously. Administration of GABA resulted in clear-cut inhibition. The predominant response to glutamate was facilitation, in several cases, however, clear inhibitory responses occurred. More interestingly, NMDA application resulted in facilitation or inhibition, or biphasic responses were observed. The inhibitory responses could not be due to cell injury or any neurotoxic effect (Meldrum and Garthwaite, 1990), since they were replicable and dose-dependent, and because the original firing rates recovered. The very same neuron responded to other drugs and its activity could be recorded for a long time (1-2 h) after glutamate or NMDA applications. Although, the possibility that an inhibitory interneuron was excited by glutamate or NMDA cannot be excluded, it is rather unlikely. Namely, the shape of action potentials was monitored continuously, the tip diameter of each pipette was between 0.2-0.4 μm, and drugs ejected from the neighbouring pipettes evoked the "appropriate" responses. The contradictory results with NMDA and glutamate can probably be explained by different NMDA receptor subtypes, by the availability of Mg-ions, or by other, so far unknown mechanisms. With reference to the latter, it was reported recently (Inoue et al., 1992), that clear hyperpolarizing responses could be induced by glutamate in cerebellar Purkinje cells. Further experiments with intra- and extracellular methods are needed to study the question. Nevertheless, from practical point of view it is clear, that GABA application in the GP can be a good method to check the neurochemical response of neurons when the microelectrophoretic technique is used.

The main result of the present experiments is, however, the discovery of GS neurons in the rat and monkey pallidum. Previously, GS neurons were described in different brain sites including the area postrema, the NTS, the LH and the amygdala (Oomura et al., 1969; Aou et al., 1984; Oomura and Yoshimatsu, 1984; Adachi and Kobashi, 1986; Nakano et al., 1986; Karádi et al., 1988, 1990, 1992b). In fact, the existence of a hierarchically organized glucose-monitoring system, located along the central neuraxis, has been postulated (Oomura and Yoshimatsu, 1984). A large amount of experimental data suggest that this system is basically involved in the regulation of feeding (Oomura, 1980; Aou et al., 1984; Nakano et al., 1986; Nishino et al., 1988; Karádi et al., 1988, 1990, 1992b). This neural network receives inputs from the internal environment, may control the local availability of glucose and is influenced by various metabolites and neurochemical agents. In the monkey LH certain GS neurons respond to external feeding related cues, taste and odor stimuli, and electrophoretic application of catecholamines (Nishino et al., 1988; Karádi et al., 1988, 1990, 1992b).

Embryological studies show (Kuhlenbeck and Haymaker, 1949) that some parts of the GP and the LH have common origin and the existence of intimate two-way interconnections are indicated between these two structures (Oomura et al., 1975). In rats, metabolic disturbances and significant increase of blood glucose concentration can be observed after bilateral lesions of the GP (Lénárd et al., 1975; Lénárd, 1977; Hahn et al., 1979, 1988). As the present results show, both the rat and monkey GP contain GS neurons. Since several GS neurons responded to catecholamines, one may suppose that their activity can also be influenced by the ascending catecholaminergic pathways. It was found recently in our experiments, that pallidal GS cells show distinct responsiveness to taste and odor stimuli and exhibit characteristic activity changes during execution of feeding related alimentary responses (Karádi et al., 1992a, Lénárd et al., 1992). Although, the exact details of their functional roles are yet to be defined, it is obvious, that GS neurons of the GP belong to the glucose-monitoring system and may play an essential role in the regulation of feeding.

Acknowledgements

This work was supported by the Hungarian National Research Found (OTKA/1404; L. Lénárd) and the Ministry of People's Welfare of Hungary (ETT T-565/1990; L. Lénárd).

REFERENCES

Adachi, A., and Kobashi, M., 1986, Consequence of hepatic gluco- and osmosensitive inputs on chemosensitive units in the medulla oblongata of rat, in: "Emotions: Neuronal and Chemical Control," Y. Oomura, ed., Jap. Sci. Soc. Press, S. Karger, AG, Tokyo/Basel, pp. 103-113.

Aou, S., Oomura, Y., Lénárd, L., Nishino, H., Inokuchi, A., Minami, T., and Misaki, H., 1984, Behavioral significance of monkey hypothalamic glucose-sensitive neurons, Brain Res. 302:69-74.

De Groot, J., 1959, The rat forebrain in stereotaxic coordinates, Verh. K. Ned. Acad. Wetensch. 52:1-40.

Faludi, B., Czurkó, A., Niedetzky, Cs., Vida, I., Hajnal, A., Karádi, Z., and Lénárd, L., 1992, Neuronal discharge rate changes in the globus pallidus during microelectrophoretic application of N-methyl-D(orDL)-aspartate, IBAGS Abstr. 4:26.

Hahn, Z., Karádi, Z., and Lénárd, L., 1988, Sex-dependent increase of blood glucose concentration after bilateral pallidal lesion in the rat, Acta Physiol. Hung. 72:99-101.

Hahn, Z., Lénárd, L., and Ruppert, F., 1979, The connection of plasma triiodothyronine levels with the sex-dependent body weight loss after bilateral pallidal lesion in rats, Acta Physiol. Hung. 53:17-21.

Inoue, T., Miyakawa, H., Ito, K., Mikoshiba, K., and Kato, H., 1992, A hyperpolarizing response induced by glutamate in mouse cerebellar Purkinje cells, Neurosci. Res. 15:265-271.

Karádi, Z., Czurkó, A., Faludi, B., Niedetzky, Cs., Vida, I., Hajnal, A., and Lénárd, L., 1992a, Functional attributes of feeding-related pallidal neurons in the rhesus monkey, ISSIB Abstr. 1:50.

Karádi, Z., Oomura, Y., Nishino, H., Scott, T.R., Lénárd, L. and Aou, S., 1988, Lateral hypothalamic and amygdaloid neuronal responses to chemical stimuli in the rhesus monkey, in: "Proc. JATS," H. Morita, ed., Ashai Univ. Press, Gifu, Japan, pp. 121-124.

Karádi, Z., Oomura, Y., Nishino, H., Scott, T.R., Lénárd, L., and Aou, S., 1990, Complex attributes of lateral hypothalamic neurons in the regulation of feeding of alert rhesus monkeys, *Brain Res. Bull.* 25:933-939.

Karádi, Z., Oomura, Y., Nishino, H., Scott, T.R., Lénárd, L., and Aou, S., 1992b, Responses of lateral hypothalamic glucose-sensitive and glucose-insensitive neurons to chemical stimuli in behaving rhesus monkeys, *J. Neurophysiol.* 67:389-400.

Kent, E.W., and Grossmann, S.P., 1973, Elimination of learned behaviors after transsection of fibers crossing the lateral border of the hypothalamus, *Physiol. Behav.* 10:953-963.

Kuhlenbeck, H., and Haymaker, W., 1949, The derivates of the hypothalamus in the human brain: Their relation to the extrapyramidal and autonomic system, *Milit. Surg.* 105:26-52.

Lénárd, L., 1977, Sex-dependent body weight loss after bilateral 6-Hydroxydopamine injection into the globus pallidus, *Brain Res.* 128:559-568.

Lénárd, L., Czurkó, A., Karádi, Z., Faludi, B., Vida, I., Niedetzky, Cs., and Hajnal, A. 1992, Functional significance of pallidal neurons: effects of electrophoretically applied glucose and neurochemicals, *IBAGS Abstr.* 4:49.

Lénárd, L., Jandó, G., Karádi, Z., Hajnal, A., and Sándor, P., 1988, Lateral hypothalamic feeding mechanisms: Iontophoretic effects of kainic acid, ibotenic acid and 6-Hydroxydopamine, *Brain Res. Bull.* 20:847-856.

Lénárd, L., Sarkisian, J., and Szabó, I., 1975, Sex-dependent survival of rats after bilateral pallidal lesions, *Physiol. Behav.* 15:389-397.

Lénárd, L., Sándor, P., Hajnal, A., Jandó, G., Karádi, Z., and Kai Y., 1991, Sex-dependent body weight changes after iontophoretic application of kainic acid into the LH or VMH, *Brain Res. Bull.* 26:141-148.

Lénárd, L., Szabó, I., Karádi, Z., and Hahn, Z., 1981, Pallidal mechanisms and feeding behavior, *in:* "Brain and Behavior, Adv. Physiol. Sci., Vol 17," G. Ádám, I. Mészáros and É. Bányai, eds., Akadémiai Kiadó, Budapest, pp. 331-341.

Marshall, J.F., Richardson, J.S., and Teitelbaum, Ph., 1974, Nigrostriatal bundle damage and the lateral hypothalamic syndrome, *J. Comp. Physiol. Psychol.* 87:808-830.

Meldrum, B., and Garthwaite, J., 1990, Excitatory amino acid neurotoxicity and neurodegenerative disease, *TiPS* 11:379.

Nakano, Y., Oomura, Y., Lénárd, L., Nishino, H., Aou, S., Yamamoto, T., and Aoyagi, K., 1986, Feeding-related activity of glucose- and morphine-sensitive neurons in the monkey amygdala, *Brain Res.* 399:167-172.

Neill, D.B., and Linn, C.L., 1975, Deficits in consummatory responses to regulatory challenges following basal ganglia lesions in rats, *Physiol. Behav.* 14:617-624.

Nishino, H., Oomura, Y., Karádi, Z., Aou, S., Lénárd, L., Kai, Y., Fukuda, A., Ito, C., Min, B.I., and Salaman, C.P., 1988, Internal and external information processing by lateral hypothalamic glucose-sensitive and insensitive neurons during bar press feeding in the monkey, *Brain Res. Bull.* 20:839-846.

Oomura, Y., 1980, Input-output organization in the hypothalamus relating feeding behavior, *in:* "Handbook of the Hypothalamus, Vol. 2," P. Morgane and and J. Panksepp, eds., Dekker, New York, pp. 557-620.

Oomura, Y., Nakamura, T., and Manchanda, S.K., 1975, Excitatory and inhibitory effects of globus pallidus and substantia nigra on the lateral hypothalamic activity in the rat, *Pharmacol. Biochem. Behav.* 3, S1:23-36.

Oomura, Y., Ono, T., Ooyama, H., and Wayner, M.J., 1969, Glucose and osmosensitive neurons of the rat lateral hypothalamus, *Nature Lond.* 222:282-284.

Oomura, Y., and Yoshimatsu, H., 1984, Neural network of glucose monitoring system, *J. Autonom. Nerv. Syst.* 10:359-372.

Sándor, P., Hajnal, A., Jandó, G., Karádi, Z., and Lénárd, L., 1992, Microelectrophoretic application of kainic acid into the globus pallidus: Disturbances in feeding behavior, *Brain Res. Bull.* 28:751-756.

Snider, S., and Lee, J.C., 1961, "A Stereotaxic Atlas of the Monkey Brain (Macaca Mulatta)," Univ. of Chicago Press, Chicago.

Szabó, I., Sarkisian, J., Lénárd, L., and Németh, L., 1977, Pallidal stimulation in rats: Facilitation of stimulus bound chewing by food deprivation, *Physiol. Behav.* 18:361-368.

LONG-LASTING CHANGES IN EXCITABILITY OF CORTICOSTRIATAL TERMINALS FOLLOWING TETANIC STIMULATION

Marianela Garcia-Munoz, Stephen J. Young, Patricia Patino
and Philip M. Groves

Department of Psychiatry, School of Medicine
University of California, San Diego
La Jolla CA 92093-0603, USA

INTRODUCTION

The presence of presynaptic auto- and heteroreceptors located on striatal afferents (e.g. Kalsner and Westfall, 1990) provides a locus for feedback regulation of transmitter release and for presynaptic interaction between afferent projections. While axo-axonic contacts are extremely rare in the striatum (Kornhuber and Kornhuber, 1983) the close proximity of cortical glutamate (GLU), nigral dopamine (DA) and other afferent terminals as they form synapses on spiny neurons (Freund et al., 1984) provides the potential for presynaptic interactions resulting from the activation of presynaptic receptors by diffusion of transmitter from synaptic sites (Cuello, 1966; Fuxe and Agnati, 1991).

To date it is not possible to use intracellular recording methods to examine physiological events related to the activation of presynaptic receptors on mammalian nerve terminals due to their small size. However, measurements of changes in the electrical excitability of the axonal terminal field may be used to infer these electrophysiological events. Excitability is evaluated by determining the stimulating current necessary to elicit an antidromic action potential from the terminal field of the neuron. Terminal excitability has been found to vary systematically with activation or blockade of presynaptic auto- and heteroreceptors on cortical, accumbens and striatal afferents (Groves et al., 1981; Mereu et al., 1985; Tepper et al., 1985; Chavez-Noriega et al., 1986; Yang and Mogenson, 1986) Increases in striatal DA induced by amphetamine or resulting from an increase in spontaneous or elicited dopaminergic firing rate have been shown to act presynaptically on nigrostriatal DA autoreceptors to decrease excitability (Tepper et al., 1985) and DA release (Kalsner and Westfall, 1990) Evidence for presynaptic interactions between nigral and cortical terminals in the striatum has also been obtained using the excitability method. Increased stimulation of DA receptors on corticostriatal terminals resulting from an increase in DA levels decreases GLU release (Earle and Davies, 1991) and excitability (Garcia-Munoz et al., 1991a).

We have recently found that the excitability of the striatal terminal field of a cortical neuron is positively correlated with its spontaneous firing rate. A finer analysis indicated that after the arrival of an action potential, the corticostriatal terminal field increased in excitability for a period of 20-40 ms. This impulse-induced facilitation of terminal excitability could be attenuated by local application of the GLU antagonists 2-amino-7-phosphonoheptanoate (AP-7) or 6,7-dinitro-quinoxaline-2,3-dione (CNQX) (Garcia-Munoz et al., 1991b). Since these effects were observed in animals with striatal lesions induced by kainic acid (Garcia-Munoz et al., 1991b), it was concluded that they were mediated by GLU autoreceptors located on corticostriatal terminals (Biziere and Coyle, 1978; Greenamyre and Young, 1989)

These results appeared consistent with a positive feedback mechanism: an increase in firing rate increased GLU release which in turn increased the activation of presynaptic autoreceptors to locally modify terminal membrane conductance and/or polarization and thereby increase electrical excitability. In support of this hypothesis, there is evidence that presynaptic activation of kainic acid, N-methyl-D-aspartate (NMDA) and *trans*-1-amino-cyclopentyl-1,3-dicarboxylate (*t*-ACPD) autoreceptors increases GLU release (Collins et al., 1983; Ferkany and Coyle, 1983; Martin et al., 1991). The possibility of a positive feedback mechanism involving GLU autoreceptors suggested that long-term changes in terminal excitability might be induced by brief high frequency stimulation of the corticostriatal pathway similar to the enduring effects observed in the phenomena of long-term potentiation (LTP) or depression (LTD). Indeed, we have recently reported that long-lasting changes in terminal excitability can be induced (Garcia-Munoz et al., 1992b). In the following sections, we review these results with additional data elucidating the receptors involved and discuss the possible functional significance of these long-term presynaptic changes.

Enduring Modifications in Postsynaptic and Presynaptic Excitability

Long-Term Potentiation and Long-Term Depression. Brief repetitive stimulation of the afferents to a neuron can initiate a long-lasting change in the amplitude of the postsynaptic response to a test stimulus. The postsynaptic response may be either increased (LTP) or decreased (LTD). Both LTP and LTD have been observed in the cerebellum (Crepel and Jaillard, 1991), in visual, somatosensory, motor and prefrontal cortical areas (Bindman et al., 1988; Artola et al., 1990; Hirsch and Crepel, 1992) and in the hippocampus (Dunwiddie and Lynch, 1978; Levy and Steward, 1983). These long-term modifications in synaptic efficacy occur in neurons receiving synapses employing GLU as a neurotransmitter, and their induction appears related to stimulation of specific GLU receptor types. The level of membrane polarization determined by the activity of the various classes of inputs to the cell during tetanization appears to establish the conditions for the degree of response of NMDA receptors and therefore, the induction of LTP. Artola et al. (1990) have suggested that cortical LTD is obtained if postsynaptic depolarization exceeds a certain critical level, but remains below the threshold for the activation of the NMDA receptors, whereas LTP is induced if this second threshold is reached.

Calcium is critical for the induction and maintenance of both LTP and LTD. A calcium chelator has been shown to prevent the induction of LTP and LTD in neurons of CA1 (Lynch et al., 1983; Christofi et al., 1991) and visual and prefrontal cortex (Hirsch and Crepel, 1992). It has been suggested that a rise in cytosolic calcium due to NMDA receptor activation leads to LTP, whereas an increase through voltage-dependent calcium channels or release from intracellular pools via activation of another type of GLU receptor leads to LTD (Hirsch and Crepel, 1992). Accordingly, blockade of NMDA receptors prevents the induction of LTP, but does not suppress LTD (Artola et al., 1990). Whether calcium enters NMDA and/or voltage dependent calcium channels appears to be determined by the level of depolarization reached during the tetanus. The level of polarization depends on the mix of convergent excitatory and inhibitory inputs activated, and on tetanus parameters. For example, in the hippocampus under conditions in which GABA transmission is not suppressed, i.e., during low frequency stimulation, the magnitude of the LTP is smaller or a short-lasting form of LTD may be observed (Barrionuevo et al., 1980; Grover and Teyler, 1990).

Enduring Modifications in Presynaptic Excitability. We have recently reported that brief high frequency stimulation of the corticostriatal pathway can induce long-lasting changes in the excitability of the striatal terminal field (Garcia-Munoz et al., 1992b). The methods employed in these experiments are briefly described in Figure 1. As illustrated in Figure 2A, we found that a long-lasting decrease in excitability could be induced following delivery of a tetanizing stimulus to the cortex (CTS). When CTS was administered at twice threshold current, a brief increase in excitability occurred prior to a long-lasting decrease. However if a second, threshold tetanus was administered during this transient elevation in excitability, a long-term increase in erminal excitability was observed (Fig. 3A).

Figure 1. Methods: The methods used in these studies have been previously described (Garcia-Munoz et al., 1992b). Briefly, the experiments were performed on rats under urethane anesthesia. A sagital view of the rat brain (after Paxinos and Watson, 1982) llustrates the electrode placements. An extracellular recording electrode and nearby bipolar stimulating electrode for delivery of a tetanizing stimulus were placed in the anteromedial prefrontal cortex. A second bipolar stimulating electrode for antidromic activation and striatal tetanic stimulation was placed alongside drug infusion cannulae in the contralateral dorsomedial striatum. The contralateral striatum was chosen to avoid stimulation of fibers of passage since crossed cortical projections do not send axons to the contralateral brainstem (Wilson, 1987). Antidromic action potentials, confirmed by collision testing, were recorded from a cortical neuron in response to stimulation of the corticostriatal terminal field. The

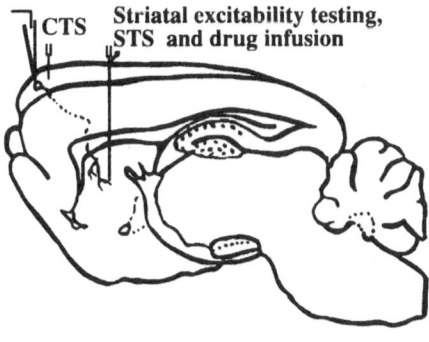

Cortical recording

CTS — Striatal excitability testing, STS and drug infusion

stimulating current was adjusted to a threshold level, defined as the minimum current necessary to evoke an antidromic response on 95-100% of non-collision trials. The threshold excitability level was periodically reassessed following an experimental condition. The effect of a treatment on excitability was evaluated in terms of the percent change in threshold current relative to control, that is, [100 (control threshold - threshold after treatment) / control threshold]. Thus, positive and negative changes represent increases and decreases in excitability, respectively. Drugs were dissolved in normal saline and delivered intrastriatally via 32 gauge stainless steel injectors using an infusion pump (0.3 µl over 5 min). Control animals received an intrastriatal infusion of saline. Tetanic stimulation was applied either through the cortical electrode (cortical tetanic stimulation, CTS) or through the striatal electrode (striatal tetanic stimulation, STS). In experiments with drug pretreatments, the tetanus was delivered 10 min after drug administration. The tetanizing stimulus consisted of five, 220 ms trains (22, 0.5ms pulses per train at 100 Hz) with a 6s intertrain interval. Tetanus pulses were presented at a threshold current just sufficient to elicit action potentials from the recorded neuron or in some cases a suprathreshold (2X threshold) level. The experiment usually continued until the recording was lost. To avoid residual drug effects, only one set of infusions, and one cell was typically examined per animal.

A long-term increase in excitability could also be consistently induced by CTS following treatments which disrupted both striatal DA and GABA transmission. In these experiments, GABAergic transmission was disrupted with a kainate lesion of the striatum or by local administration of the receptor antagonist bicuculline. Dopamine transmission was prevented by inhibiting synthesis with α-methyl-p-tyrosine or administration of the DA receptor antagonist haloperidol. In animals in which only one of these transmitter systems was impaired, CTS produced decreases or increases in almost equal numbers. Depletion of both neurotransmitters appeared necessary to produce a consistent increase in excitability following CTS. Tetanic stimulation of the striatal terminal field (STS) also initiated an enduring increase in excitability (Fig. 3I). These results were obtained in both intact animals and in rats in which postsynaptic neurons had been destroyed with kainate-induced lesions. Thus, the induction and maintenance of these long-lasting effects does not appear to depend on postsynaptic actions, but on alterations occurring in the presynaptic membrane. This is in contrast to the studies on LTP and LTD suggesting that long-term changes in synaptic efficacy involve a modification of the postsynaptic neuron resulting in the production of a retrograde message which acts on the presynaptic terminal (for references see Stevens, 1993). Our findings suggest that, similar to the induction of LTP or LTD, whether an increase or a decrease in presynaptic excitability is produced may depend on the level of depolarization of the terminal membrane achieved during the tetanization. The level of depolarization is presumed to depend on the balance of excitatory and inhibitory influences stimulating auto- and heteroreceptors on the terminal. According to this view, the glutamatergic positive feedback is fully expressed under conditions in which levels of extracellular DA and GABA are decreased and therefore, tetanization initiates a long-lasting increase in excitability. In postsynaptic studies of striatal neurons, LTD has been the most commonly observed long-term modification in synaptic efficacy (Walsh, 1991; Calabresi et al., 1992a; Tyler et al., 1992). LTD may predominate in the striatum due

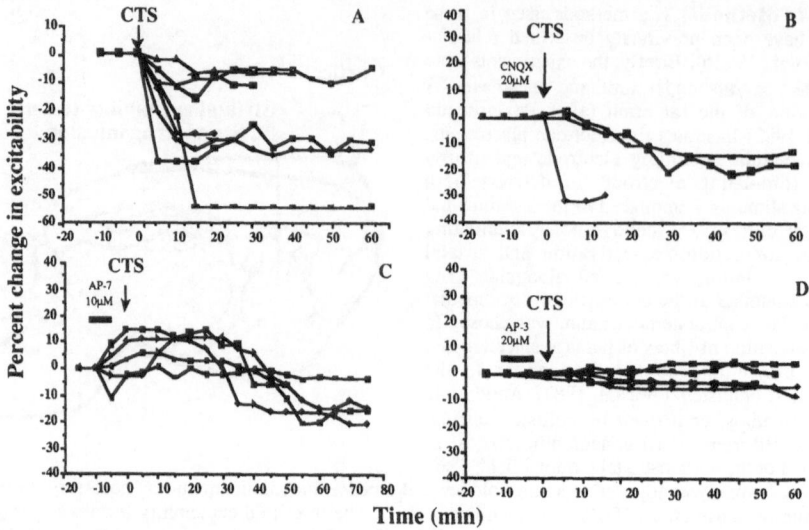

Figure 2. The effect of intrastriatal administration of GLU receptor antagonists on the excitability of the corticostriatal terminal field following CTS: (A) Threshold CTS induced a long-lasting decrease in excitability. (B) CNQX did not affect the response to CTS. (C) Administration of AP-7 modified excitability and delayed the appearance of a decrease in excitability to CTS for approximately 30 min. (D) AP-3 prevented the decrease in excitability to CTS. Each line represents an experiment on a single cell.

to the influence of DA and GABA on pre- and postsynaptic elements. LTP has been observed using *in vitro* intracellular recording of striatal spiny neurons following high frequency stimulation of the corpus callosum under conditions favoring disinhibition, i.e. under low magnesium and/or the presence of a $GABA_A$ receptor antagonist (Walsh, 1991; Calabresi et al., 1992b).

In the hippocampus a decrease in presynaptic excitability has been observed to parallel the time course of the postsynaptic LTP (Sastry, 1982). It is tempting to suggest that the initial release of GLU following the tetanus stimulated postsynaptic GLU receptors to induce LTP but only activated presynaptic GLU receptors to a level sufficient to achieve a long-term decrease in excitability. Stimulation of presynaptic heteroreceptors might contribute to a condition in which presynaptic depolarization was insufficient to activate NMDA receptors on the terminal, and therefore tetanic stimulation induced a long-term decrease in presynaptic excitability.

Glutamate Receptor Subtypes Mediating the Enduring Changes in Presynaptic Excitability

We have begun examination of the role of specific GLU receptors in long-term changes in terminal excitability. Our initial results are presented in Figures 2 and 3. GLU NMDA and non-NMDA receptors appear to participate in the induction of a long-term increase in presynaptic excitability, whereas the metabotropic receptor appears critical in the induction of a long-term decrease.

Blockade of NMDA or non-NMDA GLU receptors prevented the induction of the long-term increase in excitability that typically follows double CTS or STS (Fig. 3). These results suggest that the level of depolarization produced by the combined activation of GLU receptors is critical for the induction of an enduring increase in excitability. As mentioned above, an increase in the concentration of calcium appears to be a key factor in the induction of LTP. Depolarization opens NMDA and voltage-dependent calcium channels. Both mechanisms may be required to increase the internal concentration of calcium to sufficient levels. The mechanisms producing an increase in terminal excitability

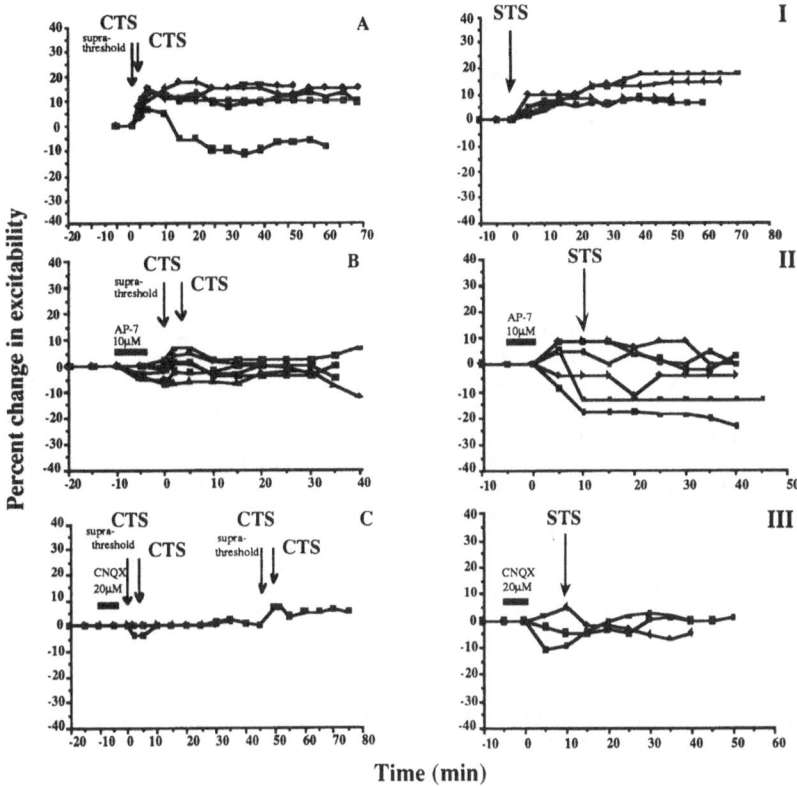

Figure 3. The effect of intrastriatal administration of GLU receptor antagonists on long-term changes in excitability induced by double CTS (A-C) and STS (I-III). (A) double CTS typically induced an enduring increase in excitability. (B) AP-7 did not affect the brief increase in excitability seen following suprathreshold CTS, but prevented the long-term increase in excitability. (C) CNQX blocked both the initial depolarization and the enduring modification of excitability. One hour after infusion, CNQX was less potent but sufficient to reduce the effect of the tetanus. (I) STS induced a long-lasting increase in excitability. (II) Following AP-7 STS produced either no effect or a decrease in excitability. (III) CNQX blocked the STS effect.

that is, a decrease in the threshold stimulating current required for eliciting an antidromic action potential, are a matter of speculation. Recent studies obtained from invertebrate neuromuscular junction with the use of calcium-sensitive fluorescent dyes support the hypothesis that the action potential-dependent enhancemen of synaptic transmission depends on the level of intracellular calcium (Zucker and Lara-Estrella, 1983). A variety of evidence indicates that calcium-dependent protein phosphorylation modulates the release of neurotransmitter. Cyclic AMP and calcium-calmodulin dependent protein kinases have been proposed to remove a constraint on neurotransmitter release and mobilize synaptic vesicles for exocytosis (Castelluci et al., 1980; Akers and Routtenberg, 1985; Nichols et al., 1990). If impulse-related GLU release is increased following tetanic stimulation of the corticostriatal pathway, increased autoreceptor stimulation could maintain the membrane at a less polarized state and, therefore, lower the threshold for the production of action potentials in the induction stage of the enduring change in excitability. Subsequently, a third messenger (possibly, phospholipase C or phospholipase A2) could induce an increase in the affinity or the conductance of GLU autoreceptors to maintain the excitability change as has been observed postsynaptically (Kauer et al., 1988).

The long-term decrease in excitability triggered by CTS (Fig. 2A) was not affected by CNQX (Fig 2B) and, therefore, does not appear to involve the non-NMDA GLU receptors. AP-7 appeared only to delay the expression, but not the induction of the enduring decrease in excitability (Fig. 2C). The long-term decrease in excitability to CTS was only prevented in the presence of AP-3, the *t*-ACPD receptor antagonist (Fig 2D). Similarly, this receptor appears critical for striatal LTD (Calabresi et al., 1992a). Stimulation of the *t*-ACPD receptor, leads to the activation of a pertusis toxin-sensitive G protein resulting in two classes of second messengers: diacylglycerol, a protein kinase C activator, and IP3 which releases intracellular calcium. An increase in intracellular calcium via IP3 may be sufficient to trigger the long-term decrease in excitability. Longer lasting modifications in excitability could be a consequence of calcium-activated second and third messenger cascades resulting in a selective GLU receptor desensitization (Ito, 1989). The dependence of the CTS induction of a long-term excitability decrease on GLU was tested using the GLU receptor agonist AP-4 (L-2-amino-4-phosphonobutyrate). In contrast to other GLU agonists, AP-4 has been reported to decrease GLU release via activation of an autoreceptor (Forsythe and Clements, 1990). Local striatal administration of AP-4 (50 μM) prevented the induction of a long-term decrease in excitability to CTS and produced a short-lasting (15 min) decrease in excitability (results not shown). The transient reduction in terminal excitability suggests that these terminals are tonically stimulated by GLU. Tetanic stimulation did not induce a long-term change in excitability under a condition in which the corticostriatal release of GLU was presumably impaired by the administration of AP-4.

Considering that LTP and LTD have been obtained in vitro in striatal cells following tetanic stimulation of corticostriatal fibers, the degree to which these postsynaptic long-term effects reflect presynaptic changes needs to be explored. The induction and maintenance of the long-lasting presynaptic effects we have observed do not appear to depend on postsynaptic actions since they can be induced in animals with kainate-induced striatal lesions. Therefore, it is possible that pre- and postsynaptic membranes may be modified concurrently by similar mechanisms to alter receptor affinity and/or conductance or to change the morphology of the synapse to maintain an altered level of excitability.

Functional Considerations

Dorsolateral, central and ventromedial sectors of the striatum are innervated by different cortical regions resulting in an anatomical and functional parcellation of the nucleus. This parcellation lacks discrete boundaries and shows considerable overlap and interdigitation (Goldman-Rakic and Selemon, 1987; Divac and Öberg, 1992; Ebrahimi et al., 1992). Superimposed on this organization are parallel input-output pathways represented by the patch/matrix compartmentalization of the striatum through which segregated cortical inputs differentially affect striatal output stations (Gerfen, 1992). Moreover, in the primate striatal matrix, cortical inputs and pallidal outputs appear to be organized in discrete zones called matrisomes. Matrisomes have been proposed to integrate information from different cortical maps to form somatotopically modular input/output systems (Flaherty and Graybiel, 1992). It has been hypothesized that the striatum serves to integrate and process through these parallel systems spatial and temporal patterns of neuronal activity from diverse cortical inputs concerned with the initiation and sequencing of motor behavior (Groves, 1983; Alexander et al., 1986; Fallon and Loughlin, 1987; Koob and Swerdlow, 1988; Divac and Öberg, 1992; Ebrahimi et al., 1992)

Motor disturbances plus signs of cognitive deficits occur in Parkinson and Huntington disease, two major clinical syndromes related to striatal dysfunction. Animal models of Parkinson disease based on the depletion of DA produce a failure to orient to stimuli and to initiate movements (Marshall et al., 1974; Hamilton et al., 1985; Sabol et al., 1985; Pisa and Cyr, 1990). Electrophysiological experiments have shown that most striatal neurons respond during the preparation for the initiation of movements, in particular those in response to a specific environmental stimulus (Rolls et al., 1983; Johnstone and Rolls, 1990). The hypothesis that the striatum may be involved in behavioral planning or sequencing has also received clinical support. Parkinson patients have difficulty with temporal sequencing and one of the earliest cognitive complaints of patients suffering from Huntington disease is difficulty in planning the day's activities (Brandt and Butters, 1986; Sagar et al., 1988).

LTP and LTD have been proposed as neural mechanisms underlying learning and memory mediated by hippocampal and cerebellar activity (Morris et al., 1988; Ito, 1989). Even though a hippocampal lesion greatly impairs the ability to store new information (Mishkin and Appenseller, 1987), learning of certain tasks is still possible. For example, amnesic patients can acquire and perform pattern-analyzing skills necessary for mirror reading, sensory-motor skills required for rotary pursuit and bimanual tracking tasks and some cognitive skills necessary to solve puzzles. The striatum appears to play a critical role in these kinds of procedural learning. (Martone et al., 1984; Cohen et al., 1985). Striatal lesions in rats impair the acquisition of several operant tasks, the learning of which is typically unaffected by hippocampal lesions (O'Keefe and Nadel, 1978). These include various avoidance tasks (Prado-Alcalá et al., 1975; Giordano and Prado-Alcalá, 1986) and visual/olfactory conditioned emotional responses (Whishaw et al., 1987; Packard and White, 1991; Packard and McGaugh, 1992). The retention of a conditioned emotional response (Viaud and White, 1989) and a win-stay visual discrimination (Packard and White, 1991) is facillitated by intrastriatal administration of amphetamine given shortly after training, but not at longer delays. These findings suggest that increased DA levels enhance memory consolidation processes within the striatum involved in the retention of these tasks. It is interesting that, according to our results and other studies, increased striatal DA levels appear to affect the probabilities for obtaining enduring decreases and increases in synaptic efficacy (Walsh, 1991; Calabresi et al., 1992b; Garcia-Munoz et al., 1992b).

Long-term presynaptic changes in corticostriatal excitability may represent an important mechanism involved in the learning of motor tasks and performance of cognitive skills. An enduring modification in excitability may reflect a long-term increase or decrease in the synaptic efficacy of the affected terminal. This may provide a "switch" to allow the passage of preferential information. Thus, a selective decrease in presynaptic excitability in one class of afferents could reflect a reduction in the influence of this class on the response of the neuron while an increase in presynaptic excitability may represent enhanced transmission in another set of inputs. Such a focusing mechanism would not only allow the facilitated input to produce the strongest modulation, but also permit, if needed, a substitution between alternative types of information. Our results would suggest that the particular pattern of increases and decreases in synaptic efficacy occurring at the synapses of a neuron or within a striatal module would depend not only on the pattern of cortical activity but also on the activity of dopaminergic, GABAergic and other afferents operating in the local environment of synapses. Activity in these afferent systems would affect the likelihood that cortical activity would institute a long-term change and whether a long-term increase or decrease in synaptic efficacy would be established.

Acknowledgements

This work was supported in part by grants from the National Institute on Drug Abuse and the Office of Naval Research.

REFERENCES

Akers, R.,and Routtenberg, A., 1985, Protein kinase C phosphorylates a 47Mr protein (F1) directly related to synaptic plasticity, *Brain Res.* 334:147-151.

Alexander, G.E., DeLong, M.R., and Strick, P.L., 1986, Parallel organization of functionally segregated circuits linking basal ganglia and cortex, *Ann. Rev. Neurosci.* 9:357-381.

Artola, A., Bröcher, S., and Singer, W., 1990, Different voltage-dependent thresholds for inducing long-term depression and long-term potentiation in slices of rat visual cortex, *Nature* 347:69-72.

Barrionuevo, G., Schottler, F., and Lynch, G., 1980, The effects of repetitive low frequency stimulation on control and "potentiated" synaptic responses in the hippocampus, *Life Sci.* 27:2385-2391.

Bindman, L.J., Murphy, K.P.S.J., and Pockett, S., 1988, Postsynaptic control of the induction of long-term changes in efficacy of transmission at neocortical synapses in slices of rat brain, *J. Neurophysiol.* 60(3):1053-1065.

Biziere, K., and Coyle, J.T., 1978, Influence of cortico-striatal afferents on striatal kainic acid neurotoxicity, *Neurosci. Lett.* 8:303-310.

Brandt, J., and Butters, N., 1986, The neuropsychology of Huntington's disease, TINS. 9:118-120.

Calabresi, P., Maj, R., Pisani, A., Mercuri, N.B., and Bernardi, G., 1992a, Long-term synaptic depression in the striatum: Physiological and pharmacological characterization, *J. Neuroscience* 12:4224-4233.

Calabresi, P., Pisani, A., Mercuri, N.B., and Bernardi, G., 1992b, Long-term potentiation in the striatum is unmasked by removing the voltage-dependent magnesium block of NMDA receptor channels, *Eur. J. Neurosci.* 4:929-935.

Castelluci, V.F., Kandel, E.R., Schwartz, J.H., Wilson, F.D., Nairn, A.C., and Greengard, P., 1980, Intracellular injection of the catalytic subunit of cyclic AMP-dependent protein kinase stimulates facilitation of transmitter release underlying behavioral sensitization in Aplysia, *Proc. Natl. Acad. Sci. U.S.A.* 77:7492-7496.

Chavez-Noriega, L., Patino, P., and Garcia-Munoz, M., 1986, Excitability changes induced in the striatal dopamine-containing terminals following frontal cortex stimulation, *Brain Res.* 379:300-306.

Christofi, G., Novicky, A.V., and Bindman, L.J., 1991, The postsynaptic induction of long-term depression (LTD) of synaptic transmission in isolated rat hippocampal slices requires extracellular calcium, *J. Physiol. (London)* 438:257P.

Cohen, N.J., Eichenbaum, H., De Acedo, H., and Corkin, S., 1985, Different memory systems underlying acquisition of procedural and declarative knowledge, *Ann. N. Y. Acad. Sci.* 444:54-71.

Collins, G.G.S., Anson, J. and Surtees, L., 1983, Presynaptic kainate and N-methyl-D-aspartate receptors regulate excitatory amino acid release in the olfactory cortex, *Brain Res.* 265:157-159.

Crepel, F., and Jaillard, D., 1991, Pairing of pre- and postsynaptic activities in cerebellar purkinje cells induces long-term changes in synaptic efficacy in vitro, *J. Physiol. (London)* 432:123-141.

Cuello, C.A., 1966, Nonclassical neuronal communications, *Frd. Proc. Fed. Soc. Exptl. Biol.* 42:2912-2922.

Divac, I., and Öberg, G.E., 1992, Subcortical mechanisms in cognition, *in:* "Neuropsychological disorders associated with subcortical lesions," G. Vallar, S. F. Cappa and C.-W. Wallesch, eds., Oxford University Press, Oxford, pp. 42-60.

Dunwiddie, T., and Lynch, G., 1978, Long-term potentiation and depression of synaptic responses in the rat hippocampus: Localization and frequency dependency, *J. Physiol. (London)* 276:353-367.

Earle, M.L., and Davies, J.A., 1991, The effect of methamphetamine on the release of glutamate from striatal slices, *J. Neural Transm.* 86:217-222.

Ebrahimi, A., Pochet , R., and Roger, M., 1992, Topographical organization of the projections from physiologically identified areas of the motor cortex to the striatum in the rat, *Neurosci. Res.* 14:39-60.

Fallon, J.H., and Loughlin, S.E., 1987, Monoamine innervation of cerebral cortex and a theory of the role of monoamines in cerebral cortex and basal ganglia, *in:* "Cerebral Cortex. Volume 6. Further aspects of cortical function, including hippocampus," E. G. Jones and A. Peters eds., Plenum Press, New York, pp. 41-127.

Ferkany, J.W., and Coyle, J.T., 1983, Kainic acid selectively stimulates the release of endogenous excitatory acidic amino acids, *J. Pharmacol. Exp. Ther.* 225:399-406.

Flaherty, A.W., and Graybiel, A.M., 1992, Multiple stages of sensorimotor processing in the primate basal ganglia, *Fourth triennial Meet. Intern. Basal Ganglia Soc.* abstr. :29.

Forsythe, I.D., and Clements, J.D., 1990, Presynaptic glutamate receptors depress excitatory monosynaptic transmission between mouse hippocampal neurones, *J. Physiol. (London)* 429:1-16.

Freund, T.F., Powell, J.F., and Smith, A.D., 1984, Tyrosine hydroxylase-immunoreactive boutons in synaptic contact with identified striatonigral neurons, with particular reference to dendritic spines, *Neurosci.* 13:1189-1215.

Fuxe, K., and Agnati, L.F., 1991, "Volume Transmission in the Brain: Novel Mechanisms for Neural Transmission," Raven Press, New York.

Garcia-Munoz, M., Young, S.J., and Groves, P.M., 1991a, Terminal excitability of the corticostriatal pathway. I. Regulation by dopamine receptor stimulation, *Brain Res.* 551:195-206.

Garcia-Munoz, M., Young, S.J., and Groves, P.M., 1991b, Terminal excitability of the corticostriatal pathway. II. Regulation by glutamate receptor stimulation, *Brain Res.* 551:207-215.

Garcia-Munoz, M., Young, S.J., and Groves, P.M., 1992a, Long-lasting changes in excitability of corticostriatal terminals following tetanic stimulation, *Fourth triennial Meet. Intern. Basal Ganglia Soc.* abst. :32.

Garcia-Munoz, M., Young, S.J., and Groves, P.M., 1992b, Presynaptic long-term changes in excitability of the corticostriatal pathway, *NeuroReport.* 3:357-360.

Gerfen, C.R., 1992, The neostriatal mosaic: multiple levels of compartmental organization, *TINS* 15:133-139.

Giordano, M., and Prado-Alcalá, R.A., 1986, Retrograde amnesia induced by post-trial injection into caudate-putamen. Protective effect of the negative reinforcer, *Pharmacol. Biochem. Behav.* 24:905-909.

Goldman-Rakic, S., and Selemon, L.D., 1987, Topography of corticostriatal projections in nonhuman primates and implications for functional parcellation of the neostriatum, *in:* "Cerebral Cortex. Volume 6. Further aspects of cortical function, including hippocampus," E. G. Jones and A. Peters, eds., Plenum Press, New York, pp. 447-466.

Greenamyre, J.T., and Young, A.B., 1989, Synaptic localization of striatal NMDA, quisqualate and kainate receptors, *Neurosci. Lett.* 101:133-137.

Grover, L.M., and Teyler, T.J., 1990, Two components of long-term potentiation induced by different patterns of afferent activation, *Nature* 347:477-479.

Groves, P.M., 1983, A theory of functional organization of the neostriatum and the neostriatal control of voluntary movement, *Brain Res. Rev.* 5:109-132.

Groves, P.M., Fenster, G.A., Tepper, J.M., Nakamura, S., and Young, S.J., 1981, Changes in dopaminergic terminal excitability induced by amphetamine and haloperidol, *Brain Res.* 221:425-431.

Hamilton, M.H., Garcia-Munoz, M., and Arbuthnott, G.W.,1985, Separation of the motor consequences from other actions of unilateral 6-hydroxydopamine lesions in the nigrostriatal neurones of rat brain, *Brain Res.* 348:220-228.

Hirsch, J.C., and Crepel, F., 1992, Postsynaptic calcium is necessary for the induction of LTP and LTD of monosynaptic EPSPs in prefrontal neurons: An in vitro study in the rat, *Hippocampus* 10:173-175.

Ito, M., 1989, Long-term depression, *Ann. Rev. Neurosci.* 12:85-102.

Johnstone, S., and Rolls, E.T., 1990, Delay, dicriminatory, and modality specific neurons in striatum and pallidum during short-term memory tasks, *Brain Res.* 522:147-151.

Kalsner, S., and Westfall, T.C., 1990, Presynaptic receptors and the question of autoregulation of neurotransmitter release, *Ann. N.Y. Acad. Sci.* 604:652-655.

Kauer, J.A., Malenka, R.C., and Nicoll, R.A., 1988, A persistent postsynaptic modification mediates long-term potentiation in the hippocampus, *Neuron* 1:911-917.

Koob, G.F., and Swerdlow, N.R., 1988, The functional output of the mesolimbic dopamine system, *Ann. NY Acad. Sci.* 537:216-227.

Kornhuber, J., and Kornhuber, M.E., 1983, Axo-axonic synapses in the rat striatum, *Eur. J. Neurol.* 22:433-436.

Levy, W.B., and Steward, O., 1983, Temporal contiguity requirements for long-term associative potentiation/depression in the hippocampus, *Neurosci.* 8:791-797.

Lynch, G., Larson, J., Kelso, S., Barrionuevo, G., and Schottler, F., 1983, Intracellular injections of EGTA block induction of hippocampal long-term potentiation, *Nature* 305:719-721.

Marshall, J.P., Richardson, J.S., and Teitelbaum, P., 1974, Nigrostriatal bundle damage and the lateral hypothalamic syndrome, *J. Comp. Physiol. Psychol.* 87:808-830.

Martin, D., Bustos, G.A., Bowe, M.A., Bray, S.D., and Nadler, J.V., 1991, Autoreceptor regulation of glutamate and aspartate release from slices of the hippocampal CA1 area, *J. Neurochem.* 56:1647-1655.

Martone, M., Butters, N., Payne, M., Becker, J.T., and Sax, D.S., 1984, Dissociations between skill learning and verbal recognition in amnesia and dementia, *Arch. Neurol.* 41:965-970.

Mereu, G., Westfall, T.C., and Wang, R.Y., 1985, Modulation of terminal excitability of mesolimbic dopaminergic neurons by d-amphetamine and haloperidol, *Brain Res.* 359:88-96.

Mishkin, M., and Appenseller, T.,1987, The anatomy of memory, *Scientific American* 256(6):80-89.

Morris, R.G.M., Kandel, E.R., and Squire, L.R., 1988, The neuroscience of learning and memory: cells, neural circuits and behaviour, *TINS* 11:125-127.

Nichols, R.A., Sihra, T.S., Czernik, A.J., Nairn, A.C., and Greengard, P., 1990, Calcium/calmodulin-dependent protein kinase II increases glutamate and noradrenaline release from synaptosomes, *Nature* 343:647-651.

O'Keefe, J., and Nadel, L., 1978, "The Hippocampus as a Cognitive Map," Oxford University Press, New York.

Packard, M.G., and McGaugh, J.L., 1992, Double dissociation of fornix and caudate nucleus lesions on acquisition of two water maze tasks: further evidence for multiple memory systems, *Behav. Neurosci.* 106:439-446.

Packard, M.G., and White, N.M., 1991, Dissociation of hippocampus and caudate nucleus memory systems by posttraining intracerebral injection of dopamine agonists, *Behav. Neurosci.* 105:295-306.

Paxinos, G., and Watson, C., 1982, "The Rat Brain in Stereotaxic Coordinates," Academic Press, New York.

Pisa, M., and Cyr, J., 1990, Regionally selective roles of the rat's striatum in modality-specific discrimination learning and forelimb reaching, *Behav. Brain Res.* 37:281-292.

Prado-Alcalá, R.A., Grinberg, J.Z., Arditti, L., Garcia-Munoz, M., Prieto, G.H., and Brust-Carmona, H., 1975, Learning deficits produced by chronic and reversible lesions of the corpus striatum in rats, *Physiol. Behav.* 15:283-287.

Rolls, E.T., Thorpe, S.J., and Maddison, S.P., 1983, Responses of striatal neurons in the behaving monkey. 1. Head of the caudate nucleus, *Behav. Brain Res.* 7:179-210.

Sabol, K.E., Neill, D.B., Wages, S.A., Church, W.H., and Justice, J.B., 1985, Dopamine depletion in a striatal subregion disrupts performance of a skilled motor task in the rat, *Brain Res.* 335:33-43.

Sagar, H.J., Sullivan, E.V., Gabrieli, J.D.E., Corkin, S., and Growden, J.H., 1988, Temporal ordering and short-term memory deficits in Parkinson's disease, *Brain* 111:525-539.

Sastry, B.R., 1982, Presynaptic change associated with long-term potentiation in hippocampus, *Life Sci.* 30:2003-2008.

Tepper, J.M., Groves, P.M., and Young, S.J., 1985, The neuropharmacology of the autoinhibition of monoamine release, *TIPS* 6:251-256.

Tyler, E.C., Lovinger, D.M., and Merritt, A., 1992, Short and long-term synaptic depression in neostriatum, *Soc. Neurosci. Abst.* 18:1351.

Viaud, M.C., White, and N.M., 1989, Dissociation of visual and olfactory conditioning in the neostriatum of rats, *Beh. Brain Res.* 32:31-42.

Walsh, J.P., 1991, Long-term potentiation (LTP) of excitatory synaptic input to medium spiny neurons of the rat striatum, *Soc. Neurosci. Abst.* 17:852.

Whishaw, I.Q., Mittleman, G., Bunch, S.T., and Dunnet, S.B., 1987, Impairments in the acquisition, retention, and selection of spatial navigation strategies after medial caudate-putamen lesions in rats, *Behav. Brain Res.* 24:125-138.

Wilson, C.J., 1987, Morphology and synaptic connections of crossed corticostriatal neurons in the rat, *J. Comp. Neurol.* 263:567-580.

Yang, C.R., and Mogenson, G.J., 1986, Dopamine enhances terminal excitability of hippocampal-accumbens neurons via D2 receptor: role of dopamine in presynaptic inhibition, *J. Neurosci.* 6:2470-2478.

Zucker, R.S., and Lara-Estrella, L.O., 1983, Post-tetanic decay of evoked and spontaneous transmitter release and a residual-calcium model of synaptic facilitation at crayfish neuromuscular junctions, *J. Gen. Physiol.* 81:355-372.

BURST FIRING INDUCED BY N-METHYL-D-ASPARTATE REQUIRES ACTIVATION OF AN ELECTROGENIC SODIUM PUMP IN RAT DOPAMINE NEURONS

Steven W. Johnson, Vincent Seutin and R. Alan North

Department of Neurology and the Vollum Institute
Oregon Health Sciences University
Portland, OR 97201, USA

INTRODUCTION

In vitro, dopamine neurons in the ventral tegmental area (VTA) and substantia nigra zona compacta (SNC) fire action potentials in a regular pacemaker pattern (Sanghera et al., 1984; Johnson and North, 1992; Grace and Onn, 1989; Lacey et al., 1989). However, *in vivo*, dopamine neurons fire in bursts (Freeman et al., 1985; Sanghera et al., 1984). The fact that burst firing occurs *in vivo* but not *in vitro* implies that afferent connections essential for burst firing are severed in the process of cutting the brain slice (Sanghera et al., 1984). Svensson and Tung (1989) showed that cooling the pre-frontal cortex in the anesthetized rat reversibly blocked burst firing in VTA dopamine neurons. Because prefrontal cortex sends axons which contain excitatory amino acids (EAA's) to the VTA (Christie et al., 1985), it was thought that cooling the cortex diminished the synaptic release of EAA's onto dopamine neurons. This conclusion is supported by the findings that glutamate receptor antagonist drugs such as kynurenate, 4-(3-phosphonopropyl)-2-piperazine carboxylic acid (CPP), and DL-aminophosphonovaleric acid (APV) reversibly blocked burst firing in dopamine VTA neurons recorded extracellularly in the rat (Charlety et al., 1991; Overton and Clark, 1992; Chergui et al., 1991). In this communication, we report on the mechanism of action underlying burst firing induced by N-methyl-D-aspartate (NMDA) in dopamine neurons *in vitro*.

METHODS

Detailed methods have been reported previously (Lacey et al., 1988; Johnson and North, 1992). Briefly, male Sprague-Dawley rats (150-250 g) were anesthetized with halothane and killed by severing major thoracic vessels. A block of tissue containing midbrain was immersed in artificial spinal fluid at 4°C; a horizontal slice (300 μm) was cut by a vibratome and placed on a nylon mesh situated in a recording chamber (500 μl) Artificial spinal fluid (36°C) entered the chamber at a rate of 2 ml/min and was drained from the top so that the slice was immersed. The slice was immobilized by placing an electron microscope grid on top of its surface, held in place by small platinum weights. VTA and SNC were identified by their relationships to white matter tracts (Paxinos and Watson, 1986). The perfusion solution was saturated with 95% O_2 and 5% CO_2; it contained (mM): NaCl 126, KCl 2.5, NaH_2PO_4 1.2, $MgCl_2$ 1.2, $CaCl_2$ 2.4, glucose 11, and $NaHCO_3$ 18 (pH 7.4). Drugs were added to the superfusate which entered the recording chamber within

30 s of turning a tap. Glass microelectrodes were filled with potassium chloride (2 M; 40 - 80 MΩ) or potassium acetate (2 M; 60 - 120 MΩ). Only the lower resistance potassium chloride electrodes were used for voltage-clamp. Membrane potentials and currents were amplified with an Axoclamp-2A amplifier. Single-electrode voltage-clamp was performed using a switching frequency of 2 - 5 kHz (33% duty cycle). Dopamine-containing "principal" cells were identified by their relatively broad action potentials, spontaneous pacemaker firing pattern, the "sag" in membrane potential in response to hyperpolarizing current pulses, and the hyperpolarization by dopamine (30 µM) (Lacey et al., 1989; Johnson and North, 1992). These cells contain dopamine because they are labeled with antibodies against tyrosine hydroxylase (Grace and Onn, 1989; Johnson and North, 1992).

RESULTS

As we have reported previously (Johnson et al., 1992), burst firing is produced by NMDA (10 - 30 µM) and consists of a burst of 3 - 20 spikes followed by a hyperpolarization (5 - 30 mV) lasting 0.5 - 2 s, repeating every 0.5 to 5 s (Figure 1A). No difference was observed between cells in the VTA and SNC, so results were pooled. The most effective voltages for burst firing occurred between -45 mV (near firing threshold) and -75 mV (point of maximum inter-burst hyperpolarization). In most cells, hyperpolarizing current (-100 to -300 pA) was applied to keep the membrane potential within this voltage range. Burst firing was not blocked by bicuculline (30 µM; n = 1), 2-hydroxysaclofen (300 µM; n = 1) or sulpiride (1 µM; n = 2), suggesting that burst firing does not depend upon release of GABA or dopamine. Tetrodotoxin (TTX; 1 µM) blocked bursts of action potentials but it did not block the membrane oscillations which underlie burst firing (n = 7). Apamin (30 - 1000 nM), a peptide from bee venom which blocks the small conductance Ca^{2+}-dependent K^+ channel (Blatz and Magleby, 1986), did not block burst firing. In fact, apamin potentiated burst firing produced by NMDA but did not cause burst firing alone. Burst firing was reversibly blocked by the NMDA-receptor antagonist APV (50 µM; n = 3) and by superfusate containing no-added Mg^{2+} (n = 3), but not by the non-NMDA- receptor antagonist CNQX (10 µM; n = 1) (Figure 1B). Neither kainate (1 - 10 µM; n = 2) nor quisqualate (1 - 10 µM; n = 2) produced burst firing, despite a 3 - 10 fold increase in spontaneous firing rates. Aspartate (500 µM) produced burst firing in 4 of 8 cells, but glutamate (100 - 1000 µM) failed to produce burst firing even in bicuculline and CNQX (n = 6). Burst firing was not blocked by TEA (1 - 30 mM; n = 4), 4-aminopyridine (1 mM; n = 1), or charybdotoxin (100 nM), which blocks large conductance Ca^{2+}-activated K^+ channels (Miller et al., 1985) (n = 2).

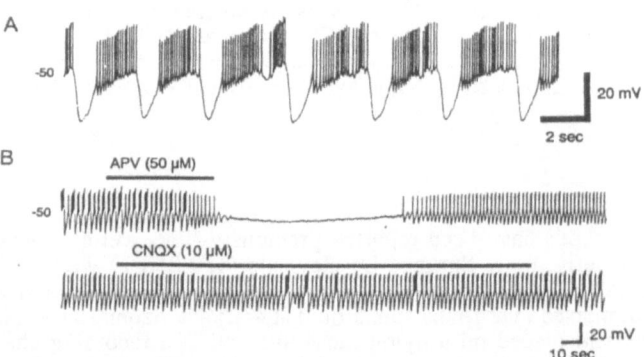

Figure 1. NMDA produces burst firing in dopamine neurons. **A**, Continuous superfusion with NMDA (10 µM) produces bursts of action potentials separated by hyperpolarizations. **B**, AVP but not CNQX blocks burst firing produced by NMDA (20 µM).

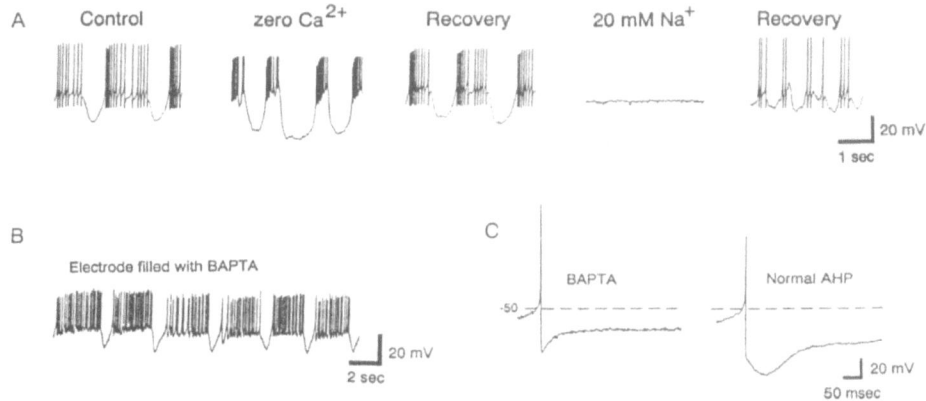

Figure 2. Burst firing is Na+- but not Ca²+-dependent. **A,** Burst firing in NMDA (20 µM) is blocked by 20 mM Na+ (with 126 mM TRIS) but not by no-added Ca²+. The entire experiment was in 3 mM Cs+ and slightly increased Mg²+ (3.2 mM) in order to reduce leakage conductance and preserve extracellular concentration of divalent cations. **B,** Burst firing in NMDA (20 µM) was recorded with electrodes filled with the Ca²+ chelating agent BAPTA (200 mM); this record was 40 min after impalement. **C,** Neurons recorded with BAPTA-filled electrodes had loss of the Ca²+-dependent slow after hyperpolarization.

Superfusate containing low Na+ (20 mM Na+, 126 mM choline or TRIS) reversibly blocked membrane oscillations underlying burst firing (n = 3) (Figure 2A). A 10 - 20 mV hyperpolarization was frequently observed in low Na+, but burst firing could not be re-established by injecting depolarizing current. In contrast, superfusate containing no-added Ca²+ failed to block burst firing even after 15 min (n = 6). Zero Ca²+ increased the number of spikes per burst and increased the magnitude and duration of the inter-burst hyperpolarization. The lack of a role of Ca²+ in NMDA-induced burst firing was also demonstrated by recording with micropipettes containing BAPTA which failed to prevent NMDA-induced burst firing (n = 3) (Figure 2B). This is despite the fact that after hyperpolarizations following spikes were abolished (Figure 2C), which suggests that BAPTA effectively reduced intracellular levels of Ca²+ (Lancaster and Nicoll, 1987).

As reported previously (Johnson et al., 1992), strophanthidin (2 - 10 µM; n = 5) and ouabain (2 - 10 µM; n = 6) blocked burst firing. Each produced a large depolarization in NMDA which required an increase in holding current of -550 to -850 pA to maintain a constant voltage. Low K+ (0.5 mM) also reversibly blocked burst firing (n = 2) (Figure 3A), as well as superfusate containing zero glucose (n = 2) (Figure 3B) and dinitrophenol (30 - 100 µM) (n = 5). No-added glucose depolarized membranes 10 - 20 mV after a delay of about 10 min, whereas dinitrophenol hyperpolarized the membrane by about 20 mV. In raised K+ (10.5 mM), the number of spikes per burst and the magnitude of the inter-burst hyperpolarization were increased (n = 2).

In voltage-clamp, a 1 s voltage step from -60 to -40 mV (thus mimicking a burst) produced an outward tail current in NMDA which was blocked by APV (n = 2). This tail current was present in NMDA but not in quisqualate (1 - 10 µM; n = 3) or kainate (1 - 10 µM; n = 3) despite evoking relatively large inward currents (400 to 1000 pA) (Figure 4A). Low Na+ also blocked the tail current in voltage-clamp (n = 2) (Figure 4B). In contrast, superfusate with no-added Ca²+ failed to block the outward tail current even after superfusing for 10 - 15 min (n = 5) (Figure 4B). Low K+ (0.5 mM) blocked the tail current in all cells tested (n = S) (Figure 4C). Hyperpolarizing test potentials reduced the outward tail current, but a reversal potential was not demonstrated. In superfusate containing raised K+ (10.5 mM), the tail current was increased; this is the opposite result predicted by the Nernst equation if the current was mediated by K+. Strophanthidin (at 3 and 10 µM; n = 2) and ouabain (1 - 3 µM; n = 3) reversibly blocked the outward tail current and produced an

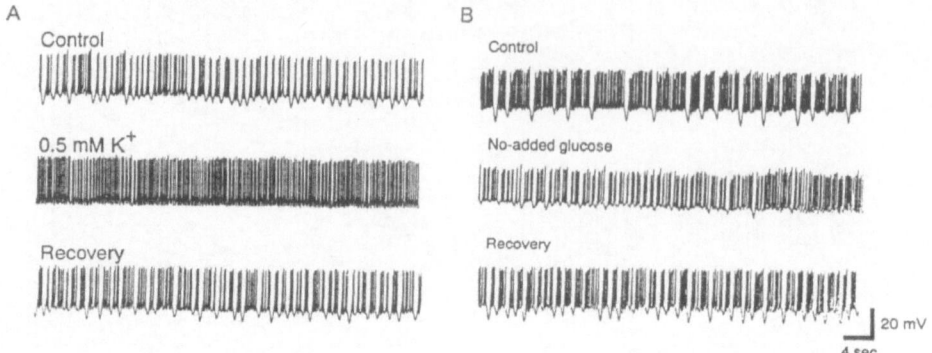

Figure 3. Inhibitors of the sodium pump block burst firing. **A**, Low K⁺ in the superfusate (0.5 mM) reversibly blocks burst firing induced by NMDA (20 μM). **B**, Superfusate containing no-added glucose also reduces burst firing produced by NMDA (20 μM).

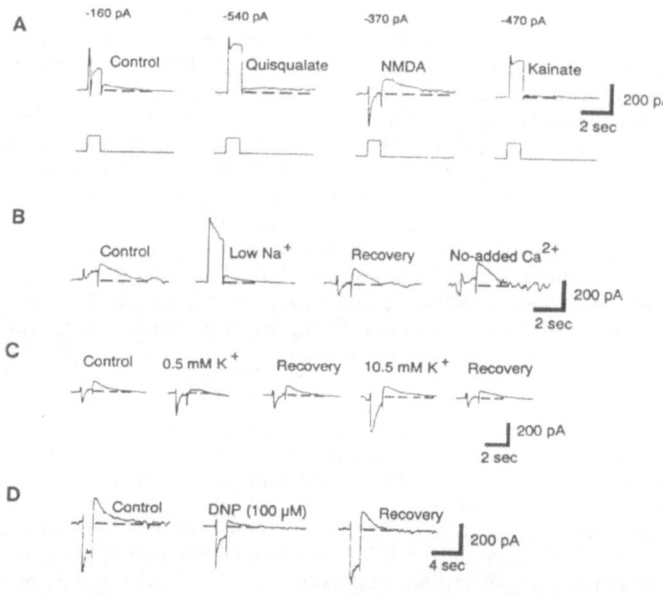

Figure 4. Outward tail currents, evoked under voltage-clamp by a 1 s depolarizing step from -60 to -40 mV, are generated by the sodium pump when superfusate contains NMDA. **A**, A relatively large outward tail current is evoked in NMDA (20 μM) but not in quisqualate (5 μM) or kainate (10 μM). A small tail current is evoked without NMDA (control) which was usually abolished by TEA (10 mM). The numbers above each record indicates magnitude of holding current (-60 mV). **B**, The tail current in NMDA (20 μM) is blocked by superfusate containing low Na⁺ (20 mM, with 126 mM TRIS) but not by no-added Ca²⁺ (15 min superfusion). **C**, The tail current (in NMDA 20 μM) is reduced in low K⁺ (0.5 mM) and increased in high K⁺ (10.5 mM). **D**, Dinitrophenol reduces the outward tail current in NMDA (20 μM). In all recordings (A - D), superfusate contained TTX (1 μM) and apamin (100 nM).

increase in inward holding current (up to -1400 pA) which was reversible with strophanthidin and partially reversible with ouabain. Furthermore, both zero glucose (n = 2) and dinitrophenol (30 and 100 μM; n = 2) blocked the tail current (Figure 4D). Zero glucose produced an inward current similar to the cardiac glycosides, but dinitrophenol (in NMDA) reduced the holding current (from about -350 to -200 pA) and also reduced the inward current evoked during the voltage step (Figure 4D); this suggests dinitrophenol may have effects other than simply inhibiting the sodium pump. The tail current was blocked by no-added Mg^{2+} (3 of 4 cells after superfusing 2 - 3 min), but not by TEA (1 - 30 mM; n = 3), 4-aminopyridine (1 mM; n = 2), or Ba^{2+} (100 - 300 μM; n = 2).

DISCUSSION

We propose that burst firing is produced by the following sequence of events: 1) Na^+ enters the cell through NMDA-gated ion channels; this current would be regenerative because of the negative slope conductance at these potentials (Mayer and Westbrook, 1987), 2) the rapid increase in intracellular Na^+ concentration activates a ouabain-sensitive pump, 3) the outward pump current ends the burst by hyperpolarizing the neuron, which thereby leads to Mg^{2+} block of the NMDA current, and 4) as the intracellular Na^+ concentration falls, the pump activity declines, and unopposed NMDA current initiates the next cycle of burst firing. It should be pointed out that this mechanism of burst firing has been described previously in an *Aplysia* neuron (Willis et al., 1974), but our report is the first in a vertebrate central neuron (Johnson et al., 1992).

Although our results clearly show that NMDA-induced burst firing does not depend upon Ca^{2+}, we do not rule out a role for Ca^{2+} in other types of burst firing. For example, Grace and Bunney (1984) showed that intracellular injection of Ca^{2+} facilitated burst firing of dopamine neurons *in vivo*. Furthermore, dopamine neurons have low threshold Ca^{2+} currents (Llinas et al., 1984) and Ca^{2+}-activated K^+ current (Silva et al., 1990) which underlie burst firing in other central neurons (Jahnsen and Llinas, 1984; McCormick and Pape, 1990; Hu and Bourque, 1992). We therefore acknowledge that Ca^{2+} may mediate some types of burst firing in dopamine neurons, but not when produced by NMDA-receptor activation.

We have consistently found that superfusate containing no-added Ca^{2+} exaggerates burst firing in NMDA, with bursts containing higher frequencies of spikes and the inter-burst hyperpolarizations more robust (Johnson et al., 1992). One possible explanation for this finding is that removal of Ca^{2+} permits more entry of Na^+ through NMDA-gated channels (Mayer and Westbrook, 1987), and the increased Na^+ makes the pump current larger. A second explanation is that removal of Ca^{2+} suppresses Ca^{2+}-activated K^+ current which interferes with burst firing (Shepard and Bunney, 1991). A third explanation is that removal of Ca^{2+} relieves the inhibitory influence this divalent cation has on Na^+/K^+ ATPase (Fukuda and Prince, 1992). The actual explanation is not known, but it is possibly due to a combination of the above.

In behaving rats and primates, single-spike firing in dopamine neurons is thought to have a permissive role in initiating movement, whereas burst firing is correlated with behavioral arousal and motivation (Miller et al., 1981; DeLong et al., 1983; Romo and Schultz, 1990; Schultz and Romo, 1990). Burst firing releases more dopamine from nerve terminals than does regularly spaced action potentials (Gonon and Buda, 1985; Gonon, 1988), which may underlie the association of burst-firing with behavioral arousal. Because schizophrenia has been linked to increased stimulation of dopamine receptors (Seeman and Lee, 1975; Creese et al., 1976), it is possible that excessive burst-firing in dopamine neurons may exacerbate this disease (Grace, 1991). Burst-firing may also facilitate release of neuropeptides such as cholecystokinin and neurotensin (Dutton and Dyball, 1979; Bean and Roth, 1991) which are co-localized with dopamine in the VTA (Hokfelt et al., 1980; Hokfelt et al., 1984). This may have important consequences for the action of dopamine because both cholecystokinin (Crawley, 1991) and neurotensin (Bissette and Nemeroff, 1988) may modulate the effect of dopamine. Therefore, burst-firing may convey a different "message" than does single-spike firing.

Excessive Na^+ influx may potentiate Ca^{2+}-induced neurotoxicity (Choi, 1988; Lees, 1991) because it reduces the rate of Ca^{2+} extrusion by the Na^+/Ca^{2+} exchanger (Baker et al., 1969; Stys et al., 1990). We hypothesize that glutamate antagonists, by decreasing Na^+

influx, are neuroprotective (Sonsalla et al., 1989; Greenamyre and O'Brien, 1991; Turski et al., 1991) because they decrease the work of the ATP-dependent sodium pump. It is interesting to note that several studies have shown that dopamine neurons in Parkinson's disease are deficient in mitochondrial enzymes (complex I) necessary for the manufacture of ATP (Parker et al., 1989; Schapira et al., 1990). Therefore, it is possible that excitatory amino acids are especially toxic to dopamine neurons in Parkinson's disease because these cells cannot keep pace with the energy requirements of the sodium pump.

Acknowledgements

This work was supported by US Department of Health and Human Services grants DA03161, MH40416, and NS01423, the National Parkinson Foundation, and the North Atlantic Treaty Organization.

REFERENCES

Baker, P.F., Blaustein, M.P., Hodgkin, A.L., and Steinhardt, R.A., 1969, The influence of calcium on sodium efflux in squid axons, *J. Physiol (Lond.)* 200:431-458.

Bean, A.J., and Roth, R.H., 1991, Extracellular dopamine and neurotensin in rat prefrontal cortex in vivo: effects of median forebrain bundle stimulation frequency, stimulation pattern, and dopamine autoreceptors, *J. Neurosci.* 11:2694-2702.

Bissette, G., and Nemeroff, C.B., 1988, Neurotensin and the mesocorticolimbic dopamine system, *Ann. N.Y. Acad. Sci.* 537:397-404.

Blatz, A., and Magleby, K.L., 1986, Single apamin-blocked Ca^{2+}-activated K^+ channels of small conductance in cultured rat skeletal muscle, *Nature* 323:718-720.

Charlety, P.J., Grenhoff, J., Chergui, K., De La Chapelle, B., Buda, M., Svensson, T.H., and Chouvet, G., 1991, Burst firing of mesencephalic dopamine neurons is inhibited by somatodendritic application of kynurenate, *Acta Physiol. Scand.* 142:105-112.

Chergui, K., Charlety, P.J., Akaoka, H., Brunet, J.L., Saunier, C.F., Buda, M., Svensson, T.H., and Chouvet, G., 1991, NMDA receptors are involved in spontaneous burst firing of dopaminergic mesencephalic neurons, *C.R. Acad. Sci. (Paris)* 313:139.

Choi, D.W., 1988, Glutamate neurotoxicity and diseases of the nervous system, *Neuron* 1:623-634.

Christie, M.J., Bridge, S., James, L.B., and Beart, P.M., 1985, Excitotoxin lesions suggest an aspartatergic projection from rat medial prefrontal cortex to ventral tegmental area, *Brain Res.* 333:169-172.

Crawley, J.N., 1991, Cholecystokinin-dopamine interactions, *Trend. Pharmacol. Sci.* 12:232-236.

Creese, I., Burt, D.R., and Snyder, S.H., 1976, Dopamine receptor binding predicts clinical and pharmacological potencies of antischizophrenic drugs, *Science* 192:481-483.

DeLong, M.R., Crutcher, M.D., and Georgopoulos, A.P., 1983, Relations between movement and single cell discharge in the substantia nigra of the behaving monkey, *J. Neurosci.* 3:1599-1606.

Dutton, A., and Dyball, R.E.J., 1979, Phasic firing enhances vasopressin release from the rat neurohypophysis, *J. Physiol. (Lond.)* 290:433-440.

Freeman, A.S., Meltzer, L.T., and Bunney, B.S., 1985, Firing properties of substantia nigra dopaminergic neurons in freely moving rats, *Life Sci.* 36:1983-1994.

Fukuda, A., and Prince, D.A., 1992, Excessive intracellular Ca^{2+} inhibits glutamate-induced Na^+-K^+ pump activation in rat hippocampal neurons, *J. Neurophysiol.* 68:28-35.

Gonon, F.G., 1988, Nonlinear relationship between impulse flow and dopamine released by rat rnidbrain dopaminergic neurons as studied by in vivo electrochemistry, *Neurosci.* 24:19-28.

Gonon, F.G., and Buda, M.J., 1985, Regulation of dopamine release by impulse flow and by autoreceptors as studied by in vivo voltammetry in the rat striatum, *Neurosci.* 14:765-774.

Grace, A.A., 1991, Phasic versus tonic dopamine release and the modulation of dopamine system responsivity: a hypothesis for the etiology of schizophrenia, *Neurosci.* 41:1-24.

Grace, A.A., and Bunney, B.S., 1984, The control of firing pattern in nigral dopamine neurons: burst firing, *J. Neurosci.* 4:2877-2890.

Grace, A.A., and Onn, S.-P., 1989, Morphology and electrophysiological properties of immunocytochemically identified rat dopamine neurons recorded in vitro, *J. Neurosci.* 9:3463-3481.

Greenamyre, J.T., and O'Brien, C.F., 1991, N-methyl-D-aspartate antagonists in the treatment of Parkinson's disease, *Arch. Neurol.* 48:977-981.

Hokfelt, T., Everitt, B.H., Thedorsson-Norheim, E., and Goldstein, M., 1984, Occurrence of neurotensin-like immunoreactivity in subpopulations of hypothalamic, mesencephalic and medullary catecholamine neurons, *J. Comp. Neurol.* 222:543-560.

Hokfelt, T., Rehfeld, J.F., Skirboll, L., Ivemark, B., Goldstein, M., and Markey, K., 1980, Evidence for coexistence of dopamine and CCK in mesolimbic neurons, *Nature* 285:476-478.

Hu, B., and Bourque, C.W., 1992, NMDA receptor-mediated rhythmic bursting activity in rat supraoptic nucleus neurones in vitro, *J. Physiol. (Lond.)* 458:667-687.

Jahnsen, H., and Llinas, R., 1984, Ionic basis for the electroresponsiveness and oscillatory properties of guinea-pig thalamic neurones in vitro, *J. Physiol. (Lond.)* 349:227-247.

Johnson, S.W., and North, R.A., 1992, Two types of neurone in the rat ventral tegmental area and their synaptic inputs, *J. Physiol. (Lond.)* 450:491 -502.

Johnson, S.W., Seutin, V., and North, R.A., 1992, Burst-firing in dopamine neurons induced by N-methyl-D- aspartate: Role of electrogenic sodium pump, *Science* 258:665-667.

Lacey, M.G., Mercuri, N.B., and North, R.A., 1988, On the potassium conductance increase activated by GABA-B and dopamine D-2 receptors in rat substantia nigra neurones, *J. Physiol. (Lond.)* 401:437-453.

Lacey, M.G., Mercuri, N.B., and North, R.A., 1989, Two cell types in rat substantia nigra zona compacta distinguished by membrane properties and the actions of dopamine and opioids, *J. Neurosci.* 9:1233-1241.

Lancaster, B., and Nicoll, R.A., 1987, Properties of two calcium-activated hyperpolarizations in rat hippocampal neurones, *J. Physiol. (Lond.)* 389:187-203.

Lees, G.J., 1991, Inhibition of sodium-potassium-ATPase: a potentially ubiquitous mechanism contributing to central nervous system neuropathology, *Brain Res. Rev.* 16:283-300.

Llinas, R., Greenfield, S.A, and Jahnsen, H., 1984, Electrophysiology of pars compacta cells in the in vitro substantia nigra - a possible mechanism for dendritic release, *Brain Res.* 294:127-132.

Mayer, M.L., and Westbrook, G L .,1987, Permeation and block of N-methyl-D-aspartic acid receptor channels by divalent cations in mouse cultured central neurones, *J. Physiol. (Lond.)* 394:501-527.

McCormick, D.A., and Pape, H.-C., 1990, Properties of a hyperpolarization-activated cation current and its role in rhythmic oscillation in thalamic relay neurones, *J. Physiol. (Lond.)* 431:291-318.

Miller, C., Moczydlowski, E., Latorre, R., and Phillips, M., 1985, Charybdotoxin, a protein inhibitor of single Ca^{2+}-activated K^+ channels from mammalian skeletal muscle, *Nature* 313:316-318.

Miller, J.D., Sanghera, M.K., and German, D.C., 1981, Mesencephalic dopaminergic unit activity in the behaviorally conditioned rat, *Life Sci.* 29:1255-1263.

Overton, P., and Clark, D., 1992, Iontophoretically administered drugs acting at the N-methyl-D-aspartate receptor modulate burst firing in A9 dopamine neurons in the rat, *Synapse* 10:131-140.

Parker, W.D., Boyson, S.J., and Parks, J.K., 1989, Abnormalities of the electron transport chain in idiopathic Parkinson's disease, *Ann. Neurol.* 26:719-723.

Paxinos, G., and Watson, C.,1986, "The Rat Brain in Stereotaxic Coordinates," Second edition, Academic Press Inc., San Diego.

Romo, R., and Schultz, W., 1990, Dopamine neurons of the monkey midbrain: contingencies of responses to active touch during self-initiated arm movements, *J. Neurophysiol.* 63:592-606.

Sanghera, M.K., Trulson, M.E., and German, D.C., 1984, Electrophysiological properties of mouse dopamine neurons: in vivo and in vitro studies, *Neurosci.* 12:793-801.

Schapira, A.H.V., Cooper, J.M., Dexter, D., Clark, J.B., Jenner, P., and Marsden, C.D., 1990, Mitochondrial complex I deficiency in Parkinson's disease, *J. Neurochem.* 54:823-827.

Schultz, W., and Romo, R., 1990, Dopamine neurons of the monkey midbrain: contingencies of responses to stimuli eliciting immediate behavioral reactions, *J. Neurophysiol.* 63:607-624.

Seeman, P., and Lee, T., 1975, Antipsychotic drugs: direct correlation between clinical potency and presynaptic action of dopamine neurons, *Science* 188:1217-1219.

Shepard, P.D., and Bunney, B.S., 1991, Repetitive firing properties of putative dopamine-containing neurons in vitro: regulation by an apamin-sensitive Ca^+-activated K^+ conductance, *Exp. Brain Res.* 86:141-150.

Silva, N.L., Pechura, C.M., and Barker, J.L., 1990, Postnatal rat nigrostriatal dopaminergic neurons exhibit five types of potassium conductances, *J. Neurophysiol.* 64:262-272.

Sonsalla, P.K., Nicklas, W.J., and Heikkila, R.E., 1989, Role for excitatory amino acids in methamphetamine-induced nigrostriatal dopaminergic toxicity, *Science* 243:398-400.

Stys, P.K., Ransom, B.R., Waxman, S.G., and Davis, P.K., 1990, Role of extracellular calcium in anoxic injury of mammalian central white matter, *Proc. Natl. Acad. Sci. (USA)* 87:4212-4216.

Svensson, T.H., and Tung, C.-S., 1989, Local cooling of pre-frontal cortex induces pacemaker-like firing of dopamine neurons in rat ventral tegmental area in vivo, *Acta Physiol. Scand.* 136:135-136.

Turski, L., Bressler, K., Rettig, K.-J., Loschmann, P.-A., and Wachtel, H., l991, Protection of substantia nigra from MPP+ neurotoxicity by N-methyl-D-aspartate antagonists, *Nature* 349:414-418.

Willis, J.A., Gaubatz, G.L., and Carpenter, D.O., 1974, The role of the electrogenic sodium pump in modulation of pacemaker discharge of aplysia neurons, *J. Cell Physiol.* 84:463-472.

PHYSIOLOGY OF TWO DISYNAPTIC PATHWAYS FROM THE SENSORI-MOTOR CORTEX TO THE BASAL GANGLIA OUTPUT NUCLEI

Hitoshi Kita

The University of Tennessee Memphis
College of Medicine
Department of Anatomy and Neurobiology
875 Monroe Avenue, Memphis, TN 38163 U.S.A.

INTRODUCTION

The neuronal connections of the basal ganglia and related nuclei have been studied extensively. The sensorimotor cortex projects to the neostriatum and the subthalamic nucleus. Because extensive anatomical and physiological studies on the cortico-striatal projection has been presented at the International Basal Ganglia Meeting at Giens, France, I will not discuss details on this projection. The subthalamic nucleus is a relatively small nucleus and is located ventrally to the zona incerta and dorsally to the cerebral peduncle. The subthalamic nucleus contains dense medium size neurons having a radiating, sparsely spined dendrites (Kita et al., 1983b; Chang et al., 1983). Both the neostriatum and the subthalamic nucleus project to the basal ganglia output nuclei, the entopeduncular nucleus (homologous to the internal segment of the pallidum in primates) and the substantia nigra pars reticulata. The output of the neostriatum is inhibitory while the subthalamic nucleus is excitatory. Thus, the signals arise in the sensorimotor cortex can be transmitted to the basal ganglia output nuclei via two parallel pathways having opposite output signals (i.e., excitatory vs. inhibitory).

The globus pallidus (homologous to the external segment of the pallidum in primates) of the rat is relatively large and is located caudomedially to the neostriatum. Anatomical and physiological studies indicated that the globus pallidus also receives parallel disynaptic projections through the neostriatum and the subthalamic nucleus (Kitai et al., 1976; Perkins and Stone, 1980; Kitai and Kita, 1987; Kita and Kitai, 1987, 1991; Robledo and Féger, 1990; Ryan and Clark, 1991). The connections are very similar to those of the entopedun-cular nucleus and the substantia nigra. The globus pallidus projects to most of the basal ganglia and related nuclei, including the neostriatum, entopeduncular nucleus, subthalamic nucleus, substantia nigra, and the pedunculopontine tegmentum (van der Kooy et al., 1981; Schmued et al.,1989; Smith and Bolam, 1989; Kita et al., 1991; Kita and Kitai, 1991). These pallidal outputs are GABAergic and inhibitory (Kita et al., 1983a; Mugnaini and Oertel, 1985; Schmued et al.,1989; Smith and Bolam, 1989; Kita et al., 1991). The neuronal connections suggest 1) that the activity of pallidal neurons is controlled by the converging inputs through the two pathways which are mediated through the neostriatum and the subthalamic nucleus and 2) that pallidal inhibitory outputs regulate the entire basal ganglia activity.

This paper is to review our recent physiological studies on the nature of these two pathways and the role of the globus pallidus in the activities of the basal ganglia The questions of interest specifically are: 1) Whether or not signals through the two pathways

The Basal Ganglia IV, Edited by
G. Percheron *et al.*, Plenum Press, New York, 1994

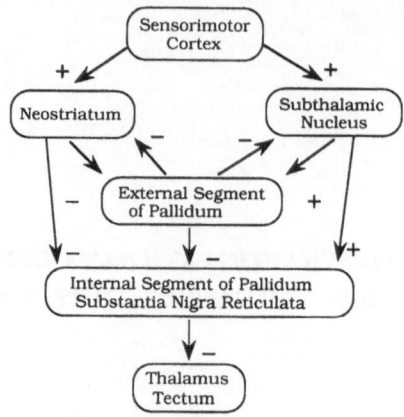

Figure 1. A schematic drawing of two parallel pathways from the sensorimotor cortex to the basal ganglia output nuclei the entopeduncular nucleus and the substantia nigra pars reticulata.

converge on single neurons in the entopeduncular nucleus and substantia nigra pars reticulata. 2) How the signals conveyed through two pathways converge on single neurons, if in fact they converge. 3) What is the role of the globus pallidus in the control of the signal transmission through the two pathways from the sensorimotor cortex to the basal ganglia output nuclei.

PHYSIOLOGICAL STUDIES

Two disynaptic pathways

Cortico-striato-entopeduncular/nigral pathways Stimulation of the sensorimotor cortex induces a short, about 3 msec, latency EPSPs followed by a long lasting disfacilitation of the cortical inputs in striatal spiny efferent neurons (Wilson, 1986). The EPSPs are short in duration because of the feed forward inhibition, exerted through GABAergic inhibitory interneurons in the neostriatum (Wilson et al., 1989; Cowan et al., 1990; Kita et al., 1990) and also probably due to the feed back inhibition mediated through the GABAergic axon collaterals in the neostriatum. When cortical stimulation evoked a large EPSP, it would trigger one or two spikes on the crest of the EPSP as shown in Figure 2. The latency of the spikes was 7-15 msec after cortical stimulation.

Stimulation of the neostriatum induced a long latency (i.e., 7-12 msec) GABAergic inhibitions in the entopeduncular nucleus and the substantia nigra pars reticulata (Yoshida and Precht, 1971; Yoshida et al., 1972; Levine et al., 1974; Ohye et al., 1976). Based on the latencies of the cortico-striatal excitations and the striato-entopeduncular/nigral inhibi-

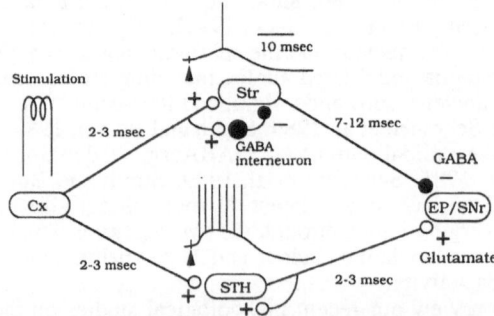

Figure 2. The neurotransmitters and the conduction times of two parallel pathways from the sensorimotor cortex to the entopeduncular nucleus and the substantia nigra pars reticulata. Responses of neostriatum and subthalamic neurons to cortical stimulation are schematically illustrated.

tions, the total conduction time of the cortical excitation to entopeduncular/nigral inhibition can be estimated to be a minimum of 14 msec.

Cortico-subthalamic-entopeduncular/nigral pathways. Most of subthalamic neurons exhibited an irregular spontaneous firing with an average rate of about 11/sec. Spontaneous burst firing was often seen in these neurons. The firing patterns and firing rates of subthalamic neurons are similar both in the urethane anesthetized rat and the unanesthetized monkey (Matsumura et al., 1992; Ryan and Clark, 1992; Fujimoto and Kita, 1993). Interestingly, subthalamic neurons in the brain slice preparation exhibit highly regular spontaneous firing (Nakanishi et al., 1987a). The observations imply that the irregular firing of subthalamic neurons *in vivo* is due to extrinsic or intrinsic synaptic inputs.

Electrical stimulation of the sensorimotor cortex evoked a short (i.e., 1.8-3.5 msec, mean=2.5 msec) latency and relatively long duration excitation followed by a long lasting inhibition in subthalamic neurons (Fig. 2, 3). An intracellular recording study indicated that the initial excitation is due to the monosynaptic input from the sensorimotor cortex (Kitai and Deniau, 1981) The long duration excitation in the subthalamic nucleus was frequently interrupted by a brief period of inhibition (Kitai and Deniau, 1981; Rouzaire-Dubois and Scarnati, 1985; Ryan and Clark, 1992; Fujimoto and Kita, 1993). Our study indicated that this brief inhibition is due to IPSPs (Fujimoto and Kita, 1993) The subthalamic neuron shown in Figure 3 exhibited two excitatory peaks in which the latency of the first excitation was approximately 3 msec and the second excitation was 15 msec. The excitations were followed by inhibition of firing, lasting about 200 msec (Fig. 3B). The long inhibition is due to the disfacilitation of cortical inputs (Fujimoto and Kita, 1993).

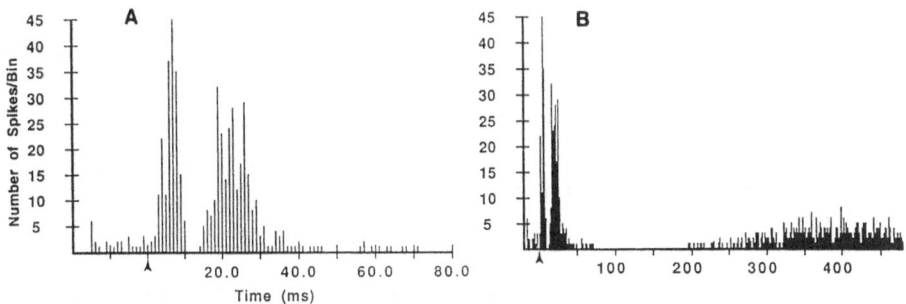

Figure 3. Peristimulus time histograms (PSTHs) of a subthalamic neuron responding to cortical stimulation of 250μA. The responses in A and B were the same, but shown with different time scale. Note the excitation with two peaks (A) and a long lasting inhibition after the second excitation in (B). One hundred trials were analyzed. Bin size=1 msec. Triangles mark the time of stimulation.

The long duration excitation observed in the subthalamic nucleus is clearly contrasted to the brief (one or two spikes) excitation observed in the neostriatum after cortical stimula-tion. There are several possible explanations for the prolonged excitations in subthalamic neurons. 1) Intracellular staining studies revealed that subthalamic efferent neurons emit axon collaterals within the nucleus (Kita et al., 1983b). The intranuclear collateral axons would spread excitations within the subthalamic nucleus resulting in a long duration excitation in each of subthalamic neurons. In fact, stimulation of a single site of the sensorimotor cortex excited neurons in a large area of the subthalamic nucleus (Fujimoto and Kita, 1993). 2) Intracellular study using brain slice preparations indicated that stimulation of the internal capsule induces NMDA responses in subthalamic neurons (Fig. 4). The duration of NMDA responses was 100-200 msec which is longer than other glutamatergic responses (Nakanishi et al., 1988). 3) subthalamic neurons have a high input resistance and a short refractory period (Kita et al., 1983a; Nakanishi et al., 1987a). Intracellular recordings in anesthetized rats and brain slice preparations indicated the membrane potential of subthalamic neurons is only slightly negative than the spike threshold (Kita et al., 1983a; Nakanishi et al., 1987a). These membrane characteristics are suited for generating multiple spikes in the subthalamic nucleus even when small EPSPs are evoked in the neurons.

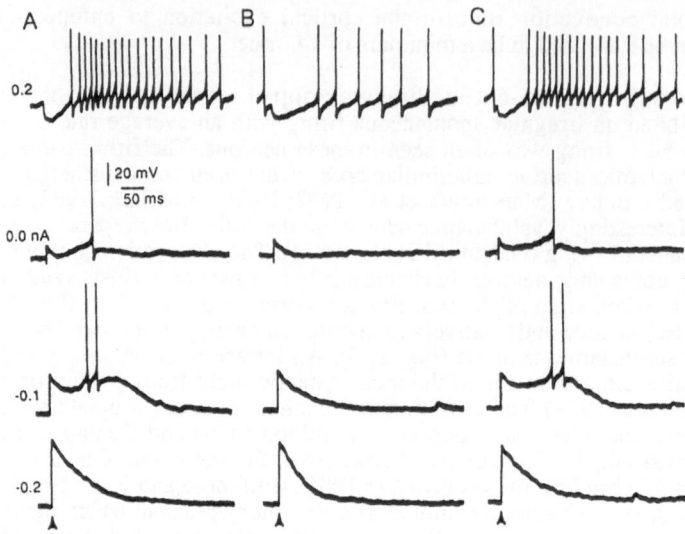

Figure 4. Responses of subthalamic neurons to stimulation of the internal capsule were studied using brain slice preparations. Stimulation of the internal capsule evoked a short duration EPSP followed by a slow EPSP (duration 100-200 msec) in subthalamic neurons. Both duration and amplitude of the initial EPSP increased with hyperpolarizing current injections. The initial EPSP could be reversed by injecting depolarizing current (A). The slow EPSP that follows the initial short EPSP displayed rather complex changes upon intracellular current injections. Depolarizing current injection clearly increased the duration of the slow EPSE and triggered repetitive action potentials (A). On the other hand, an increase in the amplitude was observed when a low intensity hyperpolarizing current (e.g., 0.1 nA), which shifted the membrane potential to about -60 mV, was injected. The slow EPSP, however, was gradually decreased upon hyperpolarizing the membrane potential more negative than -60 mV by an injection of stronger hyperpolarizing currents (e.g., 0.2 nA). Since the blockade of the response upon strong membrane hyperpolarization suggested an involvement of NMDA receptor activation, effects of APV, an NMDA antagonist, on the slow EPSP were examined. Bath application of APV with a concentration of 50 μM suppressed the slow EPSP (B). The APV effect was reversed by washout (C).

The nature of subthalamo-entopeduncular and subthalamo-nigral projections was studied using brain slice preparations. Stimulation of the subthalamic nucleus evoked short (1.2-2.4 msec) latency EPSPs in both the entopeduncular nucleus and the substantia nigra pars reticulata. The EPSPs were considered to be glutamatergic because the application of a glutamate antagonist, kynurenic acid, blocked the EPSPs (Nakanishi et al., 1990), and the neurons in the subthalamic nucleus are immunoreactive for glutamate (Kitai and Kita, 1987). Based on this data, the disynaptic pathway arising from the sensorimotor cortex and mediated through the subthalamic nucleus should convey an excitation to the entopeduncular nucleus and substantia nigra pars reticulata. The time required for an excitation to travel the cortico-subthalamo-entopeduncular and nigral pathways can be estimated to be a minimum of 3 msec.

Convergence of two disynaptic pathways

Responses of substantia nigra pars reticulata neurons to cortical stimulation. In order to determine whether or not the signals, which arise in the sensorimotor cortex and send through the two parallel pathways, can converge on single nigral neurons, recording of nigral responses to cortical stimulation were performed in anesthetized rats. Stimulation of the sensorimotor cortex, with less than 50 μA, resulted in an excitation in some of the substantia nigra pars reticulata neurons (Fig. 5A). More intense stimulation resulted in responses with multiple components of excitation and inhibition (Fig. B, C). Approximately a half of the substantia nigra neurons exhibited clear responses when the cortex was stimulated with 250 μA. In a majority of the neurons, responses began with an excitation. The excitation was often followed by one or two successions of inhibition and excitation.

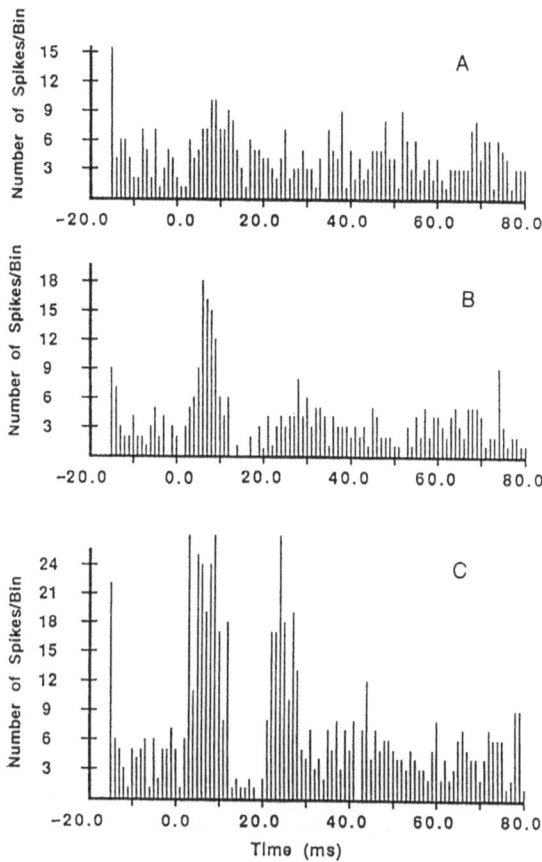

Figure 5. Peristimulus time histo-grams showing responses of a nigral neuron to sensorimotor cortex in a rat. The intensity of cortical stimulation was 50 μA (A), 100 μA (B), and 250 μA (C). Stimulation with 50 μA evoked only a small excitation (A). The same neuron exhibited multiple excitation- inhibition sequences as the stimulus intensity was increased (B and C). Each histogram consists of 100 sweeps of events. Bin size=1 msec. Triangles mark the time of stimulation.

The excitation-inhibition oscillations occurred approximately 25 msec intervals (i.e., 40 Hz). A small number of neurons responded with a long latency inhibition or inhibition-excitation sequences. Neostriatal neurons did not exhibit oscillatory responses to cortical stimulation (Wilson, 1986). We have found that the neurons in the subthalamic nucleus show two peaks of EPSPs, separated by 20-30 msec of inhibition, after cortical stimulation (Fujimoto and Kita, 1993). That observation suggests that the subthalamic nucleus participates in the oscillatory actions of the substantia nigra pars reticulata. The latency of the responses to cortical stimulation differed depending on the area of the sensorimotor cortex stimulated. When stimulation was applied to cortical site, which was similar to the aforementioned cortico-subthalamic responses were obtained, the mean latency of the initial excitation and following inhibition was 3.1 and 12 msec, respectively.

Our experiments clearly indicated, the first time, that cortical stimulation indeed produces a short latency excitation in nigral neurons. The latency of the excitation coincide very well with that expected from the data on the latency of cortico-subthalamic and subthalamo-nigral excitatory responses. In order to verify that the disynaptic cortico-subthalamo-nigral pathway is responsible for the early excitation of the cortex, experiments were performed in rats which had received a chronic ibotenic acid (an excitotoxin) lesion of the subthalamic nucleus. Stimulation of the sensorimotor cortex in subthalamic nucleus lesioned rats failed to evoke a strong, short latency excitation (Fig. 6).

This result indicated that the early excitation is mediated mainly through disynaptic projections via the subthalamic nucleus.

The long latency inhibition is most likely due to an activation of cortico-striato-nigral projection (Yoshida and Precht 1971; Dray et al., 1976). The threshold of cortical stimulation for evoking inhibition in the substantia nigra pars reticulata was higher than that for evoking excitation. The difference in the stimulus threshold may be due to a difference in the responsiveness of striatal and subthalamic neurons to cortical stimulation. Stimulation of the sensorimotor cortex evokes powerful EPSPs with multiple spikes in the subthalamic neurons (Kitai and Deniau, 1981). However, striatal neurons exhibited a short duration EPSPs with one or two spikes even when a relatively strong stimulus current was applied to the cortex. The results of our study indicate that signals derived from the sensorimotor cortex reach the substantia nigra pars reticulata by two parallel pathways which convey a short latency excitation and a long latency inhibition. Signals carried by these pathways converge on many nigral neurons.

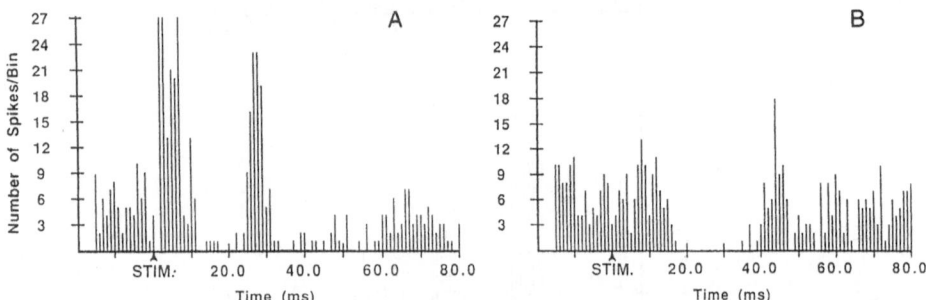

Figure 6. Peristimulus time histograms showing responses of nigral neurons to stimulation of the sensorimotor cortex in a non-lesioned rat (A) and a rat which received ibotenic acid lesion of the subthalamic nucleus (B). One hundred trials were analyzed. Bin size=1 msec. Triangles mark the time of stimulation.

Organization of cortico-nigral pathways. Neuronal responses induced by cortical stimulation were recorded from many regions of the substantia nigra pars reticulata in order to investigate the functional organization of the disynaptic cortico-nigral pathways. All nigral neurons encountered on the recording electrode tracks separated by 0.25 mm were tested by stimulation of three sites within the sensorimotor cortex, and the threshold intensity sufficient to evoke responses was noted. The stereotaxic coordinates of the three stimulus sites were for site-A: AP (from the bregma) =4.2, L=1.5 and 2.2 (for two pins); for site-B: AP=3.0, L=2.0 and 2.7; and for site-C: AP= 1.8, L=2.0 and 2.7, with penetration approximately 1 mm below the surface of the cortex.

Figure 7 shows results obtained from an experiment in which the threshold stimulus intensity and the initial response (i.e., excitation or inhibition) were mapped in the substantia nigra pars reticulata. The results indicated that the area containing the responding neurons was not restricted to a small area of the substantia nigra even when relatively low intensity stimulation (e.g., less than 100 μA) was applied. When strong stimulation was used, neurons responding to each site of stimulation were found in all regions, except in the caudomedial part of the substantia nigra. This observation can be explained by a spread of excitations from the stimulus site to other cortical areas through cortico-cortical projections, as well as, an induction of excitations in a large area of the subthalamic nucleus. This phenomenon was discussed earlier in this paper. Although neurons with initial excitation and those with initial inhibition appeared to be segregated, there were no other clear patterns in the distribution of these neurons. Further, it was evident that each neuron responded differently in terms of its threshold and its initial response (i.e., either

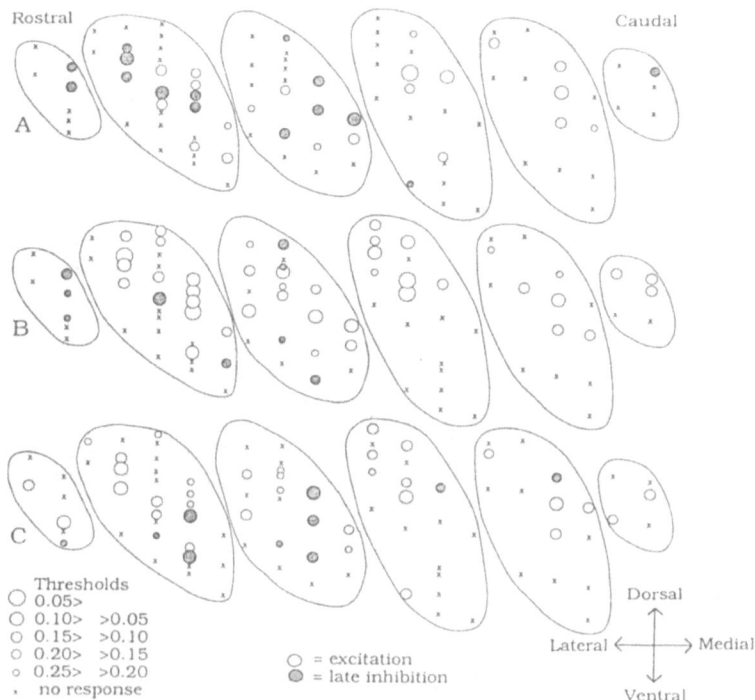

Figure 7. Distribution of recorded neurons in the substantia nigra pars reticulata. The upper, middle, and lower rows represent results of stimulation of three different cortical sites A, B, and C, respectively. Drawings in each row represent coronal sections separated by 0.25 mm, and tracts in the drawing are separated by 0.25 mm.

excitatory or inhibitory) to stimulation of the three sites, which were relatively close (i.e., about 1.2 mm apart) to each other. Anatomical studies have described the topographical organization of every one of the projections involved in the disynaptic cortico-nigral pathways. However, it is difficult to predict point to point topographical correlation between the sensorimotor cortex and the substantia nigra from anatomical knowledge. The results of this study suggest that topographical organization between the cortex and the substantia nigra pars reticulata is rather complex.

Physiology of the globus pallidus

Two Parallel pathways from the sensorimotor cortex to the globus pallidus. The sensorimotor cortex provides a similar disynaptic projections to the globus pallidus as it does to the entopeduncular nucleus and the substantia nigra. The nature of each projections involved in the cortico-pallidal pathways has been studied previously. Cortical inputs to the neostriatum and the subthalamic nucleus are excitatory with relatively short latency (i.e., approximately 3 msec)(Kitai et al., 1976; Kitai and Deniau, 1981; Fujimoto and Kita, 1993). Subthalamic input to the globus pallidus is also a short latency excitation (Robledo and Féger, 1990; Kita and Kitai, 1991). Striatal input to the globus pallidus is GABAergic inhibition with a long latency (i.e., about 7 msec) and short duration (i.e., 10-30 msec) (Noda et al., 1968; Yoshida et al., 1972; Levine et al., 1974; Park et al., 1982; Kita and Kitai,1988; Tremblay and Filion, 1989).

According to the information available on the synaptic inputs, activation of the senso-rimotor cortex should either excite the globus pallidus (due to the disynaptic excitatory pathway via through the subthalamic nucleus) or inhibit the globus pallidus (due the inhibi-tory pathway via through the neostriatum). The excitatory pathway mediated through the subthalamic nucleus should have a shorter conduction time compared to the inhibitory pathway through the neostriatum because of the long latency striato-pallidal projection. In order to determine whether or not the signals transmitted through the parallel pathways are converged on individual globus pallidus neurons, responses of pallidal neurons to cortical stimulation were examined.

In anesthetized rats, many pallidal neurons exhibited spontaneous and repetitive firing. Stimulation of the sensorimotor cortex evoked a sequence of postsynaptic responses in these neurons. The responses included: 1) an initial EPSP with approximately 4 msec latency and 10 msec duration; 2) an IPSP of 10-20 msec latency with 20-30 msec duration; and 3) a late EPSP with multiple spikes (Fig. 8A). The late EPSPs had long durations ranging from 50-100 msec and were powerful in generating spikes on their rising phase. The latency of the initial and the late EPSPs varied with changes in the stimulus intensity. This indicated that they were polysynaptically induced. The disynaptic pathway through the subthalamic nucleus is considered responsible for these EPSPs because of the following reasons: 1) both the initial and late EPSPs were greatly diminished by the ibotenic acid lesion of the subthalamic nucleus (Ryan and Clark, 1991; Kita, 1992); 2) the knife cut caudal to the SN preserved these responses; and 3) previous studies indicate that cortical stimulation results in early and late excitations in the subthalamic nucleus (Ryan and Clark, 1991; Fujimoto and Kita, 1993).

The IPSPs observed after cortical stimulation are most likely due to activation of striato-pallidal projection, since the projection is known as long latency inhibition (Yoshida et al., 1972; Park et al., 1982). Intra-venus injection of picrotoxin abolished the IPSPs and consequently left a depolarization in pallidal neurons after cortical stimulation (Fig. 8 C-E). The amplitude and the duration of the depolarization varied with the intensity

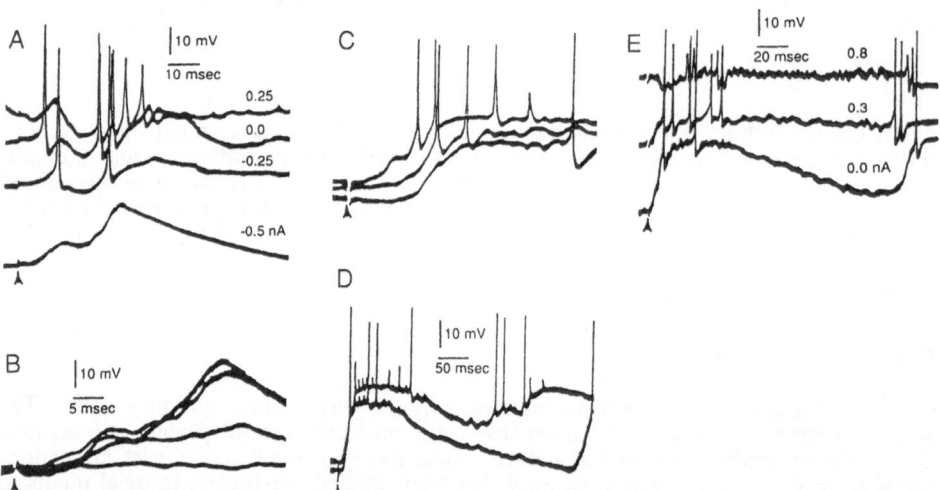

Figure 8. Responses of pallidal neurons to stimulation of the sensorimotor cortex. A: Cortical stimulation evoked a sequence of responses including initial EPSPs, IPSPs, and late EPSPs. Intracellular current injec-tions changed the amplitude of IPSPs and EPSPs. B: Responses to cortical stimulation with various inten-sities. The neuron was continuously hyperpolarized by 0.5 nA to a level close to the reversal potential of the IPSPs. The responses include multiple EPSPs with variable latencies. C-E: Responses to cortical stimulation after injection of picrotoxin. C: Responses to three stimulations with varying intensities Calibrations in A also apply to C and D: Slow sweep recordings show that a single cortical stimulation could evoke more than one depolarization. E: Injection of depolarizing currents decreases the amplitude of the depolarizations. Only the initial part of the depolarization was reversed by the application of the depolarizing current of 0.8 nA.

of the stimulation Strong stimulation often evoked large amplitude (i.e., often exceeding 20 mV) plateau depolarizations lasting more than 200 msec. Frequently, more than one plateau potentials occurred repeatedly after a single cortical stimulation (e.g., Fig. 8 D,E). The recordings with picrotoxin indicate that the disynaptic pathway through the subthalamic nucleus can produce large EPSPs in pallidal neurons when overlapping IPSPs are blocked.

The responses of pallidal neurons to cortical stimulation were always EPSPs and overlapping IPSPs. This observation indicates that the cortically derived inputs conveyed via the neostriatum and the subthalamic nucleus converge on the majority of the pallidal neurons. The results further indicate that the initial EPSPs mediated through the subthalamic nucleus arrive to the globus pallidus before the IPSPs mediated through the neostriatum. This observation is very similar to that made in the substantia nigra pars reticulata. However, a difference was that the initial excitation in the globus pallidus was much shorter than that observed in the substantia nigra. The initial EPSPs recorded in pallidal neurons are small and triggered one or no spike while one or two spikes were seen at the initial excitatory period in the substantia nigra.

Pallidal efferent sites and the conduction time of pallidal efferent axons. The pallidal efferent sites and the conduction time of their axons were studied using antidromic stimulation. Approximately 75% and 54% of the repetitively firing pallidal neurons evoked antidromic spikes in response to subthalamic and nigral stimulation, respectively. It is possible that subthalamic stimulation activates not only the fibers projecting to the subthalamic nucleus but also those which pass through the nucleus and project caudally to the substantia nigra and the pedunculopontine tegmentum. Our results suggest that some fibers end at the subthalamic nucleus and do not project further into caudal brain areas. About one-third of the continuously firing pallidal neurons responded antidromically to striatal stimulation. It is possible that more than one-third of the pallidal neurons project to the neostriatum, because the striatal stimulation might have been applied to only a portion of the pallido-striatal projection which is topographically organized (Staines et al., 1981, Kita et al., 1991). The current results also indicate that about a half of the neurons projecting to the neostriatum also project axons into the subthalamic nucleus. The collateral projection of the pallidal neurons to the neostriatum and to the substantia nigra has been reported in a double retrograde labeling study (Staines and Fibiger, 1984).

The mean antidromic latencies of the striatal, subthalamic, and nigral stimulation were very similar to each other, about 1 msec. The antidromic latency correlates well with the latency of the IPSPs recorded in the subthalamic nucleus after stimulation of the globus pallidus (Kita et al., 1983a). These results suggest that the globus pallidus is capable of very rapidly and simultaneously inhibiting not only the subthalamic nucleus but also the neostriatum and the substantia nigra pars reticulata.

Alteration of neuronal activities in the subthalamic nucleus and substantia nigra pars reticulata after a lesion of the globus pallidus. The spontaneous firing rate of subthalamic and nigral neurons are reported to increase after a lesion of the globus pallidus (Ryan and Clark, 1992). This result can be explained by the observations that both the subthalamic nucleus and substantia nigra pars reticulata receive heavy projections from the globus pallidus (Rinvik et al., 1979; Carpenter et al., 1981; Canteras et al., 1990); that pallidal neurons exhibit relatively high frequency repetitive firing; and that the pallidal output is GABAergic and inhibitory (Rouzaire-Dubois et al., 1980; van der Kooy et al., 1981; Kita et al., 1983a; Mugnaini and Oertel, 1985). Interest-ingly, the increased firing rate appears to be temporary. The increased firing of subtha-lamic and nigral neurons was no longer observed when the rats were examined more than 10 days after the lesion of the globus pallidus (Robledo and Féger, 1990; Fujimoto and Kita, 1993). It is possible that chronically lesioned rats develop some mechanisms to compensate for the activity of the subthalamic nucleus and the substantia nigra pars reticulata.

A recent study in the rat indicated that a lesion of the globus pallidus increases the magnitude and the duration of excitatory responses in the subthalamic nucleus to cortical stimulation (Ryan and Clark, 1992). We made observations similar to this in rats of which the globus pallidus had been lesioned by injections of ibotenic acid. In the globus pallidus lesioned rats, a vast majority of subthalamic neurons evoked a short latency and a long (i.e., about 20 msec) duration excitation. The response pattern clearly differed from that

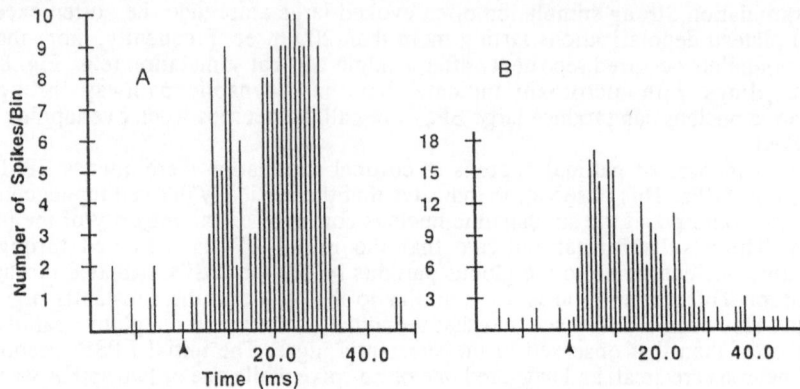

Figure 9. Peristimulus time histograms of subthalamic neurons responding to cortical stimulation in a non-lesioned rat (A) and in a rat which received a lesion of the globus pallidus (B). The neuron response obtained in intact rat shows a brief inhibition while the response from lesioned rat lacks the inhibition. Note the difference of the scale of the ordinate. Bin size was 1 msec and 50 trials were analyzed. Triangles mark the time of stimulation.

recorded in non-lesioned rats. The response recorded from non-lesioned rats consisted of two peak excitations which are separated by an intervening inhibition (Fig. 9). These observations indicated that globus pallidus can effectively control the firing activity of the subthalamic nucleus.

FUNCTIONAL IMPLICATIONS

Cortico-subthalamo-entopeduncular/nigral pathways. The physiological studies summarized in this paper indicate that an activation of the direct cortico-subthalamic projection produces a powerful and prolonged excitation in the subthalamic nucleus and that the excitation is then transmitted to the entopeduncular nucleus and the substantia nigra pars reticulata. Studies also indicate that the conduction time of these disynaptic excitatory pathways is several milliseconds shorter than the conduction time of disynaptic inhibitory pathways which are relayed through the neostriatum.

The role of the subthalamic nucleus in the movement control has been implicated by recordings in awake animals and clinicopathological and experimental lesion studies. Only a few reports are available for recordings from the subthalamic nucleus of animals perform-ing behavioral tasks (Georgopoulos et al., 1983; DeLong et al., 1985; Matsumura et al., 1992). These studies in monkeys indicated that the increase in firing, which is related to limb and saccadic eye movements, occur mainly after the initiation of the movements in the subthalamic nucleus. It has been reported that, under similar experimental conditions, some neurons in the neostriatum increase firing well ahead of movements (Crutcher and Delong, 1984; Hikosaka et al., 1989). The observations indicate, if the sensorimotor cortex is an origin of premovement activities (i.e., related to the initiation of movements), the activity is transmitted to the neostriatum but not to the subthalamic nucleus. The implication is that the neostriatum and the subthalamic nucleus receive afferent inputs from different sets of neurons in the sensorimotor cortex. Anatomical studies indicated that the neostriatum receives projections from various cortical areas (most of the projection arose from the layer III). The subthalamic nucleus receives major projections from motor related cortices (mostly from the layer Vb).

On the other hand, Matsumura and his associates have also found a great similarity in the neuronal activities between the neostriatum and the subthalamic nucleus. Activities of subthalamic neurons are not only related to movement, but are also related to other events which include visual stimuli and reward (Matumura et al., 1992). DeLong et al. reported somatosensory, especially joint movement, stimulus related neurons in the subthalamic nucleus (DeLong et al., 1985). Anatomical studies indicated that the neostriatum and the subthalamic nucleus receive afferent inputs from similar brain sites including the cortex, the thalamus, and the globus pallidus. It is possible that the activity of the subthalamic nucleus is involved in a wide range of motor controls, similar to those of the neostriatum.

Our observations suggest that the movement and sensory related activities of the neostriatum and the subthalamic nucleus might be conveyed to and converged on neurons in the entopeduncular nucleus and substantia nigra pars reticulata. The subthalamic activity may exert an excitation to entopeduncular and nigral neurons which otherwise over react to striatal inhibitory inputs. The inhibitory response of nigral neurons to cortical stimulation was relatively short in duration in non-lesioned animals, while the inhibition lasted a long period in subthalamic nucleus lesioned animals.

Pathological or experimental lesion studies of the subthalamic nucleus have indicated that this nucleus is involved in ballisms and tremors (Whittier and Mettler, 1949; Carpenter, 1961; Hammond et al., 1979; Bregman et al., 1992). Also, it has been reported recently that a lesion of the subthalamic nucleus can remedy experimental parkinsonism that is produced by a lesion of the nigro-striatal dopaminergic pathway (Bregman et al., 1990). The results can be, at least in part, explained by the idea: 1) that both the neostriatum and the subthalamic nucleus are involved (even though, each nuclei may not process same information) in performing multiple stages of appropriate and purposeful movements; 2) that the outputs of the neostriatum and the subthalamic nucleus converge on single neurons in the entopeduncular nucleus and the substantia nigra; and 3) that overall func-tions of the neostriatum and the subthalamic nucleus are positive and negative controls, respectively, on movements.

The globus pallidus. Activities of pallidal neurons in awake animals have been studied by a number of investigators using a variety of behavioral paradigms (Travis and Sparks, 1968; DeLong, 1971; Neafsey et al., 1978; Iansek and Porter, 1980; Anderson and Horak, 1985; Mink and Thach, 1987; Hamada et al., 1990; Nambu et al., 1990; Gardiner and Kitai, 1992; Kato and Hikosaka, 1992). These studies appear to indicate that pallidal neurons change their activities in relation to a variety of behavioral events and sensory stimuli. Many of the authors also reported that single neurons show activity changes in relation to various events in the tasks for which the animals performed. Thus, it appears that relatively specific sensory and motor related activities occurred in the neostriatum and the subthalamic nucleus are integrated into the activities of pallidal neurons. Anatomical and physiological studies indicated pallidal neurons receive converged inputs from a various regions of the neostriatum and the subthalamic nucleus.

All the studies performed in monkeys indicated striking similarities in the activities between the external (i.e., homologus to the globus pallidus) and the internal (i.e., homologus to the entopeduncular nucleus) segments of the pallidum. This may be due to the similarity of afferent connections to these nuclei. Despite their similarities in neuronal activities and their afferent projections, the functions of the external and the internal segments of the pallidum may differ according to the dissimilarities in their efferent sites. Experimental lesion of the internal segment of the pallidum, which project to the thalamus, generally results in a depression of contralateral muscle activation (Horak and Anderson, 1984). The hyperkinasia induced by a lesion of the subthalamic nucleus is ameliorated by an additional lesion of the internal segment of the pallidum (Carpenter, 1961). Behavioral studies pretaining to specific lesions of the external segment are scarce. Horak and Anderson (1984) indicated that lesion of the external segment of the pallidum generally increases contralateral muscle activity. An injection of a GABA antagonist, bicuculline, in the external segment induced chorea in monkeys (Crossman et al., 1988). Based on the neuronal activities of the pallidal neurons recorded in awake animals, and a wide efferent sites of the globus pallidus which include most of the basal ganglia nuclei, it is possible that the function of the globus pallidus is to regulate the entire basal ganglia system simulta-neously on the basis of converged inputs from ongoing activities in the neostriatum and the subthalamic nucleus.

Acknowledgements

I am most grateful to Prof. S. T. Kitai for his guidance and support in conducting these studies. I am also grateful to Drs. K. Fujimoto and H. Nakanishi for excellent studies they have performed in my laboratory. They are responsible for most of the data presented in this paper. This research was supported by the USPHS Grants NS-25783 and NS-26473, and the Human Frontier Science Program Grant.

REFERENCES

Anderson, M.E., and Horak, F.B., 1985, Influence of the globus pallidus on arm movements in monkeys. III. Timing of movement-related information, *J. Neurophysiol.* 54:433-448.

Bergman, H., Wichmann, T., and DeLong, M.R., 1990, Reversal of experimental Parkinsonism by lesions of the subthalamic nucleus, *Science* 249:1436-1438.

Bergman, H.T., Wichmann, B.C., and DeLong, M.R., 1992, Tremor in MPTP treated primates is associated with low frequency neural oscillations in the basal ganglia, *Internatl. Basal Ganglia Society IV Abstr.* 7.

Canteras, N.S., Shammah-Lagnado, S.J., Sliva, B.A., and Ricardo, J.A., 1990, Afferent connections of the subthalamic nucleus: a combined retrograde and anterograde horseradish peroxidase study in the rat, *Brain Res.* 513:43-59.

Carpenter, M.B., 1961, Brain stem and infratentorial neuraxis in experimental dyskinesia, *Arch. Neurol.* 5:504-524.

Carpenter, M.B., Carleton, S.C., Keller, J.T., and Conte, P., 1981, Connections of the subthalamic nucleus in the monkey, *Brain Res.* 224:1-29.

Chang, H.T., Kita, H., and Kitai, S.T., 1983, The fine structure of the rat subthalamic nucleus: an electron microscopic study, *J. Comp. Neurol.* 221:113-123.

Crossman, A.R., Mitchell, I.J., Sambrook, M.A., and Jackson, A., 1988, Chorea and myoclonus in the monkey induced by gamma-aminobutyric acid antagonism in the lentiform complex, *Brain* 111:1211-1233.

Crutcher, M.D., and DeLong, M.R., 1984, Single cell studies of the primate putamen. II. relations to direction of movement and pattern of muscular activity, *Exp. Brain Res.* 53:244-258.

Cowan, R.L., Wilson, C.J., Emson, P.C., and Heizmann, C.W., 1990, Parvalbumin-containing GABAergic interneurons in the rat neostriatum, *J. Comp. Neurol.* 302:198-205.

Dray, A., Gonye, T.J. and Oakley, N.R., 1976, Caudate stimulation and substantia nigra activity in the rat, *J. Physiol.* 259:825-849.

DeLong, M.R., 1971, Activity of pallidal neurons during movement, *J. Neurophysiol.* 34:414-427.

DeLong, M.R., Crutcher, M.D., and Georgopoulos, A.P., 1985, Primate globus pallidus and subthalamic nucleus: functional organization, *J. Neurophysiol.* 53:530-543.

Fujimoto, K., and Kita, H., 1992, Responses of rat substantia nigra pars reticulata units to cortical stimulation, *Neurosci. Let.* 142:105-109.

Fujimoto, K., and Kita, H., 1993, Response characteristics of subthalamic neurons to the stimulation of the sensorimotor cortex in the rat, *Brain Res.* In press.

Gardiner, T.W., and Kitai, S.T., 1992, Single-unit activity in the globus pallidus and neostriatum of the rat during performance of a trained head movement, *Exp. Brain Res.* 88:517-530.

Georgopoulos, A.P., DeLong, M.R., and Crutcher, M.D., 1983, Relations between parameters of step-tracking movements and single cell discharge in the globus pallidus and subthalamic nucleus of the behaving monkey, *J. Neurosci.* 3:1586-1598.

Hamada, I., DeLong, M.R., and Mano, N., 1990, Activity of identified wrist-related pallidal neurons during step and ramp wrist movements in the monkey, *J. Neurophysiol.* 64:1892-1906.

Hammond, C., Féger, J., Bioulac, B., and Souteyrand, J.P., 1979, Experimental hemiballism in the monkey produced by unilateral kainic acid lesion in corpus Luysii, *Brain Res.* 171:577-580.

Hikosaka, O., Sakamoto, M., and Usui, S., 1989, Functional properties of monkey caudate neurons I. Activities related to saccadic eye movements, *J. Neurophysiol.* 61:780-798.

Horak, F.B., and Anderson, M.E., 1984, Influence of globus pallidus on arm movements in monkeys I. Effect of kainic acid-induced lesions, *J. Neurophysiol.* 52:290-304.

Iansek, R., and Porter, R., 1980, The monkey globus pallidus: neuronal discharge properties in relation to movement, *J. Physiol.* 301:439-455.

Kato, M., and Hikosaka, O., 1992, Saccadic responses of external pallidal neurons in monkey, *Internatl. Basal Ganglia Society IV Abstr.* 44.

Kita, H., 1992, Responses of globus pallidus neurons to cortical stimulation; intracellular study in the rat. *Brain Res.* 589:84-90.

Kita, H., Chang, H.T., and Fujimoto, K., 1991, Pallido-neostriatal projections of the rat, *Soc. Neurosci. Abstr.* 17:453.

Kita, H., Chang, H.T., and Kitai, S.T., 1983a, Pallidal inputs to subthalamus: intracellular analysis, *Brain Res.* 264:255-265.

Kita, H., Chang, H.T., and Kitai, S.T., 1983b, The morphology of intracellularly labeled rat subthalamic neurons: a light microscopic analysis, *J. Comp. Neurol.* 215:245-257.

Kita, H., and Kitai, S.T., 1987, Efferent projections of the subthalamic nucleus in the rat: light and electron microscopic analysis with the PHA-L method, *J. Comp. Neurol.* 260:435-452.

Kita, H., and Kitai, S.T., 1988, Glutamate decarboxylase immunoreactive neurons in rat neostriatum: their morphological types and populations, *Brain Res.* 447:346-352.

Kita, H., and Kitai, S.T., 1991, Intracellular study of rat globus pallidus neurons: membrane properties and responses to neostriatal, subthalamic and nigral stimulation, *Brain Res.* 564:296-305.

Kita, H., Kosaka, T., and Heizmann, C.W., 1990, Parvalbumin-immunoreactive neurons in the rat neostriatum: a light electron microscopic study, *Brain Res.* 536:1-15.

Kitai, S.T., and Deniau, J.M., 1981, Cortical inputs to the subthalamus: intracellular analysis, *Brain Res.* 214:411-415.

Kitai, S.T., and Kita, H., 1987, Anatomy and physiology of the subthalamic nucleus: a driving force of the basal ganglia, in:"The Basal Ganglia II," M.B. Carpenter and A. Jayaraman, eds., Plenum Press, New York. pp. 257-273.

Kitai, S.T., Kocsis, J.D., Preston, R.J., and Sugimori, M., 1976, Monosynaptic inputs to caudate neurons identified by intracellular injection of horseradish peroxidase, *Brain Res.* 124:601-606.

Levine, M.S., Hull, C.D., and Buchwald, N.A., 1974, Pallidal and entopeduncular intracellular responses to striatal, cortical, thalamic, and sensory inputs, *Exp. Neurol.* 44:448-460.

Matsumura, M., Kojima, J., Gardiner, T.W., and Hikosaka, O., 1992, Visual and oculomotor functions of monkey subthalamic nucleus, *J. Neurophysiol.* 67:1615-1632.

Mink, J.W., and Thach, T., 1987, Preferential relation of pallidal neurons to ballistic movements, *Brain Res.* 417:393-398.

Mugnaini, E., and Oertel, W.H., 1985, An atlas of the distribution of GABAergic neurons and terminals in the rat CNS as revealed by GAD immunohistochemistry, *in*: "Handbook of Chemical Neuroanatomy, GABA and Neuropeptides in the CNS," A. Björklund and T. Hokfelt, eds., Elsevier, Amsterdam, pp. 436-595.

Nakanishi, H., Kita, H., and Kitai, S.T., 1987a, Electrical membrane properties of rat subthalamic neurons in an in vitro slice preparation, *Brain Res.* 437:35-44.

Nakanishi, H., Kita, H., and Kitai, S. T., 1987b, Intracellular study of rat substantia nigra pars reticulata neurons in an in vitro slice preparation: electrical membrane properties and response characteristics to subthalamic stimulation, *Brain Res.* 437:45-55.

Nakanishi, H., Kita, H., and Kitai, S.T., 1988, An N-Methyl-D-Aspartate receptor mediated excitatory postsynaptic potential evoked in subthalamic neurons in an in vitro slice preparation of the rat, *Neurosci. Let.* 95:130-136.

Nakanishi, H., Kita, H., and Kitai, S.T., 1990, Intracellular study of rat entopeduncular nucleus neurons in an in vitro slice preparation: electrical membrane properties, *Brain Res.* 527:81-88.

Nambu, A., Yoshida, S., and Jinnai, K., 1990, Discharge patterns of pallidal neurons with input from various cortical areas during movement in the monkey, *Brain Res.* 519:183-191.

Neafsey, E.J., Hull, C.D., and Buchwald, N.A., 1978, Preparation for movement in the cat. II. unit activity in the basal ganglia and thalamus, *Electroencephal. and Clinical Neurophysiol.* 44:714-723.

Noda, H., Manohar, S. and Adey, W.R., 1968, Responses of cat pallidal neurons to cortical and subcortical stimuli, *Exp. Neurol.* 20:585-610.

Ohye, C., LeGuyader, C., and Féger, J., 1976, Responses of subthalamic and pallidal neurons to striatal stimulation: an extracellular study on awake monkeys, *Brain Res.* 111:241-252.

Park, M.R., Falls, W.M., and Kitai, S.T., 1982, An intracellular HRP study of the rat globus pallidus. I. Responses and light microscopic analysis, *J. Comp. Neurol.* 211:284-294.

Perkins, M.N., and Stone, T.W., 1980, Subthalamic projections to the globus pallidus: an electrophysiological study in the rat, *Exp. Neurol.* 68:500-511.

Rinvik, E., Grofova, I., Hammond, C., Féger, J., and Deniau, J.M., 1979, A study of the afferent connections to the subthalamic nucleus in the monkey and the cat using the horseradish peroxidase technique, *Adv. Neurol.* 24:53-69.

Robledo, P., and Féger, J., 1990, Excitatory influence of rat subthalamic nucleus to substantia nigra pars reticulata and the pallidal complex: electrophysiological data, *Brain Res.* 518:47-54.

Rouzaire-Dubois, B., Hammond, C., Hamn, B., and Féger, J., 1980, Pharmacological blockade of the globus pallidus-induced inhibitory response of subthalamic cells in the rat, *Brain Res.* 200:321-329.

Rouzaire-Dubois, B., and Scarnati, E., 1985, Bilateral corticosubthalamic nucleus projections; an electrophysiological study in rats with chronic cerebral lesions, *Neurosci.* 15:69-79.

Ryan, L.J., and Clark, K.B., 1991, The role of the subthalamic nucleus in the response of globus pallidus neurons to stimulation of the prelimbic and agranular frontal corticies in rats, *Exp. Brain Res.* 86:641-651.

Ryan, L.J., and Clark, K.B., 1992, Alteration of neuronal responses in the subthalamic nucleus following globus pallidus and neostriatal lesion in rats, *Brain Res. Bull.* 29:319-327.

Schmued, L., Phermsangngam, P., Lee, H., Thio, S., Chen, E., Truong, P., Colton, E., and Fallon, J., 1989, Collateralization and GAD immunoreactivity of descending pallidal efferents, *Brain Res.* 487:131-142.

Smith, Y., and Bolam, J.P., 1989, Neurons of the substantia nigra reticulata receive a dense GABA-containing input from the globus pallidus in the rat, *Brain Res.* 493:160-167.

Staines, W.A., Atmadja, S., and Fibiger, H.C., 1981, Demonstration of a pallidostriatal pathway by retrograde transport of HRP-labeled lectin, *Brain Res.* 206:446-450.

Staines, W.A., and Fibiger, H.C., 1984, Collateral projections of neurons of the rat globus pallidus to the striatum and substantia nigra, *Exp. Brain Res.* 56:217-220.

Travis, R.P., and Sparks, D.L., 1968, Unitary responses and discrimination learing in the squirrel monkey: the globus pallidus, *Physiol. and Behav.* 3:187-196.

Tremblay, L., and Filion, M., 1989, Responses of pallidal neurons to striatal stimulation in intact waking monkeys, *Brain Res.* 498:1-16.

van der Kooy, D., Hattori, T., Shannak, K., and Hornykiewicz, O., 1981, The pallido-subthalamic projection in rat; anatomical and biochemical studies, *Brain Res.* 204:253-268.

Whittier, J.R., and Mettler, F.A., 1949, Studies on the subthalamus of the rhesus monkey. II. Hyperkinesia and other physiologic effects of subthalamic nucleus, with special references to the subthalamic nucleus of Luys, *J. Comp. Neurol.* 90:319-372.

Wilson, C.J., Kita, H., and Kawaguchi, Y., 1989, GABAergic interneurons, rather than spiny cell axon collaterals, are responsible for the IPSP responses to afferent stimulation in neostriatal aspiny neurons, *Soc. Neurosci. Abstr.* 15:907.

Wilson, C., 1986, Postsynaptic potentials evoked in spiny neostriatal projection neurons by stimulation of ipsilateral and contralateral neocortex, *Brain Res.* 367:201-213.

Yoshida, M., and Precht, W., 1971, Monosynaptic inhibition of neurons of the substantia nigra by caudato-nigral fibers, *Brain Res.* 32:225-228.

Yoshida, M., Rabin, A., and Anderson, M., 1972, Monosynaptic inhibition of pallidal neurons by axon collaterals of caudato-nigral fibers, *Exp. Brain Res.* 15:333-347.

REGULATORY ACTION OF THE DOPAMINERGIC NIGROSTRIATAL

PATHWAY ON THE CORTICOSTRIATAL TRANSMISSION

Eugenio Scarnati[1], Tiziana Florio[1], Franca Cerrito[2]
and Silvia Di Loreto[3]

[1]Department of Biomedical Technology
[2]Department of Experimental Medicine
  University School of Medicine
[3]Tissue Typing Institute CNR
  I-67100 L'Aquila, Italy

INTRODUCTION

The issue of integration versus segregation of information within the basal ganglia has originated a lively controversy that has been largely discussed elsewhere (DeLong, 1990; Alexander and Crutcher, 1990, 1991; Percheron and Filion, 1991; Selemon and Goldman-Rakic, 1991). Briefly, if we consider the organization of the structures through which information is conveyed from the cerebral cortex to the striatum, globus pallidus and substantia nigra, two neuronal systems where transmission of information might be differently carried out can be identified. The first is the corticostriatal system, where signals fro different cortical areas have been claimed to reach specific neurons, while the second is the striatopallidonigral system, where evidence for the occurrence of both integration and segregation has been provided.

In the monkey, the separation within the striatum of fibers arising from sensorimotor and association cortices (Percheron et al., 1984) as well as a certain degree of segregation of terminal fields from various areas of association cortex (Selemon and Goldman-Rakic, 1985) caused the impression that separate subsets of striatal neurons could subserve different functions. This idea could seem also to apply to the rat striatum since some studies reported that fibers originating from the anterior frontal lobe, and in particular those arising from the prelimbic cortex, innervate specific striatal regions, which form the so-called patch compartment, whereas most neocortical areas, including the sensorimotor cortex, were shown to project to the matrix compartment (Herkenham and Pert, 1981; Graybiel, 1984; Gerfen, 1985; Donoghue and Herkenham, 1986). However, it should be remembered that before the above-mentioned controversy began, convergence of electrophysiologicals signals was already known to be a common event occurring in the striatum. Indeed, several subcortical structures were described to send signals converging on striatal neurons in association with those originating from the cortex. Among these structures there were the substantia nigra, the amygdaloid nuclear complex, and the intralaminar thalamus (Dafny, 1975; Kocsis et al., 1977). In addition a careful inspection of responses of monkey striatal neurons during the execution of well-defined motor tasks, reveals that at least some units modulate their discharge during the expectation of impending behavioral events as well as during movement preparation (Hikosaka et al., 1989; Apicella et al., 1992). These modulations mimic those that can be recorded from the associational and motor cortical areas which innervate the striatum and, therefore, could be indicative of mixed cortical inputs.

Thus, our contribution to this book will review and discuss some data recently obtained in our laboratories concerning both the occurrence in the rat striatum of signals originating from functionally unrelated cortical regions and the function of dopaminergic fibers within this nucleus.

CONVERGENCE OF LIMBIC AND SENSORIMOTOR INPUTS OCCURS IN STRIATAL NEURONS AND THE DOPAMINERGIC NIGROSTRIATAL PATHWAY FOCUSES THE ACTION OF THESE INPUTS

This conclusion has been reached after studying the effects of cortical stimulation in adult Sprague-Dawley rats either intact or bearing a large 6-hydroxydopamine (6-OHDA) lesion of the nigrostriatal dopaminergic pathway. This pathway was destroyed by injecting stereotaxically into the substantia nigra 2.5 μl of sterile saline containing 10 μg of 6-OHDA. After fifteen days the rats were screened for circling behavior in response to apomorphine (1 mg/kg i.p.) and animals which made at least 3-4 turns/min for 45-60 min following apomorphine administration were selected for further electrophysiological investigation. The electrophysiological experiments were performed by means of conventional extracellular recording and stimulation techniques under chloral hydrate (400mg/kg i.p.) anesthesia. In short, two pairs of glass-coated stimulating electrodes were positioned: one in the prelimbic region of the anterior frontal cortex, the other in the region of the sensorimotor cortex corresponding to the forelimb area. When an active neuron was encountered its spontaneous activity was recorded and the effects of stimulation of the

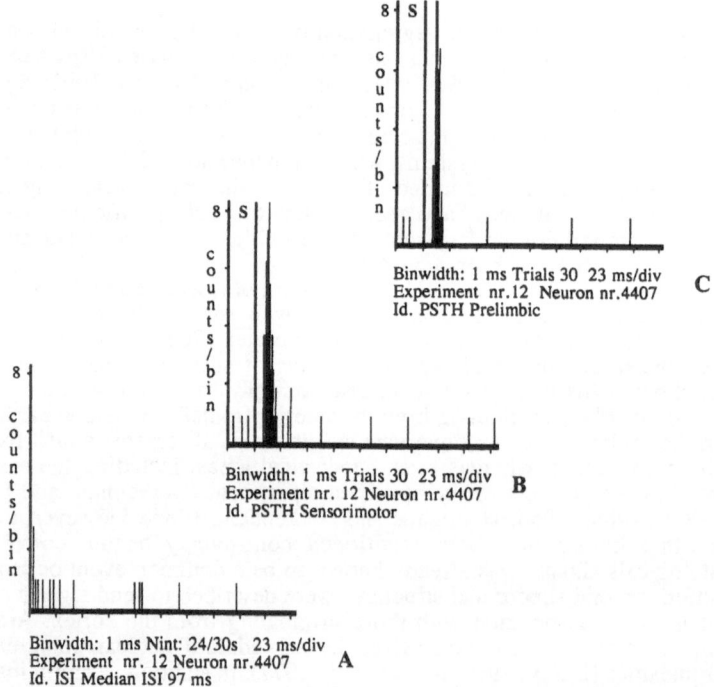

Figure 1. Example of a neuron receiving convergent excitatory inputs from cortex. This unit, as shown in the interval interspike histogram (ISI) (A), discharged at a frequency of 0.8 imp/s. It responded with a burst of impulses to stimulation of both the sensorimotor (B) and prelimbic cortex (C). The vertical bar (S) in the peristimulus time histograms (PSTH) indicates the stimulus artefact. The PSTHs were constructed from 30 consecutive stimulations whereas the ISI refers to the impulses collected in a 30 sec epoch.

cerebral cortices were studied. At the end of the experiments the high-affinity uptake of ^{3}H-DA in striatal synaptosomal pellet (P2) fractions was measured to quantify the nigral lesion and only the electrophysiological data coming from animals which presented a 70-85% reduction of the ^{3}H-DA high affinity uptake with respect to the contralateral unlesioned side were taken into account.

Of the 74 units which responded to stimulation of the cerebral cortex, 39 (52.7%) in normal rats could be orthodromically activated by stimulating both prelimbic and sensorimotor cortices (Fig. 1). Both responses consisted of a burst of 2-5 spikes in response to a single shock and their high frequency following was 300-350 Hz. The latency of the response evoked from the prelimbic cortex ranged from 2.0 to 12.0 ms (mean 7.6 ± 2.4 ms) whereas the duration varied from 3.0 to 10 ms (mean 7.4 ± 1.8 ms). For the sensorimotor cortex the latency ranged from 3.0 to 16.0 (mean 8.3 ± 3.1 ms) and the duration from 3.0 to 12.0 ms (mean 7.0 ± 2.5 ms). A mean threshold current of 350 ± 130 µA was required to evoke a response from the prelimbic and of 650 ± 236 µA from the sensorimotor cortex.

In rats with a large lesion of the nigrostriatal pathway the proportion of neurons orthodromically activated by stimulating both cortices (55/61 90.2%) was significantly higher than in normal rats (P<0.001 chi-square=22.2). In addition, the threshold intensities required to evoke the cortical responses were lower than those required by the normal rats, being of 200 ± 80 µA and 400 ± 148 µA for the prelimbic and the sensorimotor cortex, respectively. Most neurons showing convergent responses to cortical stimulation were located in the dorsolateral striatum.

TETANIC STIMULATION OF THE SUBSTANTIA NIGRA DEPRESSES THE TERMINAL EXCITABILITY OF PRELIMBIC CORTICOSTRIATAL NEURONS

Recordings were made in intact rats from neurons located in medial and lateral prefrontal cortex. These neurons were identified as corticostriatal on the basis of their antidromic activation from the striatum and collision of the antidromically-evoked impulse

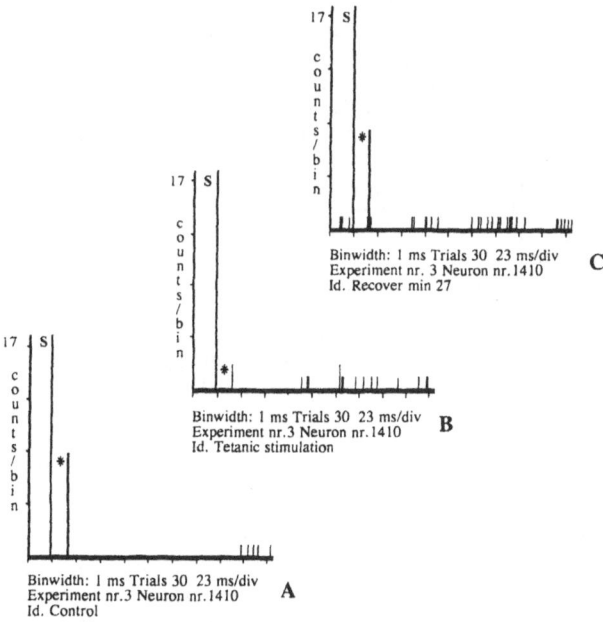

Figure 2. Example of a prelimbic cortical neuron antidromically activated by stimulating the striatum. The vertical bar (asterisk) in each peristimulus time histogram illustrates the occurrence of the antidromic impulse in 50% of trials at fixed latency following stimulation of the striatum (A). Soon after the application of a tetanic stimulation to the substantia nigra, the occurrence of the antidromic impulse was reduced (B), to recover 27 min later (C).

279

with a spontaneous one. A total of 65 neurons were recorded. The latency of the antidromic impulse was 8.4 ± 4.5 ms and a threshold current having a mean intensity of 125.3 ± 77.4 μA was required. After antidromic identification, the stimulating current was set to a value giving an antidromic response on 50 ± 5% of trials. Thereafter, a tetanic stimulation was applied to the substantia nigra by using one or three trains of stimuli - (1-3s duration, 10-100 Hz frequency, at 20 sec intervals). All of the 65 neurons tested showed a decreased terminal excitability as demonstrated either by a quite suppression of the antidromically-evoked impulse (22/65 33.9%) or by a decrease of its occurrence (43/65 66.1%) (Fig. 2). The duration of the decreased terminal excitability ranged from 4 to 70 min, with a mean value of 33.3 ± 21.6 min. The tetanic stimulation did not affect the latency of the antidromic impulse. The occurrence of the antidromic impulse was not affected by single shock stimulation of the substantia nigra during two-three minutes of continuous stimulation at frequencies of either 0.5 or 1 Hz.

WHAT DO CONVERGENCE OF CORTICOSTRIATAL SIGNALS AND FOCUSING ACTION OF DOPAMINE IN THE STRIATUM REVEAL?

As pointed out in the introduction, fibers originating from the anterior frontal lobe and in particular those arising from the prelimbic cortex were described to innervate specific regions of the striatum, which form the so-called patch compartment where a high opiate-receptor density exists. On the contrary, most neocortical areas examined, including the sensorimotor cortex, were shown to project to the matrix which is characterized by its own neurochemical markers such as calbindin immunoreactive neurons and somatostatin containing fibers.

The convergent excitatory responses that we recorded in striatal neurons by stimulating prelimbic and sensorimotor cortices challenge the view of a complete segregation between the prelimbic input to the patch compartment and the sensorimotor input to the matrix compartment. If such a segregation existed we should not have recorded convergent responses on striatal neurons by stimulating the two cortices. Our conclusion is also supported by data of Ebrahimi et al. (1992). In describing the topographical organization of motor cortex projections to the rat striatum, these authors have provided anatomical evidence to the fact that the dorsolateral striatum is reached by fibers originating from both the medial agranular cortex and the granular areas corresponding to the limb representation. Accordingly, the majority of convergent responses that we have recorded were located in this region of the striatum. Most recently, each cortical area has been shown to project to both the patch and matrix of the striatum and a laminar origin of these inputs has been demonstrated. Thus, while deep layer V and layer VI corticostriatal neurons project principally to the patches, superficial layer V and layer III and II corticostriatal neurons project principally to the matrix (Gerfen, 1989, 1992). Therefore, our data are in agreement with and extend Gerfen's current hypothesis of a rather widespread projection of corticostriatal projections to both patch and matrix compartment.

The electrophysiological characteristics of the cortically-evoked responses are in favour of their monosynaptic nature. The high frequency following of the responses as well as their latencies, which are in the range of the fast EPSPs evoked by stimulating both frontal (Wilson and Groves, 1981) and sensorimotor cortex (Calabresi et al., 1990), strongly suggest that we have recorded from the direct corticostriatal pathway. The possibility of activation of corticocortical fibers seems unlikely since the frontal cortex is not a source of afferents to the forelimb area of the sensorimotor cortex (see review in Zilles and Wree, 1985). In any case, however, even if oligosynaptic pathways could have contributed to the recorded responses, the result is always the same, i.e. a high degree of convergence of signals originating from functionally unrelated cortical areas.

There is also anatomical evidence of intrastriatal overlapping of projections from functionally related cortical areas. Indeed, the somatotopic organization in the rat striatum investigated by means of 2-Deoxy-D-[14C]-glucose autoradiography (Brown, 1992) has revealed that stimulation of hindlimb, trunk and forelimb activated discrete regions of the primary sensory cortex as well as of the sensorimotor striatum. The pattern of activation of the latter was reminiscent of the anatomic arrangement of corticostriatal projections, however, at certain anteroposterior levels, hindlimb, trunk and forelimb representations were juxtaposed in different combinations, forming a combinational map in which convergence of functionally-related cortical regions was also present. Overlapping projections from the sensorimotor cortex to the putamen have been described in the monkey as well (Flaherty and Graybiel, 1991). Taken together, these observations indicate that the extent at which the striatum receives convergent information from the cerebral cortex is larger than it was, even recently, thought.

A major factor which seems to merge the influence of various regions of the cerebral cortex on striatal neurons is the dopaminergic innervation. In particular, the amount of dopaminergic nigrostriatal fibers reaching the striatum has a key role in this process. It has long been known that the cerebral cortex is linked to the striatum through an excitatory monosynaptic pathway utilizing glutamate as a neurotransmitter (Herrling, 1985). Complex interactions between glutamate, released by corticostriatal fibers, and dopamine, released from nigrostriatal fibers, occur in the striatum (Romo et al., 1986) and it is also known that iontophoretically applied dopamine, as well as nigral stimulation, suppresses the cortically-evoked excitation of striatal neurons (Brown and Arbuthnott, 1983; Hirata et al., 1984). This effect might be brought through a reduction of terminal excitability of the corticostriatal pathway, which is under the control of DA receptors located on the terminals. Tetanic stimuli applied to the substantia nigra may induce in the striatum a long-lasting depression of terminal excitability of corticostriatal fibers whereas single shock stimulation is ineffective, as in our study, or cause a depression strictly confined to the stimulation epoch (Garcia-Munoz et al., 1991). The different effect of the two kinds of stimulation might be related to a greater release of dopamine induced by bursting stimulations than that caused by regularly spaced stimuli (Gonon, 1988).

The nigrostriatal dopaminergic pathway appears to act as a device which focuses the action of corticostriatal fibers in discrete regions of the striatum and its integrity is a requisite for the most appropriate functioning of the corticostriatal transmission. This is also suggested by the absence in 6-OHDA denervated rats, of the long-term synaptic depression induced in the striatum by tetanic stimulation of the corticostriatal pathway (Calabresi et al., 1992). It would be interesting, in the future, to investigate corticostriatal transmission in long-term studies, to ascertain if adaptive changes occurring in denervated striatum, restore the normal function of this pathway.

As far as the focusing action of dopamine is concerned, there are several reports showing that this mechanism is widely diffused within the basal ganglia. Brown (1988) reported that when tactile stimuli were applied to forelimb in the rat, a metabolic focus of activation, indexed by ^{14}C-deoxyglucose uptake, could be observed in the portions of the cerebral cortex and the striatum related to the forelimb representation. These foci were surrounded by an inhibitory ring. After destroying the dopamine input to the striatum by intranigral administration of 6-OHDA, the foci became larger, thus suggesting that in the absence of dopamine the outlines of sensory representation in the striatum are lost. Likewise, in monkeys that were made parkinsonian with the neurotoxin 1-methyl-4-phenyl-1,2,3,6-tetrahydropyridine (MPTP) many pallidal neurons had abnormally large somatosensory receptive fields (Tremblay et al., 1989). In this study an increased proportion of neurons responding to striatal stimulation was observed and many neurons also showed convergent response to both caudate and putamen stimulations. Still, neuroleptic drugs augmented whereas stimulation of the substantia nigra depressed the responsiveness of pallidal neurons to cortical stimulation (Toan and Schultz, 1985). Hence, besides the striatum, the pallidum also is a target for the focusing action of dopamine and this action would be a pivotal factor controlling the functional capacity of anatomically established relationships among the cerebral cortex and these nuclei. Probably, dopamine released from the nigrostriatal pathway causes a suppression of signal transmission at synaptic level and, as suggested by Toan and Schultz (1985), only the most active cortical inputs would still be able to act on their targets. A degeneration of at least 80-90 % of the dopaminergic nigral fibers innervating the striatum is required for the loss of the focusing action of dopamine (Florio et al., 1993), being partial lesions of the substantia nigra ineffective at all. As in other parts of the CNS like the retina and the gray matter of the spinal cord, also in the striatum there might be local neurons or short neuronal chains, which would restrict the spread of incoming signals to discrete neuronal pools. Dopamine, if in sufficient amount, might act on these mechanisms, so preventing a widespread diffusion of neural transmission. It is conceivable that a lack of dopamine in the striatum and pallidum also may cause a dissipation of descending cortical motor commands which could also be involved in causing the peculiar motor deficits of Parkinson's disease.

In conclusion, a functional segregation of limbic and sensorimotor inputs does not seem to exist in the rat striatum. In this way the striatum does not simply provide channels for cortical loops (Alexander and Crutcher, 1990) but is actively involved in the integrative processing of descending motor, behavioral and cognitive cortical information.

Acknowledgements

The present work has been partly supported by grants from Consiglio Nazionale delle Ricerche and Ministero dell'Universita e della Ricerca Scientifica e Tecnologica (Quota 40%).

REFERENCES

Alexander, G.E., and Crutcher, M.D., 1990, Functional architecture of basal ganglia circuits: neural substrates of parallel processing, *Trends Neurosci.* 13:266-271.

Alexander, G.E., and Crutcher, M.D., 1991, Parallel processing in the basal ganglia: up to a point, *Trends Neurosci.* 14:56-57.

Apicella, P., Scarnati, E., Ljungberg, T., and Schultz, W., 1992, Neuronal activity in monkey striatum related to the expectation of predictable environmental events, *J. Neurophysiol.* 68:945-960.

Brown, J.R., and Arbuthnott, G.W., 1983, The electrophysiology of dopamine (D2) receptors: a study of the actions of dopamine on corticostriatal transmission, *Neuroscience* 10:349-355.

Brown, L.L., 1988, The function of dopamine in the striatum: a sharpener of sensory information, *Neurosci. Abstr.* 14:718.

Brown, L.L., 1992, Somatotopic organization in rat striatum: evidence for a combinational map, *Proc. Natl. Acad. Sci. USA* 89:7403-7407.

Calabresi, P., Maj, R., Mercuri, N.B., and Bernardi, G., 1992, Coactivation of D1 and D2 dopamine receptors is require for long-term synaptic depression in the striatum, *Neurosci.Lett.* 142:95-99.

Calabresi, P., Mercuri, N.B., Stefani, A., and Bernardi, G., 1990, Synaptic and intrinsic control of membrane excitability of neostriatal neurons. I. An in vivo analysis, *J. Neurophysiol.* 63:651-661.

Dafny, N., 1975, Electrophysiological properties of caudate neurons following substantia nigra, motor cortex, and amygdaloid nuclear complex stimulation of the rat, *Appl. Neurophysiol.* 38:259-272.

DeLong, M.R., 1990, Primate models of movement disorders of basal ganglia origin, *Trends Neurosci.* 13:281-285.

Donoghue, J.P., and Herkenham, M., 1986, Neostriatal projections from individual cortical fields conform to histochemically distinct striatal compartments in the rat, *Brain Res.* 365:397-403.

Ebrahimi, A., Pochet, R., and Roger, M., 1992, Topographical organization of the projections from physiologically identified areas of the motor cortex to the striatum in the rat, *Neurosci. Res.* 14:39-60.

Flaherty, A.W., and Graybiel, A.M., 1991, Corticostriatal transformations in the primate somatosensory system. Projections from physiologically mapped body-part representations, *J. Neurophysiol.* 66:1249-1263.

Florio, T., Di Loreto, S., Cerrito, F., and Scarnati, E., 1993, Influence of prelimbic and sensorimotor cortices on striatal neurons in the rat: electrophysiological evidence for converging inputs and the effects of 6-OHDA-induced degeneration of the substantia nigra, in press.

Garcia-Munoz, M., Young, S.J., and Groves, P.M., 1991, Terminal excitability of the corticostriatal pathway. I .Regulation by dopamine receptor stimulation, *Brain Res.* 551:195-206.

Gerfen, C.R., 1985 The neostriatal mosaic. I. Compartmental organization of projections from the striatum to the substantia nigra in the rat, *J. Comp. Neurol.* 236:454-476

Gerfen, C.R., 1989, The neostriatal mosaic: striatal patch-matrix organization is related to cortical lamination, *Science* 246:385-388

Gerfen, C.R., 1992, The neostriatal mosaic: multiple levels of compartmental organization, *Trends Neurosci.* 15:133-139.

Gonon, F.G., 1988, Non-linear relationship between impulse flow and dopamine released by rat midbrain dopaminergic neurons as studied by in vivo electrochemistry, *Neuroscience* 24:19-28.

Graybiel, A.M., 1984, Correspondence between the dopamine islands and striosomes of the mammalian striatum, *Neuroscience* 13:1157-1187.

Herkenham, M., and Pert, C.B., 1981, Mosaic distribution of opiate receptors, parafascicular projections and acetylcholinesterase in rat striatum, *Nature* 291:415-418.

Herrling, P. L., 1985, Pharmacology of the corticocaudate excitatory postsynaptic potential in the cat: evidence for its mediation by quisqualate- or kainate-receptors, *Neuroscience* 14:417-426.

Hikosaka, O., Sakamoto, M., and Usui, S., 1989, Functional properties of monkey caudate neurons. III.Activities related to expectation of target and reward, *J. Neurophysiol.* 61:814-832.

Hirata, K., Yim, C.Y., Mogenson, G.J., 1984, Excitatory input from sensory motor cortex to neostriatum and its modification by conditioning stimulation of the substantia nigra, *Brain Res.* 321:1-8.

Kocsis, J.D., Sugimori, M., and Kitai, S.T., 1977, Convergence of excitatory synaptic inputs to caudate spiny neurons, *Brain Res*. 124:403-413.

Percheron, G., Yelnik, J., and François, C., 1984, The primate striato-pallido-nigral system: an integrative system for cortical information, *in*:"The Basal Ganglia," J.S. McKenzie, R.E. Kemm and L.N. Wilcock, eds., Plenum Press, New York, pp. 87-105.

Percheron, G., and Filion, M., 1991, Parallel processing in the basal ganglia: up to a point, *Trends Neurosci*. 14:55-56.

Romo, R., Chéramy, A., Godeheu, G., and Glowinski, J., 1986, In vivo presynaptic control of dopamine release in the cat caudate nucleus-III. Further evidence for the implication of corticostriatal glutamatergic neurons, *Neuroscience* 19:1091-1099.

Selemon, L.D., and Goldmann-Rakic, P.S., 1985, Longitudinal topography and interdigitation of corticostriatal projections in the rhesus monkey, *J. Neurosci*. 5:776-794.

Selemon, L.D., and Goldman-Rakic, P.S., 1991, Parallel processing in the basal ganglia: up to a point, *Trends Neurosci*. 14:58-59.

Toan, D.L., and Schultz, W., 1985, Responses of rat pallidum cells to cortex stimulation and effects of altered dopaminergic activity, *Neuroscience* 15:683-694.

Tremblay, L., Filion, M., and Bédard, P.J., 1989, Responses of pallidal neurons to striatal stimulation in monkeys with MPTP-induced parkinsonism., *Brain Res*. 498:17-33.

Wilson, C. J., and Groves, P.M., 1981, Spontaneous firing patterns of identified spiny neurons in the rat neostriatum, *Brain Res*. 220:67-80.

Zilles, K., and Wree, A., 1985, Cortex: areal and laminar structure, *in*: "The Rat Nervous System," G. Paxinos, ed., Academic Press, Sydney, pp. 375-415.

OLFACTORY BULB INFLUENCE ON NEURONS OF VENTRAL STRIATUM

John S. McKenzie, Antonio G. Paolini and Wolgang A.A. Kunze

Department of Physiology
University of Melbourne
Parkville
Australia, 3052

INTRODUCTION

Odours And Action

Olfactory signals can motivate action in a range of animals, as innate responses or as learned patterns of behaviour, a distinction often difficult to make. For example, sexual responses of male rats to female odours and other odours can be enhanced by prior sexual experience (cited by West et al., 1992) with a parallel effect on ventral striatal (accumbens) neuron responses to experimentally associated odours (*ibid*). The detection of food and skilled reaching for it by rats is based on olfactory input (Whishaw and Tomie, 1989). Olfactory discrimination is often investigated by association of odours with particular instrumental actions (Slotnik, 1990). That odours can influence human mood and behaviour seems obvious to subjective introspection and casual observation, although it is not clear how much is innate (Doty, 1991; Engen, 1982, Schmidt and Beauchamp, 1992). The commercial importance of the perfume and fragrance industry is objective evidence for olfactory influences on human behaviour at many levels.

If odours can motivate action, how do odour signals gain access to motor systems?

Ventral Striatum

It has been suggested (e.g. Graybiel, 1976 cited by Mogenson et al., 1980) that the nucleus accumbens (Acc) of the ventral striatum provides an interface for limbic system access to the basal ganglia and thereby to motor control systems. This model was based on evidence for medial forebrain bundle pathways from preoptic regions to the ventral tegmental area, and subsequent mesolimbic dopaminergic projections to Acc. Limbic inputs to Acc also derive from the hippocampal formation and could be influenced by olfaction via oligosynaptic pathways through the entorhinal cortex or following long-circuit processing in association cortices. The Acc receives anatomic projections from the hippocampal formation, amygdala, perirhinal and primary olfactory cortices, but does not receive direct olfactory bulb input, nor any apparently from the anterior olfactory nucleus (Newman and Winans, 1980a).

On the other hand, the olfactory tubercle region, now considered as the most ventral, superficial division of the ventral striatum (reviewed in Heimer and Alheid 1991) receives direct input from the olfactory bulb in a range of species (Heimer 1978; Newman and Winans 1990b), including monkeys (Turner et al., 1978), although its relative importance as an olfactory target appears to diminish in primate and especially human brains (Stephan and Andy 1982; Alheid et al., 1990; Price , 1990). This input suggests that the olfactory

The Basal Ganglia IV, Edited by
G. Percheron *et al.*, Plen. m Press, New York, 1994

tubercle may act as a ventral striatal interface for olfactory influences onto the basal ganglia and thus on the preparation and regulation of motor activity.

According to anatomical findings, the olfactory tubercle (OT) receives input both from the main olfactory bulb and from a number of other olfactory structures to which the bulb projects (Newman and Winans 1980b). Like Acc, the OT projects to the ventral pallidum and reticulate substantia nigra (Newman and Winans, 1980b). The output is carried by axons mainly of its medium spiny neurons (Heimer et al., 1987). Some ventral pallidal target cells are located within the deep multiform layer and may occur more superficially in the tubercle (Young et al., 1984; Walaas and Ouimet, 1989).

Both Acc and OT regions of the ventral striatum contain a preponderance of medium spiny neurons (Chronister et al., 1981; Hedreen, 1981;Millhouse and Heimer, 1984), with sparse large aspiny neurons and various small neuron types particularly in OT of rodents and carnivores (Adrianov and Mering, 1959; Meyer and Wahle, 1986a; Meyer et al. 1989). Most medium-sized output cells in OT are arranged in a dense-cell layer interrupted by granule islands and by dwarf spiny cells which are probably also output neurons (Meyer and Wahle 1986). Various types of small, medium and large spiny, and large aspiny cells occur in the deeper multiform layer.

Present Investigation

We have recorded from OT and deep Acc neurons with extracellular and intracellular methods to investigate (a) whether olfactory bulb input could drive identified ventral striatum output to the ventral pallidum, (b) to what extent olfactory bulb influences on OT or Acc output cells are monosynaptic, (c) the morphology of such output cells using intracellular dye-filling.

METHODS

All experiments were performed on male hooded rats anaesthetized with intraperitoneal urethane (1.3g/Kg) and breathing spontaneously.

Extracellular Studies

Micropipettes containing 1M KCl or NaCl (10-30Mohm) were inserted stereotaxically, with vertical location of recorded single units determined by a subsequent pial-touch change of impedance. Units within 0.3 mm of the ventral pial surface were

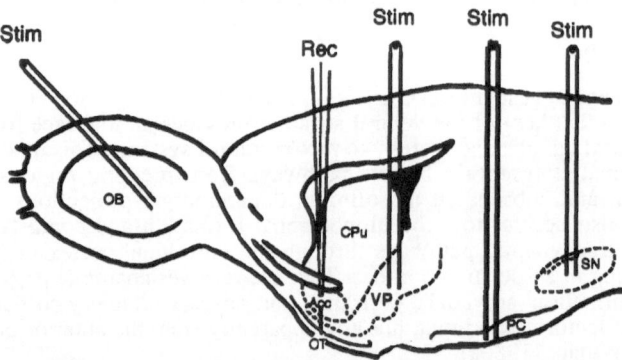

Figure 1. Schematic diagram of experiments to investigate responses of ventral striatum neurons in anaesthetized rats. Recording micropipette traverses caudate-putamen (CPu), accumbens and olfactory tubercle. Stimulating electrodes insulated but for tips inserted stereotaxically in ventral pallidum (VP), primary olfactory (piriform) cortex (PC), at times into substantia nigra (SN), and parafascicular thalamus (not shown); and into caudal olfactory bulb (OB) under visual control.

taken as being in OT. The search strategy was to advance the micropipette in 2μm steps while stimulating the ventral pallidum with brief pulses, seeking antidromic or spontaneous unit discharge in OT. On isolating a unit, its possible antidromic response to ventral pallidium, and effects of stimulating OB and olfactory cortex, and in some cases parafascicular thalamus or substantia nigra, were determined (Fig. 1). Positions of the stainless steel stimulating cathodes, either at the tips of fine coaxial pairs or in parallel assemblies, were established histologically.

Intracellular Studies

Intracellular recording electrodes for ventral striatum were aimed at stereotaxic coordinates 10.7 mm anterior to the inter-aural plane and 1.0-2.5 mm lateral to the midline. Stimulating electrodes for ventral pallidum were inserted at anterior 8.7 mm, lateral 2.6 mm, 2.0 mm above the inter-aural plane.

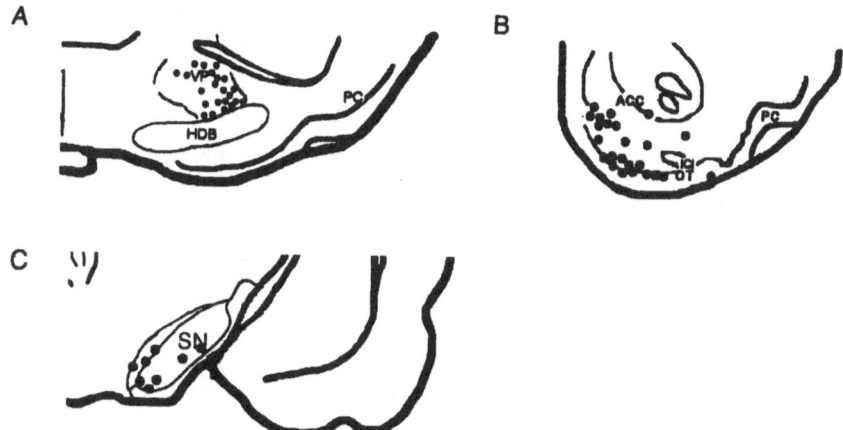

Figure 2. A, Composite diagrams of stimulating positions; B, sites of impaled lucifer yellow filled neurons in ventral striatum. OB and PC positions not shown. HDB: nucleus of the horizontal limb of diagonal band. IC: Island of Calleja: other abbrevations in Figure1 legend.

Micropipettes for cell impalement were filled with IM KAcetate (70-120Mohm) or 4% Lucifer Yellow (LY; 130-280 Mohm). They were stepped from positions 7mm below the dorsal brain surface through the ventral Acc and OT to the ventral pial surface. Upon stabilised cell impalement, stimuli to the OB and ventral pallidum, and sometimes to olfactory cortex or substantia nigra (Fig. 2), were used to classify the cell response as antidromic, synaptic (EPSP or IPSP) or absent. Antidromicity was determined by constant latency of action potential response with no discernible pre-potential, ability to follow high frequency stimulus bursts, and collision with spontaneous discharge where available. Synaptic potentials were taken as monosynaptic if latency was constant on changing stimulus strength over a wide range (approx. 0.5-2.0 mA). Cells were filled with LY by passing approximately 1.3nA of hyperpolarizing current through the micropipette, for at least 3 min. and optimally up to 15 min.

Sections of 120 μm through the ventral striatum were examined for LY-filled cells using a fluorescence microscope set up for FITC. Computer digitizing methods allowed reconstruction using a software paint package.

Positions of other impaled cells were determined as for extracellular units (see above).

RESULTS

Extracellular Units

Most of the units recorded were in OT , in which the search was concentrated. In both OT and Acc, the units were divided between two distinct classes - spontaneously discharging, or silent. None of 21 OT and 11 Acc spontaneously discharging units was evoked by ventral pallidal stumuli. The 21 silent units were evoked by ventral pallidum, so that their detection depended on the search strategy of 0.6Hz shocks to ventral pallidum. Of these, 13 in OT and 2 in Acc were considered to respond antidromically because of invariant latency and ability to follow high frequency volleys. Two OT units and 4 Acc units which failed the antidromic criteria were evoked trans-synaptically by ventral pallidal stimuli. They were tentatively identified as intrinsic but could have had axons projecting outside the area of effective stimuli.

Only the spontaneous units responded to stimulation of sites other than in ventral pallidum. While 8 of 20 in OT responded to olfactory bulb stimulation with facilitation of discharge or biphasic facilitation then suppression, no Acc unit was influenced. Some spontaneous OT units were excited from piriform cortex, parafascicular thalamus, or both; but ventral pallidal stimulation had no effect on them. No antidromically identified output unit responded to these anatomically known sources of input to OT.

Intracellular Recording

Upon impalement of cells in OT or Acc (ventral region), and removal of membrane-stabilising negative current, stimuli were applied to the ventral pallidum to test for antidromic activation, and to the olfactory bulb. Results were obtained from 43 cells in OT and 79 in Acc (Fig. 3).

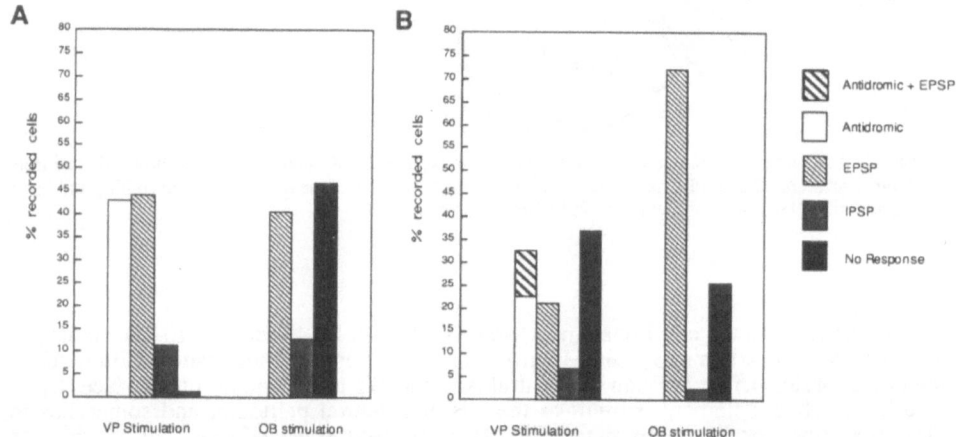

Figure 3. The percentage response to ventral Pallidal and Olfactory Bulb stimulation in A, Accumbens (n=79) and B, Olfactory tubercle (n=43) cells.

Olfactory Tubercle. Three-quarters of all impaled OT cells responded with excitatory post-synaptic potentials (EPSPs) to bulb stimulation, the great majority monosynaptically, including 80 percent of those antidromically identified as outputneurons. A similar proportion of those not responding either antidromically or synaptically to ventral pallidum also gave EPSPs to olfactory bulb volleys; most of these cells were spontaneously active. Neurons that responded to ventral pallidal stimuli with only EPSPs, mostly monosynaptic, were almost as common as identified output cells. A few responded with inhibitory postsynaptic potentials (IPSPs). Examples of intracellular OT responses are shown in figures 4 and 5.

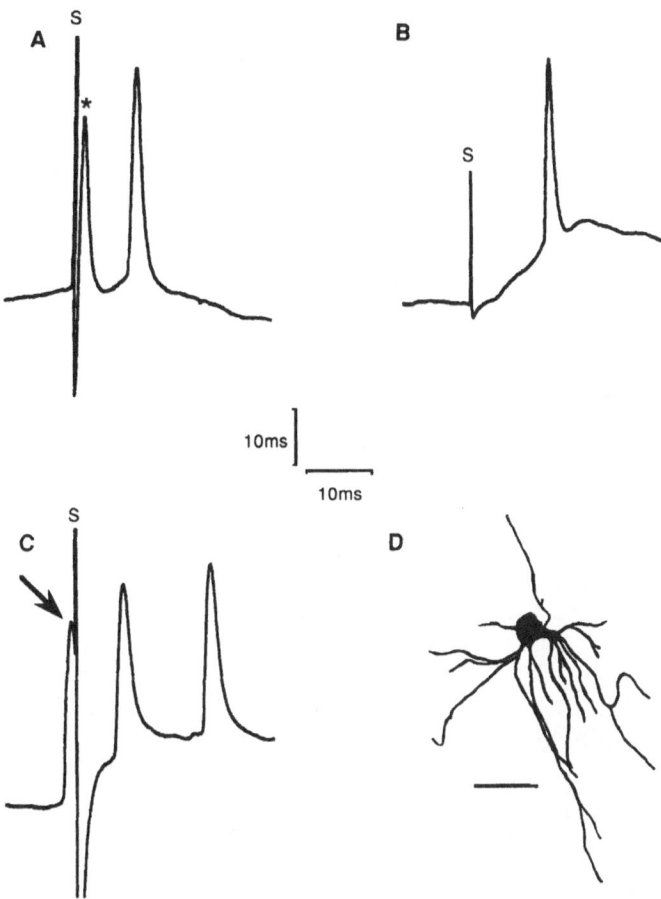

Figure 4. Computer print-outs of intracellular recordings from LY-filled spiny crescent cell in OT, deep to dense-cell layer. A, antidromic spike response (asterisk) to ventral pallidum stimulation (S); B, Monosynaptic response to olfactory bulb stimulation (S); C, Collision extinction of antidromic spike on ventral pallidum stimulation (S) following spontaneous discharge (arrow); D, Reconstruction of filled cell from computer digitized images (dendritic spines not resolved), calibration 40μm. Resting potential: -50 mV.

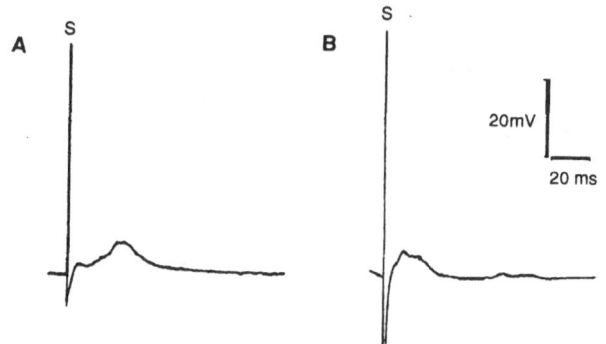

Figure 5. Intracellular records from large aspiny neurone in OT multiform layer (Fig.6a), showing monosynaptic EPSP responses to A, olfactory bulb; B, ventral pallidum stimulations (S). Resting potential: -55 mV

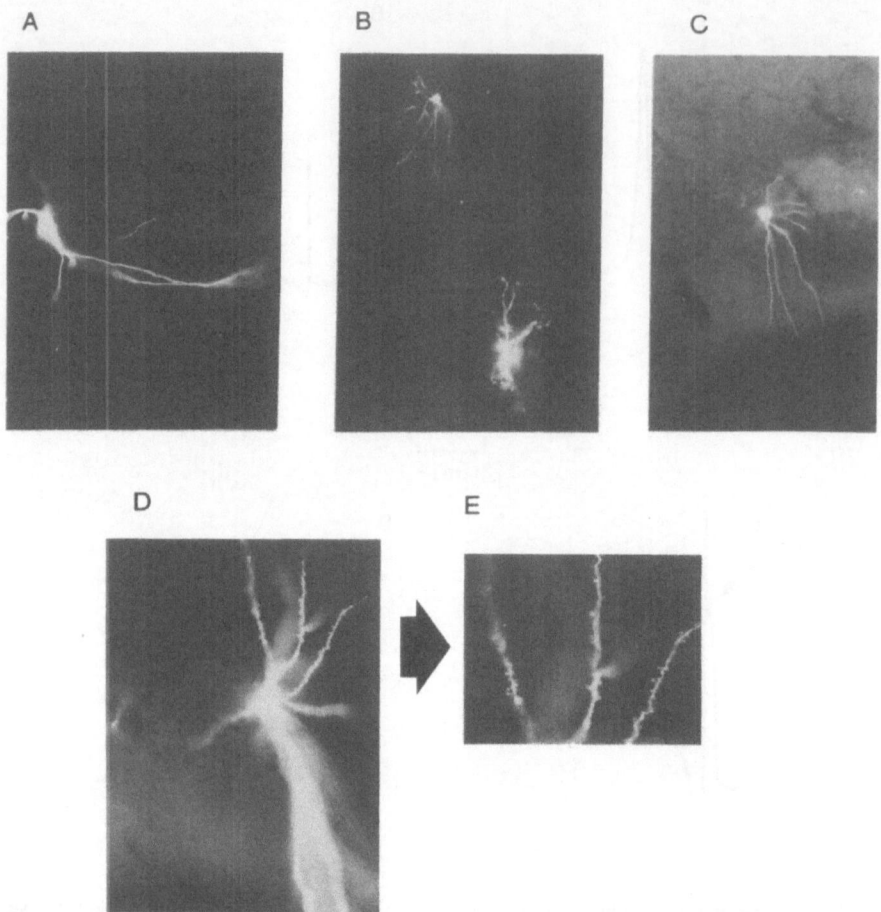

Figure 6. Photomicrographs of LY-filled neurones in the ventral striatum. A, large aspiny cell in OT multiform layer. B, medium spiny cells in Acc (upper) and OT (lower). C, medium spiny cell in OT dense-cell layer, dendrites reaching to pial surface. D, OT medium spiny neurone (some dendrites out of focus) showing spine morphology. E, Enlargement from indicated zone in D. Calibration: A,B,C, 200μm; D, 80μm; E, 40μm.

All LY-filled neurons recorded responded with monosynaptic EPSPs to olfactory bulb. The majority were located in the multiform or dense-cell layers (Fig. 2B) and were medium-sized (up to 20μm somata) with spiny dendrites (Fig. 6). About half of those in the multiform layer were identified as output cells, the remainder not responding to ventral pallidum. Two large aspiny neurons in the multiform layer, not fired antidromically from ventral pallidum, responded to bulb stimulation with monosynaptic EPSPs. Most of the medium spiny cells and one crescent (spiny) cell in the dense - cell layer, as well as one spiny dwarf cell in the moleculer layer, were identified as output cells; most also gave monosynaptic EPSPs on ventral pallidal stimulation. Dendrites of dense-cell layer neurons often extended to the pial surface. A minority had spontaneous discharge at resting membrane potential.

Substantia nigra stimulation elicited an antidromic spike in one (filled), and EPSPs in 3 (one filled) dense-cell layer neurons, with no response in one (filled) in the multiform layer. The first cell also responded antidromically to ventral pallidum. Filled cells were medium-sized spiny.

Nucleus Accumbens. All impaled cells were in, or on the margin of, the shell region of Acc (Alheid and Heimer., 1988). Two thirds of identified output cells were spontaneously active. An equally numerous group responded with EPSPs to ventral pallidum, but few of these were spontaneous. A smaller group responded with IPSPs. Most synaptic responses to ventral pallidum appeared to be monosynaptic. About half of the output cells responded to olfactory bulb stimulation with polysynaptic EPSPs or, in a few, IPSPs. Equal numbers of neurons had only EPSP responses to ventral pallidum and to bulb. Three cells appeared to have monosynaptic excitatory responses to the bulb. All except one of the LY-filled cells were in the shell region. Half of them were medium spiny cells (Fig. 6); one third of these were identified as output neurons, and the majority

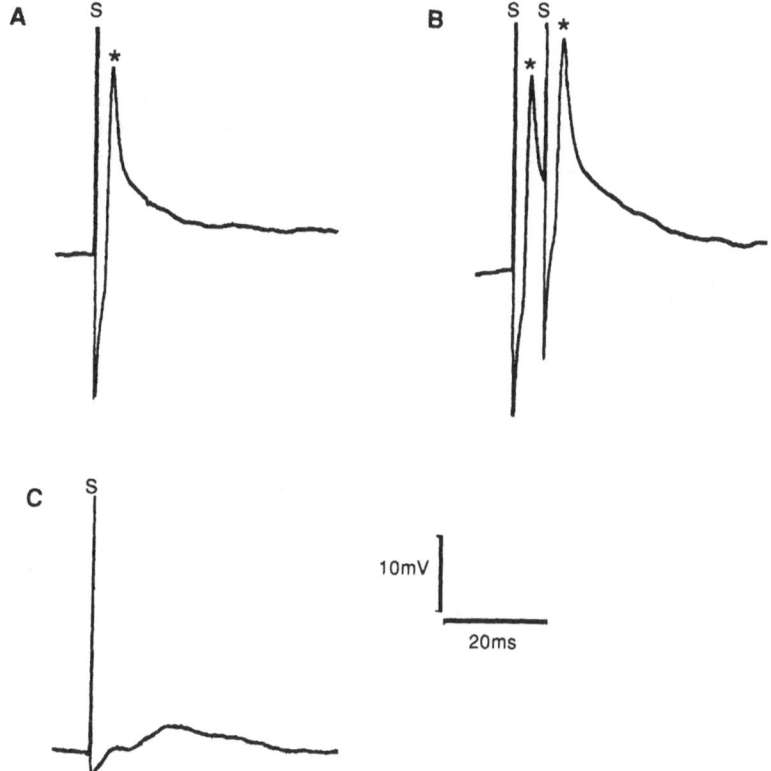

Figure 7. Intracellular records from medium spiny neurone in Acc. Antidromic responses (asterisk) to single (A) and closely paired (B) ventral pallidal stimuli (S). C, polysynaptic EPSP on olfactory bulb stimulation (S). Resting potential: -35 mV.

responded to olfactory bulb with polysynaptic EPSPs (Fig. 7). The small, medium, or large aspiny or spine poor cells often gave monosynaptic EPSP responses to ventral pallidum, although one of spindle shape was antidromically activated by both pallidal and nigral stimulation, and one large aspiny cell responded with monosynaptic IPSP to pallidum, polysynaptic EPSP to bulb and monosynaptic EPSP to nigra. One small spine-poor cell had EPSP responses to pallidum (monosynaptic) and bulb (polysynaptic) but an IPSP to nigra. Twelve of 14 Acc cells tested responded to nigra with monosynaptic EPSPs, but only 5 were identified antidromically as projecting to ventral pallidum, including the spine-poor spindle cell.

DISCUSSION

Neurophysiological Properties

While units recorded by extracellular pipettes in OT were either spontaneously active with no observed extrinsic projections, or silent unless activated from ventral pallidum, a number of impaled OT neurons (although a minority) were both spontaneously active and projected to ventral pallidal territory. It is curious that only spontaneously active units in the extracellular study responded synaptically to olfactory bulb, olfactory cortex and parafasicular thalamic sources of input whereas a large proportion of impaled cells, whether displaying spontaneous activity or not, had excitatory synaptic responses to the bulb (and to other sources less frequently tested during the intracellular experiments), often leading to action potential discharge. While it is possible that the spontaneous *extracellular* discharges were produced by ventral pallidal neurons in the multiform or more superficial layers (Heimer and Alheid, 1991 for review; Walaas and Ouimet, 1989), in which spontaneous activity would be expected by analogy with dorsal pallidal cells (eg. Everett et al., 1984), some *impaled*, medium spiny, identified projection neurons in OT were spontaneously active during intracellular recording, and these were certainly not typical pallidal cells. The difference between extracellular and intracellular findings in spontaneous activity and in responsiveness to synaptic input of OT output cells, is possibly due to effects of impalement increasing neuronal excitability. Further, the intracellular OT recordings displayed considerably fewer cases of spontaneous discharge than those in Acc, where nearly 70 percent of output cells were spontaneously active. It would hence appear that OT output cells are less excitable, or receive less effective excitatory input, than those located more dorsally in the Acc region of the ventral striatum.

The incidence of antidromic responses to our ventral pallidum stimulation was roughly comparable for impaled neurons in OT (35%) and Acc (43%). Therefore the ventral pallidal projections of OT neurons are not restricted to zones in the closely dorsal vicinity of the OT (Heimer et al., 1987; Zahm and Heimer, 1987), but extend to more caudal regions of Acc projections.

The finding of predominantly monosynaptic EPSP responses to olfactory bulb in OT, compared to practically total polysynaptic responses in Acc, is consistent with anatomical data in the hamster showing that OT receives direct projections from the bulb, but Acc does not (Newman and Winans, 1980 a,b). There was no antidromic evidence that either OT or ventral Acc projects to the bulb, again in accord with anatomical findings (Newman and Winans 1980 a, b, Macrides et al., 1981).

The IPSP responses in Acc to bulb stimuli are explicable as resulting from activation of GABAergic interneurons, or of output cells with intrinsic collaterals, by way of olfactory cortex.

The preponderance of monosynaptic EPSPs over antidromic responses to nigral stimulation, in OT and particularly Acc neurons, indicates that mesolimbic projections were activated by the stimuli more easily than were ventral striato-nigral preterminal axons. Synaptic excitation from the nigra is consistent with reports that D_1 dopamine receptors outweigh D_2 receptors in OT and Acc (Boyson et al., 1986; Dubois et al. 1986; Rev. Gerfen, 1992; but see Wamsley et al., 1989, for reported equivalence in occurrence).

Morphological properties and neurophysiology

The finding in the OT multiform layer of medium-sized spiny cells with identified output to pallidal regions, as well as a large proportion in the dense-cell layer, is concordant with descriptions based on Golgi preparations (Millhouse and Heimer, 1984), although tract-tracing studies appear to have concentrated on output from the more superficial zone (Heimer et al., 1987, Zahm and Heimer, 1987). A dwarf cell, a crescent cell, and a superficial medium spiny cell were the only filled, identified output neurons with spontaneous activity. A number of medium spiny output cells in the dense-cell and multiform layers were silent, although all morphologically identified spiny cells responded with monosynaptic EPSPs to olfactory bulb. The large aspiny neurons recorded in the multiform layer that were excited by bulb stimulation without projecting to pallidum may represent the class of large interneurons, sparsely distributed as in the dorsal striatum (Chang et al., 1982; Tagaki et al., 1984).

In the Acc shell region, half of the filled cells were medium spiny, the remainder varying through small to large aspiny or spine-poor classes. While on anatomical grounds this appears to be a region of mixed or transitional properties between the striatum and the extended amygdala (Alheid and Heimer, 1988; Alheid et al., 1990; Heimer and Alheid, 1991), one-third of cells filled in Acc were antidromically invaded from the ventral pallidum. It is of course not possible to exclude antidromic activation of extended amygdala axons projecting to the lateral hypothalamus more caudally.

Synaptic Responses To Ventral Pallidal Simulation

Synaptic responses in the ventral striatum to stimulation of ventral pallidum would be expected to be inhibitory, from activation either of ventral pallido-striatal axons (Kuo and Chang, 1992) or of the GABAergic output axons with intrinsic striatal collaterals. There were a few IPSP responses, but most were EPSPs in both OT and Acc. No cholinergic ventral pallido-striatal projections were found by Kuo and Chang (1992), and only rare asymmetric synapses. While the possibility of a few excitatory neurons in the ventral striatum with intrinsic collaterals cannot be excluded, containing for example substance- P or neurotensin (Zahm and Heimer, 1988), on statistical grounds of GABAergic neurone abundance most such responses should be inhibitory. A more likely explanation for the EPSP responses is the activation of excitatory pathways to the ventral striatum from more caudal regions such as the non-specific thalamus or posterior perirhinal and hippocampal regions (Newman and Winans, 1990).

Olfactory Striatum?

As summarized in figure 8, intracellular recording in the rat has here provided neurophysiological evidence that the OT directly mediates olfactory bulb signal transfer to the basal ganglia. In this the OT differs from the immediately more dorsal zone of the ventral striatum, the Acc shell, which receives no monosynaptic input from the bulb but only indirect transfer of odour information from primary olfactory cortex (Newman and Winans, 1980a). Its output to ventral pallidum, the frequent "double pyramid" conformation of its principal spiny neurons (Millhouse and Heimer, 1984; Millhouse 1987, fig.6), and its capacity of both receiving and transferring olfactory information (results), suggest the OT to be a specialized part of the basal ganglia with hybrid properties of striatum and olfactory cortex. It cannot be regarded as merely an extension to the basal brain surface of the Acc component of the ventral striatum.

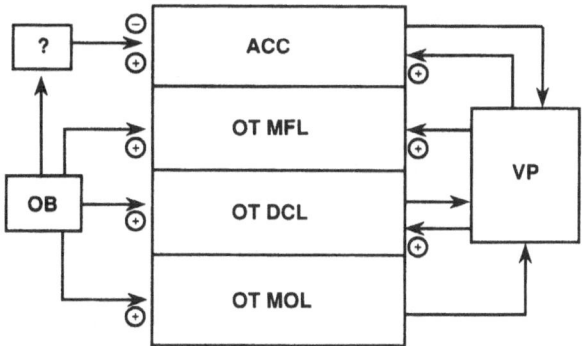

Figure 8. Connections of ventral striatum observed. Abbreviations: Acc, nucleus accumbens; OB, olfactory bulb; OT, olfactory tubercle; DCL, MFL, MOL, dense cell, multiform, molecular layers; VP, ventral pallidum; plus and minus, excitation and inhibition.

In rodents (Heimer, 1978) and carnivores (Meyer and Wahle, 1986) the laminar appearance of the OT with olfactory tract fibres covering the molecular layer is more marked laterally than in the medial, more "striatal" zone. In the tubercle of the cat there isa sequence of "cortical parts" through this latero-medial gradation, with segments of dense-cell layer neurones. These, called "pyramidal" cells by Meyer and Wahle (1986), are equivalent in Golgi morphology to the medium spiny cells with dendrites often reaching to the pial surface of the rat OT (Millhouse and Heimer, 1984; present results). These alternate with a series of "cap" regions in which dwarf spiny cells lie superficial to granule cells in lateral areas, but are displaced by granule cells clustered in "islands" reaching the pial surface in more medial OT where there is no longer an overlying fibre plexus. Wahle and Meyer (1986) proposed that the dwarf cells are the main target of olfactory bulb projections to OT, as suggested for the rat by Millhouse and Heimer (1984). Our intracellular recordings show the medium spiny cells to be the most frequent target of bulb projections, though dwarf cell impalement would admittedly be less favoured statistically. It seems likely that in both rat and cat there would be a continuum of medium, small, and dwarf spiny cells with basically similar connections, including projection to more caudal ventral pallidal neurones (Wahle and Meyer, 1986, Fig. 6).

The granule cell clusters of rat and cat OT appear to have no counterpart in the dorsal striatum (Millhouse, 1987), which supports the view that the tubercle is more than a component of the ventral striato-pallidal complex.

The monkey OT has a more restricted lateral laminate zone, although all three of its divisions receive olfactory afferents in the superficial molecular layer (Turner et al., 1978). Direct evidence regarding distribution of olfactory tract fibres to the human olfactory tubercle is unavailable. Little subsists of a dense-cell layer (named "pyramidal" layer in the detailed drawings of Crosby and Humphrey, 1941), and nearly all tract fibres on the surface of the tubercle have disappeared (Heimer et al., 1977; Alheid et al., 1990; Price, 1990). This suggests a tendency in primates for olfactory influences on the ventral striatum to be increasingly mediated by primary olfactory cortices, which would facilitate integration with neocortical association processes up-stream of striatal involvement, an integration in concordance with the cognitive and cultural influences that dominate much of at least conscious human reaction to bodily odours and social olfactory signals (Stoddart, 1990).

Nonetheless, poverty of laminar structure and of olfactory tract cover does not impel the conclusion that the OT in primates has little or no olfactory relevance. It is interesting in this context that olfactory bulb-dependent odour detection in the rat, using an instrumental task, was considerably preserved after transection of the olfactory cortex together with the lateral OT, but abolished if the lesions included the more medial "striatal" region of the OT (Slotnik, 1990).

The most compelling evidence against an essential olfactory function of the OT comes from the brains of anosmatic dolphins, porpoises and whales which have a prominent OT as a surface feature (Morgane et al, 1980) but no vestige of olfactory bulb, tract, or nucleus. In Nissl-stained sections through the basal forebrain of the porpoise *Phocaena phocena* by Breathnach (1953) the caudate nucleus reaches the ventral surface anteriorly, and more caudally the structure shows the characteristic layers of OT present in macrosmatic mammals - plexiform, dense cell, and multiform layers, particularly more laterally. In the dolphin *Tursiops truncatus,* Jacobs et al (1971) describe the OT as being continuous with the fundus striati and caudate nucleus, illustrating very restricted plexiform, dense-cell and multiform layers.

In both these toothed cetaceans, there is a piriform cortex external to the OT, of typical primary olfactory cortex structure, through severely limited in extent compared to macrosmatic mammals.

Olfactory peduncles have been found in all mysticetes (baleen whales) studied (Jacobs and Jensen, 1964; Morgane, 1980), but there appear to be no available illustrations of the internal structure of the olfactory centres, including the prominent OT which admits relatively large olfactory tracts. This is perhaps attributable to technical histological difficulties with tissue from great whales, yet detailed fibre analyses of their cranial nerves has been possible (Jacobs and Jensen, 1964). Internal structural data for the OT of the mystaceti is needed for comparison with the anosmatic cetaceans, especially regarding the contribution of olfactory tract fibres to the plexiform layer, the relative prominence of dense-cell layer, and extent of piriform cortex. Indeed, the presence of what appears structurally to be primary olfactory cortex in the odontoceti may raise the suggestion of possible alternative chemosensory inputs to the OT in these animals. One such possibility is the nervus terminalis, regarded as a chemosensitive input contributing to mating behaviour in rodents (cited in Wysocki and Meredith, 1987). The nervus terminalis has been described in the porpoise brain (Johnston, 1914) as fanning caudally on the base of the forebrain from a corticomedial point, with lateral strands entering the anterior perforated space, the OT The existence of such a chemical sense despite the absence of regular olfactory epithelium, bulbs, and tracts, could underlie anecdotal accounts of olfactory-type behaviour in cetaceans, explain the paradox of piriform cortex and restore some chemosensory function to the dominantly striatal OT.

CONCLUSION

The olfactory tubercle in the rat represents an "olfactory" striatum, analogous to the "limbic" striatum of nucleus accumbens. It provides direct synaptic access of olfactory bulb output to the basal ganglia, and thus may be capable of mediating motor responses to odour signals, as in feeding and mating. The impact of olfactory bulb signals on basal ganglia activities, and possibly thereby on reactions to odour that are below the threshold of consciousness, would seem rewarding areas for further investigation.

REFERENCES

Adrianov, O.S., and Mering, T.A., 1959, "Atlas of the Dog Brain," State Publishing of Medical Literature, Moscow.

Alheid, G.F., and Heimer, L., 1988, New perspectives in basal forebrain organization of special relevance for neuropsychiatric disorders: The striatopallidal, amygdala, and corticopetal components of substantia innominata, *Neurosci.* 27:1-39.

Alheid, G.F., Heimer, L., and Switzer, R.C., 1990, Basal ganglia, in: "The Human Nervous System," G. Paxinos, ed., Academic Press, San Diego. pp. 483-582

Breathnach, A.S. , 1953, The olfactory tubercle, prepyriform cortex and precommissural region of the porpoise (*Phocaena phocena*), *J. Anat.* 87:96-113.

Boyson, S.J., McGonigle, P., and Molinoff, P.B., 1986, Quantitative autoradiographic localization of the D_1 and D_2 subtypes of dopamine receptors in rat brain, *J. Neurosci.* 6:3177-3188.

Chang, H.T., Wilson, C.J., and Kitai, S.T., 1982, A Golgi study of rat neostriatal neurons: light micoscopic analysis, *J. Comp. Neurol.* 208:107-126.

Crosby, E.C., and Humphrey, T., 1941, Studies of the vertebrate telecephalon II. The nuclear pattern of the anterior olfactory nucleus, tuberculum olfactorium and the amygdaloid complex in adult man, *J. Comp. Neurol.* 74:309-352.

Chronister, R.B., Sikes, R.W., Trow, T.W., and DeFrance, J.F., 1981, The organisation of nucleus accumbens, in: "The Neurobiology of the Nucleus Accumbens," R.B. Chronister and J.F. De France, eds., Haer Institute for Electrophysiological Reasearch, Brunswick, pp. 97-146

Doty, R.L., 1991, Olfactory function in neonates, in: "The Human Sense of Smell," D.G. Laing, R.L. Doty, W. Breipohl, eds, Springer-Verlag, Berlin Heidelberg, pp. 155-161.

Dubois, A., Savasta, M., Curet, O., and Scatton, B., 1986, Autoradiographic distribution of the D1 agonist [3H] SKF 38393 in the rat brain and spinal cord. Comparison with the distribution of D2 dopamine receptors, *Neurosci.* 19:125-137.

Engen, T., 1982, "The Perception of Odors," Academic Press, New York.

Everett, P.W., Kemm, R.E., and McKenzie, J.S. , 1984, Neuronal activity in basal ganglia output nuclei and induced hypermotility, in: "The Basal Ganglia, Structure and Function," J.S. McKenzie, R.E. Kemm and L.N. Wilcock, eds., Plenum Press, New York, pp. 235-245.

Gerfen, C.R., 1992, The neostriatal mosaic: multiple levels of compartmental organization in the basal ganglia, *Ann. Rev. Neurosci.* 15:285-320.

Hedreen, J.C. , 1981, Neurons of the nucleus accumbens and other striatal regions in the rat, in: "The Neurobiology of the Nucleus Accumbens," R.B. Chronister, J.F. De France, eds, Haer Institute for Electrophysiological Research, Brunswick pp. 82-96.

Heimer, L., VanHoesen, G.W., and Rosene, D., 1977, The olfactory pathway and the anterior perforated substance in the primate brain, *Internat. J. Neurol.* 12:42-52.

Heimer, L. ,1978, The olfactory cortex and the ventral striatum, in: "Limbic Mechanisms," K.E. Livington and O. Hornykiewiiz, eds., Plenum Press, New York, pp. 95-187.

Heimer, L., and Alheid, G.F. , 1991, Piecing together the puzzle of basal forebrain anatomy, in: "The Basal Forebrain, Anatomy to Function," T.C. Napier, P.W. Kalwas and I. Hanin, eds., Plenum Press, New York, pp. 1-42.

Heimer, L., Zaborsky, L., Zahm, D.S., and Alheid, G.F. , 1987, The ventral striatopallido-thalamic projection I: The striatopallidal link oringinating in the striatal parts of the olfactory tubercle, *J. Comp. Neurol.* 255:571-591.

Jacobs, M.S., and Jensen, A.V. , 1964, Gross aspects of the brain and a fiber analysis of cranial nerves in the great whale, *J. Comp. Neurol.* 123:55-72.

Jacobs, M.S., Morgane, P.J., and McFarland, W.L., 1971, The anatomy of the brain of the bottlenose dolphine (Tursiops truncatus). Rhinic lobe (*Rhinencephalon*), *J. Comp. Neurol.* 141:205-272.

Johnston, J.B., 1914, The nervus terminalis in man and mammals, *Anat. Record*, 8:185-198.

Kuo, H., and Chang, H.T., 1992, Ventral pallido-striatal pathway in the rat brain: a light and electron microscopic study, *J. Comp. Neurol.* 321:626-636.

Macrides, F., Davis, B.J., Young, W.M., Nadi, N.S., and Margolis, F.L., 1981, Cholinergic and catecholaminergic afferents to the olfactory bulb in the hamster: A neuroanatomical, biochemical, and histochemical investigation, *J. Comp. Neurol.*, 203:495-514.

Meyer, G., Gonzales-Hernandez, T., Carsillo-Padilla, F., and Ferres-Torres, R., 1989, Aggregations of granule cells in the basal forebrain (islands of Calleja): Golgi and cytoarchitectonic study in different mammals including Man, *J. Comp. Neurol.* 284:405-428.

Meyer, G., and Wahle P., 1986, The olfactory tubercle of the cat I. Morphological components, *Exp. Brain Res.* 62:515-527.

Millhouse, O.E. , 1987, Granule cells of the olfactory tubercle and the question of the islands of Calleja, *J. Comp. Neurol.* 265:1-24 .

Millhouse, O.E., and Heimer, L, 1984, Cell configuration in the olfactory tubercle of the rat, *J. Comp. Neurol.* 228:571-597.

Morgane, P.J., Jacobs, M.S., and McFarland, W.L., 1980, The anatomy of the brain of bottlenose dolphin (*Tursiops truncatus*). Surface configurations of the telencephalon of the bottlenose dolphin with comparative anatomical observations in four other cetacean species, *Brain Res. Bull.* 5, Suppl. 3:1-107.

Mogenson, G.J., Jones, D.L., and Yim, C.Y., 1980, From motivation to action: functional interface between the limbic system and the motor system, *Progress in Neurobiol.* 14:69-97.

Newman, R., and Winans, S.S., 1980a, An experimental study of the ventral striatum of the golden hamster. I. Neuronal connections of the nucleus accumbens, *J. Comp. Neurol.* 191:167-192.

Newman, R., andWinans, S.S., 1980b, An experimental study of the ventral striatum of the golden hamster. II. Neuronal connections of the olfactory tubercle, *J. Comp. Neurol.* 191:193-212.

Price, J.L., 1990, Olfactory System, *in:* "The Human Nervous System," G. Paxinos, ed., Academic Press, San Diego, pp. 979-998.

Schmidt, H.J., and Beauchamp, G.K., 1992, Human olfaction in infancy and early childhood, *in:* "Science of Olfaction," M.J. Serby and K.L.Chodor, eds., Springer-Verby, New York, pp. 378-395.

Slotnik, B.M., 1990, Olfactory perception, *in:* "Comparative Perception Vol I Basic Mechanisms," M.A. Berkley and W.C. Stebbins, eds., John Wiley, New York, pp. 155-214.

Stephan, H., and Andy, O.J., 1982, Anatomy of the limbic system, *in:* "Stereotaxy of the Human Brain," G. Schaltenbrand and A.E. Walker, eds., Thieme-Stratton, New York, pp. 269-292.

Stoddart, D.M., 1990, "The Scented Ape. The Biology and Culture of Human Odour," Cambridge University Press, Cambridge.

Tagaki, H., Somogyi, P., and Smith, A.D., 1984, Aspiny neurons and their local axons in the neostriatum of the rat: a correlated light and electron microscopic study of Golgi-impregnated material, *J. Neurocytol.* 13:239-265.

Turner, B., Gupta, K.C., and Mishkin, M., 1978,. The locus and cytoarchitecture of the projection areas of the olfactory bulb in Macaca mulatta, *J. Comp. Neurol.* 177:381-396.

Walaas, S.I., and Onimet, C.C., 1989, The ventral striatopallidal comple: an immunocytochemical analysis of medium sized striatal neurons and striatopallidal fibres in the basal forebrain of the rat, *Neurosci.* 28:663-672.

Wahle, P., and Meyer, G., 1986, The olfactory tubercle of the cat II. Immunohistochemical compartmentation, *Exp. Brain Res.* 62:528-540.

West, C.H.K., Clancy, A.N., and Michael, R.P., 1992, Enhanced responses of the nucleus accumbens neurons in male rats to normal odors associated with sexually receptive females, *Brain Res.* 585:49-55.

Wamsley, J.K., Gehlert, D.R., Filloux, F.M., and Dawson, T.M., 1989, Comparison of the distribution of D_1 and D_2 receptors in the rat brain, *J. Chem. Neuroanat.* 2:119-137.

Whishaw, I.Q., and Tomie, J-A., 1989, Olfaction directs skilled forelimb reaching in the rat, *Behav. Brain Res.* 32:11-21.

Wysocki, C.J., and Meredith , M., 1987, The vomeronasal system, *in:* "The Neurobiology of Taste and Smell," T.E. Finger, W.L. Silver, eds., Wiley, New York, pp. 125-150.

Young, W.S., Alheid, G., and Heimer, L., 1984, The ventral pallidal projection to the mediodorsal thalamus: a study with fluorescent retrograde tracers and immunohistofluorescence, *J. Neurosci.*, 4:1626-1638.

Zahm, D.S., and Heimer, L., 1987, The ventral striatopallidothalamic projection. III. Striatal cells of the olfactory tubercle establish direct synaptic contact with ventral pallidal cells projecting to mediodorsal thalamus, *Brain Res.* 404:327-331.

Zahm, D.S., and Heimer, L., 1988, Ventral striatopallidal parts of the basal ganglia in the rat: I. Neurochemical compartmentation as reflected by the distribution of neurotensin and substance P immunoreactivity, *J. Comp. Neurol.* 272:516-535.

FROM PREPARATION TO ACTION:
INVOLVEMENT OF THE VENTRAL STRIATUM

Michèle Fabre-Thorpe and Marie-Françoise Montaron

Institut des Neurosciences (CNRS-UPMC)
Département des Neurosciences de la Vision Active
9 Quai Saint-Bernard
75005 Paris - France

INTRODUCTION

In animals, rhythmic electrocorticographic patterns (ECoG) have been used to characterize a variety of attentive states. Rougeul-Buser and her colleagues have extensively investigated the phenomenology of these ECoG rhythms, their possible behavioral and cognitive value, and the cerebral structures involved in their generation and control (Bouyer et al., 1981; Bouyer et al., 1992). They found that periods of rhythmic cortical activity were produced in a variety of situations that involved focused attention, such as when the animal sees something that is new or significant. For example, placing a live mouse in a transparent box in front of a cat so that the cat can see the mouse but not reach it, is especially effective in producing sustained of focused attention. Alternatively, focused attention can be produced by giving the cat a behavioral task to perform. These periods are associated 35-45 Hz rhythms developing in two neocortical foci of restricted spatial extension: in the frontal motor and premotor cortex (area 4γ and 6aβ) and in the posterior parietal area 5a (Bouyer et al., 1987). By analogy with rhythms recorded in humans studied in similar situations, these 35-45 Hz ECoG rhythms have been called Beta Rhythms.

Both the focused attentive behavior and the concomitant beta rhythms have been shown to be, at least partly, under dopaminergic control of a complex circuit involving the ventral tegmental area (Montaron et al., 1982) and one of its main targets, namely the ventral striatum or accumbens nucleus (N Acc). For instance, after bilateral kainic lesions of N Acc, cats displayed longer periods of attentive behavior in natural situations requiring focused attention (Bouyer et al., 1986); they had difficulties in disengaging their attention once a specific target or a specific task had been selected.

The potential importance of the N Acc was stressed by Mogenson who partly on the basis of its anatomical position considered that it could play an important role as functional interface between limbic and motor structures (Mogenson et al., 1980; Swanson and Mogenson, 1981). The strategic importance of its anatomical position has been largely documented in various animals such as the rat, the cat and the primate. If we consider more specifically the connections in the cat, the ventral striatum receives indeed projections from various limbic structures such as the amygdala (Royce, 1978; Groenewegen et al., 1980), and hippocampus (Groenewegen et al., 1982; Jayaraman, 1985), and sends information to various extrapyramidal motor structures such as the globus pallidus, the entopeduncular nucleus, the substantia nigra and the mesencephalic reticular formation (Groenewegen and Russchen, 1984).

A possible role of the ventral striatum in directing attention towards pertinent cues has also been supported by unit recording studies. For example, some neurons in N Acc were responsive to cues that enabled the animals to prepare for the task performance; whereas others had responses associated with reward (Rolls and Williams, 1987; Williams, 1989; Apicella et al., 1991; Rolls and Johnstone, 1992). Thus a lesion of the nucleus accumbens should interact with the preparation of a task and therefore with its execution. It was thus interesting to study with precision how such a lesion could be reflected in the performance of a task motor outcome.

In the present study we have analyzed the effect of a bilateral neurotoxic lesion of the N Acc on a period of attentive motor preparation and on the execution of the associated motor performance. The cats had to catch a target that appeared for a brief period of time after a variable delay. To catch the target and be rewarded, the cat had to focus its attention throughout the delay period in order to perform its reaching movement as quickly and accurately as possible when the target appeared. Beta rhythms were used as an attentional ECoG index to assess the quality of the focused attention displayed by the cat during the preparatory period. Motor performance was assessed in terms of speed and accuracy.

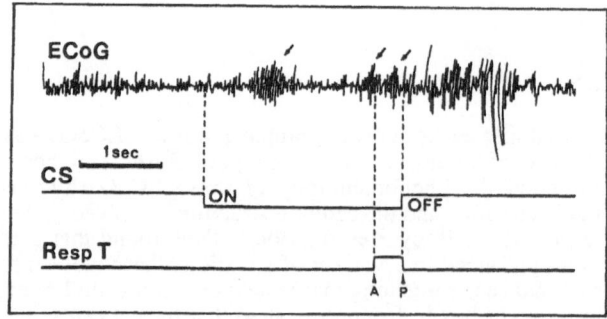

Figure 1. Recordings made during testing sessions. The upper trace represents the ECoG recorded from one frontal motor electrode showing bursts of beta rhythms (arrows) during both the preparation and the execution phase. Middle trace shows the delivery of the cue sound (CS). Lower trace shows the apparition of the target (A) and the accurate pointing movement of the cat (P). The time elapsed between A and P was considered as the response time (Resp T).

METHODS

The cats were first implanted, under deep anesthesia, with an array of electrodes within the sensorimotor cortical areas to record their ECoG. They were then overtrained on a task in which they had to point towards a moving lever visible and accessible for a short period of time (800 ms) in an aperture located in front of the animal. A trial (Figure 1) was divided into two phases: (i) preparation and (ii) execution. The phase of preparation started with the delivery of a cue sound (Figure 1, CS, ON). After a variable delay (1-5 sec), the phase of execution started with the target-lever appearing (Figure 1, A) and crossing the aperture. To be rewarded, the cat had to reach and press on the lever (Figure 1, P). If the cat missed the target, it disappeared from view and the cue sound was turned off (Figure 1, CS, OFF).

During testing, continuous ECoG signals were monitored as well as task related events. The behavioral outcome of the trial (success, missing of the target, inattention etc...) was also coded. The occurrence of beta rhythms was analyzed by the experimenter for every second during the preparation phase. The evaluation of the cat's visuo-motor performance included measurements of both accuracy (number of trials for which the target had been reached) and overall response time (Resp T). Although Resp T includes both reaction time and movement time, it is nevertheless a good index of the animal's performance.

When visuo-motor scores had stabilized over a two week period, the cats underwent a second surgical intervention under deep anesthesia, in which the N Acc was bilaterally lesioned using kaïnic acid. A glass-micropipette was stereotaxically lowered in the N Acc (using Jasper and Ajmone-Marsan atlas, 1954) and 0.3 µl of a phostphate buffered kaïnic acid solution (0.2 µg/µl) injected slowly over a period of 10mn. Side effects of brain injection of kaïnic acid were avoided by injecting the cat with a high dose of valium (7mg).

Two weeks of recovery were allowed before postoperative recording of ECoG and visuo-motor scores started. They were analyzed and statistically compared to preoperative ones.

RESULTS

A bilateral neurotoxic lesion of N Acc was carried out in four male adult animals. In two cats (S1 and S2), the lesions extended from anteriority Fr+15 to Fr+17.5, between L2 and L4 and between H+1 and H-2 (coordinates based on Jasper and Ajmone-Marsan, 1954). These lesions were almost total and nearly symmetrical. The third cat (S3) underwent a larger symmetrical lesion extending more anteriorly to Fr+19. In the last cat (S4), the lesion was more restricted on both sides and slightly asymmetrical along the rostrocaudal axis. In the present study we did not find any clear differences between the different subjects and the postoperative behavior will be considered for the overall group of cats (refer to Table 1 for individual performances).

After lesion, the animal's general behavior was recorded, but no obvious deficits could be noticed on a variety of simple clinical tasks. In contrast, we found evidence that the lesion actually led to a paradoxical improvement in the studied performance. We will first consider the ECoG and the preparatory period and second the performance of the pointing movement itself.

Nucleus Accumbens Lesions and Beta Rhythms

The normal response of the cats to the cue sound was an orientation towards the "target action field". They then stayed immobile, focusing their attention on the aperture

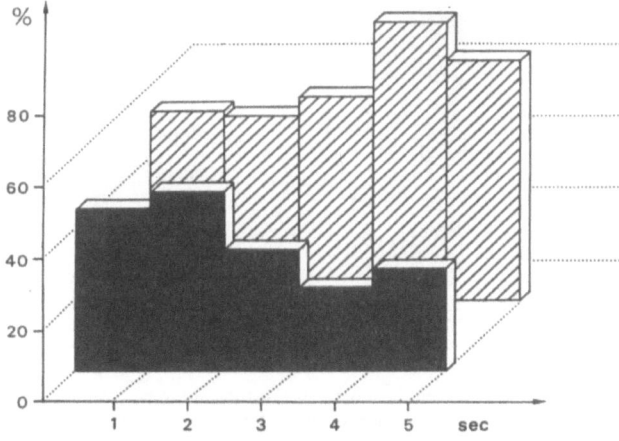

Figure 2. Effect of Nucleus Accumbens lesion on the occurrence of beta rhythms in one subject. The percentage of trials where beta rhythms were present is expressed preoperatively (black) and postoperatively (hatched) for each second of the preparatory period. Note the strong increase of beta rhythms after lesion especially during the two last seconds.

where the lever was to appear. During this preparatory period, beta rhythms were clearly seen on the frontal cortical lead. Bursts of beta rhythms were always present when the cat displayed attentive behavior toward the action field. On the other hand, the ECoG was mostly desynchronized when the cat was inattentive to the task.

After lesion (Figure 2), the first observation was a clear increase in the probability of beta rhythms. This increase was not uniform during the preparatory period; it was especially pronounced for most cats during the last seconds of expectation. The cats all appeared able to maintain sustained focused attention for longer immobile periods (e.g. more than 5sec) without disengaging from the task-goal.

Nucleus Accumbens Lesions and Visuo-Motor Performance

Trials were divided into two groups: (i) successful reaching trials (when the cat reached the target accurately), and (ii) unsuccessful reaching trials. For the overall number of successful trials, the corresponding Resp T distribution was computed. For the group involving unsuccessful reaching two cases were separately considered, the first involved those trials where the cat was totally inattentive to the task, whereas for the second, the cat was involved in the task but missed the target. The percentage of the total number of trials falling into each of these groups was evaluated pre- and post-operatively, and statistical comparisons were made using a paired Wilcoxon t-test to analyze the effect of the lesion. The statistical comparison of the pre- and post-operative Resp T distributions was done using a two-tailed Mann Whitney U test.

Global Performance. In all cats (Figure 3 and Table 1), the percentage of successful reaching trials increased after lesion. Relatively to that type of trials, the mean increase was about 15% ranging from 4% to 41%. The one-tailed paired Wilcoxon t-test showed that this increase was significant (n=4, p<.05). The percentage of unsuccessful trials corresponding to inattentive behavioral episodes (Table 1, UA) fell for three of the four cats. The fourth animal had such a low preoperatively level of inattentive trials (about 1%) that any change would have been impossible to detect. Considering the three first animals,

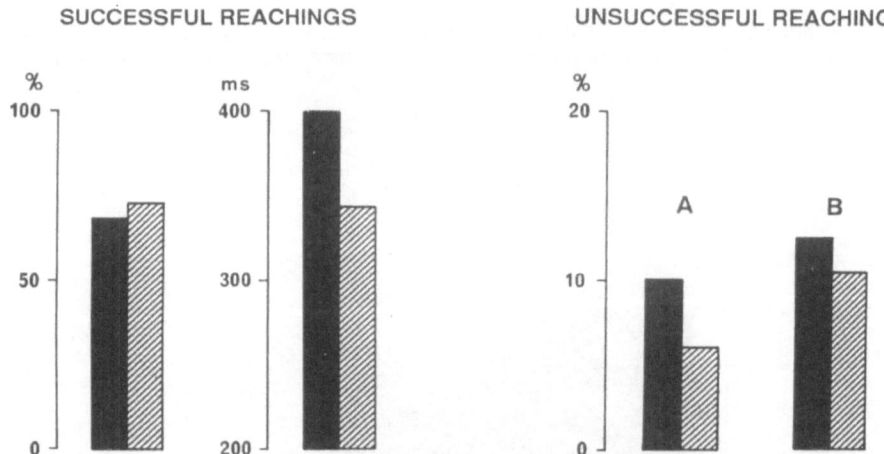

Figure 3. Effect of Nucleus Accumbens lesion on the visuo-motor performance in one subject. The preoperative performance is shown in black whereas the postoperative performance is shown with hatching. On the left is shown the percentage of successful trials relatively to the total number of trials and the mean of the corresponding Resp T distribution in milliseconds. On the right for the group of unsuccessful trials is shown, in A, the percentage of inattentive trials and in B the percentage of trials where the target was missed, both relatively to the total number of trials.

300

the proportion of inattentive trials was more than halved (decrease of 57%) after lesion. Finally for the percentage of unsuccessful trials during which the cat missed the target (Table 1, UB) the lesion had no clear effect.

Speed of Execution. The statistical comparison of the pre and postoperative Resp T distributions, revealed a clear and reliable shift of the response times towards shorter values. The mean response time decreased in all cats (Figure 3 and Table 1), indicating that the movement was either triggered earlier or performed more quickly. The statistical comparison of the pre- and post-operative Resp T distributions using a two-tailed Mann Whitney U test showed that this postoperative shift was significant for all cats (ranging from z< .01 to z< .0001).

Table 1. ECoG and Visuomotor Performance: pre- and post-operaative values for each subject (S1-S4). (1) β rhythms: % of trials in which beta rhythms were present during thetwo last seconds of the preparatory period. (2) S Trials: % of successful trials. (3) Resp T: mean and corresponding standard error of the response time distribution for successful trials. (4) UA Trials: % of unsuccessful trials during which the cat was inattentive. (5) UB Trials: % of unsuccessful trials during which the cat missed the target. All % are given relatively to the total number of trials.

	β Rhythms		S Trials		Resp T (msec)		UA Trials		UB Trials	
	Pre	Post	Pre	Post	Pre	Post	Pre	Post	Pre	Post
S1	23%	74%	68%	72%	398 ±13	342 ±13	10%	6%	12%	10%
S2	31%	66%	79%	82%	362 ±7	311 ±6	1%	2%	3%	8%
S3	16%	46%	58%	82%	384 ±16	358 ±13	20%	9%	16%	3%
S4	51%	56%	76%	82%	535 ±10	510 ±10	8%	2%	3%	5%

DISCUSSION

After bilateral neurotoxic lesion of the nucleus accumbens, the animals showed an increase in the amount of beta rhythms during the preparatory period that was associated with an increase in their ability to sustain episodes of focused attention. In parallel, there was a remarkable improvement in both the accuracy and speed of performance in the visuo-motor task.

Most studies have reported an absence of simple motor or sensori-motor deficits in learned tasks after ventral striatal lesions (Amalric and Koob, 1987; Robbins et al., 1990). Indeed, there has been at least one report describing an improvement in discriminative accuracy when using visual stimuli of low brightness (Cole and Robbins, 1989), which the authors interpreted in terms of the speed-accuracy trade-off (Fitts, 1954; Pew, 1969) or on the basis of the animals hypoactivity. According to their first interpretation, an increase of movement velocity would induce an increase of spatial errors, whereas an increase in accuracy would be associated with a slowing down of the movement execution. In the present experiment, however, the improvement of the visuo-motor scores cannot be explained in this way, as all four subjects showed postoperative improvement in both speed and accuracy, despite a long period of preoperative training. Likewise, hypoactivity cannot be the answer as, following the lesion, the proportion of trials inducing no movement was decreased and the speed of performance of the successful movements was increased.

Our interpretation of these results takes in account the involvement of the ventral striatum in the pathway underlying focused attention (Bouyer et al., 1986) together with its position as an interface between limbic and motor brain structures. The ECoG showed an increase of beta rhythms throughout the period where the cat was waiting for the target to appear, but especially towards the end of this preparatory period. Thus, whereas normal cats displayed a transient attentive period, lesioned subjects could maintain higher level of attention for longer periods of time. The improvement of motor scores is probably not directly interpretable in terms of motor execution but may be linked to a better state of preparation of the subject to perform the task related response. However, this apparent "improvement" may well be specific to the testing situation and may have to be considered as pathological because of its counterpart, that is an impairment in switching attention from the ongoing task to other behavioral sequences. Such an inability would be very disabling for a free animal in its natural environment especially in predator-prey relationships. This finding can be compared, in a way, with the difficulty in suppressing ongoing behavior and in switching to another that was shown after 6-hydroxydopamine lesions in rats (Robbins and Koob, 1980; Taghzouti et al., 1985). Although this hypothesis needs to be objectively tested (e.g. with a precise pre- and post-lesional investigation of the cat's behavioral response to the sudden introduction of unexpected stimuli during completion of the task), the marked decrease of the proportion of unsuccessful trials due to inattention showed, in our testing situation, that lesioned animals were less easily diverted from their ongoing behavioral goal.

One can wonder about the role played by the mesolimbic dopaminergic system innervating the ventral striatum. The kainic acid lesions used in the present study damaged the intrinsic cellular organization of the nucleus accumbens and disrupted the overall output from this structure. Most other studies have used 6-OHDA that lesioned the main N Acc dopaminergic afferents from the ventral tegmental area. Thus the results obtained using these two techniques are difficult to compare. Many tests used to study the role of the ventral striatum are widely considered as tests of limbic system dysfunction; they require choice between two sequences or at least behavioral flexibility in switching from one response to another. In these conditions, dopaminergic lesions induced deficits in switching behavior that have been described as a form of perseveration (Louilot et al., 1987). Recently, NMDA-receptors in N Acc have been reported to be involved in this ability to switch between cue-directed behaviors (van den Bos et al., 1992). Perseveration can also be used to explain the inability of dopaminergic or excitotoxic lesioned animals to extinguish unrewarded responding. The omission of reward appears much less efficient in such subjects (Annett et al., 1989; Robbins et al., 1990).

The deficits induced by ventral striatal lesions have been interpreted in terms of a weakening of the degree of attentional control exerted by the appropriate stimuli (Carli et al., 1989; Reading et al., 1991). Our view is that the ventral striatum is more specifically involved in balancing the amount of attention allocated to the performance of a task and the amount available for processing any other external stimuli. Such balance is of primary importance to decide whether it is appropriate to pursue the ongoing task or to generate an appropriate behavioral response to an unexpected stimulus. In our study, the task response requires the processing of one stimulus and involves only a forepaw movement. In such a task, the lack of distractibility would be an advantage, the subject would allocate more attention to the stimulus; this would improve the processing of the sensori-motor information and ultimately the motor execution itself.

REFERENCES

Amalric, M., and Koob, G.F., 1987, Depletion of dopamine in the caudate nucleus but not in nucleus accumbens impairs reaction-time performance in rats, *J. Neurosci.* 7:2129-2134.

Annett, L.E., McGregor, A., and Robbins, T.W., 1989, The effects of ibotenic acid lesions of the nucleus accumbens on spatial learning and extinction in the rat, *Behav. Brain Res.* 31:231-242.

Apicella, P., Ljungberg, T., Scarnati, E., and Schultz, W., 1991, Responses to reward in monkey dorsal and ventral striatum, *Exp. Brain Res.* 85(3):491-500.

Bouyer, J., Montaron, M., Buser, P., Durand, C., and Rougeul, A., 1992, Effects of mediodorsalis thalamic nucleus lesions on vigilance and attentive behaviour in cats, *Behav. Brain Res.* 51:51-60.

Bouyer, J., Montaron, M., and Rougeul, A., 1981, Fast fronto-parietal rhythms during combined focused attentive behaviour and immobility in cat: cortical and thalamic localizations, *EEG Clin. Neurophysiol.* 51:244-252.

Bouyer, J.J., Montaron, M.F., Fabre-Thorpe, M., and Rougeul, A., 1986, Compulsive attentive behavior after lesion of the ventral striatum in the cat: A behavioural and electrophysiological study, *Exp Neurol.* 92:698-712.

Bouyer, J.J., Montaron, M.F., Vahné, J.M., Albert, M.P., and Rougeul, A., 1987, Anatomical localization of cortical beta rhythms in cats, *Neuroscience* 22:863-870.

Carli, M., Jones, G.H., and Robbins, T.W., 1989, Effects of unilateral dopamine depletion from the dorsal and ventral striatum on visual neglect in the rat: a neural and behavioural analysis, *Neuroscience* 29:309-327.

Cole, B.J., and Robbins, T.W., 1989, Effects of 6-hydroxydopamine lesions of the nucleus accumbens septi on performance of a 5-choice serial reaction time task in rats: implications for theories of selective attention and arousal, *Behav. Brain Res.* 33:165-179.

Fitts, P.M., 1954, The information capacity of the human motor system in controlling the amplitude of movement, *J. Exp. Psychol.* 67:103-112.

Groenewegen, H.J., Becker, N.E.H.M., and Lohman, A.H.M., 1980, Subcortical afferents of the nucleus accumbens speti in the cat, studied with retrograde axonal transport of horseradish peroxidase and bisbenzimid, *Neuroscience* 5:1903-1916.

Groenewegen, H.J., Room, P., Witter, M.P., and Lohman, A.H.M., 1982, Cortical afferents of the nucleus accumbens in the cat studied with anterograde and retrograde transport techniques, *Neuroscience* 7:977-995.

Groenewegen, H.J., and Russchen, F.T., 1984, Organization of the efferent projections of the nucleus accumbens to pallidal, hippothalamic, and mesencephalic structures: a tracing and immunohistochemical study in the cat, *J. Comp. Neurol.* 223:347-367.

Jayaraman, A., 1985, Organization of thalamic projections in the nucleus accumbens and the caudate nucleus in cats and its relation with hippocampal and other subcortical afferents, *J. Comp. Neurol.* 231:396-420.

Louilot, A., Taghzouti, K., Deminiere, J.M., Simon, H., and Le Moal, M., 1987, Dopamine and behavior: functional and theoretical considerations, *in*: "Neurotransmitter interactions in the basal ganglia," M. Sandler et al., eds., Raven Press, New-York, pp. 193-204.

Mogenson, G.J., Jones, D.L., and Yim, C.Y., 1980, From motivation to action: functional interface between the limbic system and the motor system, *Prog. Neurobiol.* 14:69-97.

Montaron, M., Bouyer, J., Rougeul, A., and Buser, P., 1982, Ventral mesencephalic tegmentum (VMT) controls electrocortical beta rhythms and associated attentive behaviour in the cat, *Behav. Brain Res.* 6:129-145.

Pew, R.W., 1969, The speed-accuracy operating characteristic, *Acta Psychol.* 30:16-26.

Reading, P.J., Dunnett, S.B., and Robbins, T.W., 1991, Dissociable roles of the ventral, medial and lateral striatum on the acquisition and performance of a complex visual stimulus-response habit, *Behav. Brain Res.* 45:147-161.

Robbins, T.W., Giardini, V., Jones, G.H., Reading, P., and Sahakian, B.J., 1990, Effects of dopamine depletion from the caudate-putamen and nucleus accumbens septi on the acquisition and performance of a conditional discrimination task, *Behav. Brain Res.* 38:243-261.

Robbins, T.W., and Koob, G.F., 1980, Selective disruption of displacement behaviour by lesions of the mesolimbic dopamine system, *Nature* 285:409-412.

Rolls, E.T., and Johnstone, S., 1992, Neurophysiological analysis of striatal function., *in*: "Neuropsychological disorders associated with subcortical lesions," C. Wallesch and G. Vallar, eds., Oxford University Press, Oxford, pp. 61-97.

Rolls, E.T., and Williams, G.V., 1987, Neuronal activity in the ventral striatum of the primate, *in*: " The basal Ganglia II - Structure and function - current concepts," M.B. Carpenter and A. Jayaraman, eds., Plenum, New York, pp. 349-356.

Royce, G.J., 1978, Cells of origin of subcortical afferents to the caudate nucleus: a horseradish peroxidase study in the cat, *Brain Res.* 153:465-475.

Swanson, L.W., and Mogenson, G.J., 1981, Neural mechanisms for the functional coupling of autonomic, endocrine and somatomotor responses in adaptive behavior, *Brain Res. Rev.* 3:1-34.

Taghzouti, K., Simon, H., Louilot, A., Herman, J.P., and Le Moal, M., 1985, Behavioral study after local injection of 6-hydroxydopamine into the nucleus accumbens in the rat, *Brain Res.* 344:9-20.

van den Bos, R., Charria Ortiz, G.A., and Cools, A.R., 1992, Injections of the NMDA-antagonist D-2-amino-7-phosphonoheptanoic acid (AP-7) into the nucleus accumbens of rats enhance switching between cue-directed behaviours in a swimming test procedure, *Behav. Brain Res.* 48:165-170.

Williams, G.V., 1989, Neuronal activity in the primate caudate nucleus and ventral striatum reflects the association between stimuli determining behavior, *in*: "Neuronal mechanisms in disorders of movement," A. R. Crossman and M. A. Sambrook, eds., John Libbey, London, pp. 63-73.

ACTIVITY OF MONKEY STRIATAL AND DOPAMINE NEURONS DURING THE PERFORMANCE OF DELAYED RESPONSE TASKS

Wolfram Schultz, Paul Apicella, Tomas Ljungberg and Ranulfo Romo

Institut de Physiologie
Université de Fribourg
CH-1700 Fribourg
Switzerland

INTRODUCTION

The striatum (caudate nucleus and putamen) lies at the crossroads of major forebrain centers involved in the organization of behavioral output. It receives afferents from limbic structures and from association and motor cortical areas dealing with the cognitive and motivational control of behavior, the preparation of behavioral acts and the control of ongoing movements. Since a major part of basal ganglia output is directed back to the frontal lobe, the functions of the striatum are intimately linked to those of prefrontal, premotor and motor cortex. The striatum is also innervated by midbrain dopamine neurons which are involved in basic behavioral activating processes and goal-directed behavior. The study of behavioral deficits in Parkinsonian patients and experimentally lesioned animals demonstrates that the intact striatal dopamine innervation is important for the expression of striatal behavioral control functions.

A widely used paradigm for probing the temporal organization of behavior by the frontal cortex consists of various forms of tasks involving temporal delays (Jacobsen and Nissen, 1937; Passingham, 1975; Mishkin and Manning, 1978; Sawaguchi et al., 1989). In delayed response tasks, often considered as prototypic behavior planning paradigms, subjects are presented with external stimuli indicating different response requirements and must remember the information contained in those stimuli for some delay (usually a few seconds) before carrying out the appropriate response. Extensive experience in these tasks renders the individual components predictable to the subject, such that a particular task event serves as reliable predictor for subsequent components. In this way, subjects are able to prepare individual behavioral reactions and expect external stimuli, including reward.

Many neurons in the prefrontal, premotor, supplementary motor and even the primary motor cortex show sustained activity during the delay period during which task specific information is retained in working memory (Fuster, 1973; Kubota et al., 1974; Tanji and Evarts, 1976; Tanji et al., 1980; Weinrich and Wise, 1982). The heavy and highly specific projections linking the frontal cortex to different regions of the basal ganglia (Goldman and Nauta, 1977; Yeterian and Van Hoesen, 1978; Percheron et al., 1984; Selemon and Goldman-Rakic, 1985; Arikuni and Kubota, 1986) make it likely that these nuclei are engaged in processes underlying delayed response behavior. Accordingly, lesions to the head of the caudate nucleus lead to a deficit in spatial delayed responding (Bättig et al., 1960), and neurons in caudate and putamen show sustained activity during delays in various forms of delayed response tasks (Alexander, 1987; Schultz and Romo, 1988; Hikosaka et al., 1989), similar to those seen in the frontal cortex.

An intact dopaminergic innervation of the striatum and the prefrontal cortex is a prerequisite for the correct performance of delayed response tasks in monkeys, as shown by the effects of MPTP-induced striatal dopamine depletions (Schneider and Kovelowski, 1990) and local interference with prefrontal dopaminergic neurotransmission (Brozoski et al., 1979; Sawaguchi and Goldman-Rakic, 1991). However, preliminary evidence suggests that midbrain dopamine neurons may not show the sustained activity during delays typically found in the striatum and frontal cortex (Schultz and Romo, 1990), suggesting a different role in delay tasks as compared to these structures. The present report describes certain characteristics of neuronal activity in the basal ganglia in order to evaluate the relative contributions of striatal and dopamine neurons in the performance of delay tasks and assess how they may contribute to the temporal organization of behavior.

METHODS AND RESULTS

Using methods described previously (Schultz, 1986; Romo and Schultz, 1990; Apicella et al., 1992), *Macaca fascicularis* monkeys are seated in a primate chair for a few hours each weekday and trained in spatial and conditional go-nogo delayed response tasks. Extracellular activity of single dopamine neurons and of neurons in caudate and putamen is recorded with moveable microelectrodes during task performance. Electromyographic activity and eye movements are monitored during neuronal recordings through chronically implanted electrodes. Recording sites are reconstructed on histological sections of the brain after all electrophysiological data have been collected.

Neurons in the Striatum

The basic temporal structure of a delay task is defined by a brief initial instruction cue containing task-specific information, a subsequent delay period of several seconds during which explicit external information about the specific behavioral reaction is absent, followed by a trigger signal which elicits the behavioral reaction without indicating how to react. In our experiments, an instruction light either determines whether the animal needs to move its arm or refrain from moving after the delay (go-nogo task), or indicates which one of two levers needs to be touched after the delay (spatial task) (Apicella et al., 1992; Schultz and Romo, 1992). The trigger signal is a small light that is the same for the different behavioral reactions. It thus determines the time of the behavioral response without indicating which one of two possible reactions should be performed. Because the instruction cue remains on for only 1 s, the information contained in it needs to be stored in working memory until the trigger signal comes up and the subject is allowed to react. The correct reaction leads to the delivery of a drop of liquid or a morsel of food which may be delayed by a few seconds without degrading the animal's performance. Thus, the task contains at least two time intervals: from instruction to trigger, and from trigger to reward (Figure 1). Through the training, the individual task-related signals become good predictors for subsequent signals, behavioral reactions and the delivery of reward. Because the different task events are separated by time intervals of several seconds in each trial, the expectation of behaviorally significant external signals, the preparation for execution vs. inhibition of movement (go-nogo task), the preparation of movements towards different spatial positions (spatial task), and the expectation of reward should occur at different time periods and separated from each other.

Different neurons in caudate and putamen are activated separately during those time periods in which information is stored in working memory, behavioral reactions are prepared, and stimuli and reward are expected. Figure 1 shows how two neurons are activated during the intervals between the instruction and the trigger, and between the trigger and the reward, respectively. In some instances the activity between the instruction and trigger stimuli develops independently of the forthcoming behavioral reaction (Figure 2), suggesting that it is related to the expectation of the trigger stimulus that is common to the different behavioral situations. Other striatal neurons show a specificity for the forthcoming behavioral reaction by being activated selectively before the execution or the inhibition of movement (go-nogo task), or selectively before a movement towards a particular spatial target (Figure 3). This activity may be related to the memorized information contained in the instruction cue or to the preparation of the upcoming behavioral reaction.

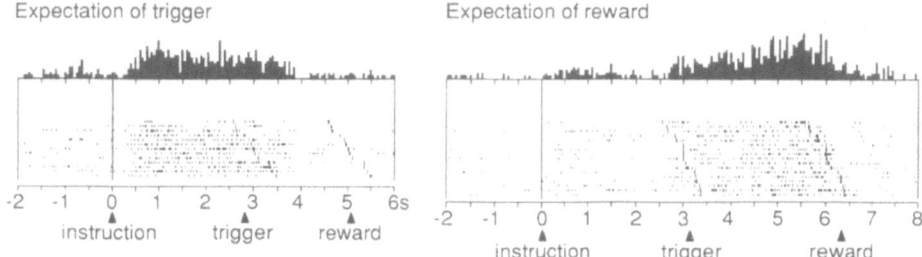

Figure 1. Sustained activity of two striatal neurons occurring differentially in the intervals between different task events of a delayed go-nogo task. **Left:** activity in a putamen neuron is increased between the instruction and trigger stimuli. **Right:** activity in a ventral striatal neuron is increased between the trigger signal and the delivery of a drop of liquid reward. Each dot in the rasters indicates the occurrence of a neuronal impulse whose distance to the vertical instruction lines corresponds to their real time intervals. Horizontal lines of dots correspond to individual trials which are displayed rank-ordered for instruction-trigger intervals. The temporal occurrences of the trigger and reward vary randomly in respect to the instruction and are shown by small vertical lines in the rasters. Histograms above the rasters are composed of the dots shown below them.

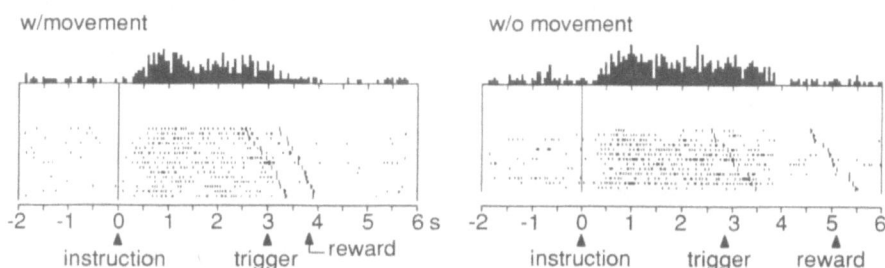

Figure 2. Activity of a putamen neuron related to the expectation of the trigger signal during performance of a delayed go-nogo task. The sustained activity occurs independently of arm movements, thus indicating a relationship to the trigger signal as common event following the instruction cue. Trials to the left and right alternate randomly during the experiment and are separated for analysis.

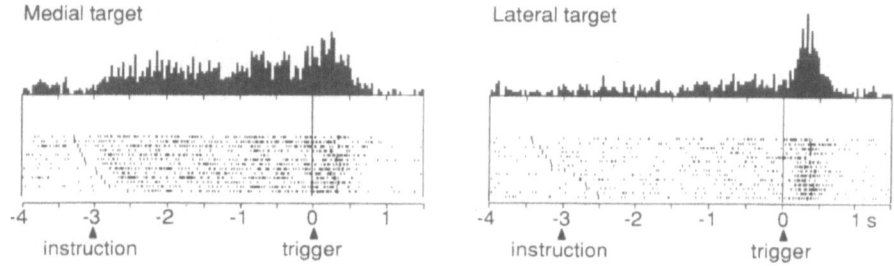

Figure 3. Activity of a caudate neuron related to the preparation of arm movement in a spatial delayed response task. The sustained activity occurs only when the instruction cue prepares for a reaching movement towards a particular target (left), whereas delay activity is absent before reaching to another target (right). The neuron in addition shows movement execution-related activity independent of the target (to the right of the trigger lines). Both targets are located contralateral to the side of neuronal recording, at 18° and 36° lateral to the midline and at eye level. Trials to the left and right alternate randomly during the experiment and are separated for analysis. The difference to the neuronal activity of the preceding figure consists in the selectivity of delay activity for a particular target, suggesting a relationship to the movement rather than to the common trigger signal.

Another class of striatal neurons is activated before the delivery of reward. These activations usually begin after the last signal preceding the reward, namely the trigger stimulus, and end immediately after reward is delivered (Figure 4). The activity often develops slowly and may even begin before the trigger stimulus (Figure 4, right), indicating that it may not be directly induced by a particular stimulus but evolves as a result of the predictions coming up while the trial advances. By contrast, the offset of this activity closely follows the delivery of reward, even when reward is delivered earlier or later during individual trials. This reward expectation-related activity is found more frequently in the ventral striatum, as opposed to the more dorsal parts of caudate and putamen. About three quarters of expectation- and preparation-related activity in the ventral striatum are related to the expectation of reward (Schultz et al., 1992). (The ventral striatum here is defined as the nucleus accumbens and the neighboring common projection fields of the amygdala and the orbitofrontal and cingulate cortex.) This particular result is interesting in the light of the supposed role of the ventral striatum in incentive learning (Fibiger and Phillips, 1986). An expectation of reward underlying the approach behavior towards conditioned, reward-predicting stimuli constitutes a central tenet of incentive learning theories (Bindra, 1968; Dickinson, 1980).

This short description shows that already in the relatively simple delay tasks a considerable variety of sustained neuronal activities is found in both caudate and putamen which is differentially related to the expectation of various predictable events, the preparation of movements, and working memory. Each individual class of activity comprises relatively small fractions (2-10% of >2000 quantitatively tested neurons), suggesting a high degree of functional specialization of sustained activity in the striatum.

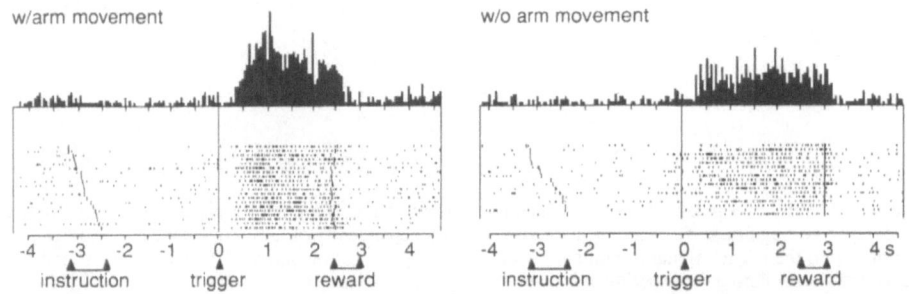

Figure 4. Activity of a putamen neuron related to the expectation of reward during performance of a delayed go-nogo task. The sustained activity occurs independently of an arm movement and thus appears to be related to the two situations, namely the delivery of a drop of liquid reward. Note the relatively slow onset of activity in the absence of arm movements (right), which contrasts with the reliable offset of activity immediately after reward in both types of trials. Supplementary tests not shown here reveal that a temporal displacement of reward results in a concordant displacement of offset of activity, further indicating a relationship to reward. Trials to the left and right alternate randomly during the experiment and are separated for analysis.

Midbrain Dopamine Neurons

Dopamine neurons of the substantia nigra (A9) and adjoining groups A8 and A10 discharge initially negative or positive impulses at low frequencies (0.5-8.5 imp/s) and with polyphasic wave forms of relatively long durations (1.8-5.5 ms). In these characteristics, DA neurons contrast with reticulata neurons of the substantia nigra discharging impulses of <1.1 ms duration at median rates of 70-90 impulses/s, with a few neurons discharging short impulses (<1.0 ms) at low rates, and with presumptive fibers discharging very short impulses (0.1-0.3 ms; all measures with 100 Hz lower cut off filtering). Dopamine neurons with these electrophysiological characteristics are antidromically activated from caudate or putamen (Schultz and Romo, 1987) and are depressed by low systemic doses of the dopamine receptor agonist apomorphine (Schultz, 1986).

In simple behavioral tasks, dopamine neurons respond to reward delivered unexpectedly or without explicit predictive signals. This occurs for example when a morsel of food is encountered during self-initiated hand movements into a covered box (Romo and Schultz, 1990) or when a drop of liquid reward is delivered for a correct behavioral reaction during the learning of a new task (Ljungberg et al., 1992). When task performance is better established, dopamine neurons respond to learned reward-predicting stimuli, for example a small light eliciting an arm movement in a reaction time task, but the response to the reward is lost. After extensive overtraining of the task, the responsiveness of dopamine neurons to the various task stimuli is considerably reduced, both in terms of numbers of responding neurons and of response magnitudes in individual neurons.

Dopamine neurons are studied in the same go-nogo and spatial delay tasks as striatal neurons (Schultz and Romo, 1990; Schultz et al., 1993). They respond to both the

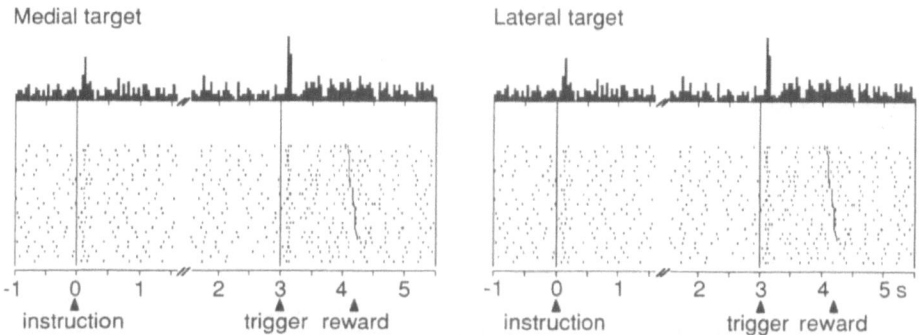

Figure 5. Responses of a dopamine neuron of the pars compacta of substantia nigra (group A9) to instruction and trigger signals in a spatial delayed response task. In contrast to striatal neurons, these responses are time-locked to the preceding stimulus, are phasic, occur to both signals and are unselective for the position of the target of reaching. Consistent sustained changes of activity during the instruction-trigger and trigger-reward intervals are absent. Both targets are located contralateral to the side of neuronal recording, at 18° and 36° lateral to the midline and at eye level. Trials to the left and right alternate randomly during the experiment and are separated for analysis.

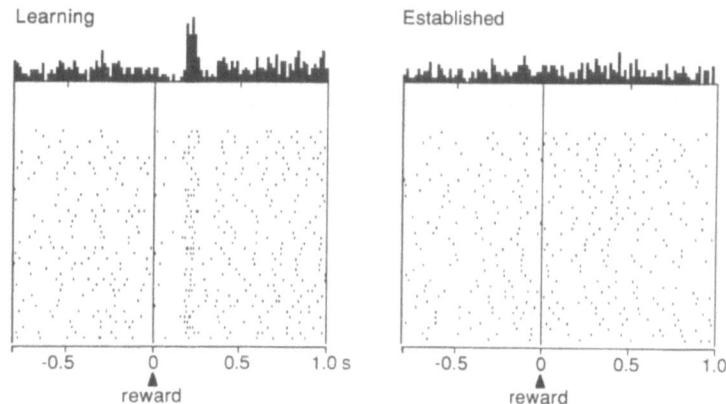

Figure 6. Activity of two A10 dopamine neurons following the delivery of reward during and after learning the last step for a spatial delayed response task (left and right, respectively). Trials are pooled over the medial and lateral targets, as described for the preceding figure.

instruction and the trigger stimuli with phasic responses occurring at latencies of 80-100 ms and lasting <200 ms. Responses are unselective for execution vs. inhibition of movement in the go-nogo task and for the direction of movement in the spatial task (Figure 5). Dopamine neurons also respond to reward while animals learn the delay task in consecutive steps, but loose the reward response as soon as each step is well established (Figure 6). Like in the other tasks, the responses are rather homogeneous: the different stimuli elicit very similar responses, and different dopamine neurons either respond in the same way or not at all. The fractions of responding neurons in the delay tasks comprise 25-49% of the A8, A9 and A10 catecholamine cells and thus are considerably lower than for simple reaction time tasks (50-85%). Although not studied during extensive overtraining in delay tasks, responses to the conditioned stimuli, particularly to the instruction cue, have a tendency to become smaller and less frequent with progressing experience in the task. Responsive neurons are found with approximately equal proportions in the different dopamine neuron groups, except for reward responses showing a slightly higher occurrence in group A10 during learning. In contrast to the reproducible phasic responses, dopamine neurons do not show consistent sustained activity during the time periods between the different task events that would resemble that found in the striatum (Figure 5).

DISCUSSION

These studies reveal several important differences between dopamine and striatal neurons during delay tasks (Figure 7). (1) Consistent sustained delay activity is frequent in striatal neurons but is absent in dopamine neurons. (2) Instead, a large fraction of dopamine neurons shows homogeneous, phasic responses to conditioned stimuli. The responses are not specific for spatial components of movements and are reduced with extended task experience. Although phasic responses are also found in striatal neurons, they constitute only a small fraction of striatal activity, are often selective for the behavioral significance of the stimulus and do not usually occur to different stimuli in the same neuron (Schultz and Romo, 1992). (3) The majority of dopamine neurons respond to reward during learning or when it occurs unexpectedly, but the response is lost when the task is acquired and the reward is predicted by a stimulus. By contrast, only few striatal neurons respond to reward, and these responses are observed during well established task performance (Apicella et al., 1991). Thus, the described sustained activities of striatal neurons and the phasic responses of dopamine neurons play considerably different roles in delay tasks. It is suggested that striatal neurons process information related to the representation of task events specifically related to the performance of delay tasks, whereas dopamine neurons respond to the salient stimuli important for task performance. In the present delay tasks, salient stimuli are the conditioned instruction and trigger signals because they precede and predict reward, and the delivery of reward during learning. Thus, dopamine neurons contribute major motivational and attentional components that are crucial for task performance, and possibly even more so for task learning, but are not specifically encoding delay tasks.

The sustained activities of striatal neurons occur in the intervals between individual task events, notably between the instruction cue and the trigger signal or behavioral reaction, and between the trigger signal and reward. There are two possible interpretations for these activities. The sustained activity may reflect a working memory process by which information about the event preceding the activity is kept in a temporary store until it is used for the subsequent behavioral reaction. This pertains particularly to the go-nogo-specific and the direction-specific activity between the instruction cue and the behavioral reaction to the trigger stimulus. This interpretation has been proposed for similar activities in the prefrontal and the inferotemporal cortex (Fuster, 1973; Kubota et al., 1974; Fuster and Jervey, 1982; Miyashita and Chang, 1988; Funahashi et al., 1989). In contrast to these activities developing directly in response to a particular signal to be memorized, many sustained activities in the striatum appear to be related to the occurrence of a forthcoming task event. This is suggested by their often sluggish onset and their better temporal relationship to that event, in particular by the offset of activity that covaries closely with temporal displacement of the event (Apicella et al., 1992; Schultz et al., 1992). It is suggested that the animal, through its experience in the task, has memorized the individual task components and the reactions to be performed in each situation, such that the complete task is internally represented in the animal's brain. A particular signal that comes up in such a situation evokes a central representation of a subsequent task event, thus allowing the animal to prepare for or expect that task event. The presently described sustained activities may constitute neuronal correlates of expectation and preparation of the different task events, suggesting that striatal neurons have access to central representations of individual task components. In analogy to working memory engaged in the short term storage, use and deletion of information, the representations of individual task events in these tasks are evoked for short time spans of a few seconds and then deleted from current processing. However, both processes are based upon rules that are stored through experience over a longer time. Thus, the observed striatal activity appears to reflect both the short term storage of information and the short term evocation of long term stored information.

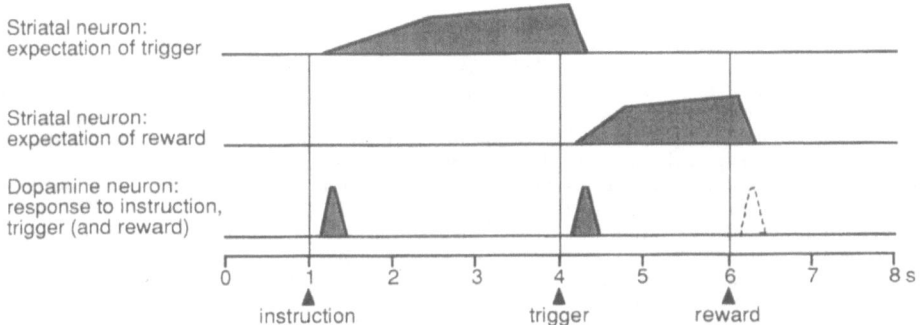

Figure 7. Schematic comparison of neuronal activities between striatal and dopamine neurons during performance of delayed response tasks. Whereas striatal neurons display sustained activity related selectively to the occurrence of predictable forthcoming events (top), dopamine neurons respond phasically to several learned stimuli that are crucial for task performance and, particularly, predict reward. During learning of delayed response tasks, dopamine neurons respond to the delivery of liquid reward (interrupted line). This response is lost when task performance is established. Thus, striatal neurons process specific task-related information and have access to central representations of task events, whereas dopamine neurons signal the motivational consequences of behavioral performance.

Dopamine neurons apparently play a dual role in the performance of cognitive tasks. (1) Unlike striatal neurons, dopamine neurons do not show changes of activity in relation to working memory, expectation of events and preparation of reactions. However, the performance of delay tasks engaging these processes is deficient when dopaminergic neurotransmission is impaired (Brozoski et al., 1979; Sawaguchi and Goldman-Rakic, 1991), and activity in prefrontal neurons related to these processes is crucially dependent upon an intact dopaminergic neurotransmission (Sawaguchi et al., 1990a, b). This suggests that dopamine released by spontaneous impulses or via local presynaptic interactions in the innervated structures exerts an enabling effect that allows striatal and prefrontal neurons to process information specific for the performance of delay tasks. (2) The phasic responses to reward-related stimuli transmit information to postsynaptic structures indicating the presence of these objects of central importance. During learning, the observed responses to reward may participate in labelling arbitrary stimuli with incentive properties. They may do this by inducing plastic processes in the postsynaptic structures, thereby contributing to the acquisition of the task. With established task performance, the responses to conditioned, reward-predicting stimuli may serve to consolidate the plastic changes. With overtraining, when task performance becomes automatic and is centrally represented in a fixed, procedural form, informations about reward and reward-predicting stimuli loose their central importance and the neuronal responses fade away. Thus, phasic responses of dopamine neurons may play a role in modifying neuronal circuits in the striatum and the frontal cortex during the learning of tasks directed at the obtainment of reward. By contrast, tonically released dopamine appears to be necessary for maintaining the functions of circuits underlying these tasks. This dual role of dopaminergic influences is not limited to delay tasks and could apply to other kinds of behavior depending on the postsynaptic structure innervated, such as the learning and execution of movements in the case of the sensorimotor putamen.

Acknowledgements

We thank B. Aebischer, J. Corpataux, A. Gaillard, A. Pisani, A. Schwarz and F. Tinguely for technical assistance. The study was supported by the Swiss NSF (Grants 3.533-0.83, 3.473-0.86, 31-28591.90), the United Parkinson Foundation (Chicago), and by postdoctoral fellowships to P. A. by the Fyssen Foundation (Paris) and the Fondation pour la Recherche Médicale (Paris).

REFERENCES

Alexander, G.E., 1987, Selective neuronal discharge in monkey putamen reflects intended direction of planned limb movements, *Exp. Brain Res.* 67:623-634.

Apicella, P., Ljungberg, T., Scarnati, E., and Schultz, W., 1991, Responses to reward in monkey dorsal and ventral striatum, *Exp. Brain Res.* 85:491-500.

Apicella, P., Scarnati, E., Ljungberg, T., and Schultz, W.,1992, Neuronal activity in monkey striatum related to the expectation of predictable environmental events, *J. Neurophysiol.* 68:945-960.

Arikuni, T., and Kubota, K.,1986, The organization of prefrontocaudate projections and their laminar origin in the macaque monkey: A retrograde study using HRP-gel, *J. Comp. Neurol.* 244:492-510.

Bättig, K., Rosvold, H.E., and Mishkin, M., 1960, Comparison of the effects of frontl and caudate lesions on delayed response and alternation in monkeys, *J. Comp. physiol. Psychol.* 53:400-404.

Bindra, D., 1968, Neuropsychological interpretation of the effects of drive and incentive-motivation on general activity and instrumental behavior, *Psychol. Rev.* 75:1-22.

Brozoski, T.J., Brown, R.M., Rosvold, H.E., and Goldman, P.S., 1979, Cognitive deficit caused by regional depletion of dopamine in prefrontal cortex of rhesus monkey, *Science* 205:929-932.

Dickinson, A., 1980, "Contemporary animal learning theory," Cambridge University Press, Cambridge.

Fibiger, H.C., and Phillips, A.G., 1986, Reward, motivation, cognition: psychobiology of mesotelencephalic dopamine systems, *in:* "Handbook of Physiology-The Nervous System IV," American Physiological Society, Bethesda, pp 647-675.

Funahashi, S., Bruce, C.J., and Goldman-Rakic, P.S., 1989, Mnemonic coding of visual space in the monkey's dorsolateral prefrontal cortex, *J. Neurophysiol.* 61:331-349.

Fuster, J.M., 1973, Unit activity of prefrontal cortex during delayed-response performance: Neuronal correlates of transient memory, *J. Neurophysiol.* 36:61-78.

Fuster, J.M., and Jervey, J.P., 1982, Neuronal firing in the inferotemporal cortex of the monkey in a visual memory task, *J. Neurosci.* 2:361-375.

Goldman, P.S., and Nauta, W.J.H., 1977, An intricately patterned prefronto-caudate projection in the rhesus monkey, *J. comp. Neurol.* 171:369-386.

Hikosaka, O., Sakamoto, M., and Usui, S., 1989, Functional properties of monkey caudate neurons. III. Activities related to expectation of target and reward, *J. Neurophysiol.* 61:814-832.

Jacobsen, C.F., and Nissen, H.W., 1937, Studies of cerebral function in primates: IV. The effects of frontal lobe lesions on the delayed alternation habit in monkeys, *J. Comp. Physiol. Psychol.* 23:101-112.

Kubota, K., Iwamoto, T., and Suzuki, H., 1974, Visuokinetic activities of primate prefrontal neurons during delayed-response performance, *J. Neurophysiol.* 37:1197-1212.

Ljungberg, T., Apicella, P., and Schultz, W., 1992, Responses of monkey dopamine neurons during learning of behavioral reactions, *J. Neurophysiol.* 67:145-163.

Mishkin, M., and Manning, F.J., 1978, Non-spatial memory after selective prefrontal lesions in monkeys, *Brain Res.* 143:313-323.

Miyashita, Y., and Chang, H.S., 1988, Neuronal correlate of short-term memory in the primate temporal cortex, *Nature* 331:68-70.

Passingham, R., 1975, Delayed matching after selective prefrontal lesions in monkeys (*Macaca mulatta*), *Brain Res.* 92:89-102.

Percheron, G., Yelnik, J., and Francois, C., 1984, A Golgi analysis of the primate globus pallidus. III. Spatial organization of the striopallidal complex, *J. comp. Neurol.* 227:214-227.

Romo, R., and Schultz, W., 1990, Dopamine neurons of the monkey midbrain: Contingencies of responses to active touch during self-initiated arm movements, *J. Neurophysiol.* 63:592-606.

Sawaguchi, T., and Goldman-Rakic, P.S., 1991, D1 Dopamine receptors in prefrontal cortex: involvement in working memory, *Science* 251:947-950.

Sawaguchi, T., Matsumura, M., and Kubota, K., 1989, Delayed response deficits produced by local injection of bicuculline into the dorsolateral prefrontal cortex in Japanese macaque monkeys, *Exp. Brain Res.* 75:457-469.

Sawaguchi, T., Matsumura, M., and Kubota, K., 1990a, Catecholaminergic effects on neuronal activity related to a delayed response task in monkey prefrontal cortex, *J. Neurophysiol.* 63:1385-1400.

Sawaguchi, T., Matsumura, M., and Kubota, K., 1990b, Effects of dopamine antagonists on neuronal activity related to a delayed response task in monkey prefrontal cortex, *J. Neurophysiol.* 63:1401-1412.

Schneider, J.S., and Kovelowski, C.J., 1990, Chronic exposure to low doses of MPTP: I. Cognitive deficits in motor asymptomatic monkeys, *Brain Res.* 519:122-128.

Schultz, W., 1986, Responses of midbrain dopamine neurons to behavioral trigger stimuli in the monkey, *J. Neurophysiol.* 56:1439-1462.

Schultz, W., Apicella, P., and Ljungberg, T., 1993, Responses of monkey dopamine neurons during performance of a delayed response task, *J. Neurosci.* 13:900-913

Schultz, W., Apicella, P., Scarnati, E., and Ljungberg, T., 1992, Neuronal activity in monkey ventral striatum related to the expectation of reward, *J. Neurosci.* 12:4595-4610.

Schultz, W., and Romo, R., 1987, Responses of nigrostriatal dopamine neurons to high intensity somatosensory stimulation in the anesthetized monkey, *J. Neurophysiol.* 57:201-217.

Schultz, W., and Romo, R., 1988, Neuronal activity in the monkey striatum during the initiation of movements, *Exp. Brain Res.* 71:431-436.

Schultz, W., and Romo, R., 1990, Dopamine neurons of the monkey midbrain: Contingencies of responses to stimuli eliciting immediate behavioral reactions, *J. Neurophysiol.* 63:607-624.

Schultz, W., and Romo, R., 1992, Role of primate basal ganglia and frontal cortex in the internal generation of movements: comparison with instruction-induced preparatory activity in striatal neurons, *Exp. Brain Res.* 91:363-384.

Selemon, L.D., and Goldman-Rakic, P.S., 1985, Longitudinal topography and interdigitation of corticostriatal projections in the rhesus monkey, *J. Neurosci.* 5:776-794.

Tanji, J., and Evarts, E.V., 1976, Anticipatory activity of motor cortex neurons in relation to direction of an intended movement, *J. Neurophysiol.* 48:633-653.

Tanji, J., Taniguchi, K., and Saga, T., 1980, Supplementary motor area: Neuronal responses to motor instructions, *J. Neurophysiol.* 43:60-68.

Weinrich, M., and Wise, S.P., 1982, The premotor cortex of the monkey, *J. Neurosci.* 2:1329-1345.

Yeterian, E.H., and Van Hoesen, G.W., 1978, Cortico-striate projections in the rhesus monkey: The organization of certain cortico-caudate connections, *Brain Res.* 139:43-63.

PATHOPHYSIOLOGY OF THE BASAL GANGLIA
AND RELATED DISORDERS

PARKINSONIAN TREMOR IS ASSOCIATED WITH LOW FREQUENCY NEURONAL OSCILLATIONS IN SELECTIVE LOOPS OF THE BASAL GANGLIA

Hagai Bergman[1], Thomas Wichmann[2], Benny Karmon[1] and Mahlon R. DeLong[2]

[1]Department of Physiology
  The Hebrew University - Hadassah Medical School, Jerusalem, Israel 91010
[2]Department of Neurology
  Emory University, Atlanta, Georgia, USA 30322

INTRODUCTION

Parkinsonian tremor (Lance et al., 1963; Findley et al., 1981; Elble and Koller, 1990), has most often been attributed to neuronal oscillatory activity in cerebello-thalamocortical, rather than in basal ganglia-thalamocortical circuits. This notion is supported by several lines of evidence. Lesions of the substantia nigra produce tremor in monkeys only in combination with damage to cerebellorubral, and cerebellothalamic pathways (Poirier, 1960). Similarly, systemic treatment of monkeys with the neurotoxin 1-methyl-4-phenyl-1,2,3,6-tetrahydropyridine (MPTP), which causes clinical and pathological changes closely resembling human parkinsonism (e.g., DeLong, 1990), does not result in low frequency tremor in most primate species (Burns et al., 1983; Langston et al., 1984; Jenner et al., 1986). In addition, periodic oscillatory neuronal activity was found in the cerebello-thalamocortical circuits in animal models of tremor in *in-vitro* (Llinas and Yarom, 1981a,b; Jahnsen and Llinas, 1984a,b) and *in-vivo* studies (de Montigny and Lamarre, 1973; Llinas and Volkind, 1973; Lamarre and Joffroy, 1979). Furthermore, parkinsonian tremor in humans is effectively reduced by thalamic lesions, especially when placed in the Vim nucleus, the cerebellar receiving area of the ventrolateral thalamus. In fact, electrophysiological recording during lesioning surgeries revealed rhythmic neuronal discharge in Vim, in synchrony with tremor (Ohye et al., 1974; Lenz et al., 1988; Narabayashi, 1990).

Recently, however, evidence has accumulated indicating that parkinsonian tremor may be more closely related to alterations in the activity of the basal ganglia than previously suggested. In both man and the African green monkey, MPTP induces degeneration of the nigrostriatal pathway and results in parkinsonian tremor (Redmond et al., 1985; Elsworth et al., 1987; Bergman et al., 1990). The MPTP treated African green monkey is therefore to date the most accurate model of parkinsonian tremor. In addition, there is now good evidence that pallidal lesions in parkinsonian patients are at least as effective in alleviating tremor as the thalamic lesions mentioned above (Svennilson et al., 1960; Laitinen et al., 1992). Furthermore, damage of the STN, experimentally in parkinsonian monkeys, or secondary to hemorrhage in parkinsonian patients ameliorates tremor (Bergman et al., 1990; Aziz et al., 1991; Sellal et al., 1992).

The abnormalities of the neuronal activity in the basal ganglia that may be related to tremor have not been well established. Previous studies of the neuronal activity in the basal ganglia of MPTP-treated macaques, which did not express tremor, have concentrated on changes in "tonic" discharge rates, and the neuronal responses to peripheral input (Miller and DeLong, 1987; Filion and Tremblay, 1991), although oscillatory neuronal activity mainly at frequencies higher than the parkinsonian tremor frequency (10-15 Hz as opposed to 3-7 Hz) has also been described in the basal ganglia of these monkeys. The present study focused on the relationship between tremor and neuronal activity in the STN and the globus pallidus in MPTP treated African green monkeys, expressing overt parkinsonian tremor.

METHODS

In three African green monkeys, metal chambers permitting microelectrode recording were positioned over the cerebral hemisphere. The cylinder was tilted 36° anteriorly to allow penetrations in parasagittal planes to the subthalamic nucleus (target stereotaxic coordinates A = 7, H = 1, L = 5, Winters et al., 1969; Contreras et al., 1981), as well as to large parts of the internal segment of the globus pallidus (GPi). This arrangement also allowed penetrations to the rostro-medial parts of the external segment of the globus pallidus (GPe), although more lateral and caudal parts of GPe could not be sampled.

After recording in the normal state (Wichmann et al., 1993a), the monkeys were treated with MPTP (Bergman et al., 1990). Spontaneous neuronal discharge was recorded from the basal ganglia nuclei with conventional electrophysiological methods. Autocorrelation functions were calculated for delays of 500 ms (bin equal 1 ms) and 2500 ms (bin equal 5 ms) for all neurons with acceptable isolation quality. A computerized feature extracting algorithm (Karmon and Bergman, 1993) was used for detection and grading of periodic oscillatory activity. Neurons with oscillatory discharge were then grouped into cells with low (< 8 Hz) and cells with high frequency oscillations (> 8 Hz). The average wave form of extracellularly recorded action potentials (spike) of a given neuron was calculated from the last 100-256 spikes recorded from that neuron.

After completion of the experiment, the monkeys were sacrificed with an overdose of pentobarbital, and perfused transcardially with normal saline and 10% neutral formalin. The brains were then blocked, frozen and sectioned in the parasagittal plane. Alternate 50 μm sections were stained with cresyl violet, and tyrosine hydroxylase immunohistochemistry.

RESULTS

Behavioral effects of the MPTP treatment

During the course of injections with MPTP (0.3-0.4 mg/kg/day IM) and the first days thereafter, the monkeys became severely akinetic. Several days after the development of akinesia, rigidity and a low frequency (< 8 Hz) tremor developed (Figure 1). The tremor was more prominent in the proximal limbs and the trunk. It occurred usually during postural adjustments or states of increased arousal. Daily computer assisted measurements of the amount of time during which the monkeys exhibited tremor in their home cages revealed that on the average, tremor was present in more than one third of the observation periods (see Bergman et al., 1990, for technical details).

Neuronal oscillations at low frequencies

In the untreated (control) state, less than 2.5% of the total number of recorded neurons in all three structures (total number of neurons: 291 in STN, 175 in GPi, 208 in GPe) demonstrated periodic activity at low frequencies. In each case, the periodic oscillations were very weak.

Following MPTP, and the development of an intermittent low frequency tremor, 16% (53/326) of neurons in the STN and 25% (39/154) of cells in GPi showed strong 4-8 Hz rhythmic bursting (Figures 2A, 3, 4). There was no significant change in the number of neurons oscillating at low frequencies in the sampled areas of GPe (5/208 before, 8/209 after MPTP treatment). The occurrence of bursts of neuronal activity was highly correlated with tremor (Figures 3, 4), when present, but often appeared even without overt tremor.

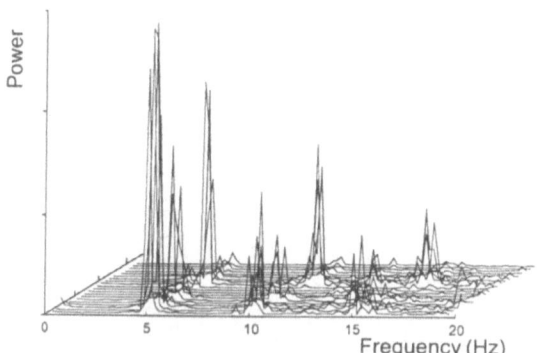

Figure 1. Power spectra of successive trains of tremor recorded by an accelerometer attached to the wrist of an African green monkey treated with MPTP.

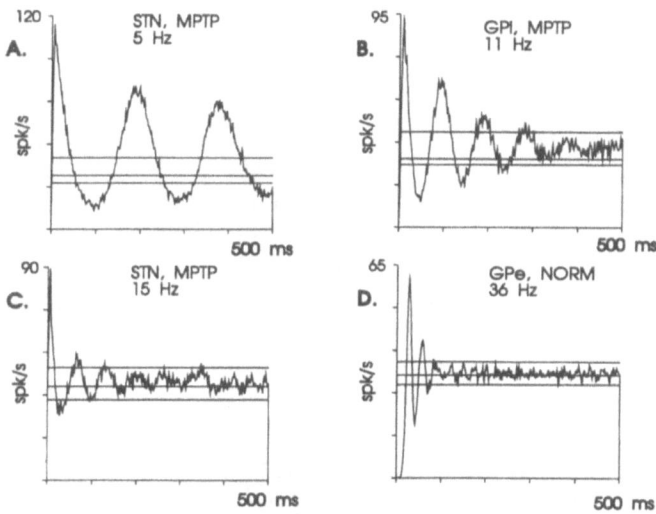

Figure 2. Examples of autocorrelograms of basal ganglia neurons with periodic oscilatory activity.

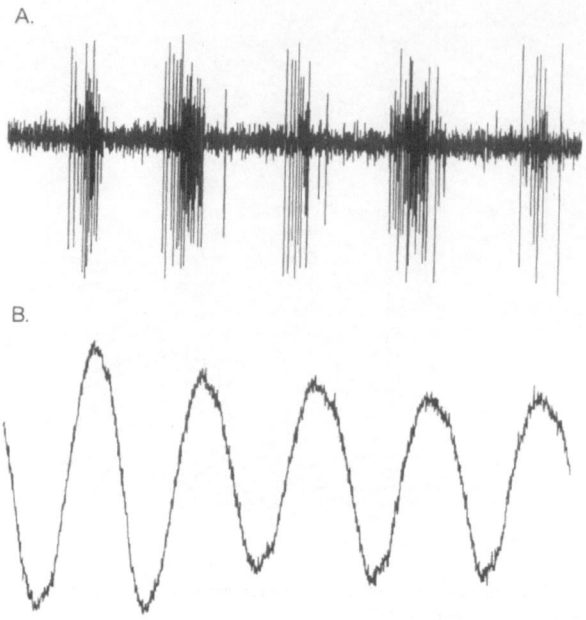

Figure 3. Example of one second of simultaneously recorded tremor (lower trace, B) and STN mulltiunit neuronal activity (upper trace, A).

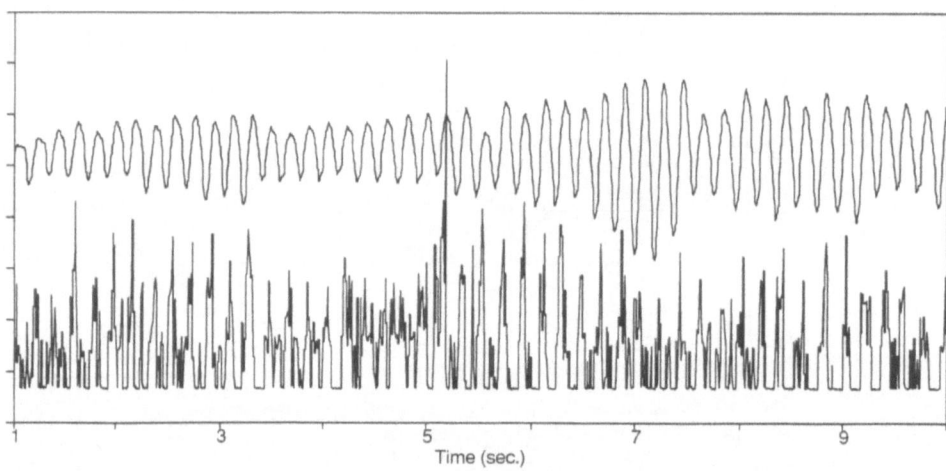

Time (sec.)

Figure 4. Continuous recording of tremor (lower trace) and integrated neuronal activity (upper trace) in GPI. Neuronal activity was passed through sample and hold integrator.

Neuronal oscillations at high frequencies

Neuronal oscillatory activity at frequencies above 8 Hz (mostly > 20 Hz) was occasionally found in the basal ganglia of the untreated monkeys (1.7% of cells in STN, 4.6% in GPi, 6.7% in GPe, Figure 2D).

After MPTP, 11% of the recorded neurons in STN demonstrated periodic oscillatory activity at frequencies above 8 Hz (median 15.7 Hz, Figure 2C). Similarly, 15% of cells in

GPi and 10% of cells in GPe discharged with periodic bursts at higher frequency (median 11.0 Hz and 10.9 Hz, respectively; see Figure 2B).

Shape of spikes

Following the MPTP treatment, there was no significant change in the wave forms of extracellularly recorded action potentials in all three structures. The duration of action potentials, however, was significantly increased in the STN and GPi, but not in GPe (Figure 5).

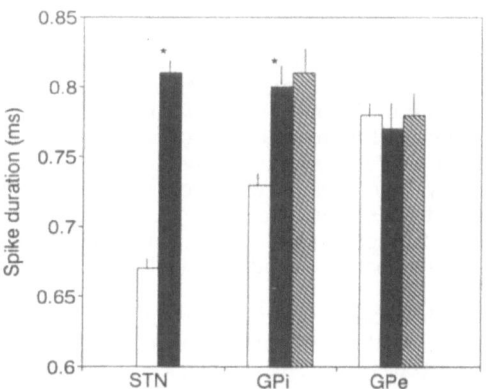

Figure 5. Spike's duration in the STN and pallidal complex. Empty bars represent duration in the normal state, black bars after MPTP treatment, and striped bars represent spike duration after STN lesion.

Histology

The substantia nigra pars compacta, the striatum, GPe, GPi, and the STN were virtually devoid of dopaminergic cells and terminals, as revealed by tyrosine hydroxylase immunohistochemistry.

DISCUSSION

The fact that periodic oscillatory activity at frequencies close to the rate of peripheral tremor was detected in the basal ganglia of MPTP treated monkeys, suggests the possibility that the basal ganglia may play a role in the development of parkinsonian tremor. Conceivably, oscillatory activity in the basal ganglia may result from instability within portions of the basal ganglia-thalamocortical "motor" circuit. On the basal ganglia level, this circuitry is grossly divided into a "direct" and an "indirect" pathway, linking the putamen with the principal basal ganglia output structures, the internal pallidal segment (GPi) and the substantia nigra, pars reticulata (Albin et al., 1989; Alexander et al., 1990). The "direct" pathway consists of an inhibitory monosynaptic connection, whereas the excitatory "indirect" pathway is made up of a series of connections, starting with an inhibitory pathway between the putamen and the external pallidal segment (GPe), and then branching into an inhibitory pathway linking GPe and GPi monosynaptically, and a pathway linking both structures via the intercalated subthalamic nucleus (STN), which gives rise to the only excitatory pathway in this circuitry. STN is also the recipient of a strong cortico-subthalamic projection (Hartmann von-Monakow et al., 1978; Kita, 1992). Given this scheme, the basal ganglia-thalamocortical "motor" circuit can be seen as a neuronal feedback system with many ancillary loops. Such feedback systems have a tendency to develop spontaneous oscillations, if gains or delays in the system are increased (Glass and Mackey, 1988), especially, when intrinsic electrophysiological properties of cells in the loop favor responses at certain frequencies (Llinas, 1988). It has previously been demonstrated that the gain of the basal ganglia-thalamocortical "motor" circuit is increased in MPTP treated animals (Miller and DeLong, 1987; Filion et al., 1988), thus creating conditions under which spontaneous oscillations may develop. Theoretically, several loops involving the basal ganglia could generate oscillatory activity, including the striato-nigral feedback loop, the cortex-striatum-GPi-thalamic circuit, and the cortex-striatum-GPe-STN-GPi-thalamic pathway.

Alternatively, oscillatory activity in the basal ganglia may result from changed intrinsic properties of individual neurons, leading them to assume pacemaker activity after treatment with MPTP. Previous studies of neurons in pallidal brain slice preparations from rat and guinea-pig revealed periodic firing at frequencies of 5 and 10 Hz (Nakanishi et al., 1985; Nambu and Llinas, 1990). GPi neurons oscillating at 12-15 Hz have occasionally been observed in Rhesus monkeys, following MPTP treatment (Miller and DeLong, 1987; Filion and Tremblay, 1991) or after lesions of the ventromedial tegmentum (Filion, 1979). This has lead to the proposal that a thalamic filter mechanism, might operate to transform the 12-15 Hz rhythmic pacemaker activity in GPi into the 3-6 Hz rhythmic activity in thalamic neurons (Paré et al., 1990). The present finding of oscillatory activity at 4-8 Hz in the basal ganglia makes such a mechanism unnecessary, since obviously the low frequency oscillations in STN and GPi could directly drive rhythmic oscillatory activity in the thalamus. The presumed increased tonic inhibition from GPi upon cells in VL thalamus (Miller and DeLong, 1987; Filion et al., 1988) may further strengthen the tendency of thalamic cells to develop oscillatory activity under parkinsonian conditions (Jahnsen and Llinas, 1984a,b; Deschênes et al., 1984).

The presence of high frequency oscillations in the basal ganglia of our monkeys, occurring in addition to the oscillatory activity around 5 Hz, suggests that MPTP treatment may unveil basal ganglia mechanisms capable of generating neuronal oscillations at distinctly different frequencies. Similar properties have previously been demonstrated in cells in the inferior olive (Llinas and Yarom, 1986), the thalamus (Jahnsen and Llinas, 1984a,b; Deschênes et al., 1984), and in neocortical pyramidal cells (Silva et al., 1991; Llinas et al., 1991). In these experiments, the expression of a specific oscillatory frequency in a given cell was dependent on its membrane potential. The functional consequences of the oscillations at higher frequencies remain unclear.

Our sample of GPe cells included mainly neurons in the "non-motor territories" of GPe (DeLong et al., 1985), and may therefore be not entirely comparable to the sample obtained in GPi and STN, which was taken mostly from the "motor" territories (Bergman et al., 1993). Previous studies in MPTP treated macaques, however, have also found periodic high frequency oscillatory activity in GPi, but not in GPe (Miller, 1988; Tremblay et al., 1989). An interesting possibility is that the difference between GPe and GPi neurons, and the corresponding changes in the duration of spikes, may be due to direct toxic effects of MPTP on the dopaminergic innervation of GPi and possibly of STN (but see Lavoie et al., 1989), rather than secondary to the well-known reduction of the dopamine content in the striatum. This has previously been suggested by studies in humans, in which the severity of parkinsonian tremor correlated with the degree of dopaminergic denervation in GPi (Bernheimer et al., 1974). A differential dopaminergic innervation of the two pallidal segments has previously been reported in squirrel monkeys (Parent and Smith, 1987). Conceivably, reduction of the dopamine content in GPi may induce oscillatory activity in this nucleus. Dopamine loss in GPi, and possibly in STN, may be more prominent in African green monkeys than in macaques (Parent et al., 1990; Schneider and Dacko, 1991), making the former species more prone to develop tremor than the latter. The degree of reduction of dopaminergic input to individual neurons may determine whether a cell expresses low- or high-frequency oscillations. A severe reduction may result in low frequency oscillations, whereas less severe dopamine loss may lead to periodic oscillations at higher frequencies (as seen previously in macaques, and in 10-15% of the cells in this study), which may not be effective in synchronizing the whole basal ganglia-thalamocortical loop, and therefore may not be sufficient to induce tremor. Our finding that oscillatory activity at low frequency can exist without leading to tremor emphasizes that intrinsic oscillations in the output nuclei of the basal ganglia may be necessary, but not sufficient for the development of tremor. A "critical mass" of synchronized oscillatory activity, possibly as the result of excessive gain in the cortex-STN-GPi-thalamic loop, may be needed before visible tremor can develop.

Acknowledgements

This research was supported in part by NIH grant 5-20231, and by an endowment fund for basic research in the life sciences: Charles A. Revson foundation, administrated by the Israeli academy of sciences and humanities.

REFERENCES

Albin, R.L., Young, A.B., and Penney, J.B., 1989, The functional anatomy of basal ganglia disorders, *TINS* 12:366-375.

Alexander, G.E., Crutcher, M.D., and DeLong, M.R., 1990, Basal ganglia-thalamocortical circuits: parallel substrates for motor, oculomotor, `prefrontal' and `limbic' functions, *Prog. Brain Res* . 85:119-146.

Aziz, T.Z, Peggs, D., Sambrook, M.A., and Crossman, A.R., 1991, Lesion of the subthalamic nucleus for the alleviation of 1-methyl-4-phenyl-1,2,3,6-tetrahydropyridine (MPTP)-induced parkinsonism in the primates, *Movement disorders* 6:288-293.

Bergman, H., Wichmann, T., and DeLong, M.R., 1990, Reversal of experimental parkinsonism by lesions of the subthalamic nucleus, *Science* 249:1436-1438.

Bergman, H., Karmon, B., Wichmann, T., and DeLong, M.R., 1993, The primate subthalamic nucleus: II. Neural activity in the subthalamic nucleus and pallidum in the MPTP model of parkinsonism, *J. Neurophysiology* submitted

Bernheimer, H., Birkmayer, W., Hornykiewicz, O., Jellinger, K., and Seitelberger, F., 1974, Brain dopamine and the syndromes of Parkinson and Huntington, *J. Neurol. Sci.* 20:415-455.

Burns, R.S., Chiueh, C.C., Markey, S.P., Ebert, M.H., Jacobowitz, D.M., and Kopin, J.J., 1983, A primate model of parkinsonism: selective destruction of dopaminergic neurons in the pars compacta of the substantia nigra by N-methyl-4-phenyl-1,2,3,6-tetrahydropyridine, *Proc. Natl. Acad. Sci. USA* 80:4546-4550.

Contreras, C.M., Mexicano, G., and Guzman-Flores, C., 1981, A stereotaxic brain atlas of the green monkey (Cercopithecus Aethiops Aethiops), *Bol. Est. Med. Biol., Mex.* 31:383-428.

DeLong, M.R., Crutcher, M.D., and Georgopoulos, A.P., 1985, Primate globus pallidus and subthalamic nucleus: Functional organization, *J. Neurophysiol.* 53:530-543.

DeLong, M.R., 1990, Primate models of movement disorder of basal ganglia origin, *TINS* 13:281-285.

de Montigny, C., and Lamarre, Y., 1973, Rhythmic activity induced by harmaline in the olivo-cerebello-bulbar system of the cat, *Brain Res.* 53:81-95.

Deschênes, M., Paradis, M., Roy, J.P., and Steriade, N., 1984, Electrophysiology of neurons of lateral thalamic nuclei in cat: resting properties and burst discharge, *J. Neurophysiology* 51:1196-1219.

Elble, R.J., and Koller, W.C., 1990, "Tremor," The John Hopkins University Press, Baltimore and London.

Elsworth, J.D., Deutch, A.Y., Redmond, D.E., Sladek, J.R., and Roth, R.H., 1987, Differential responsiveness to 1-methyl-4-phenyl-1,2,5,6-tetrahydropyridine toxicity in subregions of the primate substantia nigra and striatum, *Life Sciences* 40:193-202.

Filion, M., 1979, Effects of interruption of the nigrostriatal pathway and dopaminergic agents on the spontaneous activity of globus pallidus neurons in a awake monkey, *Brain Res.* 178:425-441.

Filion, M., Tremblay, L., and Bédard, P.J., 1988, Abnormal influences of passive limb movement on the activity of globus pallidus neurons in parkinsonian monkeys, *Brain Res.* 444:165-176.

Filion, M., and Tremblay, L., 1991, Abnormal spontaneous activity of the globus pallidus neurons in monkeys with MPTP-induced parkinsonism, *Brain Res.* 547:142-151.

Findley, L.J., Gresty, M.A., and Halmagyi, G.M., 1981, Tremor, the cogwheel phenomena and clonus in Parkinson's disease, *J. Neurol. Neurosurg. Psychiatry* 44:534-546.

Glass, L., and Mackey, M.C., 1988, "From Clocks to Chaos," Princeton University Press, N.J.

Hartmann von-Monakow, K., Akert, K., and Kunzle, H., 1978, Projections of the precentral motor cortex and other cortical areas of the frontal lobe to the subthalamic nucleus in the monkey, *Exp. Brain Res.* 33:395-403.

Jahnsen, H., and Llinas, R., 1984a, Electrophysiological properties of guinea-pig thalamic neurones: an in vitro study, *J. Physiol.* 349:205-226.

Jahnsen, H., and Llinas, R., 1984b, Ionic basis for the electroresponsiveness and oscillatory properties of guinea-pig thalamic neurones in vitro, *J. Physiol.* 349:227-247.

Jenner, P., Rose, S., Nomto, M., and Marsden, C.D., 1986, MPTP-induced parkinsonism in the common marmoset: behavioral and biochemical effects *in*: "Advances in Neurology, Vol 45," M.F. Yahr and K.J. Bergmann, eds., Raven Press, New York, pp. 183-190.

Karmon, B., and Bergman, H., 1993, Detection of neuronal periodic oscillations in the basal ganglia of normal and parkinsonian monkeys, *Is. J. Med. Sci.* submitted.

Kita, H., 1992, Responses of globus pallidus neurons to cortical stimulation: intracellular study in rat, *Brain Res.* 589:84-90.

Laitinen, L.V., Bergenheim, A.T., and Hariz, M.I., 1992, Leksell's posteroventral pallidotomy in the treatment of Parkinson disease, *J. Neurosurg.* 76:53-61.

Lamarre, Y., and Joffroy, A.J., 1979, Experimental tremor in monkey: Activity of thalamic and precentral cortical neurons in the absence of peripheral feedback, *in*: "Advances in Neurology, Vol. 24," L.J. Poirier, T.L. Sourkes and P.J. Bédard, eds, Raven Press, New York, pp. 109-122.

Lance, J.W., Schwab, R.S., and Peterson, E.A., 1963, Action tremor and the cogwheel phenomena in Parkinson's disease, *Brain* 86:95-110.

Langston, J.W., Forno, L.S., Rebert, C.S., and Irwin, I., 1984, Selective nigral toxicity after systemic administration of 1-methyl-4-phenyl-1,2,5,6-tetrahydropyridine (MPTP) in the squirrel monkey, *Brain Res.* 292:390-394.

Lavoie, B., Smith, Y., and Parent, A., 1989, Dopaminergic innervation of the basal ganglia in the squirrel monkey as revealed by tyrosine hydroxylase immunohistochemistry, *J. Comp. Neurol.* 289:36-52.

Lenz, F.A., Tasker, R.R., Kwan, H.C., Schider, S., Kwong, R., Dorsrovsky, J.O., and Murphy, J.T., 1988, Single unit analysis of the human ventral thalamic nuclear group: correlation of thalamic "tremor cells" with the 3-6 Hz component of parkinsonian tremor, *J. Neurosci.* 8:754-764.

Llinas, R.R., 1988, The intrinsic electrophysiological properties of mammalian neurons: Insights into central nervous system function, *Science* 242:1654-1664.

Llinas, R.R., and Volkind, R.A., 1973, The olivo-cerebello system: functional properties as revealed by harmaline induced tremor, *Exp. Brain Res.* 18:69-87.

Llinas, R.R., and Yarom, Y., 1981a, Electrophysiology of mammalian inferior olivary neurons in vitro. Different types of voltage dependent ionic conductance, *J. Physiol.* 315:549-567.

Llinas, R.R., and Yarom, Y., 1981b, Properties and distribution of ionic conductances generating electroresponsiveness of mammalian inferior olivary neurons in vitro, *J. Physiol.* 315:569-584.

Llinas, R.R., and Yarom, Y., 1986, Oscillatory properties of guinea-pig inferior olivary neurons and their pharmacological modulation: an in vitro study, *J. Physiol.* 376:163-182.

Llinas, R.R., Grace, A.A., and Yarom, Y., 1991, In vitro neurons in mammalian cortical layer 4 exhibit intrinsic oscillatory activity in the 10- to 50 Hz frequency range, *Proc. Natl. Acad. Sci. USA* 88:897-901.

Miller, W.C., 1988, "Examination of the motor abnormalities and changes in neuronal activity in the globus pallidus in the primate MPTP model of parkinsonism," Ph.D. thesis, Baltimore.

Miller, W.C., and DeLong, M.R., 1987, Altered tonic activity of neurons in the globus pallidus and subthalamic nucleus in the primate MPTP model of parkinsonism, *in*: "The Basal Ganglia II," M.B. Carpenter and A. Jayaraman, eds., Plenum Press, New York, pp. 415-427.

Nakanishi, H., Hori N., and Katsuda, N., 1985, Neostriatal evoked inhibition and the effects of dopamine on globus pallidus neurons in rat slice preparations, *Brain Res.* 358:282-286.

Nambu, A., and Llinas, R.R., 1990, Electrophysiology of the globus pallidus neurons: an in vitro study in guinea pig brain slices, *Soc. Neurosci. Abstr.* 16:428.

Narabayashi, H., 1990, Surgical Treatment in the Levodopa era, *in:* "Parkinson's disease," G. Stern, ed., Chapamn & Hall, London, pp. 597-646

Ohye, C., Saito, U., and Fukamachi, A., 1974, An analysis of the spontaneous rhythmic and non-rhythmic burst discharge in the human thalamic, *J. Neurol. Sci.* 22:245-259.

Paré, D., Curro'Dossi, R., and Steriade, M., 1990, Neuronal basis of the parkinsonian resting tremor: A hypothesis and its implications for treatment, *Neuroscience* 35:217-226.

Parent, A., and Smith, Y., 1987, Differential dopaminergic innervation of the two pallidal segments in the squirrel monkey (Saimiri sciureus), *Brain Res.* 426:397-400.

Parent, A., Lavoie, B., Smith, Y., and Bédard, P., 1990, The dopaminergic nigropallidal projection in primates: Distinct cellular origin and relative sparing in MPTP treated monkeys, *in*: "Advances in Neurology, Vol 53," M.B. Streifler, A.D. Korczyn, E. Melamed and M.B.H. Youdim, eds., Raven Press, New York, pp. 111-116.

Poirier, L.J., 1960, Experimental and histological study of midbrain dyskinesia, *J. Neurophysiol.* 23:534-551.

Redmond, D.E., Roth, R.H., and Sladek, J.R., 1985, MPTP produces classic parkinsonian syndrome in African green monkeys, *Soc. Neurosci. Abstr.* 11:166.

Schneider, J.S., and Dacko, S., 1991, Relative sparing of the dopaminergic innervation of the globus pallidus in monkeys made hemi-parkinsonian by intracarotid MPTP infusion, *Brain Res.* 556:292-296.

Sellal, F., Hirsch, E., Lisovoski, F., Mutschler, V., Collard, M., and Marescaux, C., 1992, Contralateral disappearance of parkinsonian signs after subthalamic hematoma, *Neurology* 42:255-256.

Silva, L.R., Amitai, Y., and Connors, B.W., 1991, Intrinsic oscillations of neocortex generated by layer 5 pyramidal neurons, *Science* 251:432-435.

Svennilson, E., Torvik, A., Lowe, R., and Leksell, L., 1960, Treatment of parkinsonism by stereotactic thermolesions in the pallidal region. A clinical evaluation of 81 cases, *Acta Psychiat. Neurol. Scand.* 35:358-377.

Tremblay, L., Filion, M., and Bédard, P.J., 1989, Responses of pallidal neurons to striatal stimulation in monkeys with MPTP induced parkinsonism, *Brain Res.* 498:17-33.

Wichmann, T., Bergman, H., and DeLong, M.R., 1993a, The primate subthalamic nucleus I. Functional properties in intact animals, *J. Neurophysiology*.submitted.

Wichmann, T., Bergman H., and DeLong, M.R., 1993b, The primate subthalamic nucleus III. Changes in behavior and pallidal neuronal activities induced by subthalamic inactivation in the model of parkinsonism, *J. Neurophysiology* submitted.

Winters, W.D., Kado, R.T., and Adey, W.R., 1969, "A Stereotaxic Brain Atlas for Macaca nemestrina," Univ of California Press, Berkeley and Los Angeles.

POTENTIAL CHANGES WITHIN THE BASAL GANGLIA DURING PUTAMEN-INDUCED DYSTONIA AND CAUDATE-INDUCED LOCOMOTOR HYPERACTIVITY IN THE FREELY MOVING CAT

Ken-ichi Fujimoto, Hiroshi Yamada, Eiji Nakamura, Satoshi Nakamura, Shinichi Muramatsu and Mitsuo Yoshida

Department of Neurology
Jichi Medical School, Minamikawachi-machi
Tochigi-ken, 329-04, Japan

INTRODUCTION

Although the striatum is a key structure in the extrapyramidal system, the function of the striatum still remains to be understood. The striatum consists of two distinctly separate structures, i.e., the caudate (Cd) and the putamen (Put). The Cd and Put hold the same place in the basal ganglia; efferent fibers of both Cd and Put supply inhibitory synapses to neurons in the substantia nigra pars reticulata (SNr) (Yoshida and Precht, 1971; Yoshida et al., 1971; Yoshida et al., 1972; Yoshida et al., 1981) and this inhibition is mediated by γ-amino butyric acid (GABA) (Precht and Yoshida, 1971; Yoshida, 1981). From clinical observations, however, different roles of Cd and Put would be expected. In Huntington's disease, which is characterized by degenerative cell loss in the Cd (Lange et al., 1976), choreiform involuntary movements of muscle groups including those of face, fingers, hands, and trunk are prominent, whereas in striatonigral degeneration, which is characterized by cell loss in the Put (Adams et al., 1964), symptoms resemble those of Parkinson's disease with rigidity and akinesia.

Extensive investigations have been carried out in rodents. Behaviors such as circling, postural asymmetry, locomotor hyperactivity, and dyskinesia have been induced by stimulation and lesioning of the caudoputamen ensemble (CPU) (Ungerstedt and Arbuthnott, 1970; Marsden et al., 1975; McKenzie and Viik, 1975; Kelly and Moore, 1976). However, since the Cd and Put are situated together (Von Bonin and Shariff, 1951) no functional differences between the Cd and Put have been revealed by use of rodents. In the rabbit, cat, dog, and primates, the Cd and Put are anatomically distinctly separate. In cats, specific roles of the Cd in motor behavior have been investigated with electrical or pharmacological stimulation (Forman and Ward, 1957; Stevens et al., 1961; Laursen, 1963; McLennan et al., 1964; Cools, 1973; Ohno and Tsubokawa, 1987; Kitama et al., 1991), however the Put has barely been studied due to the fact that the Put is technically difficult to stimulate separately from surrounding structures.

MATERIALS AND METHODS

Cooperative adult cats, weighting 2.8-3.5 kg, were anesthetized with pentobarbital sodium (30mg/kg) and fixed in a stereotaxic apparatus. After the cranium was exposed by removal of the scalp, burr holes were made through the cranium. Recording electrodes,

The Basal Ganglia IV, Edited by
G. Percheron *et al.*, Plenum Press, New York, 1994

consisting of stainless-steel insect pins (No. 000) insulated except for the tip, were inserted stereotaxically. Their coordinates were A (from ear bar) =18.5, L=4.0, Z=+6.0 for the anterior Cd; A=15.5, L=5.0, Z=+5.5 for the posterior Cd; A=13.5, L=10.5, Z=+1.5 for the anterior Put; A=10.5, L=11.8, Z=0 for the posterior Put; A=11.5, L=9.5, Z=-0.5 for the globus pallidus (GP); A=11.0, L=6.5, Z=-1.8 for the entopeduncular nucleus (Ent); A=8.0, L=4.5, Z=-3.0 for the subthalamic nucleus (STH); A=4.5, L=5.0, Z=-4.0 for the SNr. In order to record potentials in the cortex (Cx), a silver-ball electrode was placed on the dura above the motor cortex. Electrodes were placed bilaterally. All wires from electrodes were soldered to a small connector attached to the skull with orthodontic resin. In order to provide bicuculline (BIC)-injection sites at coordinates of A=17.0, L=5.0, Z=+6.0 for the Cd and A=12.0, L=11.0, Z=+1.5 for the Put, stainless-steel tubes with a length of 1.2 mm and outer diameter of 0.8 mm were inserted vertically through the burr holes. The bottom of the pipe touched, but did not penetrate, the dura. All electrodes and guide tubes were fixed to the cranium with orthodontic resin. The distances from the top of guide tubes to corresponding injection sites were calculated for later injections.

BIC (bicuculline methobromide, Cambridge Research, U.K.) was dissolved in saline, and $3.0\mu g/1.5\mu l$ of the solution was drawn into a stainless-steel needle (outer diameter of 0.5 mm) connected to a Hamilton syringe by a polyethylene tube. Marks in millimeters were put on the stainless-steel needle so that the distance from the top of the guide tube to each structure could be determined. The needle was inserted and BIC was injected into the corresponding structure for 2 min, and the needle was kept in situ for 1 min after the injection allowing the solution to diffuse into each structure. In the same animal, successive injections of BIC were given at an interval of at least 2 days. In order to observe general behavior, the cat was left on a 2x2 m floor and the behavior before and after the BIC injection was recorded by a conventional video camera. In order to record the head movement, a cat was placed in a small cage made of acrylic plate and the position of a small light attached to the forehead of a cat was traced by an X-Y recorder (Muromachi Kikai, Tokyo). Field potentials were recorded by an EEG machine (EEG-5214, Nihon Kohden, Tokyo).

At the end of the experiment, electrolytic lesions were made with negative current of 3V, 30sec in order to identify recording sites, and then $1.5\mu l$ of a mixed solution of 2 mg of BIC and 20 mg of the dye fast green FCF (FCF) in 1.0 ml of saline was injected into each structure under anesthesia with pentobarbital sodium (30 mg/kg). Since the extent of diffusion of FCF was confirmed as almost the same as that of BIC in a preliminary experiment using radioactive BIC (Yoshida et al., 1991), the extent of BIC diffusion was determined by sketching the extent of FCF diffusion on a brain atlas. Histological examination of Cresyl violet stained sections verified the placement of the recording electrodes.

RESULTS

Effects of BIC Injection into the Caudate

When the cats were left alone on a 2x2 m floor before BIC injection, they walked around slowly or crouched in the corners. Usually the amount of walking was almost zero. From each electrode, background field potentials with fast waves were recorded at this time. Since the injection took 2 min, it was difficult to determine the exact onset of the injection. For convenience, time zero was determined 1 min after the injection was completed. When BIC was injected into the Cd, spikes appeared within 1 or 2 min and lasted for 15 to 20 min. Spikes seen in the Cd were sporadic without bursting discharge (Figure 1). At the same time, the amount of walking increased gradually. The floor of 2x2 m was divided into 16 squares of 0.5x0.5 m each. The amount of walking was measured by counting the number of squares crossed by the cat. Figure 2 shows a change of locomotor activity which was obtained in 5 cats with injection of BIC into the Cd. The locomotor hyperactivity began with the appearance of spikes, reaching a maximum within 5 min and recovering to the state before the injection within another 10 min. The maximum rate of locomotor hyperactivity was 38±13.0 (mean±SD) square cross/min, which corresponded to approximately 19 m/min. Although BIC injection into the Cd was unilateral, the locomotor hyperactivity was always forward; neither postural asymmetry nor any circling tendency was observed.

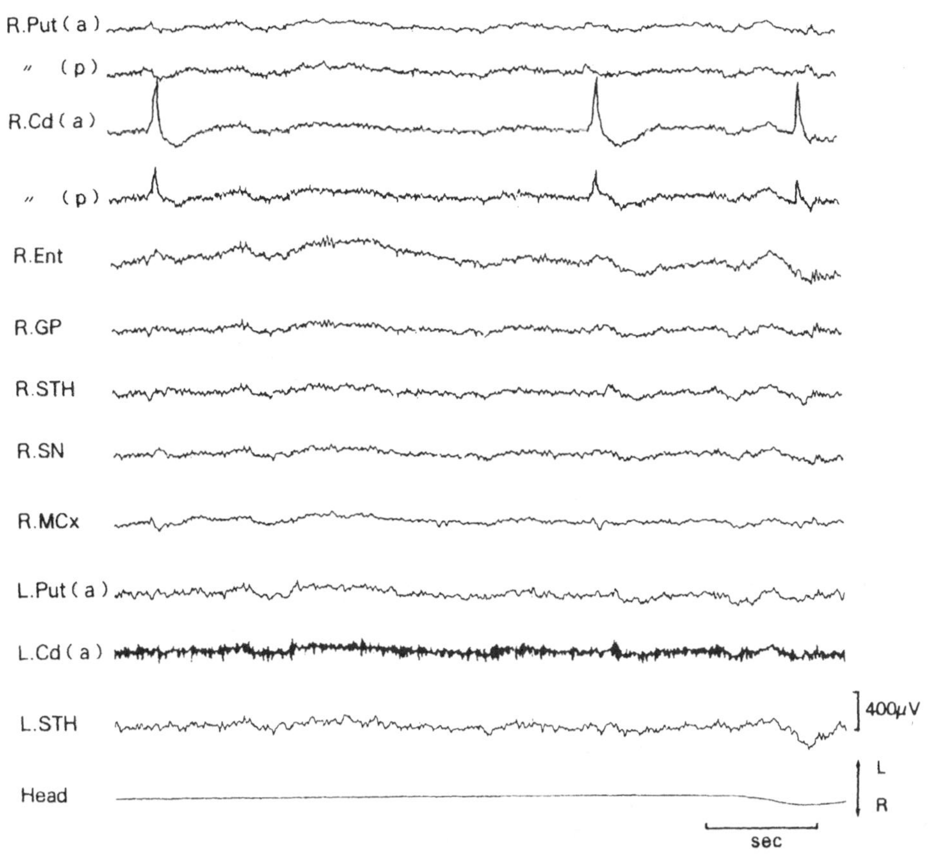

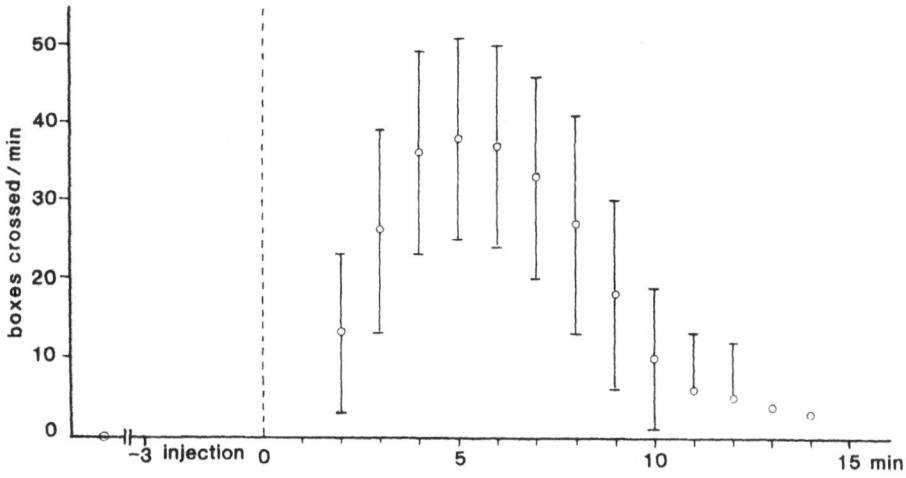

Figure 1. Field potentials of a freely moving cat about 10 min after the injection of BIC into the right (R) caudate (Cd). Single spikes were recorded in the anterior (a) and posterior (p) part of the R.Cd. Put: putamen, Ent: entopeduncular nucleus, GP: globus pallidus, STH: subthalamic nucleus, SN: substantia nigra, MCx: motor cortex, L: left.

Figure 2. Locomotor hyperactivity produced by unilateral injection of bicuculline into the caudate. The 2x2 m floor was divided into 16 squares, and number of squares crossed by cats within 1 min was counted. Data were mean±SD obtained from 5 cats. (from Yoshida et al., 1991).

Figure 3. Dystonic movements of neck and trunk about 10 min after the injection of bicuculline into the right putamen. (from Yoshida et al., 1991).

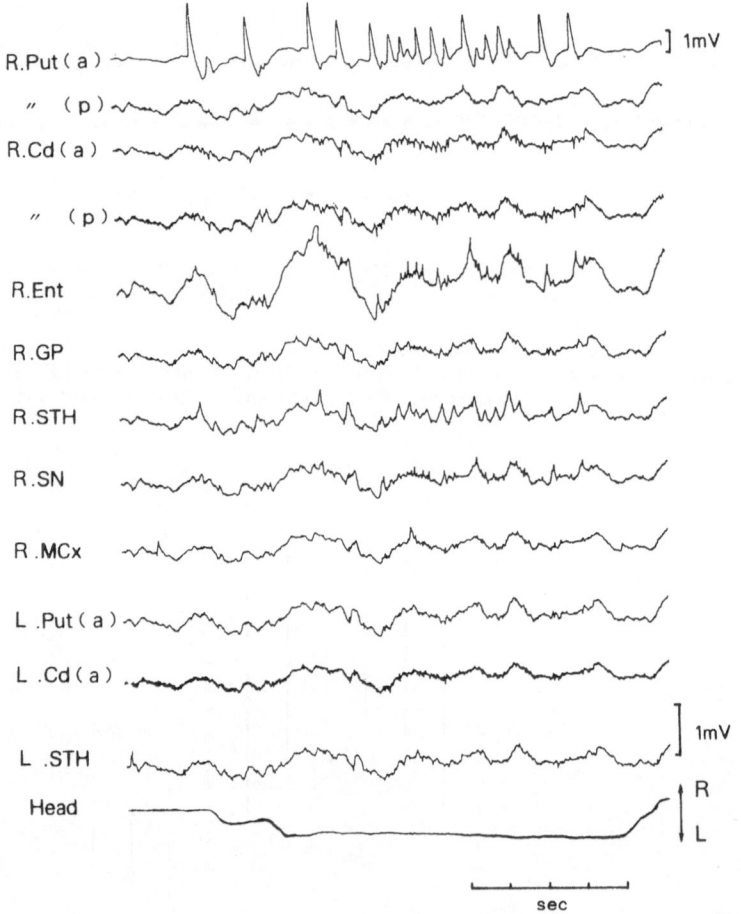

Figure 4. Field potentials and head movement of a cat about 10 min after the injection of bicuculline into the right putamen. High amplitude spikes were recorded in the anterior part of the right putamen. Note the difference of the calibration. Just after the start of consecutive putaminal spikes, the head turned stepwise to the left and it returned abruptly on interruption of the consecutive spikes. Abbreviations; See Figure 1.

Effects of BIC Injection into the Putamen

When BIC was injected into the Put, single spikes usually appeared within one min in the Put. At this time, twitch movement of the ear on the contralateral side was observed synchronized with the spikes. After about 5 min, bursts of spikes appeared in the Put and dystonic movements of the head and trunk (dystonia) toward the contralateral side of the injection occurred in a dramatic way. As shown in Figure 3, just after the start of consecutive putaminal spikes, the head turned stepwise almost 180°. Figure 4 shows burst discharges and the head movement recorded by an X-Y recorder. Mean latency of the head movement after the first spike was 134.2±52.1 msec. When the consecutive spikes were interrupted, the head returned abruptly. Locomotor hyperactivity never occurred when BIC was injected into the Put and the cat was sitting during the entire period in which dystonia was observed.

Diffusion of BIC within the Caudate and Putamen

The shape of the volume of injected BIC was spherical within Cd or Put. Figure 5 is a sketch of the extent of FCF diffusion in the case whose field potentials were presented in Figures 1 and 4. The relationship between the extent of FCF diffusion and that of BIC had been correlated precisely (Yoshida et al. 1991). In order to approach the Cd, an injecting needle must penetrate the lateral ventricle, and if the needle pass right through the Cd, it must be in the nucleus accumbens. Since locomotor hyperactivity occurred even when BIC did not enter the lateral ventricle, and it did not occur even when BIC entered the lateral ventricle, it was not due to the effect of leakage of BIC into the lateral ventricle. Since locomotor hyperactivity occurred in those cases where the extent of BIC diffusion was far dorsal or caudal to the nucleus accumbens, it was not due to the spread of BIC into the nucleus accumbens (Yoshida et al., 1991).

In case of injection aimed at the Put, interpretation of the results is a little difficult. The Put is a long nucleus surrounded by other structures, e.g. the claustrum, the globus pallidus, and the anterior sylvian gyrus. Possible involvement of these structures in the BIC-induced effects should be considered. From precise examination, statistically close parallelism was observed between the involvement of the Put in the BIC-injected area and the occurrence of dystonia, while, parallelism was not observed between the involvement of the surrounding structures and the occurrence of dystonia (Yoshida et al., 1991). We therefore conclude that the locomotor hyperactivity was an effect of BIC covering a substantial part (either the head or body) of the Cd, and dystonia represents genuine effects of BIC injected into the Put.

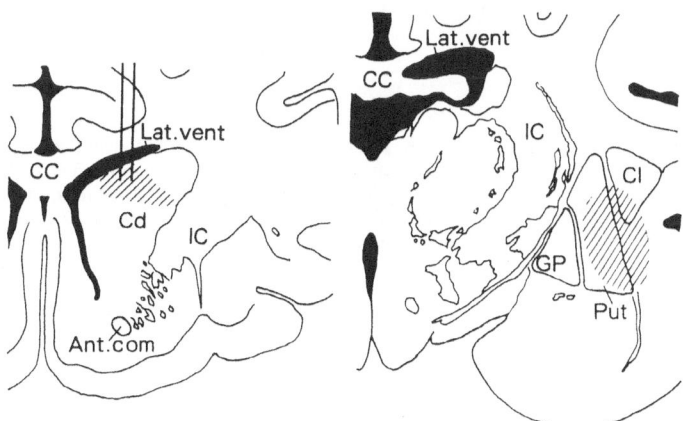

Figure 5. Extent of bicuculline spreading within the caudate (left) and the putamen (Right). Shaded areas indicate the extent of fast green FCF diffusion. CC: corpus callosum, Cd: caudate, IC: internal capsule, Lat. vent: lateral ventricle, Ant. com: anterior commissure, Put: putamen, GP: globus pallidus, Cl: claustrum.

331

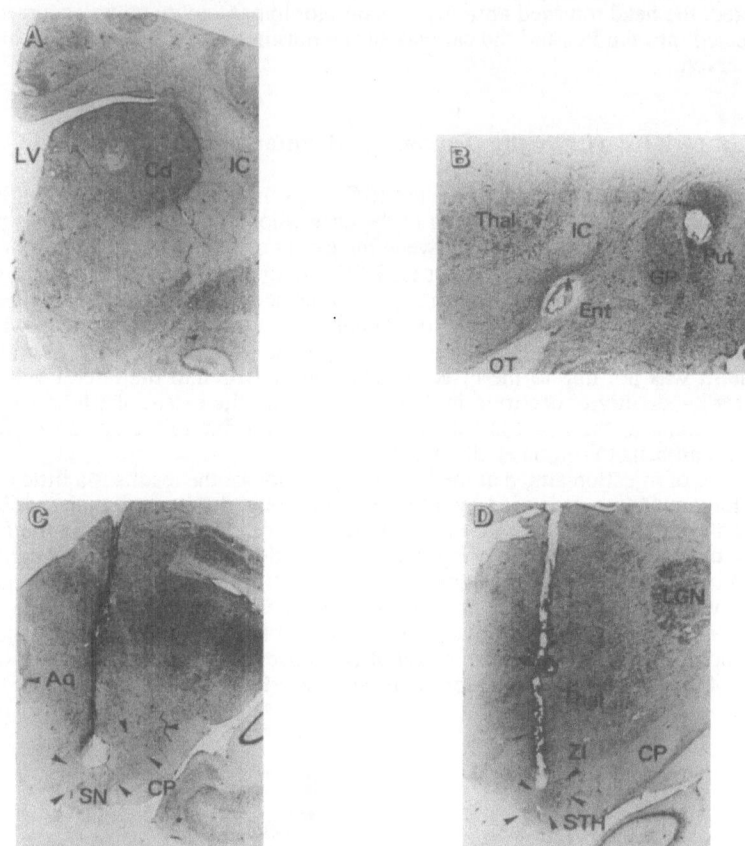

Figure 6. Coronal sections of a cat brain show the location of the recording electrodes in the caudate (**A**), putamen and entopeduncular nucleus (**B**), substantia nigra (**C**), and subthalamic nucleus (**D**). Cd: caudate, LV: lateral ventricle, IC: internal capsule, Put: putamen, Ent: entopeduncular nucleus, Thal: thalamus, OT: optic tract, GP: globus pallidus, STH: subthalamic nucleus, CP: cerebral peduncle, ZI: zona incerta, LGN: lateral geniculate nucleus, SN: substantia nigra, Aq: aqueduct.

Location of Recording Electrodes

Figure 6 shows sections of the case whose field potentials were presented in Figure 1 and 4. Electrolytic lesions in the Cd, Put, Ent, SN, and STH were clearly recognized. Since the recording elctrode in the GP was situated anteroposterior very close to the slice plane used for sketching the extent of FCF diffusion within the Put, the electrolytic lesion in the GP was not verified histologically.

DISCUSSION

Effects of BIC within the Striatum

BIC is known to block the effect of GABA (Kravitz, 1967; Obata, 1972). Efferent fibers of both Cd (Yoshida et al., 1972) and Put (Yoshida et al., 1981) are predominantly inhibitory in nature and their transmitter substance is GABA (Precht and Yoshida, 1971; Yoshida, 1981). Inhibitory axon collaterals with GABAergic synapses on the efferent neurons of the neostriatum (Delgado, 1979; Park et al., 1980) and GABAergic interneurons within the neostriatum (McGeer and McGeer, 1975) have been reported. We have already reported that administration of BIC into these nuclei would therefore effectively block these GABAergic inhibitions and thereby result in powerful activation of the neostriatal efferent neurons (Muramatsu et al., 1990). Since the potentials were generated not continuously but periodically, some inhibitory process would also be involved. Firstly inhibitory influences from peripheral areas not reached by BIC may be exerted at the site activated by BIC. Secondly, excessive firing of neurons may have produced summation of after-hyperpolarization, resulting in decreased firing of the neurons themselves. This pharmacological stimulation would have much stronger activating effects than electrical stimulation, since the latter would also activate GABAergic inhibitory synapses existing abundantly within the striatum. The extent of the area of injected BIC appeared to be spherical within the Cd or Put in the present study and the volume affected increased with time. We therefore assume that relatively wider areas of Cd and Put were activated by BIC so that clear behavioral changes resulted from this procedure. Especially in the Put, a single spike was not enough to produce dystonia. Bursting of spikes, which was brought on by the excitation of neurons in wider areas, was necessary to produce dystonic movement. The BIC-induced behavioral changes would thus suggest a functional role of the structures into which BIC was injected.

Caudate and Locomotor Hyperactivity

Injection of BIC into the head and perhaps also body of the Cd resulted in locomotor hyperactivity. The locomotor hyperactivity correlated with spikes seen in the Cd. When picrotoxin (PTX), another selective antagonist of GABA, was injected into the CPU in a rat, spikes were seen in the cortex as well as the CPU (Muramatsu et al., 1990), whereas spikes were not seen in the cortex when BIC was injected into the Cd in a cat. Though, potential changes were sometimes seen in some other areas including the Put, STH, and cortex, they were very weak compared with the spikes in the Cd. Since locomotor hyperactivity was observed earlier than the appearance of potential changes in these areas other than the Cd, it is reasonable to conclude that activation of the Cd per se could produce locomotor hyperactivity in cats.

In previous electrophysiological studies using rats (Muramatsu et al., 1990), injections of PTX into the CPU produced spikes and SNr cells showed marked inhibition in correspondence with the spikes. In contrast, VM neurons showed excitation in relation to the CPU spikes. Since striato-nigral fibers are inhibitory in nature (Yoshida and Precht, 1971; Yoshida et al., 1971; Yoshida et al., 1972; Yoshida et al., 1981), it is reasonable that the SNr neurons were strongly inhibited in relation to the CPU spikes. The SNr neuron exerts monosynaptic GABAergic inhibitory effects on neurons in the ventromedial thalamic nucleus (VM) (Yoshida and Omata, 1979; Ueki, 1983), and the latter neurons generally receive tonic inhibition from the inhibitory SNr (Chevalier et al., 1985). Thus, transient removal of the SNr inhibition could lead to excessive excitation of VM neurons (Deniau and

333

Chevalier 1985). The VM neurons then exert monosynaptic excitatory effects on neurons in the cortex and locomotor hyperactivity is generated.

When PTX was injected into the CPU of a rat, rhythmical jaw movements or dyskinesia consisting of rapid flexion and extension of the contralateral forelimb and occasional flexion of the neck were observed (Muramatsu et al., 1990; Nakamura et al., 1990), whereas injection of BIC into the Cd of a cat produced locomotor hyperactivity. In contrast, electrical or pharmacological stimulation of the nucleus accumbens is reported to produce locomotor hyperactivity in rodents (Von Voigtlander and Moore, 1973; Kelly and Moore, 1976; Pycock and Marsden, 1978; Swerdlow and Koob, 1984). These differences must be due to the difference of species. Energy metabolism studied by the 2-deoxy[14C]glucose (2-DG) method increased not in the nucleus accumbens but in the Cd of cats walking on a moving treadmill (Schwartzman et al., 1986). It is therefore reasonable that activation of the Cd by BIC per se could produce locomotor hyperactivity in cats.

The presently observed locomotor hyperactivity occurred in a symmetrical manner without accompanying circling behavior, in spite of unilateral injection of BIC into the Cd. In rodents, unilateral destruction of the nigrostriatal dopamine neuron with 6-hydroxydopamine resulted in vigorous rotatory movements or asymmetrical postural changes (Ungerstedt and Arbuthnott, 1970). Since the Cd and Put are not separated in rodents (Von Bonin and Shariff, 1951), pharmacological activation of the striatum must result in mixed effects of these 2 nuclei.

Putamen and Dystonia

After the injection of BIC into the Put, contralateral dystonic movements of the head and trunk were observed. A single spike discharge was not enough to produce dystonia. Burst discharges were necessary for dystonic movement. Though the Cd and Put share similar phylogenetic origins, neuronal morphologies, synaptic ultrastructures, and afferent and efferent connections, responses of neurons to the injection with BIC were different between the Cd and Put; a neuron in the Cd responded with a single discharge and that in the Put with burst discharges. So far, no anatomical or electrophysiological evidence to explain this phenomenon has been reported. This difference must be due to a difference of excitability of cellular membranes or neuronal networks including recurrent axon collaterals.

The relations of behavior to spikes were substantially different between the Cd and Put. In the Cd, though frequency of spikes is correlated with locomotor hyperactivity, movements of the limbs did not occur in correspondence with the spikes. In contrast, dystonic movements of the head and trunk were directly correlated with bursts of spikes in the Put. The same phenomenon has been described in monkeys (Alexander and DeLong, 1985a; Alexander and DeLong, 1985b). Microstimulation of the Put resulted in movements of individual body parts including legs, arms, and a face. In contrast, microstimulation of the Cd failed to produce motor effects. The Put is more directly involved in motor functions, whereas the Cd is involved in more complex behavioral functions.

These results were unexpected from classical studies which indicated dystonia to be produced by electrical stimulation of the Cd (Forman and Ward, 1957; Laursen, 1963). Recent careful studies also confirmed the classical conclusion (Ohno and Tsubokawa, 1987; Kitama et al., 1991). How can the discrepancy between our results and theirs be explained? Though Ohno and Tsubokawa (1987) ruled out the possibility of current spread to the corticofugal fibers by chronic ablation of the sensorimotor cortex, the following possibilities of activating putaminofugal fibers remained. (1) Dopaminergic nigroputaminal neurons could be activated via axon reflex by stimulating nigrocaudate fibers at the Cd. (2) Putaminofugal fibers that ran in the internal capsule close to the lateral edge of the Cd (Szabo, 1962) could be stimulated electrically. (3) Electrical stimulation of the caudal portion of the head of the Cd could stimulate putaminofugal fibers by spread of current. Cools (1973) demonstrated head-turning in cats by injecting dopamine into the head of the Cd. His volume (10µl) of dopamine was relatively large compared with that (1.5µl) of the BIC used in our experiment and diffusion of dopamine into other structures including the Put might have occurred.

Despite these controversial results of animal studies, a much clearer correlation between dystonia and the Put has been clinically presented. Computed tomography (CT) and magnetic resonance imaging (MRI) revealed a localized lesion due to stroke in the unilateral Put of an adult patient who showed contralateral hemi-dystonia (Burton et al., 1984) and also bilateral putaminal lesions occured in 2 children showing generalized dystonia probably due to Leigh's disease (Burton et al., 1984) or some other disease (Marsden et al., 1986). Further, localized lesions within the Put in association with dystonia were confirmed by positron emission tomography (PET) (Fross et al., 1987). Also, a tumor located in the Put resulted in manifestation of contralateral hemi-dystonia (Narbona et al., 1984). Three patients showing generalized dystonia perhaps due to mitochondrial encephalopathy also showed bilateral putaminal lesions upon CT (Berkovic et al., 1987). Though torticollis has been related to lesion in the head of the Cd in 3 patients, interpretation of these findings is compromised by the fact that the etiology of the lesion was a vascular malformation in all cases, raising the possibility of ischemia in other basal ganglia or brainstem nuclei (Marsden et al., 1985; Obeso and Gimenez-Roldán, 1988), Patients with striatonigral degeneration in which the Put was rather selectively degenerated showed marked hypertonia with proximal dominance (Adams et al., 1964). All these lines of clinical evidence strongly support our results, indicating that the Put is the structure responsible for the manifestation of dystonia.

CONCLUSION

Though the Cd and Put share common properties concerning fiber connections, they play different roles in the basal ganglia. The Cd is involved in the manifestation or driving of locomotion, while the Put is related to regulation of tonus in contralateral muscles. After the injection of BIC into the Cd, single spikes were recorded and frequency of the spikes correlated with locomotor hyperactivity. After the injection of BIC into the Put, strong bursting of spikes, which was necessary to produce dystonic movements of the head and trunk, was recorded.

REFERENCES

Adams, R.D., Van Bogaert, L., and Eecken, H.V., 1964, Striatonigral degeneration, *J. Neuropathol. Exp. Neurol.* 23:584-608.

Alexander, G.E., and DeLong, M.R., 1985a, Microstimulation of the primate neostriatum. I. Physiological properties of striatal microexcitable zones, *J. Neurophysiol.* 53:1401-1416.

Alexander, G.E., and DeLong, M.R., 1985b, Microstimulation of the primate neostriatum. II. Somatotopic organization of striatal microexcitable zones and their relation to neuronal response properties, *J. Neurophysiol.* 53:1417-1430.

Berkovic, S.F., Karpati, G., Carpenter, S., and Lang, A.E., 1987, Progressive dystonia with bilateral putaminal hypodensities, *Arch. Neurol.* 44:1184-1187.

Burton, K., Farrell, K., Li, D., and Calne, D.B., 1984, Lesions of the putamen and dystonia: CT and magnetic resonance imaging, *Neurology* 34:962-965.

Chevalier, G., Vacher, S., Deniau, J.M., and Desban, M., 1985, Disinhibition as a basic process in the expression of striatal functions. I. The striato-nigral influence on tecto-spinal/tecto-diencephalic neurons, *Brain Res.* 334:215-226.

Cools, A.R., 1973, Chemical and electrical stimulation of the caudate nucleus in freely moving cats: the role of dopamine, *Brain Res.* 58:437-451.

Delgado, J.M.R., 1979, Inhibitory functions of the neostriatum, *in:* "The neostriatum," I. Divac and R.G.E. Oberg, eds., Pergamon, Oxford, pp. 241-261.

Deniau, J.M., and Chevalier, G., 1985, Disinhibition as a basic process in the expression of striatal functions. II. The striato-nigral influence on thalamocortical cells of the ventromedial thalamic nucleus, *Brain Res.* 334:227-233.

Forman, D., and Ward, J.W., 1957, Responses to electrical stimulation of caudate nucleus in cats in chronic experiments, *J. Neurophysiol.* 20:230-244.

Fross, R.D., Martin, W.R.W., Li, D., Stoessl, A.J., Adam, M.J., Ruth,T.J., Pate, B.D., Burton, K., and Calne, D.B., 1987, Lesions of the putamen: their relevance to dystonia, *Neurology* 37:1125-1129.

Kelly, P.H., and Moore, K.E., 1976, Mesolimbic dopaminergic neurons in the rotational model of nigrostriatal function, *Nature* 263:695-696.

Kitama, T., Ohno, T., Tanaka, M., Tsubokawa, H., and Yoshida, K., 1991, Stimulation of the caudate nucleus induces contraversive saccadic eye movements as well as head turning in the cat, *Neurosci. Res.* 12:287-292.

Kravitz, E.A., 1967, Acetylcholine, γ-aminobutyric acid and glutamic acid: physiological and chemical studies related to their roles as neurotransmitter agents, *in:* "The Neuroscience," G.G. Quarton, T. Melnechuk and F.O. Schmitt, eds., Rockefeller University Press, New York, pp. 433-444.

Lange, H., Thorner, G., Hope, A., and Schroder, K.F., 1976, Morphometric studies of the neuropathological changes in choreatic disease, *J. Neurol. Sci.* 28:401-425.

Laursen, A.M., 1963, Corpus striatum, *Acta Physiol. Scand.* 59, Suppl. 211:42-73.

Marsden, C.D., Lang, A.E., Quinn, N.P., McDonald, W.I., Abdallat, A., and Nimri, S., 1986, Familial dystonia and visual failure with striatal CT lucencies, *J. Neurol. Neurosurg. Psychiatry* 49:500-509.

Marsden, C.D., Meldrum, B.S., Pycock, C., and Tarsy, D., 1975, Focal myoclonus produced by injection of picrotoxin into the caudate nucleus of the rat, *J. Physiol. (Lond.)* 246/2:96.

Marsden, C.D., Obeso, J.A., Zarranz, J.J., and Lang, A.E., 1985, The anatomical basis of symptomatic hemidystonia, *Brain* 108:463-483.

McGeer, P.L., and McGeer, E.G., 1975, Evidence for glutamic acid decarboxylase-containing interneurons in the neostriatum, *Brain Res.* 91:331-335.

McKenzie, G.M., and Viik, K., 1975, Chemically induced choreiform activity: antagonism by GABA and EEG patterns, *Exp. Neurol.* 46:229-243.

McLennan, H., Emmons, P.R., and Plummer, P.M., 1964, Some behavioral effects of stimulation of the caudate nucleus in unrestrained cats, *Can. J. Physiol. Pharmacol.* 42:329-339.

Muramatsu, S., Yoshida, M., and Nakamura, S., 1990, Electrophysiological study of dyskinesia produced by microinjection of picrotoxin into the striatum of the rat, *Neurosci. Res.* 7:369-380.

Nakamura, S., Muramatsu, S., and Yoshida, M., 1990, Role of the basal ganglia in manifestation of rhythmical jaw movement in rats, *Brain Res.* 535:335-338.

Narbona, J., Obeso, J., Tunon,T., Martinez-Lage, J.M., and Marsden, C.D., 1984, Hemi-dystonia secondary to localized basal ganglia tumor, *J. Neurol. Neurosurg. Psychiatry* 47:704-709.

Obata, K., 1972, The inhibitory action of γ-aminobutyric acid, a probable synaptic transmitter, *Int. Rev. Neurobiol.* 15:167-187.

Obeso, J.A., and Giménez-Roldán, S., 1988, Clinicopathological correlation in symptomatic dystonia, *Advances in Neurol.* 50:113-122.

Ohno, T., and Tsubokawa, H., 1987, Regional differences in the cat caudate nucleus as to the effectiveness in inducing contraversive head-turning by electrical stimulation, *Neurosci. Res.* 4:497-516.

Park, M.R., Lighthall, J.W., and Kitai, S.T., 1980, Recurrent inhibition in the rat neostriatum, *Brain Res.* 194:359-369.

Precht, W., and Yoshida, M., 1971, Blockage of caudate-evoked inhibition of neurons in the substantia nigra by picrotoxin, *Brain Res.* 32:229-233.

Pycock, C.J., and Marsden, C.D., 1978, The rotating rodent: a two-component system?, *Eur. J. Pharmacol.* 47:167-175.

Schwartzman, R.J., Eidelberg, E., and Alexander, G.M., 1986, Asymmetrical regional changes in energy metabolism of the central nervous system during walking, *Brain Res.* 398:113-120.

Stevens, J.R., Kim, C., and MacLean, P.D., 1961, Stimulation of caudate nucleus. Behavioral effects of chemical and electrical excitation, *Arch. Neurol.* 4:47-54.

Swerdlow, N.R., and Koob, G.F., 1984, The neural substrates of apomorphine-stimulated locomotor activity following denervation of the nucleus accumbens, *Life Sci.* 35:2537-2544.

Szabo, J., 1962, Topical distribution of the striatal efferents in the monkey, *Exp. Neurol.* 5:21-36.

Ueki, A., 1983, The mode of nigro-thalamic transmission investigated with intracellular recording in the cat, *Exp. Brain Res.* 49:116-124.

Ungerstedt, U., and Arbuthnott, G.W., 1970, Quantitative recording of rotational behavior in rats after 6-hydroxydopmamine lesions of the nigrostriatal dopmanine system, *Brain Res.* 24:485-493.

Von Bonin, G., and Shariff, G.A., 1951, Extrapyramidal nuclei among mammals; a quantitative study, *J. Comp. Neurol.* 94:427-438.

Von Voigtlander, P.F., and Moore, K.E., 1973, Turning behavior of mice with unilateral 6-hydroxydopamine lesions in the striatum: effects of apomorphine, L-DOPA, amanthadine, amphetamine and other psychomotor stimulants, *Neuropharmacology* 12:451-462.

Yoshida, M., 1981, The GABA-ergic systems and the role of basal ganglia in motor control, *in:* "GABA and the Basal Ganglia, Advance in Biochemical Psychopharmacology, Vol. 30," G. Di Chiara, and G.L. Gessa, eds., Raven Press, New York, pp. 37-52.

Yoshida, M., Nagatsuka, Y., Muramatsu, S., and Niijima, K., 1991, Differential roles of the caudate nucleus and putamen in motor behavior of the cat as investigated by local injection of GABA antagonists, *Neurosci. Res.* 10:34-51.

Yoshida, M., Nakajima, N., and Niijima, K., 1981, Effect of stimulation of the putamen on the substantia nigra in the cat, *Brain Res.* 217:169-174.

Yoshida, M., and Omata, S., 1979, Blocking by picrotoxin of nigra-evoked inhibition of neurons of ventromedial nucleus of the thalamus, *Experientia* 35:794.

Yoshida, M., and Precht, W., 1971, Monosynaptic inhibition of neurons of the substantia nigra by caudate-nigral fibers, *Brain Res.* 32:225-228.

Yoshida, M., Rabin, A., and Anderson, M.E., 1971, Two types of monosynaptic inhibition of pallidal neurons produced by stimulation of the diencephalon and substantia nigra, *Brain Res.* 30:235-239.

Yoshida, M., Rabin, A., and Anderson, M.E., 1972, Monosynaptic inhibition of pallidal neurons by axon collaterals of caudate-nigral fibers, *Exp. Brain Res.* 15:333-347.

PALLIDAL AND NIGRAL HYPOKINESIA: AN EXPERIMENTAL ANALYSIS IN THE MONKEY

Elisabeth Trouche, François Viallet, Paul Apicella, Meryem Alamy, Jean-Claude Pons and Eric Legallet

Laboratoire de Neurobiologie Cellulaire et Fonctionnelle
Centre National de la Recherche Scientifique
31, chemin Joseph Aiguier
F 13402 Marseille Cedex 20 France

INTRODUCTION

Hypokinesia is certainly the deficit which is most frequently observed when the basal ganglia (BG) have been impaired either experimentally or for pathological reasons. The term hypokinesia is classically used to describe motor deficits resulting in a weakening and a general slowing down of voluntary movements. A distinction has recently been made between two aspects of hypokinesia, namely bradykinesia and akinesia (Hallett, 1990). Bradykinesia can be defined as the gradual slowing down of a movement as it is being performed, and akinesia as difficulty in initiating movement, or the loss of this ability. The type of hypokinesia encountered in parkinsonian patients is generally thought to provide an excellent model for BG dysfunction in general. Numerous studies have focused on this topic over the last few years. It has emerged from all these data that parkinsonian bradykinesia and akinesia actually reflect a whole range of multiple deficits affecting a large number of processes, and that these processes are involved not only in motor activity but also in complex aspects of behaviour such as motivation, memorizing, and the subject's awareness of the surrounding context. There exist moreover other variables which have to be taken into account when attempting to interpret this already complex clinical picture, such as whether or not the subjects are undergoing treatment, and the individual compensatory strategies used to compensate the deficient function. The multiplicity of the factors involved in parkinsonian hypokinesia has been pointed out by Hallett (1990). This is what makes it so difficult to investigate the topic experimentally and to usefully extrapolate information about the working of the intact BG from the data obtained on subjects with lesions. It therefore seemed worth developing an experimental method of investigating hypokinesia in animals involving both the use of standardized techniques for assessing the subjects' performance levels and procedures for performing specific, restricted lesions on the nuclei constituting the BG.

The first precise evaluation of experimentally-induced nigral hypokinesia in monkeys originated from our laboratory (Viallet et al., 1983). It was observed at that time that unilateral electrolytic destruction of the substantia nigra (SN) resulted in contralateral deficits consisting of delayed movement initiation and considerable slowing down of the movement execution, without effect on the movement accuracy. This hypokinesia is always accompanied by postural abnormalities consisting of persistent semi-flexion of the contralateral limbs and bending of the head and body towards the side contralateral to the lesion (Viallet et al., 1983). Hypokinetic states have also been induced by performing intranigral 6-OHDA (Viallet et al., 1984; Apicella et al., 1990) and systemic MPTP injections

The Basal Ganglia IV, Edited by
G. Percheron *et al.*, Plenum Press, New York, 1994

(Doudet et al., 1985; Schultz et al., 1989b). The akinesia observed subsequent to nigral lesions has been attributed both to deficits affecting motor initiation (Viallet et al., 1983; Apicella et al., 1990) and to the existence of sensory neglect (Apicella et al., 1991). This dysfunction is known to involve the nigrostriatal dopaminergic system in particular. Another feature which should be taken into account when describing nigral akinesia is the existence of oculomotor deficits, since it has been reported that the latency of subjects' oculomotor saccades lengthened after MPTP injections (Brooks et al., 1986; Schultz et al., 1989a). As far as nigral bradykinesia is concerned, the picture built up on the basis of kinematic movement analyses showed a slowing down of both the overall velocity of movements and that of the initial acceleration phase (Viallet et al., 1983).

For globus pallidus (GP) lesion, only bilateral lesions were found to induce marked hypokinesia, accompanied by characteristic bending of the neck and limbs (Denny-Brown and Yanagisawa, 1976), which mimicked the human pallidal syndrome (Martin, 1967) consisting of dystonic flexion. A similar flexion dystonia has been induced by cooling either the region including the GP (Hore et al., 1977) or the GP itself (Beaubaton et al., 1981; Trouche et al., 1979), or by performing kaïnic acid (DeLong and Coyle, 1979) or quisqualic acid injections (Alamy et al., 1990, 1991). There seems to exist a general agreement among most of the recent data on the effects of pallidal damage on the performance of voluntary motor activities (Beaubaton et al., 1981; Horak and Anderson, 1984; Mink and Thach, 1991b) as to the fact that unilateral GP lesions result in the lengthening of the movement execution time performed with the contralateral limb without affecting its latency: in other words, GP lesions result in bradykinesia but not in akinesia. It has been suggested by Hore and Vilis (1980) that this bradykinesia might be due to co-contractions between groups of agonist and antagonist muscles. GP bradykinesia is accompanied by deterioration of the movement accuracy (Beaubaton et al., 1981). When the visual information about the ongoing movement is abolished, the deterioration is further accentuated (Beaubaton et al., 1981; Alamy et al., 1991). Data obtained at our laboratory (Beaubaton et al., 1981) have suggested that overreliance on visual afferents to control their movements might explain the GP bradykinesia observed in subjects with pallidectomy.

Since 1976, our group has focused on systematically exploring the changes in the latency and the duration of movements of various kinds (goal-directed movements involving a target and non-directed movements; tasks involving speed and/or accuracy constraints) occurring as the result of specific lesions to one of the structures belonging to the BG: the putamen (Beaubaton et al., 1980), the GP (Amato et al., 1978; Trouche et al., 1979; Beaubaton et al., 1981; Alamy et al., 1990; 1991), the SN (Viallet et al., 1983), and the nigro-striatal dopaminergic system (Viallet et al., 1984; Apicella et al., 1990) .

In what follows, it is proposed to present some of the data which seem to be most liable to contribute to a better understanding of hypokinesia, particularly that induced by nigral or pallidal lesion.

EXPERIMENT 1: SENSORY NEGLECT AS A COMPONENT OF NIGRAL HYPOKINESIA

In order to address the hypothesis that impaired dopaminergic transmission in the striatum may result in the neglect of stimuli presented to the side contralateral to the dopamine (DA) depletion, monkeys with unilateral 6-OHDA-induced lesion of the nigrostriatal dopaminergic pathway were tested in a behavioral task specifically designed to assess the latency of the movement initiation and to exclude the possibility that a movement direction bias might occur.

Methods

The subjects were two baboon monkeys (*Papio papio*) trained to perform a visual reaction time task involving hand movements which were not spatially oriented. The monkey sat in a special cage described in earlier publications (Trouche and Beaubaton, 1980). It was required to keep its hand motionless on a metal bar situated centrally at the bottom of a semi-circular panel placed in front of it, on which five red light emitting diodes (LEDs) were arranged horizontally at eye level and at arm's length. Four LEDs were positioned at angles of 30 and 60° on each side and the fifth one was positioned centrally (left part of Figure 1).

The trial was initiated when the monkey's hand contacted the bar. After an unpredictable time interval (0.5 to 2 s), one LED was switched on at a point which varied from trial to trial. In response to this signal, the monkey had to withdraw its hand from the bar within less than 500 ms in order to receive a liquid reward. We measured the reaction time (RT) from the onset of the trigger light to the withdrawal of the hand from the resting bar. A failure to release the bar within 500 ms after the signal onset led to the light being switched off and to no reward being delivered.

We analyzed the RTs and the percentage of "incorrect" responses (RTs > 500ms) with each trigger location. The data were subjected to a three-factor analysis of variance (arm x period x stimulus location) as previously described (Apicella et al., 1991). After training to stable performance levels, lesions of the nigral dopaminergic neurons were performed by infusing 6-OHDA locally into the left SN (Apicella et al., 1990). The monkeys were tested over a 4-month period after 6-OHDA injection, at the end of which they were sacrificed, and their brains where removed and processed for histological and biochemical examination (Apicella et al., 1990).

Results

Biochemistry. The results of the biochemical analysis showed that the DA levels decreased significantly (35 to 71%) in the striatum ipsilateral to the 6-OHDA injection, as compared with the intact side, in both monkeys.

Reaction time performance. Comparisons between the pre- and post-operative performances showed that the 6-OHDA injection induced a significant increase in the RT (P<0.01), whichever arm was used. In both monkeys, the significant interaction (P<0.01) between arms and periods confirmed that the contralateral arm was more severely affected by the lesion than the ipsilateral one. A significant interaction (P<0.01) was also observed between periods and stimulus locations when the monkeys performed the task with the arm ipsilateral to the 6-OHDA injection, so that impaired dopaminergic transmission was associated with increased RT only when the stimulus was presented in the hemispace contralateral to the lesion (right part of Figure 1). This interaction was not significant in one monkey when it performed the task with the arm contralateral to the 6-OHDA injection, while in the other monkey, an interaction of this kind existed but at a lower level of significance (P < 0.05).

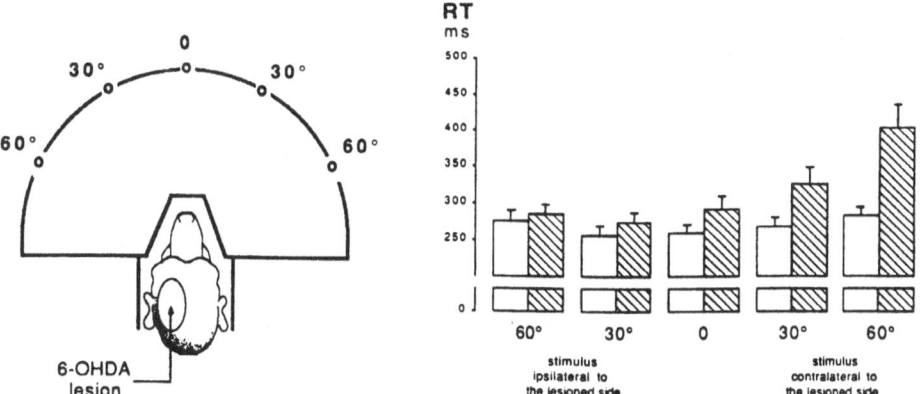

Figure 1. Effects of unilateral lesion of the nigral dopaminergic neurons on reaction time performance in a monkey performing movements with the arm ipsilateral to the lesion. Histograms indicate mean reaction times calculated on 16 to 18 sessions (256-288 trials for each of the five stimulus location), with vertical bars showing standard deviation, before (white) and after (hatched) lesion. RT: reaction time.

In addition, when both monkeys used the arm ipsilateral to the lesion, the selective RT lengthening which occurred in response to right-sided stimuli was paralleled by a significant increase ($P<0.01$) in the number of "incorrect" responses, as compared to the pre-operative data.

Discussion

In the present study, behavioral reactivity to visual stimuli presented to the hemispace either ipsilateral or contralateral to an unilateral striatal DA depletion was assessed in monkeys by measuring the latency of nonspatially oriented arm movements. Although slowness in initiating visually-triggered arm movements has been reported in monkeys after striatal DA depletion (Viallet et al., 1983; Schneider et al., 1988; Schultz et al., 1989b; Apicella et al., 1990), there have been few attempts (Apicella et al., 1991; Bankiewicz et al., 1991; Schneider et al., 1992) to quantify the effects of varying the location of the trigger stimulus in the visual field on this slowing of the movement. The results of our experiments show that the monkeys exhibited a severe increase in the latency of responses performed with the arm contralateral to the DA-depleted side, which was little if at all influenced by the location of the trigger stimulus in either hemispace. Conversely, when the monkeys responded with the ipsilateral arm, the RTs showed no change when the responses were triggered by stimuli located in the ipsilateral hemispace, while an increase in the RT was observed when the stimuli were located in the contralateral hemispace. This delay in the movement initiation can be said to be a behavioral correlate of a lack of reactivity to stimuli presented in the hemispace opposite the striatal DA depletion. The latter effect can be compared with the signs of unilateral "sensory neglect" observed in rats (Ljungberg and Ungerstedt, 1976) and cats (Feeney and Wier, 1979), which failed to respond or orient to external stimuli applied on the side contralateral to the damaged nigrostriatal system.

The main point which requires to be discussed as regards the origin of the neglect reported in our monkeys concerns the respective contributions of attentional deficits and initiation difficulties. Recent studies performed on rats have suggested that the neglect induced by unilateral 6-OHDA injection into the striatum may result from defective initiation of actions towards the contralateral hemispace, regardless of the location of the stimulus (Carli et al., 1989). The visual RT task used by the latter authors involved movements of the whole body towards or away from the trigger stimuli, while in our RT procedure the same motor response to the stimuli was required, irrespective of the stimulus location. The stimuli contained no inherent directional information about motor responding, but supplied only timing signals about the forthcoming movement onset. From the results obtained with our procedure it can be concluded that the neglect was specific to the stimulus location and not to the orientation of the response in space. This suggests a difficulty in perceptual aspects of the task, as the monkeys showed an impaired ability to attend to the stimuli emanating from the contralateral hemispace when they performed the task with the ipsilateral arm. There is a possibility however that the motor initiation may have been delayed when the trigger stimulus was located in one specific hemispace. In this case, it is the coupling between stimuli emanating from the contralateral hemispace and the motor response which may have been deficient. On the other hand, it remains to be established whether the position of the stimuli may affect the oculomotor processes involved in behavioral reactivity, since impairments in gaze orientation and in generating saccades have been reported after marked bilateral DA depletion in the striatum (Brooks et al., 1986; Schultz et al., 1989a).

It has been demonstrated that the nigrostriatal dopaminergic system is involved in attention and motor initiation, two closely related processes which when disturbed lead to an impaired ability to react to external stimuli. Destruction of dopaminergic terminals may disrupt sensory information processing within the striatum (Schneider, 1991), thus causing abnormal behavioral reactions to environmental stimuli. Further studies are needed to specify which levels of sensorimotor processing are impaired, in order to be able to explain the lack of responsiveness observed when the stimuli were located in the hemispace contralateral to the striatal DA depletion.

EXPERIMENT 2: KINESTHETIC REAFFERENTS AND NIGRAL HYPOKINESIA

In experiment 2, we investigated the effects on a pointing movement parameters of withdrawing the visual reafferents relating to the arm movements, in order to determine whether visual feed-back actually plays a compensatory role when the BG are impaired.

Methods

Behavioral task. The subjects were four baboons (*Papio papio*), and the experimental set-up has been fully described in an earlier publication (Trouche and Beaubaton, 1980; Viallet et al., 1983). Two pointing conditions were used that differed in the availability of visual feedback: 1) under the visual closed-loop condition, the visual cues were available over the entire hand -trajectory. 2) under the visual open-loop condition, a special set-up made it possible to present an image of the target without involving any change in the experimental environment. The animal's hand was hidden by a mirror, and the movement was performed without any visual control being possible.

The following experimental variables were recorded: the reaction time (RT), corresponding to the interval between the onset of the luminous target and the release of the handle; and the movement time (MT), corresponding to the interval between the release of the handle and the first finger contact on the screen. The pointing error (E) was expressed in terms of the distance between the computed mean of the pointing XY coordinates and the actual target. Film recordings made it possible to determine the mean length of the finger trajectories (L).

During the preoperative period, a preliminary training phase was devoted to the learning of the pointing task under the visual closed-loop condition. When the performances had become stabilized, two animals (RI, DE) underwent SN lesions. The two others (MA, AL) were trained under the visual open-loop condition before the SN lesion, which was again performed only after stabilization of the performance levels. The effects of SN lesion on the pointing movement were therefore studied separately under visual control in two animals (RI, DE) and without visual control in the two others (MA, AL).

Lesion. The SN lesions were performed using the stereotaxic approach described in an earlier publication (Viallet et al., 1983).

Results

The effects of SN lesion on the pointing movement variables are summarized in Table 1.

Table 1. Effects of unilateral SN lesion on the contralateral limb in visual closed-loop and open-loop conditions. Pointing movement variables: RT, reaction time; MT, movement time; e, pointing error; L, length index trajectory. Values are means ± SD. *, $p < 0.005$ (Student's t - test).

	VISUAL CLOSED-LOOP				VISUAL OPEN-LOOP			
	RI		DE		MA		AL	
	Preop	Postop	Preop	Postop	Preop	Postop	Preop	Postop
RT	258±28	288±30	258±47	296±50	239±36	304±48	255±40	302±62
(ms)	13.54*		10.27		20.13*		10.5*	
MT	167±23	258±46	138±29	236±73	167±24	203±39	164±23	218±41
(ms)	32.27*		22.62		14.51*		18.72*	
E	4.7±7.5	4.9±10.1	11.7±7.5	12.2±14	49±12.2	44±17.5	39.8±15	30.7±23
(mm)	0.29		0.54		4.8*		5.4*	
L	22.5±0.7	22.3±0.6	22.8±0.6	23±1	25.7±0.6	21.6±1.1	24.9±0.9	21.4±1.7
(cm)	0.97		0.76		14.72*		8.11*	

Under the visual closed-loop condition, the SN lesion led to an increase in both the RTs and the MTs of the arm contralateral to the lesion without causing any changes in the pointing error or the trajectory length.

Under the visual open-loop condition, an increase in both the reaction and movement times of the arm contralateral to the lesion was also observed after SN lesion. Pointing error data were more complex. During the preoperative period, the mean pointing error values were actually much larger than those obtained under the visual closed-loop condition. This defective terminal accuracy when visual cues were lacking was due to a lengthening of the index trajectories, i. e., the actual movement was hypermetric. As shown in figure 2, when visual feedback was not available, a marked rise was observed in the spatial spread of the contact points on the screen, along with a shift in the mean point of contact to above the actual target. During the early postoperative period, the mean pointing error values decreased slightly, although the real changes in pointing accuracy consisted of a significant shortening of the index trajectories, i. e., the actual movement was hypometric, and the spatial spread of contact points remained large, while the mean point shifted to slightly below the actual target.

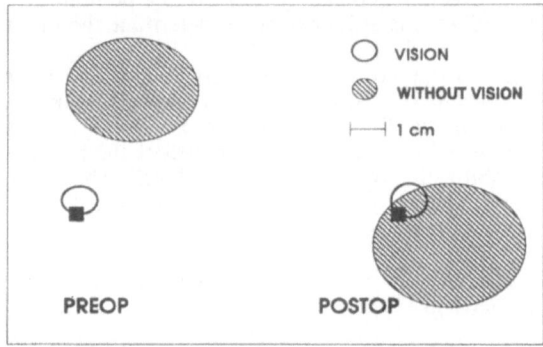

Figure 2. Spatial distribution of pointing before and after SN lesion.

Discussion

In pointing-movement experiments carried out on animals after SN lesion, no conspicuous changes in the amplitude of the spatial errors was found to accompany increase in the reaction time and movement time. Any spatial errors observed were in fact always very slight. When visual feed-back was abolished during a movement, however, the spatial errors increased considerably in amplitude due to a decrease in the trajectory amplitude. These results suggest that the functional exclusion of the BG leads to an increased use of corrective visual feed-back, which becomes responsible for controlling the movement and ensures that terminal accuracy is maintained. It can thus be assumed that when the feed-forward system of movement control is put out of order by lesion of the BG, it is replaced by a feed-back control system, and the latter system relies on both visual and proprioceptive information.

The mechanism whereby a hand trajectory is corrected on the basis of visual cues is thought to be a relatively slow one. The lengthening of the movement time might then result from an increased use of visual feedback, which is compatible with the slowing down in the performance of the movement observed, whereas the terminal accuracy was not affected (Hore et al., 1977).

When visual reafferents are lacking, the feedback control of movement will operate on the basis of proprioceptive information. The modalities underlying this mechanism are thought to be two-fold: a reflex one involving a servo-mechanism (Marsden et al., 1977) that is thought to regulate the physical state of the muscle with respect to the constraints imposed by the movement, and a kinesthetic one, involving sensory information contributing to the reorganization of the motor program (Bizzi, 1980). Long-latency reflexes recordings obtained on parkinsonian patients (Tatton and Lee, 1975) strongly suggest that the BG are involved in the control of the above proprioceptive afferent mechanisms. Impairment of the BG seems to disturb gain regulation in the transmission of proprioceptive information and might well affect error-detection mechanisms in the absence of compensatory visual information. During a pointing movement performed with no visual feed-back available, the impairment of the BG might explain the insufficiency of the hand trajectory amplitude. Complementary experimental data suggest that the disturbed proprioceptive processing resulting from impairment of the BG may also be reflected in defective adaptive movement control. Actually the study on the recovery of contralateral lesion-induced deficits under open-loop conditions in monkeys with unilateral SN lesions has suggested that subjects used intact proprioceptive information originating from the ipsilateral side to compensate for defective proprioceptive feed-back from the contralateral side (Viallet et al., 1987).

EXPERIMENT 3: SOME EVIDENCE FOR THE EXISTENCE OF PALLIDAL AKINESIA

The role of the globus pallidus in motor initiation has been giving rise to some speculation. In order to analyze the role of the GP in this phase, monkeys with unilateral GP lesion were tested in two behavioral tasks involving two different constraints (accuracy or speed).

Methods

Behavioral tasks. The subjects were six baboons (*Papio papio*) trained to perform reaction time tasks of two types: the one (Fig. 3A) with an accuracy constraint, requiring accurate index pointing at a spatially defined target and thus depending strongly on visual guidance of the limb. The material and methods used for this task were the same as those described previously (Trouche and Beaubaton, 1980). The other task (Fig. 3B) consisted of withdrawing the hand, with a speed constraint instead of the accuracy constraint: this was a non spatially oriented movement, requiring minimal visual guidance.

The reaction time (RT) was measured in the two tasks and we also analyzed the percentage of "incorrect responses" (RT>500 ms) in the reaction time task with a speed constraint. The methods of statistical analysis used were as described in a previous article (Trouche and Beaubaton, 1980).

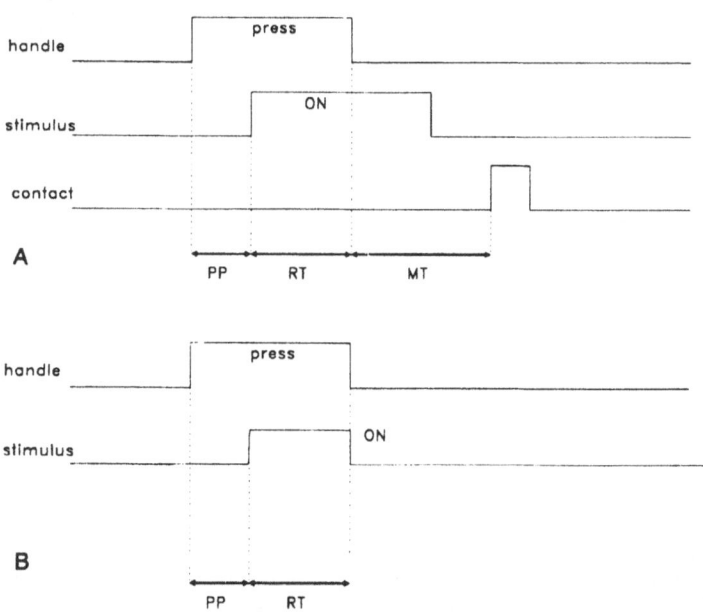

Figure 3 . Schematic diagram of the experimental procedure in reaction time tasks with accuracy constraint (A) and with speed constraint (B). MT: movement time, PP: preparatory period, RT: reaction time.

The RTs could be decomposed into two phases on the basis of the changes in the force exerted by the subject's hand on the starting platform, which were measured by means of a system of strain gauges attached to the platform. The following variables were therefore measured:
- the premotor phase t1, corresponding to the latency of the force change;
- and the motor phase t2, corresponding to the duration of the force change.
   The monkeys were tested during a 6-months period after GP lesion. At te end of the experiments the subjects were sacrificed and the brains were cut serially in frontal 50 μm thick sections and stained in order to determine the extent of the lesion.

   Lesion. The GP was located using the stereotaxic coordinates given in the atlas by Davis and Huffman (1968). These coordinates were then corrected using ventriculographic methods and by performing electrophysiological recordings. In order to perform a specific GP lesion, a neurotoxic aminoacid analogue, quisqualic acid (QUIS), was used. The lesion was induced by unilaterally injecting 0.6μl of quisqualic acid at a concentration of 0.18M at 4 points on four antero-posterior planes (A15, A16, A17, A18). The injections were performed using a Hamilton syringe implanted obliquely at an angle of 70° to the horizontal in order to spare the fibers of the internal capsule.

Results

   Histology. The results of the histological analysis and the comparisons between the neuronal densities on the side of the QUIS injection and the contralateral side show that in practically all the subjects the lesion encompassed the dorso-median part of the two pallidal segments.
   Reaction time performances.1) Reaction time task with an accuracy constraint after GP lesion: GP lesion did not bring about any statistically significant changes in the RT in any of the animals. In subject N, for example, the data did not show any change in the RT (Fig. 4A). During the premotor (t1) and motor phases (t2), the data showed the existence of no statistically significant changes in t1, but a significant increase was observed in t2 (Alamy, 1992).

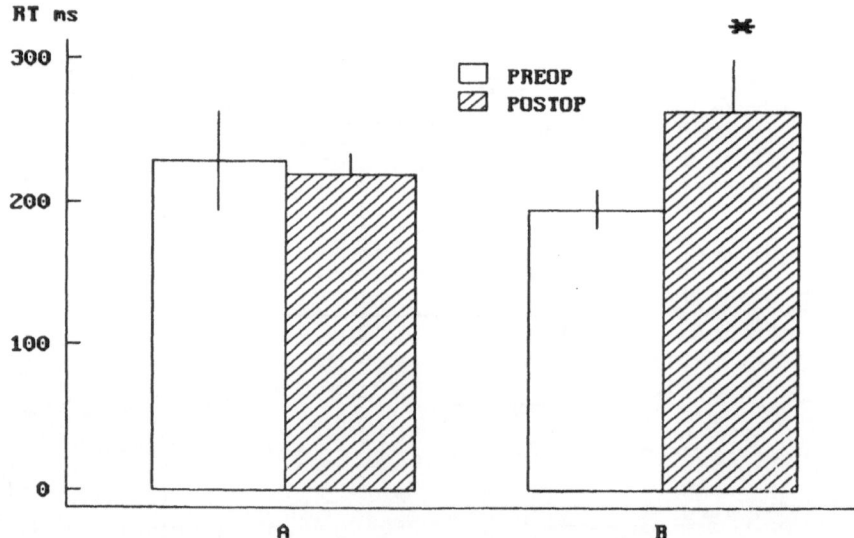

Figure 4. Effect of GP lesion on reaction time (RT) in two behavioural tasks, in reaction time tasks with accuracy constraint (A) and with speed constraint (B).
   Each histogram represent mean value of 750 trials recorded in 15 sessions before (PREOP) and after (POSTOP) GP lesion. Comparison between these performances showed a significant lengthening in the RT in the case of non spatially aiming movement ($p < 0.05$) and no change for the spatially aiming movement.

2) Reaction time task with a speed constraint after GP lesion: after an unilateral quisqualic acid lesion to the pallidum, a significant lengthening of the RT was observed in all the animals on the side opposite the lesion. In subject P for example, the RT increased by 34% (Fig. 4B). This is a statistically significant increase ($p < 0.05$). During the RT, the latency of the change in the force (t1) and the duration of the isometric force (t2) were measured separately. Comparisons between the t1 and t2 values before and after GP lesion showed that both values increased significantly (Alamy et al., 1990). In addition, a significant increase in the percentage of "incorrect responses" was observed in the contralateral limb (Alamy et al., 1990).

Discussion

These data clearly show that pallidal akinesia can be detected using well defined reaction time procedures such as a reaction task with speed constraints. In this particular case, both the RT and the percentage of "incorrect" responses increased after the GP lesion. The lengthening of the RT and the increase in the percentage of "incorrect" responses observed in our study after GP lesion are two signs of akinesia which are in line with the idea that the GP may be involved in the control of motor initiation.

Our findings are in agreement with electrophysiological data suggesting that the GP is involved in the control of the motor initiation phase of movements (Neafsey et al., 1978; Nambu et al., 1990).

It remains however to be explained why no akinesia was observed when reaction time tasks with accuracy constraints were used. Similar findings have been reported by other authors: no change in the movement latency was observed in subjects with GP lesions performing a RT task with accuracy requirements resembling that used here (Horak and Anderson, 1984) or in a step-tracking task (Hore and Vilis, 1980; Mink and Thach, 1991b).

Upon analysing the RT data more closely, a lengthening of the motor phase was observed however in the present study in the case of the RT task with accuracy constraints. A similar lengthening was observed in the triggering of RT task with speed constraint, which seems to indicate that the GP may be involved in the control of the movement initiation phase (Alamy et al., 1990).

Whether or not the premotor phase of the RT lengthened after GP lesions was found to depend on which behavioural task was used for this purpose: the premotor phase of the RT showed an impairment only when the RT test with speed constraints involving the movement initiation was used (Alamy et al., 1990). The speed requirements may have had two effects which would account for this deficit: first they may have caused a faster triggering of the movement. The lengthening of the RT after GP lesion may show that the GP is involved in controlling the speed of movement initiation. Secondly, the speed constraint may have given the subjects less time to use visual cues, in which case the lengthening of the RT observed shows that visual cues are used to compensate for pallidal akinesia (Alamy, 1992).

Previous data (Hamada et al., 1990) have shown that although changes in the activity of GP cells accompany both fast and slow movements, the discharge frequency is higher in the case of fast movements. On similar lines, Mink and Thach (1991a) have also reported that the discharge pattern of the GP neurons showed more conspicuous changes when the subjects performed fast movements. Other studies have shown on the other hand that the GP is involved in the control of speed and amplitude during the execution phase (Georgopoulos et al., 1983) and during motor preparation (Trouche et al., 1991).

As regards the second hypothesis, the fact that visual feedback is known to play an important compensatory role after GP lesion in the control of an ongoing movement (Hore et al., 1977; Alamy et al., 1991) may explain why the effects of GP lesion were more obvious in the reaction time task with a speed constraint, which was designed to reduce the use of visual information (Alamy et al., 1990). Moreover, in this framework, Mink and Thach (1991a) have reported that in both segments of the GP, a higher proportion of neurons discharged during a ballistic movement corresponding to a visual open loop movement than during a visual closed loop movement.

On the basis of all the above data, it can therefore be conclude that the GP lesions may have affected the initiation of fast movements, which seem to be controlled on a feedforward basis, rather than that of slow movements, which depend more strongly on the use of peripheral afferents (Alamy, 1992).

CONCLUSION

Nigral and pallidal lesions both result in deficits specifically affecting the preparatory phase of motor activites. It is generally recognized that the preparatory state includes several stages which are all necessary for movements to be properly performed. These include perception of the stimulus signal, and the planning, programming and triggering of the movement. Each stage involves its own functional processes, such as those mediating signal recognition, and the mnesic, attentional and motivation processes. Our investigations on nigral and pallidal akinesia have shown that the mechanisms which are impaired are not the same with both types of lesion. The akinesia resulting from destruction of the nigro striatal dopaminergic system may be due to an overall lowering of some aspect of the subject's reactivity. The specific type of reactivity involved might be said to be a particular attentional state, an alertness to stimuli which are behaviorally relevant. Pallidal akinesia on the other hand might lead to a disorder focusing on the parametric control of specific aspects of movements, such as their amplitude or their velocity.

Although as we have seen, the factors contributing to nigral and pallidal akinesia are quite different, they have one main common feature: in both cases, it is the feedforward motor control aspects which are preponderantly impaired. By making greater use of feedback control during movement execution, i. e. by enhancing the processing of visual information about ongoing movements, it might be possible for these subjects to compensate for their feedforward deficits, but this strategy in turn would contribute to inducing pallidal or nigral bradykinesia (Trouche et al., 1983).

From the point of view of the functional anatomy of the BG, the destruction of the nigrostriatal dopaminergic system due to either experimental or pathological causes can be said to result in hyperactivity at the level of the internal GP segment. This gives rise to the paradoxical situation mentioned by Percheron et al. (1990), where hypokinesia can result either from hyperactivity of the GP segment or from the fact that the activity of this segment has been interrupted.

Acknowledgements

The authors wish to thank for technical assistance D. Casanova, A. Christolomme, M. Gavioli, R. Massarino and G. Povéda.

The English version of this manuscript was translated and revised by Dr Jessica Blanc.

REFERENCES

Alamy, M., Legallet, E., Pons, J.C., and Trouche, E., 1990, Globus pallidus and motor preparation: Reaction time after unilateral neurochemical lesion in the monkey, *Europ. J. Neurosci. Suppl.* 3:296.

Alamy, M., Legallet, E., Povéda, G., and Trouche, E., 1991, The compensatory role of visual feed-back after neurotoxic globus pallidus lesion: effect of the suppression of visual afferents on pointing movement variables in monkeys, *Europ. J. Neurosci. Suppl.* 4:152.

Alamy, M., 1992, Rôle du pallidum dans la régulation de la motricité volontaire chez le singe: effet d'une lésion excitotoxique unilatérale sur la posture, la préparation et l'exécution du mouvement. Implication des informations visuelles dans la compensation des déficits moteurs, Doctoral thesis, Aix-Marseille II University.

Amato, G., Trouche, E., Beaubaton, D., and Grangetto, A., 1978, The role of the internal pallidal segment on the initiation of a goal-directed movement, *Neurosci. Lett.* 9:159-163.

Apicella, P., Trouche, E., Nieoullon, A., Legallet, E., and Dusticier, N., 1990, Motor impairments and neurochemical changes after unilateral 6-hydroxydopamine lesion of the nigrostriatal dopaminergic system in monkeys, *Neuroscience* 38:655-666.

Apicella, P., Legallet, E., Nieoullon, A., and Trouche, E., 1991, Neglect of contralateral visual stimuli in monkeys with unilateral striatal dopamine depletion, *Behav. Brain Res.* 46:187-195.

Bankiewicz, K.S., Oldfield, E.H., Plunkett, R.J., Schuette, W.H., Cogan, D.G., Hogan, N., Zuddas, A., and Kopin, I.J., 1991, Apparent unilateral visual neglect in MPTP-hemiparkinsonian monkeys is due to delayed initiation of motion, Brain Res. 541:98-102.

Beaubaton, D., Amato, G., Trouche, E., and Legallet, E., 1980, Effect of putamen cooling on the latency, speed and accuracy of a pointing movement in the baboon, Brain Res. 196:572-576.

Beaubaton, D., Trouche, E., Amato, G., and Legallet, E., 1981, Perturbation de l'incitation et de l'exécution d'un mouvement visuellement guidé chez le babouin, après exclusion temporaire ou définitive du segment interne du globus pallidus, *J. Physiol. (Paris)* 77:107-118.

Bizzi, E., 1980, Central and peripheral mechanisms in motor control, *in*: "Tutorials in motor behavior," G.E. Stelmach, and J. Requin, eds., Amsterdam, North Holland, pp. 131-143.

Brooks, B.A., Fuchs, A.F., and Finocchio, D., 1986, Saccadic eye movement deficits in the MPTP monkey model of Parkinson's disease, *Brain Res.* 383:402-407.

Carli, M., Jones, G.H., and Robbins, T.W., 1989, Effects of unilateral dorsal and ventral striatal dopamine depletion on visual neglect in the rat: a neural and behavioural analysis, *Neuroscience* 29:309-327.

Davis, R., and Huffman, R.D., 1968, "A stereotaxic Atlas of the Brain of the Baboon (Papio), The southwest Foundation for Research and Education," ed., University of Texas Press, Austin and London,

DeLong, M.R., and Coyle, J.T., 1979, Globus pallidus lesions in the monkey produced by kainic acid: histologic and behavioral effects, *Appl. Neurophysiol.* 42:95-107.

Denny-Brown, D., and Yanagisawa, N., 1976, The role of the basal ganglia in the initiation of movement, *in*: "The basal ganglia," M.D. Yahr, ed., Raven Press, New-York, pp. 115-148.

Doudet, D., Gross, C., Lebrun-Grandie, P., and Bioulac, B., 1985, MPTP primate model of Parkinson's disease: a mechanographic and electromyographic study, *Brain Res.* 335:194-199.

Feeney, D.M., and Wier, C.S., 1979, Sensory neglect after lesions of substantia nigra or lateral hypothalamus: differential severity and recovery of function, *Brain Res.* 178:329-346.

Georgopoulos, A.P., DeLong, M.R., and Crutcher, M.D., 1983, Relations between parameters of step-tracking movements and single cell discharge in the globus pallidus and subthalamic nucleus of the behaving monkey, *J. Neurosci.* 8:1586-1598.

Hallett, M., 1990, Neurophysiologie clinique de l'akinésie, *Rev. Neurol.* 146(10):585-590.

Hamada, I., DeLong, M.R., and Mano, N., 1990, Activity of identified wrist-related pallidal neurons during step and ramp wrist movements in the monkey, *J. Neurophysiol.* 64:1892-1906.

Horak, F.B., and Anderson, M.E., 1984, Influence of globus pallidus on arm movements in monkeys. I. Effects of kaïnic acid-induced lesions, *J. Neurophysiol.* 52:290-304.

Hore, J., Meyer-Lohmann J., and Brooks, V.B., 1977, Basal ganglia cooling disables learned arm movements of monkeys in the absence of visual guidance, *Science* 195:585-586.

Hore, J., and Vilis, T., 1980, Arm movement performance during reversible basal ganglia lesions in the monkey, Exp. *Brain Res.* 39:217-228.

Ljungberg, T., and Ungerstedt, U., 1976, Sensory inattention produced by 6-OHDA-induced degeneration of ascending DA neurons in the brain, *Exp. Neurol.* 53:585-609.

Marsden, C.D., Merton, P.A., and Morton, H.B., 1977, The sensory mechanism of servo-action in human muscle, *J. Physiol. (Lond.)* 265:521-535.

Martin, J.P., 1967, "The Basal Ganglia and Posture," Pitman, London.

Mink, J.W., and Thach, W.T., 1991a, Basal ganglia motor control. I. Non exclusive relation of pallidal discharge to five movement modes, *J. Neurophysiol.* 65:273-300.

Mink, J.W., and Thach, W.T., 1991b, Basal ganglia motor control. III. Pallidal ablation: normal reaction time, muscle cocontraction, and slow movement, *J. Neurophysiol.* 65:330-350.

Nambu, A., Yoshida, S.I., and Jinnai, K., 1990, Discharge patterns of pallidal neurons with input from various cortical areas during movement in the monkey, *Brain Res.* 519:183-191.

Neafsey, E.J., Hull, C.D., and Buchwald, N.A., 1978, Preparation for movement in the cat. II. Unit activity in the basal ganglia and thalamus, *Electroenceph. Clin. Neurophysiol.* 44:714-723.

Percheron, G., Parent, A., Crossman, A.R., Filion, M., Mitchell, I.J., Bédard, P.J., François, C., Yelnik, J., and Fénelon, G., 1990, Substrat anatomo-physiologique de l'akinésie chez le primate, *Rev. Neurol.* 146(10):572-584.

Schneider, J.S., 1991, Responses of striatal neurons to peripheral sensory stimulation in symptomatic MPTP-exposed cats, *Brain Res.* 544:297-302.

Schneider, J.S., McLaughlin, W.W., and Roeltgen, D.P., 1992, Motor and nonmotor behavioral deficits in monkeys made hemiparkinsonian by intracarotid MPTP infusion, *Neurology* 42:1565-1572.

Schneider, J.S., Unguez, U., Yuwiler, A., Berg, S.C., and Markham, C.H., 1988, Deficits in operant behaviour in monkeys treated with N-methyl-4-phenyl-1,2,3,6-tetrahydropyridine (MPTP), *Brain* 111:1265-1285.

Schultz, W., Romo, R., Scarnati, E., Sundström, E., Jonsson, G., and Studer, A., 1989a, Saccadic reaction times, eye-arm coordination and spontaneous eye movements in normal and MPTP-treated monkeys, *Exp. Brain Res.* 78:253-267.

Schultz, W., Studer, A., Romo, R., Sundström, E., Jonsson, G., and Scarnati, E., 1989b, Deficits in reaction times and movement times as correlates of hypokinesia in monkeys with MPTP-induced striatal dopamine depletion, *J. Neurophysiol.* 61:651-668.

Tatton, W.G., and Lee, R.G., 1975, Evidence for abnormal long-loop reflexes in rigid parkinsonian patients, *Brain Res.* 100:671-676.

Trouche E., Alamy M., and Viallet F., 1991, Initiation impairment after globus pallidus lesion depend on the reaction time task, *Soc. Neurosci. Abstr.* vol 16:483.8.

Trouche, E., and Beaubaton, D., 1980, Initiation of a goal-directed movement in the monkey. Role of the cerebellar dentate nucleus, *Exp. Brain Res.* 40:311-322.

Trouche, E., Beaubaton, D., Amato, G., Legallet, E., and Zenatti, A., 1979, The role of the internal pallidal segment on the execution of a goal-directed movement, *Brain Res.* 175:362-365.

Trouche, E., Beaubaton, D., Viallet,. F., and Legallet, E., 1983, Contrôle proactif et rétroactif d'un mouvement de pointage: participation du noyau dentelé et de la substance noire, *in* : "Space Physiology-Physiologie Spatiale," A. Berthoz and F. Lestienne, eds., Cepadue, Toulouse, pp. 183-194.

Viallet, F., Trouche, E., Beaubaton, D., and Legallet E., 1984, Unilateral electrolytic and 6-OHDA lesions of the substantia nigra in baboons: Behavioural and biochemical data, *in*: "The basal ganglia: Structure and function," J.S. McKenzie, R.E. Kemm and L.N. Wilcock, eds., Plenum, New-York and London, pp. 373-391.

Viallet, F., Trouche, E., Beaubaton, D., and Legallet, E., 1987, The role of visual reafferents during a goal directed movement. A comparative study between open-loop and closed-loop performances in monkeys after unilateral electrolytic lesion of the substantia nigra, *Exp. Brain Res.* 65:399-410.

Viallet, F., Trouche, E., Beaubaton, D., Nieoullon, A., and Legallet, E., 1983, Motor impairments after unilateral electrolytic lesions of the substantia nigra in baboons: behavioural data with quantitative and kinematic analysis of a pointing movement, *Brain Res.* 279:193-206.

THE ROLE OF THE INTERNAL SEGMENT OF THE GLOBUS PALLIDUS IN MEDIATING DYSKINESIAS

Wendy C. Graham, Robert G. Robertson, Tipu Z. Aziz, David Peggs, Ian J. Mitchell, Michael A. Sambrook and Alan R.Crossman

Experimental Neurology and Myology Group, Department of Cell and Structural Biology, University of Manchester, M13 9PT, UK

INTRODUCTION

Changes in neuronal activity within the internal segment of the globus pallidus (GPi), which subsequently influence thalamo-cortical pathways, are believed to reflect movement-related information processed within the basal ganglia. Previous work from this laboratory would support the view that in the hypokinetic movement disorder of Parkinson's disease, the GPi is overactive, being driven by a disinhibited subthalamic nucleus (STN) mediating excitatory transmission. In contrast it is well documented that lesion or damage within the STN in man gives rise to choreic or ballistic dyskinesias (hyperkinesias), which, if STN output is reduced, would imply the presence of an underactivity of GPi neurones in the production of these movements. Therefore the current view is that an overall increase or decrease in GPi neuronal activity explains the two extremes of the movement spectrum. In this chapter we examine the sufficiency of this view by discussing more specifically the role of the GPi in mediating dyskinesias (chorea, ballism and dystonia), by drawing upon results from our own and other related studies.

REDUCTION OF EXCITATORY TRANSMISSION WITHIN THE GPI

Pharmacological Manipulations

Early studies involving intracerebral injections of the GABA antagonist bicuculline within the ventral lentiform nucleus in the monkey, proposed that the resulting contralateral limb choreiform dyskinesias were due to decreased input to the GPi from the STN. The reduction in STN activity was believed to result from an overactivity of the inhibitory pathway from the external pallidal segment (GPe) to the STN. Thus analysis of 2-deoxyglucose (2-DG) autoradiographs from these monkeys revealed decreased 2-DG uptake in both the GPi and the GPe, to which the STN projects, reflecting its underactivity (Mitchell et al., 1989). With the acceptance of the glutamatergic nature of STN transmission, it was proposed that the GPi was receiving less excitatory input during choreiform dyskinesia.

To confirm this more directly, the excitatory amino acid antagonist kynurenic acid (50-100ug) was injected within the GPi itself, to block STN-mediated neurotransmission (Robertson et al., 1989). Again the result was choreiform and ballistic dyskinesias. Analysis of 2-DG autoradiographs from monkeys rendered unilaterally dyskinetic in this way was also carried out (see Table 1). Interestingly the autoradiographs again revealed an involvement of the STN in the neural mechanisms mediating chorea/ballism, where there was a significant ipsilateral increase in 2-DG uptake (see Fig. 1). However, by what means could the STN, presynaptic to the pharmacological manipulation within the GPi, become involved?

Table 1. 2-DG results expressed as percentage change in optical density (O.D.) between injected and control sides following unilateral kynurenic acid injection in the GPi.

| | PERCENTAGE CHANGE IN O.D. | | | |
AREA	1	2	3	4
GPe	3.81 ± 0.96	0.23 ± 0.71	-1.07 ± 3.05	-3.99 ± 0.94
GPi	0.67 ± 1.45	2.22 ± 1.69	-7.69 ± 2.39	-2.99 ± 0.64
CN	5.41 ± 1.08	-0.46 ± 0.46	-3.21 ± 0.59	6.36 ± 0.99
Put d	2.27 ± 0.65	-1.48 ± 0.53	-2.64 ± 0.45	-4.94 ± 0.90
Put v	-4.31 ± 1.02	-1.55 ± 0.79	-1.02 ± 0.71	-7.78 ± 0.69
VA	1.72 ± 0.96	-2.08 ± 0.60	0.10 ± 0.42	-4.70 ± 0.45
VL	2.36 ± 0.80	-1.80 ± 0.65	-0.43 ± 0.85	-3.57 ± 1.01
STN[a]	8.98 ± 1.66	8.14 ± 1.78	3.82 ± 0.59	3.03 ± 1.05
Pf	3.17 ± 0.68	-1.78 ± 0.81	-2.64 ± 0.77	-0.27 ± 1.05
CM	1.55 ± 0.97	-4.08 ± 1.34	-4.38 ± 0.80	-2.14 ± 1.42
PPN	-0.08 ± 1.22	-0.40 ± 1.06	2.60 ± 0.86	-1.08 ± 1.15

Values derived by comparing O.D. of a given structure on the injected side with that from uninjected (control) side for each monkey (1 - 4), expressing the difference as a percentage change. Positive percentages denote an increase and negative percentages a decrease in 2-DG uptake. [a]: $p < 0.05$, matched paired t-test of O.D. values. GPe, GPi: external and internal pallidal segments; CN: caudate nucleus; Put d,v: dorsal and ventral putamen; VA,VL: anterior and lateral ventral thalamic nuclei; STN: subthalamic nucleus; Pf: parafascicular thalamic nucleus; CM: centromedian thalamic nucleus; PPN: pedunculopontine nucleus.

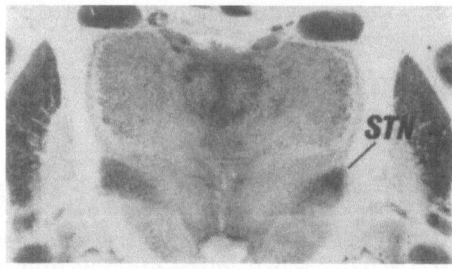

Figure 1. Autoradiograph showing increase in 2-DG uptake in right STN during choreiform dyskinesia elicited by injection of kynurenic acid in left GPi.

The answer may lie in the existence, between the two pallidal segments, of reciprocal, presumed GABAergic pathways, where GPi axons are believed to influence several GPe neurones (Hazrati et al., 1990). If GPi activity was reduced, following kynurenic acid blockade of excitatory input, the GPe would be disinhibited and thus GPe-STN transmission may increase. Indeed the area of the STN showing the greatest increase in 2-DG uptake was the dorsolateral area, where GPe efferents are known to terminate. STN activity would then be reduced thus introducing a positive feedback to the already underactive GPi. Although it appears that for the induction of chorea/ballism the underactivity of the STN is crucial, and thus increased inhibitory input to the STN seems most probable in this model, it is not possible to rule out an indirect recruitment or involvement of other pathways to the STN (e.g. cortical, intralaminar thalamic).

If GPi activity was reduced by blockade of glutamatergic transmission, it seemed reasonable to predict reductions in 2-DG uptake in target areas of GPi neurones, specifically the ventral anterior/ventral lateral and centromedian (VA/VL and CM) thalamic nuclei, and the pedunculopontine nucleus (PPN). However, no significant changes in 2-DG uptake were revealed in these areas. This might be explained in the light of the recent demonstration of contralateral pallido-thalamic and pallido-tegmental projections (Hazrati and Parent, 1991). Although only possibly 10-20% of GPi neurones projected bilaterally to these areas, there was profuse arborization of the contralateral GPi terminals, particularly in the thalamus. Thus the bilaterality of GPi projections may obscure small or diffuse changes in ipsilateral terminal activity. However it should be stressed that the very nature of the induced movement disorder, that is unilateral hyperkinesia, would very obviously indicate disordered or disrupted pallido-thalamic activity unilaterally.

Recently it has been suggested that changes in glucose utilization may be relatively small as compared to changes in neuronal firing rates, and that the 2-DG technique, as well as not providing information on the pattern of neuronal activity i.e. phasic changes, may only reveal large and uniform changes in that activity (Féger and Robledo, 1991). This may suggest that underlying hyperkinetic movements there is only a moderately reduced or disordered GPi output. Alternatively there may be a major change in the pattern of activity within the GPi neuronal population, which is not revealed by the 2-DG technique, i.e. phasic rather than tonic changes in firing.

The preceding discussion has dealt with reducing excitatory input to the GPi in a normal monkey, where prior to the pharmacological manipulations it is assumed the presence of "normo"-active neural circuitry, helping to sustain normal movement. In Parkinson's disease it is well accepted that the STN to GPi pathway is overactive, thus there is heightened excitatory influence within the GPi. It might therefore be predicted that introduction of kynurenic acid into the GPi in this situation might reduce its activity to more physiological levels, sufficient to restore normal movement. However, although there was reversal of parkinsonian symptoms following GPi injection of kynurenic acid and MK-801 (NMDA antagonist) in the MPTP-exposed parkinsonian monkey, there were also accompanying choreiform dyskinesias (Graham et al., 1990a).

Silencing the STN

Choreic and ballistic dyskinesias have also been apparent in the parkinsonian primate, where reversible or irreversible silencing of the STN has been carried out.

With much experimental evidence pointing to the overactivity of the STN in parkinsonism in the primate, a major step forward in research was made with the decision to see if removal of this driving force would alleviate parkinsonian symptoms. Initially it was shown that in two MPTP-exposed primates injection of the GABA agonist muscimol into the STN resulted in the alleviation of akinesia, tremor and rigidity contralaterally, as did the subsequent ablation of the nucleus with ibotenic acid (Bergman et al., 1990a and b). The reversal of symptoms however was again accompanied by contralateral limb dyskinesia, described as transient, disappearing after one week in one monkey, and lessened but not fully disappeared at three weeks in the other monkey.

In our own studies the STN has been lesioned in parkinsonian primates by a thermocoagulative technique (Aziz et al., 1992). MPTP-exposed monkeys received either unilateral or bilateral lesions of the STN, with dramatic effect. Parkinsonian symptoms were reversed bilaterally following unilateral lesion. Following STN ablation several of these animals exhibited contralateral limb choreic and ballistic movements. These dyskinesias remitted with time, yet could be induced by stress or excitement. Histologically it was revealed (after a post-lesion follow-up time of at least 4 months), that in the monkeys where bilateral reversal followed a unilateral lesion, approximately 40-50% of the STN was destroyed. For dyskinesias to have remitted in the presence of such a substantial lesion, some compensatory mechanism may be implicated. Indeed in man such movements become a lesser problem with time, when the STN has been involved in vascular accident or following surgery.

ENHANCEMENT OF INHIBITORY TRANSMISSION WITHIN THE GPI

In our behavioural studies in normal monkeys, having confirmed that reducing the excitatory influence within the GPi induced hyperkinetic movements, subsequent experiments involved increasing the inhibitory influence within the GPi, by directly injecting the GABA agonist muscimol (5μg). When injected into sites which had previously given kynurenate-induced chorea, we saw again choreiform dyskinesias. Interestingly we never elicited ballism following muscimol within the GPi. The duration of action, as measured by the visible dyskinesias, was also significantly shorter than that with kynurenic acid (1 v. 3-4 hours). This may be due to muscimol diffusing less within the GPi, thus affecting a smaller population of neurones.

Thus it was also possible to elicit hyperkinesia by imitating an increase in GABAergic striato-pallidal or intra-pallidal pathway activity.

REDUCTION OF INHIBITORY TRANSMISSION WITHIN THE GPI

It has also been possible to antagonise the GABAergic input to the GPi (from the striatum and GPe), by directly injecting bicuculline (10-25μg). In this case the elicited involuntary movements were classified as dystonic rather than choreic or ballistic.

Bicuculline has been used on several previous occasions in our studies where injections have been centred either in the ventral lentiform nucleus (Mitchell et al., 1989) inducing choreiform dyskinesias, or within the STN (Mitchell et al., 1985) inducing choreiform and ballistic movements. In the first case the effects of bicuculline were attributed simply to its blockade of GABA, ultimately leading to a reduction in activity of the pathway from the STN to the GPi. For the injection of bicuculline within the STN to produce similar dyskinesias a state of depolarization blockade subsequent to GABA antagonism was believed to have been induced, again resulting in less excitatory drive to the GPi. It is therefore possible that different neuronal populations may vary in their sensitivity to excessive excitatory activity during GABA blockade. Thus the question is raised as to what may happen to GPi neurones following bicuculline application.

If bicuculline was acting simply by blocking the GABAergic input, so that GPi activity was regulated predominantly by the STN, the expected overall change would be an increase in pallido-thalamic activity. So in dystonia the neural mechanisms would be more akin to those found in parkinsonism, particularly if the increased GPi activity led to inhibition of the GPe. On the other hand if bicuculline did cause a state of depolarization blockade in the GPi then this would lead to cessation of activity of GPi neurones. It could be argued that this would be more akin to lesion or ablation of the GPi.

A whole spectrum of movement disorders have been reported to occur following lesion, damage or inactivation of the pallidal segments. It seems very likely that these are closely related to the extent of the lesion and the involvement of associated pathways and structures. Akinesia, rigidity, choreoathetosis and flexor deviation have all been reported, and in addition to these signs torsion dystonia where pallidal degeneration was the major pathological finding (see Discussion in Aizawa et al., 1991 and Mink and Thach, 1991). Very recently, slowed movements, but with no rigidity, have been reported in man in a case of pure pallidal degeneration, with no other pathological abnormalities (Aizawa et al., 1991). The slowness of movement was distinct from parkinsonian akinesia (and did not respond to L-DOPA treatment). In addition dystonic posturing was present in the neck and fingers.

More relevant to our discussion are results from studies where lesions have been made within the boundaries of the GPi. Following kainic acid lesion of the GPi flexed posturing with contralateral head deviation, not dissimilar to that observed following the bicuculline injections within the GPi, was seen (Delong and Coyle, 1979). This would equate cessation of GPi activity, and presumably a major disinhibition of the motor thalamus, with flexed postures. It may also support further the idea that in inducing hyperkinetic movements, kynurenic acid and muscimol have led to a moderate reduction or have induced some phasic changes in pallido-thalamic activity and that to induce dystonia a major reduction or cessation of GPi activity must occur. It may also suggest that the neural mechanisms underlying rigidity in parkinsonism are very different to those responsible for dystonia.

INVOLVEMENT OF THE GPI IN DYSKINESIAS FOLLOWING CHRONIC L-DOPA/DOPAMINE AGONIST TREATMENT IN MPTP-EXPOSED PRIMATES

Further information regarding the role of the GPi in mediating dyskinesias, comes from a 2-DG study involving parkinsonian monkeys exhibiting peak-dose dyskinesias subsequent to chronic L-DOPA/dopamine agonist treatment (Mitchell et al., 1992). 2-DG uptake was compared in animals exhibiting peak-dose chorea or dystonia, with that from parkinsonian monkeys treated acutely with L-DOPA or apomorphine to render them asymptomatic during the period of 2-DG uptake, but where no dyskinesias were present.

Results showed that in chorea and dystonia, as compared with the non-dyskinetic state, there was an increase in 2-DG uptake in the STN. This was interpreted as suggesting increased inhibitory pallidal input, presumably due to the inhibitory effect of dopaminergic stimulation on the striatal efferents to the GPe, so causing disinhibition of GPe-STN transmission. STN activity would therefore be reduced.

There was also increased 2-DG uptake in the GPi in the choreic and dystonic animals, which if the STN was underactive, would suggest that another major afferent of the GPi was overactive. The main candidate for this would be the GABAergic input from the striatum. Taken together this implies that in L-DOPA/dopamine agonist-induced dyskinesias, there is an increased inhibitory striatal input to the GPi in the presence of a reduced excitatory subthalamic input.

Further evidence for this overactivity in the striato-GPi pathway was provided by receptor autoradiography experiments, where the D1 and D2 dopamine receptors were visualized in these same animals. It is well documented that in the untreated parkinsonian state the striatal D2 receptor population is increased (Graham et al., 1990b and references therein). In contrast we favour the view that the D1 receptor population is not affected similarly. Following chronic anti-parkinsonian treatment the D2 receptor population was found to be down-regulated towards more control receptor levels, whilst in the same animals in the caudal striatum there appeared to be a trend towards an increase in striatal D1 receptors (Graham et al., 1993). Similarly, in a hemi-parkinsonian monkey chronically treated with apomorphine, and exhibiting peak-dose dystonia, there was an asymmetry of D1 receptor binding, where the greater binding density was on the MPTP-exposed side of the brain. No such asymmetry was apparent in an untreated hemi-parkinsonian animal (see Fig. 2).

The relevance of this becomes apparent when the location of the D1 and D2 receptors is considered. Following initial studies in the rodent brain, there is now growing

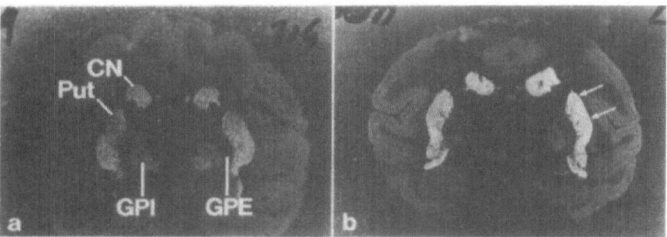

Figure 2. Autoradiographs of ^{3}H-SCH 23390 binding to D1 receptors in hemi-parkinsonian monkeys. a: not treated and b: treated longterm with dopamine agonists. Note increased binding (b: arrows) on MPTP-exposed side of the brain. CN: caudate nucleus, GPE, GPI: external and internal pallidum, Put: putamen. Reproduced by permission of Elsevier Science Publishers.

acceptance of the idea that D1 and D2 receptors are differentially localized on striatal efferents (Gerfen et al., 1990). Thus it is proposed that the majority of D2 receptors are located on striato-GPe projection neurones, with D1 receptors appearing predominantly on striato-GPi and striato-SNR (substantia nigra pars reticulata) neurones.

Following longterm L-DOPA/dopamine agonist treatment, when it is believed the STN is contributing less excitatory drive to the GPi, the direct inhibitory striatal pathway to the GPi may therefore play a crucial role in mediating dyskinesias. The increased D1 receptor population may produce enhanced sensitivity to dopaminergic stimulation, and thus enhanced GABAergic transmission may follow.

With reference again to the 2-DG study, one other observation also deserves mention. Although the result was not statistically significant, in dystonia the motor thalamus (VA/VL) showed a tendency of reduced 2-DG uptake, possibly implying a more severe reduction in pallido-thalamic activity. In support of this were observations made from 2-DG autoradiographs from the hemi-parkinsonian monkey exhibiting dystonia after chronic treatment with apomorphine (Mitchell et al., 1990). Here again there was an apparent increase in 2-DG uptake in the STN and GPi and a reduction in 2-DG uptake in the VA/VL. This would add further weight to the suggestion that for dystonia, rather than chorea, to occur there must be a greater decrease in GPi activity.

PHENOMENON OF "ON" AND "OFF" DYSTONIA

The possibility of increased striato-GPi activity, and hence decreased GPi-thalamic activity, underlying the dystonia seen "on-dose" after longterm treatment with L-DOPA/dopamine agonists in parkinsonism, raises the question as to how it is possible for "off" dystonia to occur simultaneously with parkinsonian symptoms. We have stated already that increased GPi-thalamic activity is believed to sustain parkinsonian symptoms, and that rigidity and dystonia would appear to be supported by different neural mechanisms. However is it not possible that more than one pattern of neural activity may lead to dystonia? In "off" dystonia in parkinsonism the increased excitatory input to the GPi, with the believed reduced striatal inhibitory input, may leave the GPi more susceptible to depolarization blockade. Thus superimposed upon a general overactivity of GPi neurones, there may be phasic periods of depolarization blockade causing disinhibition of the motor thalamus. It could be envisaged that if this was to occur in small somatotopically distinct areas then it would be possible for dystonia to be present as well as parkinsonian symptoms. As was stated above in "on" dystonia, a swing in the opposite direction may occur such that there is an excessive inhibitory striatal input with reduced excitatory activity in the GPi.

Interestingly, observations (unpublished) we have made from primates which have been more severely affected by MPTP exposure may support this speculation. In these animals there appears to be a greater probability of "off" dystonia being apparent. This may represent a greater intensity of STN-GPi activation and hence increased likelihood of depolarization blockade occurring. As off-dose chorea has not been observed, it would seem unlikely that there is a phase of decreased excitatory input or increased inhibitory input to the GPi preceding or underlying neural mechanisms leading to off-dose dystonia.

CONCLUSIONS

In conclusion it appears that a decrease in GPi neuronal activity will lead to dyskinesias. However, for this statement to apply to chorea, ballism and dystonia, it is likely that there is a range of activity patterns, both phasic and tonic, that are induced as the excitatory input to the GPi is reduced or the inhibitory input is increased. Initially this imbalance between the dual inputs to the GPi may give rise to a pattern of activity resulting in choreiform movements, whereas we have proposed that with a more severe reduction, or even cessation of GPi activity, dystonia may become apparent. Undoubtedly direct recordings of pallidal and thalamic activity occurring during dyskinesias induced by these various means will shed more light on these theories. These electrophysiological studies are awaited with interest.

REFERENCES

Aizawa, H., Kwak, S., Shimizu, T., Goto, J., Nakano, I., Mannen, T., and Shibasaki, H., 1991, A case of adult onset pure pallidal degeneration, *J. Neurol. Sci.* 102:76-82.

Aziz, T.Z., Peggs, D., Agarwal, E., Sambrook, M.A., and Crossman, A.R., 1992, Subthalamic nucleotomy alleviates parkinsonism in the 1-methyl-4-phenyl-1,2,3,6-tetrahydropyridine (MPTP)-exposed primate, *Brit. J. Neurosurg.* 6:575-582.

Bergman, H., Wichmann, T., and DeLong, M., 1990a, Amelioration of parkinsonian symptoms by inactivation of the subthalamic nucleus (STN) in MPTP-treated monkeys, *Movement Disorders* 5 supp 1:79.

Bergman, H., Wichmann, T., and DeLong, M., 1990b, Reversal of experimental parkinsonism by lesions of the subthalamic nucleus, *Science* 249:1436-1438.

DeLong, M.R., and Coyle, J.T., 1979, Globus pallidus lesions in the monkey produced by kainic acid: histologic and behavioural effects, *Appl. Neurophysiol.* 42:95-97.

Féger, J., and Robledo, P., 1991, The effects of activation or inhibition of the subthalamic nucleus on the metabolic and electrophysiological activities within the pallidal complex and substantia nigra in the rat, *Eur. J. Neurosci.* 3:947-952.

Gerfen, C.R., Engber, T.M., Mahan, L.C., Susel, Z., Chase, T.N., Monsma, F.J., and Sibley, D.R., 1990, D1 and D2 dopamine receptor-regulated gene expression of striatonigral and striatopallidal neurons, *Science* 250:1429-1432.

Graham, W.C., Robertson, R.G., Sambrook, M.A., and Crossman, A.R., 1990a, Injection of excitatory amino acid antagonist into the medial pallidal segment of a 1-methyl-4-phenyl-1,2,3,6-tetrahydropyridine (MPTP) treated primate reverses motor symptoms of parkinsonism, *Life Sci.* 47:PL91-97.

Graham, W.C., Clarke, C.E., Boyce, S., Sambrook, M.A., Crossman, A.R., and Woodruff, G.N., 1990b, Autoradiographic studies in animal models of hemi-parkinsonism reveal dopamine D2 but not D1 receptor supersensitivity. II. Unilateral intra-carotid infusion of MPTP in the monkey (Macaca fascicularis), *Brain Res.* 514:103-110.

Graham, W.C., Sambrook, M.A., and Crossman, A.R., 1993, Differential effect of chronic dopaminergic treatment on dopamine D1 and D2 receptors in the monkey brain in MPTP-induced parkinsonism, *Brain Res.* 602:290-303.

Hazrati, L.-N., Parent, A., Mitchell, S., and Haber, S.N., 1990, Evidence for interconnections between the two segments of the globus pallidus in primates: a PHA-L anterograde tracing study, *Brain Res.* 533:171-175.

Hazrati, L.-N., and Parent, A., 1991, Contralateral pallidothalamic and pallidotegmental projections in primates: an anterograde and retrograde labelling study, *Brain Res.* 567:212-223.

Mink, J.W., and Thach, W.T., 1991, Basal ganglia motor control. III. Pallidal ablation: normal reaction time, muscle cocontradtion, and slow movement, *J. Neuropysiol.* 65:330-351.

Mitchell, I.J., Sambrook, M.A., and Crossman, A.R., 1985, Subcortical changes in the regional uptake of ^{3}H-2-DG in the brain of the monkey during experimental choreiform dyskinesia elicited by injection of a GABA antagonist into the subthalamic nucleus, *Brain* 108:421-438.

Mitchell, I.J., Jackson, A., Sambrook, M.A., and Crossman, A.R., 1989, The role of the subthalamic nucleus in experimental chorea, *Brain* 112:1533-1548.

Mitchell, I.J., Luquin, R., Boyce, S., Clarke, C.E., Robertson, R.G., Sambrook, M.A., and Crossman, A.R., 1990, Neural mechanisms of dystonia: evidence from a 2-deoxyglucose uptake study in a primate moαel of dopamine agonist-induced dystonia, *Movement Disorders* 5:49-54.

Mitchell, I.J., Boyce, S., Sambrook, M.A., and Crossman, A.R., 1992, A 2-deoxyglucose study of the effects of dopaminergic agonists on the parkinsonian primate brain, *Brain* 115:809-824.

Robertson, R.G., Farmery, S.M., Sambrook, M.A., and Crossman, A.R., 1989, Dyskinesia in the primate following injection of an excitatory amino acid antagonist into the medial segment of the globus pallidus, *Brain Res.* 476:317-322.

LOCAL INACTIVATION OF THE SENSORIMOTOR TERRITORIES OF THE INTERNAL SEGMENT OF THE GLOBUS PALLIDUS AND THE SUBTHALAMIC NUCLEUS ALLEVIATES PARKINSONIAN MOTOR SIGNS IN MPTP TREATED MONKEYS

Thomas Wichmann, Mark.S. Baron and Mahlon R. DeLong

Dept. Neurology, Emory University School of Medicine, Woodruff Memorial Building, 1639 Pierce Drive, Atlanta, GA 30322, U.S.A.

INTRODUCTION

Recent physiological and anatomical studies have yielded considerable insight into the role of the basal ganglia in the pathophysiology of Parkinson's disease. Experiments in monkeys before and after treatment with the neurotoxin 1-methyl-4-phenyl-1,2,3,6-tetrahydropyridine (MPTP) have revealed that in parkinsonism, the inhibitory basal ganglia output, emanating from the internal segment of the globus pallidus (GPi), and directed at the ventrolateral thalamus (VL), as well as the brainstem, is increased (see Albin et al., 1989, DeLong, 1990). Both phenomena have been directly linked to the development of parkinsonian motor signs. These observations have allowed the formulation of a unified scheme of parkinsonian pathophysiology, in which all parkinsonian motor signs result from abnormalities of basal ganglia output.

An important aspect of the current circuit model is that the neuronal basal ganglia-thalamocortical loops stay largely segregated, dividing the basal ganglia into "motor" and "non-motor" territories. Parkinsonian motor signs may arise in large part from abnormalities of discharge specifically affecting the "motor" territory of the basal ganglia, whereas similar abnormalities in the "non-motor" territory may have little bearing on the development of parkinsonism. This model is supported by the finding that localized inactivation of the "motor" territory of the subthalamic nucleus (STN) strikingly reduces parkinsonian motor signs, presumably by reducing the activity of the "motor" territory of GPi, counteracting increased output of the motor part of GPi to VL (Bergman et al., 1990, Aziz et al., 1991). Strong direct support for the hypothesis that overactivity of neurons in the posterior and ventral "motor" portion of GPi is involved in the development of parkinsonism comes from studies exploring the effects of stereotaxic lesions of this area in parkinsonian patients (Svennilson et al., 1960, Laitinen et al., 1992). These lesions have marked and prolonged beneficial effects on all parkinsonian motor signs, including akinesia, rigidity and tremor. The ideal target for these lesions, however, remains uncertain.

The goal of our experiments was to test the above mentioned hypothesis further by exploring the effects of transient inactivation of discrete portions of GPi and STN in a parkinsonian monkey. A clear definition of the region responsible for the development of parkinsonian motor signs may help to reduce the size of stereotaxic lesions needed to treat parkinsonism in humans, and to decrease the rate of unwanted side effects of these procedures.

METHODS

Animals, behavioral conditioning, MPTP treatment and surgery

A Rhesus monkey (Macaca mulatta, 4 kg) was trained to sit in a primate chair. The animal then received intracarotid injections of MPTP (MPTP HCL, Sigma, St. Louis, MO; 0.4 mg/kg) on both sides. The two surgeries were carried out two weeks apart with sterile technique under general anesthesia with isoflurane.

The monkey was observed for nineteen weeks after the MPTP treatment until the parkinsonian motor signs had stabilized. At this time, another surgery was performed to place a permanent recording chamber. A 20 mm diameter hole was made in the monkey's skull with a trephine over the right hemisphere, and a cylindrical stainless steel chamber was cemented in place with dental acrylic. The cylinder was positioned so as to permit recording of the activity of neurons in GPi and in the STN. It was tilted 25 degrees anteriorly in the sagittal plane, with its axis aimed at the stereotaxic coordinates A = 12, H = 3 and L = -7. Several screws were also imbedded into the dental acrylic cap to allow head fixation during subsequent recording and injection sessions.

Pharmacological treatment

Injections into GPi were carried out with a combined recording/injection device (Hamada and DeLong, 1992), consisting of a 30-ga injection canula through which a thin coated tungsten wire (50 µm base diameter, O.D. 75 µm) was threaded. The cut end of the wire protruded 0.3 - 0.5 mm from the tip of the canula, permitting recording of multiunit activity close to the injection site with standard electrophysiological techniques. Pressure injections were done with a Hamilton syringe that was connected to the injection canula with a Teflon tube. The injection canula and the recording wire were protected by a guide tube. The whole assembly was lowered into the brain with a microdrive (MO-95, Narishige, Tokyo), coupled to a linear potentiometer for depth measurements. Once inside the brain, the injection canula was lowered through the guide tube, and further advanced (18-25 mm from the tip of the guide tube) toward the injection target. Muscimol (Sigma, St. Louis, MO, 1 µg/µl in normal saline) or normal saline was applied in 0.1 µl boluses, in thirty second intervals to a total volume of 1 µl. After injections, the canula was left at the injection site for an additional period of five minutes to prevent backflow along the injection track. Injections of saline were carried out as controls. Only one injection was carried out per day.

Data acquisition and analysis

Daily the monkey was placed into a custom made Plexiglas cage and filmed for three twenty-minute periods (a baseline preinjection period, and 20 - 40 minutes and 60 - 80 minutes postinjection). The tapes were later reviewed by an observer blinded to the treatment and to the location of the injection. The severity of the monkey's parkinsonian signs and possible side effects of the injections were rated.

The behavioral scoring data was used to determine the degree of change from the baseline at a given injection site. Since the monkey did not express tremor, and assessment of muscle tone was found to be unreliable, the scoring data reported here was based on behavioral observations only, and therefore reflects largely changes in akinesia and bradykinesia in the side contralateral to injections. The akinesia was rated on a scale of 0 to 5 (0 - no akinesia, 1 - mild akinesia, 2 - moderate akinesia, 3 - severe akinesia with only 3 to 4 movements-/observation period, 4 - severe akinesia with 2 or less movements- /observation period, and 5 - no movement throughout the observation period). For each injection, a summary baseline score was established by averaging baseline scores over five days prior to the injection. Post-injection scores were obtained using the same rating scale. The final score of the effect of a given injection was computed by subtracting post-injection scores from baseline scores. Effect scores ranged between 0 and 3.

Histological analysis

At the completion of the experiment, the monkey was sacrificed with an overdose of pentobarbital (100 mg/kg), and perfused transcardially with normal saline, followed by 10% neutral formalin. The brain was then blocked and frozen, followed by sectioning in the sagittal plane. Alternate 50 μm sections were stained for cresyl violet and tyrosine hydroxylase.

Injection canula tracks were easily seen in the cresyl-violet sections of GPi. The histological information, with the recording data and measurements of the penetration depths during the actual recording/injection sessions were used to reconstruct the injection sites.

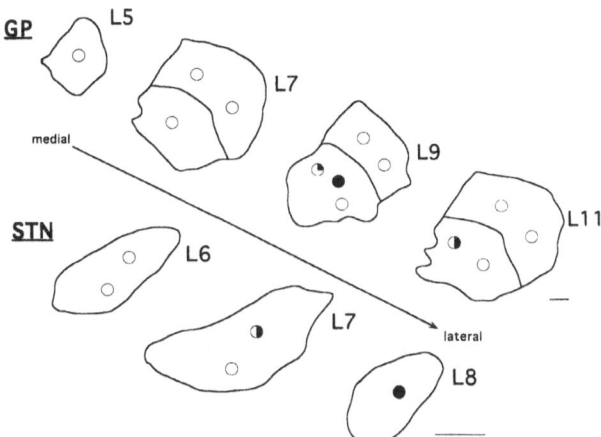

Figure 1. Improvement in contralateral arm motility following muscimol injections into the globus pallidus and into the subthalamic nucleus. Results of muscimol injections (1 μg/μl) are shown in sagittal planes. The most medial plane (LS) includes only GPi. Open circle = no effect (effect score = 1), quarter filled circle = mild effect (effect score = 1), half filled circle = moderate effect (effect score = 2), filled circle = marked effect (effect score = 3). Bars represent 1 mm.

RESULTS

Effect of the MPTP injections

The monkey received bilateral intracarotid injections of equal amounts of MPTP. The first of the two injections resulted in intermittent arm dystonia, starting almost immediately after the injection, and giving way to initially severe akinesia and rigidity after several days, which gradually improved over the following weeks. Injection of MPTP into the contralateral carotid artery two weeks later had the same effects, except that the parkinsonian motor signs (akinesia, rigidity), affecting mainly upper and lower extremities, remained stable after this injection. After several weeks, the monkey was essentially hemiparkinsonian. Throughout the experiment, the monkey was always able to maintain adequate food and water intake with his less affected upper extremity. Behavioral effects of the injection were assessed only on the animal's more affected side.

Muscimol injections into the two segments of the globus pallidus

Injections were carried out in a 2 mm by 2 mm grid into both nuclei of the globus pallidus, throughout their entire anterior-posterior and the medio-lateral extent. Some of the results of this series of experiments are shown in figure 1. None of the injections into GPe had any effect on the monkey's general behavior or the severity of the contralateral parkinsonian motor signs.

Injections into GPi were, in contrast, often very effective in ameliorating parkinsonism. Beginning 10 minutes after the injections, the monkey began to move its upper and lower extremity on the contralateral side in a purposeful way. This was usually first evident as scratching movements, followed by the monkey's ability to retrieve food objects and feed himself with the hand contralateral to the injection. Furthermore, the monkey's overall mobility was markedly improved by increased voluntary movements of his previously parkinsonian leg. The maximal effect of the injections was usually seen about 20 - 40 minutes after the injection, with the monkey returning to his baseline parkinsonian state about 90 - 120 minutes after the injection. The effectiveness of the GPi injections was on the location of the injection. Injections into the central and lateral portions of the nucleus were most effective in alleviating parkinsonian motor signs, whereas injections into the medial parts of the nucleus were virtually without effect. Control injections with normal saline into GPe or GPi had no effect.

Muscimol injections into the STN

The size of the STN mandated that these injections were more closely spaced (using a 1 mm grid) than those into GPi. Injections into particular areas within the STN produced defects similar to those into the GPi (see figure). Injections into the most lateral areas of the nucleus had the most prominent effects on parkinsonian motor signs, with almost complete reversal of parkinsonian akinesia and rigidity on the side contralateral to the inactivation. In addition, a moderate amount of dyskinetic movements was noted after the STN injections. The time course of the development and disappearance of the dyskinesias coincided with the amelioration of parkinsonism. Control injections with normal saline into the STN had no effect.

DISCUSSION

The findings of this study support the view that increased output from the basal ganglia is important in the development of parkinsonian motor signs, and that local inactivation of the output nuclei of the "motor" circuit is sufficient to markedly ameliorate parkinsonian motor signs, including akinesia.

The hypothesis that altered basal ganglia output is essential for the development of parkinsonian motor signs is based on a series of studies, indicating that the neuronal activity in GPi is increased in parkinsonian monkeys (Pan et al., 1983; 1985, Crossman et al., 1985; Miller and DeLong, 1987; Porrino et al., 1987; Filion et al., 1988; Schwartzman et al., 1988; Mitchell et al., 1989; Gerfen et al., 1990). Although these changes in discharge in GPi are certainly suggestive of a role of the basal ganglia in the development of parkinsonian motor signs, they could still be only an epiphenomenon. Direct evidence for a causative involvement of altered activity in the basal ganglia comes from experiments, in which the STN was permanently lesioned (Bergman et al., 1990; Aziz et al., 1991). Inactivation of the STN in these experiments resulted in a long lasting amelioration of parkinsonian motor signs, presumably by reducing the excitatory drive of GPi from the STN. The relevance of these findings for the understanding of the pathophysiology of Parkinson's disease in humans is underscored by the recent description of a patient whose parkinsonian motor signs were strongly reduced after a hemorrhage into the STN (Sellal et al., 1992).

Inactivation of GPi as done in this study in a parkinsonian monkey, has also been shown empirically to be a highly effective means of ameliorating parkinsonian signs in humans (Svennilson et al., 1960, Laitinen et al., 1992). Contrary to the widely held view that lesions of the cerebellar receiving area in the thalamus (Vim in humans) produce better amelioration of parkinsonian signs than do pallidal lesions (Hassler et al., 1960), the above mentioned studies on the effects of pallidal lesions in parkinsonian patients have demonstrated impressive benefits from GPi lesions on all parkinsonian motor signs. The key to further improving the outcome of pallidal lesions in parkinsonism lies in the exact localization of the lesioning target. In the study by Svennilson et al. (1960), systematic variation of the lesion site resulted in dramatic improvement of the overall success rate of these procedures when the lesion was placed in the ventral posterior portion of GPi. A follow-up study, using the same target, confirmed this result (Laitinen et al., 1992). Our series of experiments with electrophysiologically guided and histologically verified temporary inactivations of regions of GPi showed the same phenomenon in a more precise manner than currently achievable in human parkinsonian patients.

The fact that inactivation of a relatively small area in the STN or in GPi reverses parkinsonism, whereas inactivation of neighboring regions is ineffective, can be explained by the concept that only a small portion of the basal ganglia is actually involved in the production of parkinsonian motor signs, supporting the "segregated circuit" hypothesis of the basal ganglia-thalamocortical loops. According to this hypothesis, the basal ganglia are subdivided into multiple independent regions with little interaction (Alexander et al., 1986). "Motor" portions of the basal ganglia have been mostly defined by the responses of neurons in these areas to behavioral events (e.g. DeLong et al., 1985; Hamada et al., 1990) and by anatomical methods (e.g. Szabo, 1962, 1967; Cowan and Powell, 1966; Tokuno et al., 1992; Hoover and Strick, 1993). The regions which most effectively ameliorated parkinsonian motor signs in our experiments were found within the previously described "motor" areas. Recording of neuronal activity in GPi and in the STN of MPTP treated monkeys has not revealed major regional differences in the activity changes of pallidal neurons (Wichmann et al., unpublished). Although this suggests, that multiple basal ganglia circuits change their activity after MPTP, the effectiveness of local inactivation of "motor" areas demonstrates that the characteristic motor phenomena of parkinsonism are explainable by abnormalities confined to portions of the "motor" circuit.

These considerations have relevance for new attempts to treat Parkinson's disease with ablative surgery. Although more advanced neurosurgical approaches have been geared Svennilson et al., 1960), the lesion size was still substantial, leading to potentially serious side effects. For instance, 14% of Laitinen's patients (Laitinen et al., 1992) suffered visual impairment secondary to an inadvertent extension of the GPi lesion into the optic tract. If local inactivation of the "motor" areas in GPi is sufficient to ameliorate parkinsonian signs, the lesions could be made much smaller, thus avoiding potential damage to areas such as the optic tract and the internal capsule. Efforts to identify the "motor" portion within GPi with electrophysiologic mapping of the nucleus and selective lesioning of this area should help to minimize lesion size.

Although lesions in the "subthalamic area", adjacent to the subthalamic nucleus, have been performed as treatment of parkinsonism (Struppler et al., 1993); lesions of the STN itself have not been carried out intentionally in humans, partly for fear of inducing long lasting dyskinesias. The experiments presented here show that small lesions of the "motor" portion of the STN are a highly effective means to ameliorate parkinsonism, and that dyskinesias after such a procedure may be of only moderate severity. Previous experience with permanent damage of the STN in animals and in humans (Bergman et al., 1990; Aziz et al., 1991; Sellal et al., 1992) had long-lasting effects on parkinsonian signs, whereas dyskinesias were reportedly short-lived. If this result is confirmed in more extended animal experiments, it may be worthwhile to consider the STN as another possible lesioning target parkinsonian patients. The biggest practical advantage of this type of lesion over lesions in GPi is that the target identification is much easier in the STN than it is in GPi, since the neuronal activity in the STN is very distinct, and neuronal responses to somatosensory input are much more robust than in GPi (e.g. Wichmann et al., 1993). In addition, from a theoretical standpoint, lesioning of the STN would seem less likely to induce side effects it leaves the "direct" pathway and the pallidofugal fibers intact. Lesions of the STN would also be likely to affect larger areas of basal ganglia output, including the substantia nigra, pars reticulata, and the midbrain tegmentum, and may therefore be more effective than GPi lesions. Potential disadvantages of this procedure obviously include the proximity of the lesion to the internal capsule and the greater potential for such lesions to induce bleeding in the richly vascularized STN. Neither of these issues has been a problem in the above-mentioned animal studies.

Although the findings of our study provide strong evidence for the role of altered activity in "motor" portions of the basal ganglia in the development of parkinsonian motor signs, they do not necessarily permit conclusions about the function of the basal ganglia under normal conditions. The proposed role of the basal ganglia in motor control is unfortunately largely based on observations in parkinsonian or dyskinetic patients, or in profound changes in motor performance associated with alterations in basal ganglia output cannot be taken as evidence for a similarly profound role of the basal ganglia in voluntary movement under normal conditions.

REFERENCES

Albin, RL, Young, A.B., and Penney, J.B., 1989, The functional anatomy of basal ganglia disorders, *Trends Neurosci.* 12:366-375.

Alexander, G.E., DeLong, M.R., and Strick, P.L., 1986, Parallel organization of functionally segregated circuits linking basal ganglia and cortex, *Ann. Rev. Neurosci.* 9: 357-381.

Aziz, T.Z., Peggs, D., Sambrook, M.A., and Crossman, A.R., 1991, Lesion of the subthalamic nucleus for the alleviation of l-methyl-4-phenyl-1,2,3,6-tetrahydropyridine (MPTP)-induced parkinsonism in the primates, *Mov. disorders* 6: 288-293.

Bergman, H., Wichmann, T., and DeLong, M.R., 1990, Reversal of experimental parkinsonism by lesions of the subthalamic nucleus, *Science* 249: 1436-1438.

Cowan, W.M., and Powell, TP.S., 1966, Strio-pallidal projection in the monkey, *J. Neurol. Neurosurg. Psychiatry* 29:426-439.

Crossman, A.R., Mitchell, I.J., and Sambrook, M.A., 1985, Regional brain uptake of 2-deoxyglucose in N-methyl-4-phenyl-1,2,3,6-tetrahydropyridine (MPTP)-induced parkinsonism in the macaque monkey, *Neuropharmacol.* 24:587-591.

DeLong, M.R., 1990, Primate models of movement disorders of basal ganglia origin, *Trends Neurosci.* 13:281-285.

DeLong, M.R., Crutcher, M.D., and Georgopoulos, A.P., 1985, Primate globus pallidus and subthalamic nucleus: functional organisation, *J. Neurophysiol.* 53:53-543.

Filion, M., Tremblay, L., and Bédard, P.J., 1988, Abnormal influences of passive limb movement on the activity of globus pallidus neurons in parkinsonian monkeys, *Brain Res.* 444:165-176.

Gerfen, C.R., Engber, T.M., Mahan, L.C., Susel, Z., Chase, T.N., Monsma, F.J.Jr., and Sibley, D.R., 1990, D1 and D2 dopamine receptor-regulated gene expression of striatonigral and striatopallidal neurons, *Science* 250:1429-1432.

Hamada, I. and DeLong, M.R., 1992, Excitotoxic acid lesions of the primate subthalamic nucleus results in reduced pallidal neuronal activity during active holding, *J. Neurophysiol.* 68:1859-1866.

Hamada, I., DeLong, M.R., and Mano, N.-I., 1990, Activity of identified wrist-related pallidal neurons during step and ramp wrist movements in the monkey, *J. Neurophysiol.* 64:1892-1906.

Hassler, R., Reichert, T., Mundinger, F., Umbach, W., and Gangleberger, J.A., 1960, Physiological observations in stereotaxic operations in extrapyramidal motor disturbances, *Brain* 8:337.

Hoover, J.E., and Strick, P.L., 1993, Multiple output channels in the basal ganglia, *Science* 259:819-821.

Laitinen, L.V., Bergenheim, A.T., and Hariz, M.I., 1992, Leksell's posteroventral pallidotomy in the treatment of Parkinson disease, *J. Neurosurg.* 76:53-61.

Miller, W.C., and DeLong, MR., 1987, Altered tonic activity of neurons in the globus pallidus and subthalamic nucleus in the primate MPTP model of parkinsonism, in: "The Basal Ganglia II," M.B. Carpenter and A. Jayaraman, eds., Plenum Press, New.York, pp.415-427.

Mitchell, I.J., Clarke, C.E., Boyce, S., Robertson, R.G., Peggs, D., Sambrook, M.A., Crossman, A.R., 1989, Neural mechanisms underlying parkinsonian symptoms based upon regional uptake of 2-deoxyglucose in monkeys exposed to l-methyl-4-phenyl-1,2,3,6-tetrahydropyridine, *Neuroscience* 32:213-226.

Pan, H.S., Frey, K.A., Young, A.B., and Penney, J.B. Jr., 1983, Changes in [3H]muscimol binding in substantia nigra, entopeduncular nucleus, globus pallidus and thalamus after striatal lesions as demonstrated by quantitative receptor autoradiography, *J. Neurosci.* 3:1189-1198.

Pan, H.S., Penney, J.B., and Young, A.B., 1985, Gamma-aminobutyric acid and benzodiazepine receptor changes induced by unilateral 6-hydroxydopamine lesions of the medial forebrain bundle, *J. Neurochem.* 45:1396-1404.

Porrino, L.J., Burns, R.S., Crane, A.M., Palombo, E., Kopin, I.J., and Sokoloff, L., 1987, Changes in local cerebral glucose utilization associated with Parkinson's syndrome induced by l-methyl-4-phenyl-1,-2,3,6-tetrahydropyridine (MPTP) in the primate, *Life Sciences* 40:1657-1664.

Schwartzman, R.J., Alexander, G.M., Ferraro, T.N., Grothusen, J.R., and Stahl, S.M., 1988, Cerebral metabolism of parkinsonian primates 21 days after MPTP, *Exp. Neurol.* 102:307-313.

Sellal, F., Hirsch, E., Lisovoski, F., Mutschler, V., Collard, M., and Marescaux, C., 1992, Contralateral disappearance of parkinsonian signs after subthalamic hematoma, *Neurology* 42:255-256.

Struppler, A., Gurfinkel, V., Mathis, J., and Max, T., 1993, Motor performance in parkinsonism following stereotactic thalamotomy, *Adv. Neurol.* 60:403-407.

Svennilson, E., Torvik, A., Lowe, R., and Leksell, L., 1960, Treatment of parkinsonism by stereotaxic thermolesions in the pallidal region. A clinical evaluation of 81 cases, *Acta Psychiat. Neurol. Scand.* 35:358-377.

Szabo, J., 1962, Topical distribution of the striatal efferents in the monkey, *Exp. Neurol.* 5:21.

Szabo, J., 1967, The efferent projections of the putamen in the monkey, *Exp. Neurol.* 19:463-476.

Tokuno, H., Kimura, M., and Tanji, J., 1992, Pallidal inputs to thalamocortical neurons projecting to the supplementary motor area: an anterograde and retrograde double labeling study in the macaque monkey, *Exp. Brain. Res.* 90:635-638.

Wichmann, T., Bergman, H., and DeLong, M.R., 1993, The primate subthalamic nucleus I. Functional properties in intact animals, *J. Neurophysiol.* submitted.

ROLE OF THE SUBTHALAMIC NUCLEUS
IN NORMAL AND PATHOLOGICAL CONDITIONS

José A. Obeso,[1] Jorge Guridi[1,2] and María-Trinidad Herrero[1]

[1]Experimental Neurology Group and Movement Disorders Unit
  Department of Neurology, Clínica Universitaria, University of Navarra
[2]Neurosurgery Service, Hospital de Navarra, Pamplona, Spain

INTRODUCTION

Unilateral lesion of the subthalamic nucleus (STN) produces both in humans and animals a violent and continuous hemidyskinesia (hemiballism) which may interfere with normal movement (Whittier and Mettler, 1949; Carpenter et al., 1950; Dierssen and Gioino, 1961). This well established observation has traditionally been used to support the crucial role of the STN in motor control. The prevailing view for many years was that the STN exerted a powerful inhibitory influence upon the medial globus pallidus (GPM) (Whittier and Mettler, 1949; Martin, 1967). However, as it is often the case, such clinicopathological deduction upon the normal physiology of the STN was not entirely correct.

Appropiate understanding of the role of the STN is now more needed than ever before. Recent findings in parkinsonian monkeys (Bergman et al., 1990; Aziz et al., 1991; Benazzouz et al., 1992; Guridi et al., 1993a) indicate that inactivation of the STN may induce a marked motor improvement. The STN could, therefore, become a new target for stereotaxic surgery of Parkinson's disease (Guridi et al., 1993b).

NORMAL FUNCTION OF THE STN

The STN is a small and ovoidal structure with an approximate volume of 20-30 mm^3 in humans. In primates, the major afferent projections to the STN arise from: 1) The lateral globus pallidus (GPL) forming an inhibitory (GABA) connection; 2) The sensorimotor cortex forming an excitatory connection; 3) The pedunculopontine nucleus (PPN -pars compacta-) forming a cholinergic excitatory pathway. The STN projects to: 1) The lateral and medial segments of the globus pallidus; 2) The substantia nigra pars reticulata (SNpr); 3) Striatum and PPN, but much less densely than the pallidal and reticulata pathways (Parent, 1990).

The STN is somatotopically organized according to a complex pattern. The 80% dorsolateral "sensorimotor" zone projecting to the globus pallidus (GP) and the putamen (P). The 20% ventromedial "associative" zone projecting to the caudate nucleus and the SNpr, and the 10% "overlapping" zone projecting both to the lenticular (GP and P) and the SNpr (Parent, 1990).

Recent findings clearly indicate that STN output pathways are excitatory. Thus, glutamatergic neurons in the STN have been stained by immunocytochemical methods (Albin et al., 1989; Parent et al., 1989), microstimulation of the STN produces excitation of the SNpr and GPL (Kitai and Kita, 1987) as well as the entopeduncular nucleus (Robledo and Féger, 1990). In monkeys the work of Crossman and his group has extensively shown that inactivation of the STN leads to metabolic reduction of the GPM activity (Mitchell et al., 1989a, b). Hamada and DeLong (1992a) have recently described how chemical lesion of the STN decreases the firing rate of GPM neurons during active holding. The exact chemical nature of the amino acid mediating the excitatory activity of the STN is not determined yet, but it seems to be glutamatergic (Robledo and Féger, 1990).

Electrophysiological studies upon the STN firing characteristics are not very abundant. DeLong et al. (1985) described a mean basal firing rate of 24 impulses/s in monkeys. In that study, STN neurons were observed to discharge in doublets or triplets, giving a "bursting" quality to the recording. Active movement and joint rotation were the two manouvres that consistently activated STN neurons. However, the increase in STN firing during a movement generally begins after those in the motor cortex and too late to account for movement initiation (Georgopoulos et al., 1983). The situation equally applies to GPM, but STN firing antedated GPM discharges.

Summary of basal ganglia role in movement

The major contribution of the STN to cortically mediated movements in primates is probably mediated by its efferent projection to the GPM. It is important, therefore, to summarize some physiological aspects of this major basal ganglia output structure. Cells in GPM have a very high discharge rate (70 Hz) and their firing is distinctively regular under basal conditions. The GPM exerts therefore a tonic inhibition of the motor thalamus. Loss of that inhibition, following local muscimol administration or kainic acid excitotoxic lesion, caused cocontraction of arm muscles and slowness of movement (Mink and Thach, 1991). Brotchie et al. (1991a) found that pallidal neurons were more active when the movements were predictable and well practiced. In keeping with this possible inhibition of the motor thalamo-cortical circuit, Aizawa et al. (1991) showed in the monkey that neurons in the supplementary motor area discharge very little when the animal performed overlearned movements.

The general, albeit simplified, idea from these and many other studies is that the striatopallidal complex is mainly concerned with the execution of predictable and automatic movements. This function is mainly undertaken by inhibiting unwanted muscle activity, which is achieved by means of the high tonic activity of the GPM. It is likely that when unusual circumstances occur or novel external signals are received, the striatum uses its powerful inhibitory projection to GPM in order to reduce the GPM-thalamic inhibition, thus interrupting the automatic sequences run by the premotor cortices and facilitating the new action (Marsden and Obeso, 1993). The recent demonstration of three different, topographically organized neuronal regions in GPM, each separately projecting to the primary motor area, premotor area and supplementary motor area (Hoover and Strick, 1993) provides the anatomical basis for this dual inhibitory/facilitatory action of the basal ganglia.

Contribution of the subthalamus to the motor control

Recent detailed anatomical studies by Parent and his collaborators have shown that the subthalamic efferent fibers establish numerous synaptic contacts on GPM neurons, forming a diffuse plexus within the pallidal neuropil (Parent and Hazrati, 1993). This arrangement contrasts with the much higher specificity of the synaptic contacts of the striatopallidal projection. During movement the STN increases its activity, showing a firing pattern which follows almost exactly that in the primary motor cortex, but antedating by a few milliseconds the changes in GPM activity (Georgopoulos et al., 1983). Cortical stimulation produces a powerful excitatory response in the STN with a mean latency in the rat of 2.5 ms. This response is thought to be mediated by a monosynaptic connection (Kitai and Deniau, 1981).

The overall picture is that the STN contributes to keep the GPM high activity rate in the resting condition, thus exerting a general inhibitory effect of unwanted movements. Under activation, the phasic changes in STN activity are probably related with the ending of an action (when performing a single movement) and the end/begining of movements when executing sequential tasks (Brotchie et al., 1991a).

There are still many problems to be solved about the physiological role of the STN. For instance, one would like to know which is the functional significance of the somatotopic arrangement of the STN. At present, experimental data (Hamada and DeLong, 1992b) and clinical findings suggest that the STN exerts upon the GPM an "all or nothing" type of control. Against this interpretation are the clustering of neurons discharging in relation with movement of a certain body part (DeLong et al., 1985), the finding of abnormal metabolic activity (2-deoxyglucose uptake), during experimentally-induced chorea in monkeys (Mitchell et al., 1989b), mainly on the dorsolateral third of the STN, and the different output organization of STN efferents. Thus, the GPM projection stems mainly form the dorsolateral region while the SNpr projection arise in the ventral 20% of the nucleus (Parent, 1990). Also, the role plays by the ventromedial region of the STN, more related with the limbic system, is very important to understand.

PATHOPHYSIOLOGY OF THE SUBTHALAMIC NUCLEUS

The better understanding of the role of the basal ganglia in movement control, summarized above, and a large serie of experiments in monkeys done by A. Crossman and his group have provided a well founded basis to value the extreme importance of the STN in the origin of many movement disorders (Mitchell et al., 1989a,b; Aziz et al., 1991). The crucial finding is that <u>increased</u> activity of the STN is associated with <u>parkinsonism</u>, while <u>reduced</u> activity leads to <u>chorea-ballism</u>.

The anatomical basis sustaining these findings are progressively better understood (Figure 1). In animals with lesion of the substantia nigra pars compacta, the dopaminergic loss removes the excitatory drive on the GABA-SP-Dynorphin output neurons projecting to

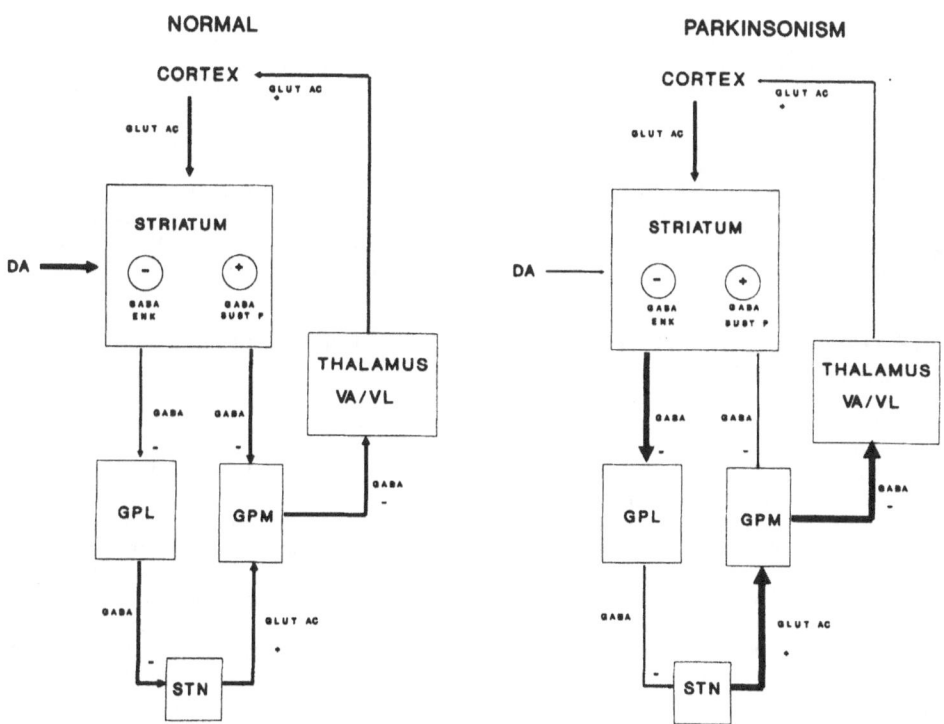

Figure 1. Schematic diagram of the main basal ganglia circuits and their activity under normal conditions and following dopaminergic lesion in Parkinson's Disease. Abbreviations: DA, dopamine; ENK, enkephalin; GABA, γ-aminobutyric acid; GLUT AC, glutamic acid; GPL, lateral segment of the globus pallidus; GPM medial segment of the globus pallidus; STN, subthalamic nucleus; SUBST P, substance-P; VA/VL, ventral anterior and ventral lateral thalamic nuclei.

GPM; this leaves the "direct" striato-GPM circuit underactive and leads to increased activity in GPM neurons. The nigrostriatal lesion also reduces the inhibitory activity of dopamine on the GABA-enkephalin neurons projecting to GPL (Figure). This produces excessive inhibitory activity in the striato-GPL circuit resulting in overinhibition of GPL and, consequently, reduction in the normal inhibitory tone that GPL exerts upon the STN. As a result, neuronal activity in the STN-GPM circuit is abnormally high (Mitchell et al., 1989a). This abnormality is likely to be quantitatively more important than the reduction in the GABA inhibition by the "direct" circuit, since the latter is physiologically underactive during routine behavior (Marsden and Obeso, 1993). Electrophysiological recording "in vivo" of neuronal activity both in the rat and monkey have confirmed the abnormally high firing of STN and GPM neurons after dopaminergic depletion (Miller and DeLong, 1987; Filion et al., 1988; Hollerman and Grace, 1992). The opposite state characterises chorea-ballism. In this situation, lesion of the STN or accentuation of the inhibitory tone in the GPL projection reduces the excitatory activity upon GPM and results in decrease inhibitory drive on the motor thalamus. Filion and Tremblay (1991) have elegantly shown in the parkinsonian monkey that the administration of apomorphine drastically reduces GPM discharges and how this is associated with choreic dyskinesias.

The subthalamic-GPM pathway plays, therefore, a paramount role in the origin of both parkinsonism and dyskinesias. It is likely that disfunction of the STN is also involved in the origin of drug-induced dyskinesias, like the ones induced by dopaminergic drugs in Parkinson's disease (Crossman, 1990; Luquin et al., 1992) and the syndrome of tardive dyskinesias provoked by chronic exposure to neuroleptic drugs (Mitchell et al., 1992).

Regarding the role of the STN in the pathophysiology of parkinsonism, the conventional explanation is that the excessive inhibition of the motor thalamus by the GPM, this in turn abnormally driven by the STN, leads to inhibition of the motor cortical areas (Albin et al., 1989). However, this is a very simplistic view. Both in monkeys and humans parkinsonism is a complex constellation of symptoms and signs with different pathophysiological mechanisms. For instance, rigidity is probably related with hyperexcitability of the primary motor cortex, which produces enhancement of the long latency stretch reflex. Slowness of movement initiation and the concatenation of sequential movements is probably related with desactivation of the supplementary motor area (Obeso et al., 1991). We believe that the heterogeneous GPM projections to the different cortical motor areas provides the basis for the origin of different signs under the influence of the same basic disturbance in the STN-GPM circuit. On the other hand, the reduction in blinking frequency and arm swinging or the short steps of the typical parkinsonian gait are probably mediated by subcortical pathways. Here, the connections of the STN/GPM with the PPN and other brainstem nuclei may be of the greatest importance (Moon Edley and Graybiel, 1983).

THE SUBTHALAMIC NUCLEUS AS A SURGICAL TARGET FOR PARKINSON'S DISEASE

According with the above, reduction of the exagerated excitatory drive or increasing the GABA inhibition of GPM should lead to improvement of akinesia and rigidity. In two parkinsonian (by MPTP) monkeys, DeLong's group demonstrated for the first time that lesion (with ibotenic acid) of the STN can produce a marked motor improvement (Bergman et al., 1990). More recently, Aziz et al. (1991) reported successful relief of parkinsonism in 6 parkinsonian macaques. The lesion (made by radiofrequency) was bilateral in 2 and unilateral in 4. The latter group of animals showed clear cut improvement on both sides. Persistent hemichorea was reported in out of the 8 monkeys operated by both groups. We have performed subthalamotomy (kainic acid) in 7 parkinsonian monkeys. One was very severe (Grade V), three were severe but capable of feeding themselves (Grade III) and two were mild (Grade II). One of the animal in Grade III had received l-dopa/carbidopa (25-50 mg daily) for several weeks and has developed mild choreiform dyskinesias in the lower limbs. The most severe monkey showed mild choreiform movements in the leg contralateral to the lesion but no improvement and had to be killed a few days after surgery. The other 6 animals improved strikingly. This effect was noticeable after 2-3 days in terms of the facial expression and spontaneous mobility, but actually continue to improve for several weeks. Fine motor tasks with the hands improved bilaterally. However, the side ipsilateral to the lesion remained parkinsonian. In the other side (contralateral to STN lesion) there was a persistent hemidyskinesia. This was particularly violent in one monkey, who was fairly mild. In the monkey with l-dopa-induced dyskinesias, there was no accentuation of the dyskinesia induced by subthalomotomy when rechallenged with levodopa.

These results are very encouraging and suggest that marked improvement could be achieved by producing a small lesion of the STN in humans. Indeed, two cases of parkinsonian patients (Sellal et al., 1992; Burnett and Jankovic, 1992) in whom lesion of the STN occurred fortitously and produced improvement, support this possibility. The major problem with the proposal of surgery of the STN in Parkinson's disease is the concern with dyskinesias (Guridi et al., 1993b). On this regard, the following points should be taken into account: 1. Lesion of the STN has a much greater probability of provoking an intense hemiballism in a normal or only mildly parkinsonian person than in a severe case. This is so because the pathological increment in GPM activity associated with parkinsonism depends not only of the STN-GPM circuit, but also of the reduced GABA inhibition by the direct striato-GPM projection. In this sense, the latter abnormality would act as an opposing force to the development of hemiballism. 2. In normal monkeys, a small lesion of the STN reduced GPM activity without producing dyskinesias during a manual activity (Hamada and DeLong, 1992b). A similar effect could be achieved in patients. 3. Even if severe hemidyskinesia would developed, it would always be possible to eliminate this by thalamotomy as demonstrated in patients for levodopa-induced dyskinesias (Narabayashi et al., 1984) and parkinsonian monkeys (Page 1992).

The alternative to subthalamotomy is pallidotomy. However, precise definition of the GPM zones more closely associated with the capital signs of parkinsonism is lacking. The precise anatomical organization described by Hoover and Strick (1993) in the monkey should also be worked out in humans to know more precisely what is the right target. This is of fundamental importance because it is difficult to understand how a parkinsonian patient is going to move better after interruption of the major basal ganglia output to the cortex.

REFERENCES

Aizawa, H., Inase, M., Mushiake, H., Shima, K., and Tanji, J., 1991, Reorganization of activity in the supplementary motor area associated with motor learning and functional recovery, *Exp. Brain Res.* 84:668-671.

Albin, R., Young, A.B., and Penney, J.B., 1989, The functional anatomy of basal ganglia disorders, *Trends Neurosci.* 12:366-375.

Aziz, T.Z., Peggs, D., Sambrook, M.A., and Crossman, A.R., 1991, Lesion of the subthalamic nucleus for the alleviation of 1-methy-4-phenyl-1,2,3,6-tetrahydropyridine (MPTP) induced parkinsonism in the primates, *Mov. Disord.* 6:288-292.

Benazzouz, A., Gross, C.H., Féger, J., Boraud, I., and Bioulac, B., 1992, Reversal of rigidity and improvement in motor performance by subthalamic high frequency stimulation in MPTP-treated monkeys, *Eur. J. Neurosci.* 5:382-389.

Bergman, M., Wichmann, T., and DeLong, M.R., 1990, Reversal of experimental parkinsonism by lesion of subthalamic nucleus, *Science.* 249:1436-1438.

Brotchie, P., Iansek, R., and Horne, M.K., 1991a, Motor function of the monkey globus pallidus. 1. Neuronal discharge and parameters of movement, *Brain.* 114:1667-1683.

Brotchie, P., Iansek, R., and Horne, M.K., 1991b, Motor function of the monkeys globus pallidus. 2. Cognitive aspects of the movement and phasic neuronal activity, *Brain.* 114:1685-1702.

Burnett, L., and Jankovic, J., 1992, Subthalamotomy and Parkinson's disease, *Mov. Disord.* 7:(Suppl. 1), 160.

Carpenter, M.B., Whittier, J.R., and Mettler, F.A., 1950, Analysis of choreoid hyperkinesia in the rhesus monkey, *J. Comp. Neurol.* 92:293-322.

Crossman, A.R., 1990, A hypothesis on the pathophysiological mechanisms that underlie levodopa of dopamine agonist-induced dyskinesia in Parkinson's disease: implications for future strategies in treatment, *Mov. Disord.* 5:100-108.

DeLong, M.R., Crutcher, M.D., and Georgopoulos, A.P., 1985, Primate globus pallidus and subthalamic nucleus: functional organization, *J. Neurophysiol.* 53:540-543.

Dierssen, G., and Gioino, G., 1961, Correlación anatómica del hemibalismo, *Rev. Clin . Esp.* 82:280-305.

Filion, M., and Tremblay, L., 1991, Abnormal spontaneous activity of globus pallidus neurons in monkeys with MPTP-induced parkinsonism, *Brain Res.* 547:142-151.

Filion, M., Tremblay, L., and Bédard, P.J., 1988, Abnormal influences of passive limb movement on the activity of globus pallidus neurons in parkinsonian monkeys, *Brain Res.* 444:165-176.

Georgopoulos, A.P., DeLong, A.R., and Crutcher, M.D., 1983, Relations between parameters of step-tracking movements and single cell discharge with globus pallidus and subthalamic nucleus of the behaving monkey, *J. Neurosci.* 3:1586-1598.

Guridi, J., Luquin, M.R., Guillén, J., Herrero, M.T., and Obeso, J.A. 1993a, Antiparkinsonian effect of subthalamotomy in MPTP-monkeys of different severity, *Mov. Disord.* 9:(in press).

Guridi, J., Luquin, M.R., Herrero, M.T., and Obeso, J.A., 1993b, The subthalamic nucleus: a possible target for stereotaxic surgery in Parkinson's disease, *Mov. Disord.* in press.

Hamada, I., and DeLong, M.R., 1992a, Excitotoxic acid lesions of the primate subthalamic nucleus result in transient dyskinesias of the contralateral limbs, *J. Neurophysiol.* 68:1850-1858.

Hamada, I., and DeLong, M.R., 1992b, Excitotoxic acid lesions of the primate subthalamic nucleus result in reduced pallidal neuronal activity during active holding, *J. Neurophysiol.* 68:1859-1866.

Hollerman, J.R., and Grace, R.R., 1992, Subthalamic nucleus cell firing in the 6-OHDA-treated rat: basal activity and response to haloperidol, *Brain Res.* 590:29-2991.

Hoover, J., and Strick, P.L., 1993, Multiple output channels in the basal ganglia, *Science.* 259:819-821.

Kitai, S.J., and Deniau, J.M., 1981, Cortical inputs to the subthalamus: intracellular analysis, *Brain Res.* 214:411-415.

Kitai, S.J., and Kita, H., 1987, Anatomy and physiology of the subthalamic nucleus: a driving force of the basal ganglia, *Adv. Behav. Biol.* 32:357-373.

Luquin, M.R., Laguna, J., and Obeso, J.A., 1992, Selective D2 receptor stimulation induces dyskinesia in parkinsonian monkeys, *Ann. Neurol.* 31:551-554.

Marsden, C.D., and Obeso, J.A., 1993, The paradox of stereotaxic surgery in Parkinson's disease for contemporary theories of the functions of the basal ganglia: leading to a dual concept of their mysterious perceptuo-motor functions, *Brain* in press.

Martin, J.P., 1967, "The Basal Ganglia and Posture," Pitman Medical, London.

Miller, W.C., and DeLong, M.R., 1987, Altered tonic activity of neurons in the globus pallidus and subthalamic nucleus in the primate MPTP model of parkinsonisn, *Adv. Behav. Biol.* 32:415-427.

Mink, J.W., and Thach, W.T., 1991, Basal ganglia motor control. III. Pallidal ablation: normal reaction time, muscle contraction and slow movement, *J. Neurophysiol.* 65:330-351.

Mitchell, I.J., Clarke, C.E., Boyce, S., Robertson, R.G., Peggs, D., Sambrook, M.A., and Crossman, A.R., 1989a, Neural mechanisms underlying parkinsonian symptoms based upon regional uptake of 2-deoxyglucose in monkeys exposed to 1-methyl-4-phenyl-1,2,3,6-tetrahydropyridine (MPTP), *Neurosci.* 32:213-226.

Mitchell, I.J., Jackson, A., Sambrook, M.A., and Crossman, A.R., 1989b, The role of the subthalamic nucleus in experimental chorea, *Brain.* 112:1533-1548.

Mitchell, I.J., Boyce, S., Sambrook, M.A., and Crossman, A.R., 1992, A 2-deoxyglucose study of the effects of dopamine agonists on the parkinsonian primate brain, *Brain.* 115:809-824.

Moon Edley, S., and Graybiel, A.M., 1983, The afferent and efferent connections of the feline nucleus tegmenti pedunculopontinus pars compacta, *J. Comp. Neurol.* 217:187-215.

Narabayashi, H., Yokochi, F., and Nakajima, Y., 1984, Levodopa induced dyskinesia and thalamotomy, *J. Neurol. Neurosurg. Psych.* 47:831-839.

Obeso, J.A., Artieda, J., Oliver, A., Muruzábal, J., and Barraquer-Bordás, L., 1991, Motor performance following the bilateral lesion of the supplementary motor area (SMA) in man, *J. Physiol.* 438: 31P.

Page, R.D., 1992, The use of thalamotomy in the treatment of levodopa-induced dyskinesia, *Acta Neurochir.* 114:77-117.

Parent, A., 1990, Extrinsic connections of the basal ganglia, *Trends Neurosci.* 13:254-258.

Parent, A., and Hazrati, L.N., 1993, Anatomical aspects of information processing in primate basal ganglia, *Trends Neurosci.* 16:111-116.

Parent, A., Hazrati, L.N., and Smith, Y., 1989, The subthalamic nucleus in primates. A neuroanatomical and immunohistochemical study, *in*: "Neural mechanisms in disorders of movement," A.R. Crossman and M.A. Sambrook, eds., John Libbey, London, pp. 29-35

Robledo, P., and Féger, J., 1990, Excitatory influence of rat subthalamic nucleus to substantia nigra pars reticulata and the pallidal complex: electrophysiological data, *Brain Res.* 518:47-54.

Sellal, F., Hirsch, E., Lisovoski, F., Mutschler, V., Collard, M., and Marescaux, C., 1992, Contralateral disappearance of parkinsonian signs after subthalamic hematoma, *Neurology* 42:255-256.

Whittier, J.R., and Mettler, F.A., 1949, Studies on the subthalamus of the rhesus monkey, *J. Comp. Neurol.* 90:281-317.

THE SUBTHALAMIC NUCLEUS: A MORE COMPLEX STRUCTURE THAN EXPECTED

Jean Féger[1], Mireille Mouroux[1], Abdelhamid Benazzouz[2], Thomas Boraud[2], Christian Gross[2] and Alan R. Crossman[3]

[1] Laboratoire de Pharmacologie, Faculté des Sciences Pharmaceutiques Biologiques, Université R. Descartes, 75006, Paris, France
[2] Laboratoire de Neurophysiologie, CNRS URA 1200, Université de Bordeaux II, France
[3] Experimental Neurology and Myology group, Department of Cell and Structural Biology, University of Manchester Medical School, Manchester, Great Britain

INTRODUCTION

Almost all the papers on the subthalamic nucleus, after the usual review of its anatomical connections, summarize the function of this structure with the classical reference to hemiballism. This type of anatomopathological approach is still widely used in neurobiology, despite the fact that it is a little strange to define subthalamic nucleus functions by reference to its dysfunction. The physiological functions have only recently been freed from the impressive and restrictive clinical description of hemiballism. Currently, facts and hypotheses stress the relationship between hypo- or hyperactivity of the subthalamic neurons and two opposite motor dysfunctions, hyperkinesia as in hemiballism and akinesia as in parkinsonism. Moreover, the importance of subthalamic inputs arising from associative or limbic structures should not be overlooked since they are an integral part of this concept of the function of the subthalamic nucleus and suggest that this structure has broader and more complex involvement in a large range of activities which may or may not have a motor component.

The function of the subthalamic nucleus must be re-evaluated in the light of a new model of the functional organization of the basal ganglia (Albin et al., 1989; Alexander and Crutcher, 1990). As became clear and amazing to remark at this last basal ganglia meeting, this new model, founded on the duality of the striatal efferents with a direct and an indirect circuit, is now widely accepted. The model implies a balance, at the level of the internal pallidum and the pars reticulata of the substantia nigra, between an inhibitory effect of the direct striatal output and an excitatory effect provided by the last step of the indirect circuit arising from the striatum. This indirect pathway involves the external pallidum and the subthalamic nucleus. The subthalamic efferents drive the neuronal activity of the pallidal and nigral neurons through glutamatergic mediation (Kitai and Kita, 1987; Robledo et al., 1988; Smith and Parent, 1988; Robledo and Féger, 1990). Since the striato-pallidal and the pallido-subthalamic projections are both GABAergic and inhibitory, it is assumed that changes in the discharge rate of the subthalamic neurons result from the release of inhibition.

At least apparently, this new design of the basal ganglia relationships appears to provide an easier and more coherent explanation of opposing movement disorders, roughly akinesia or hyperkinesia, in terms of the failure of a particular part of the basal ganglia. The experimental data obtained with models of Parkinson's disease and of hemiballism are reviewed in the first part of this paper. They support this concept and emphasize the contribution of the subthalamic nucleus to the expression of this motor disorder primarily linked to striatal dopaminergic depletion. However, as is discussed in the second part of this paper, the subthalamic nucleus cannot be thought of simply as a relay converting a disinhibition from the external pallidum into excitation of basal ganglia output: the internal pallidum and the pars reticulata of the substantia nigra. The convergence of cortical and thalamic afferents in the subthalamic nucleus, in addition to dopaminergic innervation arising from substantia nigra and ventro-tegmental area, suggests that the subthalamic nucleus should be involved in a larger range of functions within the basal ganglia.

SUBTHALAMIC NUCLEUS, HYPO- AND HYPERKINETICS DISORDERS

According to this new model of basal ganglia organization, the loss of striatal dopamine in Parkinsonism leads to underactivity of striatal neurons in the direct circuit and overactivity of the striatal projection to the external segment of the globus pallidus, the first step of the indirect circuit. This results in increased firing of the GABAergic neurons arising from the internal segment of the globus pallidus and from the pars reticulata of the substantia nigra (DeLong, 1990). This conclusion is supported by experimental data obtained with various techniques in animals having impaired striatal dopaminergic transmission (Crossman, 1987; Miller and DeLong, 1987; Mitchell et al., 1989; Gerfen, 1992; Robertson et al., 1992).The reversal of parkinsonian bradykinesia following a lesion of the subthalamic nucleus is believed to provide clear verification of this model.

Actually, an alleviation of the clinical signs of Parkinsonism have been observed in MPTP-treated monkeys following lesion or functional impairment of the subthalamic nucleus or of its pallidal efferents (Bergman et al., 1990; Graham et al., 1990; Aziz et al., 1991; Brotchie et al., 1991; Benazzouz et al., 1993). There is also an anecdotal clinical case report on the improvement of a Parkinsonian patient that is attributed to a superposed transient hemiballismus following a subthalamic hematoma, documented with computerized tomography (Sellal et al., 1992). The pioneering experiments of Bergman, Wichman and DeLong (1990) performed on African green monkeys showed the three main features of Parkinsonism following MPTP-induced lesion of the dopaminergic nigro-striatal neurons. Akinesia, rigidity and tremor were reduced in the contralateral limbs by a unilateral lesion of the subthalamic nucleus. In two animals, dyskinesia of the contralateral arm and leg also began a few minutes after the injection of the excitotoxic compound, but unlike the long lasting alleviation of the parkinsonism, this dyskinesia was transient. Similarly, unilateral radiofrequency subthalamotomy dramatically improved the motor abilities in MPTP-treated monkeys (Aziz et al., 1991). An unexpected result was also obtained in this last experiment. Unlike the strictly contralateral location of the hemiballism that affects the arm and the leg, unilateral subthalamotomy provided bilateral alleviation of the Parkinsonism. Further lesioning of the remaining subthalamic nucleus in one animal induced only a small additional change. In these experiments, the contralateral hemiballism was transient and disappeared without change in the stable and unremitting alleviation of the MPTP-induced Parkinsonism.

The improvement of Parkinsonism was completely dissociated from the production of hemiballistic or dyskinetic movement in experiments performed on hemi-Parkinsonian monkeys by Benazzouz et al. (1993). In these studies, a unilateral lesion of the nigral dopaminergic neurons was obtained by a slow injection of MPTP through the internal carotid artery. These animals had been previously trained to carry out conditioned extension or flexion of the forearm contralateral to the lesioned side. The subsequent subthalamic impairment was obtained by repeated stimulation of the subthalamic nucleus ipsilateral to the lesioned substantia nigra at a frequency of 100-130 hz. This experimental procedure was directly derived from an empirical method used as a substitute for thalamotomy in the treatment of Parkinsonian tremor in ambulatory patients with an intrathalamic electrode connected to an implanted stimulator (Benabid et al., 1989). In contrast to the experiments summarised above, the changes in rigidity and motor activity were more precisely determined. Moreover, since the blockade of the subthalamic nucleus was only temporary, motor performances could be compared more analytically. On the same day, the monkey's motor ability could be switched from an impaired to an alleviated state, allowing comparisons free from bias due to such factors as progressive recovery from the parkinsonian state observed in the MPTP model. In these animals, the rigidity was correlated with the rectified and integrated electromyogram recorded from the biceps and triceps brachii while the animal was not moving the arm, and motor ability was quantified using computerised analysis of the recorded mechanogram in a classical conditioned movement (fast ballistic flexion and extension of the forearm). Maximal speed and acceleration were calculated from the recorded mechanogram. The values obtained in the normal state, after MPTP-treatment and during subthalamic high frequency stimulation were compared.

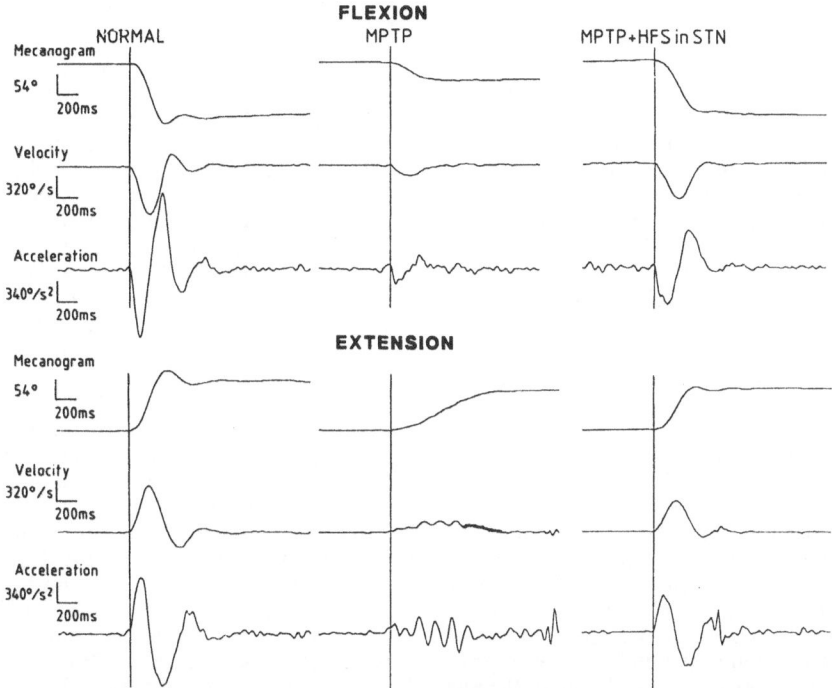

Figure 1. Representation of the successive mechanogram and the derived velocity and acceleration diagrams for flexion and extension movements of the forearm in monkey before, after MPTP-treatment and during a high frequency stimulation applied to the subthalamic nucleus ipsilateral to the lesioned substantia nigra. The vertical line marks the onset of movement (Benazzouz et al., 1993).

Qualitatively, this high frequency stimulation induced a decrease in rigidity, and the monkey began to move the contralateral limbs. This improved state remained noticeable at the end of stimulation and disappeared slowly at a rate that depended on the duration of the stimulation. Neither stereotypy nor dyskinesia were seen during or after the stimulation.

Quantitatively, the electromyographic activities recorded in the disabled arm decreased when stimulation began. Analysis of the movement parameters in flexion or extension indicated a return to the characteristics found in the normal monkey. The electromyographic activities for the movement changed from disorganised to time locked. Moreover, the increase in electromyographic activity at the end of stimulation was progressive, due to a prolonged effect of this high frequency stimulation. Further experiments are planned to determine whether unilateral stimulation produces a bilateral relief of rigidity and akinesia in animals with Parkinsonism.

The mechanism underlying the disappearance of tremor after thalamic stimulation performed on Parkinsonian patients (Benabid et al., 1989), or the alleviation of rigidity and akinesia by subthalamic stimulation in MPTP-treated monkeys remains unknown. There are several possibilities, including suppression of spiking activity by a depolarisation block, activation of a recurrent inhibitory circuit or impairment of synaptic transmission due to transmitter depletion, that may explain how high frequency stimulation could reproduce the effects obtained with a lesion.

In all these experiments where Parkinson's syndrome has been alleviated by an action on the subthalamic nucleus (Bergman et al., 1990; Aziz et al., 1991; Benazzouz et al., 1993) and obviously in the experiments using a high frequency electrical stimulation, the disappearance of akinesia cannot be attributed simply to a superimposed hyperkinesia. Clinical or experimentally-induced hemiballism is frequently transient. This is generally explained by invoking a compensatory or regulatory mechanism, acting in both normal and in MPTP-treated monkeys. But the same lesion improves the Parkinsonism, without any compensatory mechanism, since the relief of the akinesia seems permanent. Despite the lack of sufficient experimental data, the conflicting nature of that which is available, provides the basis for certain conclusions to be drawn.

First, the two classes of motor activity released after subthalamic lesion, the unwanted and erratic movements characteristic of hemiballism and the voluntary movements previously blocked in Parkinsonism, are obviously different. It could be speculated that changes in cortical excitability following disinhibition of thalamo-cortical neurons do not have the same repercussions on unwanted movements, probably due to poor motivation or poorly organised premotor preparation, as they do on voluntary movements driven by greater motivation and skilled coordination between the activities in premotor and motor areas. It, therefore, seems important to continue the attempts, initiated by Purdon Martin (1967), to investigate the nature of the trigger of these unwanted movements, or conversely, the origin of the rest periods which give hemiballism its intermittent character. This feature was observed in our experiments (Hammond et al., 1979) and in those reported by Hamada and DeLong (1992a).

Second, as proposed in our previous reviews (Féger et al., 1989, 1991b) and discussed in a recent report on pallidal activity recorded in monkeys following a subthalamic lesion (Hamada and DeLong, 1992b), the link between the subthalamic lesion, the decrease in discharge rate of pallidal neurons and dyskinesia is probably more complex than is currently assumed. It is probably necessary to take into account the way the subthalamic lesion is produced and the relative size of the lesion, the chronological changes in neuronal activity in the subthalamic cells close to the lesion and in the structures receiving subthalamic efferents. These various factors could perhaps explain the published discrepancies in the delay and duration of experimental hemiballism (Whittier and Mettler, 1949; Carpenter and Mettler, 1951; Hammond et al., 1979; Crossman et al., 1984; Hamada and Delong, 1992a). Moreover, blockage of synaptic activation related to subthalamic input to the neurons in the internal segment of the globus pallidus produces dyskinesia (Robertson et al., 1989), but this result is not obtained with a lesion of the internal pallidum (Horak and Anderson, 1984; Mink and Thach, 1991). It is difficult to explain this discrepancy since our electrophysiological experiments (Robledo and Féger, 1990) showed how the firing rate of the entopeduncular neurons was strongly dependant on tonic subthalamic activity. An important limitation, however, is that these experiments were performed in rats. The changes in neuronal activities, with special attention to the pattern of firing in the pallidum, the thalamic relay and in the different motor areas during dyskinesia must also be demonstrated in primates.

Third, MPTP-treated monkeys, at least in the experiments with radiofrequency lesions (Aziz et al., 1991), show bilateral changes following unilateral subthalamic lesion. This result is not reproduced if the synaptic effect of the subthalamic projection to the neurons located in the internal pallidum is blocked, as with unilateral microinjection of an excitatory amino acid antagonist in the medial segment of the globus pallidus (Graham et al., 1990). In this study, the changes in motor activity following a unilateral injection remained strictly unilateral. If the results could be confirmed suggesting a bilateral influence, it could be due to the involvement of the nigral efferents or the bilateralized projections to the pedunculopontine nucleus.

It is also difficult to explain the discrepancy between the somatotopic organisation described in the subthalamic nucleus (Carpenter and Carpenter, 1951; Hartmann-Von Monakow et al., 1978; DeLong et al., 1985) and the relatively stereotyped features of the hemiballism, affecting mainly the limbs, obtained after small lesions in different parts of the subthalamic nucleus. This might be explained by recent anatomical data showing that the subthalamo-pallidal fibers supply numerous axon collaterals that run perpendicular to the parent axon and loosely ensheathe the dendrites of the pallidal neurons (Smith et al., 1990).

In the internal segment of the globus pallidus, the subthalamic axonal varicosities on the soma and the proximal part of the dendrites are more numerous and larger as on the distal dendrites. The size and shape of the subthalamopallidal varicosities remain homogeneous in the external segment of the globus pallidus (Parent and Hazrati, 1993). With this organisation, subthalamic neurons could be connected to several groups of neurons lying in separate parts of the internal segment of the globus pallidus. The differences between the synaptic organization of the striatopallidal and subthalamopallidal axons suggest that the subthalamic nucleus could exert a more diffuse control providing a generalized activation on the pallidal neurons and the striatal input a more focalized effect. The circuits from the pallidum to the cortex through the thalamic relay are also distributed in separated channels ending in different cortical motor areas (Hoover and Strick, 1993). In experiments performed with transneuronal tracers, it was shown that the arm representations in the supplementary motor area, in the ventral premotor area and in primary motor cortex are correlated with different parts of the internal segment of the globus pallidus. Thus, a dysfunction of separate parts of the subthalamic nucleus that are connected to the different pallidal regions related to the representations of the limbs in the three cortical motor area, could induce identical abnormal movements.

In conclusion, there is good evidence for involvement of the subthalamic nucleus in the clinical expression of both hypo-and hyperkinetic disorders. However, the relationship between Parkinsonism and hemiballism or hyper- and hypoactivity of subthalamic neurons is more complex than that described by the present model of basal ganglia organisation.

THALAMIC PARAFASCICULAR AFFERENTS AND SUBTHALAMIC ACTIVITY

In the dual circuit model, the discharge rate of pallidal and nigral neurons depends on the ratio between two opposite driving forces: the inhibitory synaptic effect due to the direct striatal output and the excitation provided by the last relay in the indirect striatal output, the subthalamic nucleus. This raises the question of how the tonic and phasic activities of subthalamic neurons are adjusted? Various factors could induce changes in the discharge rate of these cells. These include intrinsic controls, such as the characteristics of ionic channels governing the electrophysiological properties of subthalamic neurons, the influence of local ionic homeostasis with crosstalk between neurons and glial cells, plus extrinsic influences due to the various afferents to the subthalamic nucleus that contribute greatly to the control of the neuronal activity. The most completely described subthalamic inputs are the inhibitory pallidal afferents (Rouzaire-Dubois et al., 1980) and the excitatory projections onto the subthalamic nucleus from the cerebral cortex (Hammond et al., 1978; Kitai and Deniau, 1981; Rouzaire-Dubois and Scarnati, 1985) and from the pedunculopontine nucleus (Hammond et al., 1983). However, chronic lesions of a large part of the cortex induce only small decreases in the mean firing rate of subthalamic neurons (Rouzaire-Dubois and Scarnati, 1985), whereas lesion of the globus pallidus slightly increases the rate of discharge and changes its pattern (Ryan and Clark, 1992 and our unpublished results). There is also a projection to the subthalamic nucleus from the centre médian-parafascicular complex, a part of the thalamic intralaminar structure. This last pathway is poorly documented, since anatomical data only were available (Sugimoto and Hattori, 1983; Sugimoto et al., 1983; Groenewegen and Berendse, 1990), when electrophysiological and pharmacological experiments were begun in our laboratory and completed with an anatomical study using a double labelling method (Bentivoglio et al., 1980; Féger and Crossman, 1984).

Anatomical Results

The main purpose of these experiments was to describe the location of the intralaminar neurons projecting to the subthalamic nucleus and to determine if this projection arises from a specific thalamic population or is a collateral of the massive thalamo-striatal projection (Césaro et al., 1985).

We found that the thalamic projection to the subthalamic nucleus arises only from the parafascicular part of the centre médian-parafascicular complex. In this structure, the neurons that are retrogradely labelled from the subthalamic nucleus lie in the posterior part around the fasciculus retroflexus, mainly between the fasciculus and the wall of the third ventricle.

Unilateral injection into one subthalamic nucleus indicated that the labelled neurons lie in the ispsilateral parafascicular thalamic nucleus only. The neurons projecting to the striatum arise from different subdivisions of the thalamus. Their density decreases from the anterior to the posterior part in the parafascicular nucleus. Rostrally, these neurons form a large ring encircling the fasciculus and are also present in a dorso-lateral extension. Caudally the bulk of the thalamic neurons projecting to the striatum lie around the upper half of the fasciculus. Following simultaneous injections of two dyes, true blue and nuclear yellow, into the subthalamic nucleus and striatum respectively, the neurons in this thalamic area were labelled with only one fluorescent dye. Double-labelled neurons were never observed in the thalamus.

These experiments also provided additional data on the relationship between the cortical projections to the striatum and to the subthalamic nucleus. Cortical neurons projecting to the subthalamic nucleus were found in the prefrontal, fronto-parietal area and in the anterior and medial cingulate cortex. These projections appear to be strictly ipsilateral. There were also numerous double labelled neurons at this level. The percentage of the total number of cortical neurons projecting to the subthalamic nucleus which were double labelled varied with different cortical areas. They were more numerous in the prefrontal cortex (38 ± 6,3 %) than in the cingulate (15,5 ± 2,5 %) or in the somatomotor cortex (9 ± 2,6 %).

A supplementary anatomical experiment examined the topographic organisation of this thalamic projection using an anterogradely transported marker. PHA-L was injected either between the wall of the third ventricle and the fasciculus retroflexus or lateral to this structure. It confirmed a previous report by Berendse and Groenewegen, (1990, 1991b), showing that the distribution of the terminals at subthalamic level and in the striatum reproduced the same mediolateral arrangement. Injection into the medial part of the parafascicular nucleus showed terminals in the medial and central third of the subthalamic nucleus and in the nucleus accumbens and the antero-medial striatum. Injection into the lateral part of the parafascicular nucleus revealed terminals mainly in the lateral part of the subthalamic nucleus and in the most caudo-lateral part of the striatum.

Electrophysiological and Pharmacological Results

The synaptic effect and the pharmacological characteristics of the cortical projection to subthalamic neurons have been comprehensively described (Hammond et al., 1978; Kitai and Deniau, 1981; Rouzaire-Dubois and Scarnati, 1985). They provide an excitatory input through glutamatergic transmission (Rouzaire-Dubois and Scarnati, 1987). The excitation obtained after direct electrical stimulation of various cortical areas, or following somesthesic activation relayed through the somatosensory cortex, is marked by a burst of action potentials. This feature is always found in experiments using extracellular recording techniques, corresponds with the large depolarization seen in intracellular recordings of this response (Kitai and Deniau, 1981), and is due to specific properties of subthalamic neuron membranes (Kitai and Kita, 1987).

As described in our preliminary reports (Féger and Mouroux, 1991a; Féger et al., 1992), electrical stimulation within the parafascicular nucleus induces a response with a reproducible but complex pattern. Extracellular recordings with a glass micropipette show a short-latency initial excitation with a single spike only, followed by a 5 to 7 ms silent period, a delayed excitation with a burst of action potentials, and a long period in which the spontaneous activity is blocked. During the recording of one neuron, there is only slight variation in the latency of the initial response to successive stimulations, as is usually observed when a monosynaptically driven action potential is recorded extracellularly. However, this discharge of subthalamic neurons could also be attributed to a disynaptic pathway via a cortical relay receiving thalamic afferents and at the origin of the cortico-subthalamic projection. This seems unlikely, however, because the short latency action potential evoked by thalamic stimulation is always recordable in animals with a large transection of the ipsilateral internal capsule. These results and the anatomical evidence for a direct projection from the parafascicular nucleus to subthalamic neurons provide strong evidence that this stimulation is mediated by a direct parafascicular-subthalamic pathway. The second phase of activation with a burst of action potentials could be attributed to a polysynaptic pathway, including a thalamo-striato-pallido-subthalamic circuit. The burst of discharge may be due to disinhibition. In direct support of this hypothesis, it has been shown that when the globus pallidus is chronically lesioned with a microinjection of the excitotoxic agent, quinolinic acid, the delayed burst of action potentials is not observed, whereas the initial excitation in response to thalamic stimulation is present. Extracellular recordings cannot be used to analyse the two phases marked by a depression of the spontaneous activity. If the blocking of spiking activity is due to an IPSP, pallido-subthalamic inhibition could be involved. This is also in good agreement with the recent description of a projection from the parafascicular nucleus to the globus pallidus (Kincaid et al., 1991).

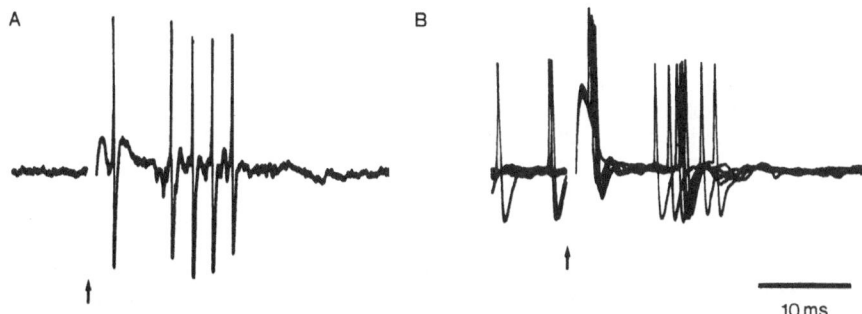

Figure 2. Extracellular recordings of a subthalamic neuron response after single shock parafascicular stimulation (arrows): A- single oscilloscopic record, B- superimposition of five successive responses showing the very small latency variability of the earlier response (Féger and Mouroux, 1991).

In spite of their importance, the pharmacology of the thalamic efferents to the basal ganglia is poorly documented. Thus, the anatomical organization of the thalamo-striatal projection and morphological aspects of the synaptic junctions have been comprehensively described (see Parent, 1986; Dubé et al., 1988; Lapper and Bolam, 1992), but the identity of the neurotransmitters responsible for their excitatory nature (Kocsis et al. 1977; Wilson et al., 1990) remains undetermined. After early reports of cholinergic mediation (Simke and Saelens, 1977; Saelens et al. 1977), glutamatergic transmission has been suggested on the basis of neurochemical and pharmacological arguments (Lehman and Scatton, 1982; Scatton and Lehmann, 1982; Nieoullon et al., 1985; Alberch et al., 1990). Whether thalamic neurons connected to the subthalamic nucleus have the same neurochemical characteristics as those projecting to the striatum, has been examined experimentally. First, the pharmacological effect of kynurenic acid, a nonselective antagonist of excitatory amino acid receptors, on the excitatory response induced with an electrical stimulation of the parafascicular nucleus was checked. This drug was injected within the subthalamic nucleus through a micropipette connected to a mechanically driven microsyringe. The antagonist blocked the initial spike, but not the secondary burst, in a dose (0.035, 0.1 and 0.3 µg in 200 nl) dependent manner. The receptors involved in this glutamatergic transmission were identified using two other drugs, APV and CNQX, selective antagonists respectively of the NMDA and Kainate/AMPA receptors. Again, both antagonists suppressed only the short-latency excitation in a dose dependent manner, and had no effect on the second phase of activation (Mouroux and Féger, 1993).

As our anatomical data provide strong evidence for a separate channel linking the parafascicular nucleus and the subthalamic nucleus, another set of experiments were performed to determine the balance between the monosynaptically driven excitation and the indirect inhibitory and disinhibitory effects upon subthalamic neurons, following a tonic activation or inhibition of the parafascicular nucleus as a whole. The unit activity of single subthalamic neurons was recorded either after activation with a thalamic microinjection of carbachol, a stable cholinergic agonist, or depression with muscimol, or after a lesion of the parafascicular nucleus with an excitotoxic drug. The preliminary results suggest that the discharge rate of subthalamic neurons depends on this excitatory input from the parafascicular nucleus. A less pronounced effect was observed after removal of the cortical excitatory input by decortication (Rouzaire-Dubois and Scarnati, 1987), acute freezing of the somatosensorimotor area (Hammond et al., 1978), or a transection of the internal capsule (our recent results to be published).

These experimental studies on the thalamic projection to the subthalamic nucleus have revealed a new glutamatergic excitatory input acting on NMDA and non-NMDA receptors. The thalamic neurons providing these efferents to the subthalamic nucleus lie in the parafascicular nucleus and are separate from those projecting to the striatum. It is noticeable that this glutamatergic excitatory projection produces an excitation of subthalamic neurons with a pattern, a single discharge, that is unlike the burst of action potential so distinctive for the cortically evoked response. Is this discrepancy related to the dendritic localisation of the excitatory input ? An explanation should be provided by an anatomical study in progress aimed at identification of the thalamic terminals on subthalamic neurons. This is the first pharmacological identification of thalamic input to a component of the basal ganglia. Glutamatergic transmission in the thalamic input to subthalamic neurons could perhaps be extrapolated to the excitatory thalamo-striatal projection. Further experiments are planned to determine the synaptic effect of another thalamic projection to the basal ganglia, the thalamo-pallidal pathway (Kincaid et al., 1991).

Functional aspects

There are too few data for a detailled discussion on the functional aspects of this projection arising from part of the intralaminar thalamic nuclei. However, the topographical organization of the parafascicular efferents to both the subthalamic nucleus and to the striatum, is medio-lateral (Berendse and Groenewegen, 1990). The medial part of the parafascicular nucleus is connected to the medio-ventral part of the striatum, including the nucleus accumbens, and to the medial part of the subthalamic nucleus, whereas the lateral part is connected to more caudal and lateral parts of the striatum and to the lateral part of the subthalamic nucleus. The cortico-subthalamic projection shows striking parallels with the organisation of the the the cortico-striatal projection (Berendse and Groenewegen, 1990). The associative and/or limbic cortical areas project mainly to the medial part of the subthalamic nucleus, whereas the somatomotor and somatosensory-cortical areas project mainly to the dorsolateral part of the subthalamic nucleus (Afsharpour, 1985; Berendse and Groenewegen, 1991b; Canteras et al., 1988). Consequently, the thalamic and cortical projections to the subthalamic nucleus are matched with associative or motor attributes. The position of their terminals within the subthalamic nucleus suggests that there are functionally different sectors. The subdivision of this structure into largely associative and motor sectors is correlated also with the distribution of the subthalamic efferents to the internal segment of the globus pallidus, since the medial part of the subthalamic nucleus sends fibers that terminate in the medial part of the entopeduncular nucleus and in the subcommissural part of the globus pallidus in the rat (Groenewegen and Berendse, 1990). A contingent of subthalamofugal fibers ending in the subcommissural part of the globus pallidus and in the substantia inominata have also been described in the squirrel monkey (Smith et al., 1990). All these data support the notion of a subthalamic nucleus with various functions as a corollary to the anatomical subdivision described in rodents and more precisely in primates. Is there cross talk between the motor and the so-called associative parts? The finding of numerous terminals in the dorsomedial part of the subthalamic nucleus following injection of an anterogradely transported marker (PHA-L) into the subcommissural ventral pallidum (Groenewegen and Berendse, 1990) would suggest that this is so. This anatomical observation could be correlated with the motor hyperactivity and an increase in labelled 2-deoxyglucose accumulation in the subthalamic nucleus in response to microinjection of a dopaminergic agonist into the nucleus accumbens. Lesioning the ventral pallidum and substantia innominata depresses both the motor activity and the labelling of the subthalamic nucleus (Patel et al., 1985). Further electrophysiological and behavioural experiments are required to verify the suggestion of Groenewegen and Berendse (1990) that there is a transfer of associative or limbic influences to the somatomotor system through the subthalamic nucleus.

Another set of anatomical, electrophysiological and behavioural data may indicate an associative or cognitive function of the subthalamic nucleus. An important contingent of cortical projections to the subthalamic nucleus in the primate arises from the frontal eye field and from the supplementary frontal eye field (Akert and Hartmann-Von Monakow, 1980). These projections terminate in the rostromedial part of the subthalamic nucleus. Curiously, a recent report stresses the importance of the reciprocal connections between the parafascicular nucleus and these areas in the frontal cortex which are involved in oculomotor control (see ref. in Sadikot et al., 1992). In addition, visual and oculomotor functions have been attributed to the subthalamic nucleus on the basis of single-unit recordings from the subthalamic nucleus in monkeys trained to perform visuo-oculomotor tasks (Matsumara et al., 1992). A change in the activity of numerous neurons was related to the fixation of the eye on a target, visually-guided or memory-guided saccades, or to a visual stimulus. In addition, spiking activity of some neurons was correlated with the expectation of a reward associated with a visual object. The proportion of subthalamic neurons with visuo-oculomotor properties (36%) is greater than the proportion of neurons with skelotomotor activities evoked from an active movement or passive manipulation of the body (25 % of all the recorded units). The bulk of subthalamic neurons with visuo-oculomotor properties were found in the ventromedial part of this structure (Matsumara et al., 1992). This location fits well with the subregion of the subthalamic nucleus containing neurons which project to the substantia nigra (Parent et al., 1989). This important involvement of the subthalamic nucleus in visuo-oculomotor activities is not anticipated from clinical observations, which stress the unwanted movements of the limbs. Similarly, abnormal ocular movement have not been described in reports on experimental hemiballism in monkeys. In recent experiments (Hamada and Delong, 1992a) small lesions were localized in the dorsal part of the subthalamic nucleus, a part where the neurons project to the globus pallidus (Parent et al., 1989). In our experiment with a large lesion involving all the subthalamic nucleus (Hammond et al., 1979), a permanent deviation of the head to the opposite side appeared early. This dystonia was attributed, maybe erroneously, to a partial lesion of nigral neurons. In contrast, paucity of eye movements is characteristic of patients with Parkinson's disease. This could be due to the lack of facial movements and blinking, features that are independant of visuo-oculomotor related activities. Matsumara et al. (1992) suggest that the direct cortico-subthalamic projection arising from the prefrontal cortex and the indirect circuit from the caudate nucleus may be involved in the transmission of visuomotor information to the subthalamic nucleus, since in this part of the cortex and in the caudate nucleus, the visuo-oculomotor related activities of some neurons have features similar to those of subthalamic neurons. Considering the importance of the reciprocal relationships between the parafascicular nucleus and the cortical frontal eye field and the supplementary frontal eye field, the parafascicular projection to the subthalamic nucleus could be involved in these visuo-oculomotor attributes of the subthalamic nucleus.

Any further discussions on the function of the subthalamic nucleus must take into account the results obtained in our double-labelling anatomical studies, showing the distinct identities of the parafascicular neurons connected to either the subthalamic nucleus or the striatum, with a large proportion of cortical neurons providing a separate projection to these two structures. This identification means that the information provided directly to subthalamic neurons could be qualitatively different from that going to the striatum and reaching the subthalamic nucleus indirectly.

CONCLUSION

The subthalamic nucleus is at the crossroads of a complex neuronal network. At the simplest level, the subthalamic nucleus is the last relay in the indirect circuit linking the striatum to the internal segment of the globus pallidus and the pars reticulata of the substantia nigra. At another level, the subthalamic nucleus also provides an interface between the striatal subdivisons characterized by more associative or more motor attributes. The duality of the dopaminergic innervation of the subthalamic nucleus, with one component arising from the ventrotegmental area and another from the pars compacta of the substantia nigra, reinforces this ambivalence. The greatest complexity is provided by the evidence on the organisation of the thalamic and a large part of the cortical projections, to the subthalamic nucleus via channels that are separate from the projections to the striatum. Could this complexity be correlated with connectionist models of motor processing discussed by Alexander et al. (1992) ?

Functionally, the transfer of cortical information directly to the subthalamic nucleus, or indirectly by two circuits through the striato-pallidal pathway or the parafascicular thalamic nucleus, allows adjustment in the timing and patterns of subthalamic neuronal activity, and finally in the outputs of the basal ganglia. The qualitative differentiation of information transmitted through the specific projections from the parafascicular nucleus and from the cerebral cortex to the subthalamic nucleus or to the striatum could allow the selection of neural signals related to various behaviours with a motor expression.

Acknowledgements

The anatomical and electrophysiological experiments reported were made possible in part by research grants from INSERM (CRE 910804) and from the European Science Foundation. The skilful collaboration of Mark Bevan in the anatomical experiments is gratefully acknowledged.

REFERENCES

Afsharpour, S., 1985, Topographical projections of the cerebral cortex to the subthalamic nucleus, *J. Comp. Neurol.* 236:14-28.

Akert, K., and Hartmann-Von Monakow, K., 1980, Relationships of precentral, premotor and prefrontal cortex to the mediodorsal and intralaminar nuclei of the monkey thalamus, *Acta Neurobiol. Exp.* 40:7-25.

Alberch, A., Arenas, E., Sanchez Arroyes, R., and Marsal, J., 1990, Excitatory amino acids release endogenous acetylcholine from rat striatal slices: regulation by GABA, *Neurochem. Int.* 17:107-116.

Albin, R.L., Young, A.B. and Penney, J.B., 1989, The functionnal anatomy of basal ganglia disorders, *Trends Neurosci.* 12:366-375.

Alexander, G.E., and Crutcher, M.D., 1990, Functionnal architecture of basal ganglia circuits: neural substrates of parallel processing, *Trends Neurosci.* 13:266-271.

Alexander, G.E., Delong, M.R., and Crutcher, M.D., 1992, Do cortical and basal ganglionic motor areas use motor programs to control movement, *Behav Brain Sci.* 15:656-665.

Aziz, T.Z., Peggs, D., Sambrook, M.A., and Crossman, A.R., 1991, Lesion of the subthalamic nucleus for alleviation of 1-methyl-4-phenyl-1,2,3,6-tetrahydro-pyridine (MPTP)-induced parkinsonism in the primate, *Mov. Disord.* 6:288-292.

Benabid, A.L., Pollak, P., Louveau, A., Hommel, M., Perret, J., and De Rougemont, J., 1989, Chronic Vim-thalamic stimulation in movement disorders, *in*: "Neural mechanisms in disorders of movement," A.R. Crossman and M.A. Sambrook, eds., Plenum Press, New York, pp. 413-415.

Benazzouz, A., Gross, C., Féger, J., Boraud, T., and Bioulac, B, 1993, Reversal of rigidity and improvement in motor performance by subthalamic high frequency stimulation in MPTP-treated monkeys, *Eur. J. Neurosci.* 5:382-389.

Bentivoglio, M., Kuypers, H.G.M., Catsman-Berrevoets, C.E., Loewe, H., and Dann, O., 1980, Two new fluorescent retrograde neuronal tracers which are transported over long distances, *Neurosci. Lett.* 18:25-30.

Berendse, H.W., and Groenewegen, H.J., 1990, Organization of the thalamostriatal projections in the rat, with special emphasis on the ventral striatum, *J. Comp. .Neurol.* 299:187-228.

Berendse, H.W., and Groenewegen, H.J., 1991a, Restricted cortical termination fields of the midline and intralaminar thalamic nuclei in the rat, *Neuroscience* 42:73-102.

Berendse, H.W., and Groenewegen, H.J., 1991b, The connections of the medial part of the subthalamic nucleus in the rat: evidence for a parallel organization, *in* " The Basal Ganglia III," G. Bernardi, M. B. Carpenter, G. D. Chiara, M. Morelli and P. Stanzione, eds., Plenum Press, New York, pp. 89-98.

Bergman, H., Wichman, T., and DeLong, M.R., 1990, Reversal of experimental parkinsonism by lesions of the subthalamic nucleus, *Science* 249:1436-1438.

Brotchie, J.M., Mitchell, I.J., Sambrook, M.A., and Crossman, A.R., 1991, Allievation of parkinsonism by antagonism of excitatory amino-acid transmission in the medial segment of the globus pallidus in rat and primate, *Mov. Disord.* 6:133-138.

Canteras, N.S., Shammah-Lagnado, S.J., Silva, B.A., and Ricardo, J.A., 1988, Somatosensory inputs to the subthalamic nculeus: a combined retrograde and anterograde horseradish peroxidase study in the rat, *Brain Res.* 458:53-64.

Carpenter, M.B., and Mettler, F.A., 1951, Analysis of subthalamic hyperkinesia in the monkey with special reference to ablation of agranular cortex, *J. Comp. Neurol.* 95:125-158.

Carpenter, M.B., and Carpenter, C.S., 1951, Analysis of somatotopic relations of the corpus Luysi in man and monkey, *J. Comp. Neurol.* 95:349-370.

Césaro, P., Nguyen-Legros, J., Polin, B., and Laplante, S., 1985, Single intralaminar thalamic neurons project to cerebral cortex, striatum and nucleus reticualaris thalami: A retrograde anatomical tracing study in the rat, *Brain Res.* 325:29-37.

Crossman, A.R., 1987, Primate models of dyskinesia: the experimental approach to the study of basal-ganglia related involuntary movements disorders, *Neuroscience* 21:1-40.

Crossman, A.R., Sambrook, M.A., and Jackson, A., 1984, Experimental hemichorea/hemiballismus in the the monkey. Studies of the intracerebral site of action in a drug induced dyskinesia, *Brain* 107:579-596.

DeLong, M.R., 1990, Primate models of movement disorders of basal ganglia origin, *Trends Neurosci.* 13:281-285.

DeLong, M.R., Crutcher, M.D., and Georgopoulos, A.P., 1985, Primate globus pallidus and subthalamic nucleus: functional organization, *J. Neurophysiol.* 53:530-543.

Dubé, L., Smith, A.D., and Bolam, J.P., 1988, Identification of synaptic terminals of thalamic or cortical origin in contact with distinct medium size spiny neurons in the rat neostriatum, *J. Comp. Neurol.* 267:455-471.

Féger, J., and Crossman, A.R., 1984, Identification of different subpopulations of neostriatal neurones projecting to globus pallidus or substantia nigra in the monkey: a retrograde fluorescence double-labelling study, *Neurosci. Lett.* 49:7-12.

Féger, J., and Mouroux, M., 1991, Mise en évidence de l'effet excitateur de l'efférence thalamo-subthalamique issue du noyau parafasciculaire, *C. R. Acad. Sci. Paris*, 313 série III:447-452.

Féger, J., Robledo, P., and Renwart, N., 1991, The subthalamic nucleus: new data, new questions, *in*: "The Basal Ganglia III," G. Bernardi, M.B. Carpenter, G.D. Chiara, M. Morelli, and P. Stanzione, eds., Plenum Press, New York, pp. 99-108.

Féger, J., Mraovitch, S., and Mouroux, M., 1992, The thalamic parafascicular projection to the subthalamic nucleus : a glutamatergic excitatory pathway, *Neuroscience Abstract* 18:136-5.

Féger, J., Vezole, I., Renwart, N., and Robledo, P., 1989, The rat subthalamic nucleus: electrophysiological and behavioural data, *in*: "Neural mechanisms in disorders of movement," A.R. Crossman, and M.A. Sambrook, eds., John Libbey, London, pp. 37-43

Gerfen, C.R., 1992, The neostriatal mosaic: multiple levels of compartmental organization, *Trends Neurosci.* 15:133-139.

Graham, W.C., Robertson, R.G., Sambrook, M.A., and Crossman, A.R., 1990, Injection of excitatory amino-acid antagonists into the pallidal segment of a 1-methyl-4 phenyl-1,2,3,6-tetrahydropyridine (MPTP) treated primate reverses motor symptoms of parkinsonism, *Life Sci.* 47:91-97.

Groenewegen, H.J., and Berendse, H.W., 1990, Connections of the subthalamic nucleus with ventral striatopallidal parts of the basal ganglia in the rat, *J. Comp. Neurol.* 294:607-622.

Hamada, I., and Delong, M.R., 1992a, Excitotoxic acid lesions of the primate subthalamic nucleus result in transient dyskinesias of the contralateral limbs, *J. Neurophysiol.* 68:1850-1858.

Hamada, I., and Delong, M.R., 1992b, Excitotoxic acid lesions of the primate subthalamic nucleus result in reduced pallidal neuronal activity during active holding, *J. Neurophysiol.* 68:1859-1866.

Hammond, C., Deniau, J.M., Rouzaire-Dubois, B., and Féger, J., 1978, Peripheral input to the rat subthalamic nucleus, an electrophysiological study, *Neurosci. Lett.* 9:171-176.

Hammond, C., Féger, J., Bioulac, B., and Souteyrand, J.P., 1979, Experimental hemiballism in the monkey produced by unilateral kainic acid lesion in corpus luysi, *Brain Res.* 171 577-580.

Hammond, C., Rouzaire-Dubois, B., Féger, J., Jackson, A., and Crossman, A.R., 1983, Anatomical and electrophysiological studies on the reciprocal projections between the subthalamic nucleus and the nucleus tegmenti pedonculopontinus in the rat nucleus, *Neuroscience* 9:41-55.

Hartmann-Von Monakow, K., Akert, K., and Künzle, H., 1978, Projections of the precentral motor cortex and other cortical areas of the frontal lobe to the subthalamic nucleus in the monkey, *Exp. Brain Res.* 33:395-403.

Hoover, J.E., and Strick, P.L., 1993, Multiple output channels in the basal ganglia, *Science* 259:819-821.

Horak, F.B., and Anderson, M.E., 1984, Influence of globus pallidus on arm movements in monkeys, *J. Neurophysiol.* 52:290-304.

Kincaid, A.E., Penney, J.B., Young, A.B., and Newman, S.W., 1991, The globus pallidus receives a projection from the parafascicular nucleus in the rat, *Brain Res.* 553:18-26.

Kitai, S.T., and Deniau, J.M., 1981, Cortical inputs to the subthalamic nucleus: intracellular analysis, *Brain Res.* 214:411-415.

Kitai, S.T., and Kita, H.,1987, Anatomy and physiology of the subthalamic nucleus: a driving force of the basal ganglia, *in*: "The Basal Ganglia II, Structure and Function - Current concepts,", M.B. Carpenter and A. Jayaraman, eds., Plenum Press, New York, pp. 357-373.

Kocsis, J.D., Sugimori, M. and Kitai, S.T., 1977, Convergence of excitatory synaptic inputs to caudate spiny neurones, *Brain Res.* 124:403-413.

Lapper, S.R., and Bolam, J.P., 1992, Input from the frontal cortex and the parafascicular nucleus to cholinergic interneurons in the dorsal striatum of the rat, *Neuroscience* 51:533-545.

Lehmann, J., and Scatton, B., 1982, Characterization of the excitatory amino acid receptor-mediated release of [^{3}H]acetylcholine from rat striatal slices, *Brain Res.* 252:77-89.

Martin, J.P., 1967, "The basal ganglia and Posture," Pitman Medical, London.

Matsumara, M., Kojima, J., Gardiner, T. W., and Hikosaka, O., 1992, Visual and oculomotor functions of monkey subthalamic nucleus, *J. Neurophysiol.* 67:1615-1632.

Miller, W.C., and DeLong, M.R., 1987, Altered tonic activity of neurons in the globus pallidus and subthalamic nucleus in the primate MPTP model of Parkinsonism, *in*: "The Basal Ganglia II, Structure and Function - Current concepts," M.B. Carpenter and A. Jayaraman, eds., Plenum Press, New York, pp. 415-427.

Mink, J.W., and Thach, W.T., 1991, Basal ganglia motor control: III Pallidal ablation: normal reaction time, muscle contraction and slow movement, *J. Neurophysiol.* 65:330-351.

Mitchell, I.J., Clarke, C.E., Boyce, S., Robertson, R.G., Peggs, D., Sambrook, M.A., and Crossman, A.R., 1989, Neural mechanisms underlying parkinsonian symptoms based upon regional uptake of 2-deoxyglucose in monkeys exposed to 1-methyl-4-phenyl-1,23,6-tetrahydropyridine (MPTP), *Neuroscience* 32:213-226.

Mouroux, M. and Féger, J., 1993, Evidence that the parafascicular projection to the subthalamic nucleus is glutamatergic, *Neuroreport* 4: in press.

Nieoullon, A., Scarfone, E., Kerkerian, L., Errami, M., and Dusticier, N., 1985, Changes in choline acetyltransferase, glutamate decarboxylase, high affinity glutamate uptake and dopaminergic activity induced by kainic lesion of the the thalamostriatal neurons, *Neurosci. Lett.* 58:299-304.

Parent, A. ,1986, "Comparative Neurobiology of the Basal ganglia," J. Wiley, New York.

Parent, A., Hazrati, L.-N., and Smith, Y., 1989, The subthalamic nucleus in primates. A neuroanatomical and immunohistochemical study, *in*: "Neural mechanisms in disorders of movement," A.R. Crossman, and M.A. Sambrook, eds., John Libbey, London, pp. 29-35.

Parent, A. and Hazrati, L.N., 1993, Anatomical aspects of information processing in primate basal ganglia, *Trends Neurosci.* 16:111-116.

Patel, S., Slater, P., and Crossman, A.R., 1985, A lesioning and 2-deoxyglucose study of the hyperactivity produced by an intra accumbens agonist, *Naunyn Schmiedbergs Arch. Pharmacol.* 331:334-340.

Robertson, G.S., Vincent, S.R., and Fibiger, H.C., 1992, D1 and D2 dopamine receptors differentially regulate *c-fos* expression in striato nigral and striatopallidal neurons, *Neuroscience* 49:285-296.

Robertson, R.G., Farmery, S.M., Sambrook, M.A., and Crossman, A.R., 1989, Dyskinesia in the primate following injection of an excitatory amino acid antagonist into the medial segment of the globus pallidus, *Brain Res.* 476:317-322.

Robledo, P., and Féger, J., 1990, Excitatory influence of rat subthalamic nucleus to subtantia nigra pars reticulata and the pallidal complex: electrophysiological data, *Brain Res.* 518:47-54.

Robledo, P., Vezolle, I., and Féger, J., 1988, Mise en évidence d'un effet excitateur des efférences subthalamo-nigrales et subthalamo-pallidales chez le rat, *C. R. Ac. Sci.* 307:133-138.

Rouzaire-Dubois, B., Hammond, C., Hamon, B., and Féger, J., 1980, Pharmacological blockade of the globus pallidus-induced inhibitory responses of the subthalamic cells in the rat, *Brain Res.* 200:321-329.

Rouzaire-Dubois, B., and Scarnati, E., 1985, Bilateral corticosubthalamic nucleus projections: an electrophysiological study in rats with chronic cerebral lesions, *Neuroscience* 15:69-79.

Rouzaire-Dubois, B., and Scarnati, E., 1987, Pharmacological study of the cortical-induced excitation of subthalamic nucleus neurones in the rat: evidence for amino acids as putative neurotransmitters, *Neuroscience* 21:429-440.

Ryan, L. J., and Clark, K. B., 1992, Alteration of neuronal responses in the subthalamic nucleus following globus pallidus and neostriatal lesions in rats, *Brain Res. Bull.* 29:319-327.

Sadikot, A. F., Parent, A., and François, C., 1992, Efferent connections of the centromédian and parafascicular thalamic nuclei in the squirrel monkey: A PHA-L study of subcortical projections, *J. Comp. .Neurol.* 315:137-159.

Saelens, J.K., Edwards-Neak, S., and Simke, J.P., 1977, Further evidence for cholinergic thalamostriatal neurons, *J. Neurochem.* 32:1093-1094.

Scatton, B., and Lehmann, J., 1982, N-methyl-d-aspartate-type receptors mediate [^{3}H] acetylcholine release evoked by excitatory amino acids, *Nature* 297:422-424.

Sellal, F., Hirsch, E., Lisovoski, F., Collard, M., and Marescaux, C., 1992, Contralateral disappearance of parkinsonian signs after subthalamic hematoma, *Neurology* 42:255-256.

Simke, J.P., and Saelens, J.K., 1977, Evidence for a cholinergic tract connecting the thalamus with the head of the striatum in the rat, *Brain Res.* 126:487-495.

Smith, Y., Hazrati, L.-N., and Parent, A., 1990, Efferent projections of the subthalamic nucleus in the squirrel monkey as studied by the PHA-L anterograde tracing method, *J. Comp. Neurol.* 294:306-323.

Smith, Y., and Parent, A., 1988, Neurons of the subthalamic nucleus in primates display glutamate but not GABA immunoreactivity, *Brain Res.* 453:353-356.

Sugimoto, T., and Hattori, T., 1983, Confirmation of thalamosubthalamic projections by electron-microscopic autoradiography, *Brain Res.* 264:335-339.

Sugimoto, T., Hattori, T., Mizuno, N., Itoh, K., and Sato, M., 1983, Direct projection from the centre médian-parafascicular complex to the subthalamic nucleus in the cat and rat, *J. Comp. .Neurol.* 214:209-216.

Whittier, J.R., and Mettler, F.A., 1949, Studies of the subthalamus of the rhesus monkey. II. Hyperkinesia and other physiologic effects of subthalamic lesions, with special references to the subthalamic nucleus of Luys, *J. Comp. Neurol.* 90:319-372.

Wilson, C.J., Chang, H.T., and Kitai, S.T., 1990, Firing patterns and synaptic potentials of identified giant aspiny interneurons in the rat neostriatum, *J. Neuroscience* 10:508-519.

NEURONAL ACTIVITY OF THE HUMAN BASAL GANGLIA IN PARKINSONISM COMPARED TO OTHER MOTOR DISORDERS

Chihiro Ohye, Masafumi Hirato, Yasuhiro Kawashima,
Nobuaki Hayase, Kenji Satake and Akio Takahashi

Department of Neurosurgery
Gunma University School of Medicine
Maebashi, Gunma, Japan

INTRODUCTION

To elucidate the function of the basal ganglia in human disease states, we have recorded their spontaneous neuronal activity during the course of stereotactic thalamotomy and more recently pallidotomy for Parkinson's disease and other kinds of involuntary movement. The activity of the basal ganglia was correlated with that of the lateral oral thalamus to which the internal segment of the globus pallidus conveying information from the striatum is known to project. The results will be interpreted taking into account recent knowledge about this pallido-striato-thalamic connection.

In an earlier series, the neuronal activity of the basal ganglia was recorded mainly from the caudate nucleus (Ohye et al., 1991; Ohye, 1993a). The direction of our standard tracking for the surgery of tremor was towards the thalamic nucleus ventralis intermedius, usually passing through the head of the caudate nucleus (Ohye, 1988). Therefore, when we examined the functional relation between the basal ganglia and the thalamus, it was almost necessarily the relation between the caudate nucleus (via the pallidonigral complex) and the the thalamus that we observed. According to Alexander and Crutcher (1990) the neuronal circuits involving the basal ganglia and the thalamus would be the following. The caudate nucleus would project indirectly to the thalamus via the substantia nigra (pars reticulata). The putamen would project to the lateral oral thalamus through the internal segment of the globus pallidus (GPi). We thought that direct information could be obtained if the activity of the putamen and GPi could be recorded and correlated with that of the part of the thalamus receiving pallidal afferences. It happened that on several occasions, our electrode passed through the GPi in its way between the caudate nucleus and the thalamus.

There is a recent tendency for revival of stereotactic pallidotomy in cases of parkinsonian akinesia (Laitinen et al., 1992), and on such occasions, we tried also by the usual microrecording technics, to record directly the activity of the putamen and the GPi. Although the number of pallidotomy cases is still limited, some preliminary results will be considered together with previous ones.

It will be shown that in cases with parkinsonian rigidity and akinesia the internal segment of the globus pallidus (GPi) is hyperactive.

SUBJECTS AND METHODS

Some ten cases of Parkinson's disease (PA), four cases with essential tremor, two cases with dystonia and two with chorea were the subjects of this study. On all of them, stereotactic selective thalamotomy was performed in an awake state under local anesthesia by our microrecording technics. In Parkinson's disease, cases with tremor type and rigid type were separately considered as we previously found that metabolic state and electrical activity in deep structures were somewhat different between these two types of Parkinsonian syndrome (Ohye, 1991; Ohye et al., 1991). The details of the stereotactic operations with microrecording were already described in previous papers (Ohye, 1988, Ohye et al., 1990); only the part relevant to this study will be briefly reported again.

Before the operation, stereotactic magnetic resonance imaging (MRI) is done in order to get the necessary brain image from which the electrode track during the operation is determined (Kawashima et al., 1992). This is made easy because the anterior and posterior commissures are clearly visualized in mid sagittal view. The brain slices were made about 45° to the intercommissual line. The slice passing 5-6 mm anterior to the posterior commissure contains the nucleus ventralis intermedius which is the target of surgery for tremor (Ohye, 1988). The intersection of this plane of section and the skull is the burr hole point, 3-4 mm lateral to the midline. Using Leksell's stereotactic apparatus, an electrode introduced through this burr hole automatically passes through the plane of the selected MRI picture. The electrode is a bipolar concentric type, with outer diameter of 0.3 mm, tip about 10 μm, and interpolar distance about 0.3 mm. Electrical resistance is less than 100 Kohm. We usually use a pair of electrodes set rostrocaudally wih a 3 mm interval parallel to the midsagittal plane. The anterior electrode is always aimed at the tentative target point, but sometimes the introduction of this electrode is omitted during recording in order to avoid undue damage to the brain until neurosurgical coagulation is finally to be performed. The electrodes are introduced gently into the brain with the aid of a motor-driven, stepping micromanipulator.

As shown in Figure 1, the electrode almost always passes through the caudate nucleus before entering the thalamus in our anterolateral to posteromedial directional approach. If carefully observed, there are also chances of passing through the internal segment of the globus pallidus, on the way between the caudate nucleus and the thalamus, especially when the electrode passes closer to the intercommissural line, and is slightly more laterally angled, which happened in several cases. According to our long experience of depth recording, it is easy to identify caudate and thalamic activities by their respective characteristic spontaneous discharge patterns. Therefore, if some other particular electrical activity is intercalated between them, it would be, most probably, the activity of the internal segment of the globus pallidus. In this respect, the internal capsule between the caudate nucleus and the thalamus exhibits just a low background activity with small positive spikes.

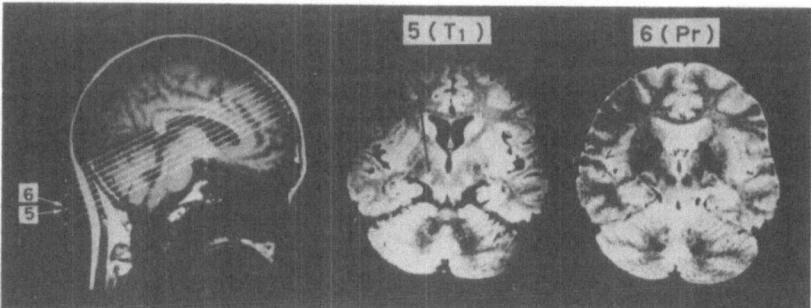

Figure 1. Magnetic resonance images showing the basal ganglia, thalamus and an electrode trajectory passing through these structures. Left: A sagittal scout view to show levels of cutting planes (white lines). Middle: Selected fifth T1 weighted image (marked by 5 and an arrow) with a trajectory (a black line) toward the thalamic nucleus ventralis intermedius. Right: Sixth proton image (marked 6 and an arrow). In this image, the globus pallidus is shown by dark, low-signal area.

Spontaneous electrical activity in the caudate nucleus, GPi, and the pallidal projection zone of the dorsal thalamus is continuously recorded and is electrically integrated on line, so that it is possible to measure quantitatively the basic activity of each structure. The integrated value during a given interval of time (usually for one second) is shown on a digital meter, by an arbitrary unit. For example, according to this system, the minimum value is about 30 in white matter, and the maximum value is about 300 in the thalamic nucleus ventralis intermedius. We have shown previously that the value changes almost parallel to the characteristics of depth structure, mostly depending on respective cytoarchitectonic nature and spontaneous activity level (Ohye, 1993a). If necessary, a bar histogram of the background activity along a track is made with the aid of a personal computer. Attention is paid also to spontaneous discharge pattern and to the sensory response to natural stimuli on the controlateral extremities, if any.

RESULTS

Spontaneous activity in the caudate nucleus, internal segment of the globus pallidus and pallidal territory of the thalamus

When the electrode reached the caudate nucleus from overlying white matter, the background activity increased suddenly, with fast oscillation of around 30 Hz, superimposed

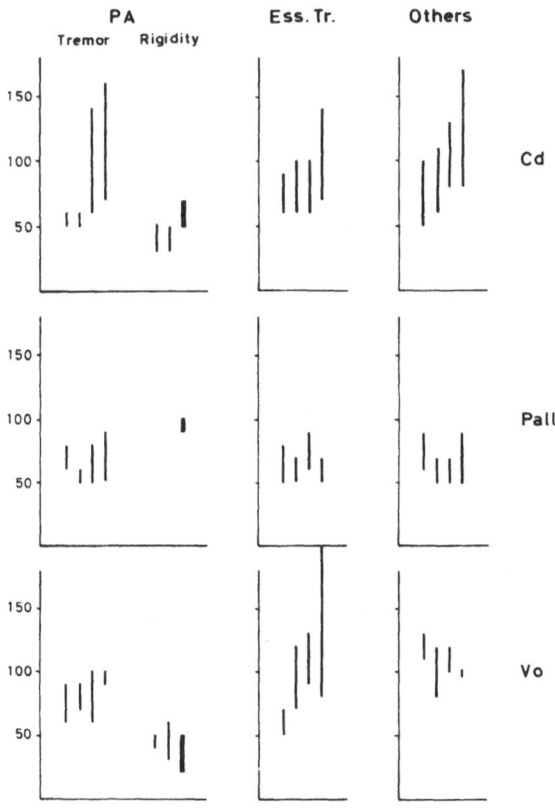

Figure 2. Basic background activity in Cd, GPi (Pall) and anterodorsal thalamus (VO) in cases with different involuntary movements. PA: Parkinson's disease. Ess Tr: Essential tremor. Others: Other kinds of involuntary movement. In this bar graph, range of spontaneous activity (maximum and minimum values) in each case at Cd, GPi and VO respectively are aligned on the same vertical position and put in the order of low to high, from left to right, in each disease group. Activity represented by thick line was recorded from a case with rigid and akinetic case.

on low to medium amplitude spike discharges. In a stable state, spike discharge showed either a sporadic pattern of 1-3 Hz, or long irregular burst patterns. Background activity was high in general, but if carefully compared, as shown in Figure 2 and in previous papers, the level of caudate activity was different in the individual cases. It was higher in the tremor type Parkinsonian syndrome than in the rigid type. In cases with essential tremor and other involuntary movements, caudate activity was also high. Therefore, in this regard, reduced activity in rigid type Parkinsonian syndrom seemed to be a specific profile.

In some cases the electrode passed through the medial edge of the internal segment of the globus pallidus before entering the thalamus. Activity of GPi at this point was not marked in cases tested irrespective of the types of Parkinson's disease (Fig. 2) except in the rigid type, which will be referred to in a later section. In two cases with choreic movement, many irregular burst discharges were found in GPi (Kawashima et al., 1992). It was not necessarily time locked with the patient's involuntary movement, but some coincidence was noticed.

At the level of the thalamus, corresponding to the nucleus ventralis oralis (VO), the spontaneous activity was high in the tremor type of parkinsonian syndrome and in essential tremor, being almost parallel to the level of the activity in the caudate nucleus. In contrast, in the rigid type of Parkinsonism, it was low. It was also noticed that in the other involuntary movement group, thalamic activity was inversely related to that of the caudate nucleus (Fig. 2), as already pointed out in previous papers (Ohye, 1993b; Ohye et al., 1991).

Neuronal activity of the Putamen

In one case of mixed rigid and tremor types of Parkinsonism, a pair of recording electrodes was introduced accidentally, into the anterior part of the putamen due to a small miscalculation of the lateral coordinate. General spontaneous activity of the putamen was quite similar to that of the caudate nucleus. There were fast oscillations with sporadic spike discharges. Retrospective analysis of the activity revealed a heterogeneous distribution of high and low activity zones. Several examples of the activity of the putamen along the trajectory are shown in Figure 3. In this case, the background spontaneous activity was high in the dorsal part, decreased during next 10 mm, and increased again in further ventral part. Recordings were stopped at the midpoint of the putamen.

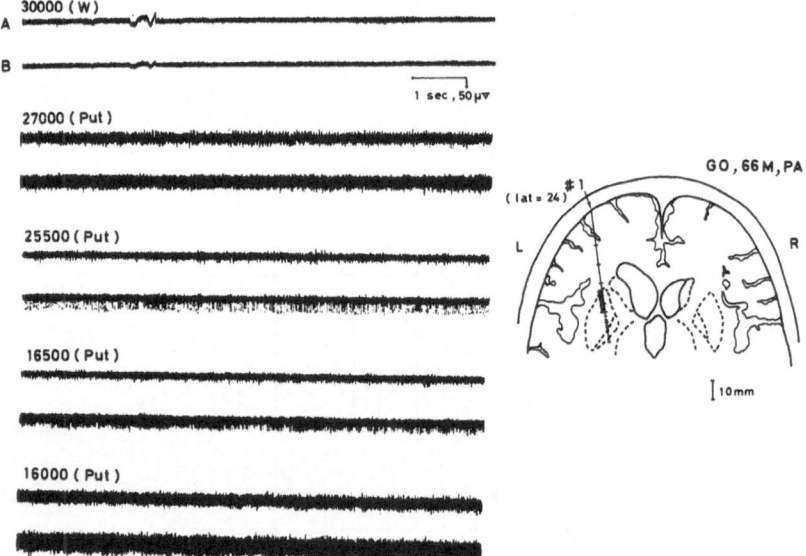

Figure 3. Examples of the spontaneous activity in different levels of the putamen (Put). A pair of traces (A and B) is the simultaneous recording by a pair of electrodes set in parallel rostrocaudally at 3 mm interval. Numbers indicate distance in microns from tentative zero (end of trajectory). Image on the right is a coronal brain section traced from a MRI image showing electrode trajectory through the left Put. In this case, only dorsal Put (thick line) was explored.

In another case of akinetic and rigid type of Parkinsonism, on the occasion of a pallidotomy, an electrode passed through the putamen. The distribution of spontaneous activity was again heterogeneous, as shown in Figure 3.

Neuronal activity of the internal segment of the globus pallidus (GPi)

In a case of the rigid type of Parkinsonism, a recording electrode was intentionally introduced into the internal segment (GPi) in order to study its spontaneous activity. It was found that the activity was very much elevated. In this case, after fixing the electrode in the GPi, the second electrode was inserted, as usually, toward the thalamic Vim nucleus passing through the caudate nucleus. Comparing the activity in GPi and thalamic VO, it was clearly shown that GPi activity was much higher than that of VO as shown in Figure 4. Before entering the thalamus, the activity of neurons of the caudate nucleus was recorded (not shown in the Figure) and was less than that in GPi.

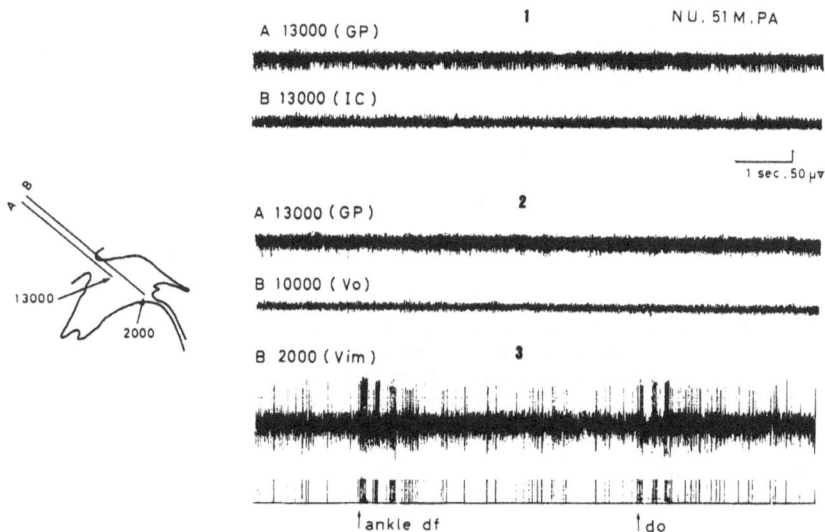

Figure 4. Examples of hyperactive GPi. On the left side is shown a sagittal view of electrode trajectory in relation to the third ventricle traced from X-ray film. In this case, A-electrode is fixed in GPi at 13000 μm, and B-electrode is advanced to Vim (kinesthetic zone), through internal capsule (IP) and dorsal thalamus (VO). Right bottom is response of a Vim neuron to ankle dorsiflexion.

In another case of Parkinsonian syndrom with akinesia, rigidity and tremor (in decreasing order of severity), stereotactic pallidotomy was planned. In this case, the anterior part of the internal segment of the globus pallidus was the target. Recordings were made by a pair of electrodes as shown in Figure 5, passing through the caudate nucleus and through the globus pallidus (first external and then internal segments). Referring to the standard atlas, this electode probably did not reached the putamen. Recording of spontaneous activity revealed that the activity of the caudate nucleus neurons was moderate, that of GPe neurons was relatively low, while it was highly exaggerated in the target zone (GPi), where homogeneous high activity was noticed. Through the whole trajectory in what was considered to be the internal segment of the globus pallidus, continuous high frequency spike discharges of more than 50 Hz were very characteristic. During the recording process, tremor was always manifested in the contralateral extremities, but there were no rhythmic discharges, nor any sensory response to natural stimuli on the contralateral extremities.

PET results in this case

It is noteworthy that in this case, the PET study (Hitachi, PCT H-1) using F18-FDG clearly exhibited a higher accumulation in the corresponding left lenticular nucleus (rigidity

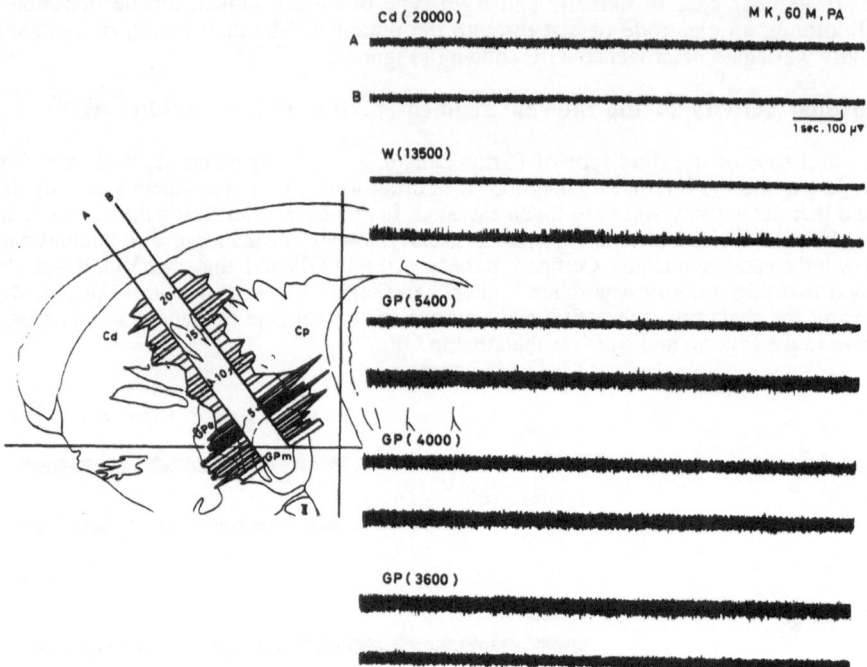

Figure 5. A sequential change of background activity along the trajectories towards GPi for pallidotomy. On the left is shown a bar histogram of the background activity on a sagittal view of the corresponding striatal zone of the standard atlas. On the right, examples of the activity at different levels are shown. A pair of recording traces are the same as in the previous figures. Note the hyperactivity of GPi neurons.

and tremor were on the right side) as shown in Figure 6. Quantitative assays revealed some 20% more accumulation of FDG on the left side than on the right side.

DISCUSSION

The present data on the spontaneous activity of the basal ganglia in Parkinsonian syndrome and in other kinds of involuntary movement in humans supports and supplements previous findings and hypotheses on the functional relation of the striatopallidothalamic connection. In the case where the neuronal activity of the internal segment of the globus pallidus (GPi) was recorded directly, to our surprise, this activity was markedly increased especially in the rigid and akinetic type of Parkinsonism. We do not have a normal control, but comparison of the activity of GPi neurons in the tremor type of Parkinsonism with that in cases of essential tremor shows that this increase is significant. On the basis of recent studies on the experimental model of Parkinsonism obtained in MPTP treated monkeys, it was postulated that the subthalamic nucleus was hyperactive (Crossman,1987; Albin et al., 1989; Mitchell et al., 1989; DeLong,1990). If so, the internal segment of the globus pallidus (GPi), which receives excitatory input from this nucleus, should be also hyperactive. The present observation corresponds to what one would expect from the data from experimental studies.

To summarize, the present observations on the activity of the neurons of the internal segment of the globus pallidus, together with our previous, somewhat fragmentary, observations on the neuronal activity in the caudate nucleus, putamen and thalamic VO nucleus, demonstrate that in the rigid and akinetic type of the Parkinsonian syndrom, the neuronal activity in the caudate nucleus is relatively low, in the putamen a mixture of low and high, in the internal segment of the globus pallidus high, and VO low. All this confirms our previous conclusion. In contrast, in the tremor type of the Parkinsonian syndrom, the relation is reversed. Neuronal activity in the caudate nucleus is high, that of GPi is low, and that of VO is high. In cases with essential tremor, the same pattern of activity was found in the different record sites as in the tremor type of the Parkinsonian syndrom. It should be noted that the previous metabolic study on Parkinsonian patients by PET scan showed different metabolic states between the tremor and the rigid types of Parkinsonian syndrome. In the present observation also, an interesting finding was that FDG metabolism in the lenticular nucleus region was elevated in the rigid and akinetic type of Parkinsonism in parallel with exaggerated spontaneous activity of GPi.

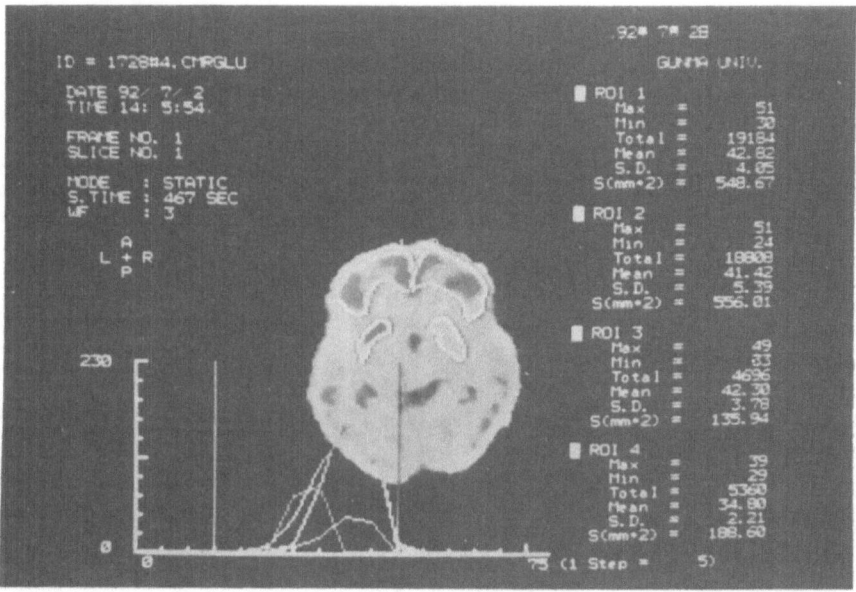

Figure 6. PET image showing glucose metabolism in the same case as illustrated in Figure 5. ROI (region of interest, circumscribed area) 1 is set in the left frontal area, ROI 2 in the right frontal, ROI 3 in the left BG and ROI 4 in the right BG. Note that FDG uptake in the left BG (controlateral to the clinically affected side, hence operated side in this case) is higher than that in the right.

Recent anatomical and physiological studies in monkeys have revealed that the internal segment of the globus pallidus receives various inputs : from striatum first, from the lateral segment (GPe) and the subthalamic nuclei. It sends fibers to the lateral and oral part of the thalamus, to the habenula, the subthalamic nucleus and to the pedunculopontine nucleus (Parent and De Bellefeuille, 1983; Hazrati and Parent, 1992). The main target of the external segment of the globus pallidus is the subthalamic nucleus. In addition to these complex input and output relations, a morphological and functional organization within the internal segment of the globus pallidus, with an associative and a sensorimotor pallidum, has been proposed (Percheron et al., 1987). In this regard, from our limited experiences of pallidotomy, it was not possible to find such an organization within the GPi. We could not find either, using our electrophysiological method, a particular place of the internal segment of the globus pallidus which could be proposed as the best target for the alleviation of parkinsonian akinesia. As far as we tested, there was no definite response to natural sensory stimuli or to voluntary action, although responses from wide peripheral areas were found in the internal segment of the globus pallidus of MPTP treated monkeys (Filion et al., 1988). There were also no rhythmic discharges related to tremor in this segment of the globus pallidus. We are awaiting the results of further studies to try to define the most appropriate target point in GPi. In any case, it seems likely that destruction of the hyperactive GPi, especially in its posterior part, could be effective for the alleviation of the akinetic state in Parkinsonism. Such a positive effect would be very important in relation to the experimental result that a lesion of the subthalamic nucleus resulted in the amelioration of akinesia in MPTP treated monkeys (Bergman et al., 1990; Aziz et al., 1991), as well as to an incidental improvement of rigidity and akinesia seen after a subthalamic hemorrhage in a case of Parkinson disease (Sellal et al., 1992). If ansotomy, pallidotomy, pallidoansotomy, already effected in the forties and fifties (Meyers, 1940; Cooper, 1956), and recently proposed again (Laitinen et al., 1992), proves really effective on a long term basis, it might replace subthalamic lesions and even striatal implantations, neither of which are easy to realize widely.

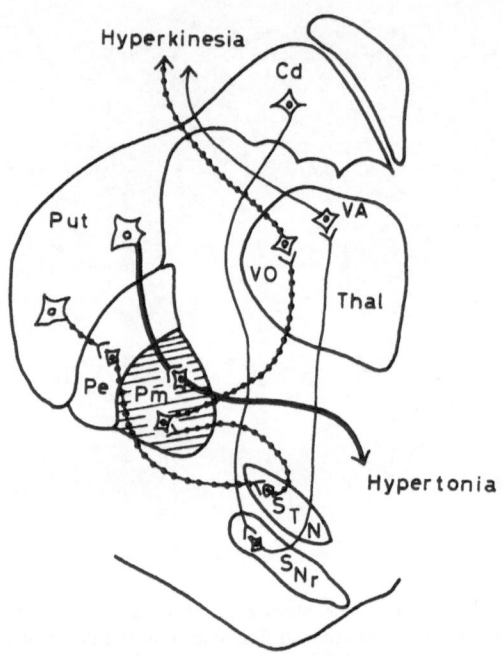

Figure 7. A simplified and hypothetical diagram of the basal ganglia in relation to the disorders of movement linked to basal ganglia disfunction. Neuronal circuits responsible for hyperkinesia (beads line) and hypertonia (thick line) may be differently involved. GPi (Pm: shaded area) is hyperactive in rigid and akinetic Parkinsonian syndroms probably reflecting the hyperactivity of the subthalamic nucleus (STN).

Finally, although oversimplified, some hypothetical neuronal circuits which may mediate different involuntary movements related to the basal ganglia are presented in Figure 7, based on clinical (Ohye, 1993a) and experimental (Page, 1992) studies. An emphasis is placed on the possibility that hyperkinetic (dyskinesia, chorea, ballism) and hypertonic (dystonia and akinesia) symptoms correspond to different circuits. In this case however, parkinsonian rigidity is somewhat difficult to explain since it is considerably ameliorated by pallidotomy as well as VO thalamotomy, hence it could be more related to the hyperkinetic circuit. An understanding of the organization of the internal segment of the globus pallidus in relation to rigidity will be helpful in solving this problem. In any case the main result of the present study, performed during neurosurgical interventions in humans, provides a clear confirmation of the changes of activity previously observed in MPTP treated monkeys. This brings strong support to the pathophysiological interpretations recently proposed, particularly those involving the internal segment of the globus pallidus.

REFERENCES

Albin, R.L., Young, A.B., and Penney, J.B., 1989, The functional anatomy of basal ganglia disorders, *TINS* 12:366-375.

Alexander, G.E., and Crutcher, M.D., 1990, Functional architecture of basal ganglia circuits; neural substrates of parallel processing, *TINS* 13:266-271.

Aziz, T.Z., Peggs, D., Sambrook, M.A., and Crossman, A.R., 1991, Lesions of the subthalamic nucleus for the alleviation of 1-methyl-4-phenyl-1,2,3,6-tetrahydropyridine (MPTP)-induced parkinsonism in the primate, *Mov. Disord.* 6:288-292.

Bergman, H., Wichmann, T., and DeLong, M.R., 1990, Reversal of experimental parkinsonism by lesions of the subthalamic nucleus, *Science* 249:1436-1438.

Cooper, I.S.,1956, "The Neurosurgical Alleviation of Parkinsonim," Thomas, Sprinfield.

Crossman, A.R., 1987, Primate models of dyskinesia: the experimental approach to the study of basal ganglia related involuntary movement disorders, *Neuroscience* 21:1-40.

DeLong, M.R., 1990, Primate model of movement disorders of basal ganglia origin, *TINS* 13:281-285.

Filion, M., Tremblay, L., and Bédard, P.J., 1988, Abnormal influence of passive limb movement on the activity of globus pallidus neurons in parkinsonian monkeys, *Brain Res.* 444:165-170.

Hazrati, L-N., and Parent, A., 1992, The striatopallidal projection displays a high degree of anatomical specificity in the primate, *Brain Res.* 592:213-227.

Kawashima, Y., Chen, H-J., Takahashi, A., Hirato, M., and Ohye, C., 1992, Application of magnetic resonance imaging in functional stereotactic thalamotomy for the evaluation of the thalamus, *Stereotact. Funct. Neurosurg.* 58:33-38.

Kawashima, Y., Takahashi, A., Hirato, M., and Ohye, C., 1991, Stereotactic Vim-Vo thalamotomy for choreic movement disorder, *Acta Neurochir.* suppl 52:103-106.

Laitinen, L.V., Bergenheim, A.T., and Hariz, M.I., 1992, Leksell's posteroventral pallidotomy in the treatment of Parkinson's disease, *J. Neurosurg.* 76:53-61.

Mitchell, I.J., Clarke, C.E., Boyce, S., Robertson, R.G., Peggs, D., Sambrook, M.A., and Crossman, A.R., 1989, Neural mechanisms underlying parkinsonian symptoms based upon regional uptake of 2-deoxyglucose in monkeys exposed to 1-methyl-4-phenyl-1,2,3,6-tetrahydro-pyridine, *Neuroscience* 32:213-226.

Meyers, R., 1940, A surgical procedure for postencephalitic tremor, with notes on the physiology of the premotor fibers, *Arch. Neurol. Psychiat.* 44:455-459.

Ohye, C., 1988, Selective thalamotomy for movement disorders: Microrecording, stimulation, techniques and results, *in*: "Modern Stereotactic Neurosurgery," L.D. Lunsford, ed., Martinus Nijhoff, Boston, pp. 318-331.

Ohye, C., 1991, Positron emission tomographic study in Parkinson's disease- Rigid vs tremor type, *in*: "Parkinson's Disease. From Clinical Aspects to Molecular Basis," T. Nagatsu et al., eds., Springer, Wien, New York, pp. 179-186.

Ohye, C., 1993a, Involuntary movement due to vascular lesion in the basal ganglia (in Japanese), *Brain and Nerve* 45:119-127.

Ohye, C., 1993b, Dynamic aspects of striatothalamic connection studied in cases with movement disorder, *in*: "Parkinson's Disease. From Basic Research to Treatment," H. Narabayashi et al., eds., Raven Press, New York, pp. 78-83.

Ohye, C., Shibazaki, T., Hirato, M., Kawashima, Y., and Matsumura, M., 1990, Strategy of selective Vim-thalamotomy guided by microrecording, *Stereotact. Funct. Neurosurg.* 54-55:186-191.

Ohye, C., Shibazaki, T., Hirato, M., Kawashima, Y., Matsumura, M., and Shibazaki, T., 1991, Neural activity of the basal ganglia in Parkinson's disease depth recording and PET scan, *in*: "Basal Ganglia III," G.Bernardi et al., eds., Plenum, New York, pp. 637-644.

Page, R.D., 1992, The use of thalamotomy in the treatment of Levodopa-induced dyskinesia, *Acta Neurochir.* 114:77-117.

Parent, A., and De Bellefeuille, L., 1983, The pallidointralaminar and pallidonigral projections in primate as studied by retrograde doublelabeling method, *Brain Res.* 278:11-27.

Percheron, G., François, C., and Yelnik, J., 1987, Spatial organization and information processing in the core of the basal ganglia, *in*: "The Basal Ganglia II," M.B. Carpenter and A. Jayaraman, eds., Plenum, New York, pp. 205-226.

Sellal, F., Hirsch, E., Lisovoski, F., Mutschler, V., Collard, M., and Marescaux, C., 1992, Contralateral disappearance of parkinsonian signs after subthalamic hematoma, *Neurology* 42:255-256.

NEUROCHEMISTRY AND NEUROPHARMACOLOGY

OF THE BASAL GANGLIA

BIOCHEMISTRY AND PHARMACOLOGY OF L-DOPA IN RELATION TO BASAL GANGLIA CIRCUITRY

Peter DeBoer, Stephen R. Wachtel and Elizabeth D. Abercrombie

Center for Molecular and Behavioral Neuroscience
Rutgers University
197 University Avenue
Newark, NJ 07104 (USA)

Few drugs make it to the movies. Although the wondrous effects of L-DOPA were made known to a wide audience by the movie "Awakenings", the actual mechanism of action of this drug remains somewhat of an enigma. L-DOPA is thought to be converted by the enzyme aromatic L-amino acid decarboxylase (AADC) into the neurotransmitter dopamine (DA) (Lloyd et al., 1975; Melamed et al., 1985), thus substituting for the loss of dopaminergic input to the striatum that is the underlying pathology of Parkinson's disease. However, the effects of the DA formed from L-DOPA on target areas of the dopaminergic system are not well described. As an example of the action of L-DOPA on postsynaptic targets, we will discuss the impact of L-DOPA on striatal cholinergic function. Another important aspect to consider in understanding the actions of L-DOPA involves the assumption that the main site of action of the DA formed from this compound is in the striatum. In this paper we also will discuss evidence that L-DOPA is converted into DA not only at the level of the striatum, but also at the level of the substantia nigra pars reticulata (SNpr) and may thus have effects on basal ganglia functioning beyond restoration of dopaminergic tone in striatum.

We have studied the biochemistry and pharmacology of L-DOPA using *in vivo* brain microdialysis, a technique allowing measurement of extracellular levels of neurotransmitters in freely moving animals. Briefly, a small diameter (250 μm) membrane that is permeable to neurotransmitters and their metabolites is surgically implanted into the brain of rats. The membrane is perfused with an artificial cerebrospinal fluid (aCSF) and small molecules diffuse down a concentration gradient into the aCSF. Transmitter levels in the aCSF samples thus obtained are analyzed by high performance liquid chromatography.

For the studies described here, we have implanted the dialysis probes into the striatum and SNpr and analyzed the levels of DA and acetylcholine (ACh) in the dialysates after systemic injections of L-DOPA. Transmitter levels were analyzed in intact rats and rats that had received a unilateral 6-hydroxydopamine (6-OHDA) lesion before the dialysis experiment. This lesion decreased the tissue DA levels in the ipsilateral striatum by more than 90%. Rats treated with 6-OHDA in this way show many biochemical and behavioral changes that are analogous to those occuring in patients with Parkinsonism (Zigmond and Stricker, 1989).

The Basal Ganglia IV, Edited by
G. Percheron *et al.*, Plenum Press, New York, 1994

EFFECT OF L-DOPA ON THE OUTPUT OF STRIATAL ACh

The clinical finding that muscarinic antagonists and dopamine agonists both are beneficial in the treatment of Parkinson's disease led to the hypothesis of a striatal "DA-ACh balance" (McGeer et al., 1961; Barbeau, 1962). In this model, an increase in the dopaminergic tone in the striatum is reflected in a decrease in the output of striatal ACh, whereas a decrease in the striatal dopaminergic tone is reflected in an increase in the output of striatal ACh. Thus, in Parkinsonian brain, the degeneration of the nigrostriatal dopaminergic system should lead to an overactivity of the striatal cholinergic system. Treatment of these patients with direct DA agonists such as bromocriptine (BROMO) or indirect DA agonists such as L-DOPA, therefore, is hypothesized to restore the balance that normally exists between striatal cholinergic and dopaminergic systems.

A large body of evidence indicates that the output of striatal ACh is tonically inhibited by DA via DA D-2 receptors. It may be assumed that these D-2 receptors are at least partly located on the cholinergic neurons since, in the presence of tetrodotoxin (TTX) *in vitro*, DA D-2 agonists have an inhibitory effect on potassium-evoked ACh efflux. Furthermore, dopaminergic and cholinergic terminals both synapse upon the medium spiny output neurons of the striatum. Thus, apart from the presence of a "DA-ACh balance" at the level of the striatal cholinergic interneuron, both systems additionally are in a position to directly modulate the activity of the striatal output neurons (for review see: Lehmann and Langer, 1983; Stoof et al., 1992). Using *in vivo* microdialysis for ACh, we and others have tested the effect of different classes of dopaminergic compounds on the output of striatal ACh (Bertorelli and Consolo, 1990; Damsma et al., 1990a, b; DeBoer et al., 1992). These studies have led to a slightly revised model of dopaminergic modulation of striatal ACh. As discussed below, we hypothesize an important role for extrastriatal DA acting upon the D-1 subclass of DA receptor.

Five different subtypes of DA receptors have been described, however pharmacological studies are based primarily on the D-1/D-2 classification. According to the model that is emerging, based largely on *in vivo* microdialysis studies, striatal ACh output is modulated by both DA D-1 and D-2 receptors in an antagonistic manner. Stimulation of DA D-1 receptors results in an increased output of striatal ACh (Bertorelli and Consolo, 1990; Damsma et al., 1990a, b), whereas stimulation of DA D-2 receptors results in a decrease in this variable (Bertorelli and Consolo, 1990; Damsma et al., 1990a). With respect to striatal ACh, indirect DA agonists that are administered systemically act more like DA D-1 receptor agonists than D-2 agonists, i.e. they increase the output of striatal ACh (Damsma et al., 1991; Consolo et al., 1992; DeBoer et al., 1993). However, local perfusion of indirect DA agonists through the microdialysis probe results in D-2-like responses, i.e. a decrease in the output of striatal ACh is observed (DeBoer et al., 1990). Thus, systemic injections of indirect DA agonists appear to activate dual influences upon striatal ACh efflux. These influences consist of an inhibitory D-2 effect that is predominantly intrastriatal and a stimulatory D-1 effect that may originate outside the striatum.

It is clear from the above that both DA D-1 and D-2 receptors regulate the output of striatal ACh. Of particular interest to the present discussion are the data of Bertorelli et al. (1992) demonstrating that whole brain DA depletion resulted in a loss of both D-1 and D-2-mediated regulation of striatal ACh, but that the DA D-2-mediated regulation was more resistant to disruption by DA-depletion than the D-1-mediated regulation (Bertorelli et al., 1992). This observation raises the possibility that, in early Parkinson's disease, the DA D-1-mediated regulation of striatal ACh is preferentially impaired, perhaps leading to a decrease in the levels of extracellular striatal ACh. A similar scenario recently was hypothesized to occur by Robertson et al. (1992). In such a model, L-DOPA influences the relation between striatal DA and ACh in a direction that is opposite to the classical concept of the "DA-ACh balance", i.e. increasing instead of decreasing striatal ACh. Additional studies with partially lesioned rats will shed further light on this possibility.

Given the newly appreciated complexity of the regulation of striatal ACh output by DA, we considered it important to determine the effects of L-DOPA administration on striatal ACh efflux. In particular, we were interested in knowing the effect of L-DOPA administered in the 6-OHDA rat model of Parkinson's disease. L-DOPA therapy of Parkinson's disease has been hypothesized to restore the "DA-ACh balance" via D-2-mediated inhibition of striatal ACh. The more recent data described above, suggest the possibility that L-DOPA may, in fact, increase the output of striatal ACh. Such a result would necessitate a reconsideration of current models of the actions of L-DOPA in Parkinsonism. We therefore have re-evaluated the effect of two anti-Parkinsonian drugs, the indirect DA agonist L-DOPA and the direct DA agonist BROMO, on the output of striatal ACh.

The tissue level of DA in the striatum of animals treated with 6-OHDA several weeks prior to the experiment was <90% of control levels. These animals had a significantly (t-test, t=3.5, p=.0035) higher basal output of striatal ACh relative to unlesioned animals (88 fmol/min vs. 52 fmol/min). This result corroborates the notion of a tonic inhibition of striatal ACh by DA in the intact striatum. DA formed from L-DOPA (50 mg/kg), however, evoked a potent stimulation of ACh output in 6-OHDA-lesioned animals. Unlesioned animals were significantly less responsive, but a significant increase in the output of striatal ACh was measured in these animals as well. L-DOPA (100 mg/kg) stimulated the output of ACh to the same extent in both groups (Figure 1). It is interesting to note, however, that this same dose of L-DOPA increased striatal DA levels significantly more (t-test, t=3.9, p=0.003) in lesioned (from 4 fmol/sample to 1290 fmol/sample) than in unlesioned rats (from 60 fmol/sample to 330 fmol/sample). BROMO (4 mg/kg) inhibited the output of striatal ACh in both groups, but moreso in lesioned than in unlesioned animals. The more potent decrease in the output of striatal ACh in lesioned vs. unlesioned animals may be attributed to sensitization of the DA D-2 receptors on the striatal cholinergic interneurons.

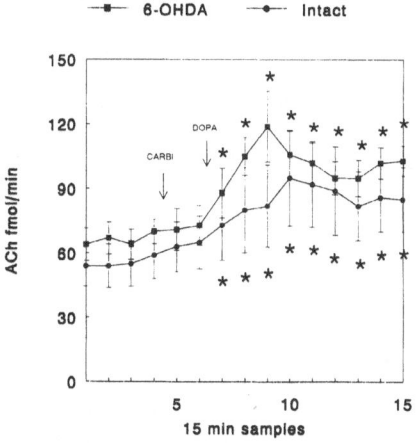

Figure 1. Effect of 100 mg/kg of the indirect DA agonist L-DOPA on the output of striatal ACh from the striatum of 6-OHDA-treated (n=5) and untreated (n=5) rats. All animals were pretreated with a peripheral decarboxylase inhibitor (carbidopa, CARBI). Numbers on the abscissa represent successive 15 min dialysate samples. Values are mean ± SEM for ACh content of the samples expressed in fmol/min. *, significantly different from pre-drug values, p ≤ 0.05. [*Reprinted with permission from DeBoer et al., 1993*]

Thus, in the 6-OHDA-treated rat model, it appears that L-DOPA produces effects on striatal ACh that more closely resemble those of a DA D-1 agonist than those of a D-2 agonist. Interestingly, both L-DOPA and BROMO are used successfully to treat Parkinson's disease. Our results may indicate that neurochemical processes in addition to the balance between DA and ACh in striatum are important for the therapeutic action of these drugs. It is therefore of interest to explore further the substrate(s) for the increase in striatal ACh output observed in response to systemic L-DOPA.

EFFECT OF L-DOPA ON EXTRACELLULAR DA IN SNpr

As the main input station of the basal ganglia, the striatum of rodents receives topographically organized input from the cortex. The SNpr and the entopeduncular nucleus (EP) are the primary basal ganglia output stations to thalamus. Analogous to the circuitry described in primates, the activity of the striatal output neurons can influence SNpr and EP via a direct as well as an indirect pathway (Gerfen, 1992). Cortical activation of the direct pathway results in an increased activity of the striatal medium-spiny GABA-ergic neurons that in turn inhibit the GABA-ergic cells of the SNpr and EP. Cortical activation of the indirect pathway through globus pallidus results in disinhibition of the subthalamic nucleus, which sends excitatory projections to the EP and SNpr.

DA can influence the flow of information through this circuitry at two different levels. First, the medium spiny neurons of striatum receive direct dopaminergic inputs from the substantia nigra pars compacta (SNpc) (Freund et al., 1984). At this level, it appears that DA-mediated regulation of the direct pathway occurs primarily via actions on DA D-1 receptors and that DA influences the indirect pathway mainly by actions on DA D-2 receptors. Second, DA released from the dendrites of SNpc DA neurons can act upon DA receptors located in the nigra. This DA regulates the activity of SNpc DA neurons through DA D-2 autoreceptors (Wachtel et al., 1989) and also appears to modulate the activity of SNpr cells via an action on DA D-1 receptors located on the terminals of the GABA-ergic striatonigral fibers (Cheramy et al., 1981; Walters et al., 1987; Beckstead, 1988).

The importance of DA actions at both striatal and extrastriatal sites is exemplified in studies showing that rotational behavior in unilaterally 6-OHDA-lesioned animals is dependent on activation of DA receptors in both the striatum and the SNpr. In particular, the importance of nigral DA D-1 receptors is indicated by the fact that when L-DOPA is systemically administered to unilaterally 6-OHDA-lesioned animals, rotational behavior is suppressed by intranigral application of the D1-selective antagonist SCH23390 or of an AADC inhibitor such as carbidopa (for review see: Robertson and Robertson, 1987; Robertson, 1992). We were therefore interested in comparing the neurochemical effects of systemic administration of L-DOPA (50 mg/kg) in striatum and SNpr of intact and 6-OHDA-lesioned animals.

In intact animals, 50 mg/kg L-DOPA increased the output of nigral DA from 5 to 41 fmol/sample, whereas in 6-OHDA-lesioned animals the output of DA increased from 1.5 to 181 fmol/sample. In contrast, the same dose of L-DOPA slightly decreased the output of striatal DA in intact animals from 43 to 39 fmol/sample and increased the output of DA in 6-OHDA-lesioned animals from 19 to 114 fmol/sample. These data are shown in Figure 2.

We hypothesize that the relative lack of an increase in extracellular DA in the intact striatum after 50 mg/kg L-DOPA can be attributed to intraneuronal metabolism of the DA formed from L-DOPA in DA terminals combined with high affinity uptake by DA terminals of extracellular DA that may derive from other sources (Abercrombie et al., 1990). Inhibition of monoamine oxidase-A, but not monoamine oxidase-B, enhances the effect of L-DOPA in intact animals by overcoming the effect of intraneuronal metabolism (Wachtel and Abercrombie, 1993). Higher doses of L-DOPA (100 and 200 mg/kg) increase extracellular levels of DA, probably by overcoming the clearing effect of the uptake system (Abercrombie et al., 1990). We have observed that pretreatment of animals with the DA uptake inhibitor GBR12909 reveals an increase in striatal DA output after 50 mg/kg L-DOPA (unpublished data). The present data on the formation of DA from L-DOPA in SNpr may indicate less efficient clearance of DA from the extracellular space by the DA-transporter and a lesser metabolic capacity relative to striatum.

It is evident that both in the SNpr and in the striatum, L-DOPA is converted into DA. In striatum of intact animals, a highly efficient uptake mechanism combined with a high capacity for metabolism of DA within DA nerve endings, keeps extracellular levels of DA at control values after low doses of L-DOPA. In the SNpr, however, an increase in the output of extracellular DA can be measured even after low doses of L-DOPA. These data are especially relevant when considering the partial depletions of DA in the human basal ganglia that exist at the point where the symptoms of Parkinson's disease begin to emerge. Based on the results summarized above, one might hypothesize that in the earlier stages of the disease L-DOPA causes a relatively greater increase in extracellular DA in the SNpr than in striatum. The therapeutic effect of L-DOPA in these earlier stages may therefore be preferentially located at the level of the SNpr. This model, however, does not exclude the possibility (probability) that a full behavioral response to L-DOPA requires dopaminergic stimulation at sites in both the striatum and SNpr. Since nigral DA most likely is modulating the release of GABA from striatonigral fibers via stimulation of DA D-1 receptors, DA formed from L-DOPA in the SNpr presumably acts to amplify the striatal influence upon nigral outputs to thalamus and tectum. Increases in extracellular DA in the striatum after L-DOPA directly affect corticostriatal and thalamostriatal transmission via the direct and indirect pathways. Thus, to understand the beneficial effect of L-DOPA in the Parkinsonian brain, it is necessary to take into account the effect of L-DOPA in both the striatum and the SNpr.

STRIATUM SUBSTANTIA NIGRA

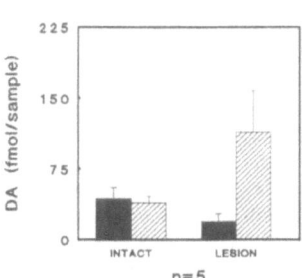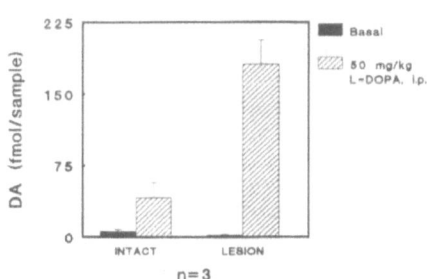

Figure 2. The peak-effect of 50 mg/kg L-DOPA on the output of striatal and SNpr DA in intact and 6-OHDA-treated animals. All animals were pretreated with a peripheral decarboxylase inhibitor. Values are mean ± SEM for the DA content of the dialysis sample expressed in fmol/min.

STRIATAL OR EXTRASTRIATAL DA D-1 RECEPTORS?

The observation that systemically administered indirect DA-agonists such as amphetamine and cocaine increase extracellular levels of striatal ACh, whereas intrastriatal application of amphetamine produces decreases in ACh, prompted the hypothesis that the DA D-1 receptors apparently responsible for the former effect were located outside the striatum (Damsma et al., 1991). These authors suggested that DA stimulation of D-1 receptors in frontal cortex might be particularly important. It was suggested that cortical DA D-1 mediated increases in striatal ACh output were caused by an increase in activity of the corticostriatal glutamatergic pathway, although the ability of cortical DA D-1 receptor stimulation to activate this pathway was not demonstrated. When the excitatory effects of these inputs were blocked at the level of the striatum by application of an NMDA-receptor antagonist, the D-1-mediated increases in striatal ACh could be prevented (Damsma et al., 1991).

An alternative hypothesis has been put forward by Consolo and coworkers, based on experiments testing the ability of the D-1 selective antagonists SCH23390 and SCH39166 injected into the frontal cortex or the striatum to block the effect of systemically administered amphetamine on striatal ACh output. The results of these studies suggested that the striatum rather than the frontal cortex, was in fact the origin of DA D-1-mediated effects on striatal ACh (Consolo et al., 1992). This model, however, has difficulty accounting for the decreases in the output of striatal ACh observed when amphetamine is directly infused into the striatum. Also, intrastriatal application of DA D-1 agonists does not produce an increase in striatal ACh output either *in vitro* or *in vivo* (DeBoer et al., 1992; Dolezal et al., 1992).

Since our neurochemical studies on the effects of L-DOPA administration showed a DA D-1-like effect of this compound on striatal ACh efflux as well as a prominent effect on extracellular DA in SNpr, we became interested in the possibility that the high density of D-1 receptors located in SNpr might be an important substrate for D-1-mediated increases in striatal ACh output. Studies examining metabolic activity of brain areas after L-DOPA show increases in this measurement in structures in addition to the striatum, especially in the SNpr (Trugman et al., 1991). Other experiments conducted our laboratory have shown that intranigral injections of AMPH activate behavior and also inhibit the activity of the SNpr output cells (Timmerman and Abercrombie, 1993). This latter effect would lead to disinhibition of thalamic projections to cortex. Thus, an increased input to corticostriatal neurons via a nigrothalamocortical circuit may be initiated by stimulation of DA D-1 receptors in SNpr (see also Cheramy et al., 1981; Besson et al., 1986; Girault et al., 1986). Activation of DA D-1 receptors along the striatonigral axis, therefore, may play a role in determining the neurochemical and behavioral effects of indirect DA-agonists such as L-DOPA.

In conclusion, we propose that DA formed from L-DOPA has significant actions at sites in addition to the striatum and that actions upon DA D-1 receptors in SNpr may be particularly relevant to the use of L-DOPA in Parkinson's disease. The stimulatory effects of L-DOPA on striatal ACh output are hypothesized to occur via a nigrothalamocortical circuit, implying a widespread effect of L-DOPA on brain activation which includes cortical arousal. In animals sustaining depletions of striatal DA that exceed 90%, resting levels of extracellular ACh in striatum are elevated as compared to controls, consistent with the classical concept of an antagonistic "DA-ACh" balance. However, the recent data indicating opposing influences of DA D-1 and D-2 receptors in the regulation of striatal ACh output and the characteristics of this modulation make another kind of interaction possible under certain conditions. One such condition of particular importance is L-DOPA administration during the initial stages of Parkinson's disease. In this case, striatal cholinergic function may be impaired due to decreased DA D-1-mediated excitation rather than decreased DA D-2-mediated inhibition. The stimulatory action of L-DOPA on striatal ACh output may therefore restore cholinergic function to more normal levels. The nature of the "DA-ACh balance" thus becomes a function of the degree of nigrostriatal DA loss. Further experimental tests of this model are necessary. It is clear, however, that the neuropharmacology of L-DOPA is far more complex than previously thought.

Acknowledgements

Supported by the Tourette's Syndrome Association, Inc. (PDB), Individual Post-doctoral Fellowship NS09206 (SRW), a NATO International Collaborative Research Grant, The American Parkinson's Disease Association, and USPHS Grant NS19608 (EDA).

REFERENCES

Abercrombie, E.D., Bonatz, A.E., and Zigmond, M.J., 1990, Effects of L-DOPA on extracellular dopamine in striatum of normal and 6-hydroxydopamine-treated rats, *Brain Res.* 525:36-44.

Barbeau, A., 1962, The pathogenesis of Parkinson's disease: A new hypothesis, *Canad. Med. Assoc. J.* 87:802-807.

Beckstead, R.M., 1988, Association of dopamine D1 and D2 receptors with specific cellular elements in the basal ganglia of the cat: The uneven topography of dopamine receptors in the striatum is determined by intrinsic striatal cells, not nigrostriatal axons, *Neurosci.* 27(3):851-863.

Bertorelli, R. and Consolo, S., 1990, D1 and D2 dopaminergic regulation of acetylcholine release from striata of freely moving animals, *J. Neurochem.* 54:2145-2148.

Bertorelli, R., Zambelli, M., DiChiara, and Consolo, S., 1992, Dopamine depletion preferentially impairs D1- over D2-receptor regulation of striatal in vivo acetylcholine release, *J. Neurochem.* 59:353-357.

Besson, M.J., Kemel, M.L., Gauchy, C., Girault, J.A., Spampinato, U., Lantin, N., Desban, M., and Glowinski, J., 1986, In vivo measurement of [3H]GABA release: An approach to the study of the regulation of GABA-containing neurons in the Basal Ganglia and associated structures in the cat and the rat, *in:* "Neurochemical Analysis of the Conscious Brain: Voltammetry and Push-Pull Perfusion," R.D. Myers and P.J. Knott, eds., *Ann. NY Acad. Sci.* 473:475-488.

Cheramy, A., Leviel, V., and Glowinski, J., 1981, Dendritic release of dopamine in the substantia nigra, *Nature* 289:537-542.

Consolo, S., Girotti, P., Russi, G., and DiChiara, G., 1992, Endogenous dopamine facilitates striatal in vivo acetylcholine release by acting on D1 receptors localized in the striatum, *J. Neurochem.* 59:1555-1557.

Damsma, G., De Boer, P., Westerink, B.H.C., and Fibiger, H.C., 1990a, Dopaminergic regulation of striatal cholinergic interneurons: An in vivo microdialysis study, *Naunyn Schmiedeberg's Arch. Pharmacol.* 342:523-527.

Damsma, G., Robertson, G.S., Tham, C-S., and Fibiger, H.C., 1991, Dopaminergic regulation of striatal acetylcholine release: Importance of D1 and NMDA receptors, *J. Exp. Pharmacol. Ther.* 259(3):1064-1072.

Damsma, G., Tham, C-S., Robertson, G., and Fibiger, H.C., 1990b, Dopamine D1 receptor stimulation increases striatal acetylcholine release in the rat, *Eur. J. Pharmacol.* 186:335-338.

DeBoer, P., Abercrombie, E.D., Heeringa, M., and Westerink, B.H.C., 1993, Differential effect of systemic administration of bromocriptine and L-DOPA on the release of acetylcholine from striatum of intact and 6-OHDA-treated rats, *Brain Res.* 608:198-203.

DeBoer, P., Damsma, G., Fibiger, H.C., Timmerman, W., de Vries, J.B., and Westerink, B.H.C., 1990, Dopaminergic-cholinergic interactions in the striatum: The critical significance of calcium concentrations in brain microdialysis, *Naunyn-Schmiedeberg's Arch. Pharmacol.* 341:528-534.

DeBoer, P., Damsma, G., Schram, Q., Stoof, J.C., Zaagsma, J., and Westerink, B.H.C., 1992, The effect of intrastriatal application of directly and indirectly acting dopamine agonists and antagonists on the in vivo release of acetylcholine measured by brain microdialysis: The importance of the post-surgery interval, *Naunyn-Schmiedeberg's Arch. Pharmacol.* 345:144-152.

Dolezal, V., Jackisch, R., Herrting, G., and Allgaier, C., 1992, Activation of dopamine D1 receptors does not affect D2 receptor-mediated inhibition of acetylcholine release in rabbit striatum, *Naunyn-Schmiedenberg's Arch. Pharmacol.* 345:16-20.

Freund, T.F., Powell, J.F., and Smith A.D., 1984, Tyrosine hydroxylase-immunoreactive boutons in synaptic contact with identified striatonigral neurons, with particular reference to dendritic spines, *Neurosci.* 13(4):1189-1215.

Gerfen, C.R., 1992, The neostriatal mosaic: Multiple levels of compartmental organization, *Trends Neurosci.* 15(4):133-139.

Girault, J.A., Spampinato, U., Desban, M., Glowinski, J., and Besson, M.J., 1986, Enhancement of glutamate release in the rat striatum following electrical stimulation of the nigrothalamic pathway, *Brain Res.* 374:362-366.

Lehmann, J. and Langer, S.Z., 1983, The striatal cholinergic interneuron: Synaptic target of dopaminergic terminals?, *Neurosci.* 10(4):1105-1120.

Lloyd, K.G., Davidson, L., and Hornykiewicz, 1975, The neurochemistry of Parkinson's disease: Effect of L-DOPA therapy, *J. Pharmacol. Exp. Ther.* 195:453-464.

McGeer, P.L., Boulding, J.E., Gibson, W.C., and Foulkes, R.G., 1961, Drug-induced extrapyramidal reactions, *J. Am. Med. Assoc.* 177(10):101-106.

Melamed, E., Globus, M., Uzzan, A., and Rosenthal, J., 1985, Is dopamine formed from exogenous L-DOPA stored within vesicles in striatal dopaminergic nerve terminals: Implications for L-DOPA's mechanism of action in Parkinson's disease, *Neurology (Suppl)* 35:118.

Robertson, G.S. and Robertson, H.A., 1987, D1 and D2 dopamine receptor agonist synergism: Separate sites of action?, *Trends Pharmacol.* 8:295-299.

Robertson, G.S., Hubert, G.W., Tham, C.-S., and Fibiger, H.C., 1992, Lesions of the mesotelencephalic dopamine system enhance the effects of selective dopamine D1 and D2 receptor agonists on striatal acetylcholine release, *Eur. J. Pharmacol.* 219:323-325.

Robertson, H.A., 1992, Dopamine receptor interactions: Some implications for the treatment of Parkinson's disease, *Trends Neurosci.* 15(6):201-206.

Stoof, J.C., Drukarch, B., DeBoer, P., Westerink, B.H.C., and Groenewegen, H.J., 1992, Regulation of the activity of striatal cholinergic neurons by dopamine, *Neurosci.* 47(4):755-770.

Timmerman, W. and Abercrombie, E.D., 1993, Behavioral and electrophysiological effects of amphetamine-induced release of dendritic dopamine, *Soc. Neurosci. Abstr.* in press.

Trugman, J.M., James, C.L., and Wooten, G.F., 1991, D1/D2 dopamine receptor stimulation by L-DOPA, *Brain* 114:1429-1440.

Wachtel, S.R., and Abercrombie, E.D., L-3,4-Dihydroxyphenylalanine-induced dopamine release in the striatum of intact and 6-hydroxydopamine-treated rats: Differential effects of monamine oxidase A and B inhibitors, *J. Neurochem.* submitted.

Wachtel, S.R., Hu, X-.T., Galloway, M.P., and White, F.J., 1989, D1 dopamine receptor stimulation enables the postsynaptic, but not autoreceptor, effects of D2 dopamine agonists in nigrostriatal and mesoaccumbens dopamine systems, *Synapse* 4:327-346.

Walters, J.R., Bergstrom, D.A., Carlson, J.H., Weick, B.G., and Pan, H.S., 1987, Stimulation of D-1 and D-2 dopamine receptors: Synergistic effect on single unit activity in basal ganglia output nuclei, *in:* "Neurophysiology of Dopaminergic Systems - Current Status and Clinical Perspectives," L.A. Chiodo and A.S. Freeman, eds., Lakeshore Publishing Company, pp. 285-316.

Zigmond, M.J., and Stricker, E.M., 1989, Animal models of parkinsonism using selective neurotoxins: Clinical and basic implications, *Int. Rev. Neurobiol.* 31:1-79

THE ROLE OF CALCIUM IONS IN DOPAMINE SYNTHESIS
AND DOPAMINE RELEASE

Vincent Leviel, Valérie Olivier and Bernard Guibert

Institut Alfred Fessard
C.N.R.S.
91198 Gif sur Yvette Cedex, France

INTRODUCTION

The release of dopamine (DA) from striatal terminals appears to be a rather complex and misunderstood phenomenon. It has been observed from many years that this release is only partially dependent on calcium ions in the extracellular space (Arbilla and Langer, 1978; Raiteri et al., 1979; Schwartz et al., 1980; Okuma et al., 1983; Woodward et al., 1988). However, it is now well known that the synthesis of DA is modulated by means of alterations of tyrosine hydroxylase (TH) activity under the control of calcium-dependent phosphorylating mechanisms (Fung and Uretsky, 1982; El Mestikawy et al., 1983; Albert et al., 1984; Chowdhury and Fillenz, 1988; Zigmond et al. 1989). In addition, the synthesis of DA is the basis of its release (Costa et al., 1972; Roth et al., 1973; Schwarz et al., 1980) probably through an increase of both the vesicular and the cytosolic stores. Thus, in the striatum calcium is clearly related to the extracellular DA concentration both by its action on synthesis and release.

In previous studies we attempted to dissociate treatments increasing extracellular DA by means of modulations of amine synthesis from those acting on the mechanism of release itself (Leviel et al., 1989, 1990). The method was based on measurement of the specific activity of the amine released during continuous superfusion of the tissue *in vivo* with radiolabelled tyrosine. Indeed, the fluctuations of DA specific activity in the extracellular space can reflect the involvement in the release of either neosynthesized DA or previously stored amine.

The role of calcium in these two complementary processes is presented here with only the spontaneous release of the amine considered. The effects of a reduction in extracellular calcium ions or a blockade of calcium channels (with cadmium ions) were investigated first and compared to the effects of a TH inhibitor (α-mpt). Secondly, a calcium ionophore was used (A-23187) to observe the consequences of an increase in calcium entry.

MATERIAL AND METHODS

Rats (Wistar male) weighing 300g were halothane anesthetized in pure oxygen and tracheotomized. They were implanted with a push-pull cannula in the anterior part of the caudate nucleus (ant: 8mm; lat: 2.75mm; ht: 6mm), the horizontal plane passing through the interaural axis and incisor bar. Cannulae (1.0 mm outer diameter) were supplied (Flow rate: 12.5 µl/min) with an artificial cerebrospinal fluid (CSF) adjusted to pH 7.4 with an O_2-CO_2 (95:5 v/v) mixture. 3,5-[3H]TYR (50 Ci/mmole, Dositek, France) was purified by

The Basal Ganglia IV, Edited by
G. Percheron *et al.*, Plenum Press, New York, 1994

high performance liquid chromatography (HPLC) on a C18 Microbondapak column (Millipore-Waters, France) using H3PO4 (1mM, pH 3) as a mobile phase. One hour after implantation, [3H]TYR was added to artificial CSF (80 µCi/ml) and superfusates collected thereafter as successive 20 minute fractions. Substances were applied locally in the striatum by addition to the artificial CSF at the following concentrations: A23187 (Tocris Neuramin) 1µM and 100µM; α-methyl-p-tyrosine (Sigma) 10µM; cadmium (Sigma) 100µM (For maintenance of osmolarity, Mg++ concentration was balanced).

The labelled and endogenous DA and dihydroxyphenylacetic acid (DOPAC) were measured in the superfusing fluid. In brief, the DA and DOPAC concentrations in CSF were determined by electrochemical detection with potential of the working electrode at 800mV (Metrohm 641/656) after HPLC analysis (column C18 Brownlee RP18, 5µm, 2.1x220 mm, maintained at 25°C; Mobile phase 100mM NaH2PO4, 0.1mM EDTA-Na2, 0.28mM sodium octyl sulfate, 6% methanol, adjusted to pH 3.3; Flow rate 0.25ml/min). The radioactivity corresponding to each HPLC peak was counted using a continuous flow scintillation detector (Radiomatic, Flo-One β A250).

At the end of each experiment, to verify the cannula location, the animals were intracardially perfused with a 3% formaldehyde solution, the brain was removed, sliced (100 µm) and stained with cresyl violet.

For an easier comparison, the results concerning DA and DOPAC release are presented after standardization to 100 % of the spontaneous release defined from the first three fractions (1 hour). In contrast, the specific activities were expressed in Ci/mmole. Statistical analysis was conducted using a two tailed Student's t test by comparing the mean of corresponding fractions of control and treated groups. Technical details and data treatment are described elsewhere (Leviel et al., 1989).

RESULTS

α-Methyl-p-tyrosine and Superfusion with Calcium Free Medium

As already described (Leviel et al., 1989), α-mpt, when added to the artificial CSF, produced a sharp decrease of DA release stabilised after about 60 min to 40% of the base value and remained unchanged during the following, two hours. DOPAC release decreased slowly, reaching 50 % of the base value after two hours superfusion with α-mpt. These effects were accompanied by a lowering of the specific activity of the two substances (Figure 1 presents the decrease observed after 80 minutes of superfusion with α-mpt).

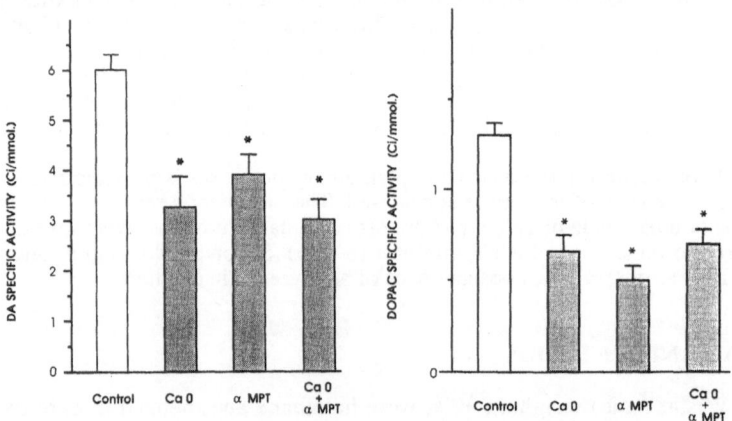

Figure 1. Effects of calcium removal (CaO) or α-mpt addition in the superfusing fluid supplying a push-pull cannula implanted in the striatum of halothane anesthetized rats. Values are expressed in Ci/mmole (Means + SEM, n=8). <u>On the left</u>, are presented the specific activity of the DA released in extracellular space 80 min. after removal of Ca++ or addition of α-mpt. The fourth bar correspond the fraction collected 100 min after Ca++ removal and 20 min after α-mpt addition. <u>On the right</u>, are presented the specific activity for DOPAC in the same experimental conditions. *: P < 0.05.

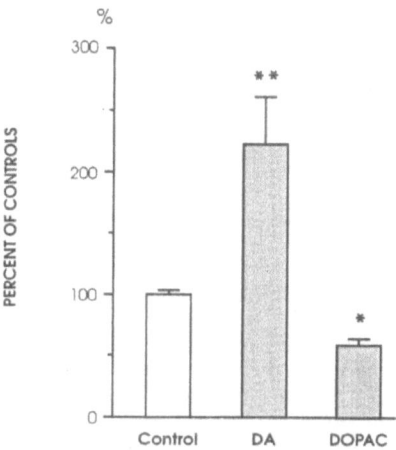

Figure 2. Effect of a 20 min 100µM cadmium application on DA and DOPAC release observed in the same conditions that in Figure 1. Values are obtained from 8 animals and expressed in percent of the spontaneous release measured on the same number of untreated animals. *: P < 0.05 ; ** : P < 0.01.

Removing calcium ions from the superfusing medium (Ca0), produced the same effect with the same amplitude as α-mpt: a decreased release and specific activity of DA and DOPAC (the specific activities for DA and DOPAC 80 min after calcium removal are presented on Figure 1).

During this treatment, when DA release was stabilised to a lower value, an additionnal application of α-mpt was unable to reduce further any of the measured parameters. Thus the effects of simultaneous superfusion with a calcium free medium and α-mpt seemed not to be additive (the effects on the DA and DOPAC specific activity are presented on Figure 1).

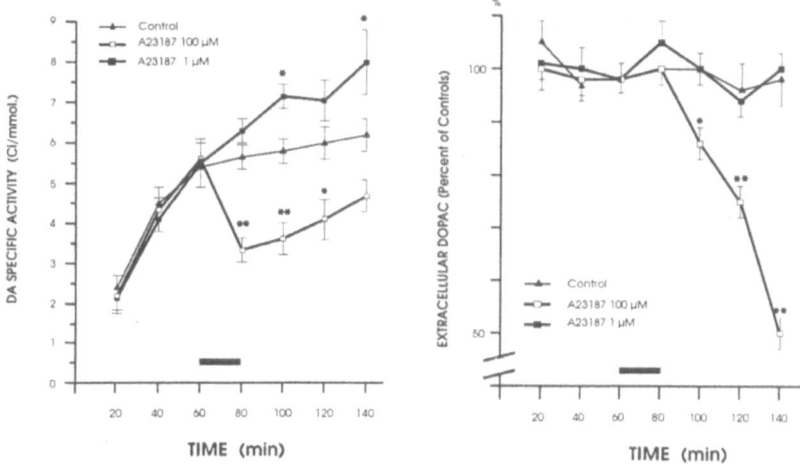

Figure 3. On the left, time dependent-alterations of extracellular DA specific activity produced by a local application in the striatum of 1µM and 100µM calcium ionophore A23187. Values are expressed in Ci/mmole and compared to corresponding fractions in untreated animals. On the right, blockade by 100µM A23187 of DOPAC efflux in the extracellular space. Values are expressed in percent of the spontaneous efflux measured on untreated rats. * : P < 0.05 ; ** : P < 0.01 ; n=8.

Effects of Local Application of Cadmiun

When cadmium was applied, a rapid increase of DA extracellular concentration was observed, reaching 225% of basal rate after 80 min (Fig. 2). The specific activity of DA decreased after such a treatment (6.5 to 3.3 Ci/mmole, data not shown). Extracellular concentrations of DOPAC decreased immediately after the addition of cadmium reaching 40% of controls (Fig. 2), but its specific activity remained unchanged.

Effects of Local Application of A23187

The Calcium ionophore A23187 was applied locally during 20 min, and the effects of two different concentrations were compared. At 1μM A23187 had no effect on DA or DOPAC release but induced an increase of their specific activities (Fig. 3). At a concentration of 100μM, A23187 induced a sharp increase of DA release (not shown) accompanied by reduced specific activity for the amine (Fig. 3). This treatment also produced a decrease of DOPAC release (Fig. 3) .

After removal of the high doses of A23187 from the superfusing fluid, the values for DA and DOPAC returned progressively to base levels while those for the tritiated amine decreased to about 50% of base release (not shown).

DISCUSSION

Previous studies confirmed by the present results show that calcium ions constitute *in vivo* a key component in the DA synthetic pathway. The reduced DA release during superfusion with calcium free medium resulted from a reduced neosynthesis of the amine. Surprisingly, a moderate calcium entry into terminals forced the synthesis of the amine and left the release unaffected. In contrast, a massive calcium entry resulted in an increased DA release simultaneously with a decreased DA metabolic path. It thus appears that some component of the calcium dependent DA release could be mediated through synthesis alterations. With respect to that concept, some receptor systems known to modulate DA release at the presynaptic level could act by modulating calcium entry into DA terminals. It is interesting to note that with a different technique, developed on synaptosomal preparations, Kapatos and Zigmond (1982) did not observe that TH activity is dependent on calcium availability. However, results comparable to the present study were obtained from synaptosomal preparations by Desce et al. (personal communication).

Calcium Ions Considered as a Synthesis Activator

The synthesis of DA in dopaminergic terminals requires tyrosine hydroxylation that for a long time has been defined as limiting the synthesizing process (Levitt et al., 1965). The enzyme responsible for this activity (TH) is controlled by multiple processes including cofactor concentration (Kettler et al., 1974; Susuki et al., 1988), the DA level (Udenfriend et al., 1965; Andersson et al., 1988) and receptors located on the terminal membranes (Zivkovic and Guidotti, 1974; El Mestikawy et al., 1986). Several studies have reported that TH activity is dependent on, at least, the phosphorylation of different serine residues (Haycock, 1990; Walaas and Greengard, 1991). Three types of kinase are involved: cAMP, Ca/phospholipids and Ca/calmodulin dependent protein kinases (Zigmond et al., 1989). The internal calcium concentration is thus related to TH phosphorylation, and many studies have correlated DA synthesis with calcium entry (Fung and Uretsky, 1982; Albert et al., 1984; El Mestikawy et al., 1986; Chowdhury and Fillenz, 1988).

In the present experiments the calcium ionophore A23187 applied at low doses remained without effect on DA release since the extracellular amine concentration was unchanged. However, we observed an increase in the specific activity of released DA and DOPAC, suggesting that synthesis was increased in the terminals. In fact, neosynthesized DA presents a higher specific activity than the stored amine. In the opposite situation, during superfusion with calcium free medium, the base release of the amine was reduced, reaching a new steady state corresponding to about 40% of the base level. In this condition the specific activity of the released amine was reduced, showing that it was the stored amine that was called up preferentially for release. The involvement of the stored amine is undoubtedly the consequence of a reduction in neosynthesized molecules. Thus we can conclude that the removal of calcium ions from the extracellular space results in a reduction of TH activity. In line with this observation, when added after calcium removal, the TH competitive inhibitor α-mpt did not further reduce the release of DA and its specific activity. This last effect confirms that TH activity was already reduced by the absence of calcium.

Calcium Ions Considered as a Synthesis Inhibitor

The local application of high doses of the calcium ionophore resulted in an increase of the base release of DA. This effect could be expected since the mechanism of neurotransmitter release is under the control of intraterminal Ca^{++} concentration and since evoked release is well known to be mediated by calcium entry (Katz and Miledi 1970; Augustine et al., 1987). Also, in the different paradigms used, increasing extracellular calcium has already been described as able to potentiate base (Westerink et al., 1989) or evoked (Zurgil et al., 1986) DA release.

A double component of the release of DA in the striatum was however defined, one being evoked by spikes conducted along dopaminergic axons and the other being *spike-independent* and probably modulated at the presynapic level (Grace, 1991). It had been proposed that the glutamate induced efflux of DA could be mediated by this *spike-independent* mechanism (Leviel et al., 1990; Grace, 1991) something which has been recently confirmed (Keefe et al., 1992). If the *spike-mediated* release is likely to be calcium dependent, the case of the *spike-independent* release remains questionable and the glutamate induced DA efflux is only partially affected by calcium blockade (Grace, 1991). In conclusion, as expected, increasing calcium entry can produce an increased release.

A reduction in the specific activity of DA was observed during application of high doses of Ca ionophore. In addition, DOPAC efflux was also reduced which suggested an impairment in amine synthesis. Finally, we also observed that after removal of the ionophore from the superfusing medium, the base release of labelled DA was persistently reduced. This probably reflects a reduced intraterminal synthesis of the amine.

The effects of cadmium ions could appear surprising, since cadmium is known to block calcium-ion entry (Llinas et al., 1981; Suszkiw et al., 1984; Yaari et al., 1987) and that a decrease in DA release was reported (Westerink et al., 1989) after higher doses (300 μM). Our results, however, confirm in vivo that cadmium is able to penetrate the terminal in the place of calcium, thus playing the same role (Nilson and Volle, 1976; Nishumura et al., 1984; Guan et al., 1987). Whatever the molecular mechanism by which calcium and cadmium compete to enter terminals, the metabolic consequences are quite similar for the two ions, in good confirmation of what has been observed in different simpler models (Asai et al., 1982; Racké et al., 1988; Molgo et al., 1989). Indeed, during cadmium application the release of DA was sharply enhanced, DA specific activity was decreased and, after removal of cadmium from the superfusing fluid, the ability to release tritiated DA was impaired.

It can be concluded that these two bivalent cations can activate, by their massive entry, the calcium dependent release of DA and simultaneously decrease amine synthesis.

Calcium Ions, a Modulator of the DA Release

The reduction of DA synthesis by massive calcium entry and the activation of DA synthesis by moderate calcium entry have to be considered in the light of the possible mechanisms of DA release. It is now established that glutamate modulates DA release even in the absence of electrical activity in dopaminergic axons (Keefe et al.,1992) confirming its possible action through a *spike-independent* mechanism. This does not mean that its action should be considered as totally *calcium-independent*. In the cat as in the rat, low doses of glutamate resulted in an increased DA release, while in contrast higher doses were responsible for an indirect inhibition at DA release (Romo et al., 1986; Leviel et al., 1990). In addition, it has been proposed that, in the rat, glutamate acts mainly by modulating DA synthesis (Leviel et al., 1990). More recently, this role of glutamate in DA synthesis was confirmed (Chowdhury and Fillenz, 1991; Arias-Montano et al., 1992). A recently described glutamate receptor, negatively coupled to adenyl cyclase (Prezeau et al., 1992), could furnish a molecular basis of this effect. However, it is not definitely known if there are two different mechanisms responsible for the activation and inhibition of DA synthesis by glutamate. Nevertheless, this hypothesis is supported by the fact that the metabotropic receptor being negatively coupled to cyclase could mediate a reduction in TH activity whereas NMDA receptors linked to a calcium channel could mediate an activation of this enzyme. Finally, receptor modulated calcium entry could be responsible for the intraterminal modulation of DA synthesis.

In conclusion, calcium ions can be implicated in various steps of DA metabolism and release. While the release is directly in relation with the intraterminal ionic concentration, calcium also interacts with synthesis through a more complex scheme: activator at low concentrations and inhibitor at high concentrations. In addition, the effects of presynaptic receptors located on DA terminals in the striatum are likely to exert their action by means of calcium entry to increase or decrease DA release.

REFERENCES

Albert, K.A., Helmer-Matyjek, E., Nairn, A.C., Muller, T.H., Haycock, J.W., Greene, L.A., Goldstein, M., and Greengard, P., 1984, Calcium/phospholipid-dependent protein kinase (protein kinase C) phosphorylates and activates tyrosine hydroxylase, *Proc. Natl. Acad. Sci. USA* 81:7713-7717.

Andersson, K.K., Cox, D.D., Que, L.Jr., Flatmark, T., and Haavik, J., 1988, Resonance raman studies on the blue-green-colored bovine adrenal tyrosine 3-monooxygenase (tyrosine hydroxylase) - Evidence that the feedback inhibitors adrenaline and noradrenaline are coordinated to iron, *J. Biol. Chem.* 263:18621-18626.

Arbilla, S., and Langer, S.Z., 1978, Morphine and β-endorphin inhibit release of noradrenaline from cerebral cotex but not of dopamine from rat striatum, *Nature* 271:559-560.

Arias-Montano, J.A., Martinez-Fong, D., and Aceves, J., 1992, Glutamate stimulation of tyrosine hydroxylase is mediated by NMDA receptors in the rat striatum, *Brain Res.* 569:317-322.

Asai, F., Nishimura, M., Satoh, E., and Urakawa, N., 1982, Mechanism of cadmium induced contraction in ileal longitudinal muscle of guinea pig, *Br. J. Pharmacol.* 75:561-567.

Augustine, G.J., Charlton, M.P., and Smith, S.J., 1987, Calcium action in synaptic transmitter release, *Annual Rev. Neurosci.* 10:633-693.

Costa, E., Gropetti, A., and Naimzada, M.K., 1972, Effect of amphetamine on the turnover rate of brain catecholamines and motor activity, *Br. J. Pharmacol.* 44:742-751.

Chowdhury, M., and Fillenz, M., 1988, K+-dependent stimulation of dopamine synthesis in striatal synaptosomes is mediated by protein kinase C, *J. Neurochem.* 50:624-629.

Chowdhury, M., and Fillenz, M., 1991, Presynaptic adenosine A₂ and N-methyl-D-aspartate receptors regulate dopamine synthesis in rat striatal synaptosomes, *J. Neurochem.* 56:1783-1788.

El Mestikawy, S., Glowinski, J., and Hamon, M., 1983, Tyrosine hydroxylase activation in depolarized dopaminergic terminals - Involvement of Ca++ dependent phosphorylation, *Nature* 302:830-832.

El Mestikawy, S., Glowinski, J., and Hamon, M., 1986, Presynaptic dopamine autoreceptors control tyrosine hydroxylase activation in depolarized striatal dopaminergic terminals, *J. Neurochem.* 46:12-22.

Fung, Y.K., and Uretsky, N.J., 1982, The importance of calcium in the amphetamine-induced stimulation of dopamine synthesis in mouse striata in vivo, *J. Pharmacol. Exp. Ther.* 223:477-482.

Grace, A.A., 1991, Phasic versus tonic dopamine release and the modulation of dopamine system responsivity : a hypothesis for the etiology of schizophrenia, *Neuroscience* 41:1-24.

Guan, Y.Y., Quastel, D.M.J., and Saint, D.A., 1987, Multiple actions of cadmium on transmitter release at the mouse neuromuscular junction, *Can. J. Physiol. Pharmacol.* 65:2131-2136.

Haycock, J.W., 1990, Phosphorylation of tyrosine hydroxylase in situ at serine 8, 19, 31 and 40, *J. Biol. Chem.* 265:11682-11691.

Kapatos, G., and Zigmond, M.J., 1982, Influence of calcium on dopamine synthesis and tyrosine hydroxylase activity in rat striatum, *J. Neurochem.* 39:327-335.

Katz, B., and Miledi, R., 1970, Further study of the role of calcium in synaptic transmission, *J. Physiol.* 207:789-801.

Keefe, K.A., Zigmond, M.J., and Abercrombie, E.D., 1992, Extracellular dopamine in striatum: influence of nerve impulse activity in medial forebrain bundle and local glutamatergic input, *Neuroscience* 47:325-332.

Kettler, R., Bartholini, G., and Pletscher, A., 1974, In vivo enhancement of tyrosine hydroxylation in rat striatum by tetrahydrobiopterin, *Nature* 249:476-478.

Leviel, V., Gobert, A., and Guibert, B., 1989, Direct observation of dopamine compartmentation in striatal nerve terminal by in vivo measurement of the specific activity of released dopamine, *Brain Res.* 499:205-213.

Leviel, V., Gobert, A. and Guibert, B., 1990, The glutamate-mediated release of dopamine in the rat striatum: further characterization of the dual excitatory-inhibitory function, *Neurosci.* 39:305-312.

Levitt, M., Spector, S., Sjoerdsma, A., and Udenfriend, S., 1965, Elucidation of the rate-limiting step in norepinephrine biosynthesis in the perfused guinea pig heart, *J. Pharmacol. Exp. Ther.* 148:1-8.

Llinas, R., Steinberg, I.Z., and Walton, K., 1981, Presynaptic calcium currents in squid giant synapse, *Biophys. J.* 33:289-322.

Molgo, J., Pécot-Dechavassine, M., and Thesleff, S., 1989, Effects of cadmium on quantal transmitter release and ultrastructure of frog motor nerve endings, *J. Neural Transm.* 77:79-91.

Nilson, R., and Volle, R.L., 1976, Blockade by cadmium ($Cd2^+$) of transmitter release at the mouse neuromuscular junction, *Fed. Proc. Fed. Am. Soc. Exp. Biol.* 35:696.

Nishimura, M., Tsutsui, I., Yagasak, O., and Yanagiya, I., 1984, Transmitter release at the mouse neuromuscular junction stimulated by cadmium ions, *Arch. Int. Pharmacodyn.* 271:106-121.

Okuma, Y., Fukuda, Y., and Osumi, Y., 1983, Neurotensin potentiates the potassium-induced release of endogenous dopamine from rat striatal slices, *Eur. J. Pharmacol.* 93:27-33.

Prezeau, L., Manzoni, O., Homburger, V., Sladeczek, F., Curry, K., and Bockaert, J., 1992, Characterization of a metabotropic glutamate receptor : direct negative coupling to adenyl cyclase and involvement of a pertussis toxin-sensitive G protein, *Proc. Natl. Acad. Sci. USA* 89:8040-8044.

Racké, K., Hering, B., and Hochgesand, U., 1988, Effects of gadolinium and cadmium on the electrically evoked release of ^{45}calcium from the isolated rat neurohypophysis, *Naunyn-Schmiedeberg's Arch Pharmacol.* 337:301-307.

Raiteri, M., Cerrito, F., Cervoni, A.H., and Levi, G., 1979, Dopamine can be released by two mechanisms differentially affected by dopamine transport inhibitor nomifensine, *J. Pharmacol. Exp. Ther.* 208:195-202.

Romo, R., Cheramy, A., Godeheu, G., and Glowinski, J., 1986, In vivo presynaptic control of dopamine release in the cat caudate nucleus. III. Further evidence for the implication of corticostriatal glutamatergic neurons, *Neuroscience* 19:1091-1099.

Roth, R.H., Walters, J.R., and Aghajanian, G.K., 1973, Effects of impulse-flow on the release and synthesis of dopamine in the rat striatum, *in*: "Frontiers in Catecholamine Research," E. Usdin. and S. Snyder, eds, Pergamon Press, New-York, pp. 457-464.

Schwarz, R.D., Uretsky, N.J., and Bianchine, J.R., 1980, The relationship between the stimulation of dopamine synthesis and release produced by amphetamine and high potassium in striatal slices, *J. Neurochem.* 35:1120-1127.

Suszkiw, J., Toth, G., Murzwsky, M., and Cooper, G.P., 1984, Effects of Pb^{2+} and Cd^{2+} on acetylcholine release and Ca^{2+} movements in synaptosomes and subcellular fractions from rat brain and *torpedo* electric organ, *Brain Research* 323:31-46.

Suzuki, S.S., Watanabe, Y., Tsubokura, S., Kagamiyama, H., and Hayaishi, O., 1988, Decrease in tetrahydrobiopterin content and neurotransmitter amine biosynthesis in rat brain by an inhibitor of guanosine triphosphate cyclohydrolase, *Brain Research* 446:1-10.

Udenfriend, S., Zaltzman-Nirenberg, P., and Nagatsu,T., 1965, Inhibitors of purified beef adrenal tyrosine hydroxylase, *Biochem. Pharmacol.* 14:837-845.

Walaas, S.I., and Greengard, P., 1991, Protein phosphorylation and neuronal function, *Pharmacol. Rev.* 43:299-349.

Westerink, B.H.C., Hofsteede, R.M., Tuntler, J., and de Vries, J.B., 1989, Use of calcium antagonism for the characterization of drug-evoked dopamine release from the brain of conscious rats determined by microdialysis, *J. Neurochem.* 52:722-729.

Woodward, J.J., Chandler, L.J., and Steven, W.L., 1988, Calcium-dependent and independent release of endogenous dopamine from rat striatal synaptosomes, *Brain Research* 473:91-98.

Yaari, Y., Hamon, B., and Lux, H.D., 1987, Development of two types of calcium channels in cultured mammalian hippocampal neurons, *Science* 235:680-682.

Zigmond, R.E., Schwarzschild, M.A., and Rittenhouse, A.R., 1989, Acute regulation of tyrosine hydroxylase by nerve activity and by neurotransmitters via phosphorylation, *Ann. Rev. Neurosci.* 12:415-461.

Zivkovic, B., and Guidotti, A., 1974, Changes of kinetic constant of striatal tyrosine hydroxylase elicited by neuroleptics that impair the function of dopamine receptors, *Brain Research* 79:505-509.

Zurgil, N., Yarom, M., and Zisapel, N., 1986, Concerted enhancement of calcium influx, neurotransmitter release and protein phosphorylation by a phorbol ester in cultured brain neurons, *Neuroscience* 19:1255-1264.

PRESYNAPTIC REGULATION OF DOPAMINE RELEASE IN STRIATAL COMPARTMENTS AND FUNCTIONAL HETEROGENEITY OF THE MATRIX

Jacques Glowinski, Christian Gauchy, Marie-Odile Krebs, Léon Tremblay, Marcel Desban and Marie-Lou Kemel

Chaire de Neuropharmacologie INSERM U 114
Collège de France
11, place Marcelin Berthelot
75231 Paris cedex 05 - France

INTRODUCTION

The regulation of dopamine (DA) release from nerve terminals of the nigrostriatal DA neurones depends not only on nerve activity but also on presynaptic processes. DA itself inhibits its own release by acting on DA autoreceptors (D2 and D3 types) present on DA nerve terminals. Other transmitters located in striatal afferent fibers (glutamate), striatal interneurones (acetylcholine-ACh, somatostatine) or collaterals of the medium-sized spiny neurones which innervate either the substantia nigra and the entopeduncular nucleus or the external globus pallidus (GABA, opioid peptides, tachykinines) can facilitate or reduce the release of DA (Chesselet, 1984; Chéramy et al., 1986; Kemel et al., 1989; Krebs et al., 1989; Gauchy et al., 1991). These effects are either direct or indirect mediated by receptors located on DA neurones or on neurones in contact with DA nerve terminals. Due to the heterogeneity of the striatum, local circuits involved in the presynaptic control of DA release could differ from one area to another and this may have functional significance. The striatum is divided indeed in two main compartments, the striosomes (or patches) and the matrix which can be distinguished by several biochemical markers and their afferent or efferent neurones (Graybiel, 1990; Gerfen, 1992). In addition, the matrix itself is heterogeneous since some efferent cells are grouped in clusters (matrisomes) (Desban et al., 1989; Graybiel, 1990). Moreover, although some nigral DA cells project to both striatal compartments, a group of nigral DA cells located in the so-called denso-cellular zone in the cat and in both the ventral part of the pars compacta and the pars reticulata in the rat innervate exclusively the striosomes while other DA cells located in the A8 and A10 DA cell groups project only to the matrix (Gerfen et al., 1987; Jimenez-Castellanos and Graybiel, 1987). Therefore, DA neurones innervating either the striosomes or the matrix could exhibit different properties and be submitted to different presynaptic regulations of DA release. In fact, DA cells which innervate exclusively the striosomes mature first during ontogenesis (Graybiel, 1990), while those responsible for the matrix innervation exhibit a faster turnover rate of DA, contain the calcium calmodulin binding protein and are more sensitive to the neurotoxic effect of MPTP (Gerfen, 1985; Graybiel et al., 1987; Turner et al., 1988; Moratalla et al., 1992).

All these informations led us to develop an *in vitro* method allowing to estimate the release of ³H-DA continuously synthesized from ³H-tyrosine in striosomal or matrix-enriched areas using push-pull cannulae vertically applied on saggital and/or coronal brain slices of the rat or the cat (Kemel et al., 1989; Krebs et al., 1991). This strategy has required the three-dimensional mapping of the striosomal network organization in the two species using acetylcholinesterase (AChE) and ³H-naloxone binding as markers of the matrix and the striosomes respectively (Desban et al., 1989; unpublished observations).

We will briefly summarized results obtained during these past few years which have allowed to demonstrate that different local circuits contribute to the presynaptic regulation of DA release in striatal compartments (rat, cat) and in subcompartments of the matrix (cat). This could be shown by examining the effects of the glutamatergic agonist N-methyl-D-aspartate (NMDA), ACh and the agonists of NK1, NK2 and NK3 tachykinin receptors on ³H-DA release in the absence or presence of bicuculline, naloxone or tetrodotoxin (TTX).

Role of NMDA and NMDA-activated Local Inhibitory Circuits in the Control of Dopamine Release in Striatal Compartments of the Rat

The cortico-striatal glutamatergic neurones innervate either the striosomes or the matrix depending on their cortical layer of origin (Gerfen, 1992). They project onto the medium-sized spiny GABAergic neurones which contribute to either the direct (striato-nigral or entopeduncular nucleus) or the indirect (striato-globus pallidus) striatal efferent pathways. Several studies performed *in vitro* or *in vivo* have indicated that glutamate stimulates presynaptically the release of DA by acting on AMPA and NMDA receptors (Barbeito et al., 1990; Leviel et al., 1990; Galli et al., 1991; Krebs et al., 1991). Some of these receptors are located on DA nerve terminals since a glutamate-evoked release of DA involving both types of receptors was observed not only on striatal slices in the presence of TTX but also, more convincingly, on purified striatal synaptosomes (Krebs et al., 1991; Desce et al., 1992). Thus, we have compared the stimulatory effects of NMDA (50 μM) on ³H-DA release in striosomes- and matrix-enriched areas of the rat striatum using mainly sagittal slices. These experiments performed in the absence of magnesium to remove the magnesium block of NMDA receptors were then repeated in the presence of either bicuculline (5 μM) or naloxone (1 μM) to determine the inhibitory influences of GABA and dynorphin released from collaterals of striatal efferent cells in the control of the NMDA-evoked release of ³H-DA.

Confirming the existence of differences in the presynaptic regulation of DA release in striatal compartments, the NMDA-evoked release of ³H-DA was found to be much more important in matrix- than in striosomal-enriched areas (Fig. 1) (Krebs et al., 1991). This difference persisted in the presence of glycine which increased the NMDA responses. Parallel experiments indicated that NMDA stimulates markedly the release of preloaded ³H-GABA in both striatal compartments, this effect being slightly more pronounced in the matrix- than in striosomal-enriched areas (Galli et al., 1992; unpublished observations). Therefore, as expected the continuous blockade of GABA A receptors with bicuculline dramatically increased the NMDA-evoked release of ³H-DA in the two striatal compartments. However, the desinhibitory effect of bicuculline on DA release was much larger in striosome- than in matrix-enriched areas (Fig. 1). Although less pronounced, similar results were obtained when naloxone was used instead of bicuculline (Fig. 1).

These results suggest that under the local application of NMDA, dynorphin is co-released with GABA from either the collaterals of the medium-sized spiny neurones projecting to the substantia nigra (and/or the entopeduncular nucleus) and/or their dendrites (as suggested by experiments made with TTX). In fact, the involvement of Kappa opiate receptors in a direct presynaptic inhibitory control of DA release has already been demonstrated using striatal synaptosomes (Werling et al., 1988). In addition, in our experiments, the co-application with naloxone of U 50488 (1 μM), the agonist of Kappa opiate receptors, totally (matrix) or partially (striosomes) reduced the desinhibitory effect of naloxone on the NMDA-evoked release of ³H-DA. Finally, complementary experiments performed in the combined presence of bicuculline and naloxone revealed that the desinhibitory effects of these antagonists on the NMDA-evoked responses were additive in matrix- but not in striosomal- enriched areas. Therefore, due to the matrix heterogeneity, more complex local circuits seem to intervene in the presynaptic control of DA release in this striatal compartment.

As illustrated in Figure 1, additional informations were obtained when similar experiments were performed in the presence of TTX, which abolishes indirect effects requiring nerve impulse flow. 1) In the matrix only, the NMDA-evoked release of ³H-D

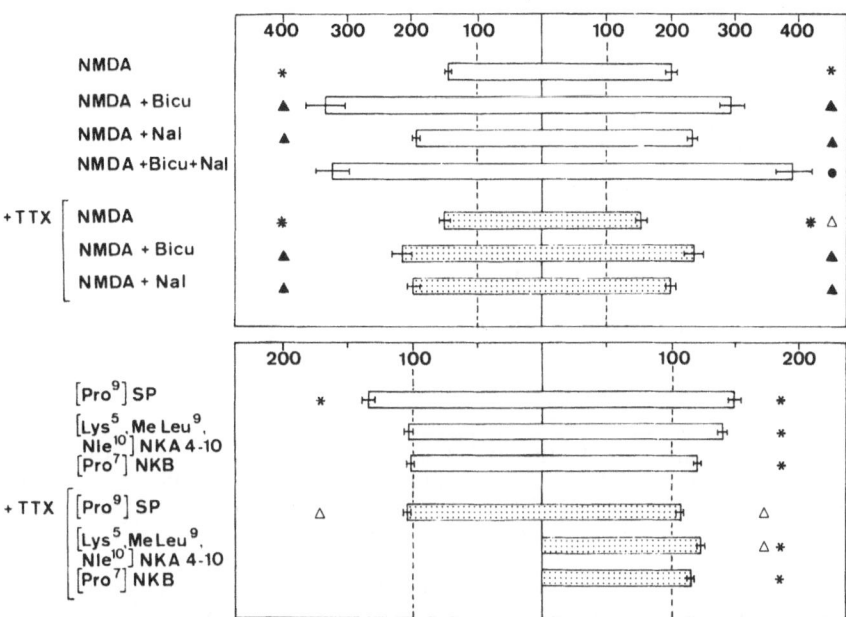

Figure 1. Presynaptic control of dopamine release in striosomal- and matrix-enriched areas of the rat striatum by NMDA and tachykinin agonists.

Superfusion experiments were performed as previously described (Krebs et al., 1991) NMDA (50 µM) and tachykinin agonists (0.1 µM) were applied for 25 min, 65 min after the beginning of the superfusion. When included, TTX (1 µM) bicuculline (Bicu, 5 µM) and/or naloxone (Nal, 1 µM) were added throughout the superfusion with ^{3}H-tyrosine. In each experiment, ^{3}H-DA recovered in successive 5 min fractions was expressed as a percentage of the mean spontaneous release of ^{3}H-DA determined during the four fractions preceding the onset of the agonist application. The average release of ^{3}H-DA during each 25 min pharmacological treatment was then calculated and expressed on a 5 min basis. Results are the mean ± SEM of data obtained in 7 to 18 experiments.

*, p<0.05 effect of NMDA or tachykinin agonists (in the absence or the presence of TTX) when compared to respective control values.

▲, p<0.05 effect of NMDA in the presence of bicuculline or naloxone in TTX-free or TTX-containing medium when compared to the effects of NMDA alone in the absence or in the presence of TTX respectively.

●, P<0.05 effect of NMDA in the presence of bicuculline and naloxone when compared to the effect of NMDA obtained in the presence of bicuculline alone.

Δ, p<0.05 effect of NMDA or tachykinin agonists in the presence of TTX when compared to the effects of these treatments in the absence of TTX.

was partially reduced indicating that, in the absence of TTX, not yet identified excitatory neurones contribute to the NMDA response in this compartment. 2) The desinhibitions by bicuculline or naloxone of the NMDA responses were still observed but each antagonist induced similar TTX-resistant effects in the two striatal compartments. This suggests that NMDA stimulates the release of GABA and dynorphin not only from collateral fibers but also from dendrites (TTX-resistant process) of the medium-sized spiny neurones and that these inhibitory transmitters reduce the release of DA by acting directly on DA nerve terminals. Supporting this statement, NMDA was also shown to stimulate the release of preloaded ^{3}H-GABA through a TTX-resistant process in both the matrix and striosome-enriched areas (Galli et al., 1992; unpublished observations). 3) In the striosomal-enriched area part of the bicuculline desinhibitory effect on the NMDA response was TTX-sensitive indicating that in this compartment nerve activity is also required in GABAergic neurones for their inhibitory control on DA release .

Role of Tachykinines in the Control of DA release in Striatal Compartments of the Rat

In both striatal compartments, the medium-sized spiny GABAergic neurones which are rich in dynorphin also contain substance P (SP) and neurokinin A (NKA), the latter peptide being present in two of the three SP precursors (Krause et al., 1987; Gerfen and Young III, 1988). In addition, in the matrix, some of the GABAergic neurones containing metenkephalin which innervate the external globus pallidus are also rich in neurokinin B (NKB) (Graybiel, 1990; Gerfen, 1992). Finally, the presence in this striatal compartment of neurones containing both SP and NKB has also been described (Burgunder and Young III, 1989). These anatomical data and results indicating that SP stimulates the release of DA through a TTX-sensitive process on whole rat striatal slices (Petit and Glowinski, 1986) led us to compare the effects of selective agonists of NK1, NK2 and NK3 receptors ((Pro⁹)SP, (Lys⁵, Meleu⁹ Nle¹⁰)NKA (4-10) and (Pro⁷)NKB respectively) on the release of DA in striatal compartments of the rat. Recently developed nonpeptide NK1 and NK2 antagonists (RP 67580 and SR 48968 respectively (Garret et al., 1991; Emonds-Alf et al., 1992)) were also used for this purpose (Tremblay et al., 1992).

Several differences were observed (Fig. 1): 1) When used in small concentrations (10^{-8} and/or 10^{-7}M), the three tachykinin agonists stimulated the release of ^{3}H-DA in the matrix-enriched area (the (Pro⁷)NKB-evoked response being of smaller amplitude) while the NK1 agonist (Pro⁹)SP was the only agonist which also enhanced ^{3}H-DA release in the striosomal-enriched area. 2) As shown in the matrix, these responses were specific since the (Pro⁹)SP-evoked response was suppressed by the NK1 antagonist (10^{-6}M) and that of (Lys⁵, Meleu⁹ Nle¹⁰)NKA (4-10) by the NK2 antagonist (10^{-6}M). In addition, both antagonists were without effect on the (Pro⁷)NKB-evoked response. 3) The presence of TTX abolished the NK1 agonist-evoked release of ^{3}H-DA in both compartments while responses induced in the matrix-enriched area by the NK2 and NK3 agonists were only partially (NK2 agonist) or not (NK3 agonist) affected. This suggests that the NK1 agonist regulates indirectly the release of DA while the effects of the NK2 and NK3 agonists are partially or totally mediated by NK2 and NK3 receptors located on DA nerve terminals. 4) The cholinergic interneurones which possess NK1 receptors (Gerfen, 1991) and are activated by NK1 agonists (as indicated by the estimation of ACh release) (Arenas et al., 1991; Petitet et al., 1991) seem to be involved in the indirect effect of the NK1 agonist on DA release in the matrix but not in the striosomes. Indeed, the pharmacological interruption of cholinergic transmission (atropine plus pempidine) reduced the (Pro⁹)SP-evoked response in the matrix but not in the striosomal-enriched area.

These results further demonstrate that the presynaptic regulation of DA release differs in the two striatal compartments and that more complex local circuits occur in the matrix. They also show that central biological responses induced by the stimulation of NK2 receptors can be demonstrated although these receptors have not been yet identified in the brain using binding studies or molecular cloning. Finally, close similarities were seen between the effects of the NK1 and NK2 agonists in the striatum and those of SP and NKA in the substantia nigra since as shown *in vivo* in either the rat or the cat these two peptides interact by different processes with nigrostriatal DA neurones (Baruch et al., 1988; Reid et al., 1990).

Role of Acetylcholine in the Control of DA Release in Striatal Compartments of the Cat : Functional Heterogeneity of the Matrix

The cholinergic interneurones could be involved in the transfer of information between the matrix and the striosomes (Graybiel et al., 1986). Indeed, although this has recently been challenged, the cell bodies of these neurones are often located around or near the striosomes and their neurites which are more abondant in the matrix are also present in the striosomes. According to biochemical and electrophysiological studies, the cholinergic interneurones are innervated by cortico-striatal glutamatergic neurones and/or thalamic excitatory neurones known to project mainly to the matrix (Scatton and Lehman, 1982; Graybiel, 1990; Sadikot et al., 1990; Wilson et al., 1990; Lapper and Bolam, 1992). Interestingly, recent ultrastructural studies have indicated that the cholinergic interneurones located in the dorso-lateral part of the rat striatum are synaptically contacted by thalamic neurones from the thalamic parafascicularis nucleus but not by those from the prefrontal cortex (Lapper and Bolam, 1992). Numerous striatal cells possess muscarinic and/or nicotinic receptors and we

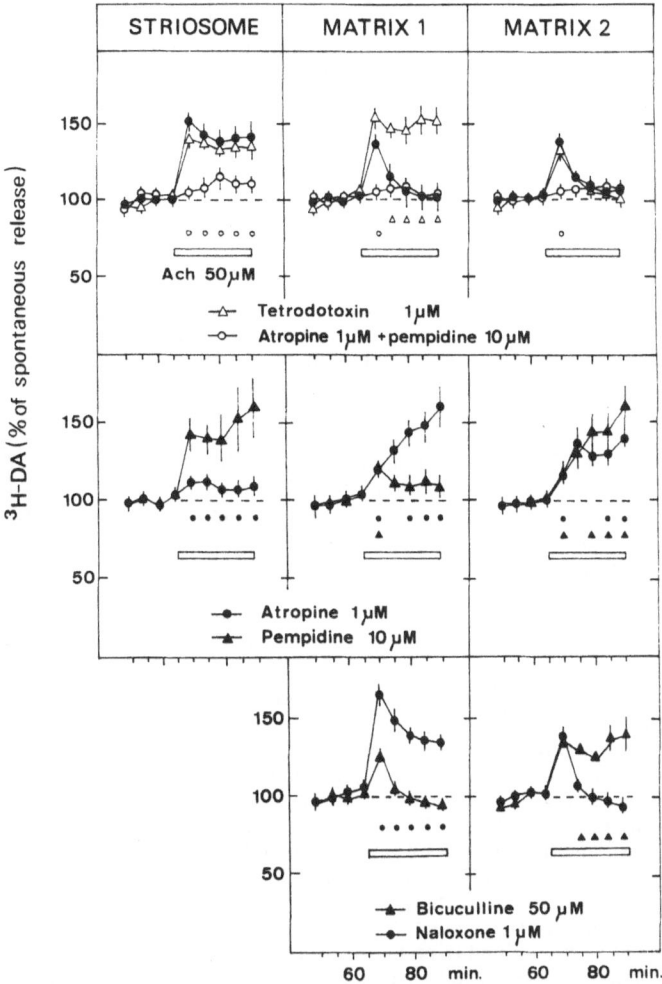

Figure 2. Cholinergic control of dopamine release in the striosomes and subcompartments of the matrix of the cat caudate nucleus.

Superfusion experiments were performed as previously described (Kemel et al., 1989). ACh (50 μM) was applied during the last 25 min of the experiment, 65 min after the onset of superfusion with ^{3}H-tyrosine. When used, TTX (1 μM), atropine (1 μM), pempidine (10 μM), bicuculline (50 μM) or naloxone (1 μM) were added throughout the superfusion with ^{3}H-tyrosine.

In each experiment ^{3}H-DA release was expressed as a percentage of the mean spontaneous release estimated in the four fractions preceding the application of ACh. Results are the mean ± SEM of data obtained in 7 to 14 experiments. During the application of ACh performed in either the absence or the presence of TTX in the striosomes, the matrix 1 or the matrix 3, the release of ^{3}H-DA was significantly increased (p<0,05) during the five (striosomes) or the first (matrix 1 and 2) fractions of the ACh application.

O, p<0.05 effect of ACh in the presence of atropine and pempidine (-O-) when compared to the ACh-evoked-response (-●-).

Δ, p<0.05 effect of ACh in the presence of TTX (-Δ-) when compared to the effect of ACh in the absence of TTX (-●-).

•, p<0.05 effect of ACh in the presence of atropine or naloxone when compared to the effect of ACh alone (as shown in the upper part).

▲, p<0.05 effect of ACh in the presence of pempidine or bicuculline when compared to the effect of ACh in the absence of the antagonist (as shown in the upper part).

have recently observed that ACh or carbachol stimulate the release of preloaded ³H-GABA in both striosomal and matrix-enriched areas. As indicated by results obtained on whole striatal slices in the presence of TTX or on striatal synaptosomes, ACh was the first neurotransmitter shown to be involved in a direct presynaptic regulation of DA release, this effect being mediated by both muscarinic and nicotinic receptors (Chesselet, 1984). All these informations led us in a serie of studies made on coronal slices of the cat to look for the effects of ACh on DA release in striosomal- and matrix-enriched areas and in anatomically defined subcompartments of the matrix (Kemel et al., 1989; Gauchy et al., 1991; Kemel et al., 1992).

Two main observations emerge from our initial investigations. Firstly, DA nerve terminals in both striatal compartments possess muscarinic receptors involved in a presynaptic facilitation of ³H-DA release. Secondly, in contrast to that observed in a striosomal- enriched area, the ACh-evoked release of ³H-DA in matrix areas, was short lasting revealing that ACh dependent inhibitory processes are also involved in the regulation of DA release in this compartment (Kemel et al., 1989). A further detailed analysis indicated that depending on the matrix area, ACh could indirectly reduce the release of ³H-DA through its releasing effect on either GABA or dynorphin and this functional heterogeneity was found to be closely associated in the limited site of the ACh application with the presence or not of clusters of cells innervating the substantia nigra (Gauchy et al., 1991; Kemel et al., 1992).

AChE staining and ³H-naloxone binding indicated that the two juxtaposed matrix areas (denominated matrix 1 and matrix 2) in the dorso-lateral part of the caudate nucleus chosen for our release study were practically devoid of striosomal tissue. In addition, retrograde tracing studies revealed that the matrix 2 area contained a higher density of efferent striatonigral cells than the matrix 1 area. Indeed, following injections of wheat germ agglutinin Horseradish peroxidase in either the substantia nigra pars reticulata or the internal pallidum, dispersed labeled cells were seen in the matrix 1 area while the matrix 2 was characterized by the presence of cells grouped into clusters retrogradely labeled only from the pars reticulata.

Several differences were found in the regulation of ³H-DA release when the effects of ACh (50 µM) were examined in these two matrix areas. Indeed, as indicated in Figure 2, the immediate and short lasting ACh-evoked release of ³H-DA seen in the two matrix areas was affected differently in matrix 1 and matrix 2 when experiments were performed in the presence of either TTX (1 µM), pempidine (50 µM), bicuculline (50 µM) or naloxone (1 µM) while similar changes occurred in the presence of atropine (50 µM). These results and additional experiments described in details elsewhere allowed us to reach several conclusions. 1) DA nerve terminals in matrix 2 possess both muscarinic and nicotinic receptors involved in a cholinergic presynaptic facilitatory control of DA release while those present in matrix 1 possess only muscarinic receptors. 2) Muscarinic receptors located on striatal neurones are also involved in an indirect cholinergic inhibitory control of DA release in both matrix areas but these inhibitory responses are TTX-sensitive in matrix 1 and TTX-insensitive in matrix 2 suggesting that in the latter case ACh exerts its effect through inhibitory fibers (containing GABA) in contact with DA nerve terminals. 3) Nicotinic receptors are also involved in the indirect cholinergic control of DA release: in matrix 1, this regulation is facilitatory and TTX-sensitive, mediated by yet unidentified neurones, while in matrix 2 this regulation is inhibitory and TTX-resistant. 4) More important, the inhibitory effect of ACh on DA release seems to be mediated by dynorphin in matrix 1 and by GABA in matrix 2 and the corresponding receptors (Kappa opiate receptors and GABA-A receptors) appear to be located on DA nerve terminals.

Therefore, the indirect ACh-evoked inhibition of ³H-DA release in matrix 1 involves dynorphin and requires nerve activity in dynorphin containing neurones while that intervening in matrix 2 involves GABA and is TTX-resistant. These differences could be related to the already described anatomical characteristics of matrix 1 and matrix 2. As already seen in the rat for responses mediated by NMDA, in the cat, ACh could release GABA from dendrites in matrix 2. This will explain why the ACh-evoked inhibitory responses are TTX-resistant. It remains to be determined why GABA is only effective in the control of DA release in matrix 2 while dynorphin intervenes only in matrix 1. In any case, these results are reminescent to those observed with NMDA in the rat since in the matrix, the desinhibitory effects of bicuculline and naloxone on the NMDA-evoked release of ³H-DA were additive suggesting the intervention of distinct local circuits.

CONCLUSIONS

Numerous studies have been dedicated to the striatal compartments. However much has still to be done in order to understand the respective roles of these compartments in the treatment and transfer of information in the basal ganglia. Undoubtedly, sophisticated electrophysiological approaches combined with anatomical techniques will be of great help for this purpose. As shown in this study, some insights on the neuronal interactions occurring in striatal compartments but also in matrix subcompartments have been obtained thanks to a new strategy based on the identification and analysis of local circuits involved in the presynaptic control of DA release.

Experiments performed in the rat or the cat using either NMDA or ACh alone or in the presence of various antagonists have revealed important differences between striosomes- and matrix-enriched areas. These differences in the NMDA- or ACh-evoked regulations of DA release are mainly attributed to indirect effects i. e. which are TTX-sensitive. In general, more complex regulations are seen in the matrix. This was further revealed by the responses obtained with NK1, NK2 and NK3 tachykinin agonists and by additional experiments performed in the cat indicating that ACh could act differently in controling the release of DA depending on the presence or not at the site of its application of neurones grouped in clusters innervating the substantia nigra pars reticulata. The different populations of medium-sized GABAergic neurones which possess NMDA and/or cholinergic receptors and which are rich in either dynorphin, met-enkephalin or distinct tachykinins play an important role in controlling locally, in discrete territories, dopaminergic transmission. Finally, two basic problems which will have still to be further elucidated emerge from these studies: they concern the dendritic release of transmitters located in medium-sized spiny neurones and the regulatory processes of co-transmission.

Acknowledgements

This research was supported by grants from Caisse Nationale de l'Assurance Maladie des Travailleurs salariés, Direction des Recherches, Etudes et techniques (90.078), Human Frontiers, Rhône Poulenc Rorer, and Fonds de la Recherche en Santé du Québec (fellowship to Léon Tremblay).

REFERENCES

Arenas, E., Alberch, J. Perez-Navarro, E., Solsona, C., and Marsal, J., 1991, Neurokinin receptors differentially mediate endogenous acetylcholine release evoked by tachykinins in the neostriatum, *J. Neurosci.* 11:2332-2338.

Barbeito, L., Chéramy, A., Godeheu, G., Desce, M., and Glowinski, J., 1990, Glutamate receptors of a quisqualate-kainate subtype are involved in the presynaptic regulation of dopamine release in the cat caudate nucleus in vivo, *Eur. J. Neurosci.* 2:304-311.

Baruch, P., Artaud, F., Godeheu, G., Barbeito, L., Glowinski, J., and Chéramy, A., 1988, Substance P and neurokinin A regulate by different mechanisms dopamine release from dendrites and nerve terminals of the nigrostriatal dopaminergic neurons, *Neuroscience* 25:889-898.

Burgunder, J.M., and Young III, W.S., 1989, Distribution, projection and dopaminergic regulation of the neurokinin B mRNA-containing neurons of the rat caudate putamen, *Neuroscience* 32:323-335.

Chéramy, A., Romo, R., Godeheu, G., Baruch, P., and Glowinski, J., 1986, In vivo presynaptic control of dopamine release in the cat caudate nucleus II. Facilitatory or inhibitory influence of L-glutamate, *Neuroscience* 19:1081-1090.

Chesselet,, M.F., 1984, Presynaptic regulations of neurotransmitter release in the brain: facts and hypothesis, *Neuroscience* 12:347-375.

Desban, M., Gauchy, C., Kemel, M.L., Besson, M.J., and Glowinski, J., 1989, Three-dimensional organization of the striosomal compartment and patchy distribution of striatonigral projections in the matrix of the cat caudate nucleus, *Neuroscience* 29:551-566.

Desce, J.M., Godeheu, G., Galli, T., Artaud, F., Chéramy, A., and Glowinski, J., 1992, L-glutamate-evoked release of dopamine from synaptosomes of the rat striatum: Involvement of AMPA and N-methyl-D-aspartate receptors, *Neuroscience* 47:333-339.

Emonds-Alt, X., Vilain, P., Goulaouic, P., Proietto, V., Van Broeck, D., Advenier, C., Naline, E., Neliat, G., Le Fur, G., and Breliere, J.C., 1992, A potent and selective non-peptide antagonist of the neurokinin A (NK2) receptor, *Life Sci.* 50:101-106.

Galli, T., Desce, J.M., Artaud, F., Kemel, M.L., Chéramy, A., and Glowinski, J., 1992, Modulation of GABA release by AMPA and NMDA receptors in matrix-enriched areas of the rat striatum, *Neuroscience* 50:769-800.

Galli, T., Godeheu, G., Artaud, F., Desce, J.M., Pittaluga,A., Barbeito, L., Glowinski, J., and Chéramy, A., 1991, Specific role of N-acetyl-aspartyl-glutamate in the in vivo regulation of dopamine release from dendrites and nerve terminals of nigrostriatal dopaminergic neurons in the cat, *Neuroscience* 42:19-28.

Garret, C., Carruette, A., Fardin, V., Moussaoui, S., Peyronel, J.F., Blanchard, J.C., and Laduron, P.M., 1991, Pharmacological properties of a potent and selective non-peptide substance P antagonist, *Proc. Natl., Acad. Sci. USA* 88:10208-10212.

Gauchy, C., Desban, M., Krebs, M.O., Glowinski, J., and Kemel, M.L., 1991, Role of dynorphin-containing neurons in the presynaptic inhibitory control of the acetylcholine-evoked release of dopamine in the striosomes and the matrix of the cat caudate nucleus, *Neuroscience* 41:449-458.

Gerfen, C.R., 1985, The neostriatal mosaic: I Compartmental organization of projections from the striatum to the substantia nigra in the rat, *J. Comp. Neurol.* 236:454-478.

Gerfen, C.R., 1991, Substance P (neurokinin-1) receptors mRNA is selectively expressed in cholinergic neurons in the striatum and basal forebrain, *Brain Res.* 556:165-170.

Gerfen, C.R., 1992, The neostriatal mosaic: multiple levels of compartmental organization, *Trends Neurosci.* 15:133-138.

Gerfen, C.R., Herkenham, M., and Thibault, J., 1987, The neostriatal mosaic: II Patch-and matrix-directed mesostriatal dopaminergic and non-dopaminergic systems, *J. Neurosci.* 7:3915-3934.

Gerfen, C.R., and Young III, W.S., 1988, Distribution of striatonigral and striatopallidal peptidergic neurons in both patch and matrix compartments: an in situ hybridization histochemistry and fluorescent retrograde tracing study, *Brain Res.* 460:161-167.

Graybiel, A.M., 1990, Neurotransmitters and neuromodulators in the basal ganglia, *Trends Neurosci.* 13:244-254.

Graybiel, A.M., Baughman, R.W., and Eckenstein, F., 1986, Cholinergic neuropil of the striatum observes striosomal boundaries, *Nature* 323:625-627.

Graybiel, A.M., Hirsch, E.C., and Agid, Y., 1987, Differences in tyrosine hydroxylase-like immunoreactivity characterize the mesostriatal innervation of striosomes and extrastriosomal matrix at maturity, *Proc. Natl. Acad. Sci. USA* 84:303-307.

Jimenez-Castellanos, J., and Graybiel, A.M., 1987, Subdivisions of the dopamine-containing A8-A9-A10 complex identified by their differential mesostriatal innervation of striosomes and extrastriosomal matrix, *Neuroscience* 23:223-242.

Kemel, M.L., Desban, M., Glowinski, J., and Gauchy, C., 1989, Distinct presynaptic control of dopamine release in striosomal and matrix areas of the cat caudate nucleus, *Proc. Natl. Acad. Sci. USA* 86:9006-9010.

Kemel, M.L., Desban, M., Glowinski, J., and Gauchy, C., 1992, Functional heterogeneity of the matrix compartment in the cat caudate nucleus as demonstrated by the cholinergic presynaptic regulation of dopamine release, *Neuroscience* 50:597-610.

Krause, J.E., Chirgwin, S.M., Carter, M.S., Xu, Z.S., and Herskey, A.D., 1987, Identification of three rat preprotachykinins encoding both substance P and neurokinin A., *Proc. Natl. Acad. Sci. USA* 84:881-885.

Krebs, M.O., Kemel, M.L., Gauchy, C., Desban, M., and Glowinski, J., 1989, Glycine potentiates the NMDA-induced release of dopamine through a strychnine-insensitive site in the rat striatum, *Eur. J. Pharmacol.* 166:567-570.

Krebs, M.O., Trovero, F., Desban, M., Gauchy, C., Glowinski, J., and Kemel, M.L., 1991, Distinct presynaptic regulation of dopamine release through NMDA receptors in striosome- and matrix-enriched areas of the rat striatum, *J. Neurosci.* 11:1256-1262.

Lapper, S.R., and Bolam, S.P., 1992, Input from the frontal cortex and the parafascicular nucleus to cholinergic interneurons in the dorsal striatum of the rat, *Neuroscience* 51:533-545.

Leviel, V., Gobert, A., and Guibert, B., 1990, The glutamate-mediated release of dopamine in the rat striatum: further characterization of the dual excitatory-inhibitory function, *Neuroscience* 39:305-312.

Moratalla, R., Quinn, B., Delanney, L.E., Irwin, I., Langston, J.W., and Graybiel, A.M., 1992, Differential vulnerability of primate caudate-putamen and striosome-matrix dopamine systems to the neurotoxic effects of 1-methyl-4-phenyl-1,2,3,6-tetrahydropyridine, *Proc. Natl. Acad. Sci. USA* 89:3859-3863.

Petit, F., and Glowinski, J., 1986, Stimulatory effect of substance P on the spontaneous release of newly synthesized ^{3}H-dopamine from rat striatal slices; a tetrodotxin-sensitive process, *Neuropharmacology* 25:1015-1021.

Petitet, F., Glowinski, J., and Beaujouan, J.C., 1991, Evoked release of acetylcholine in the rat striatum by stimulation of tachykinin NK1 receptors, *Eur. J. Pharmacol.* 192:303-204.

Reid, M.S., Herrera-Marschitz, M., Hökfelt, T., Ohlin, M., Valentino, J.L., and Ungerstedt, U., 1990, Effects of intranigral substance P and neurokinin A on striatal dopamine release. I. Interactions with substance P antagonists, *Neuroscience* 36:643-658.

Sadikot, A.F., Parent, A., and François, C., 1990, The centre median and parafascicular thalamic nuclei project respectively to the sensorimotor and associative limbic striatal territories in the squirrel monkey, *Brain Res.* 510:161-165.

Scatton, B., and Lehman, J., 1982, N-methyl-D-aspartate-type of receptors mediate striatal ^{3}H-acetylcholine release evoked by excitatory amino acids, *Nature* 297:422-424.

Tremblay, L., Kemel, M.L., Desban, M., Gauchy, C., and Glowinski, J., 1992, Distinct presynaptic control of dopamine release in striosomal- and matrix-enriched areas of the rat striatum by selective agonists of NK1, NK2 and NK3 tachykinin receptors, *Proc. Natl. Acad. Sci.,USA* 89:11214-11218.

Turner, B.H., Wilson, J.S., McKenzie J., and Richtand, N., 1988, MPTP produces a pattern of nigrostriatal degeneration which coincides with the mosaic organization of the caudate nucleus, *Brain Res.* 473:60-64.

Werling, L.L., Frattali, A., Porthogheze, P.S., Takemori, A.E., and Cox, B.M., 1988, Kappa receptor regulation of dopamine release from striatum and cortex of rats and guinea pigs, *J. Pharmac. exp Ther.* 246:282-286.

Wilson, C.J., Chang, H.T., and Kitai, S.T., 1990, Firing patterns and synaptic potentials of identified giant aspiny interneurons in the rat neostriatum, *J. Neurosci.* 10:508-519.

D₁ RECEPTOR MEDIATED TROPHIC ACTION OF DOPAMINE ON THE SYNTHESIS OF GABA AT THE TERMINALS OF STRIATAL PROJECTIONS

Jorge Aceves[1], Benjamin Floran[1] and Martha Garcia[2]

[1]Departamento de Fisiología, Biofísica y Neurociencias
CINVESTAV-IPN. México D.F., México
[2]Escuela Nacional de Ciencias Biologicas del IPN
México D.F., México

INTRODUCTION

There is no doubt that degeneration of the dopaminergic innervation of the basal ganglia affects GABAergic activity in these structures. Among the parameters of GABAergic activity affected is the activity of the enzyme in charge of the synthesis of GABA, glutamic acid decarboxylase. Several approaches have been followed to test the effect of the degeneration on the activity of the enzyme. One approach has been to see whether the activity of the enzyme is affected by Parkinson's disease (McGeer et al., 1971; Lloyd and Hornykiewicz, 1973; McGeer and McGeer, 1976; Javoy-Agid et al., 1981). There is consensus that Parkinson's disease markedly reduces GAD activity in internal globus pallidus and substantia nigra (most probably in pars reticulata). However conflicting results have been found in the striatum. Here some authors (Lloyd and Hornykiewicz, 1973) have found a significant decrease in GAD activity, but others (McGeer and McGeer, 1976; Perry et al., 1983) have not found any effect of the disease on the activity of the enzyme. Another approach has been to see whether experimentally induced degeneration of the dopaminergic nigrostriatal system affects GAD activity. Without exception, it has been found that the degeneration of the dopaminergic nigrostriatal system (induced by 6-hydroxydopamine) increases GAD activity in the rat neostriatum (Vincent et al., 1978; Segovia and Garcia-Muñoz, 1987; Vernier et al., 1988; Segovia et al., 1989, 1990, 1991). Degeneration of the dopaminergic input to neostriatum not only increases GAD activity but GAD mRNA as well (Segovia et al., 1990). From this, these authors have proposed that dopaminergic neurons down regulate the gene expression for GAD in neostriatal GABAergic neurons. In other words, the dopaminergic innervation appears to exert a tonic inhibitory action on GAD activity mediated by a down regulation in the synthesis of GAD due to a decrease in GAD mRNA. This proposition is the opposite of that of Lloyd and Hornykiewicz, who based on the observation that levodopa partially recovered the GAD activity in the basal ganglia of parkinsonian brains (Lloyd and Hornykiewicz, 1973) proposed that "GAD-containing neurons could be under a continuous "trophic" influence of the dopaminergic system", adding that "degeneration of the dopaminergic pathway might result in a biochemical "atrophy" of the GAD neurones". How to reconcile these two propositions? The conflicting results may arise, in part, from the conditions in which GAD activity was measured, that is, in the presence of saturating concentrations of substrate (glutamic acid) and cofactor (pyridoxal phosphate). It appears that these conditions do not reflect the true functional GAD activity of the GABAergic neurons (Miller et al., 1978; Itoh and Uchimura, 1981). The functional activity may the activity measured without exogenously added pyridoxal phosphate (Itoh, 1983) and in the presence of normal ADP and ATP concentrations (Miller and Walters, 1977). Because of these considerations, we decided to study the effect of the degeneration of the dopaminergic system on the synthesis of GABA in slices isolated from the output nuclei, that is, from the substantia nigra reticulata and entopeduncular nucleus, and from the caudate-putamen (neostriatum) in the rat. We assumed that in well oxygenated slices the level of all those compounds affecting GAD activity would be within a normal range, therefore more appropriately reproducing *in vivo* conditions. In addition, since it has been demonstrated that D_1 receptors are present on the terminals of striatal projections (Barone et al., 1987), we also studied whether the dopaminergic effect was mediated by these receptors, and whether the dopaminergic innervation regulated the number of the receptors present on the striatonigral terminals.

METHODS

Adult male Wistar rats (250-300 g) were used. Unilateral lesions of the dopaminergic nigrostriatal system were done by the stereotaxic injection of 8 µg of 6-hydroxydopamine (free base) into the pars compacta of anesthetized (pentobarbital) rats. The lesion was evaluated 1 week after the injection by the turning induced by dl-metamphetamine (10 mg/Kg s.c.). Rats with no less than 12 turns/min were used. The incubation chambers and procedures are described elsewhere (Góngora et al., 1988; Aceves et al., 1992). The synthesis of GABA was assessed by its accumulation following inhibition of GABA transaminase with aminooxyacetic acid (Bernasconi et al., 1982). [³H]SCH 23390 was used to label D_1 receptors in homogenates of the pars reticulata. Nonspecific binding was determined as the binding found in the presence of 1 µM cold SCH 23390. Thirty unilateral lesioned rats were required to obtain enough tissue for a binding curve. The effect of the 6-OHDA lesion on binding was estimated by comparison with sham (saline) lesions.

RESULTS

Effect of the 6-hydroxydopamine lesion on basal GABA content

As Table 1 shows, the lesion of the dopaminergic system did not affect the basal GABA content in any of the studied nuclei. Incubation times for up to 1 h slightly increase GABA content, as judged by measurements made in slices of substantia nigra pars reticulata (content at 0 time: 56.3 ± 2.8; content at 60 min: 61.6 ± 0.8 nmol/mg prot.; n=8; $P<0.05$). It is known that, as result of the reduced levels of ADP and ATP, brain GAD undergoes both a post-mortem activation and a further activation during preparative procedures (Miller et al., 1977). The changes in GAD activity are reflected in changes in GABA synthesis. The relatively small increase in GABA content observed after 1 h of incubation indicates that the slices were well oxygenated, and consequently that the levels of ATP and ADP did not change appreciably during the incubation period.

Table 1. Effect of the 6-hydroxydopamine-induced lesion of the dopaminergic nigrostriatal system on basal GABA content

condition	pars reticulata	entopeduncular nucleus	caudate-putamen
intact	60.0 ± 3.7 (12)	9.2 ± 0.5 (8)	19.5 ± 2.1 (8)
lesioned	57.7 ± 2.4 (12)	9.4 ± 0.8 (3)	21.3 ± 1.7 (8)

Contents (mean ± sem) are expressed in nmol/mg protein. In parenthesis, number of determinations

Effect of the 6-hydroxydopamine lesion on the accumulation of GABA

Even though the lesion did not affect basal GABA content, it drastically affected the accumulation of the aminoacid following the inhibition of GABA transaminase. Figure 1 shows the effect of the lesion on GABA accumulation in slices isolated from the pars reticulata and from the caudate-putamen. It can be seen that the loss of the dopaminergic input to the neostriatum produced a drastic reduction in the accumulation of GABA following inhibition of GABA transaminase with aminooxyacetic acid (AOAA). In the pars reticulata the effect of th lesion was present even 2 months after the lesion. As can be seen in the same figure, there was also a marked reduction in the accumulation of GABA in the caudate-putamen. The lesion of the dopaminergic nigrostriatal system produced also a drastic reduction of GABA accumulation in slices isolated from the entopeduncular nucleus, as judged by the lack of accumulation in lesioned slices after 20 min of incubation in the presence of AOAA (GABA content of lesioned slices: before AOAA, 9.4 ± 0.8, after AOAA, 8.6 ± 0.8 nmol/mg protein; content in intact slices: before AOAA, 9.2 ± 0.4, after AOAA, 15.0 ± 1.2 nmol/mg protein).

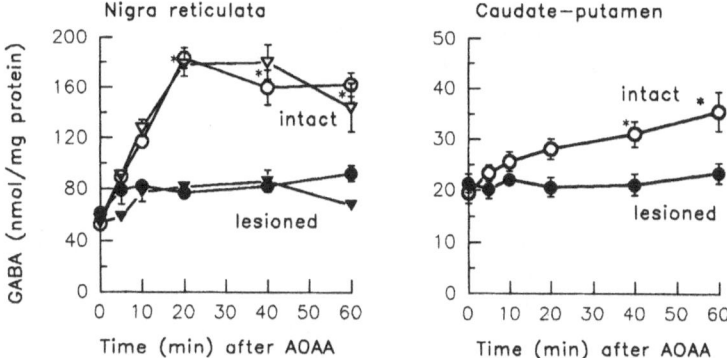

Figure 1. Effect of the unilateral lesion induced by 6-hydroxydopamine on the accumulation of GABA in slices from the pars reticulata and from the caudate-putamen following the inhibition of GABA transaminase with 10 μM aminooxyacetic acid (AOAA). In the pars reticulata, the measurements were done at 2 weeks (circles) and 2 months (triangles) after the lesion. In the caudate-putamen, the measurements were done 2 weeks after the lesion.

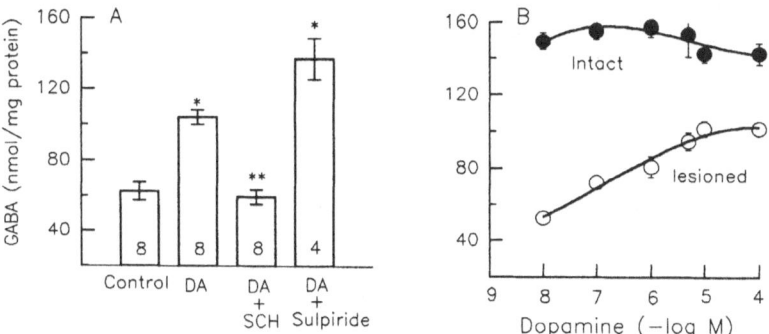

Figure 2. Effect of dopamine on the accumulation of GABA in slices of the pars reticulata. Measurements were made 10 min following addition of AOAA. The drugs were added concurrently with the inhibitor. In A, the effect of the dopaminergic compounds was tested in lesioned slices. Control means the amount of GABA found in the presence of AOAA alone. In B, a concentration-response curve was also determined in slices isolated from the intact side.

Activation of D_1 receptors partially recovers GABA accumulation

As can be seen in figure 2, addition of dopamine to the incubation medium produced a clear increase in the accumulation of GABA following inhibition of GABA transaminase. The effect of dopamine was inhibited by SCH 233390, a selective D_1 antagonist, but not by sulpiride, a selective D_2 antagonist, indicating that the effect was mediated by D_1 receptors. The EC_{50} of the dopamine effect was about 1 μM (see part B of the figure). This value is close to the EC_{50} (3.2 μM) for the stimulation by dopamine of the release of [^{3}H]GABA in 6-hydroxydopamine lesioned slices of the pars reticulata (Floran et al., 1990). However, GABA accumulation was only partially recovered by the activation of the D_1 receptors. This can be inferred comparing the accumulation observed in lesioned compared to intact slices in the presence of maximal concentrations (100 μM) of dopamine (Fig. 2).

Effect of the 6-hydroxydopamine lesion on the number and affinity of D_1 receptors in the pars reticulata

As can be seen in figure 3, the lesion produced a drastic reduction of the number of D_1 receptors present in the pars reticulata. As seen from the Scatchard plots in the figure, the lesion did not affect the affinity of the receptors for [^{3}H]SCH 23390, which was used to label the receptors.

DISCUSSION

The majority of experimental studies on the effect of the degeneration of the dopaminergic cells of the pars compacta on GAD activity have been done in striatal homogenates of the rat brain. However, the rat neostriatum contains 2 types of GABA neurons, interneurons and projection neurons (Bolam et al., 1985; Kita and Kitai, 1988). Therefore, GAD activity measured in striatal homogenates could correspond either to that of the interneurons or to that of the projection neurons. On the other hand, in mammalian brain, there are two different isoforms of GAD: GAD65 and GAD67 (Kaufman et al. 1991). The two isoforms are encoded by two different genes (Erlander et al., 1991), and have different requirements for cofactor. In the conditions in which GAD has been so far measured, namely in the presence of saturating concentrations of pyridoxal phosphate and of glutamate, one would expect that the activity measured would correspond to that of GAD67. Apparently this isoform is located mainly in the interneurons (Gonzales et al., 1991).

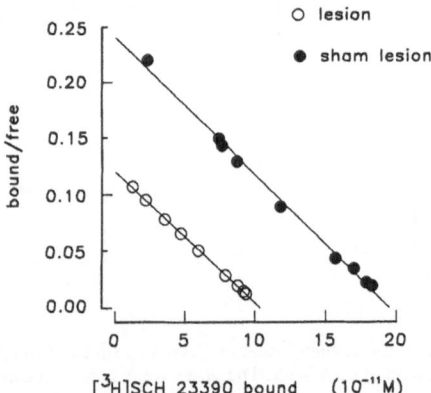

Figure 3. Effect of the 6-hydroxydopamine-induced lesion on the specific binding of [^{3}H]SCH 23390 in homogenates of the pars reticulata. Points represent the mean of triplicate determinations. In homogenates from sham lesioned slices, the binding parameters were: Bmax=650 ± 15 fM/mg prot.; Kd=0.81 nM. In homogenates from lesioned slices, the binding parameters were: Bmax=352 ± 34 fM/mg prot.; Kd=0.85 ± .01 nM.

Without exception, it has been found that the loss of the dopaminergic input to the neostriatum produces an increase in GAD activity. From this, it has been concluded that under normal conditions, dopamine exerts a tonic inhibitory action on GAD activity (Segovia et al., 1990), and consequently on GABA synthesis. However, as the present results show, when the synthesis of GABA is estimated by its accumulation following inhibition of GABA transaminase (Bernasconi et al., 1982) in well oxygenated striatal slices (Góngora et al., 1988) the result of the degeneration of the dopaminergic input is a clear reduction in the synthesis of the aminoacid. From this, it can be concluded that dopamine normally exerts a tonic excitatory action on the synthesis of GABA. How to reconcile the two conclusions? The contradiction may be explained assuming that GAD activity measured in striatal homogenates is that of the interneurons while GAD activity estimated in the present experiments in striatal slices is that of the projection neurons. The drastic reduction in GABA synthesis produced by the lesion in the pars reticulata and entopeduncular nucleus, targets of the striatal projections, support the proposition that in the present experimental conditions, the activity of GAD is that of the projection neurons. Therefore, the present results support the original proposition of Lloyd and Hornykiewicz (1973) that normally the dopaminergic system exerts a continuous trophic influence on GAD-containing neurons. However the statement should be restricted to the GABA projection neurons. The dopaminergic influence on the GABA interneurons could be the opposite to that of the projection neurons.

From the results, it is clear that the dopaminergic effect is mediated by D_1 receptors. It is known that these receptors are located both in the soma and dendrites of the GABAergic neurons in the neostriatum and in the presynaptic terminals of these neurons in the entopeduncular nucleus and in the pars reticulata (Barone et al., 1987). They most probably are located also in the terminals of the recurrent collaterals of these neurons. The D_1-mediated dopaminergic action would be exerted at membrane level and at genomic level. Activation of the D_1 receptors located in the somato-dendritic neuronal membrane in the striatum would allow the firing of the neuron induced by the glutamatergic input from the cortex (see figure 4). Activation of the receptors located in the presynaptic terminals would facilitate the release of GABA (Floran et al., 1990) and, according to the present results, would stimulate the synthesis of the aminoacid transmitter, probably coupling synthesis with release. The role of the presynaptic D_1 receptors would be, then, facilitation of the GABAergic neurotransmission at synapsis in the target nuclei of the striatal projection (see figure 4). At genomic level, there would be a tonic (trophic) D_1-mediated stimulation of the expression of the genes encoding for D_1 receptors and for GAD. The loss of the

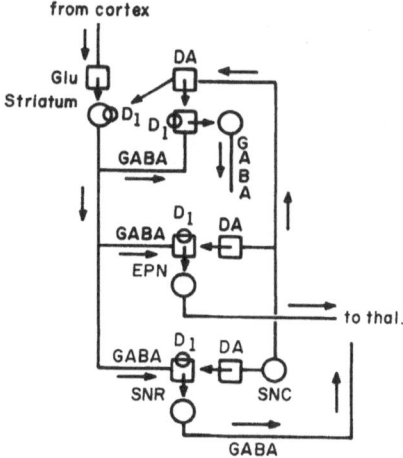

Figure 4. Diagram of the striatonigral and nigrostriatal pathways. In the first, recurrent collaterals and collaterals to the entopeduncular nucleus (EPN) are shown. In the second, collaterals to this nucleus are also shown. The dopaminergic input to the pars reticulata (SNR) is via the dendrites from the pars compacta (SNC). For clarity, only the thalamic projection from the pars reticulata and from the entopeduncular nucleus is shown. Abbreviations: DA, dopamine; D_1, D_1 DA receptor; Glu, glutamic acid; thal., thalamus.

dopaminergic influence would result in a decrease in the expression of mRNAs encoding the D_1 receptor (Gerfen et al., 1990) and presumably in the expression of mRNAs encoding GAD65 (Gonzales et al., 1991).

As the present results show, the reduced expression of D_1 receptors in the soma of the neostriatal neuron (Gerfen et al., 1990) results in a drastic reduction in the number of D_1 receptors present in the terminals of the neuron in the pars reticulata. This reduction in the number of D_1 receptors could explain why the activation of these receptors in the pars reticulata by supramaximal concentrations of dopamine (Fig. 3) only partially recovered the reduced synthesis of GABA produced by the dopaminergic denervation. If the number of receptors is less, the maximal response should also be less. However, in recent experiments we have found that the dopaminergic denervation potentiated the effect of dopamine mediated by D_1 receptors on the the release of GABA from slices of the pars reticulata of rat substantia nigra (intact slices: $+14 \pm 1\%$, n=32; lesioned slices: $+41 \pm 2\%$, n=48; measurements were made in the presence of 100 µM dopamine in both conditions). How to explain the discrepant effect of dopamine on the synthesis of GABA as compared to its effect on the release of the aminoacid in dopamine-denervated slices? Furthermore, how to explain the potentiation of the D_1-mediated effect of dopamine on GABA release in the presence of a marked reduction in the number of D_1 receptors? Missale et al. (1989) found that the dopaminergic depletion produced by reserpine resulted in a marked increase in the responsiveness of striatal adenylate cyclase to D_1 receptor stimulation while no significant changes were found in the number of D_1 receptors as estimated by specific binding techniques. These authors found that the potentiation of the activation of adenylate cyclase produced by the dopaminergic denervation was due to a marked enhancement of the coupling efficiency of Gs protein. This enhancement of the coupling efficiency might also explain the potentiation of the effect of dopamine on GABA release even in the presence of significantly less number of D_1 receptors. But why, then, there was no potentiation of the effect of dopamine on GABA synthesis? Possibly because in this case there was not only a decrease in the number of D_1 receptors but also in the number of GAD enzymes. In this case, even in the presence of enhanced coupling efficiency of Gs protein the effect of dopamine would be less.

Even though the loss of the dopaminergic input to the neostriatum markedly decreased the number of D_1 receptors in striatonigral terminals (Fig. 3), the effect on the number of D_1 receptors present in the neostriatum is very small (unpublished results; Adams et al., 1991) or nonexistent (Savasta et al., 1988). The reason for the difference is unknown.

It is interesting that even though the dopaminergic denervation produced a marked reduction in the synthesis of GABA it did not modify the basal GABA content. By decreasing release, the reduction of the activity of the nigrostriatal pathway associated with the loss of the dopaminergic influence may be the mechanism by which basal GABA content is maintained within normal values.

Acknowledgements

This work was supported by a grant (0586N-9108 from CONACYT (México).

REFERENCES

Aceves, J., Floran, B., Martinez-Fong, D., Benitez, J., Sierra, A., and Flores, G., 1992, Activation of D_1 receptors stimulates accumulation of γ-aminobutyric acid in slices of the pars reticulata of 6-hydroxydopamine-lesioned rats, *Neurosci. Lett.* 145:40-42.

Adams, C.F., Stromberg, I., Van Horne, C., Hoffer, B.J., and Boyson, S.J., 1991, Effect of human fetal ventral mesencephalic xenografts on D_1 and D_2 receptor binding in the unilaterally 6-hydroxydopamine(6-OHDA)-lesioned rat striatum, *Soc. Neurosci. Abst.* 307.3.

Barone, P., Tucci, I., Parashos, S.A., and Chase, N., 1987, D-1 dopamine receptors changes after striatal quinolinic acid lesion, *Eur. J. Pharmacol.* 138:141-145.

Bernasconi, R., Maitre, L., Martin, P., and Raschdorf, F., 1982, The use of inhibitors of GABA transaminase for the determination of GABA turnover in mouse brain regions: an evaluation of aminooxyacetic acid and gabaculline. *J. Neurochem.* 38:57-66.

Bolam, J.P., Powell, J.F., Wu, J.-Y., and Smith, A.D., 1985, Glutamate decarboxylase-immunoreactive structures in the rat neostriatum: a correlated light and electron microscopic study including a combination of Golgi impregnation with immunocytochemistry, *J. Comp. Neurol.* 237:1-20.

Erlander, M.G., Tillakaratne, N.J.K., Feldblum, S., Patel, N., Tobin, A.J., 1991, Two genes encoding distinct glutamate decarboxylases, *Neuron* 7:91-100.

Floran, B., Aceves, J., Sierra, A., and Martinez-Fong, D., 1990, Activation of D_1 dopamine receptor stimulates the release of GABA in the basal ganglia of the rat, *Neurosci. Lett.*116:136-140.

Gerfen, Ch.R., Engber, T.M., Mahan, L.C., Susel, Z., Chase, T.N., and Monsma, F.J., 1990, D_1 and D_2 dopamine receptor-regulated gene expression of striatonigral and striatopallidal neurons, *Science* 250:1429-1432.

Góngora, J.L., Sierra, A., Mariscal, S., and Aceves, J., 1988, Physostigmine stimulates phosphoinositide breakdown in the rat striatum, *Eur. J. Pharmacol.* 155:49-55.

Gonzales, C., Kaufman, D.L., Tobin, A.J., and Chesselet, M.-F., 1991, Distribution of glutamic aciddecarboxylase (Mr 67 000) in the basal ganglia of the rat: an immunohistochemical study with a selective cDNA-generated polyclonal antibody, *J. Neurocytol.* 20:953-961.

Itoh, M., 1983, Effect of haloperidol on glutamate decarboxylase activity in discrete brain regions, *Psychopharmacology* 79:169-172.

Itoh, M., and Uchimura, H., 1981, Regional differences in cofactor saturation of glutamate decarboxylase (GAD) in discrete brain nuclei of the rat, *Neurochem. Res.* 6:1283-1289.

Javoy-Agid, F., Ploska, A., Agid, Y., 1981, Microtopography of tyrosine hydroxylase, glutamic acid decarboxylase, and choline acetyltransferase in the substantia nigra and ventral tegmental area of control and parkinsonian patiens, *J. Neurochem.* 37:1218-1227.

Kaufman, D.L., Houser, C.R., and Tobin, A.J., 1991, Two forms of the γ-aminobutyric acid synthetic enzyme glutamate decarboxylase have distinct intraneuronal distributions and cofactor interactions, *J. Neurochem.* 56:720-723.

Kita, H., and Kitai, S.T., 1988, Glutamate decarboxylase immunoreactive neurons in rat neostriatum: their morphological types and populations, *Brain Res.* 447:346-352.

Lloyd, K.G., and Hornykiewicz, O., 1973, L-glutamic acid decarboxylase in Parkinson's disease: effect of L-Dopa therapy, *Nature* 243:521-523.

McGeer, P.L., and McGeer, E.G., 1976, Enzymes associated with metabolism of catecholamines, acetylcholine and GABA in human controls and patients with Parkinson's disease and Huntington's chorea, *J. Neurochem.* 26: 65-76.

McGeer, P.L., McGeer, E.G., Wada, J.A., and Jung, E., 1971, Effects of globus pallidus lesions and Parkinson's disease on brain glutamic acid decarboxylase, *Brain Res.* 32:425-431.

Miller, L.P., Martin, D.L., Mazumder, A., and Walters, J.R., 1978, Studies on the regulation ofGABA synthesis: substrate-promoted dissociation of pyridoxal-5'-phosphate from GAD, *J. Neurochem.* 30:361-369.

Miller, L.P., Walters, J.R., and Martin, D.L., 1977, Post-mortem changes implicate adenine nucleotides and pyridoxal-5'-phosphate in regulation of brain glutamate decarboxylase, *Nature* 266:847-848.

Missale, C., Nisoli, E., Liberini, P., Rizzonelly, P., Memo, M., Buonamici, M., Rossi, A., and Spano, P.F., 1989, Repeated reserpine administration up-regulates the transduction mechanisms of D_1 receptors without changing the density of [^{3}H]SCH23390 binding, *Brain Res.* 483:117-122.

Perry, T.L., Javoy-Agid, F., Agid, Y., and Fibiger, H.C., 1983, Striatal GABAergic activity is not reduced in Parkinson's disease, *J. Neurochem.* 40:1120-1123.

Savasta, M., Dubois, A., Benavides, J., and Scatton, B., 1988, Different plasticity changes in D1 and D2 receptors in rat striatal subregions following impairement of dopaminergic transmission, *Neurosci. Lett.* 85:119-124.

Segovia, J., and Garcia-Muñoz, M., 1987, Changes in the activity of GAD in the basal ganglia of the rat after striatal dopaminergic denervation, *Neuropharmacology* 26:1449-1451.

Segovia, J., Armstrong, D.M., Benzing, W.C., and Hornby, P.J., 1991, Striatal glutamic acid immunoreactivity is increased after dopaminergic deafferentation: densitometric analysis, *Neurosci. Lett.* 122:252-256.

Segovia, J., Meloni, R., and Gale, K., 1989, Effect of dopaminergic denervation and trasplant-derived reinnervation on a marker of striatal GABAergic function, *Brain Res.* 493:185-189.

Segovia, J., Tillakaratne, N.J.K., Whelan, K., Tobin, A.J., and Gale, K., 1990, Parallel increases in striatal glutamic decarboxylase activity and GAD mRNA levels in rats with lesions of the nigrostriatal pathway, *Brain Res.* 529:345-348.

Vernier, P., Julien, J-F., Rataboul, P., Fourrier, O., Feuerstein, and C., Mallet, J., 1988, Similar time course changes in striatal levels of glutamic acid decarboxylase and proenkephalin mRNA following dopaminergic deafferentation in the rat, *J. Neurochem.* 51:1375-1380.

Vincent, S.R., Nagy, J.I., and Fibiger, H.C., 1978, Increased striatal glutamate decarboxylase after lesions of the nigrostriatal pathway, *Brain* Res. 143:168-173.

CHOLINERGIC MODULATION OF GABAERGIC
EFFERENT STRIATAL NEURONS

Mihalis Mavridis, Nathalie Kayadjanian and Marie Jo Besson

Laboratoire de Neurochimie Anatomie, IDN, CNRS URA 1488
Université P.et M. Curie, 9 Quai St Bernard
75005 Paris

INTRODUCTION

From clinical and experimental data it has been postulated that the physiology of basal ganglia involves an equilibrium between dopamine (DA) and acetylcholine (ACh) in the modulation these two neurotransmitters exert. The existence of a balance between a global inhibitory effect of DA and a global excitatory effect of ACh has been proposed (Hornykiewicz, 1971) and has constituted a basis for antimuscarinic drug medication of Parkinson's disease and neuroleptic-induced extrapyramidal side effects (Adams and Victor, 1989). This interaction has been extensively studied in the neostriatum, the major receptive component of the basal ganglia, which contains high levels of DA and ACh. The effect of DA on efferent striatal neurons has been thoroughly analyzed and the results obtained support an opposing action of DA on the two main striatal output pathways: the striatopallidal and the striatonigral or entopeduncular pathways (Wooten and Collins, 1983; Young et al., 1986). The effect of ACh on striatal neurons is however less well understood. In this chapter we will consider the organization of cholinergic innervation of the striatum and of one of its output structures, the substantia nigra (SN) as well as the role of ACh in the regulation of neuronal activity in these two basal ganglia structures.

THE CHOLINERGIC INNERVATION OF THE STRIATUM

The cholinergic system of the neostriatum is mainly intrinsic (Fibiger, 1982) and consists of large aspiny interneurons with a positive choline acetyltransferase (ChAT) immunoreactivity (Bolam et al., 1984). They represent approximately 1 to 2 % of the total neuronal population of the striatum (Graybiel et al., 1986). Considering the heterogeneity of the striatum, cholinergic interneurons reside in both striatal compartments: the striosomes (patches) and the matrix (Mesulam et al., 1984; Mesulam et al., 1992). The ChAT positive neurons are mainly found at the boundaries of striosomes, but they arborize predominantly in the matrix (Graybiel et al., 1986). They are the target of afferent projections, in particular they receive direct dopaminergic inputs (Kubota et al., 1987; Chang, 1988). Although it has been suggested from various approaches that ACh neurons are under a direct glutamatergic control from the cerebral cortex, a recent study by Lapper and Bolam (1992) shows only appositions between corticostriatal terminals and ACh neurons. This study also indicates that the glutamatergic innervation of ACh neurons could have a thalamic origin. The ACh neurons are also contacted by collaterals of medium-sized spiny neurons which correspond to efferent striatal neurons (Bolam et al., 1986) and intrinsic interneurons such as those containing neuropeptide Y (NPY) (Vuillet et al., 1992). On the other hand, cholinergic terminals establish synaptic contacts with medium-sized spiny efferent neurons (Izzo and Bolam, 1988; Smith and Bolam, 1990), NPY containing interneurons (Vuillet et al., 1992) and make tight appositions with tyrosine hydroxylase positive terminals of nigrostriatal neurons (Chang, 1988). Cholinergic neurons have a high sensitivity to small depolarizing potentials and a tonic firing, properties which place them, despite their small number, in an excellent position to act as modulators of the activity of neostriatal efferent neurons and interneurons (Wilson et al., 1990).

Concerning the distribution of cholinergic receptors, the neostriatum contains one of the highest density of muscarinic receptors. Among the pharmacologically defined receptors, there is a high density of the M_1 subtype (Wamsley et al., 1984) characterized by its high affinity for the antagonist pirenzepine (Hammer et al., 1980). A much lower density has been reported for the M_2 subtype (Regenold et al., 1989; Wang et al., 1989) labelled by ligands such as AF-DX 116 which is an efficient muscarinic antagonist at cardiac M_2 receptors (Hammer et al., 1986). Recently, five muscarinic receptor genes encoding distinct receptor proteins (m1 to m5) have been cloned (Kubo et al., 1986; Peralta et al., 1986; Bonner et al., 1987, 1988). These genetically defined receptors are not yet well characterized by the available muscarinic agonists and antagonists. The pharmacological correspondence is roughly the following: M_1 = m1, m4 and m5, M_2 = m2 and M_3 = m3 (Dörje et al., 1991). In the striatum, three muscarinic receptor mRNAs (m1, m2 and m4) are expressed (Weiner et al., 1990) and their relative levels of expression fit well with the corresponding receptor protein distribution determined by immunohistochemistry with specific antibodies (Levey et al., 1991). The muscarinic receptor mRNAs expressed by striatal neurons display specific patterns of localization (Bernard et al., 1992). The m1 receptor mRNA is found in most striatal neuronal populations including all the efferent neurons and some categories of interneurons. The m4 receptor gene is also expressed in most neuronal populations, yet its distribution differs according to subpopulations of efferent neurons: it is expressed in almost all striatonigral neurons but in only 30% of striatopallidal neurons. The m2 receptor mRNA is exclusively expressed in cholinergic interneurons which also contain m1 and m4 receptor mRNAs.

Nicotinic receptors in addition have been detected in the striatum by autoradiography of ^3H-nicotine binding sites (Clarke et al., 1985) and by immunohistochemical detection of the receptor (Swanson et al., 1987). A relatively high density of nicotinic receptors has been revealed with these two approaches. The molecular cloning of the nicotinic receptor indicates the existence of a large variety of α and β subunits which constitute the neuronal nicotinic receptor. Within the striatum, only the α3 and β2 subunit mRNAs are expressed but at a low level (Wada et al., 1989). Thus, the high density of ^3H-nicotine receptors detected by autoradiography and immunodetection corresponds likely to receptors localized on afferences. In fact, a decrease of ^3H-nicotine binding has been detected after a lesion of nigral DA neurons (Clarke and Pert, 1985) as well as after a lesion of serotoninergic terminals (Schwartz et al., 1984) suggesting a localization of nicotinic receptors at least on these afferences.

INTERACTIONS ACH/DA IN THE STRIATUM

During the development of the neostriatum an overlapping compartmentalization of ACh and DA-containing elements has been observed in different mammalian species (Nastuk and Graybiel, 1985). This intimate relationship, reflected by the correlation between the topographical distribution of various DA and ACh markers (receptors, uptake sites) (Joyce, 1991), persists to some degree in the adulthood and provides an anatomical basis for the complex interactions between these two neuromodulators in the neostriatum.

The DA regulation of the striatal cholinergic activity has recently been questioned, both in terms of direction (inhibitory or excitatory) and of DA receptor subtypes involved (D_2 or D_1). The prevailing opinion, based on original *in vivo* experiments (Stadler et al., 1973; Guyenet et al., 1975), is that DA, released from terminals of nigrostriatal neurons, tonically inhibits the activity of striatal ACh interneurons (Lehmann and Langer, 1983). In line with this theory, *in vitro* experiments on striatal slices repeatedly demonstrated, both in the neostriatum and in the nucleus accumbens, that ACh release is inhibited by endogenous DA, apomorphine or D_2-DA agonists and potentiated by the blockade of D_2 receptors (Stoof et al., 1982, 1987). On the other hand, these *in vitro* experiments failed to show a clear role for the DA-D_1 receptors in the modulation of striatal ACh efflux (Wedzony et al., 1988; Tedford et al., 1992). Thus, the inhibitory influence of DA is thought to represent a direct effect mediated by D_2 receptors located on striatal ACh interneurons (Le Moine et al., 1990; Weiner et al., 1990). Recent *in vivo* microdialysis studies confirmed this direct inhibitory action, since systemically administered or locally infused D_2 agonists and antagonists decrease and increase, respectively, the striatal release of ACh (Damsma et al., 1991; Bertorelli et al., 1992; De Boer et al., 1992).

These same *in vivo* studies have revealed in addition, an opposite (i.e. facilitatory) role for DA acting on D_1 receptors in the modulation of striatal cholinergic activity (Bertorelli and Consolo, 1990; Damsma et al., 1990; Bertorelli et al., 1992). This regulation is only observed when the D_1 agonists or antagonists are given systemically but not when applied locally in the striatum (Damsma et al., 1991; De Boer et al., 1992, but see Consolo et al., 1992). It seems therefore that the control of striatal cholinergic activity through the D_1 receptor represents an indirect, polysynaptic effect implicating possibly cortical or thalamic glutamatergic striatal afferents (Damsma et al., 1991; De Boer et al., 1992). However, these studies indicate that the predominant *in vivo* effect of endogenous DA on striatal ACh neurons is the excitatory one (i.e the D_1-mediated). Thus, an increase in DA transmission produced by systemic administration of amphetamine, cocaine or nomifensine enhances striatal ACh release (Damsma et al., 1991; Consolo et al., 1992), whereas the interruption of the mesotelencephalic DA projection or a pharmacological depletion of DA (by reserpine or α-methyl-p-tyrosine) drastically reduces striatal ACh overflow (Bertorelli et al., 1992; Robertson et al., 1992).

Behavioural and biochemical studies, however, support the idea of a tonic DA inhibitory control on ACh striatal neurons. Thus, motor effects evoked by a stimulation or a blockade of neostriatal DA receptors are modulated by muscarinic drugs in a direction compatible with an inhibitory control exerted by DA on ACh neurons. For instance, the circling behaviour evoked by a systemic administration of a direct or an indirect DA agonist in rats with a unilateral destruction of the DA nigrostriatal pathway is increased by atropine and inhibited by pilocarpine, centrally active muscarinic antagonist and agonist, respectively (Kaakkola, 1981). Furthermore, the muscarinic antagonist scopolamine reduces the catalepsy induced in rats by D_1 or D_2 receptor antagonists (Ögren and Fuxe, 1988). On the other hand, chronic treatment with D_2 agonists increases muscarinic receptor density in the neostriatum, presumably as a result of reduced ACh release by cholinergic interneurons (Majocha and Baldessarini, 1984). An opposite mechanism seems to operate in the down regulation of striatal muscarinic receptors observed after an unilateral lesion of the mesostriatal DA pathway (Joyce, 1991) or a prolonged treatment with D_2 antagonists (Boyson et al., 1988).

The same complexity holds for the cholinergic regulation of striatal DA transmission. Original *in vitro* experiments on striatal slices indicated that endogenous ACh, as well as muscarinic and nicotinic agonists enhance the DA release in the striatum (Giorguieff et al., 1977; Lehmann and Langer, 1982). The existence of muscarinic and nicotinic receptors located presynaptically on striatal DA terminals was proposed because this facilitatory effect was insensitive to tetrodotoxin (Giorguieff-Chesselet et al., 1979; Raiteri et al., 1982). For the muscarinic-mediated modulation, the involvement of the M_1 (Raiteri et al., 1984) or the M_2 (Schoffelmeer et al., 1986) receptor subtype was suggested. In fact the DA neurons express the m5 receptor subtype (Weiner et al., 1990) which is not well differentiated neither by M_1 nor by M_2 antagonists (Dörje et al., 1991). It is also known now that DA neurons express specific subunits of the neuronal nicotinic receptor (Wada et al., 1989).

When the cholinergic control of DA release was further analyzed in the two neurochemical compartments of the striatum (striosomes and matrix) a much more complex pattern of modulation emerged. In the striosomes, the activation of striatal DA release through a direct stimulation of presynaptic muscarinic receptor is the prevailing mechanism. In the matrix, ACh can influence DA release directly and indirectly via polysynaptic circuits. Both muscarinic and nicotinic receptors are involved in the direct facilitatory effect. Opposing modulations, implicating muscarinic and nicotinic receptors have been described for the indirect action (Kemel et al., 1989, 1992). The global cholinergic stimulatory effect on striatal DA release reflects actually the concurrent operation of a local monosynaptic and various polysynaptic circuits.

INTERACTIONS ACH/GABA IN THE STRIATUM

Despite anatomical evidence showing a cholinergic input on cell bodies and proximal dendrites of GABA projection neurons in the rat striatum (Smith and Bolam, 1990), few studies have focused their interest on the modulation of the activity of striatal GABA neurons by ACh. A direct approach to test whether cholinergic drugs interact with GABA neurons is to study their effect on the release of GABA. Another approach is to study the regulation of glutamic acid decarboxylase (GAD the biosynthetic enzyme of GABA) through its activity or the changes in expression of mRNAs encoding the enzyme.

Cholinergic drugs can modulate the release of GABA in the striatum. However, opposite effects were described when considering the basal release of GABA or the release evoked by a depolarization produced by a high K^+ concentration. An increase of the basal GABA release produced by ACh or cholinergic drugs was reported in in vivo experiments in the rat (Van der Heyden et al., 1980; Girault et al., 1986) and in the cat (Besson et al., 1982) and in vitro in the caudate nucleus of the rabbit (Limberger et al., 1986). This latter study reported the involvement of nicotinic receptors and a persistence of the effect in the presence of tetrodotoxin, arguing in favor of a direct nicotinic action on GABA neurons. By contrast, the depolarization-evoked release of ^3H-GABA measured in rat striatal slices was reported to be inhibited by ACh (Marchi et al., 1990; Sugita et al., 1991) through an interaction with muscarinic receptors.

The cholinergic regulation of striatal GABA neurons can also be examined by measuring variations of GAD mRNA levels following treatment with cholinergic drugs. It should be noted that most GABA neurons express two isoforms of GAD, the GAD_{67} and GAD_{65} encoded by distinct genes and that the longer isoform is preferentially regulated in basal ganglia neurons (Soghomonian et al., 1992). The regulation of GAD_{67} mRNA was investigated by in situ hybridization histochemistry (ISHH) in rats subjected to a chronic treatment with an antagonist acting on M_1 receptors, trihexyphenidyl (TRI; Artane^R). Because this drug is commonly used in the treatment of Parkinson's disease, the influence of chronic TRI administration on striatal GAD_{67} mRNA levels was investigated also in the rat model of unilateral parkinsonism (injection of 6-OHDA in the SN pars compacta). Moreover, with this paradigm we can get an insight into the functional balance between ACh and DA in the regulation of the activity of striatal efferents GABA neurons. In an attempt to differentiate the effect of TRI on striatal efferent neurons, variations in the levels of striatal preproenkephalin (PPE) mRNA, encoding the precursor for enkephalin, were measured in parallel. In fact PPE mRNA is coexpressed with GAD mRNAs predominantly in striatopallidal neurons (Gerfen et al., 1990) and the cholinergic control on this neuronal population could be examined.

Chronic TRI administration in sham-lesioned rats led to a small (18 %) but significant increase in striatal GAD_{67} mRNA expression, while PPE mRNA levels were in the control range value (Fig. 1). In agreement with previous reports, the unilateral striatal DA deafferentation resulted in a significant increase in PPE mRNA (Young et al., 1986; Gerfen et al., 1991) and GAD_{67} expression (Vernier et al., 1988; Soghomonian et al., 1992) in the ipsilateral vs the contralateral striatum. However, only the ipsilateral rise in PPE mRNA levels was significant when compared to values from control (sham-lesioned) rats. The chronic TRI administration in 6-OHDA injected rats **did not** counterbalance the effects of the lesion neither on PPE nor on GAD_{67} mRNAs expression. In fact, in the DA-denervated striatum the TRI treatment produced a further increase in both PPE and GAD_{67} mRNA levels. These increases were significantly different when compared with values from both TRI-treated sham-lesioned rats and those from the denervated striatum of vehicle-treated 6-OHDA injected rats. Hence, after chronic TRI administration, the ipsilateral vs contralateral differences observed in lesioned rats persisted for GAD_{67} and was amplified for PPE mRNA. These effects found after chronic TRI administration on PPE mRNA expression in 6-OHDA injected rats contrast with those reported after chronic scopolamine treatment which attenuates the lesion-induced increase in striatal PPE mRNA expression examined by dot-blot hybridization (Pollack and Wooten, 1992). This discrepancy resides probably on methodological differences between the two studies: the scopolamine treatment was instituted the day following the 6-OHDA lesion, lasted one week and, importantly, was intermittent. The difference in the drug schedule has already been shown to play an important role in the regulation of striatal efferent neurons by dopaminergic agonists (Gerfen et al., 1990).

STRIATAL GAD$_{67}$ mRNA

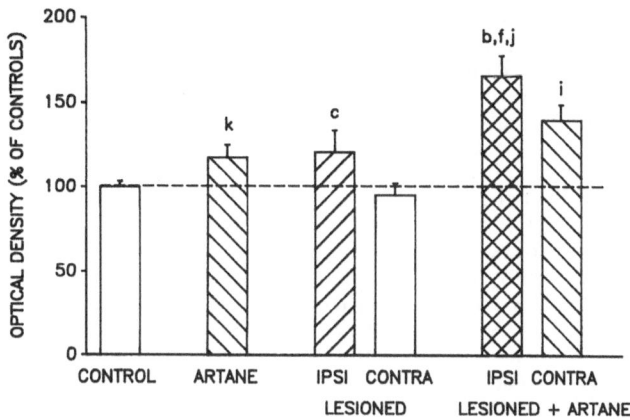

STRIATAL PPE mRNA

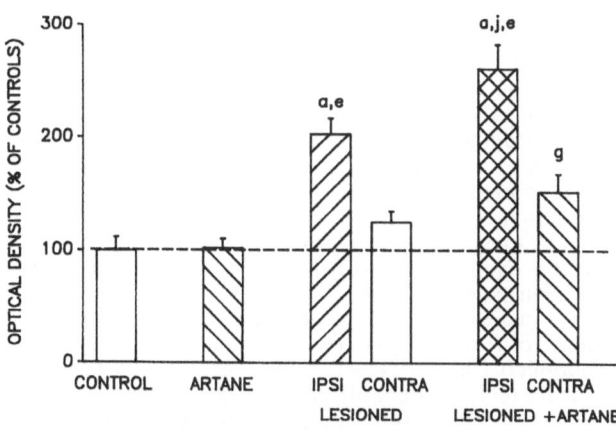

Figure 1. Effects of a muscarinic antagonist, trihexyphenidyl (Artane), on the expression of GAD$_{67}$ and PPE mRNA in sham-operated (control) and 6-OHDA lesioned rats. Rats were pretreated with desipramine (25mg/kg i.p., 30 min before) to protect the ascending NA pathway from the 6-OHDA lesion. Half of the rats were injected with 6-OHDA (8μg in 4μl) into the left SN pars compacta while the others were subjected to a sham-lesioning procedure. The efficacy of the lesion was tested two weeks later by examining the rotational behaviour to a low dose (0,05 mg/kg s.c.) of apomorphine (administered also to sham-lesioned rats). On the same day, an osmotic minipump (Alzet 2ML2) was subcutaneously implanted. The pump was filled with Artane (TRI, daily delivery: 5 mg/kg) for half of the 6-OHDA injected and sham-lesioned rats and with vehicle for the other half. Animals were sacrificed 15 days later. The expression of mRNAs encoding PPE and GAD$_{67}$ was assayed on neostriatal sections by ISHH using oligonucleotide probes labelled with ^{35}S-dATP. Sections were apposed to x-ray films for 8-10 days and the optical densities of the developed images were quantified by densitometric analysis to measure changes in mRNA levels. For sham-lesioned rats, treated either with vehicle or with Artane, values from the left and right striatum were pooled since the sham operation did not produce *per se* any significant ipsi vs contralateral difference for PPE or GAD$_{67}$ mRNA levels. The paired t-test was applied for comparisons between the two sides in the same group of lesioned rats [a: P<0.001, b: P<0.01, c: P<0.05]. The unpaired t-test was applied for comparisons 1) between either side in lesioned groups and their corresponding unlesioned group [e: P<0.001, f: P<0.01, g: P<0.05] 2) between corresponding striata in vehicle- and Artane-treated lesioned rats [i: P<0.01, j: P<0.05] 3) between Artane and control group [k: P<0.05].

The unilateral 6-OHDA injection did also produce a small **contralateral** effect, an increase in PPE and a decrease in GAD_{67} mRNA expression, compared to control values. Thus, the difference in GAD_{67} mRNA levels between the ipsilateral and the contralateral striatum was significant partly because of this contralateral effect. On the other hand, the chronic administration of TRI produced, in the intact side of lesioned rats, increases of both GAD_{67} and PPE mRNA levels which were significant vs control values. Furthermore, the increase of PPE mRNA levels was significant when compared to values obtained in sham-lesioned TRI-treated rats. Such bilateral effects following unilateral manipulations of basal ganglia nuclei, and particularly a lesion of the nigrostriatal pathway are commonly reported in behavioural (McKenzie et al., 1991), in *in vivo* release (Robinson and Whishaw, 1988) and recently, in ISHH studies (Cadet et al., 1992) making mandatory, in unilateraly lesioned animals, the comparison of both sides with a proper control group (Robinson, 1991).

Because the drug was systemically administered, it cannot be excluded that the effects of TRI on striatal GAD_{67} and PPE mRNA levels, in sham-lesioned as well as on the contralateral side of 6-OHDA-injected rats, are partly mediated through an interaction of the drug with extrastriatal muscarinic receptors. In fact, the increase in striatal PPE and GAD_{67} mRNA expression can be explained by a blockade of muscarinic M1 receptors at the level of SN pars compacta DA neurons (see below) resulting in a decrease in striatal DA transmission (Góngora-Alfaro et al., 1991).

INTERACTIONS ACH/GABA IN THE SUBSTANTIA NIGRA: CHOLINERGIC MODULATION OF THE SPONTANEOUS GABA RELEASE

The presence of cholinergic terminals and somata in the SN has been well established in the rat (Gould and Butcher, 1985; Martinez-Murillo et al., 1989a). Electrophysiological studies and neuroanatomical tracing experiments indicate that the pedonculopontine tegmental nucleus (PPN) is the probable source of cholinergic afferents to the SN (Beninato and Spencer, 1987; Clarke et al., 1987; Gould et al., 1989), although the existence of a glutamatergic projection from the PPN to the SN has been also shown (Scarnati et al., 1986; Clements and Grant, 1990).

Most studies concerning the effect of ACh on the activity of the nigral neurons have been centered on the action of ACh on DA neurons. Biochemical (Javoy et al., 1974) and behavioural studies (James and Massey, 1978; Parker et al., 1991) indicate that the action of ACh in the SN is conveyed by an ACh-DA link which is substantiated by recent ultrastructural data (Martinez-Murillo et al., 1989b).

Initial biochemical studies showing a nicotine-evoked DA release in the striatum which persisted in the presence of tetrodotoxin have suggested the existence of nicotinic receptors on DA terminals (Giorguieff-Chesselet et al., 1979). These results were further corroborated by electrophysiological studies (Grenhoff et al., 1986) indicating an activation of DA neurons by nicotine and by biochemical studies showing a decrease (30%) of nicotinic receptors in the striatum after a nigral 6-OHDA lesion (Clarke and Pert, 1985; Deutch et al., 1987). More recently the ISHH technique has allowed the detection of mRNAs encoding the α4 and β2 subunits of the neuronal nicotinic receptor in the SN pars compacta neurons as well as in neurons of the pars reticulata (Wada et al., 1989).

Concerning the other cholinergic receptor family, the pharmacologically defined M_1 and M_2 muscarinic receptors are present in both pars reticulata and compacta of the SN, as indicated by binding studies using specific ligands (Cortes and Palacios, 1986; Nastuk and Graybiel, 1991). However, with the ISHH detection, a single mRNA subtype, the m5, (pharmacologically close to the M_1 receptor) was found to be expressed in SN neurons and only in the pars compacta DA neurons (Weiner et al., 1990). Thus, part of the muscarinic receptors detected in the pars reticulata by radioligand binding could be localized on dendrites of DA neurons. However, the abundant GABA terminals from striatonigral neurons which express m1 and m4 muscarinic subtype mRNAs (Bernard et al., 1992) may significantly contribute to the amount of muscarinic binding sites found in the SN pars reticulata.

To investigate a possible interaction between ACh and these GABA terminals the cholinergic modulation of the spontaneous release of ^{3}H-GABA was analyzed *in vitro* on rat SN slices.

The superfusion of ACh (5×10^{-5} M and 5×10^{-4} M) in the presence of 5×10^{-5} M eserine (an inhibitor of acetylcholinesterase which by itself had no effect) increased the spontaneous release of ^{3}H-GABA by +13% and +18%, respectively. The pharmacological characterization of the ACh-enhanced ^{3}H-GABA release was examined by analyzing the effects of muscarinic and nicotinic agonists and antagonists. Carbachol (5×10^{-4} M), a muscarinic agonist induced a 10% increase of ^{3}H-GABA release (Fig. 2). This effect was Ca^{2+}-dependent since the carbachol-induced effect was augmented by 19% when the external Ca^{2+} concentration was increased from 1.3 mM to 2.4 mM Ca^{2+} and totally abolished in the presence of 0.4 mM Ca^{2+}. The carbachol-induced ^{3}H-GABA release was completely antagonized by atropine (10^{-6} M) indicating that this effect involves the stimulation of muscarinic receptors (Fig. 2). Furthermore, the carbachol-induced effect was partially abolished by 10^{-4} M pirenzepine, a M_1 receptor antagonist and completely inhibited by AF-DX 384MS (10^{-6} M), a M_2 antagonist. The involvement of the genetically defined muscarinic m4 receptor in this response could be possible since m4 receptors have a high affinity for pirenzepine and for AF-DX 384MS (Miller et al., 1991).

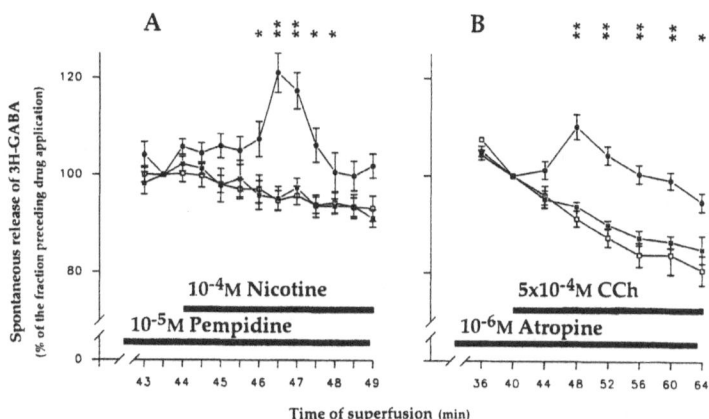

Figure 2. Effects of nicotine and carbachol on the spontaneous release of ^{3}H-GABA in the rat substantia nigra. SN slices, dissected from vibratome-cut coronal sections (300mm), were incubated for 20 min in 200 nM ^{3}H-GABA. They were then introduced in superfusion chambers and continuously superfused with a Krebs' medium (containing 2.4 mM Ca^{2+}) saturated with a mixture of 95% O_2 - 5% CO_2. After 32 min of superfusion, fractions were collected every 30 seconds (for the examination of the nicotinic effect) (A) or every 4 min (for the muscarinic effect) (B). (-) Nicotine (10^{-4} M) or carbachol (CCh) (5×10^{-4} M) was added to the superfusion medium 44 min after the beginning of the superfusion and the antagonists pempidine (10^{-5} M, A) or atropine (10^{-6} M, B) 8 min before the agonist addition. The amount of ^{3}H-GABA released was expressed in % of the fractional release measured in the fraction preceding the agonist application nicotine (-- •-- in A) and CCh (--•-- in B) or in the corresponding fraction in antagonist-treated slices (controls) and in agonist-treated slices in the presence of the antagonist. Note the different time-scale in the nicotine- and the CCh-induced effects on the ^{3}H-GABA release. Comparisons between corresponding fractions in agonist-treated and control slices (--□-- pempidine in A or atropine in B) or agonist-treated slices in the presence of the antagonist (--■--) were established by the Student's t test. * P<0.05, ** P<0.01 when compared to control slices or to agonist-treated slices in the presence of the antagonist pempidine or atropine.

The superfusion of nicotine (10^{-4} M) produced a transient increase of the ^{3}H-GABA release reaching a value of 20% at the maximun of the effect (Fig. 2). The nicotine-induced effect was totally abolished when the Ca^{2+} concentration in the superfusion medium was lowered from 2.4 mM to 0.4 mM Ca^{2+}. The nicotine-induced release of ^{3}H-GABA was totally inhibited in the presence of pempidine (10^{-5} M), a nicotinic antagonist.

It is noteworthy that the time-course of the nicotine-induced effect is very different from the carbachol-induced effect. The former was transient lasting no more than 2 min whereas the carbachol-induced ^{3}H-GABA release was building up progressively reaching a maximum value 8 min after the drug addition and faded out progressively. The transient nicotinic effect probably reflects a desensitization process well described in the literature related to the ligand-gated channel receptors and contrasts with the more prolonged increase of ^{3}H-GABA release produced by stimulation of muscarinic receptors. In fact, receptors which belong to the family of G protein-coupled receptors are less susceptible to fast desensitization.

Thus, in the SN of the rat the spontaneous ^{3}H-GABA release can be modulated through the activation of muscarinic and nicotinic receptors. Although both types of receptors are present in the SN, their localization on GABA terminals of striatonigral neurons is still uncertain and remains to be elucidated. It is known however, that DA neurons express mRNAs encoding both muscarinic and nicotinic receptors. Thus, changes in ^{3}H-GABA release observed in this study can be attributed to an indirect effect which involves DA neurons. Preliminary data indicate that in the presence of a DA-D$_1$ antagonist (SCH 23390) (10^{-6} M) the nicotine-induced effect is blocked whereas the muscarinic effect is not. These results suggest that the nicotine-induced effect is indirect and mediated by an action of nicotine on DA neurons. On the other hand the increase release of ^{3}H-GABA produced by carbachol could be a direct effect via an activation of muscarinic receptors localized on GABA terminals of striatonigral neurons. The DA-independence of the muscarinic effect on the GABA release does not exclude, however, an indirect effect via nigral elements other than DA neurons.

Acknowledgments

We thank M. Rogard for her skillfull assistance for ISSH and A. Ménetrey for her technical help in release experiments. The original studies reported here were supported by the Université P. and M. Curie, CNRS and the Human Frontiers Program.

REFERENCES

Adams, R. D., and Victor, M., 1989. "Principles of Neurology", 4th edition. McGraw-Hill, New York.

Beninato, M., and Spencer, R.F., 1987, A cholinergic projection to the rat substantia nigra from the pedunculopontine tegmental nucleus, *Brain Res.* 412:169-174

Bernard, V., Normand, E., and Bloch, B., 1992, Phenotypical characterization of the rat striatal neurons expressing muscarinic receptor genes, *J. Neurosci.* 12:3591-3600.

Bertorelli, R., and Consolo, S., 1990, D1 and D2 dopaminergic regulation of acetylcholine release from striata of freely moving rats, *J. Neurochem.* 54:2145-2148.

Bertorelli, R., Zambelli, M., Di Chiara, G., and Consolo, S., 1992, Dopamine depletion preferentially impairs D1- over D2-receptor regulation of striatal in vivo acetylcholine release, *J. Neurochem.* 59:353-357.

Besson, M.J., Kemel, M.L., Gauchy, C., and Glowinski, J., 1982, Bilateral asymmetrical changes in the nigral release of 3H-GABA induced by unilateral application of acetylcholine in the cat caudate nucleus, *Brain Res.* 241:241-248.

Bolam, J.P., Ingham, C.A., Izzo, P.N., Levey, A.I., Rye, D.B., Smith, A.D., and Wainer, B.H., 1986, Substance P-containing terminals in synaptic contact with cholinergic neurons in the neostriatum and basal forebrain: a double immunocytochemical study in the rat, *Brain Res.* 397:279-289.

Bolam, J.P., Wainer, B.H., and Smith, A.D., 1984, Characterization of cholinergic neurons in the rat neostriatum. A combination of choline acetyltransferase immunocytochemistry, Golgi-impregnation and electron microscopy, *Neuroscience* 12:711-718.

Bonner, T.I., Buckley, N.J., Young, A.C., and Brann, M.R., 1987, Identification of a family of muscarinic acetylcholine receptor genes, *Science* 237:527-532.

Bonner, T.I., Young, A.C., Brann, M.R., and Buckley, N.J., 1988, Cloning and expression of the human and rat m5 muscarinic acetylcholine receptor genes, *Neuron* 1:403-410.

Boyson, S. J., McGonigle, P., Luthin, G. R., Wolfe, B. B., and Molinoff, P. B., 1988, Effects of chronic administration of neuroleptic and anticholinergic agents on densities of D2 dopamine and muscarinic cholinergic receptors in rat striatum, *J. Pharmacol. Exp. Ther.* 244:987-993.

Cadet, J.L., Zhu, S.M., and Angulo, J.A., 1992, Quantitative in situ hybridization evidence for differential regulation of proenkephalin and dopamine D2 receptor mRNA levels in the rat striatum: effects of unilateral injections of 6-hydroxydopamine, *Mol. Brain Res.* 12:59-67.

Chang, H. T., 1988, Dopamine-acetylcholine interaction in the rat striatum: a dual-labeling immunocytochemical study, *Brain Res. Bull.* 21:295-304.

Clarke, P.B.S., and Pert, A., 1985, Autoradiographic evidence for nicotine receptors on nigrostriatal and mesolimbic dopaminergic neurons, *Brain Res.* 348:355-358.

Clarke, P.B.S., Hommer, D.W, Pert, A., and Skirboll, L.R., 1987, Innervation of substantia nigra neurons by cholinergic afferents from pedunculo-pontine nucleus in the rat: neuroanatomical and electrophysiological evidence, *Neuroscience* 23:1011-1019.

Clarke, P.B.S., Schwartz, R.D., Paul, S.M., Pert, C.B., and Pert, A., 1985, Nicotinic binding in rat brain: autoradiographic comparison of 3H acetylcholine, 3H nicotine, and 125I-α-bungarotoxin, *J. Neurosci.* 5:1307-1315.

Clements, J.R., and Grant, S., 1990, Glutamate-like immunoreactivity in neurons of the laterodorsal tegmental and pedunculopontine nuclei in the rat, *Neurosci. Lett.* 120:70-73.

Consolo, S., Girotti, P., Russi, G., and Di Chiara, G., 1992, Endogenous dopamine facilitates striatal in vivo acetylcholine release by acting on D1 receptors localized in the striatum, *J. Neurochem.* 59:1555-1557.

Cortes, R., and Palacios, J.M., 1986, Muscarinic cholinergic receptor subtypes in the rat brain. I. Quantitative autoradiographic studies, *Brain Res.* 362:227-238.

Damsma, G., Robertson, G. S., Tham, C.S., and Fibiger, H. C., 1991, Dopaminergic regulation of striatal acetylcholine release: importance of D1 and N-methyl-D-aspartate receptors, *J. Pharmacol. Exp. Ther.* 259:1064-1072.

Damsma, G., Tham, C.S., Robertson, G. S., and Fibiger, H. C., 1990, Dopamine D1 receptor stimulation increases striatal acetylcholine release in the rat, *Eur. J. Pharmacol.* 186:335-338.

De Boer, P., Damsma, G., Schram, Q., Stoof, J. C., Zaagsma, J., and Westerink, B. H. C., 1992, The effect of intrastriatal application of directly and indirectly acting dopamine agonists and antagonists on the in vivo release of acetylcholine measured by brain microdialysis: the importance of the post-surgery interval, *Naunyn-Schmiedeberg's Arch. Pharmacol.* 345:144-152.

Deutch, A.Y., Holliday, J., Roth, R.H., Chun, L.L.Y., and Hawrot, E., 1987, Immunohistochemical localization of a neuronal nicotinic acetylcholine receptor in mammalian brain. *Proc. Natl. Acad. Sci. USA* 84:8697-8701.

Dörje, F., Wess, J., Lambrecht, G., Tacke, R., Mutschler, E., and Brann, M. R., 1991, Antagonistic binding profiles of five cloned human muscarinic receptor subtypes, *J. Pharmacol. Exp. Ther.* 256:727-733.

Fibiger, H.C., 1982, The organization and some projections of cholinergic neurons of the mammalian forebrain, *Brain Res. Rev.* 4:327-388.

Gerfen, C. R., Engber, T. M., Mahan, L. C., Susel, Z., Chase, T. N., Monsma, F. J. Jr., and Sibley, D. R., 1990, D1 and D2 dopamine receptor-regulated gene expression of striatonigral and striatopallidal neurons, *Science.* 250:1429-1432.

Gerfen, C.R., McGinty, J.F., and Young, W.S. III., 1991, Dopamine differentially regulates dynorphin, substance P and enkephalin expression in striatal neurons: *in situ* hybridization histochemical analysis, *J. Neurosci.* 11:1016-1031.

Giorguieff, M.F., Le Floc'h, M.L., Glowinski, J. and Besson, M.J., 1977, Involvement of cholinergic presynaptic receptors of nicotinic and muscarinic types in the control of the spontaneous release of dopamine from striatal dopaminergic terminals in the rat, *J. Pharmacol. Exp. Ther.* 200:535-544.

Giorguieff-Chesselet, M.F., Kemel, M.L., Wandscheer, D., and Glowinski, J., 1979, Regulation of dopamine release by presynaptic nicotinic receptors in rat striatal slices: effect of nicotine in a low concentration, *Life Sci.* 25:1257-1262.

Girault, J.A., Spampinato, U., Savaki, H.E., Glowinski, J., and Besson, M.J., 1986, In vivo release of 3H γ-aminobutyric acid in the rat neostriatum - I. Characterization and topographical heterogeneity of the effects of dopamine and cholinergic agents, *Neuroscience* 19:1101-1108.

Góngora-Alfaro, JL., Hernández-López, S., Martínez-Fong, D., Brassart, J.-L., and Aceves, J., 1991, Activation of nigral M1 and M2 muscarinic receptors produces opposing effects on striatal 3,4-dihydro-xyphenylacetic acid measured by in vivo voltametry, *Brain Res.* 554:329-332.

Gould, E., and Butcher, L.L., 1985, Cholinergic neurons in the rat substantia nigra, *Neurosci. Lett.* 63:315-319.

Gould, E., Woolf, N.J., and Butcher, L.L., 1989, Cholinergic projections to the substantia nigra from the pedunculo pontine and laterodorsal tegmental nuclei, *Neuroscience* 28:611-623.

Graybiel, A.M., Baughman, R.W., and Eckenstein, F., 1986, Cholinergic neuropil of the striatum observes striosomal boundaries, *Nature* 323:625-628.

Grenhoff, J., Aston-Jones, G., and Svensson, T.H., 1986, Nicotinic effects on the firing pattern of midbrain dopamine neurons, *Acta Physiol. Scand.* 128:351-358.

Guyenet, P., Agid, Y., Javoy, F., Beaujouan, J. C., Rossier, J., and Glowinski, J., 1975, Effects of dopaminergic receptor agonists and antagonists on the activity of the neostriatal cholinergic system, *Brain Res.* 84:227-244.

Hammer, R., Berrie, C.P., Birshall, N.J.M., Burgen, A.S.V., and Hulme, E.C., 1980, Pirenzepine distinguishes between different subclasses of muscarinic receptors, *Nature* 283:90-92.

Hammer, R., Giraldo, E., Schiavi, G.B., Monferini, E., and Ladinsky, H., 1986, Binding profile of a novel cardio selective muscarine receptor antagonist, AF-DX 116, to membranes of peripheral tissues and brain in the rat, *Life Sci.* 38:1653-1662.

Hornykiewicz, O., 1971, Neurochemical pathology and pharmacology of brain dopamine and acetylcholine: rational basis for the current drug treatment of Parkinsonism, *in*: "Contemporary Neurology Series: Recent advances in Parkinson's disease, vol. 8," F.H. McDowell and C.H. Markham, eds.,Davis, Philadelphia, pp. 33-65.

Izzo, P.N., and Bolam, J.P., 1988, Cholinergic synaptic input to different parts of spiny striatonigral neurons in the rat, *J. Comp. Neurol.* 269:219-234.

James, T.A., and Massey, S., 1978, Evidence for a possible dopaminergic link in the action of acetylcholine in the rat substantia nigra, *Neuropharmacology* 17:687-690.

Javoy, F., Agid, Y., Bouvet, D., and Glowinski J., 1974, Changes in neostriatal DA metabolism after carbachol or atropine microinjections into the substantia nigra, *Brain Res.* 68:253-260.

Joyce, J. N., 1991, Differential response of striatal dopamine and muscarinic cholinergic receptor subtypes to the loss of dopamine. I. Effects of intranigral or intracerebroventricular 6-hydroxydopamine lesions of the mesostriatal dopamine system, *Exp. Neurol.* 113:261-276.

Kaakkola, S., 1981, Effect of nicotinic and muscarinic drugs on amphetamine- and apomorphine-induced circling behaviour in rats, *Acta Pharmacol. Toxicol.* 48:162-167.

Kemel, M.-L., Desban, M., Glowinski, J., and Gauchy, C., 1989, Distinct presynaptic control of dopamine release in striosomal and matrix areas of the cat caudate nucleus, *Proc. Natl. Acad. Sci. USA* 86:9006-9010.

Kemel, M.-L., Desban, M., Glowinski, J., and Gauchy, C., 1992, Functional heterogeneity of the matrix compartment in the cat caudate nucleus as demonstrated by the cholinergic presynaptic regulation of dopamine release, *Neuroscience* 50:597-610.

Kubo, T., Fukuda, K., Mikami, A., Maeda, A., Takahashi, H., Mishina, M., Haga, T., Haga, K., Ichiyama, A., Kangawa, K., Kojima, M., Matsuo, H., Hirose, T., and Numa, S., 1986, Cloning, sequencing and expression of complementary DNA encoding the muscarinic acetylcholine receptor, *Nature* 232:411-416.

Kubota, Y., Inagaki, S., Shimada, S., Kito, S., Eckenstein, F., and Tohyama, M., 1987, Neostriatal cholinergic neurons receive direct synaptic inputs from dopaminergic axons, *Brain Res.* 413:179-184.

Lapper, S.R., and Bolam, J.P., 1992, Input from the frontal cortex and the parafascicular nucleus to cholinergic interneurons in the dorsal striatum of the rat, *Neuroscience* 31:533-545.

Le Moine, C., Tison, F. and Bloch, B., 1990, D2 dopamine receptor gene expression by cholinergic neurons in the rat striatum, *Neurosci. Lett.* 117:248-252.

Lehmann, J., and Langer, S.Z., 1982, Muscarinic receptors on dopamine terminals in the cat caudate nucleus: neuromodulation of [3H]dopamine release in vitro by endogenous acetylcholine, *Brain Res.* 248:61-69.

Lehmann, J., and Langer, S.Z., 1983, The striatal cholinergic interneuron: synaptic target of dopaminergic terminals ?, *Neuroscience* 10:1105-1120.

Levey, A.I., Kitt, C.A., Simonds, W.F., Price, D.L., and Brann, M.R., 1991, Identification and localization of muscarinic acetylcholine receptor proteins in brain with subtype-specific antibodies, *J. Neurosci.* 11:3218-3226.

Limberger, N., Spath, L., and Starke, K., 1986, A search for receptors modulating the release of g-(^{3}H)aminobutyric acid in rabbit caudate nucleus slices, *J. Neurochem.* 46:1109-1117.

Majocha, R., and Baldessarini, R. J., 1984, Tolerance to an anticholinergic agent is paralleled by increased binding to muscarinic receptors in rat brain and increased behavioral response to a centrally active cholinomimetic, *Life Sci.* 35:2247-2255.

Marchi, M., Sanguineti, P., and Raiteri, M., 1990, Muscarinic receptors mediate direct inhibition of GABA release from rat striatal nerve terminals, *Neurosci. Lett.* 116:347-351.

Martinez-Murillo, R., Villalba, R., Montero-Caballero, M.I., and Rodrigo, J., 1989a, Cholinergic somata and terminals in the rat substantia nigra: An immunocytochemical study with optical and electron microscopic techniques, *J. Comp. Neurol.* 281:397-415.

438

Martinez-Murillo, R., Villalba, R.M., and Rodrigo, J., 1989b, Electron microscopic localization of cholinergic terminals in the rat substantia nigra: an immunocytochemical study, *Neurosci. Lett.* 96:121-126.

McKenzie, J.S., Shafton, A.D., and Stewart, C.A., 1991, Intrastriatal dopaminergic agents, muscarinic stimulation, and GABA antagonism compared for rotation responses in rats, *Behav. Brain Res.* 45:163-170.

Mesulam, M.-M., Mash, D., Hersh, L., Bothwell, M. and Geula, C., 1992, Cholinergic innervation of the human striatum, globus pallidus, subthalamic nucleus, substantia nigra and red nucleus, *J. Comp. Neurol.* 323:252-268.

Mesulam, M.M., Mufson, E.I., Levey, A. I., and Wainer, B.H., 1984, Atlas of cholinergic neurons in the forebrain and upper brainstem of the macaque based on monoclonal choline acetyl transferase immunoreactivity, *Neuroscience* 12:669-686.

Miller, J.H., Gibson, V.A., and McKinney, M., 1991, Binding of (3H) AF-DX 384 to cloned and native muscarinic receptors, *J. Pharmacol. Exp. Ther.* 259:601-607.

Nastuk, M.A., and Graybiel, A.M., 1985, Patterns of muscarinic cholinergic binding in the striatum and their relation to dopamine islands and striosomes, *J. Comp. Neurol.* 237:176-194.

Nastuk, M.A., and Graybiel, A.M., 1991, Pharmacologically defined M1 and M2 muscarinic cholinergic binding sites in the cat's substantia nigra: development and maturity, *Dev. Brain Res.* 61:1-10.

Ögren, S.O., and Fuxe, K., 1988, D1- and D2-receptor antagonists induce catalepsy via different efferent striatal pathways, *Neurosci. Lett.* 85:333-338.

Parker, G.C., Rugg, E.L., and Winn, P., 1991, Cholinergic stimulation of substantia nigra: abolition of carbachol-induced eating by unilateral 6-hydroxydopamine lesion of nigrostriatal dopamine neurons, *Exp. Brain Res.* 87:597-603.

Peralta, E.G., Winslow, J.W., Peterson, G.L., Smith, D.H., Ashkenazi, A., Ramachandran, J., Schimerlik, M.I., and Capon, D.J., 1986, Primary structure and biochemical properties of an M2 muscarinic receptor, *Science* 236:600-605.

Pollack, A. E., and Wooten, G. F., 1992, D2 dopaminergic regulation of striatal preproenkephalin mRNA levels is mediated at least in part through cholinergic interneurons, *Mol. Brain Res.* 13:35-41.

Raiteri, M., Leardi, R., and Marchi, M., 1984, Heterogeneity of presynaptic muscarinic receptors regulating neurotransmitter release in the rat brain, *J. Pharmacol. Exp. Ther.* 228:209-214.

Raiteri, M., Marchi, M., and Maura, G., 1982, Presynaptic muscarinic receptors increase striatal dopamine release evoked by "quasi-physiological" depolarization, *Eur. J. Pharmacol.* 83:127-129.

Regenold, W., Araujo, D.M., and Quirion, R., 1989, Quantitative autoradiographic distribution of 3H AF-DX 116 muscarinic-M2 receptor binding sites in rat brain, *Synapse* 4:115-125.

Robertson, G.S., Hubert, G.W., Tham, C.-S., and Fibiger, H.C., 1992, Lesions of the mesotelencephalic dopamine system enhance the effects of selective dopamine D1 and D2 receptor agonists on striatal acetylcholine release, *Eur. J. Pharmacol.* 219:323-325.

Robinson, T.E., 1991, Controls for lesions of the nigrostriatal dopamine system, *Science* 253:332.

Robinson, T.E., and Whishaw, I.Q., 1988, Normalization of extracellular dopamine in striatum following recovery from a partial unilateral 6-OHDA lesion of the substantia nigra: a microdialysis study in freely moving rats, *Brain Res.* 450:209-224.

Scarnati, E., Proia, A., Campana, E., and Pacitti, C., 1986, A microiontophoretic study on the nature of the putative synaptic neurotransmitter involved in the pedunculopontine-substantia nigra pars compacta excitatory pathway of the rat, *Exp. Brain Res.* 62:470-478.

Schoffelmeer, A.N.M., Van Vliet, BJ., Wardeh, G., and Mulder, AH., 1986, Muscarinic receptor-mediated modulation of [3H]-dopamine and [14C]-acetylcholine release from rat neostriatal slices: selective antagonism by gallamine but not pirenzepine, *Eur. J. Pharmacol.* 128:291-294.

Schwartz, R.D., Lehmann, J., and Kellar, K.J., 1984, Presynaptic nicotinic cholinergic receptors labeled by 3H acetylcholine on catecholamine and serotonin axons in brain, *J. Neurochem.* 42:1495-1498.

Smith, A.D., and Bolam, J.P., 1990, The neural network of the basal ganglia as revealed by the study of synaptic connections of identified neurones, *Trends Neurosci.* 13:259-265.

Soghomonian, J.-J., Gonzales, C., and Chesselet, M.-F., 1992, Messenger RNAs encoding glutamate-decarboxylase are differentially affected by nigrostriatal lesions in subpopulations of striatal neurons, *Brain Res.* 576:68-79.

Stadler, H., Lloyd, K., Gadea-Ciria, K.G., and Bartholini, G., 1973, Enhanced striatal acetylcholine release by chlorpromazine and its reversal by apomorphine, *Brain Res.* 55:476-480.

Stoof, J.C., De Boer, T., Sminia, P., and Mulder, A.H., 1982, Stimulation of D2-dopamine receptors in rat neostriatum inhibits the release of acetylcholine and dopamine but does not affect the release of gamma-aminobutyric acid, glutamate or serotonin, *Eur. J.Pharmacol.* 84:211-214.

Stoof, J.C., Verheijden, P.F.H.M., and Leysen, J.E., 1987, Stimulation of D2-receptors in rat nucleus accumbens slices inhibits dopamine and acetylcholine release but not cyclic AMP formation, *Brain Res.* 423:364-368.

Sugita, S., Uchimura, N., Jiang, Z.G., and North, R.A., 1991, Distinct muscarinic receptors inhibit release of γ-aminobutyric acid and excitatory amino acids in mammalian brain, *Proc. Natl. Acad. Sci. USA* 88:2608-2611.

Swanson, L.W., Simmons, D.M., Whiting, P.J., and Lindstrom., J., 1987, Immunohistochemical localization of neuronal nicotinic receptors in the rodent central nervous system, *J. Neurosci.* 7:3334-3342.

Tedford, C.E., Crosby, G. Jr., Iorio, L.C., and Chipkin, R.E., 1992, Effect of SCH 39166, a novel dopamine D1 receptor antagonist, on [3H]acetylcholine release in rat striatal slices, *Eur. J. Pharmacol.* 211:169-176.

Van der Heyden, J.A.M., Venema, K., and Korf, J., 1980, *In vivo* release of endogenous gamma-aminobutyric acid from rat striatum: effects of muscimol, oxotremorine and morphine, *J. Neurochem.* 34:1648-1653.

Vernier, P., Julien, J.-F., Rataboul, P., Fourrier, O., Feuerstein, C., and Mallet, J., 1988, Similar time course changes in striatal levels of glutamic acid decarboxylase and proenkephalin mRNA following dopaminergic deafferentation in the rat, *J. Neurochem.* 51:1375-1380.

Vuillet, J., Dimova, R., Nieoullon, A., and Kerkerian-Le Goff, L., 1992, Ultrastructural relationships between choline acetyltransferase- and neuropeptide Y-containing neurons in the rat striatum, *Neuroscience* 46:351-360.

Wada, E., Wada, K., Boulter, J., Deneris, E., Heinemann, S., Patrick, J., and Swanson, L.W., 1989, Distribution of Alpha 2, Alpha 3, Alpha 4, and Beta 2 neuronal nicotinic receptor subunit mRNAs in the central nervous system: A hybridization histochemical study in the rat, *J. Comp. Neurol.* 284:314-335.

Wamsley, J.K., Gehlert, D.R., Roeske, W.R., and Yamamura, H.I., 1984, Muscarinic antagonist binding heterogeneity as evidenced by autoradiography after direct labeling with [3H]-QNB and [3H]-pirenzepine, *Life Sci.* 34:1395-1402.

Wang, J.-X., Roeske, W.R., Hawkins, K.N., Gehlert, D.R., and Yammamura, H.I., 1989, Quantitative autoradiography of M2 muscarinic receptors in the rat brain identified by using a selective radioligand [3H]AF-DX 116, *Brain Res.* 477:322-326.

Wedzony, K., Limberger, N., Späth, L., Wichmann, T., and Starke, K., 1988, Acetylcholine release in rat nucleus accumbens is regulated through dopamine D2-receptors, *Naunyn-Schmiedeberg's Arch. Pharmacol.* 338:250-255.

Weiner, D.M., Levey, A.I., and Brann, M.R., 1990, Expression of muscarinic acetylcholine and dopamine receptor mRNAs in rat basal ganglia, *Proc. Natl. Acad. Sci. USA.* 87:7050-7054.

Wilson, C.J., Chang, H.T., and Kitai, S.T., 1990, Firing patterns and synaptic potentials of identified giant aspiny interneurons in the neostriatum, *J. Neurosci.* 10:508-519.

Wooten, G.F., and Collins, R.C., 1983, Effects of dopaminergic stimulation on functional brain metabolism in rats with unilateral substantia nigra lesions, *Brain Res.* 263:267-275.

Young, W.S. III, Bonner, T.I., and Brann, M.R., 1986, Mesencephalic dopamine neurons regulate the expression of neuropeptide mRNAs in the rat forebrain, *Proc. Natl. Acad. Sci. USA* 83:9827-9831.

INTERACTIONS OF DOPAMINE, EXCITATORY AMINO ACIDS, AND INHIBITORY AMINO ACIDS IN THE BASAL GANGLIA OF THE CONSCIOUS RAT

Francisco Mora and Alberto Porras

Department of Physiology
Faculty of Medicine
University Complutense of Madrid
28040 Madrid

INTRODUCTION

In the last few years, our understanding of how different types of neurotransmitters operate in the brain has advanced considerably (Boustfield, 1985; Meldrum et al., 1991). More recent studies are concentrated on the interaction of different types of neurotransmitters in specified areas of the Central Nervous System. The basal ganglia, has been one such area in the brain (Chiodo and Berger, 1986; Tanganelli et al., 1989; Carlsson and Carlsson, 1990; Consolo et al., 1992; Expósito et al., 1993).

DOPAMINE AND EXCITATORY AMINO ACID NEUROTRANSMITTERS INTERACTION IN THE NEOSTRIATUM

After the discovery that a deficit of dopamine (DA) in the basal ganglia was at the basis of Parkinson's disease, a balance or interaction between DA and acetylcholine was soon proposed as the basis of the regulation of motor functions (Barbeau, 1962). However, this type of interaction, although still generally accepted, has been revised after the finding that glutamate/aspartate (GLU/ASP) also seems to play an important role in motor functions (Carlsson and Carlsson, 1990).

Since the proposal that GLU/ASP could be the mediators of the corticostriatal pathway (Fonnum et al. 1981) and also that this pathway could interact with DA (Giorguieff et al., 1977), research has been dedicated to study this problem. Moreover, during the past 4-5 years there has been much debate and controversy, probably as a result of much interest, in this type of interaction in the basal ganglia. Most of this interest has arisen because of the implication of this type of interaction in Parkinson's disease and also schizophrenia (Carlsson and Carlsson, 1990).

Initial studies suggested that dopaminergic inhibitory fibers from the substantia nigra form axo-axonic synapses on glutamatergic terminals arising from the cerebral cortex (Mitchell and Doggett, 1980; Rowlands and Roberts, 1980). However this hypothesis has been questioned (Carlsson and Carlsson, 1990). In fact, it has been shown that glutamatergic cortical afferents and dopaminergic nigral afferents converge through axo-dendritic synapses on a common neuron in the neostriatum, probably of GABA nature (Bouyer et al., 1984). Based on these findings two new hypothesis have emerged to explain the GLU/ASP-DA interaction in the neostriatum. The first postulates a direct GLU/ASP-DA

The Basal Ganglia IV, Edited by
G. Percheron *et al.*, Plenum Press, New York, 1994

interaction within the neostriatum. The second proposes a GLU/ASP-DA interaction involving extrastriatal connections.

The GLU/ASP-DA volumetric interaction hypothesis

Although glutamatergic cortical and dopaminergic nigral afferents seem to converge through axo-dendritic synapses on a common neuron in the neostriatum (Bouyer et al., 1984), recent studies still support the idea of a reciprocal direct interaction between GLU/ASP and DA within the neostriatum (Shimizu et al., 1990; García-Muñoz et al., 1991). Thus, recent electrophysiological findings have suggested that DA released from nigral afferents could diffuse to stimulate glutamatergic terminals through non-synaptic volumetric transmission (García-Muñoz et al., 1991).

The striato-pallido-thalamo-cortico-striatal loop hypothesis

This hypothesis proposes that DA released through the nigrostriatal pathway could activate a striato-pallido-thalamo-cortico-striatal neural loop that releases GLU/ASP in the neostriatum (Barbeito et al., 1989; Carlsson and Carlsson, 1990).

Consistent with this hypothesis are recent findings showing that amphetamine, injected into rats with cortical or thalamic ablations, do not produces the classical excitation of neurons in the neostriatum (Warenycia et al., 1987; Haracz et al., 1992). This would suggest that DA produces its excitatory effects on neostriatal neurons through the activation of a corticostriatal excitatory pathway.

Some recent experimental data

We have recently performed a series of experiments in which the effects of acute injections of amphetamine and apomorphine, a D1-D2 dopamine receptor agonist, were investigated on the release of excitatory amino acids, GLU and ASP, in the neostriatum of the conscious rat (Mora and Porras, 1993; Expósito et al., 1993). Thus, amphetamine, which produces a release of DA (Butcher et al., 1988), also produces an increase in the extracellular [GLU] and [ASP] (Figure 1A). These effects were fully blocked by previous injections of haloperidol, a D1-D2 dopamine receptor blocker (Figure 1B). Also the intrastriatal perfusion of apomorphine was effective in producing a release of GLU and ASP and also these effects were blocked by previous injections of haloperidol (Figure 2).

The results obtained with amphetamine and apomorphine would fit in well with the proposed hypothesis of a striato-pallido-thalamo-cortical feedback loop. However the possibility that a kind of local volumetric interaction between GLU/ASP and DA, which is also taken place in these effects, can not be discarded (Expósito et al., 1993).

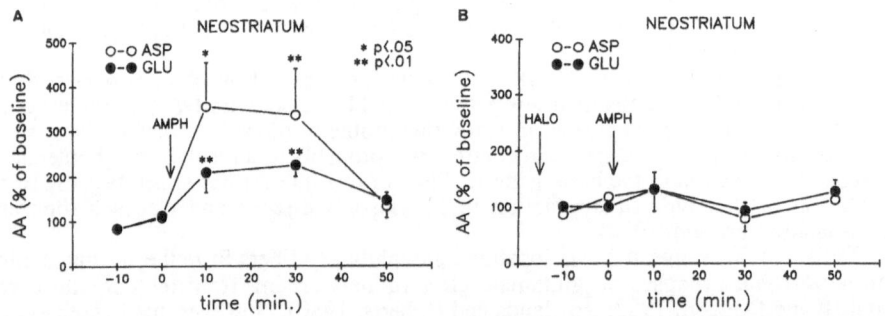

Figure 1. Extracellular [GLU] and [ASP] in the neostriatum of conscious rats obtained with a continuous perfusion system. (A) After injections of amphetamine (5 mg/kg). (B) After injections of amphetamine (5 mg/kg) plus haloperidol (3 mg/kg).

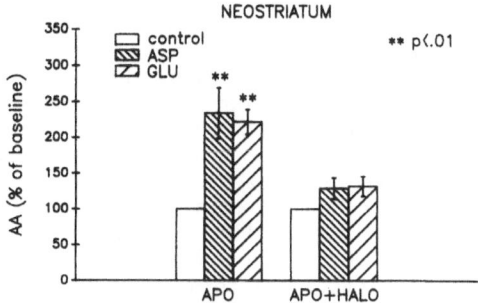

Figure 2. Effects of intrastriatal perfusion of apomorphine (10 µM) and apomorphine (10 µM) plus haloperidol (3 mg/kg i.p.) on extracellular [GLU] and [ASP] in the neostriatum of conscious rats.

DOPAMINE AND INHIBITORY AMINO ACID NEUROTRANSMITTERS INTERACTION IN THE NEOSTRIATUM

In a recent series of experiments we have shown that amphetamine, along with the release of GLU and ASP, also releases GABA, taurine (TAU), and glycine (GLY) in the neostriatum of the rat. Further, this release is fully blocked by previous injections of haloperidol (see Figure 3) suggesting that the release of inhibitory amino acid neurotransmitters could be due to a previous release of DA (Porras and Mora, 1993).

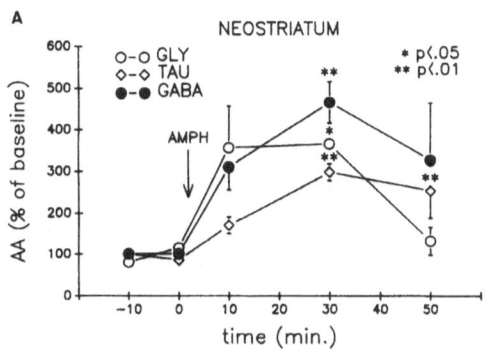

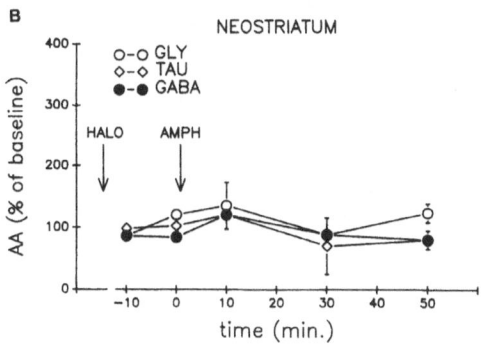

Figure 3. Extracellular [GABA], [TAU], and [GLY] in the neostriatum of conscious rats obtained with a continuous perfusion system. (A) After injections of amphetamine (1.25 mg/kg for GABA, and TAU, 5 mg/kg for GLY). (B) After injections of amphetamine (5 mg/kg) plus haloperidol (3 mg/kg).

That GABA is released by DA is supported by other studies showing an enhancement of the extracellular [GABA] in the basal ganglia under the effects of amphetamine (Besson et al., 1986). Moreover, these effects seem to be mediated by D1 dopamine receptors since intrastriatal microinjections of a specific D1 dopamine receptor agonist produces an increase of GABA (Besson et al., 1986). Since dopaminergic nigral afferents seem to synapse on GABA neurons in the neostriatum (Bouyer et al., 1984) and DA seem to have both, excitatory and inhibitory functions (Haracz et al., 1992), the possibility exists for DA releasing GABA directly from gabaergic neurons in the neostriatum.

TAU has also been shown to be released by amphetamine through the release of DA in the neostriatum. Also under the stimulation of NMDA glutamatergic receptors TAU is released in the striatum (Butcher et al., 1987). Therefore the possibility exists for TAU to be released as a consequence of a sequential release of DA and GLU/ASP.

In relation to GLY it could be speculated that because GLY has an allosteric site in the NMDA glutamatergic receptor (Johnson and Asher, 1987) this amino acid could play a complementary role to that of GLU and ASP, also released by DA. However, the extracellular [GLY] assessed by in vivo studies have been shown to be sufficient to maintain the GLY binding sites constitutively active (Johnson and Asher, 1988). Therefore this possibility looks unlikely. Since there are not reports indicating a possible relationship between DA and GLY in the basal ganglia, this type of interaction remains obscure.

DOPAMINE, EXCITATORY AMINO ACID, AND INHIBITORY AMINO ACID NEUROTRANSMITTERS INTERACTIONS IN THE NEOSTRIATUM: A HYPOTHESIS

As we have seen in the present chapter, DA released by amphetamine could be the neurotransmitter responsible for the release of excitatory amino acids and inhibitory amino acids in the neostriatum of the conscious rat. Also amphetamine and its derivates have neurotoxic effects on dopamine terminals in the neostriatum (Ellison et al., 1978; Hotchkiss and Gibb, 1980). These neurotoxic effects have been attributed to both DA and excitatory amino acids (Hotchkiss and Gibb, 1980; Johnson et al., 1989; Sonsalla et al., 1991; Muraki et al., 1992).

It is interesting that, together with amphetamine, other pathological processes which involve neuronal injury, such as hypoxia/ischemia produces a marked elevation of [DA] as well as [GLU] and [ASP] in the neostriatum (Hagberg et al., 1987; Globus et al., 1988; Damsma et al., 1990; Akiyama et al., 1991). Moreover depletion of DA and blockade of

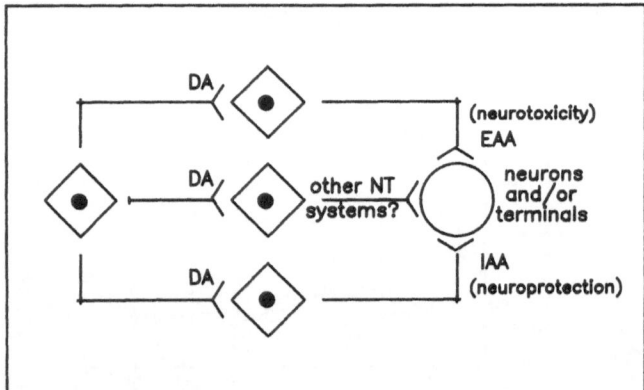

Figure 4. Schematic drawing illustrating the hypothetical role of dopamine (DA) as a potential alarm transmitter in the neostriatum. (This scheme does not take into account the multisynaptic intra- and extra-neostriatal interconnections). Release of inhibitory amino acid neurotransmitters (IAA) such as GABA, TAU, and GLY would be neuroprotective. Release of excitatory amino acid neurotransmitters (EAA) such as GLU, and ASP would be, reached certain concentrations, neurotoxic. It is hypothetized that the neurotoxicity produced by amphetamines and hypoxia/ischemia in the neostriatum could be the result of an interaction between neurotoxic excitatory and neuroprotective inhibitory neurotransmitters.

glutamatergic receptors can attenuate the neuronal death produced by hypoxia/ischemia (Weinberger et al., 1985; Gill et al., 1987; Clemens and Phebus, 1988). These findings suggest that DA may induce a secondary release of GLU/ASP which could mediate, at least in part, the neurotoxic effects produced.

Interestingly, both amphetamine and hypoxia/ischemia episodes, produce along with the release of excitatory amino acids, a release of inhibitory amino acids (Hagberg et al., 1987; Globus et al., 1988; Porras and Mora, 1993).

During hypoxia/ischemia, a protective role for TAU and GABA has been suggested (Shurr et al., 1987; Sternau et al., 1989). In the case of TAU, these suggestions have been further substantiated by the fact that significant increases of TAU are only obtained when the extracellular levels of GLU reach neurotoxic concentrations (Magnusson et al., 1991). All these considerations would suggest the possibility that the neuropathology caused by amphetamine and also ischemia/ hypoxia could be subserved by similar neurochemical mechanisms. In fact, we suggest that the end result of the neurotoxic effects produced by amphetamine and ischemia/ hypoxia could be produced by an interplay between neurotoxic excitatory and neuroprotective inhibitory amino acids. Further, and because the release of both excitatory and inhibitory amino acids are blocked by dopamine receptor antagonists, the possibility exists for DA to play the role of a general alarm neurotransmitter or amplifier of the activity of different neurochemical circuits subserving different but complementary functions (Porras and Mora, 1993) (see Figure 4). Because other types of neurotransmitters are also released by amphetamine (see Porras and Mora, 1993 for references) this hypothesis remains speculative.

Acknowledgements

We would like to thank professor Carl V. Gisolfi from Iowa University. A. Porras is a research fellow from the University Complutense of Madrid. This research has been supported by grant PB90-0252 from DGICYT.

REFERENCES

Akiyama, Y., Koshimura, K., Ohue, T., Lee, K., Miwa, S., Yamagata, S., and Kikuchi, H., 1991, Effects of hypoxia on the activity of the dopaminergic neuron system in the rat striatum as studied by in vivo brain microdialysis, *J. Neurochem.* 57:997-1002.

Barbeau, A., 1962, The pathogenesis of Parkinson's disease: a new hypothesis, *Can. Med. Association* 87:802-807.

Barbeito, L., Girault, J.A., Godeheu, G., Pittaluga, A., Glowinski, J., and Cheramy, A., 1989, Activation of the bilateral corticostriatal glutamatergic projection by infusion of GABA into thalamic motor nuclei in the cat: an in vivo release study, *Neuroscience* 28:365-374.

Besson, M., Kemel, M.L., Gauchy, C., Girault, J.A., Spampinato, U., Lautin, N., Desban, M., and Glowinski, J., 1986, In vivo measurement of [3H]GABA release: an approach to the study of the regulation of GABA-containing neurons in the basal ganglia and associated structures in the cat and in the rat, *in* : "Neurochemical Analysis of the Conscious Brain: Voltammetry and Push-Pull Perfusion," R.D. Myers and P-J. Knott, eds., Ann. N.Y. Acad. Sci., New York, pp. 475-489.

Boustfield, D., 1985, "Neurotransmitters in Action", Elsevier, Amsterdam.

Bouyer, J.J., Park, D.H., Joh, T.H., and Pickel, V.M., 1984, Chemical and structural analysis of the relation between cortical inputs and tyrosine hydroxylase-containing terminals in rat neostriatum, *Brain Res.* 302:267-275.

Butcher, S.P., Lazarewicz, J.W., and Hamberger, A., 1987, In vivo microdialysis studies on the effects of decortication and excitotoxic lesions on kainic acid-induced calcium fluxes, and endogenous amino acid release, in the striatum, *J. Neurochem.* 49:1335-1360.

Butcher, S.P., Fairbrother, I.S., Kelly, J.S., and Arbuthnott, G.W., 1988, Amphetamine-induced dopamine release in the rat striatum: an in vivo microdialysis study, *J. Neurochem.* 50:346-355.

Carlsson, M., and Carlsson, A., 1990, Interactions between glutamatergic and monoaminergic systems within the basal ganglia - implications for schizophrenia and Parkinson's disease, *Trends Neurosci.* 13:272-276.

Clemens, J.A., and Phebus, L.A., 1988, Dopamine depletion protects striatal neurons from ischemia-induced cell death, *Life Sci.* 42:707-713.

Consolo, S., Girotti, P., Russi, G., and Di Chiara, G., 1992, Endogenous dopamine facilitates striatal in vivo acetylcholine release by acting on D1 receptors localized in the striatum, *J. Neurochem.* 59:1555-1557.

Chiodo, L.A., and Berger, T.W., 1986, Interactions between dopamine and amino acid-induced excitation and inhibition in the striatum, *Brain Res.* 375:198-203.

Damsma, G., Boisvert, D.P., Mudrick, L.A., Wenkstern, D., and Fibiger, H.C., 1990, Effects of transient forebrain ischemia and pargyline on extracellular concentrations of dopamine, serotonin, and their metabolites in the rat striatum as determinated by in vivo microdialysis, *J. Neurochem.* 54:801-808.

Ellison, G., Eison, M.S., Huberman, H.S., and Daniel, F., 1978, Long-term changes in dopaminergic innervation of caudate nucleus after continuous amphetamine administration, *Science* 201:276-278.

Expósito, I., Porras, A., Sanz, B., and Mora, F., Effects of apomorphine and L-methionine sulfoximine on the release of excitatory amino acid neurotransmitters and glutamine in the basal ganglia of the conscious rat, *Eur. J. Neurosci.* submitted.

Fonnum, F., Storm-Mathisen, J., and Divac, I., 1981, Biochemical evidence for glutamate as neurotransmitter in corticostriatal and corticothalamic fibres in rat brain, *Neuroscience* 6:863-873.

García-Muñoz, M., Young, S.J., and Groves, P.M., 1991, Terminal excitability of the corticostriatal pathway. I. Regulation by dopamine receptor stimulation, *Brain Res.* 551:195-206.

Gill, R., Foster, A.C., and Woodruff, G.N., 1987, Systemic administration of MK-801 protects against ischemia-induced hippocampal neurodegeneration in the gerbil, *J. Neurosci.* 7:3343-3349.

Giorguieff, M.F., Kemel, M.L., and Glowinski, J., 1977, Presynaptic effect of L-glutamic acid on the release of dopamine in rat striatal slices, *Neurosci. Lett.* 6:73-77.

Globus, M.Y.-T., Busto, R., Dietrich, W.D., Martinez, E., Valdes, I., and Ginsberg, M.D., 1988, Effect of ischemia on the in vivo release of striatal dopamine, glutamate, and gamma-aminobutyric acid studied by intracerebral microdialysis, *J. Neurochem.* 51:1455-1464.

Haracz, J.L., Tschanz, J.T., Wang, Z., Griffith, K.E., and Rebec, G.V., 1992, Behavioral clamping of freely moving rats reveals that cortical ablation blocks amphetamine-induced excitations, but not inhibitions, of striatal neurons, *Fourth triennial Meet. Intern. Basal Ganglia Soc. Abst.*:37.

Hagberg, H., Andersson, P., Kjellmer, I., Thiringer, K., and Thordstein, M., 1987, Extracellular overflow of glutamate, aspartate, GABA, and taurine in the cortex and in the basal ganglia of fetal lambs during hypoxia-ischemia, *Neurosci. Lett.* 78:311-317.

Hotchkiss, A.J., and Gibb, J.W., 1980, Long-term effects of multiple doses of methamphetamine on tryptophan hydroxylase and tyrosine hydroxylase activity in rat brain, *J. Pharmacol. Exp. Ther.* 214:257-262.

Johnson, J.W., and Asher, P., 1987, Glycine potentiates the NMDA response in cultured mouse brain neurons, *Nature* 325:522-525.

Johnson, J.W., and Asher, P., 1988, The NMDA receptor and its channel. Modulation by magnesium and by glycine, *in*: "Excitatory Amino Acids in Health and Disease," D. Lodge, ed., John Wiley and Sons Ltd., Avon, pp. 143-164.

Johnson, M., Hanson, G.R., and Gibb, J.W., 1989, Effect of MK-801 on the decrease in tryptophan hydroxilase induced by methamphetamine and its methylenedioxy analog, *Eur. J. Pharmacol.* 165:315-318.

Magnusson, K.R., Koerner, J.F., Larson, A.A., Smullin, D.H., Skilling, S.R., and Beitz, A.J., 1991, NMDA-, kainate-, and quisqualate-stimulated release of taurine from electrophysiologically monitored rat hippocampal slices, *Brain Res.* 549:1-8.

Meldrum, B.S., Moroni, F., Simon, R.P., and Woods, J.H., 1991, "Excitatory amino acids", Raven Press, New York.

Mitchell, P.R., and Doggett, N.S., 1980, Modulation of striatal [3H]glutamic acid release by dopaminergic drugs, *Life Sci.* 26:2073-2081.

Mora, F., and Porras, A., 1993, Effects of amphetamine on the release of excitatory amino acid neurotransmitters in the basal ganglia of the conscious rat, *Can. J. Physiol. Pharmacol.* in press.

Muraki, A., Koyama, T., Nakayama, M., Ohmori, T., and Yamashita, I., 1992, MK-801, a non-competitive antagonist of NMDA receptor, prevents methamphetamine-induced decrease of striatal dopamine uptake sites in the rat striatum, *Neurosci. Lett.* 136:39-42.

Porras, A., and Mora, F., 1993, Dopamine receptor antagonist blocks the release of glycine, GABA, and taurine produced by amphetamine, *Brain Res. Bull.* 31: in press.

Rowlands, G.J., and Roberts, P.J., 1980, Activation of dopamine receptors inhibit calcium-dependent glutamate release from cortico-striatal terminals in vitro, *Eur. J. Pharmacol.* 62:241-242.

Shimizu, N., Duan, S.M., Hori, T., and Oomura, Y., 1990, Glutamate modulates dopamine release in the striatum as mesured by brain microdialysis, *Brain Res. Bull.* 25:99-102.

Shurr, A., Tseng, M.T., West, C.A., and Rigor, B.M., 1987, Taurine improves the recovery of neuronal function following cerebral hypoxia: an in vitro study, *Life Sci.* 40:2059-2066.

Sonsalla, P.K., Riordan, D.E., and Hekkila, R.E., 1991, Competitive and noncompetitive antagonists at N-methyl-D-aspartate receptors protect against methamphetamine-induced dopaminergic damage in mice, *J. Pharmacol. Exp. Ther.* 256: 506-512.

Sternau, L.L., Lust, -W.D., Ricci, A.J., and Ratcheson, R., 1989, Role for gamma-aminobutyric acid in selective vulnerability in gerbils, *Stroke* 20:281-287.

Tanganelli, S., Von Euluer, G., Fuxe, K., Agnati, L.F., and Ungerstedt, U., 1989, Neurotensin counteracts apomorphine-induced inhibition of dopamine release as studied by microdialysis in rat neostriatum, *Brain Res.* 502:319-324.

Warenycia, M.W., McKenzie, G.M., Murphy, M., and Szerb, J.C., 1987, The effects of cortical ablation on multiple unit activity in the striatum following dexamphetamine, *Neuropharmacol.* 26:1107-1114.

Weinberger, J., Nieves-Rosa, J., and Cohen, G., 1985, Nerve terminal damage in cerebral ischemia: protective effect of alpha-methyl-para-tyrosine, *Stroke* 16:864-870.

ENKEPHALIN-GABA CO-TRANSMISSION IN THE STRIATOPALLIDAL PATHWAY IN PARKINSONISM

Yannick P. Maneuf, Ian J. Mitchell, Alan R. Crossman
and Jonathan M. Brotchie

School of Biological Sciences
University of Manchester
Oxford Road
Manchester M13 9PT, U.K.

INTRODUCTION

A general consensus exists as to the important role played by the basal ganglia in the control of voluntary movements. Dysfunction of the basal ganglia is responsible for a wide variety of movement disorders, e.g. Parkinson's disease, chorea and dystonia. The basal ganglia exhibit a high level of structural complexity. Several neurotransmitter systems coexist in the basal ganglia and provide the substrate for signalling between the constituent nuclei. A characteristic of the chemical signalling within the basal ganglia is that peptides are often co-transmitted with amino acids or catecholamine. Within the basal ganglia, the striatum is a rich source of such peptidergic systems. However, very little is known regarding their functional significance in the control of movement or in the pathophysiology of movement disorders.

It is now well established that the medium spiny neurons of the striatum, which constitute more than 90% of striatal neurons, use gamma-aminobutyric acid (GABA) as their neurotransmitter. A segregation in the peptidergic content of these striatal neurons is apparent, and highly conserved across many species (Anderson and Reiner, 1990; Besson et al., 1990). Generally, striatal cells contain either enkephalin or substance P and dynorphin. Thus, in the rodent striatum, most of the cells (75%) immunoreactive for substance P also contain dynorphin and reciprocally the cells positive for dynorphin contain substance P. Nonetheless, none of the dynorphin positive cells show any immunoreactivity for enkephalin and only 5% of the neurons containing substance P are positive for enkephalin. This dichotomy is maintained with regard to the projection targets of medium spiny neurons. Thus, the GABA-enkephalinergic neurons project to the external segment of the globus pallidus (GPe). On the other hand, the internal pallidal segment (GPi, or its rodent homologue the entopeduncular nucleus) and the substantia nigra pars reticulata (SNr) are innervated by GABA co-transmitted with dynorphin and substance P (Brownstein et al., 1977; Gerfen and Young, 1988).

In the experiments presented here we have examined the role of enkephalin-GABA co-transmission in the striatopallidal pathway. This pathway illustrates a classical example of amino acid-peptide co-transmission. However, in spite of a well-defined anatomical and physiological function for the two transmitters separately, very little is known with regards to the possible functional interactions between GABA and enkephalin. In parkinsonism reduced dopaminergic transmission in the striatum causes changes in the GABA-enkephalinergic striatopallidal pathway. These changes in activity of the connection between

The Basal Ganglia IV, Edited by
G. Percheron *et al.*, Plenum Press, New York, 1994

the striatum and GPe are thought to represent a critical component of the neural mechanisms underlying the pathophysiology of Parkinson's disease.

Several lines of evidence suggest that both the GABAergic and enkephalinergic components of the connection between the striatum and GPe are overactive in parkinsonism.
1) Increased 2-deoxyglucose utilisation is seen in the GPe of MPTP-exposed primates, this suggests increased terminal activity in the inputs to the GPe (Mitchell et al., 1989).
2) Decreased levels of the $GABA_A$ receptor-ionophore complex are seen in both the 6-hydroxydopamine (Pan et al., 1985) and MPTP-exposed primate models of parkinsonism (Robertson et al., 1990). This down regulation is indicative of an overactive GABAergic input.
3) Overactive inhibitory GABAergic input from the striatum has also been suggested to account for the decreased firing rates of GPe neurons seen in MPTP-treated primates (Miller and DeLong, 1987; Filion et al., 1988).
4) *In situ* and northern hybridisation studies have demonstrated that the levels of mRNA for enkephalin precursor peptides are elevated in the striatum of MPTP-exposed primates (Frayne et al., 1991), 6-hydroxydopamine-treated rats (Gerfen et al., 1991) and reserpine-treated rats (Jaber et al., 1992).
5) Increases in mRNA are translated into increased levels of enkephalin as demonstrated by immunocytochemical studies showing that in the 6-hydroxydopamine treated rat there is increased levels of enkephalin-like immunoreactivity (Engber et al., 1991).

However, whilst illustrating the changes in both transmitters in parkinsonism, no indication of the functional role of this co-transmission is given by these studies. We have employed a behavioural pharmacological approach in a rodent model of parkinsonism to define GABA-enkephalin interactions at a functional level, with regard to the control of movement. We have subsequently used a neurochemical approach to define a potential neural mechanisms underlying such functional interaction.

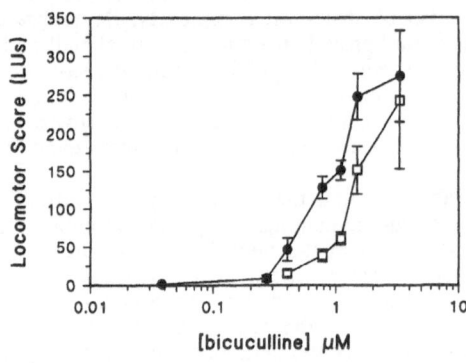

Figure 1. Dose-response curves for the locomotor effects of injections of bicuculline alone (filled circles) and bicuculline plus naloxone (0.1M) (open squares) into the globus pallidus of reserpine-treated rats. Each data point represents the mean locomotor score, in locomotor units (LU) ± SEM (n=3 to 6).

GABA-ENKEPHALIN INTERACTIONS *IN VIVO*

Given the supposition that increased GABAergic transmission in the GPe is an important component of the neural mechanisms mediating the symptoms of parkinsonism we hypothesised that antagonism of $GABA_A$ effects in the GPe would alleviate parkinsonian symptoms.

We tested this hypothesis by injecting the GABA antagonist bicuculline intracerebrally, directly into the globus pallidus (rodent homologue of GPe) of the reserpine-treated rat model of parkinsonism (5mg/kg, s.c.). Injections needles of appropriate length to allow injection into the globus pallidus were inserted down previously implanted guide cannulae (Paxinos and Watson, 1982). The animal was placed into an open field arena (50x50cm). After two minutes, a unilateral injection of bicuculline was made (0.5μl, 2.5μl/min). The needle was left in place for two minutes. The locomotion of the animals was assessed for ten minutes following drug or vehicle injection as previously described (Brotchie et al., 1991). A locomotion score, in locomotion units (LU) was obtained. This parameter provides a measure of the distance moved by the forelimb contralateral to the injection site. Bicuculline was injected at a range of concentrations (0.04μM to 3.3μM).

In reserpine-treated rats a parkinsonian syndrome characterised by akinesia, rigidity and catalepsy was observed, the locomotor score was 2.7 ± 2.16 (n=6). Injection of bicuculline into the globus pallidus resulted in a reversal of the reserpine-induced akinesia. The anti-parkinsonian effects of bicuculline were dose-dependent (Fig. 1). Injections of bicuculline resulted in a dose-dependent alleviation of the akinesia with a threshold at 0.27μM. The EC_{50} determined from the dose-response curve was 0.78μM. At this concentration the locomotor score was $128 \, LU \pm 14.5$ (n=6). The latency of onset for this anti-parkinsonian effect was typically about one minute. These anti-parkinsonian effects of bicuculline were site-specific, similar injections which were located in the striatum or the internal capsule had no effect upon locomotion (Fig. 2). Injection of vehicle into the globus pallidus elicited no noticeable behavioural effects, the locomotor score being $4.2 \, LU \pm 2$, n=5.

Given that enkephalinergic transmission in the striatopallidal pathway is also increased in parkinsonism, we hypothesised that injections of the opioid antagonist naloxone might reverse akinesia in a manner similar to bicuculline. We injected a range of concentrations of naloxone into the globus pallidus in reserpine-treated parkinsonian rats. The range of doses (0.1μM up to 0.1M) of naloxone employed is comparable to those used in similar behavioural experiments conducted in other regions of the CNS (e.g. Siegfried and Nunes de Souza, 1989). However, even at the highest concentration, 0.1M, no changes in locomotion could be seen when compared with control injections of saline. The locomotor score following injection of 0.1M naloxone was 7 ± 1.6 LU (n=5). No other behavioural changes were observed following injections of naloxone. These data suggest that the increased levels of enkephalinergic transmission observed in the parkinsonian basal ganglia are not responsible for the mediation of parkinsonian symptoms.

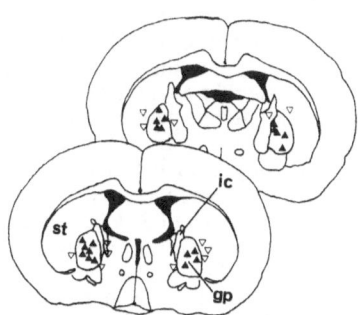

Figure 2. Injection sites in the region of the globus pallidus. Solid triangles represent injection sites at which bicuculline produced an improvement in the locomotor score. Open triangles represent sites at which bicuculline failed to reverse reserpine-induced akinesia. Abbreviations: GP, globus pallidus; IC, internal capsule; ST, striatum.

To determine the potential functional interaction between GABA and enkephalin transmission in the parkinsonian brain we conducted a series of experiments employing co-injection of the GABA and opioid antagonists in the reserpinised rat. Thus, dose-response curves for the anti-parkinsonian effects of bicuculline were constructed in the presence of naloxone.

These combined injections of naloxone and bicuculline resulted in a marked attenuation of the anti-akinetic effect of bicuculline (Fig. 3). In the presence of naloxone (0.1M) the EC_{50} for the anti-parkinsonian effects of bicuculline was found to be 1.15μM compared to 0.78μM in the absence of naloxone. Thus in the presence of naloxone the dose-response curve was shifted to the right, i.e. bicuculline was less potent as an anti-parkinsonian agent. Naloxone, thus, appears to exert an effect which increases GABAergic effects. We propose that the functional role of enkephalin in the parkinsonian globus pallidus is to attenuate GABAergic transmission. This finding of negative interaction between GABA and enkephalin is also consistent with data reported by Austin and Kalivas (1990) which suggest that in the ventral basal ganglia opioids and GABA have opposite effects on locomotor activity in normal rats.

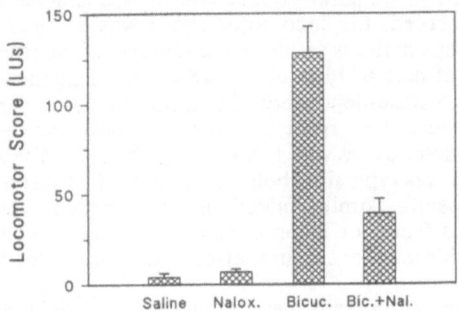

Figure 3. Comparison of the anti-parkinsonian effect of bicuculline (0.78μM) and naloxone (0.1M). Data were analysed using 2-way ANOVA and Tukey HSD. No significant difference was observed between naloxone alone and saline. Bicuculline alone significantly increased locomotor score compared to saline (p<0.001). Locomotor scores following injections of naloxone and bicuculline together were lower than those following bicuculline alone (p<0.01).

GABA-ENKEPHALIN INTERACTIONS *IN VITRO*

Enkephalin is generally considered to exerts its actions *in vivo* through interactions with mu-opioid receptors. Abou-Khalil, Penney and Young (1984) have shown that mu-opioid receptors levels in the globus pallidus are reduced following lesion of the striatum. They interpreted this finding as suggesting that in the globus pallidus, mu-opioid receptors are located pre-synaptically on the terminals of the GABAergic inputs from the striatum. We proposed that enkephalin might be attenuating GABA transmission via a pre-synaptic mechanism. We tested the hypothesis that enkephalin acts to decrease GABA release by measuring efflux of [³H]-GABA from globus pallidus slices.

Pallidal slices (400μm) were prepared from fresh rat brain. GABA terminals were loaded with [³H]-GABA by incubating in artificial CSF (aCSF) containing [³H]-GABA (1μM) for 20 minutes at 25°C. At all stages the aCSF was aerated with 95%O_2/ 5%CO_2. Following loading slices were washed in aerated aCSF for 25 minutes. GABA release was measured in each five minute period thereafter using an aerating manifold similar in design to that used to measure release of pre-loaded radioactivity from rat portal vein (Hamilton et al., 1986). [³H]-GABA release was measured in the presence of 1mM of nipecotic acid to prevent the re-uptake of GABA. To evoke release of [³H]-GABA, terminals were depolarised by adding a five minute pulse of potassium chloride (40mM) to the perfusion medium. Two such pulses of potassium were applied at 20 minutes interval (at times 15 and 35 minutes following completion of the washing stage). [³H]-GABA release was expressed as a fractional rate of release (Amoroso et al., 1990). The potassium-evoked efflux was calculated by subtracting the basal, non-stimulated release.

K+-evoked release of GABA was reduced by 77% when a modified aCSF, calcium chloride was replaced by cobalt chloride and EDTA, was used to perfuse the slice (p<0.01, Fig. 5). This calcium-dependency illustrates that the release measured by the assay probably represents GABA release from neuronal vesicular stores and not from glial or non-vesicular compartments (Adam-Vizi and Ashley, 1987). The amplitude of K+-evoked peak of release at time 15 mn was not significantly different from that at 35 mn (p>0.05, Fig. 4). Thus, a matched-pair design was employed where each slice received two pulses of depolarisation, a control pulse drug and a pulse in the presence of enkephalin. The order of these pulses was randomised. Drug effects were expressed as the ratio of the K+-evoked [³H]-GABA release in the presence of the drug divided by that observed in the absence of enkephalin.

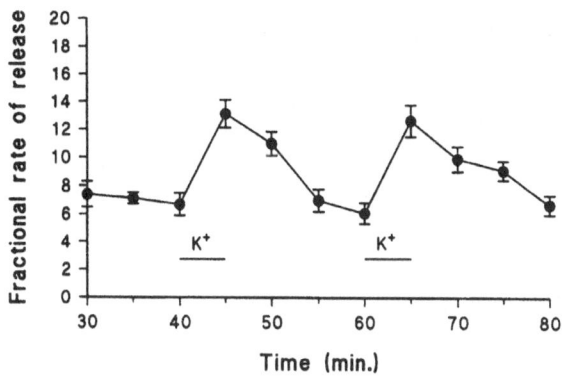

Figure 4. Depolarization-evoked release of [³H]-GABA from pallidal slices (400μm). Two five minute pulses of KCl (40mM) were applied 20 minutes apart. The amplitude of the peaks of depolarization-evoked release were not found to be significantly different (paired Student's t-test, p>0.05, n=6).

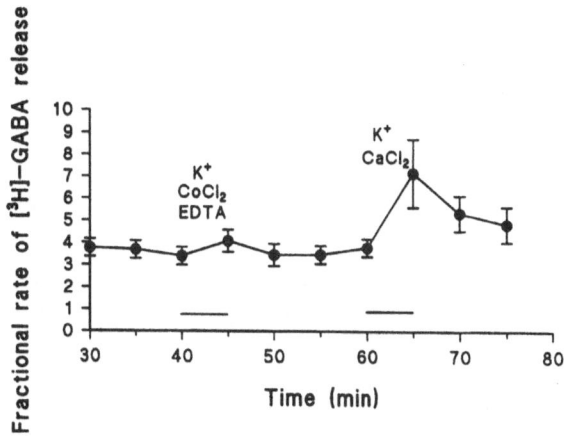

Figure 5. Calcium-dependent release of [³H]-GABA. KCl-evoked release of [³H]-GABA from pallidal slices was 77% lower when slices were perfused with aCSF in which CaCl₂ was replaced by CoCl₂ and EDTA.

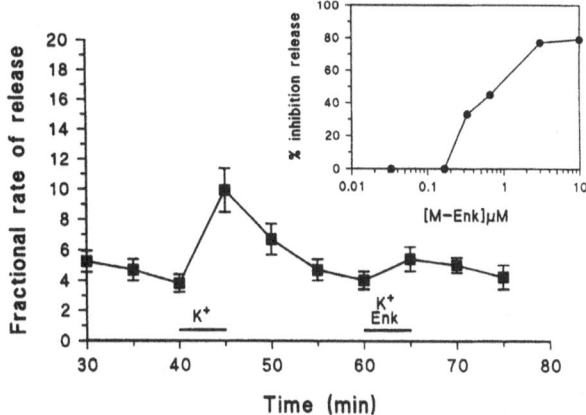

Figure 6. Effect of met-enkephalin on the KCl-evoked release of [³H]-GABA from pallidal slices. [³H]-GABA release was decreased by 75% in the presence of met-enkephalin (10μM) (p<0.05, n=4). Inset shows the dose-response curve for the reduction enkephalin-induced inhibition of evoked GABA release. (n=4 for each point).

Met-enkephalin (0.03-10μM) diminished the amplitude of K⁺-evoked GABA release in a dose-dependent manner (p<0.05, Fig. 6). The effect was maximum at a concentration of 10μM enkephalin. At this concentration the fractional rate of [³H]-GABA release was 26% of that observed in the absence of drug. The IC$_{50}$ for the effects of enkephalin was estimated as being 0.38μM. The effects of enkephalin in reducing GABA release appear to involve activation of opioid receptors as they are blocked by naloxone. For example, naloxone (5mM) blocked the reduction in GABA release elicited by of enkephalin (10μM) (Fig. 7). Similarly, other peptides, e.g. substance P (5mM), showed no effect on K⁺-evoked release of [³H]-GABA from pallidal slices.

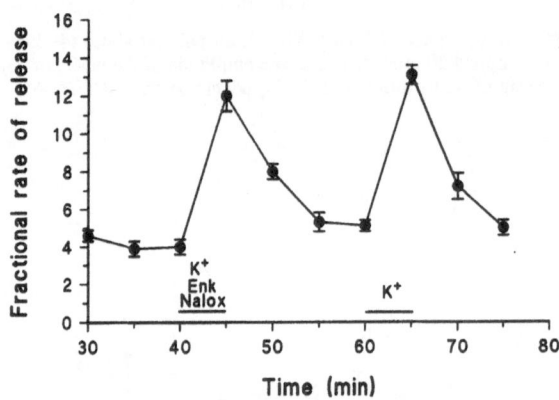

Figure 7. Naloxone (5mM) antagonises the enkephalin-induced inhibition of GABA release. KCl-evoked [³H]-GABA release in the presence of both naloxone (5mM) and met-enkephalin (10μM) was not found to be significantly different to that evoked by KCl alone (p>0.05, n=4).

This study illustrates that one of the actions of enkephalin in the basal ganglia is to reduce GABA release. In hippocampal cell cultures, enkephalin has recently been shown to induce similar reductions in the release of GABA (Cohen et al., 1992). The potency of these effects compares well with those reported here. This modulation of amino acid release may be a common feature of the physiological role of opioid-peptide co-transmission (Nicoll et al., 1980). Enkephalin appears to have similar actions to another opioid peptide dynorphin, which has been shown to reduce the release of excitatory amino acid transmitters in the hippocampus (Weisskopf et al., 1993).

CONCLUSION

The data presented in this chapter suggest that enkephalin and GABA play complementary roles in the transmission of information between the striatum and globus pallidus in the parkinsonian brain. Enkephalinergic transmission does not appear to be responsible for the generation of parkinsonian symptoms. However, enkephalin reduces the effects of GABAergic transmission. This is functional significance as we have illustrated that the increased GABAergic transmission is responsible for mediating parkinsonian symptoms. We have identified that a potential mechanism for this GABA-enkephalin interaction is a reduction in the release of GABA. However, the possibility that additional mechanisms are involved should not be overlooked, e.g. post-synaptic effects on membrane excitability (Jiang and North, 1992). Further investigation of the roles played by peptide transmitters in modulating amino acid transmission in the basal ganglia will increase our understanding of the neural mechanisms underlying both basal ganglia-related movement disorders and normal physiological motor control.

REFERENCES

Abou-Khalil, B., Penney, J.B., and Young, A.B., 1984, Evidence for the pre-synaptic localization of opiate binding sites on striatal efferent fibres, *Brain Res.* 323:21-29.

Adam-Vizi, V., and Ashley, R.H., 1987, Relation of acetylcholine release to Ca^{++} uptake and intraterminal Ca^{++} concentration in guinea pig cortex synaptosomes, *J. Neurochem.* 49:1013-1021.

Amoroso, S., Schmid-Antomarchi, H., Fosset, M., and Lazdunski, M., 1990, Glucose, sulfonylureas, and neurotransmitter release: role of ATP-sensitive K+ channels, *Science* 247:852-854.

Anderson, K.D., and Reiner, A., 1990, Extensive co-occurence of substance P and dynorphin in striatal projection neurons: an evolutionarily conserved feature of basal ganglia organization, *J. Comp. Neurol.* 295:339-369.

Austin, M.C., and Kalivas, P.W., 1990, Enkephalinergic and GABAergic modulation of motor activity in the ventral pallidum, *J. Pharmacol. Exp. Ther.* 252:1370-1377.

Besson, M.J., Graybiel, A.M., and Quinn, B., 1990, Co-expression of neuropeptides in the cat's striatum: an immunohistochemical study of substance P, dynorphin B and enkephalin, *Neuroscience* 39:33-58.

Brotchie, J.M., Mitchell, I.J., Sambrook, M.A., and Crossman, A.R., 1991, Alleviation of Parkinsonism by antagonism of EAA transmission in the medial segment of the globus pallidus in Rat and Primate, *Movement Disorders* 6:133-138.

Brownstein, M.J., Mroz, E.A., Tappaz, M.L., and Leeman, S.E., 1977, On the origin of substance P and GAD in the substantia nigra, *Brain Res.* 135:315-323.

Cohen, G.A., Doze, V.A, and Madison, D.V., 1992, Opioid inhibition of GABA release from presynaptic terminals of rat hippocampal interneurons, *Neuron* 9:325-335.

Engber, T.M., Susel, Z., Kuo, S., Gerfen,C.R., and Chase, T.N., 1991, Levodopa replacement therapy alters enzyme activities in striatum and neuropeptide content in striatal output regions of 6-hydroxydopamine lesioned rat, *Brain Res.* 552:113-118.

Filion, M., Tremblay, L., Bédard P.J., 1988, Abnormal influences of passive limb movement on the activity of globus pallidus neurons in parkinsonian monkeys, *Brain Res.* 444:165-176.

Frayne, S.E., Mitchell, I.J., Sharpe, P.T., Sambrook, M.A., and Crossman A.R., 1991, Distribution of enkephalin gene expression in the striatum of the parkinsonian primate: implications for dopamine agonist induced dystonia, *Molecular Neuropharmacol.* 1:53-58.

Gerfen, C.R., McGinty, J.F., and Young, W.S., 1991, Dopamine differentially regulates dynorphin, substance P, and enkephalin expression in striatal neurons: *in situ* hybridization histochemical analysis, *J. Neuroscience* 11:1016-1031.

Gerfen, C.R., and Young, W.S., 1988, Distribution of striatonigral and striatopallidal peptidergic neurons in both patch and matrix compartments: an *in situ* hybridization histochemistry and fluorescent retrograde tracing study, *Brain Res.* 460:161-167.

Hamilton, T.C., Weir, S.W., and Weston, A.H., 1986, Comparison of the effects of BRL34915 and verapamil on electrical and mechanical activity in rat portal vein, *Br. J. Phar.* 88:103-111.

Jaber, M., Fournier, M.C., and Bloch, B., 1992, Reserpine treatment stimulates enkephalin and D_2 dopamine receptor gene expression in the rat striatum, *Molecular Brain Res.* 15:189-194.

Jiang, Z.G., and North, R.A., 1992, Pre- and postsynaptic inhibition by opioids in rat striatum, *J. Neuroscience* 12:356-361.

Miller, W.C., and Delong, M.R., 1987, Altered tonic activity of neurons in the globus pallidus and subthalamic nucleus in the primate MPTP model of parkinsonism. *in*: "Advances in Behavioural Biology, vol.32: The Basal Ganglia II. Structure and function-Current Concepts," M.B. Carpenter and A. Jayaraman, eds., New York. Plenum Press, pp. 415-427.

Mitchell, I.J., Clarke, C.E., Boyce, S., Robertson, R.G., Peggs, D., Sambrook, M.A., and Crossman, A.R., 1989, Neural mechanisms underlying parkinsonian symptoms based upon regional uptake of 2- deoxyglucose in monkeys exposed to 1-methyl-4-phenyl-1,2,3,6-tetrahydropyridine (MPTP), *Neuroscience* 32:213-226.

Nicoll, R.A., Alger, B.E., and Jahr, C.E., 1980, Enkephalin blocks inhibitory pathways in the vertebrate CNS, *Nature* 287:22-25.

Pan, H.S., Penney, J.B., and Young, A.B., 1985, γ-amino acid benzodiazepine receptor changes induced by unilateral 6-hydroxydopamine lesions of the medial forebrain bundle, *J. Neurochemistry* 45:1396-1404.

Paxinos, G., and Watson, G., 1982, "The rat Brain in stereotaxic coordinates," Academic Press, New York.

Robertson, R.G., Clark, C.E., Boyce, S., Sambrook, M.A., and Crossman, A.R., 1990, The role of striatopallidal neurons utilising GABA in the pathophysiology of MPTP-induced parkinsonism, *Brain Res.* 531:95-104.

Siegfried, B., and Nunes de Souza, R.L., 1989, NMDA receptor blockade in the periaqueductal grey prevents stress-induced analgesia in attacked mice, *Eur. J. Pharmacol.* 168:239-242.

Weisskopf, M.G., Zalutsky, R.A., and Nicoll, R.A., 1993, The opioid peptide dynorphin mediates heterosynaptic depression of hippocampal mossy fibre synapses and modulates long-term potentiation, *Nature* 362:423-427.

CORTICAL AND DOPAMINERGIC REGULATION OF AMPHETAMINE-INDUCED CHANGES IN STRIATAL SINGLE-UNIT ACTIVITY IN AWAKE, BEHAVING RATS

George V. Rebec, Ariane Rosa-Kenig, John L. Haracz, JoAnn T. Tschanz, Karen E. Griffith, Zhongrui Wang and Jerome K. Puotz

Program in Neural Science, Department of Psychology
Indiana University
Bloomington, IN 47405 USA

INTRODUCTION

The striatum and its dopaminergic input from the ventral midbrain are known to play a critical role in the behavioral effects of amphetamine and related psychomotor stimulants. Relatively little information is available, however, on the neuronal mechanisms underlying these behavioral effects. In an initial attempt to address this issue, a series of single-unit recording experiments has been carried out in the striatum of awake, behaving animals. Early data obtained from cats revealed a tendency for amphetamine to accelerate striatal neuronal activity in parallel with the drug-induced behavioral activation (Trulson and Jacobs, 1979). Consistent with these findings, West et al. (1987) reported that amphetamine increased unit activity in virtually all striatal neurons sampled from rats trained to walk on a treadmill. In contrast, other investigators found both excitatory and inhibitory striatal responses to amphetamine in freely moving rats (Gardiner et al., 1988; Ryan et al., 1989). Subsequent research has shown that striatal neurons firing in close temporal association with movement are significantly more likely to increase activity in response to amphetamine than neurons in which neuronal activity is unrelated to movement (Haracz et al., 1989; Rebec et al., 1991). In fact, nonmotor-related neurons typically are inhibited by the drug. Although these changes in unit activity may occur secondarily to an amphetamine-induced behavioral change, data obtained from behavioral clamping techniques, which attempt to control for behavioral feedback effects by comparing neuronal activity during matched pre- and post-amphetamine behaviors, argue against this view (Haracz et al., 1993). In fact, direct infusions of amphetamine into the striatum of freely moving animals activates motor- and inhibits nonmotor-related neurons several minutes prior to the onset of overt behavioral changes (Wang and Rebec, 1992). Thus, amphetamine-induced changes in striatal activity appear to reflect a direct action of this drug on striatal neurons rather than a secondary response associated with behavioral activation. By creating a divergence in firing rate between motor- and nonmotor-related neurons, amphetamine may help to bias the striatum toward expression of the drug-induced behavioral response. In the present series of experiments, we focused on some likely mechanisms by which dopamine may regulate the changes in striatal neuronal activity produced by amphetamine.

EXPERIMENT I: EFFECTS OF D1 AND D2 ANTAGONISTS

Because amphetamine facilitates neuronal dopamine release in the striatum (Sharp et al., 1987; Kuczenski and Segal, 1989) , it seems likely that dopaminergic mechanisms play a major role in the responses of striatal neurons to amphetamine. Indeed, neuroleptic drugs, which block dopamine receptors, routinely antagonize amphetamine-induced excitations, though some neuroleptics appear to have less consistent effects on the inhibitory action of this drug (Rebec et al., 1991; Wang et al., 1992; Haracz et al., 1993). The recent development of relatively selective antagonists for D1 and D2 dopamine receptors now makes it possible to assess the relative involvement of these receptor subtypes in the neuronal and behavioral effects of amphetamine. Thus, we tested the ability of SCH-23390, a D1 antagonist, and sulpiride or eticlopride, D2 antagonists, to reverse amphetamine-induced changes in single-unit activity in the striatum of freely moving rats.

Methods

Data were obtained from male, Sprague-Dawley rats (approximately 350 g) according to procedures approved by the National Institutes of Health and the Institutional Animal Care and Use Committee at Indiana University. In preparation for surgery, animals were anesthetized (Equithesin, 0.33 ml/kg) and mounted in a stereotaxic frame. The animal was prepared for single-unit recording and a plastic hub was fixed to the skull overlying the striatum, as previously described (Wang et al., 1992). Following a 7-day recovery period, the animal was habituated to the recording chamber, which included an open-field arena (1.3 m^2) housed inside a sound-attenuating cubicle. A videotaping system recorded behavioral activity. A micromanipulator with two tungsten microelectrodes (impedances of 1-2 MΩ at 1 kHz) was inserted into the head-mounted plastic hub, and the electrodes were advanced into the striatum (Rebec et al., in press). Single-unit discharges (signal-to-noise ratio of 3:1 or more) were amplified and displayed by conventional means and stored on an audio track of the videotape. Spontaneous neuronal and behavioral activity was recorded for at least 30 min, and the animals received (sc) 1.0 mg/kg (free base) d-amphetamine sulfate (Sigma). Approximately 30 min later, during the peak of the amphetamine-induced neuronal and behavioral response, the animals were challenged (sc) with SCH-23390 (0.25, 0.5, or 1.0 mg/kg), eticlopride (0.25, 0.5, or 1.0 mg/kg), or (-)-sulpiride (10.0, 20.0 or 40.0 mg/kg), and neuronal and behavioral activity were monitored for another 30 min. Upon completion of the experiment, animals were anesthetized and current was passed through the electrode to mark the recording site. Following a transcardial perfusion with formosaline, the brain was removed, frozen, sectioned, and stained with cresyl violet for histological analysis.

Off-line analysis of neuronal data was performed in conjunction with BrainWave® Systems software, which permitted discrimination of single-unit discharges. Cases in which spike amplitude or waveform changed noticeably over time were excluded from further analysis. Independent observers rated videotaped behavioral changes according to procedures described elsewhere (Rebec et al., 1982).

Results and Discussion

Single-unit activity was recorded successfully from 49 neurons in 61 animals. In all cases, neuronal activity was slow and irregular during periods of quiet rest with a baseline rate of less than 6 spikes/sec. Consistent with previous results (see Rebec, 1992), the large majority of these neurons (n=42) more than doubled activity in close association with movement. Some neurons (n=3), however, were inhibited during movement, and others (n=4) failed to show any movement-related changes in firing rate.

All neurons excited during movement increased activity following 1.0 mg/kg D-amphetamine (sc). This response began within 10 min after injection and increased steadily to more than 200% of the quiet-baseline rate by 30 min. In contrast, amphetamine suppressed inhibitory motor-related neurons to less than 20% of the pre-drug baseline. All nonmotor-related neurons were similarly suppressed by amphetamine.

Subsequent injection (sc) of either SCH-23390 or eticlopride reversed all amphetamine-induced excitations within 15 min as shown in Table 1. This effect was evident at all doses tested. Sulpiride, in contrast, failed to alter the neuronal response to amphetamine. In every neuron challenged with 10.0, 20.0, or 40.0 mg/kg sulpiride (n=9), the amphetamine-induced increase in firing rate continued without change for the duration of the recording period. Similar results were obtained in rats that received either 0.5 mg/kg ritanserin (n=3) or vehicle (n=4) 30 min after amphetamine.

Our electrophysiological data paralleled the behavioral effects of these drugs. Thus, amphetamine increased locomotion and rearing as well as repetitive sniffing and head

Table 1. Effects of selective D1 and D2 antagonists on amphetamine-induced increases in striatal unit activity.[1]

DA antagonist	Dose (mg/kg)	n	% AMPH rate mean (±S.E.M.)	
SCH-23390	1.00	1	11	
	0.50	6	36	(±11.6)
	0.25	6	14	(±7.68)
	0.12	1	5	
Eticlopride	1.00	8	9	(±3.51)
	0.25	4	11	(±4.97)

[1]Data for the dopamine antagonists are presented as the mean (±S.E.M.) percent change from the excitatory neuronal response to amphetamine (AMPH). Firing rate was first calculated for the period just prior to injection of the dopamine antagonist (i.e., between 22 and 30 min after AMPH when the behavioral and neuronal response were at their maximum), and this value was defined as 100%. Firing rate following administration of the dopamine antagonist (i.e., between 22 and 30 min after the dopamine antagonist) was then calculated as a percent change from the 100% AMPH value. n = number of neurons included in each case.

bobbing within 10 min after injection. These responses increased steadily but were reversed shortly after administration of either SCH-23390 or eticlopride. Sulpiride, ritanserin, and vehicle had no effect on the amphetamine-induced behavioral response.

We also attempted to reverse the amphetamine-induced inhibition of striatal unit activity, even though this response occurred in relatively few neurons. In 3 motor-related neurons inhibited by amphetamine, SCH-23390 (0.5 mg/kg) and eticlopride (1.0 mg/kg), but not sulpiride (40 mg/kg), reversed the amphetamine response. Amphetamine also inhibited 4 nonmotor-related neurons, which were challenged with either SCH-23390 (0.5 mg/kg), eticlopride (1.0 mg/kg), sulpiride (10.0 m/kg), or ritanserin (0.5 mg/kg). Only eticlopride blocked the amphetamine-induced inhibition in these neurons.

Histological analysis revealed that all recording sites were confined to the striatum (sections 1.2, 1.7, and 2.4 mm anterior to bregma), according to the coordinates of Paxinos and Watson (1982). Consistent with previous results (Haracz et al., 1993; Wang et al., 1992), motor- and nonmotor-related neurons were widely scattered throughout this structure.

Our results support evidence that amphetamine accelerates the activity of excitatory motor-related neurons in the striatum of freely moving animals (see Rebec, 1992). This effect is blocked by either SCH-23390 or eticlopride, suggesting a role for both D1 and D2 dopamine receptors in this neuronal response. Similarly, either of these drugs reversed the characteristic behavioral effects of amphetamine. Although SCH-23390 has some affinity for 5-HT2 receptors, this action alone cannot explain our results because ritanserin, a relatively selective 5-HT2 antagonist, failed to alter either the neuronal or the behavioral effects of amphetamine. The failure of sulpiride in this same capacity may be related to a relatively slow rate of penetration into the central nervous system (e.g., Nishibe et al., 1982), although other interpretations, such as an action on extrastriatal D2 receptors, cannot be ruled out.

Our results are also consistent with behavioral and electrophysiological evidence that D1 or D2 receptor stimulation acts cooperatively in mediating dopamine-induced effects in the striatum (White et al., 1988; Wachtel et al., 1989). Presumably, blockade of either receptor subtype is sufficient to reverse both the excitatory actions of amphetamine on striatal neurons and the effects of this drug on open-field behavior. SCH-23390 and eticlopride also blocked the amphetamine-induced inhibition of motor-related neurons, but only eticlopride reversed the nonmotor-related inhibitory response to amphetamine. The relatively small number of amphetamine-induced inhibitions in our sample, however, precluded an adequate assessment of the role of dopaminergic mechanisms in this effect.

EXPERIMENT II: EFFECTS OF CEREBROCORTICAL ABLATIONS

The striatum receives major glutamatergic input from frontal and sensorimotor cortex. Glutamate is known to have an excitatory action on striatal neurons, and dopamine appears to enhance this effect (Chiodo and Berger, 1986). Recent models of dopamine function reflect this interaction and may shed light on the mechanisms by which amphetamine and other dopamine agonists alter striatal activity (Servan-Schreiber et al., 1990; Haracz et al., 1993). According to these models, dopamine acts as a gain-enhancing neuromodulator by facilitating or inhibiting the activity of neurons receiving, respectively, substantial or little excitatory afferent input If this view is correct, then ablation of the cerebral cortex, which serves as a primary source of excitatory striatal input, should selectively attenuate the excitatory, but not inhibitory, effects of amphetamine on striatal neurons. To test this hypothesis, we applied our behavioral clamping analysis to assess the effects of amphetamine on neuronal excitations and inhibitions in the striatum of cortical and sham lesioned rats.

Methods

Animals were prepared for single-unit recording as described above. During surgery, 15 rats sustained bilateral aspiration lesions of cerebrocortex from the frontal pole to approximately 2 mm posterior to bregma and extending 4-5 mm lateral to the midline (Tschanz et al., 1991). A separate group of 12 rats received sham lesions, and 4 control rats were left intact. Following a recovery period of 1-3 weeks, tungsten microelectrodes were lowered into the striatum, and single-unit discharges were recorded and stored on videotape (see above). Spontaneous neuronal and behavioral activity were recorded for at least 30 min, and all animals received 1.0 mg/kg d-amphetamine (sc). Drug-induced neuronal and behavioral responses were recorded for 30 min. The recording site was marked, and the brain was prepared for histological analysis (see above).

Table 2. Effect of cortical ablation on amphetamine-induced responses of striatal neurons under behavioral clamping conditions.[1]

Experimental Group	AMPH-Induced Excitations			AMPH-Induced Inhibitions		
	>	<	=	>	<	=
Control	25*	13	3	1	15**	9
Lesion	28	23	4	1	12***	10

[1]Pre- and post-amphetamine median spike counts for matched behavioral rating periods were compared for neurons excited or inhibited by amphetamine relative to pre-drug resting behavior. (>, <, and = indicate post-amphetamine firing rate is greater than, less than, or equal to behaviorally matched pre-amphetamine rate, respectively). Pooled outcomes within the lesion or control groups were analyzed with the sign test. * P = 0.02,   ** P < 0.001,   *** P < 0.004

For behavioral clamping, a rating scale was applied to videotaped 6-sec observation periods, during which similar behaviors (e.g., locomotion, rearing, and head movements) occurred before and after amphetamine. Behavioral clamping was accomplished in each case by matching pre- and post amphetamine observation periods with identical behavioral ratings. Data were obtained from a total of 32 randomly selected neurons either excited or inhibited by amphetamine relative to the pre-drug quiet baseline. Neuronal firing rates were assessed for each behavioral match. Pooled outcomes within lesioned or control groups were analyzed with the sign test.

Results and Discussion

An example of the extent of cerebrocortical ablation is shown in Figure 1. Such lesions failed to alter either basal firing in the striatum or open-field behavioral activity compared to control rats. As shown in Table 2, however, behavioral clamping revealed that cortical ablation blocked the excitatory, but not inhibitory, effects of amphetamine on striatal neurons. These differential drug-induced responses, therefore, appear to reflect different underlying mechanisms. In control rats with an intact cerebrocortex, separate subpopulations of striatal neurons were either excited or inhibited by amphetamine under behavioral

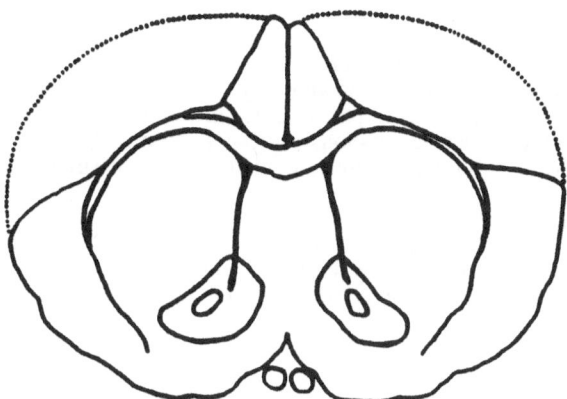

Figure 1. Representative example of cortical destruction showing the greatest extent of the lesion (dashed lines) 1.7 mm anterior to bregma (after Paxinos and Watson, 1982).

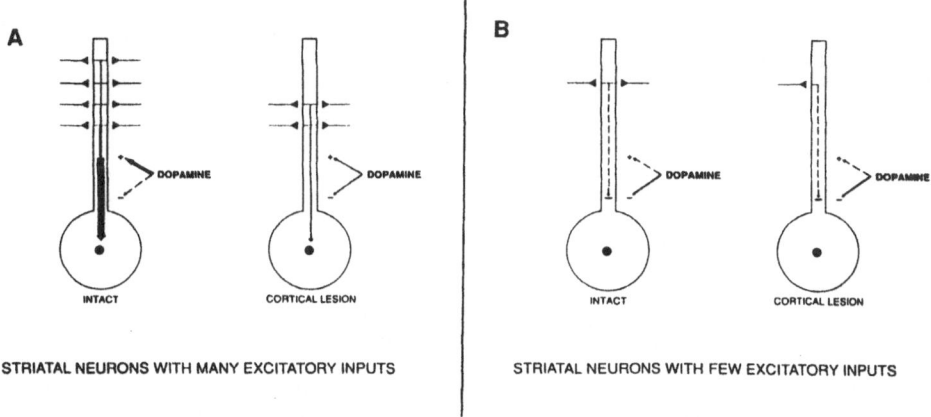

Figure 2. A model of bidirectional dopamine neuromodulation in striatum based on the view that dopamine enhances the gain of neuronal information processing by facilitating or attenuating the activity of neurons receiving substantial (A) or little (B) excitatory input. Line and arrow thickness symbolizes relative levels of signal propagation or neuromodulation, with dashed lines indicating the lowest relative levels. In neurons with many active excitatory inputs, dopamine is proposed to play a facilitatory role, but this effect is attenuated in partially denervated neurons of rats with cortical lesions (A). In contrast, the activity of neurons with a low level of excitatory input is attenuated by dopamine, regardless of whether or not the few afferents are destroyed (B).

clamping conditions. This finding suggests the divergence in firing rate produced by amphetamine is a primary drug effect independent of the secondary effects of behavior on neuronal activity.

CONCLUSION

Although the mechanisms responsible for amphetamine-induced excitations and inhibitions of striatal neurons remain to be determined, several factors suggest that dopamine is likely to play a key role in both. First, amphetamine facilitates dopamine release, and dopamine has been reported to exert both excitatory and inhibitory effects on striatal neurons (Herrling and Hull, 1980). Second, these differential actions of dopamine have been shown to modify glutamate-induced excitations (Chiodo and Berger, 1986). Third, amphetamine-induced excitations are reliably blocked by dopamine antagonists, and although further research is required to determine if these drugs reliably reverse the inhibitory neuronal response to amphetamine, clozapine (Haracz et al., 1993) has been shown to reverse at least some of these inhibitions. Finally, as shown in Fig. 2, models of bidirectional dopaminergic neuromodulation correctly predict our pattern of results, in which cortical ablation selectively impaired amphetamine-induced excitations of striatal neurons. Thus, our results support the concept of dopamine as a gain-enhancing neuromodulator in the striatum.

Acknowledgements

This research was supported by USPHS Grant DA 0241. Paul Langley provided expert technical assistance; Faye Caylor helped with the preparation of the manuscript.

REFERENCES

Chiodo, L.A., and Berger, T.W., 1986, Interactions between dopamine and amino acid-induced excitation and inhibition in the striatum, *Brain Res.* 375:198-203.

Gardiner, T.W., Iverson, D.A., and Rebec, G.V., 1988, Heterogeneous responses of neostriatal neurons to amphetamine in freely moving rats, *Brain Res.* 463:268-274.

Haracz, J.L., Tschanz, J.T., Greenberg, J., and Rebec, G.V., 1989, Amphetamine-induced excitations predominate in single neostriatal neurons showing motor-related activity, *Brain Res.* 489:363-368.

Haracz, J.L., Tschanz, J.T., Wang, Z., White, I.M., and Rebec G.V., 1993, Striatal single-unit responses to amphetamine and neuroleptics in freely moving rats, *Neurosci. Biobehav. Res.*, in press.

Herrling, P.L., and Hull, C.D., 1980, Iontophoretically applied dopamine depolarizes and hyperpolarizes the membrane of cat caudate neurons, *Brain Res.* 192:441-462.

Kuczenski, R., and Segal, D., 1989, Concomitant characterization of behavioral and striatal neurotransmitter response to amphetamine using in vivo microdialysis, *J. Neurosci.* 9:2051-2065.

Nishibe, Y., Matsuo, Y., Yoshizaki, T., Eigyo, M., Shiomi, T., and Hirose, K. ,1982, Differential effects of sulpiride and metoclopramide on brain homovanillic acid levels and shuttlebox avoidance after systemic and intracerebral administration, *Naunyn-Schmiedeberg's Arch. Pharmacol.* 321:190-194.

Paxinos, G., and Watson, C., 1982, "The Rat Brain in Stereotaxic Coordinates," Academic Press, New York

Rebec, G.V., (1992) Neuropharmacology of amphetamine: changes in the activity of striatal neurons during behavior, *in:* "Drugs of Abuse and Neurobiology, CRC Series Physiology of Drug Abuse," R. R. Watson, ed., CRC press, Boca Raton, FL, pp. 149-174.

Rebec, G.V., Haracz, J.L., Tschanz, J.T., Wang, Z., and White, I., 1991, Responses of motor- and nonmotor-related neostriatal neurons to amphetamine and neuroleptic drugs, *in:* "Basal Ganglia III," G. Bernardi, M.B. Carpenter, G. DiChiara, M. Morelli and P. Stanzione, eds., Plenum Press, New York, pp. 463-470.

Rebec, G.V., Langley, P.E., Pierce, R.C., Wang, Z., and Heidenreich, B.A., 1993, A simple micromanipulator for multiple uses in freely moving rats: electrophysiology, voltammetry, and simultaneous, *J. Neurosci. Meth.* in press.

Rebec, G.V., Peirson, E.E., McPherson, F.A., and Brugge, K., 1982, Differential sensitization to amphetamine following long-term treatment with haloperidol or clozapine, *Psychopharmacol.* 77:360-366.

Ryan, L.J., Young, S.J., Segal, D.S. and Groves, P.M., 1989, Antidromically identified striatonigral projection neurons in the chronically implanted behaving rat: relations of cell firing to amphetamine-induced behaviors, *Behav. Neurosci.* 103:3-14.

Servan-Schreiber, D., Printz, H., and Cohen, J. D., 1990, A network model of catecholamine effects: Gain, signal-to-noise ratio, and behavior, *Science* 249:892-895.

Sharp, T., Zetterstrom, T., Ljungberg, T., and Ungerstedt, U., 1987, A direct comparison of amphetamine-induced behaviours and regional brain dopamine release in the rat using intracerebral dialysis, *Brain Res.* 401:322-330.

Trulson M.E., and Jacobs B.L., 1979, Effects of d-amphetamine on striatal unit activity and behavior in freely moving cats, *Neuropharmacol.* 18:735-738.

Tschanz, J.T., Haracz, J.L., Griffith, K.E., and Rebec, G.V., 1991, Bilateral cortical ablations attenuate amphetamine-induced excitations of neostriatal motor-related neurons in freely moving rats, *Neurosci. Lett.* 134:127-130.

Wachtel, S.R., Hu, X.-T., Galloway, M.P., White, F.J., 1989, D1 dopamine receptor stimulation enables the postsynaptic, but not autoreceptor, effects of D2 dopamine agonists in nigrostriatal and mesoaccumbens dopamine systems, *Synapse* 4:327-346.

Wang, Z.R, and Rebec, G.V., 1992, Neuronal and behavioral correlates of intrastriatal infusions of amphetamine in freely moving rats. *Soc. Neurosci. Abstr.* 18:692.

Wang, Z., Haracz, J.L., and Rebec, G.V., 1992, BMY-14802, a sigma ligand and potential antipsychotic drug, reverses amphetamine-induced changes in neostriatal single-unit activity in freely moving rats, *Synapse* 12:312-321.

West, M.O., Michael, A.J., Knowles, S.E., Chapin, J.K., and Woodward, D.J., 1987, Striatal unit activity and the linkage between sensory and motor events, *in*: "Basal Ganglia and Behavior: Sensory Aspects of Motor Functioning," J.S. Schneider and T.I. Lidsky, eds., Hans Huber, Toronto, pp. 27-35.

White F.J., Bednarz L.M., Wachtel S.R., Hjorth S., and Brooderson R.J., 1988, Is stimulation of both D1 and D2 receptors necessary for the expression of dopamine-mediated behaviors? *Pharmacol. Biochem. Behav.* 30:189-193.

NICOTINE, COCAINE AND AMPHETAMINE SHARE SEVERAL COMMON MECHANISMS OF c-fos INDUCTION IN THE STRIATUM: IMPLICATIONS FOR THE FUNCTIONS OF THE STRIATUM

Hideo Kiba and A. Jayaraman

Department of Neurology
Louisiana State University School of Medicine
New Orleans, LA 70112

INTRODUCTION

The striatum (nucleus accumbens and the caudatoputamen) receives dense dopaminergic input form several subdivisions of the ventral tegmental area (VTA) and the substantia nigra pars compacta (SNpc) (Fallon, 1988). Different human neurological disorders and varieties of animals models that have manipulated the striatal dopaminergic innervation have suggested that dopamine plays a major role in integrating the various inputs and the neurotransmitter systems intrinsic to the striatum. Dysfunctions of the dopaminergic input to the striatum result in a complex spectrum of hypo- and hyperkinetic as well as brady- and hyperphrenic syndromes. Our understanding of the functions of the basal ganglia clearly depends on better definition of the role played by the dopamine in the striatum.

Nicotine, cocaine and amphetamine are some of the most commonly used and abused psychostimulant drugs (Wise and Rompre, 1989). Acute administration of these drugs results in very high levels of extracellular dopamine in the striatum. Nicotine stimulates nicotinic receptors in the soma and the terminals of the DA neurons of the midbrain and release very high levels of dopamine in the striatum (Giorguieff-Chesselet et al., 1979; Clarke and Pert, 1985; Grenhoff et al., 1986; Imperato et al., 1986). The striatum appears to be one of the most sensitive areas of the brain to respond to nicotine. Application of 1mM of nicotine into the striatum by microdialysis metho leads to two-fold increase in extracellular dopamine and administration of 100mM of nicotine into the striatum increased extracellular dopamine levels by 60-fold (Toth et al., 1992). The acute euphoric properties of cocaine are due its interactions with the DA transporter molecule and blocking of DA reuptake at the VTA-accumbens prefrontal cortex DAergic fiber terminals. Cocaine may also cause the release of DA stored in the vesicles. In vivo microdialysis studies suggest that within 10mts after acute injections of cocaine, DA levels increase even upto 300% and the elevated DA levels are correlated well with the increased locomotor activity (Hurd et al., 1988, 1989)). Amphetamine enhances release of dopamine, including the cytosolic component (McMillen, 1983), and also blocks the reuptake of dopamine (Lewander, 1977; Moore, 1978).

Nicotine, cocaine and amphetamine are powerful reinforcers (Wise and Rompre, 1989) and are excellent tools for studying the metabolic and behavioral consequences of increased dopaminergic transmission. The present report summarizes and speculates on the potential significance of the results from some of our experiments designed to study the effects of increased dopamine transmission after acute administration of these psychostimulants.

To achieve these goals we have studied the pattern of expression of the gene c-fos, a member of the immediate early gene family, in the striatum (Sheng and Greenberg, 1990;

Robertson, 1992). This molecular pathway has been studied intensely and experimentally manipulated in the central nervous system to understand issues relating to, among others, 1. neuroanatomical pathways that are specifically involved in seizure propagation (Morgan et al., 1987), and 2. to specify pharmacological properties of several dopaminergic agonists and antagonists (Robertson and Fibiger, 1992).

MATERIALS AND METHODS

To identify the area of striatum that mediate the acute effects of these psychostimulants, different groups of rats received 0.1 to 1.4mg/kg of nicotine s.c., 15 to 30mg/kg of cocaine, i.p., and 2.5 and 5.0mg/kg of amphetamine i.p. In order to identify the role played by D1 receptor in mediating the acute effects of nicotine injections, several rats received either SCH 23390 alone or SCH 23390 followed in 30 mins by injections of 1.0 or 2.0mg/kg of nicotine. In order to define the role played by glutamate on dopamine induced c-fos activation, injections of non competitive NMDA antagonist MK 801 (1mg/kg, i.p.,) and the specific and competitive antagonist CPP (4mg/kg, i.p.,), were injected 30min prior to injections of 1.0mg/kg of nicotine, 15mg/kg of cocaine and 2.5mg/kg of amphetamine.

All animals were killed 2hours after drug injections and the pattern of expression of c-fos gene within the striatum was mapped with standard ABC immunocytochemical methods using Fos antibody from Oncogene Sciences, Uniondale, NY.

ACUTE NICOTINE, COCAINE AND AMPHETAMINE INJECTIONS INDUCE c-fos IN THE STRIATUM

Results

Injections of nicotine, cocaine and amphetamine result in intense expression of Fos in different areas of the nucleus accumbens and the caudatoputamen (CPu) (Fig.1).

Significance

Acute injections of not only nicotine, cocaine and amphetamine, but also morphine (Chang et al., 1988) induce Fos activity in striatal neurons. The significance of Fos induction by acute injections of these psychostimulants remains to be explored. The c-fos gene encodes Fos, a phosphoprotein, that combines with the protein encoded by another immediate early gene, Jun, to form a potent heterodimeric transcription factor. The Fos:Jun heterodimeric complex binds to the AP-1 binding site of several late response genes to play a major role in intracellular signalling, signal transduction and transformation and direct subsequent genomic response by neurons (Sheng and Greenberg, 1990; Robertson, 1992).

The c-fos gene is activated by varieties of physiological and experimental stimuli. The seemingly nonspecific nature of c-fos activation may not be nonspecific after all. In striatal cell culture system, cocaine induces c-fos and jun B mRNAs, but not c-jun (Moratalla et al., 1993). Despite several lines of evidence showing that the SNpc cells contain nicotinic receptors, and that iontophoretic and extraneural injections of nicotine strongly stimulate the SNpc cells, these cells were totally devoid of Fos-Li even after very high doses of nicotine (Pang et al., 1993). These results may suggest that the direct and indirect stimulation of SNpc cells by nicotine may induce an immediate early gene other than c-fos. Even though c-fos can be induced by stimulation of the NMDA receptors by injecting glutamate agonists directly into the striatum (Beretta et al., 1992), or by stimulating the cerebral cortex electrically (Fu and Beckstead, 1991) or with epileptogenesis (Morgan et al., 1987), expression of yet another immediate early gene zif-268 and not c-fos is linked to NMDA receptor mediated LTP (Sheng and Greenberg, 1990; Abraham et al., 1991; Robertson, 1992). These results suggest that induction of Fos and other immediate early genes may have tissue, stimuli and function specific patterns of expression in the central nervous system. The induction of Fos by acute administration of these psychostimulants may indeed be only an "early" response to these drugs, but not a long lasting one. This early response may consist of binding to AP-1 binding sites of different genes for neuropeptides, viz., enkephalin and dynorphin genes in striatal cells, as means of directing subsequent longer lasting genomic response (Sheng and Greenberg, 1990; Robertson, 1992).

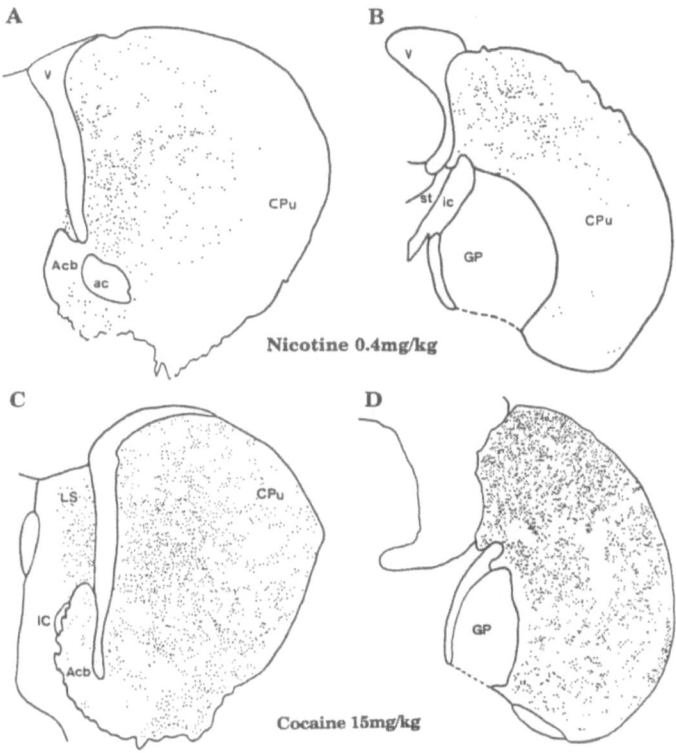

Figure 1. Camera lucida drawings of the pattern of distribution of Fos immunoreactive neurons in the striatum after nicotine and cocaine injections. Note that the distribution of Fos immunoreactive cells are more prominent in the caudatoputamen than in the nucleus accumbens.

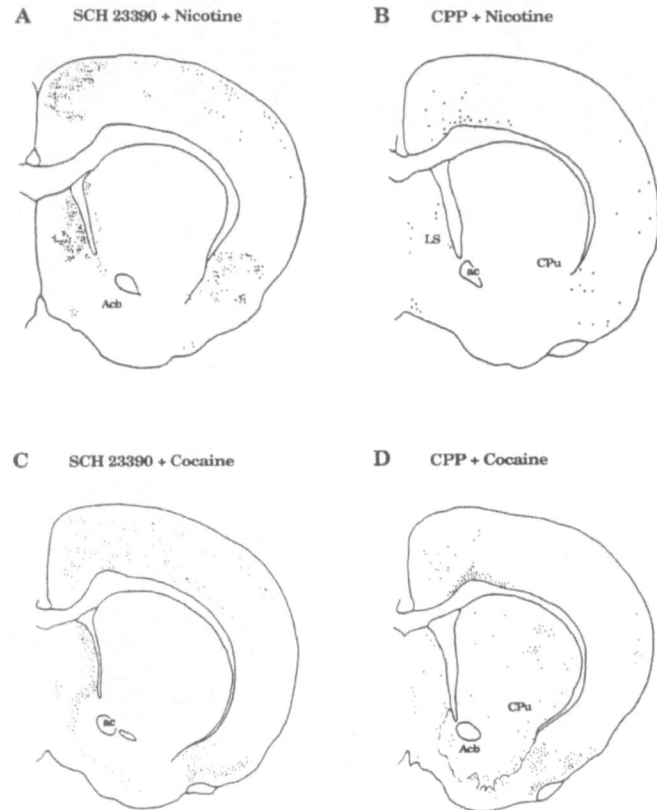

Figure 2. Camera lucida drawings of Fos reactive neurons in the forebrain in those rats receiving dopamine D1 antagonist SCH 23390 and the NMDA antagonist CPP prior to receiving nicotine and cocaine injections. Note that the striatum is relatively free of Fos labeled cells, whereas the different cortical regions continue to exhibit Fos immunoreactive cells.

NICOTINE, COCAINE AND AMPHETAMINE INDUCED c-fos EXPRESSION IS MEDIATED BY DOPAMINE D1 RECEPTORS

Results

If nicotine, cocaine and amphetamine are administered 30 min after injections of 1mg/kg of SCH 23390, the specific dopamine D1 receptor antagonist, neurons in several cortical regions were labeled prominently with Fos-Li, but Fos immunoreactive cells were totally absent in the striatum. These results suggest that expression of c-fos gene resulting from acute psychostimulant injections are mediated by stimulation of D1 receptors.

Significance

While dopamine plays a major role in intracranial self-stimulation (ICSS) (Wise and Rompre, 1989), the role played by the different dopamine receptors in ICSS is incompletely understood. The dopamine D1 and D2 receptors have both synergistic and oppositional interactions between each other (Waddington, 1989; Daly and Waddington, 1992). The synergistic effects of D1 and D2 receptors may subserve majority of dopamine mediated behaviors (Waddington and O'Boyle, 1987). Numerous studies have suggested that both D1 and D2 receptors contribute to reinforcement behavior (Corrigal and Coen, 1991; Hubner and Moreton, 1991; Witkin et al., 1991; Koob, 1992; McDougall et al., 1992). Recently it has been suggested that D1 receptors may play a more important role in the reinforcing properties of cocaine than D2 receptors (Robledo et al., 1992; Koob, 1992). Even though experimental evidence are inadequate to define the individual roles of D1 and D2 receptors in ICSS, the view that the D1 receptors (Koob, 1992; Woolverton and Johnson, 1992) may play a more important role in ICSS than D2 receptors may be supported by the fact, as shown in the current study, that stimulation of the D1 receptors appears to be a major neurochemical pathway by which psychostimulants induced acutely enhanced dopamine transmission results in intense expression of c-fos gene in the striatum. The roles of D2, D3 and D4 in ICSS remain to be studied.

PSYCHOSTIMULANTS INDUCED FOS EXPRESSION IS DEPENDENT ON NMDA RECEPTOR STIMULATION

Results

The striatum receives dense glutamatergic projections from many cerebral cortical areas and glutamate has been reported to facilitate the release of dopamine. Injections of non competitive NMDA antagonist MK 801 and the competitive antagonist CPP, were administered 30min prior to injections of 1.0mg/kg of nicotine, 15mg/kg of cocaine and 2.5mg/kg of amphetamine administration completely abolished Fos expression in the striatum and the Fos immunoreactive cells were considerably less in various cortical areas. These results suggest that blocking of NMDA receptors significantly inhibits c-fos activation in striatal neurons.

Significance

NMDA receptors play an important role in dopamine induced Fos expression by nicotine, cocaine and amphetamine. Torres and Rivier (1993) also showed that injections of ketamine, MK 801 and CPP inhibited cocaine induced Fos expression in the striatum significantly. Glutamate has been suggested to play a major role in long term potentiation (LTP) (Bliss and Collingridge, 1993). Glutamate has been demonstrated to release dopamine from the mesolimbic and mesostriatal terminals (Imperato et al., 1990; Krebs et al., 1991).

The role played by the facilitating actions of glutamate on dopamine in striatal functions remains to be studied. It is possible that glutamate-dopamine interactions in the striatum may be an example of mechanisms subserving detection of coincident signals (Bourne and Nicoll, 1993).

NICOTINE, COCAINE AND AMPHETAMINE INDUCE FOS EXPRESSION MORE PROMINENTLY IN THE CAUDATOPUTAMEN THAN IN THE NUCLEUS ACCUMBENS

Results

After 0.3 mg/kg; s.c., of nicotine significant number of Fos reactive neurons were noted in the nucleus accumbens and the CPu. Neurons with Fos like-immunoreactivity (Fos-Li) were more in the shell region than core area of the nucleus accumbens. Neurons with Fos-Li were very prominent in the ventro- and dorsomedial regions of the caudate nucleus, but the dorso- and ventrolateral CPu contained only a few scattered Fos labeled cells. These effects of nicotine were complete blocked by the central nicotinic antagonist mecamylamine.

Cocaine (15mg/kg i.p.) induced intense expression of Fos in the nucleus accumbens and the CPu. Cocaine induced a strikingly more intense activation of the c-fos gene than even high doses (1.4mg/kg) of nicotine. Fos reactive neurons were seen diffusely in the nucleus accumbens, but they were more extensively distributed in the CPu than the nucleus accumbens. The most intense expression of Fos was noted in the central and anteromedial regions of CPu, whereas dorsolateral regions of the CPu also contained significant number of Fos reactive neurons, but considerably less in intensity.

Amphetamine injections (2.5mg/kg) resulted in labeling of medium sized neurons of the striatum, but in a patchymanner, similar to the pattern reported by Graybiel et al. (1990). Fos immunoreactive cells were noted more densely in the shell regions than the core region of the nucleus accumbens. But Fos immunoreactive cells were more prominent in anteromedial and central regions of the caudate nucleus. The dorsolateral areas were relatively free of Fos reactive neurons with this dose of amphetamine.

With increasing doses of these drugs (1.4mg/kg of nicotine s.c., 30mg/kg of cocaine and 5mg/kg of amphetamine) an increased intensity of Fos expression was noted in the striatum. In these cases the neurons expressing Fos-Li were also noted in the dorsolateral and ventrolateral striatum. The dose dependent increase in distribution and the number neurons expressing Fos in the striatum appears to be, as noted by Young et al., (1991) quite characteristic of indirect dopamine enhancing agents.

Significance

The results from our studies clearly show that acute injections of these psycho-stimulants result in induction of the c-fos gene, in a dose dependent fashion, more intensely in the CPu than in the nucleus accumbens. Even with the lowest dose required to induce Fos in CPu, Fos reactive cells were noted prominently in the anteromedial and central CPu than dorso- or ventrolaterally.

The dopaminergic VTA - nucleus accumbens axis plays a major role in reinforcement behavior. The psychomotor stimulant and the positive reinforcing effects of many drugs of abuse have been proposed to be mediated by the dopaminergic neurons of the VTA and their terminals in the nucleus accumbens.

The observation that these drugs, which are powerful positive reinforcing agents, induce intense expression of Fos in the anteromedial and central CPu more intensely than in the nucleus accumbens certainly raises the question whether the medial and anteromedial CPu also plays a role in ICSS? If dopaminergic innervation in the nucleus accumbens plays a major role in ICSS, what then is the role of the dense dopaminergic terminals in the rostral, dorsomedial and central CPu, the site of convergence of dense projections from the prefrontal, frontal and ventrolateral orbital cortical regions (Jayaraman, 1985; McGeorge and Faull, 1989)? The striatum is divided into several independent (Jayaraman, 1985) but not necessarily parallel subcompartments, since the output from these subcompartments may ultimately converge in areas of the cerebral cortex that subserve attentional mechanisms (Jayaraman, 1987). Does the dopaminergic innervation of the anteromedial and central caudate, for example, play a role in selectively reinforcing, either positively or negatively, cognitive decisions derived from the orbitofrontal, prefrontal and frontal cortical areas which underlie behaviors related to positive reinforcement? Does the dopaminergic projection in the sensorimotor dorsolateral striatum have similar reinforcing effects on motor programs that are required to complete the motoric aspects of a rewarding response?

It is of interest to note that even some of the earliest studies of Routtenberg (1969) have shown that medial CPu can support ICSS. The possibility that the medial CPu may play a role in ICSS in rats,cats and monkeys have been suggested by many other earlier investigators (reviewed in Phillips et al., 1976). Using 6-OHDA lesioning of SNpc, Phillips et al. (1976) further showed that CPu, especially the head of the CPu, can support ICCS. Subsequently Phillips and Fibiger (1989) have proposed that the nucleus accumbens and the CPu may only be part of several neuroanatomical regions that may be involved in brain-stimulation-reward phenomenon.

The question whether the dense dopaminergic input to the CPu plays a role in reward and reinforcing behavior certainly remains to be answered in the future. This will require using improved techniques to stimulate the individual striatal compartments and behavior paradigms to test the motivational, cognitive and motor components of reward behavior *individually*. In this regard results from the studies of Kelley and Delfs (1991), is of great interest. Their study clearly shows that, after microinjections of dopamine enhancing psychostimulant amphetamine in several functionally different subregions of the striatum, in addition to the nucleus accumbens, the antero- and ventromedial CPu but not the dorso and ventrolateral striatum also showed significant and selective increases in the conditioned reinforcement response. These studies certainly support the view that the CPu may also play an important role in ICSS.

CONCLUSIONS

The results from these studies support the following conclusions:

1. Acute enhancement of dopaminergic transmission with psychostimulants induces expression of the immediate early gene c-fos in the striatum. The intensity of expression of Fos varies with different types of psychostimulants used in these experiments, with cocaine resulting in the most pronounced pattern of Fos expression in the striatum. The extent of Fos expression is dose dependent,

2. Fos expression in the postsynaptic striatal neurons are mediated by stimulation of dopamine D1 receptors,

3. The psychostimulant induced Fos activation is dependent on NMDA receptor stimulation. The glutamate and dopamine interactions may subserve as a mechanism of detecting coincident signals, and

4. The intense expression of Fos in the neurons of the anteromedial and central CPu suggests that *anteromedial CPu may play a role which is equally important* to that of activation of VTA-nucleus accumbens axis in the development of positive psychomotor behavior and ICSS.

Acknowledgments

Supported by The Smokeless Tobacco Research Council, NY and The Neuroscience Center of Excellence Incentive Grant from LSU. Dr. Kiba is a Visiting Scientist from Nihon University School of Dentistry at Matsudo, Matsudo, Chiba, Japan.

REFERENCES

Abraham, W.C., Dragunow, M., and Tate, W.P., 1991, The role of immediate early genes in the stabilization of long term potentiation, *Mol. Neurobiol.* 5:297-314.

Berretta, S., Robertson, H.A., and Graybiel, A.M., 1992, Dopamine and glutamate agonists stimulate neuron specific expression of fos-like protein in the striatum, *J. Neurophysiol.* 68:767-777.

Bliss, T.V.P., and Collingridge, G.L., 1993, A synaptic model of memory:long term potentiation in the hippocampus, *Nature* 361:31-39.

Bourne, H.R., and Nicoll, R., 1993, Molecular machines integrate coincident synaptic signals, *Cell* 72/*Neuron* 10:65-75.

Chang, S.L., Squinto, S.P., and Harlan, R.E., 1988, Morphine activation of c-fos expression in rat brain, *Biochem. biophys. Res. Commun.* 2:698-704.

Clarke, P.B.S., and Pert, A., 1985, Autoradiographic evidence for nicotine receptors on nigrostriatal and mesolimbic dopaminergic neurons, *Brain Res.* 348:355-358.

Corrigall, W.A., and Coen, K.M., 1991, Cocaine self-administration is increased by both D1 and D2 dopamine antagonist, *Pharmacol. Biochem. Behav.* 39:799-802.

Daly, S.A., and Waddington, J.L., 1992, Two directions of dopamine D1/D2 receptor interaction in studies of behavioural regulation: a finding generic to four new, selective dopamine D1 receptor antagonists, *Eur. J. Pharamacol.* 213:251-258.

Fallon, J.H, 1988, Topographic organization of ascending dopaminergic projections, *in:* "The Mesocorticolimbic dopmaine system," P.W. Kalivas and C.B. Nemeroff, eds., *Ann. NY Acad. Sci.* 537:1-9.

Fu, L., and Beckstead, R.M., 1991, Cortical stimulation induces fos expression in striatal neurons, *Neuroscience* 46:329-334.

Giorguieff-Chesselet, M.F., Kemel, M.L., Wandscheer, D., and Glowinski, J., 1979, Regulation of dopamine release by nicotinic receptors in rat striatal slices:Effect of nicotine in a low concentration, *Life. Sci.* 25:1257-1262.

Graybiel, A.M., Moratalla, R., and Robertson, H.A., 1990, Amphetamine and cocaine induce drug-specific activation of the c-fos gene in striosome-matrix compartments and limbic subdivisions of the striatum, *Proc. Natl. Acad. Sci. USA* 87:6912-6916.

Grenhoff, J., Ashton-Jones, G., and Svennson, T.H., 1986, Nicotinic effects on the firing pattern of midbrain dopamine neurons, *Acta Physiol. Scand.* 128:351-358.

Hubner, C.B., and Moreton, J.E., 1991, Effects of selective D1 and D2 dopamine antagonists on cocaine self-administration in the rat, *Psychopharmacology (Berl.)* 105:151-156.

Hurd, Y.L., Kehr, J., and Ungerstedt, U., 1988, In vivo microdialysis as a technique to monitor drug transport: correlation of extracellular cocaine levels and dopamine overflow, *J Neurochem.* 51:1314-1416.

Hurd, Y.L., Weiss, F., Koob, G.F., and, N.E., and Ungerstedt, U., 1989, Cocaine reinforcement and extracvellular dopamine overflow in rat nucleus accumbens:an in vivo microdialysis study, *Brain Res.* 498:199-203.

Imperato, A., Mulas, A., and DiChiara, G., 1986, Nicotine preferentially stimulates dopamine release in the limbic system of freely moving rats, *Eur. J. Pharmac.* 132:337-338.

Imperato, A., Scrocco, M.G., Bacchi, S., and Angelucci, L., 1990, NMDA receptors and in vivo dopamine release in the nucleus accumbens and caudatus, *Eur. J. Pharmacol.* 187:555-556.

Jayaraman, A., 1985, Functional subcompartments of the striatum, *Soc. Neurosci. Abst.* 11:199.

Jayaraman, A., 1987, The Basal Ganglia and Cognition: An Interpretation of Anatomical Connectivity Pattern, *in:* "Basal Ganglia and Behavior: Sensory aspects of Motor Functioning," J.S. Schneider and T. Lidsky, eds., Hans Huber Publications, Toronto, p. 149.

Kelley, A.E., and Delfs, J.M., 1991, Dopamine and conditioned reinforcement, *Psychopharmac.* 103:187-196.

Koob, G.F, 1992, Neural mechanisms of drug reinforcement, *Ann. NY. Acad. Sci.* 654:171-191.

Krebs, M.O., Trovero, F., Desban, M., Gauchy, C., Glowinski, J., and Kemel, M.L, 1991, Distinct presynaptic regulation of dopamine release through NMDA receptors in Striosome- and Matrix-enriched areas of the rat striatum, *J. Neurosci.* 11:1256-1262.

Lewander, T., 1977, Effects of amphetamines in animals, *in:* "Drug Addiction," W.R. Martin, ed., Springer-Verlag, New York, p. 33.

McDougall, S.A., Crawford, C.A., and Nonneman, A.J., 1992, Reinforced responding of the 11-day-old rat pup: synergistic interaction of D1 and D2 dopamine receptors, *Pharmacol. Biochem. Behav.* 42:163-168.

McGeorge, A.J., and Faull, R.L.M., 1989 The organization of the projection from the cerebral crotex to the striatum in the rat, *Neuroscience* 29:503-537.

McMillen, B.A., 1983, CNS stimulants: two distinct mechanisms of action for amphetamine-like drugs, *Trends Pharmacol. Sci.* 4: 429-432.

Moore, K.E., 1978, Amphetamines: Biochemical and behavioral actions in animals, *in:* "Handbook of Psychopharmacology," L.L. Iversen, S.D. Iversen, and S.H. Snyder, eds., Plenum Press, New York, p. 41.

Moratalla, R., Vickers, E.A., Robertson, H.A., Cochran, B.H., and Graybiel, A.M., 1993, Coordinate expression of c-fos and jun B is induced in the rat striatum by cocaine, *J. Neurosci.* 13:423-433.

Morgan, J.I., Cohen, D.R., Hempstead, D.L., and Curran, T, 1987, Mapping patterns of c-fos expression in the central nervous system after seizure, *Science* 237:192-197.

Pang, Y., Kiba, H., and Jayaraman, A., 1993, Acute nicotine injections induce c-fos mostly in nondopaminergic neurons of the midbrain of the rat, *Mol. Brain Res.* in press.

Phillips, A.G., and Fibiger, H.C., 1989, Neuroanatonical basis of intracranial self-stimulation: untangling the Gordian knot, *in:* "The Neuropharmacological Basis of Reward," J.M. Liebman and S.J. Cooper, eds., Clarendon Press, Oxford, pp. .

Phillips, A.G., Carter, D.A., and Fibiger, H.C., 1976, Dopaminergic substrates of intracranial self-stimulation in the caudate-putamen, *Brain Res.* 104:221-232.

Robertson, G.S., and Fibiger, H.C., 1992, Neuroleptics increase c-fos expression in the forebrain: contrasting effects of haloperidol and clozapine, *Neuroscience* 46:315-328.

Robertson, H.A., 1992, Immediate early genes, neuronal plasticity, and memory, *Biochem. Cell Biol.,* 70:729-737.

Robledo, P., Maldonado-Lopez, R., and Koob, G.F,1992 , Role of dopamine receptors in the nucleus accumbens in the rewarding properties of cocaine, *Ann. N. Acad. Sci.* 654:509-512.

Routtenberg, A., 1969, Forebrain pathways of reward in Rattus norvegicus, *Comp. Physiol. Psych.* 75:269-276.

Sheng, M., and M.E., Greenberg, M.E., 1990, The regulation and function of c-fos and other immediate early genes in the nervous system, *Neuron* 4:477-485.

Torres, G., and Rivier, C., 1993, Cocaine-induced expression of striatal c-fos in the rat is inhibited by NMDA receptor antagonists, *Brain Res. Bull.* 30:173-176.

Toth, E. Sershen, H., Hashim, A., Vizi, E.S., and Lajtha, A., 1992, Effect of Nicotine on Extracellular Levels of Neurotransmitters Assessed by Microdialysis in Various Brain Regions: Role of Glutamic Acid, *Neurochem. Res.* 17:265-271.

Waddington, J.L., 1989, Functional interactions between D-1 and D-2 dopamine receptor systems: their role in the regulation of psychomotor behaviour, putative mechanisms, and clinical relevance, *J. Psychopharmacol.* 3:54-61.

Waddington, J.L., and O'Boyle, K.M., 1987, The D-1 dopmaine receptor and the search for its functional role: from neurochemistry to behavior, *Rev. Neurosci.* 1:157-184.

Wise, R.A., and Rompre, P.P., 1989, Brain dopamine and reward, *Ann. Rev. Psychol.* 40:191-225.

Witkin, J.M., Schindler, C.W., Tella, S.R., and Goldberg, S.R, 1991, Interaction of haloperidol and SCH 23390 with cocaine and dopamine receptor subtype-selective agonists on schedule-controlled behavior of squirrel monkeys, *Psychopharmacology (Berl.)* 104:425-431.

Woolverton, W.L., and Johnson, K.M., 1992, Neurobiology of cocaine abuse, *Trends in Pharmac. Sci.* 13:193-200.

Young, S.T., Porrino, L.J., and Iadarola, M.J., 1991, Cocaine induces striatal c-Fos-immunoreactive proteins via dopaminergic D1 receptors, *Proc. Natl. Acad. Sci. USA*, 88:1291-1295.

GLUTAMATE-DOPAMINE BALANCE IN THE STRIATUM: PRE- AND POST-SYNAPTIC INTERACTIONS

Béatrice Lannes and Gabriel Micheletti

Institut de Physiologie
Faculté de Médecine
4, rue Kirschleger
F-67085 Strasbourg

INTRODUCTION

The striatum receives a major dopaminergic input, the nigro-striatal pathway originating from the substantia nigra pars compacta (SNc) (Dahlström and Fuxe, 1964; Andén et al., 1966). It receives also a massive excitatory input originating from the cortex and the thalamus (Grofova, 1979; Parent, 1990). Glutamate (GLU) is the candidate neurotransmitter of both of these pathways (Mc Geer et al., 1977; Reubi and Cuenod, 1979; Fonnum et al., 1981; Lapper and Bolam, 1992), although the neurotransmitter of the thalamo-striatal pathway is still controversial (Nieoullon et al., 1985; Nieoullon, 1986; Kilpatrick and Phillipson, 1986). During the past years, various experimental studies have established that the striatum is the site of reciprocal interactions between dopaminergic and glutamatergic neurotransmissions leading to the hypothesis that imbalance of these interactions may be involved in the pathogenesis of Parkinson's disease (Nieoullon et al., 1982) or schizophrenia (Kim et al., 1980; Carlsson and Carlsson, 1990; Grace, 1991). Our purpose here is to review these interactions and their functional implications.

A survey of the literature published since the review describing the influence of the cortico-striatal glutamatergic neurons on the striatal dopaminergic transmission (Glowinski et al., 1991) reveals three major points. 1) A concept of volume transmission has emerged lately that seems especially relevant for both glutamatergic and dopaminergic neurotransmissions. 2) Striatum is probably not a unique site for such interactions since SNc also receives excitatory glutamatergic afferents from cortical areas and from subthalamic and pedunculopontine nuclei. Excitability and firing of nigral dopaminergic neurons are controlled by these afferents through NMDA and non-NMDA glutamatergic receptors. 3) Experimental data suggest that, besides well known presynaptic striatal interactions between dopamine (DA) and GLU (Chesselet, 1984), there are interactions also at the postsynaptic level, involving NMDA receptors.

TWO MODES OF NEUROTRANSMITTER RELEASE

Classical synaptic transmission

The basic mechanisms whereby one chemical transmitter is released from presynaptic nerve terminals and acts on postsynaptic neurons are well described. The neurotransmitter is released by a propagated action potential which depolarizes axon terminals. Neurotransmitter release is actually secondary to an increased Ca^{2+} concentration in the axoplasm due to either a Ca^{2+} efflux from intracellular storage sites or to an enhanced influx of Ca^{2+} from the extracellular medium. Next, the neurotransmitter crosses the synaptic cleft and binds to its postsynaptic receptors (Katz, 1969; Augustine et al., 1987).

Volume transmission

The main concept of volume transmission is that a neurotransmitter may also be released independently of the constraints linked to synaptic transmission. This concept emphasizes that, besides the synaptic communication which is the most generally accepted structural basis for interneuronal communication, there may be an additional mechanism of neuronal interactions through non-synaptic contacts (for review, see Fuxe and Agnati, 1991; Vizi and Labos, 1991). The neurotransmitter is released from axon terminals through a "leaking" process, usually controlled by another neurotransmitter system, diffuses towards extracellular space and interacts with its receptors after travelling sometimes large distances, up to millimeters (Fuxe and Agnati, 1991). These extrasynaptic receptors present a higher sensitivity than receptors which are involved in classical synaptic transmission. Volume transmission may also imply glial cells. Regarding DA/GLU presynaptic interactions, such a glial participation may well happen since astrocytes are thought to regulate the balance between transmitter and metabolic glutamatergic pools (Barres, 1991; Martin, 1992). Of course, volume transmission coexists with classical synaptic transmission.

Phasic versus tonic release

Grace (1991) proposed that DA release depends on a dual mechanism which fits well with both synaptic and non-synaptic (volume) transmission. In the following, Grace's model is considered in order to describe striatal DA/GLU interactions.

Phasic release. One mechanism is proposed to underlie *phasic* neurotransmitter release. It is depicted by the classical model of synaptic neurotransmission and represents a spike-dependent event occuring in response to neuronal discharge elicited by either physiological or experimental (electrical/pharmacological) stimuli. This phasic release defined by *in vitro* studies is calcium-dependent and sensitive to Na^+ and Ca^{2+} channel blockers (such as tetrodotoxin (TTX) and procaine, respectively). Neurotransmitter released by this mechanism does not elicit measurable changes in its extracellular concentration since it is rapidly removed from the synaptic cleft by reuptake processes. Thus, this transient release does not induce long term changes in the responsiveness of the system. However, when non physiological stimulations are applied, (*e.g.* a neuronal stimulation with much greater frequency than a spontaneous discharge) they can actually increase extracellular neurotransmitter levels (Kuhr et al., 1984, 1987; Gonon and Buda, 1985; May et al., 1988).

Tonic release. A second mechanism proposed to be *tonic,* represents a non-synaptic process which does not imply anatomical synaptic specializations. This release results from presynaptic receptors activation, usually localized on nerve terminals although they may occupy any other place along the axon. These non-synaptic receptors usually present a higher sensitivity than synaptic ones (Vizi and Labos, 1991). Once activated, these receptors favour neurotransmitter efflux without nerve terminal depolarization. This tonic release does not depend on neither neuronal firing nor Na^+ or Ca^{2+} channels. This tonic component exhibits a longer time-course process whose activation leads to measurable modifications in extracellular concentration. These modifications are expected to induce long term compensatory adaptations.

Experimental implications. According to this model, neuronal depolarization, elicited either by electrical, pharmacological or ionic (e.g. K^+-induced) stimulation should be analyzed in its interfering with the phasic release. Conversely, experimental techniques using local applications (microiontophoresis, push-pull cannula, transstriatal dialysis or microdialysis) of a ligand, should be analyzed as interfering with the tonic release. These tonic and phasic release mechanisms may be approached by their TTX sensitivity: in the presence of TTX, indirect regulations which imply either interneurons or other nerve terminals are blocked, whereas direct regulations are TTX resistant. However, *in vivo*, the meaning of TTX sensitivity is not unanimously accepted (Westerink et al., 1992).

The glutamatergic and dopaminergic neurotransmitter systems present several characteristics which fit well with the proposed model: 1) two independent processes, corresponding to tonic and phasic release seem to regulate DA and GLU striatal release, 2) Bradford et al., (1987) have established that there is a continuous glutamatergic outward neuronal leakage, contributing to measurable glutamatergic extracellular concentrations, 3) as discussed later, most of the anatomical studies have failed to demonstrate actual axo-axonic synaptic contacts between cortico-striatal and nigro-striatal terminals. However, despite this lack of synaptic specializations, cross talk between glutamatergic and dopaminergic transmitter systems has been established that seems important for basal ganglia function. These data may suggest that, besides classical synaptic transmission, dopaminergic and glutamatergic systems could use volume transmission as a main mode for interneuronal communication, in particular for presynaptic interactions.

CORTICO-NIGRIC PROJECTIONS

We have now a fair detailed knowledge of the circuitry in which the basal ganglia are involved, including complex interconnections as well as connections with extrinsic structures. The striatum represents the core structure of the basal ganglia complex, being the main entrance for neuronal inputs: glutamatergic input originating from the cortex, the dopaminergic from the SNc. However, SNc seems also an important entrance for neuronal inputs: 1) GABAergic striato-nigric projections make synaptic contacts on dopaminergic neurons localized within both the SNc and the substantia nigra, pars reticulata (SNr) (Chéramy et al., 1981; Van den Pol et al., 1985; Bolam and Smith, 1991). 2) Afferent inputs originating from subthalamic (Hammond et al.,1983; Parent, 1990; Robledo and Féger, 1990) and pedonculopontine (Scarnati et al., 1986) nuclei have also been described. Furthermore, anatomical, electrophysiological and biochemical studies have shown that a direct cortico-nigric projection originates from the cortex (Rinvik, 1966; Afifi et al., 1974; Bunney and Aghajanian, 1976; Gariano and Groves, 1988) using glutamate as the neurotransmitter (Carter, 1982; Kerkerian et al., 1983; Fonnum et al., 1981; Kornhuber et al., 1984) and inducing burst firing in DA neurons (Kornhuber et al., 1984; Gariano and Groves, 1988). Likewise, pharmacological studies (Scarnati et al., 1986; Mereu et al., 1991) as well as degeneration studies (Tallaksen-Greene et al., 1992) indicate that NMDA and non-NMDA receptors mediate cortico-nigric connections. Thus, anatomical and experimental data suggest that GLU/DA interactions could take place in these two distinct anatomical sites and that striatal DA release could be modulated by glutamate acting at these two sites.

We now examine the reciprocal presynaptic interactions between GLU and DA, describing them, when possible, according to previously described criteria of phasic or tonic GLU/DA release. Next, we examine the receptor subtypes involved, and in particular the way glutamatergic receptor subtypes are supposed to work. We report also the pharmacological experiments that suggest that postsynaptic interactions exist besides presynaptic ones. Finally, we briefly present Grace's hypothesis for the etiology of schizophrenia, and how we could reconsider some clinical and therapeutic aspects in Parkinson's disease.

PRESYNAPTIC INTERACTIONS

Dopaminergic control on glutamate striatal release

In vitro, there are indications that glutamate striatal level is the result of both its release and uptake in cortico-striatal terminals. DA agonists inhibit the K^+-evoked, Ca^{2+}-dependent GLU release from rat striatal synaptosomes and slices whereas neuroleptics inhibit this effect (Mitchell and Doggett, 1980; Rowlands and Roberts, 1980; Maura et al., 1988). In addition, DA modulates the striatal GLU uptake: DA as well as bromocriptine or apomorphine produce marked inhibition of high affinity glutamate striatal uptake (HAGU) (Nieoullon et al., 1982). This HAGU inhibition is more important after nigro-striatal lesion because of a supersensitive response of dopaminergic receptors presynaptically involved in this uptake (Kerkerian and Nieoullon, 1988).

In vivo, DA was not found to exert identical effects on spontaneous (tonic) or evoked (phasic) GLU release. Experiments using combined push-pull cannula, drug application and electrophysiological recordings, showed that striatal application of DA agonists (apomorphine, ß-phenylethylamine) or stimulation of substantia nigra increased the basal release of L-[^{3}H]glutamate in the striatum (Godukhin et al., 1984). Conversely, apomorphine, -phenylethylamine or the D2 agonist LY171555 inhibit GLU release induced by frontal cortex stimulation or K+-induced depolarization, which is supposed to represent the phasic release of GLU. Haloperidol and sulpiride reverse these effects (Godukhin et al., 1984; Yamamoto and Davy, 1992). Moreover, DA agonists superfused into the striatum reduce the amplitude of evoked potentials produced by a frontal cortex stimulation or K+-depolarization (Godukhin et al., 1984) whereas sulpiride increases the evoked firing of the cortico-striatal pathway (Brown and Arbuthnott, 1983). Likewise, electrical stimulation of the nigro-striatal pathway as well as local striatal application of DA agonists have been described to decrease cortico-striatal excitability. This decrease is reversed by local infusion of neuroleptics and is no longer observed after dopaminergic depletion achieved by -methylparatyrosine or reserpine (Garcia-Munoz et al., 1991). Similarly, unilateral 6-hydroxydopamine lesion of the mesencephalic dopaminergic neurons induces a bilateral increase of GLU tissue and extracellular levels (Lindefors and Ungerstedt, 1990).

Dopamine receptors involved. Several studies have shown that the two major striatal dopaminergic receptor subtypes, D1 and D2, are localized postsynaptically on different striatal neuronal elements. D1 is expressed on direct striato-nigric neurons, and D2 is expressed on projecting neurons involved in the indirect striato-pallido-subthalamo-nigric pathway (Gerfen et al., 1990; Le Moine et al., 1990, 1991). However, it has been concluded from several studies that about 20 to 40% of the D2 receptors are localized on cortico-striatal terminals. Following intrastriatal injection of kainic acid that selectively destroys cell bodies and dendrites while sparing axons and terminals, only a small reduction in the D2 receptors binding could be measured (Govoni et al., 1978; Schwarcz et al., 1978; Theodorou et al., 1981; Martres et al., 1984). Conversely, decortication (Garau et al, 1978; Schwarcz et al., 1978; Theodorou et al., 1981) or selective cortical infarction (Filloux et al., 1988) have been shown to markedly reduce the number of binding sites for dopaminergic receptor ligands in striatal homogenates. These changes do not affect the DA-sensitive adenylate cyclase activity (Garau et al, 1978; Schwarcz et al., 1978), suggesting that the lost receptors are of the D2 subtype, since they are not linked to adenylate cyclase (Kebabian and Calne, 1979). However, the existence of dopaminergic D2 receptors on cortical afferents has been recently debated (Trugman et al., 1986; Joyce and Marshall, 1987; Filloux et al., 1988). Regarding the receptor subtype which could mediate these dopaminergic effects on glutamatergic transmission, pharmacological experiments emphasize the presynaptic participation of D2 rather than D1 receptors (Maura et al., 1988; Roberts and Anderson, 1979; Mitchell and Doggett, 1980; Yamamoto and Davy, 1992). However, since in some pharmacological experiments, DA agonists used to demonstrate these DA/GLU interactions are mixed D1/D2 agonists one cannot rule out participation of D1 receptors which have been recently evidenced presynaptically (Huang et al., 1992).

Anatomical considerations. Anatomically, it is unclear whether these DA/GLU interactions are direct, mediated by striatal axo-axonic synapses or indirect, via glutamatergic and dopaminergic afferents that synapse on striatal interneurons. Up to now, indirect mechanisms have not been demonstrated. Conversely, in most cases, these DA/GLU interactions were shown to be insensitive to TTX or to a previous lesion of intrinsic striatal neurons with ibotenic acid. This suggests that exogenously applied DA or DA agonist acts directly on the glutamatergic nerve endings to inhibit the release of GLU. However, the anatomical support of a direct presynaptic interaction is still a much debated question. With combined anterograde degeneration and tyrosine hydroxylase immunoreactivity, appositions between dopaminergic and cortico-striatal terminals have been described (Bouyer et al., 1984). Although controversial and rare, true axo-axonic synapses between these terminals have been observed (Kemp and Powell, 1971; Bouyer et al., 1984; Groves, 1980; Pickel et al., 1981) and quantified (Kornhuber and Kornhuber, 1983). The difficulty to demonstrate these synapses has been interpreted as a result of the bend and the large distance between the two contacts (Kornhuber and Kornhuber, 1986). Whether these relations between dopaminergic and glutamatergic terminals are appositions or true synapses, they nevertheless provide anatomical support for a direct presynaptic reciprocal interaction between dopaminergic and glutamatergic terminals. Since axo-axonic synaptic specializations are very rare in the striatum, dopamine could act through non classical synaptic transmission, a certain distance away, on dopaminergic receptors localized on glutamatergic cortico-striatal terminals by a diffusion process such as volume transmission (Fuxe and Agnati, 1991).

Glutamatergic control on dopamine release

It is well established that the cortico-striatal glutamatergic pathway is involved in the presynaptic regulation of DA release from nerve dopaminergic terminals.

In vitro, GLU activates tonic and phasic striatal DA release. It has been shown that L-glutamate stimulates striatal DA release from slices previously incubated with [^{3}H]dopamine or superfused with [^{3}H]tyrosine. This increase was dose- and Ca^{2+}-dependent and TTX resistant; it was markedly antagonized by glutamate receptor antagonists and by Mg^{2+}, suggesting the involvement of NMDA receptor subtypes presynaptically localized on dopaminergic terminals (Giorguieff et al., 1977; Roberts and Sharif, 1978; Roberts and Anderson, 1979; Marien et al., 1983; Jhamandas and Marien, 1987; Clow and Jhamandas, 1989).

In addition to its action on the spontaneous (tonic) release, GLU stimulates also K$^+$-induced (phasic) DA release (Roberts and Sharif, 1978; Rudolph et al., 1983). This stimulatory effect depends upon both the glutamatergic concentration and the degree of neuronal depolarization (Rudolph et al., 1983). The quisqualate/kainate receptor subtype is likely involved since quisqualate and kainate enhance the endogenous release (Clow and Jhamandas, 1989) although inconstantly affected by CNQX, its selective antagonist (Lonart and Zigmond, 1991). Quisqualate or kainate may act on dopaminergic terminals through both direct and indirect mechanisms since this response is partially blocked by TTX (Clow and Jhamandas, 1989). Other experiments carried out with striatal synaptosomes suggest that NMDA receptors are also involved since the stimulatory effect of NMDA is potentiated by glycine and abolished by MK 801 or the addition of Mg^{2+} (Krebs et al., 1991; Wang, 1991; Desce et al., 1992).

From experiments carried out on substantia nigra slices, it is difficult to establish the phasic or tonic nature of glutamate-evoked DA release because of the complex relationship between neuronal firing and dendritic release (Araneda and Bustos, 1989). The glutamate-evoked DA release is Ca^{2+}-dependent and antagonized by GDEE, APV and Mg^{2+} (indicating the involvement of NMDA receptor) and by TTX. The latter result suggests that GLU acts, either indirectly and includes the participation of nigric intrinsic neurons (Marien et al., 1983) or that it activates a voltage-dependent Na$^+$ channel (Araneda and Bustos, 1989).

In vivo. Unilateral electrical stimulation of motor or visual cortices activates the glutamatergic pathway and increases bilaterally striatal release of newly synthesized [^{3}H]dopamine (Nieoullon et al., 1978). Following an identical stimulation, dendritic DA release is also induced in the substantia nigra (Nieoullon et al., 1978) probably through cortico-nigric projections which involve glutamatergic receptors localized either on dendrites or on perikarya of nigric dopaminergic neurons. However, in this experimental procedure, it is impossible to determine whether striatal DA release is tonic, depending on DA release from nigro-striatal terminals or both tonic and phasic: indeed, one cannot exclude that an electrical cortical stimulation induces nigro-striatal firing, and therefore a phasic DA striatal release.

Using a push-pull technique, Romo et al., (1986) observed that the GABAergic activation of thalamic motor nuclei enhances the release of [^{3}H]dopamine in both caudate nuclei. This effect is correlated with a bilateral striatal increase of GLU and disappears after extensive lesions of the ipsilateral sensorimotor cortex. These data suggest that DA release results from a cortico-striatal presynaptic activation of dopaminergic terminals.

Using dialysis technique, it has been shown that local infusions of GLU, NMDA or kainate increase DA striatal release (Carter et al., 1988; Moghaddam et al., 1990; Imperato et al., 1990; Carrozza et al., 1991; Wedzony et al., 1991; Keefe et al., 1992; Martinez-Fong et al., 1992; Westerink et al., 1992). Other experiments suggest that GLU might be excitatory at low concentrations and inhibitory at high concentrations (Chéramy et al., 1986; Barbeito et al., 1990; Leviel et al., 1990). This latter result seems to be highly dependent on the technique, since only obtained with the push-pull cannula technique. The direct or indirect (involving striatal interneurons) nature of this control is usually defined by TTX sensitivity. However some authors suggest that *in vivo*, TTX sensitivity of the release only implies a voltage-dependent mechanism, and not necessarily an interneuron (Westerink et al., 1992). The stimulatory effect of GLU on DA release probably depends on a direct and tonic mechanism. It seems to be a direct mechanism since kainate-induced DA release remains unchanged after striatal lesions with ibotenic acid (Westerink et al., 1992). It has the property of a tonic process since this release does not satisfy classical criteria of neurotransmitter release, first and foremost because it does not depend on nigro-striatal firing: in a very elegant experiment, Keefe et al., (1992) observed that NMDA and kainate produce an increase in extracellular striatal DA despite the infusion of TTX through a second dialysis probe implanted in the medial forebrain bundle, indicating that impulse activity is not necessary for inducing DA release. Moreover, NMDA-increased release of DA was not modified in reserpine-pretreated rats, despite a massive decrease in the spontaneous release of DA (Wedzony et al., 1991).

Collectively, these results suggest that activation of the cortico-striatal glutamatergic pathway stimulates the tonic release of DA from dopaminergic nigro-striatal terminals at the presynaptic level. They suggest also that the tonic DA release proceeds from a non-vesicular pool. The previous discussion regarding the anatomical localization of DA/GLU interactions supports the hypothesis of a non-synaptic DA release since most of ultrastructural anatomical studies failed to demonstrate actual synaptic specializations between dopaminergic and glutamatergic terminals (Bouyer et al., 1984).

Striosome/matrix regulation. The modulation of DA release by the cortical glutamatergic pathway appears to differ according to the striatal compartmentalization, striosome and matrix. Using an experimental device which allows the superfusion of restricted areas, Krebs et al., (1991) found that NMDA increases the release of newly synthesized DA from the dopaminergic terminals at a higher level in the matrix. These NMDA-evoked responses were blocked by MK 801 or Mg^{2+}. Furthermore, the effect of NMDA was almost totally resistant to TTX in striosome-, whereas in matrix-enriched areas, it was diminished by more than half. These results suggest that the indirect component of the stimulatory effect of NMDA on DA release is more important in the matrix than in the striosome, whereas the direct component seems to be comparable in the two compartments. Thus, the regulation of DA release by cortico-striatal and cortico-nigric fibres seems to be more complex in the matrix where indirect regulations are more involved.

Glutamate receptor subtypes involved. In the striatum, either *in vivo* or *in vitro*, L-glutamate stimulates DA release through partially TTX-sensitive processes, suggesting a dual mechanism, direct and indirect.

In the direct process, two types of glutamatergic receptors, NMDA and non-NMDA receptors are involved. NMDA subtype is involved since NMDA stimulates DA release through a process which is potentiated by glycine and antagonized by Mg^{2+} as well as by several antagonists of NMDA-mediated glutamatergic neurotransmission: GDEE, APV, 7-Cl-Kyn, MK 801, kynurenate and CPP (Marien et al., 1983; Rudolph et al., 1983; Kabuto et al., 1989; Mount et al., 1990; Barbeito et al., 1990; Johnson and Jeng, 1991; Wang, 1991; Martinez-Fong et al., 1992). Non-NMDA subtypes are also involved since a direct stimulatory effect is observed *in vitro*, either with kainate or with quisqualate (Carter et al., 1988; Barbeito et al., 1990; Mount et al., 1990; Johnson and Jeng, 1991). This is also reported *in vivo*, where the effect is antagonized by CNQX and DNQX, two blockers of the quisqualate/kainate receptor subtype, (Imperato et al., 1990; Leviel et al., 1990; Carrozza et al., 1991). As indicated by binding studies after destruction of dopaminergic terminals by intranigral injections of 6-hydroxydopamine, receptors involved are likely presynaptically localized on nigro-striatal dopaminergic terminals (Krebs et al., 1991).

Indirect stimulatory effects have been observed with kainate and quisqualate (Clow and Jhamandas, 1989). These effects are TTX-sensitive. Other authors have proposed that NMDA receptors mediate this indirect stimulatory effect (Carter et al., 1988; Westerink et al., 1992). NMDA activity is fully antagonized by co-infusion of atropine and 2-APV, and abolished in rats with ibotenate lesions of the striatum. These data suggest that the NMDA stimulatory effects are mediated through cholinergic interneurons. Using the push-pull cannula technique, indirect inhibitory processes which are supposed to be mediated by

NMDA receptors are blocked by bicuculline. GABAergic interneurons are thus likely to mediate this indirect inhibitory effect on DA release (Chéramy et al., 1986; Leviel et al., 1990).

However, striatum is not the single site for presynaptic GLU/DA interactions. As previously described, substantia nigra receives afferent innervation which contributes to the firing of dopaminergic and non dopaminergic nigral neurons (Hammond et al., 1983; Gariano and Groves, 1988; Robledo and Féger, 1990; Smith and Grace, 1992). Activation of these afferents have been shown to evoke DA release in the striatum (Ewing et al., 1983; Imperato and Di Chiara, 1984; Gonon and Buda, 1985) and in the subtantia nigra itself (Nieoullon et al., 1977, 1978). The firing rate as well as the firing pattern modulate the DA release which reciprocally may exert autoregulatory effects on the firing of nigro-striatal neurons themselves (Nedergaard et al., 1988; Rutherford et al., 1988; Silva and Bunney, 1988; Overton and Clark, 1992; Suaud-Chagny et al., 1992). Both NMDA and non-NMDA receptors mediate glutamatergic nigral inputs: *in vivo,* activation of NMDA receptors has been shown to trigger DA release that is inhibited by NMDA receptor antagonists (Araneda and Bustos, 1989; Overton and Clark, 1992). This effect is probably indirect since it is totally TTX-dependent. Candidate interneurons involved are glycine-containing neurons since these effects are reduced by strychnine (Araneda and Bustos, 1989). *In vitro,* NMDA, AMPA, kainate and metabotropic receptor subtypes have been localized on cortico-nigric neurons by quantitative autoradiography (Tallaksen-Greene et al., 1992). Furthermore, intranigral stimulation induces postsynaptic excitatory currents whose slow and fast components present pharmacological properties of NMDA and non-NMDA receptors (Mereu et al., 1991).

Functioning of glutamate receptors. Glutamatergic receptor subtypes could specifically participate in determining the nature (tonic or phasic) of neurotransmitter release either in the striatum or in the substantia nigra. In these structures, NMDA and non-NMDA receptors might be activated either separately or simultaneously, according to relevant stimuli. They seem to play a critical role in controlling neuronal excitability: NMDA receptors increase the excitability of nerve terminals. In neuronal bodies, they act by finely tuning Ca^{2+} and other voltage-dependent conductances and modulate the coupling between dendritic synaptic inputs and somatic action potentials (Sah et al., 1989). However, NMDA receptor activation is not essential for depolarizing neurons to the threshold needed for action potential generation (Davies, 1990). In contrast, non-NMDA receptor activation seems necessary for this generation and, as suggested by Lambert and Jones, (1989) concurrent activation of NMDA and non-NMDA receptors actually induces marked synergistic actions resulting in increased depolarization and neuronal firing. By applying these data to the GLU release process, it appears that extracellular GLU concentration in the $10^{-5}M$ range (Bradford et al., 1987) is sufficient to activate extrasynaptic NMDA subtype (Sah et al., 1989; Sands et al., 1989) localized on nigro-striatal dopaminergic neurons. When relevant stimuli induce phasic GLU synaptic release, the local GLU concentration increases transiently and activates the less sensitive (Sah et al., 1989; Sands et al., 1989) kainate/quisqualate receptor subtype. Thus the concurrent activation of NMDA and kainate/quisqualate receptors in the SNc would induce dopaminergic neuronal firing, which in turn induces phasic DA release (Ewing et al., 1983; Imperato and Di Chiara, 1984; Kuhr et al., 1987; Overton and Clark, 1992). However, it has been shown *in vivo*, that in chloral hydrate-anaesthetized rats, the burst-firing of midbrain dopaminergic neurons is mediated by NMDA rather than non-NMDA receptors (Chergui et al, 1993). For tonic release we refer to the fact that, as supported by several experimental studies, DA release evoked by GLU cannot be explained by the classical model of neurotransmitter release from a vesicular pool. It has been suggested that GLU could stimulate the DA release by a mechanism involving a reversal of the DA transporter, a process which fits well with a tonic, non-synaptic release (Lonart and Zigmond, 1991).

POSTSYNAPTIC INTERACTIONS

Experimental data suggest that, besides the presynaptic interactions previously described, DA/GLU interactions could also take place at the postsynaptic level. We have demonstrated such interactions by combining behavioural, biochemical, pharmacological and molecular investigations. Adult rats were chronically treated by the specific non-competitive NMDA antagonists ketamine or MK 801. After 6 weeks, we examined their acute pharmacological responsiveness by scoring stereotypies (Costall and Naylor, 1973) induced by apomorphine (0.125mg/kg, s.c.), and by scoring the catalepsy (Schmidt and Bubser, 1989) induced by haloperidol (0.25mg/kg, i.p.). Next, we measured striatal DA and metabolite levels, D2 receptor mRNA expression and D2 receptor binding (Grigoriadis and Seeman, 1985). Animals treated with NMDA antagonists showed an increased pharmacological responsiveness to dopaminergic agents (Lannes et al., 1991), and an increased expression of striatal hybridization signals of D2 receptor mRNA. Lastly, striatal D2 binding site density (B_{max}) was significantly higher in animals treated with MK 801 without any noted difference on the affinity of the D2 receptor for its ligand [^{3}H]spiperone (Micheletti et al., 1992). No difference was found in the striatal levels of DA, 5-HT and their metabolites.

Taken together, these results indicate that chronic treatment with NMDA antagonists potentiate the behavioural, pharmacological, and molecular expression of striatal D2 receptor without any presynaptic change in striatal DA level. They suggest that the transcription of the striatal D2 receptor gene is controlled by postsynaptic NMDA-dependent events and that D2/NMDA receptor-receptor interaction could be implicated in such a postsynaptic regulation. The observed changes could take place in any of the two striatal cell types which have been shown to express the D2 receptor gene: the medium-sized GABAergic neurons which also express preproenkephalin mRNA (Gerfen et al., 1990; Le Moine et al., 1990) and the large-sized cholinergic neurons (Weiner and Brann, 1989). Anatomically, the medium-sized spiny neurons appear to be the main site for these postsynaptic DA/GLU interactions: indeed, these neurons constitute actually the majority of the striatal neurons and the major postsynaptic targets for striatal afferents. Cortical and mesencephalic afferents converge upon them: the glutamatergic input forms synapse on the top of their distal dendritic spines whereas the dopaminergic input synapses preferentially onto the distal dendritic shafts and neck of those spines. According to Smith and Bolam (1990), this anatomical arrangement could support postsynaptic DA/GLU interactions.

FUNCTIONAL IMPLICATIONS: DA/GLU IMBALANCE

In two pathological conditions DA/GLU balance in the striatum is known to be impaired, probably in opposite ways: schizophrenia and Parkinson's disease. Let us examine how the neurobiological basis of these diseases could be reconsidered in light of the phasic and tonic modes of neurotransmitter release.

Schizophrenia

In schizophrenic patients, a decreased frontal cortical activity has been evidenced (Farkas et al., 1984; Chabrol et al., 1986; Weinberger et al., 1986). It suggests that besides the classical hyperdopaminergic hypothesis of schizophrenia, a decreased release of GLU in subcortical structures (mainly the ventral striatum) could be involved in the pathogenesis of the disease (Kim et al., 1980; Carlsson and Carlsson, 1990; Wachtel and Turski, 1990). According to Grace (1991), a decreased GLU striatal release would reduce tonic DA release leading to a prolonged decrease in extracellular DA levels. Homeostatic processes would then be triggered attempting to set the responsivity of the dopaminergic system. However, these adaptative processes (including increased DA synthesis and increased number of postsynaptic dopaminergic receptors) would result in abnormally large phasic DA responses succeeding to "behaviourally relevant stimuli" (Grace, 1991) that are supposed to specifically stimulate A10 dopaminergic neurons and phasic DA release. This exaggerated response to external stimuli could be regarded as the neuronal basis for positive psychotic symptoms. Regarding the negative schizophrenic symptoms they seem correlated with DA insufficiency expressed by low cerebrospinal fluid HVA (Lindstrom, 1985) and associated with structural damage within different cortical areas (Bogerts et al., 1983) that probably results in a severe hypoactivity of glutamatergic pathways. Furthermore among various drugs which can induce psychosis-like symptoms, only NMDA antagonists like phencyclidine (PCP) are able to induce negative symptoms as well as positive ones. These negative symptoms could then result from the unique ability of PCP to reduce extracellular (tonic) DA level. Prominent negative signs could emerge after the development of positive symptoms, when the tonic DA level decreases beyond a critical point.

Furthermore, Grace proposed an hypothesis concerning neuroleptic effectiveness in schizophrenia: indeed, blockade of dopaminergic receptors cannot account for this efficacy since this blockade occurs very quickly after onset of the treatment, whereas the clinical effects are usually delayed for weeks. Neuroleptics could in fact relieve psychotic symptoms by inactivating dopaminergic neurons firing since it has been shown that chronic neuroleptic treatment induces a depolarization block in these neurons (Bunney and Grace, 1978; White and Wang, 1983). Grace emphasized that clozapine which alleviates negative symptoms may act via a dual mechanism: a decrease of the abnormal phasic dopaminergic response by eliciting depolarization block in mesolimbic neurons, and an increase in tonic extracellular DA levels (Chiodo and Bunney, 1983; Maidment and Marsden, 1987).

Parkinson's disease

Considering GLU/DA interactions, the dual tonic/phasic mode of DA release suggests an hypothesis concerning the pathogeny of akinesia in Parkinson's disease. This analysis could also provide an hypothesis for different therapeutic and side effects elicited by L-DOPA and DA agonists.

Progressive loss of dopaminergic neurons and the resulting reduction in the number of functional dopaminergic terminals probably result in a decreased tonic DA release. Classical compensatory mechanisms, like those observed after 6-hydroxydopamine lesions in animal models (sprouting of remaining terminals [Onn et al., 1986], increased DA synthesis and release from remaining neurons [Acheson et al., 1980; Stachowiak et al., 1987], increased number of postsynaptic dopaminergic receptors [Creese et al., 1977; Lisovoski et al., 1992]), progressively take place. Moreover, as previously described for nigro-striatal lesions, one can expect an increase in striatal GLU levels (Lindefors and Ungerstedt, 1990). This increase could be seen as part of the compensatory mechanisms, since GLU is known to enhance tonic DA release restoring extracellular levels of DA (*cf. supra*). The compensatory mechanisms are very efficient since in human Parkinson's disease as well as in animal models, clinical or behavioural symptoms appear only when the striatal DA depletion reaches 80 to 90% (Zigmond and Stricker, 1972; Bernheimer et al., 1973). These observations must be related to the fact that contrary to what is observed after chronic neuroleptic treatment, the firing pattern of remaining nigral dopaminergic neurons is unchanged as long as the neuronal loss remains below 80% (Hollerman and Grace, 1990). When dopaminergic loss gets more severe, the relative proportion of spontaneously active DA neurons as well as their firing rate and their burst frequency, increase. Furthermore, an even more important depletion (>96%) leads to an increased number of inactive though chronically depolarized cells, as a consequence of the emergence of a depolarization block (Hollerman and Grace, 1990). Thus, one can hypothesize that in Parkinson's disease, motor impairment could be correlated to the alteration in firing pattern rather than to the DA depletion itself: the remaining nigral neurons display reduced responsiveness to relevant stimuli because of their increased spontaneous activity.

In this context, one can view the therapeutic effects of DA agonists (e.g. apomorphine or bromocriptine) as the result of the restoration of normal firing in dopaminergic neurons rather than an activation of striatal postsynaptic receptors as usually accepted. In this hypothesis, DA agonists mimic extracellular DA. They would stimulate the autoreceptors on dopaminergic neurons, inhibiting their firing (Bernardini et al. 1990) and thus, restoring their normal responsiveness. In short, restoration of normal firing by agonists could be seen as normalizing nigro-striatal input.

There is a main difference between L-DOPA and DA agonist treatment of Parkinson's disease: dyskinesia and "on-off" phenomena only occur with L-DOPA, after the disease having evolved several years. At the onset of the disease, one can suppose that L-DOPA increases the available DA pool used for the tonic release controlled by striatal GLU. The resulting increase in extracellular DA would therefore act as described above for DA agonists. However, L-DOPA probably also increases the DA synaptic concentration. Then, once nigral loss gets more important, the compensatory mechanisms are no longer totally effective and, as previously described, the remaining neurons probably increase their firing. At that time, the large amount of DA (due to L-DOPA treatment) available for synaptic transmission might elicit abnormally large responses that could participate in the dyskinesia.

This mechanism is similar to that hypothezised by Grace (1991) to explain psychotic symptoms in schizophrenia. In a certain aspect, dyskinesia might be considered as "motor delusions".

Furthermore as it has been described above, increased electrical activity may result in depolarization block, which is a non responsive state. This depolarization block could participate to acute "off" periods as in "on-off" phenomena.

CONCLUSION

Taken together, results presented here emphasize three major points regarding DA/GLU interactions in the basal ganglia. 1) DA neurotransmission seems to depend on two modes of release, phasic and tonic. 2) Besides well known presynaptic regulations, dopaminergic and glutamatergic neurotransmission interact also through postsynaptic interactions. These latter seem to imply receptor-receptor interactions between NMDA and D2 receptors. 3) The SNc is another site for DA/GLU interactions. In this site, these interactions could be especially relevant for basal ganglia function since they might control the firing of dopaminergic neurons and therefore the phasic striatal DA release.

These phenomena and structures involved in neural plasticity might be of particular relevance for further understanding clinical and therapeutic aspects of various diseases related to basal ganglia dysfunction. Wether this kind of regulations can be observed *in vivo* between other neurotransmitter systems and structures remains to be elucidated.

Acknowledgements

This work was generously supported by the Fabriques de Tabac Réunies SA (Switzerland), the Commission Recherche de la Faculté de Médecine, Université Louis Pasteur, Strasbourg (France), the Fondation pour la Recherche Médicale, the Fondation de France and the Direction de la Recherche et des Etudes Doctorales. B.L. is a recipient of fellowships from the Association France-Parkinson and Laboratoires Biocodex. The authors thank J. Zwiller for his helpful comments.

REFERENCES

Acheson, A.L., Zigmond, M.J., and Stricker, E.M., 1980, Compensatory increase in tyrosine hydroxylase activity in rat brain after intraventricular injection of 6-hydroxydopamine, *Science* 207:537-540.

Afifi, A.K., Bahuth, N.B., Kaelber, W.W., Mikhael, E., and Nassar, S., 1974, The cortico-nigral fibre tract. An experimental Fink-Heimer study in cats, *J. Anat.* 118:469-476.

Andén, N.E., Dahlström, A., Fuxe, K., Larsson, K., Olson, L., and Ungerstedt, U., 1966, Ascending monoamine neurons to the telencephalon and diencephalon, *Acta Physiol. Scand.* 67:313-326.

Araneda, R., and Bustos, G., 1989, Modulation of dendritic release of dopamine by N-Methyl-D-Aspartate receptors in rat substantia nigra, *J. Neurochem.* 52:962-970.

Augustine, G.J., Charlton, M.P., and Smith, S.J., 1987, Calcium action in synaptic transmitter release, *Ann. Rev. Neurosci.* 10:633-693.

Barbeito, L., Chéramy, A., Godeheu, G., Desce, J.M., and Glowinski, J., 1990, Glutamate receptors of a quisqualate-kainate subtype are involved in the presynaptic regulation of dopamine release in the caudate nucleus *in vivo*, *Eur. J. Neurosci.* 2:304-311.

Barres, B.A., 1991, New roles for glia, *J. Neurosci.* 11:3685-3694.

Bernardini, G.L., Speciale, S.G., German, D.C., 1990, Increased midbrain dopaminergic cell activity following 2'CH3-MPTP-induced dopaminergic cell loss: an in vitro electrophysiological study, *Brain Res.* 527:123-129.

Bernheimer, H., Birkmayer, W., Hornykiewicz, O., Jellinger, K., and Seitelberger, F., 1973, Brain dopamine and the syndromes of Parkinson and Huntington: clinical, morphological and neurochemical correlations, *J. Neurol. Sci.* 20:415-455.

Bogerts, B., Hantsch, J., and Herzer, M., 1983, A morphometric study of the dopamine containing cell groups in the mesencephalon of normals, Parkinson patients and schizophrenics, *Biol. Psychiat.* 18:951-969.

Bolam, P., and Smith, Y., 1991, Characterization of the synaptic inputs to dopaminergic neurons in the rat substantia nigra, *in:* "The Basal Ganglia III," G. Bernardi, M.B. Carpenter, G. Di Chiara, M. Morelli, and P. Stanzione, eds., Plenum Press, New York, pp. 119-131.

Bouyer, J.J., Park, D.H., Joh, T.H., and Pickel, V.M., 1984, Chemical and structural analysis of the relation between cortical inputs and tyrosine hydroxylase-containing terminals in rat neostriatum, *Brain Res.* 302:267-275.

Bradford, H.F., Young, A.M.J., and Crowder, J.M., 1987, Continuous glutamate leakage from brain cells is balanced by compensatory high-affinity reuptake transport, *Neurosci. Lett.* 81:296-302.

Brown J.R., and Arbuthnott, G.W., 1983, The electrophysiology of dopamine (D_2) receptors: a study of the actions of dopamine on corticostriatal transmission, *Neuroscience* 10:349-355.

Bunney, B.S., and Aghajanian, G.K., 1976, The precise localization of nigral afferents in the rat as determined by a retrograde tracing technique, *Brain Res.* 117:423-435.

Bunney, B.S., and Grace, A.A., 1978, Acute and chronic haloperidol treatment: comparison of effects on nigral dopaminergic cell activity, *Life Sci.* 23:1715-1728.

Carlsson, M., and Carlsson, A., 1990, Interactions between glutamatergic and monoaminergic systems within the basal ganglia - Implications for schizophrenia and Parkinson's disease, *Trends Neurosci.* 13:272-276.

Carrozza, D.P., Ferraro, T.N., Golden, G.T., Reyes, P.F., and Hare, T.A., 1991, Partial characterization of kainic acid-induced striatal dopamine release using in vivo microdialysis, *Brain Res.* 543:69-76.

Carter, C.J., 1982, Topographical distribution of possible glutamatergic pathways from the frontal cortex to the striatum and substantia nigra in rats, *Neuropharmacology* 21:379-383.

Carter, C.J., L'Heureux, R., and Scatton, B., 1988, Differential control by N-Methyl-D-Aspartate and kainate of striatal dopamine release in vivo: a transstriatal dialysis study, *J. Neurochem.* 51:462-468.

Chabrol, H., Guell, A., Bes, A., and Moron, P., 1986, Cerebral blood flow in schizophrenic adolescents, *Am. J. Psychiat.* 143:130.

Chéramy, A., Leviel, V., and Glowinski, J., 1981, Dendritic release of dopamine in the substantia nigra, *Nature* 289:537-542.

Chéramy, A., Romo, R., Godeheu, G., Baruch, P., and Glowinski, J., 1986, *In vivo* presynaptic control of dopamine release in the cat caudate nucleus. II. Facilitatory or inhibitory influence of L-Glutamate, *Neuroscience* 19:1081-1090.

Chergui, K., Charléty, P.J., Akaoka, H., Saunier, C.F., Brunet, J.-L., Buda, M., Svensson, T.H., and Chouvet, G., 1993, Tonic activation of NMDA receptors causes spontaneous burst discharge of rat midbrain dopamine neurons in vivo, *Eur. J. Neurosci;* 5:137-144.

Chesselet, M.F., 1984, Presynaptic regulation of neurotransmitter release in the brain: facts and hypothesis, *Neuroscience* 12:347-375.

Chiodo, L.A., and Bunney, B.S., 1983, Typical and atypical neuroleptics: differential effects of chronic administration on the activity of A9 and A10 midbrain DA neurons, *J. Neurosci.* 3:1607-1619.

Clow, D.W., and Jhamandas, K., 1989, Characterization of L-Glutamate action on the release of endogenous dopamine from the rat caudate-putamen, *J. Pharm. Exp. Ther.* 248:722-728.

Costall, B., and Naylor, R.J., 1973, On the mode of action of apomorphine, *Eur. J. Pharmacol.* 21:350-361.

Creese, I., Burt, D.R., and Snyder, S.H., 1977, Dopamine receptor binding enhancement accompanies lesion-induced behavioral supersensitivity, *Science* 197:596-598.

Dahlström, A., and Fuxe, K., 1964, Evidence for the existence of monoamine-containing neurons in the central nervous system, *Acta Physiol. Scand. Suppl.*, 232:1-55.

Davies, J., 1990, NMDA receptors in synaptic pathways, in: "The NMDA Receptor," J.C. Watkins, and G.L. Collingridge, eds., IRL Press, Oxford, pp. 77-91.

Desce, J.M., Godeheu, G., Galli, T., Artaud, F., Chéramy, A., and Glowinski, J., 1992, L-Glutamate-evoked release of dopamine from synaptosomes of the rat striatum: involvement of AMPA and N-Methyl-D-Aspartate receptors, *Neuroscience* 47:333-339.

Ewing, A.G., Bigelow, J.C., and Wightman, R.M., 1983, Direct in vivo monitoring of dopamine released from two striatal compartments in the rat, *Science* 221:169-171.

Farkas, T., Wolf, A.P., Jaeger, J., Brodie, J.D., Christman, D.R., and Fowler, J.S., 1984, Regional brain glucose metabolism in chronic schizophrenia, *Arch. gen. Psychiat.* 41:293-300.

Filloux, F., Dawson, T.M., and Wamsley, J.K., 1988, Localization of nigrostriatal dopamine receptor subtypes and adenylate cyclase, *Brain Res. Bull.* 20:447-459.

Fonnum, F.F., Storm-Mathisen, J., and Divac, I., 1981, Biochemical evidence for glutamate as the neurotransmitter in corticostriatal and corticothalamic fibres in rat brain, *Neuroscience* 6:863-873.

Fuxe, K., and Agnati, L.F., 1991, Two principal modes of electrochemical communication in the brain: volume versus wiring transmission, in: "Volume Transmission in the Brain," K. Fuxe, and L.F. Agnati, eds., Raven Press, New York, pp. 1-9.

Garau, L., Govoni, S., Stefanini, E., Trabucchi, M., and Spano, P.F., 1978, Dopamine receptors: pharmacological and anatomical evidences indicate that two distinct dopamine receptor populations are present in rat striatum, *Life Sci.* 23:1745-1750.

Garcia-Munoz, M., Young, S.J., and Groves, P.M., 1991, Terminal excitability of the corticostriatal pathway. I. Regulation by dopamine receptor stimulation, *Brain Res.* 551:195-206.

Gariano, R.F., and Groves, P.M., 1988, Burst induced firing in midbrain dopamine neurons by stimulation of the medial prefrontal and anterior cingulate cortices, *Brain Res.* 462:194-198.

Gerfen, C., Engber, T.M., Mahan, L.C., Susel, Z., Chase, T.N., Monsma, F.J. and Sibley, D.R., 1990, D_1 and D_2 dopamine receptor-regulated gene expression of striatonigral and striatopallidal neurons, *Science* 250:1429-1432.

Giorguieff, M.F., Kemel, M.L., and Glowinski, J., 1977, Presynaptic effect of L-Glutamic acid on the release of dopamine in rat striatal slices, *Neurosci. Lett.* 6:73-77.

Glowinski, J., Barbeito, L., and Chéramy, A., 1991, Influence of cortico-striatal glutamatergic neurons on dopaminergic transmission in the striatum, *in*: "The Basal Ganglia III," G. Bernardi, M.B. Carpenter, G. Di Chiara, M. Morelli, and P. Stanzione, eds., Plenum Press, New York, pp. 347-355.

Godukhin, O.V., Zharikova, A.D., and Budantsev, A.Y., 1984, Role of presynaptic dopamine receptors in regulation of the glutamatergic neurotransmission in rat neostriatum, *Neuroscience* 12:377-383.

Gonon, F.G., and Buda, M.J., 1985, Regulation of dopamine release by impulse flow and by autoreceptors as studied by *in vivo* voltammetry in the rat striatum, *Neuroscience* 14:765-774.

Govoni, S., Olgiati, V.R., Trabucchi, M., Garau, L., Stefanini, E., and Spano, P.F., 1978, [^{3}H]haloperidol and [^{3}H]spiroperidol receptor binding after striatal injection of kainic acid, *Neurosci. Lett.* 8:207-210.

Grace, A.A., 1991, Phasic versus tonic dopamine release and the modulation of dopamine system responsivity: a hypothesis for the etiology of schizophrenia, *Neuroscience* 41:1-24.

Grigoriadis, D., and Seeman, P., 1985, Complete conversion of brain dopamine receptors from the high- to the low-affinity state for dopamine agonists, using sodium ions and guanine nucleotide, *J. Neurochem.* 44:1925-1935.

Grofova, I., 1979, Extrinsic connections of the neostriatum, *in*: "The Neostriatum," I. Divac, and R.G. Öberg, eds., Pergamon, Oxford, pp.37-51.

Groves, P.M., 1980, Synaptic endings and their postsynaptic targets in neostriatum: synaptic specialization revealed from analysis of serial sections, *Proc. Natl. Acad. Sci. USA* 77:6926-6929.

Hammond, C., Shibazaki, T., and Rouzaire-Dubois, B., 1983, Branched output neurons of the rat subthalamic nucleus: electrophysiological study of the synaptic effects on identified cells in the two main target nuclei, the entopeduncular nucleus and the substantia nigra, *Neuroscience* 9:511-520.

Hollerman, J.R., and Grace, A.A., 1990, The effects of dopamine-depleting brain lesions on the electrophysiological activity of rat substantia nigra dopamine neurons, *Brain Res.* 533:203-212.

Huang, Q., Zhou, D., Chase, K., Gusella, J.F., Aronin, N., and DiFiglia, M., 1992, Immunohistochemical localization of the D_1 dopamine receptor in rat brain reveals its axonal transport, pre- and postsynaptic localization, and prevalence in the basal ganglia, limbic system, and thalamic reticular nucleus, *Proc. Natl. Acad. Sci. USA* 89:11988-11992.

Imperato, A., and Di Chiara, G., 1984, Trans-striatal dialysis coupled to reverse phase high performance liquid chromatography with electrochemical detection: a new method for the study of the *in vivo* release of endogenous dopamine and metabolites, *J. Neurosci.* 4:966-977.

Imperato, A., Honoré, T., and Jensen, L.H., 1990, Dopamine release in the nucleus caudatus and in the nucleus accumbens is under glutamatergic control through non-NMDA receptors: a study in freely-moving rats, *Brain Res.* 530:223-228.

Jhamandas, K., and Marien, M., 1987, Glutamate-evoked release of endogenous brain dopamine: inhibition by an excitatory amino acid antagonist and an enkephalin analogue, *Br. J. Pharmacol.* 90:641-650.

Johnson, K.M., and Jeng, Y.-J., 1991, Pharmacological evidence for *N*-methyl-D-aspartate receptors on nigrostriatal dopaminergic nerve terminals, *Can. J. Physiol. Pharmacol.* 69:1416-1421.

Joyce, J.N., and Marshall, J.F., 1987, Quantitative autoradiography of dopamine D_2 sites in rat caudate-putamen: localization to intrinsic neurons and not to neocortical afferents, *Neuroscience* 20:773-795.

Kabuto, H., Yokoi, I., Mizukawa, K., and Mori, A., 1989, Effects of an N-methyl-D-Aspartate receptor agonist and its antagonist CPP on the levels of dopamine and serotonin metabolism in rat striatum collected in vivo by using a brain dialysis technique, *Neurochem. Res.* 14:1075-1080.

Katz, B., 1969, "The Release of Neural Transmitter Substances," Liverpool University Press, Liverpool.

Kebabian, J.W., and Calne, D.B., 1979, Multiple receptors for dopamine, *Nature* 277:93-96.

Keefe, K.A., Zigmond, M.J., and Abercrombie, E.D., 1992, Extracellular dopamine in striatum: influence of nerve impulse activity in medial forebrain bundle and local glutamatergic input, *Neuroscience* 47:325-332.

Kemp, J.M., and Powell, T.P.S., 1971, The site of termination of afferent fibers in the caudate nucleus, *Phil. Trans. R. Soc.Lond. B.* 262:403-412.

Kerkerian, L., and Nieoullon, A., 1988, Supersensitivity of presynaptic receptors involved in the dopaminergic control of striatal high affinity glutamate uptake after 6-hydroxydopamine lesion of nigrostriatal dopaminergic neurons, *Exp. Brain Res.* 62:424-430.

Kerkerian, L, Nieoullon A., and Dusticier, N., 1983, Topographic changes in high-affinity glutamate uptake in the cat red nucleus, substantia nigra, thalamus, and caudate nucleus after lesions of sensorimotor cortical areas, *Exp. Neurol.* 81:598-612.

Kilpatrick, I.C., and Phillipson, O.T., 1986, On the transmitter chemistry of thalamostriatal fibres, *Neurosci. Lett.* 67:97-98.

Kim, J.S., Kornhuber, H.H., Schmid-Burgk, W., and Holzmiller, B., 1980, Low cerebrospinal fluid glutamate in schizophrenic patients and a new hypothesis on schizophrenia, *Neurosci. Lett.* 20:379-382.

Kornhuber, J., Kim, J.S., Kornhuber, M.E., and Kornhuber, H.H., 1984, The cortico-nigral projection: reduced glutamate content in the substantia nigra following frontal cortex ablation in the rat, *Brain Res.* 322:124-126.

Kornhuber, J., and Kornhuber, M.E., 1983, Axo-axonic synapses in the rat striatum, *Eur. Neurol.* 22:433-436.

Kornhuber, J., and Kornhuber, M.E., 1986, Presynaptic dopaminergic modulation of cortical input to the striatum, *Life Sci.* 39:669-674.

Krebs, M.O., Desce, J.M., Kemel, M.L., Gauchy, C., Godeheu, G., Chéramy, A., and Glowinski, J., 1991, Glutamatergic control of dopamine release in the rat striatum: evidence for presynaptic N-Methyl-D-Aspartate receptors on dopaminergic nerve terminals, *J. Neurochem.* 56:81-85.

Kuhr, W.G., Ewing, A.G., Caudill, W.L., and Wightman, R.M., 1984, Monitoring the stimulated release of dopamine with in vivo voltammetry. I: characterization of the response observed in the caudate nucleus of the rat, *J. Neurochem.* 43:560-569.

Kuhr, W.G., Wightman, R.M., and Rebec, G.V., 1987, Dopaminergic neurons: simultaneous measurements of dopamine release and single-unit activity during stimulation of the medial forebrain bundle, *Brain Res.* 418:122-128.

Lambert, J.D.C., and Jones, R.S.G., 1989, Activation of N-methyl-D-aspartate receptors contributes to the EPSP at perforant path synapses in the rat dentate gyrus in vitro, *Neurosci. Lett.* 97:323-328.

Lannes, B., Micheletti, G., Warter, J.-M., Kempf, E., and Di Scala, G., 1991, Behavioural, pharmacological and biochemical effects of acute and chronic administration of ketamine in the rat, *Neurosci. Lett.* 128:177-181.

Lapper, S.R., and Bolam, J.P., 1992, Input from the frontal cortex and the parafascicular nucleus to cholinergic interneurons in the dorsal striatum of the rat, *Neuroscience* 51:533-545.

Le Moine, C., Normand, E., Guitteny, A.F., Fouque, B., Teoule, R., and Bloch, B., 1990, Dopamine receptor gene expression by enkephalin neurons in rat forebrain, *Proc. Natl. Acad. Sci. USA* 87:230-234.

Le Moine, C., Normand, E., and Bloch, B., 1991, Phenotypical characterization of the rat striatal neurons expressing the D1 dopamine receptor gene, *Proc. Natl. Acad. Sci. USA* 88:4205-4209.

Leviel, V., Gobert, A., and Guibert, B., 1990, The glutamate-mediated release of dopamine in the rat striatum: further characterization of the dual excitatory-inhibitory function, *Neuroscience* 39:305-312.

Lindefors, N., and Ungerstedt, U., 1990, Bilateral regulation of glutamate tissue and extracellular levels in caudate-putamen by midbrain dopamine neurons, *Neurosci. Lett.* 115:248-252.

Lindstrom, L.H., 1985, Low HVA and normal 5-HIAA CSF levels in drug-free schizophrenic patients compared to healthy volunteers: correlations to symptomatology and family history, *Psychiat. Res.* 14:265-273.

Lisovoski, F., Haby, C., Borrelli, E., Schleef, C., Revel, M.O., Hindelang, C., and Zwiller, J., 1992, Induction of D2 dopamine receptor mRNA synthesis in a 6-hydroxydopamine parkinsonian rat model, *Brain Res. Bull.* 28:697-701.

Lonart, G., and Zigmond, M.J., 1991, High glutamate concentrations evoke Ca^{++}-independent dopamine release from striatal slices: a possible role of reverse dopamine transport, *J. Pharm. Exp. Ther.* 256:1132-1138.

Maidment, N.T., and Marsden, C.A., 1987, Repeated atypical neuroleptic administration: effects on central dopamine metabolism monitored by in vivo voltammetry, *Eur. J. Pharmacol.* 136:141-149.

Marien, M., Brein, J., and Jhamandas, K., 1983, Regional release of [^{3}H]dopamine from rat brain in vitro: effects of opioids on release induced by potassium, nicotine, and L-glutamic acid, *Can. J. Physiol. Pharmacol.* 61:43-60.

Martin, D.L., 1992, Synthesis and release of neuroactive substances by glial cells, *Glia* 5:81-94.

Martinez-Fong, D., Rosales, M.G., Gongora-Alfaro, J.L., Hernandez, S., and Aceves, J., 1992, NMDA receptor mediates dopamine release in the striatum of unanesthetized rats as measured by brain microdialysis, *Brain Res.* 595:309-315.

Martres, M.P., Sokoloff, P., and Schwartz, J.C., 1984, Dopaminergic binding sites in rat striatal slices and action of guanyl nucleotides, *Naunyn Schmiedeberg's Arch. Pharmacol.* 325:116-123.

Maura, G., Giardi, A., and Raiteri, M., 1988, Release-regulating D-2 dopamine receptors are located on striatal glutamatergic nerve terminals, *J. Pharm. Exp. Ther.* 247:680-684.

May, L.J., Kuhr, W.G., and Wightman, R.M., 1988, Differentiation of dopamine overflow and uptake processes in the extracellular fluid of the rat caudate nucleus with fast-scan in vivo voltammetry, *J. Neurochem.* 51:1060-1069.

Mc Geer, P.L., Mc Geer, E.G., Scherer, V., and Singh, K., 1977, A glutamatergic corticostriatal path?, *Brain Res.* 128:369-373.

Mereu, G., Costa, E., Armstrong, D.M., and Vicini, S., 1991, Glutamate receptor subtypes mediate excitatory synaptic currents of dopamine neurons in midbrain slices, *J. Neurosci.* 11:1359-1366.

Micheletti, G., Lannes, B., Haby, C., Borrelli, E., Kempf, E., Warter, J.M., and Zwiller, J., 1992, Chronic administration of NMDA antagonists induces D2 receptor synthesis in rat striatum, *Mol. Brain Res.* 14:363-368.

Mitchell, P.R., and Doggett, N.S., 1980, Modulation of striatal [^{3}H]-glutamic acid release by dopaminergic drugs, *Life Sci.* 26:2073-2081.

Moghaddam, B., Gruen, R.J., Roth, R.H., Bunney, B.S., and Adams, R.A., 1990, Effect of L-glutamate on the release of striatal dopamine: in vivo dialysis and electrochemical studies, *Brain Res.* 518:55-60.

Mount, H., Quirion, R., Kohn-Alexander, J., and Boksa, P., 1990, Subtypes of excitatory amino acid receptors involved in the stimulation of [^{3}H]dopamine release from cell cultures of rat ventral mesencephalon, *Synapse* 5:271-280.

Nedergaard, S., Hopkins, C., and Greenfield, S.A., 1988, Do nigro-striatal neurones possess a discrete dendritic modulatory mechanism? Electrophysiological evidence from the actions of amphetamine in brain slices, *Exp. Brain Res.* 69:444-448.

Nieoullon, A., 1986, Reply to the letter to the editor by Kilpatrick and Phillipson, *Neurosci. Lett.* 67:98-99.

Nieoullon, A., Chéramy, A., and Glowinski, J., 1977, Release of dopamine in vivo from cat substantia nigra, *Nature* 266:375-377.

Nieoullon, A., Chéramy, A., and Glowinski, J., 1978, Release of dopamine evoked by electrical stimulation of the motor and visual areas of the cerebral cortex in both caudate nuclei and in the substantia nigra in the cat, *Brain Res.* 145:69-83.

Nieoullon, A., Kerkerian, L., and Dusticier, N., 1982, Inhibitory effects of dopamine on high affinity glutamate uptake from rat striatum, *Life Sci.* 30:1165-1172.

Nieoullon, A., Scarfone, E., Kerkerian, L., Errami, M., and Dusticier, N., 1985, Changes in choline acetyltransferase, glutamic acid decarboxylase, high-affinity glutamate uptake and dopaminergic activity induced by kainic acid lesion of thalamostriatal neurons, *Neurosci. Lett.* 58:299-304.

Onn, S.-P., Berger, T.W., Stricker, E.M., and Zigmond, M.J., 1986, Effects of intraventricular 6-hydroxydopamine on the dopaminergic innervation of striatum: histochemical and neurochemical analysis, *Brain Res.* 376:8-19.

Overton, P., and Clark, D., 1992, Iontophoretically administered drugs acting at the N-Methyl-D-Aspartate receptor modulate burst firing in A9 dopamine neurons in the rat, *Synapse* 10:131-140.

Parent, A., 1990, Extrinsic connections of the basal ganglia, *Trends Neurosci.* 13:254-258.

Pickel, V.M., Beckley, S.C., Joh, T.H., and Reis, D.J., 1981, Ultrastructural immunocytochemical localization of tyrosine hydroxylase in the neostriatum, *Brain Res.* 225:373-385.

Reubi, J.C., and Cuenod, M., 1979, Glutamate release in vitro from corticostriatal terminals, *Brain Res.* 176:185-188.

Rinvik, E., 1966, The cortico-nigral projection in the cat, *J. Comp. Neurol.* 126:241-254.

Roberts, P.J., and Anderson, S.D., 1979, Stimulatory effect of L-Glutamate and related amino acids on [^{3}H] dopamine release from rat striatum: an *in vitro* model for glutamate actions, *J. Neurochem.* 32:1539-1545.

Roberts, P.J., and Sharif, N.A., 1978, Effects of L-glutamate and related amino acids upon the release of [^{3}H] dopamine from rat striatal slices, *Brain Res.* 157:391-395.

Robledo, P., and Féger, J., 1990, Excitatory influence of rat subthalamic nucleus to substantia nigra pars reticulata and the pallidal complex: electrophysiological data, *Brain Res.* 518:47-54.

Romo, R., Chéramy, A., Godeheu, G., and Glowinski, J., 1986, *In vivo* presynaptic control of dopamine release in the cat caudate nucleus -I. Opposite changes in neuronal activity and release evoked from thalamic motor nuclei, *Neuroscience* 19:1067-1079.

Rowlands, G.J., and Roberts, P.J., 1980, Activation of dopamine receptors inhibits calcium-dependent glutamate release from cortico-striatal terminals in vitro, *Eur. J. Pharmacol.* 62:241-242.

Rudolph, M.I., Arqueros, L., and Bustos, G., 1983, L-Glutamic acid, a neuromodulator of dopaminergic transmission in the rat corpus striatum, *Neurochem. Intern.* 5:479-486.

Rutherford, A., Garcia-Munoz, M., and Arbuthnott, G.W., 1988, An afterhyperpolarization recording in striatal cells 'in vitro': effect of dopamine administration, *Exp. Brain Res.* 71:399-405.

Sah, P., Hestrin, S., and Nicoll, R.A., 1989, Tonic activation of NMDA receptors by ambient glutamate enhances excitability of neurons, *Science* 246:815-818.

Sands, S.B., and Barish, M.E., 1989, A quantitative description of excitatory amino acid neurotransmitter responses on cultured embryonic *Xenopus* spinal neurons, *Brain Res.* 502:375-386.

Scarnati, E., Proia, A., Campana, E., and Pacitti, C., 1986, A microiontophoretic study on the nature of the putative synaptic neurotransmitter involved in the pedonculopontine-substantia nigra pars compacta excitatory pathway of the rat, *Exp. Brain Res.* 62:470-478.

Schmidt, W.J., and Bubser, M., 1989, Anticataleptic effects of the *N*-methyl-D-aspartate antagonist MK-801 in rats, *Pharmacol. Biochem. Behav.* 32:621-623.

Schwarcz, R., Creese, I., Coyle, J.T., and Snyder, S.H., 1978, Dopamine receptors localised on cerebral cortical afferents to rat corpus striatum, *Nature* 271:766-768.

Silva, N.L., and Bunney, B.S., 1988, Intracellular studies of dopamine neurons in vitro: pacemakers modulated by dopamine, *Eur. J. Pharmacol.* 149:307-315.

Smith, A.D., and Bolam, J.P., 1990, The neural network of the basal ganglia as revealed by the study of synaptic connections of identified neurones, *Trends Neurosci.* 13:259-265.

Smith, I.D., and Grace, A.A., 1992, Role of the subthalamic nucleus in the regulation of nigral dopamine neuron activity, *Synapse* 12:287-303.

Stachowiak, M.K., Keller Jr., R.W., Stricker, E.M., and Zigmond, M.J., 1987, Increased dopamine efflux from striatal slices during development and after nigrostriatal bundle damage, *J. Neurosci.* 7:1648-1654.

Suaud-Chagny M.F., Chergui, K., Chouvet, G., and Gonon, F., 1992, Relationship between dopamine release in the rat nucleus accumbens and the discharge activity of dopaminergic neurons during local *in vivo* application of amino acids in the ventral tegmental area, *Neuroscience* 49:63-72.

Tallaksen-Greene, S.J., Wiley, R.G., and Albin, R.L., 1992, Localization of striatal excitatory amino acid binding site subtypes to striatonigral projection neurons, *Brain Res.* 594:165-170.

Theodorou, A., Reavill, C., Jenner, P., and Marsden, C.D., 1981, Kainic acid lesions of striatum and decortication reduce specific [^{3}H] sulpiride binding in rats, so D2 receptors exist postsynaptically on corticostriate afferents and striatal neurons, *J. Pharm. Pharmacol.* 33:439-444.

Trugman, J.M., Geary II, W.A., and Wooten, G.F., 1986, Localization of D2 dopaminergic receptors to intrinsic striatal neurons by quantitative autoradiography, *Nature* 323:267-269.

Van den Pol, A.N., Smith, A.D., and Powell, J.F., 1985, GABA axons in synaptic contact with dopamine neurons in the substantia nigra: double immunocytochemistry with biotin-peroxidase and protein A-colloidal gold, *Brain Res.* 348:146-154.

Vizi, E.S., and Labos, E., 1991, Non-synaptic interactions at presynaptic level, *Progr. Neurobiol.* 37:145-161.

Wang, J.K.T., 1991, Presynaptic glutamate receptors modulate dopamine release from striatal synaptosomes, *J. Neurochem.* 57:819-822.

Wachtel, H., and Turski, L., 1990, Glutamate: a new target in schizophrenia?, *Trends Pharmacol. Sci.* 11:219-220.

Wedzony, K., Golembiowska, K., and Maj, J., 1991, A search for the effects of NMDA on the release of dopamine from the rat caudate nucleus, *in*: "Monitoring Molecules in Neuroscience," H. Rollema, B.H.C. Westerink, and W.J. Drijfhout, eds., University Centre for Pharmacy, Groningen, pp. 321-324.

Weinberger, D.R., Berman, K.F., and Zec, R.F., 1986, Physiological dysfunction of dorsolateral prefrontal cortex in schizophrenia, *Arch. gen Psychiat.* 43:114-124.

Weiner, D.M., and Brann, M.R., 1989, The distribution of a dopaminergic D2 receptor mRNA in rat brain, *FEBS Lett.* 253:207-213.

Westerink, B.H.C., Santiago, M., and De Vries, J.B., 1992, The release of dopamine from nerve terminals and dendrites of nigrostriatal neurons induced by excitatory amino acids in the conscious rat, *Naunyn-Schmiedeberg's Arch. Pharmacol.* 345:523-529.

White, F.J., and Wang, R.Y., 1983, Comparison of the effect of chronic haloperidol treatment on A9 and A10 dopamine neurons in the rat, *Life Sci.* 32:983-993.

Yamamoto, B.K., and Davy, S., 1992, Dopaminergic modulation of glutamate release in striatum as measured by microdialysis, *J. Neurochem.* 58:1736-1742.

Zigmond, M.J., and Stricker, E.M., 1972, Deficits in feeding behavior after intraventricular injection of 6-hydroxydopamine in rats, *Science* 177:1211-1214.

ACETYLCHOLINE, DOPAMINE AND NMDA TRANSMISSION IN THE CAUDATE-PUTAMEN: THEIR INTERACTION AND FUNCTION AS A STRIATAL MODULATORY SYSTEM

Gaetano Di Chiara and Micaela Morelli

Department of Toxicology, University of Cagliari
Viale A.Diaz 182
09100 Cagliari, Italy

INTRODUCTION

Drugs acting on muscarinic cholinergic receptors, dopaminergic D-1, D-2, D-3, D-4 receptors and glutamatergic N-Methyl-D-Aspartate (NMDA) receptors are known to exert major influences on motor functions by an action on the Basal Ganglia. This strikingly contrasts with the elusiveness of the influences of these receptors on the function of striatal neurons as measured by electrophysiological methods that readily demonstrate the action of classical neurotransmitter receptors such as those of the glutamate/AMPA or GABA-A type. For example, only in recent times, almost 30 years after the appraisal of dopamine (DA) deficiency as a causal factor in Parkinson's disease, the electrophysiological actions of DA at the single cell level are starting to be unraveled (Calabresi et al., 1987; Lacey et al., 1987; Kitai and Surmeier, 1993). The reason for the elusive nature of dopaminergic, muscarinic and NMDA transmission might be the fact that they do not directly trigger or inhibit action potentials but rather gate or modulate their generation by "fast" neurotransmitter receptors.

Here we will review the interaction in the striatum among these modulatory receptors and their relationship with the main neuronal type of the striatum, the medium size spiny neurons. This information will be incorporated in a model of striatal modulatory functions.

CHOLINERGIC NEURONS OF THE STRIATUM: INPUT/OUTPUT RELATIONSHIPS

The ACh neuron of the striatum is a site of interaction of three most important synaptic inputs. Probably the best documented input is from medium size substance P positive spiny neurons, that make symmetrical contacts with cholinergic proximal dendrites and cell bodies (Bolam et al., 1986; Martone et al., 1992). Since in these neurons GABA is known to coexist with substance P (Chesselet et al., 1987), it is likely that both these transmitters are able to influence the activity of striatal ACh neurons although their final influence will depend also on the relative density of their specific receptors and coupling to the transduction mechanism. ACh neurons have been reported to be the only striatal neuronal type to express NK1 receptors, the physiological substance P receptors (Gerfen, 1991). These receptors mediate the stimulatory effects of NK1 receptor agonists on ACh release observed *in vitro* and *in vivo* (Arenas et al., 1991; Petitet et al., 1991). The strict relationship between substance P and ACh neurons in the striatum is also documented by the overlapping distribution of ACh neurons and substance P terminals (Martone et al., 1992). This applies not only to the dorsal but also to the ventral striatum, where the association between ACh neurons and substance P terminals survives the loss of a relationship with the striosome/patch compartment (Martone et al., 1992).

The abundance of substance P input on ACh neurons contrasts with the paucity of enkephalin input (Martone et al., 1992). Since, at least in the rat, substance P is a marker of

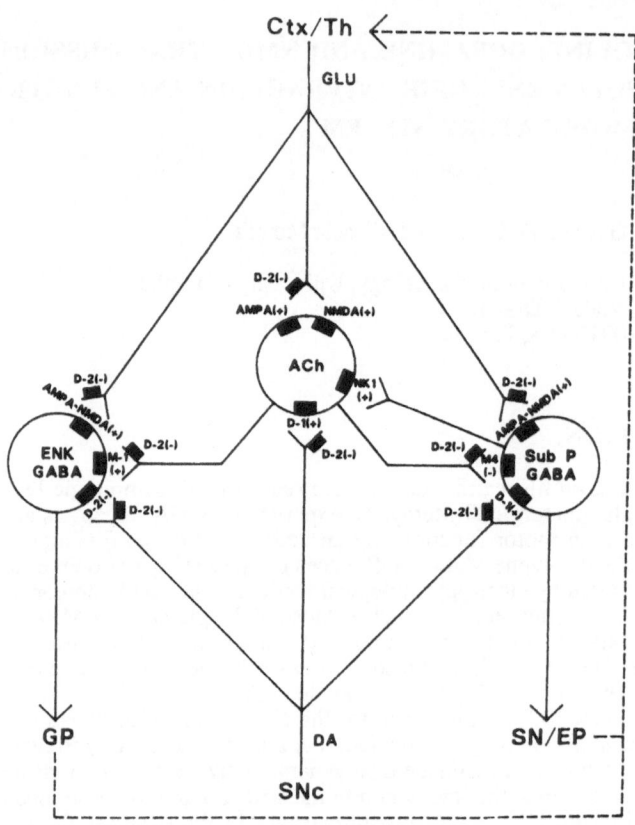

Figure 1. Diagram of the relationship between ach, da and nmda transmission in the striatum.

The model shows that dopamine (DA) input from substantia nigra pars compacta (SNc) and glutamate (GLU) input from the thalamus (Th) and cortex (Ctx) terminates on ACh interneurons and medium-size spiny neurons projecting to the globus pallidus (GP), using enkephalin (ENK) and GABA as transmitters, and to the substantia nigra (SN) and entopeduncular nucleus (EP), using substance P (Sub P) and GABA as transmitters. DA, by acting on post-synaptic D-1 receptors is postulated to facilitate while, through an action on D-2 receptors, is thought to stabilize ACh and medium-size spiny neurons to excitatory phasic input (AMPA) from Ctx and Th. DA, by acting on pre-synaptic D-2 receptors, reduce ACh release and glutamate release from their terminals (Maura et al., 1988). D-1 actions are thought to take place at the level of synaptic junctions while pre-synaptic D-2 effects would take place non-synaptically by DA diffused in the extracellular space. ACh can act on ENK neurons through facilitatory M-1 receptors and on Sub P neurons through inhibitory M-4 receptors. Therefore, Sub P and ENK neurons are under the reciprocal influence of DA and ACh through D-1 and M-4 receptors and through D-2 and M-1 receptors respectively. Combined stimulation of D-1 and blockade of M-4 receptors facilitates Sub P neurons, just like blockade of D-2 and stimulation of M-1 receptors facilitates ENK neurons. These changes are reflected in changes of c-fos induction and behavior. Sub P neurons provide a dense excitatory input, through NK1 receptors, to ACh neurons. D-1 receptors can stimulate ACh neurons in part directly (through D-1 receptors expressed by 30% of ACh neurons) but mostly, indirectly through an intra-striatal loop (mediated by Sub P neurons) and an extra-striatal loop reverberating from Ctx and Th onto ACh neurons.

striato-nigral/striato-entopeduncular neurons while enkephalin is a marker of striato-pallidal neurons (Gerfen, 1992), these observations indicate that ACh neurons receive a privileged input from striato-nigral/ striato entopeduncular neurons using substance P and GABA as transmitters. In turn ACh neurons synapse with striato-nigral medium size spiny neurons that are likely to be substance P neurons (Izzo and Bolam, 1988). In view of this, it is justified to view ACh neurons and substance P spiny neurons as part of a reciprocal intrastriatal loop (Gerfen, 1992; Martone et al., 1992) (See Fig. 1).

The second input to the ACh neurons is likely to be provided by DA neurons extrinsic to the striatum, having their cell bodies in the ventral mesencephalon (cell groups A8, A9 and A10). The issue of the existence of a dopaminergic synaptic input onto striatal ACh neurons has been uncertain for sometime (Lehmann and Langer, 1983; Pickel and Chan, 1990); evidence provided in recent years convincingly showed that DA terminals form symmetrical synaptic junctions with somata and proximal dendrites of striatal ACh neurons (Kubota et al., 1987; Chang, 1988; Dimova et al., 1993). Dopaminergic axons actually form "basket-like" pericellular arrangements around the ACh perikarya, contacting it by "en passant" synapses (Dimova et al., 1993).

The third major synaptic input to the ACh neuron is represented by asymmetric synapses originating from thalamic and cortical projections. In contrast with previous beliefs, the thalamic parafascicular nucleus (PF) rather than the cortex provides the major asymmetric input to the ACh neuron (Meredith and Wouterlood, 1990; Lapper and Bolam, 1992; Dimova et al., 1993).

An excitatory input onto proximal dendrites and somata as well as the intrinsic properties of the membrane of the ACh neuron might be the basis for two of its physiological characteristics: that of being tonically active in awake animals (in contrast with medium size spiny neurons that are mostly silent) (Kimura et al., 1984) and that of being exquisitely sensitive to afferent excitatory input (Wilson et al.. 1990). In relation to this, it has been argued that although ACh neurons receive a quantitatively minor excitatory input they are able to fire much more easily than medium size spiny neurons in response to afferent stimulation (Wilson et al., 1990). This argument can be further developed to view medium size spiny neurons on one side and ACh interneurons on the other as two different modalities by which afferent inputs can get access to the striatum and be processed within this structure (see below).

Consistent with the central position of ACh interneurons within the striatal network is the observations that cholinergic terminals make extensive symmetrical synaptic contact with medium size spiny neurons both on spines as well as on proximal dendrites, somata and even axon initial segments (Phelps et al., 1985; Izzo and Bolam, 1988).

DOPAMINE-ACETYLCHOLINE INTERACTIONS

ACh neurons have been long hypothesized to be a privileged target for DA. The early formulation of DA-ACh interactions in the striatum is that DA exerts an inhibitory influence on ACh neurons. The major impetus for this hypothesis came from in vitro and in vivo release studies showing that DA agonists and antagonists stimulate ACh release in the striatum (see Lehmann and Langer, 1983, for earlier literature).

Even earlier than the above hypothesis it was generically thought that DA and ACh exert opposite influences (the DA-ACh balance) on Basal Ganglia function, this idea deriving mainly from clinical studies showing that muscarinic antagonists were able to alleviate motor disturbances elicited by idiopathic DA deficiency or by pharmacological impairment of DA transmission (McGeer et al., 1961; Barbeau, 1962). Since the idea of the DA-ACh balance was not inconsistent with the hypothesis of the inhibitory DA-ACh interaction, it was actually lumped with the second one. However, it should be pointed out that the notion of a DA-ACh balance does not require postulating that DA is inhibitory on ACh function but simply that DA and ACh exert opposite influences on a common target, the striatal output neuron.

The hypothesis of an inhibitory DA-ACh interaction was strengthened by studies employing agonists and antagonists specific for the D-2 DA receptor subtype which demonstrated that, at least in vitro, the action of DA could be entirely attributed to stimulation of D-2 receptors (Lehmann and Langer, 1983).

On this basis DA and ACh neurons were depicted to be neurochemically in series and to interact post-synaptically (Hattori et al., 1976). However, given the uncertainties over the existence of DA synapses on ACh neurons, it was later proposed that DA acts pre-synaptically on ACh terminals to reduce ACh release by a non-synaptic mechanism (Lehmann and Langer, 1983).

At the time the inhibitory DA-ACh interaction was formulated, the D-1 receptor was considered just as a biochemical entity, being linked to stimulation of adenylate cyclase, but devoid of functional significance (Laduron, 1983). In retrospect such a restrictive position can be attributed to the lack of specific agonists and antagonists of D-1 receptors. The advent of such drugs, which started with the introduction of the first D-1 antagonist SCH 23390 (Hyttel, 1983; Iorio et al., 1983) and with the first D-1 (although partial) agonist SKF 38393 (Setler et al., 1978), opened the way to a series of studies whose results call for a **radical revision of the hypothesis of a negative DA-ACh interaction.**

SYMMETRICALLY OPPOSITE CONTROL OF ACETYLCHOLINE TRANSMISSION BY D-1 AND D-2 RECEPTORS

Measurement of extracellular ACh by brain dialysis have provided evidence that DA via D-1 receptors exerts a tonic facilitatory influence on the release of ACh in the striatum (Consolo et al., 1987; Bertorelli and Consolo, 1990; Damsma et al.,, 1990). D-1 mediated facilitation of ACh release by endogenous DA is already operative in basal conditions, as demonstrated by the ability of the D-1 antagonist SCH 23390 to reduce basal extracellular concentrations of ACh in the striatum (Consolo et al., 1987; Bertorelli and Consolo, 1990) and to increase striatal ACh levels in post-mortem homogenates (changes in tissue ACh are reciprocal to changes in extracellular ACh and ACh release) (Fage and Scatton, 1986).

In spite of this, combined blockade of D-1 and D-2 receptors or acute reduction of DA synthesis by α-methyltyrosine fails to change basal ACh release (Bertorelli et al., 1992); conversely, selective blockade of D-2 receptors increases ACh release (Bertorelli and Consolo, 1990). This indicates that in basal conditions DA reciprocally controls ACh release by a symmetric facilitatory/inhibitory influence by D-1 and D-2 receptors respectively. According to this arrangement of mutually antagonistic D-1/D-2 influences, any reduction of one input would result in an unbalance with prevalence of the reciprocal one. Thus, reduction of D-1 input would result in prevalence of D-2 mediated inhibition while reduction of D-2 input would result in prevalence of D-1 mediated facilitation of ACh-release.

Although in basal conditions the D-1/D-2 input by endogenous DA appears symmetrical, this is not so when DA-transmission is increased. Thus, treatments that increase extracellular DA like stimulation of DA release by amphetamine (Consolo et al., 1992), blockade of DA reuptake by cocaine (Consolo et al., 1992) or blockade of MAO by pargyline (Di Chiara et al., 1993), all result in stimulation of ACh release. Apparently, increase of DA transmission breaks the symmetry of the D-1/D-2 balance by favoring the facilitatory D-1 influence. The reason for this is unknown, but it is notable that the affinity of DA for D-2 receptors is reportedly one or two orders of magnitude higher than for D-1 receptors (see Table 1 in Jarvie and Caron, 1993) and this should result in large differences in the fractional occupation of the two types of DA-receptors in vivo. Apart from other factors, these differences in affinity of D-1 and D-2 receptors for DA would result in a higher fractional occupation of D-2 receptors relative to D-1 receptors in basal conditions. Stimulation of DA release might result in saturation of D-2 mediated inhibition but not of D-1 mediated facilitation so that further stimulation of DA release will result in further occupation of D-1 receptors and stimulation of ACh release.

MECHANISM OF DOPAMINERGIC INFLUENCES ON ACETYLCHOLINE RELEASE

Various possibilities exist to explain the mechanism of the D-1 mediated control of ACh release. In contrast with earlier suggestions, recent studies consistently indicate that the D-1 receptors mediating facilitation of ACh release are located in the striatum (Consolo et al., 1992). Another possible candidate for this mechanism, i.e. the substantia nigra pars reticulata, reportedly the brain area provided with the highest concentration of D-1 receptors, does not seem to be involved in this effect as local intranigral infusion of SCH 23390, fails to affect amphetamine-induced stimulation of striatal ACh release (Consolo and Di Chiara, unpublished). Similarly uninvolved appear the D-1 receptors located in the cerebral cortex (Consolo et al., 1992). In the striatum at least 30% of ACh neurons do express D-1 receptors (Guennoun and Bloch, 1992); therefore D-1 mediated facilitation of ACh release might be related in part to a direct D-1 mediated activation of ACh neurons.

However, the D-1 influence on ACh release might take place also indirectly with the interposition of substance P neurons receiving direct DA input and synapsing via intrastriatal collaterals with ACh neurons (Gerfen, 1992). DA acting on D-1 receptors would facilitate these neurons and the released substance P would activate ACh neurons (Arenas et al., 1991; Petitet et al., 1991) (Fig. 1).

A third mechanism by which D-1 receptor stimulation might control ACh release is more indirect and requires the activation, via striatal D-1 receptors, of a feedback circuit that, from the striatum and through the thalamus, reverberates onto striatal ACh neurons via thalamic and cortical excitatory projections. This possibility is further discussed below.

From these considerations it appears that there is a high degree of redundancy in the mechanisms by which DA through D-1 receptors, can facilitate ACh release. It is likely however that these different mechanisms are not functionally equivalent.

As far as the D-2 influence on ACh release, although classic views regard it as pre-synaptic via D-2 receptors located on ACh terminals (Lehmann and Langer, 1983), the possibility of a post-synaptic mechanism via somato-dendritic D-2 receptors controlling the excitability of ACh neurons should not be disregarded (Kitai and Surmeier, 1993).

ACETYLCHOLINE/DOPAMINE INTERACTION: CONVERGING INPUT ONTO STRIATAL SPINY NEURONS

Although the ACh neuron is an important target of DA input it would be quite misleading to envision the ACh neuron as the main target of DA in the striatum. In fact, the DA-ACh relationship should be viewed as only one of the mechanisms by which DA is capable of influencing striatal function. Thus, the other most important target of DA in the striatum is the medium size spiny neuron itself which constitutes as much as 95% of striatal neurons and receives monosynaptic DA input (Freund et al., 1984) as well as ACh input onto spines and dendrites (Phelps et al., 1985; Izzo and Bolam, 1988; Pickel and Chan, 1990). Medium size spiny neurons are therefore a site of converging DA and ACh input and such convergence might be in turn an additional substrate for the negative DA-ACh interaction postulated on the basis of clinical and experimental studies.

Indeed, an increasing body of evidence indicates that DA and ACh exert opposite influences on the function of medium size spiny neurons. Such reciprocal ACh/DA relationship is maintained in spite of the segregation of specific DA and muscarinic receptors on distinct populations of striatal spiny neurons. Thus, striato-nigral neurons which express D-1 but not D-2 receptors (Gerfen et al., 1990; Le Moine et al., 1991; Robertson et al., 1992), express, in addition to m_1, also m_4 receptors (Bernard et al., 1992). D-1 and m_4 receptors exert opposite influences on adenylate cyclase, D-1 stimulating and m_4 inhibiting it (Hulme et al., 1990). In striato-nigral neurons stimulation of D-1 receptors activates c-fos expression (Robertson et al., 1990) while stimulation of m_4 receptors is likely to inhibit it, as muscarinic blockade potentiates D-1 receptor mediated c-fos activation (Morelli et al., 1993a) (Fig. 3).

Quite a different picture but a similar organization and principle applies to strio-pallidal neurons. These neurons express D-2 receptors (Gerfen et al., 1990; Le Moine et al., 1990) that are inhibitory on adenylate-cyclase (Onali et al., 1985) and m_1 receptors that stimulate PI turnover (Hulme et al., 1990); only 40% of these neurons express the m_4 receptors (Bernard et al., 1992). It is interesting that in these neurons blockade of D-2 receptors results in activation of c-fos via stimulation of muscarinic (m_1?) receptors (Guo et al., 1992). It appears therefore that DA and ACh acting on reciprocally arranged receptor subtypes influence in an opposite manner the function of the two populations of medium size spiny neurons of the striatum.

The reason for such a reciprocal organization of the strio-nigral and strio-pallidal pathways as far as regards DA and ACh influences might be related to the fact that both pathways converge in their final out-put to the thalamus but are themselves organized in a reciprocal manner. In fact, the indirect strio-pallidal pathway involves an additional inhibitory synapse in respect to the direct strio-nigral / strio-entopeduncular pathway (Albin et al., 1989). With this arrangement, the direct and indirect pathways can work synergically or antagonistically depending on the fact that the input converging upon them has an opposite or, respectively, an identical influence. We interpret the reciprocal organization of DA and ACh receptor subtypes in the two subdivisions of the striatal out-put as suggesting that, at least for what concerns the tonic influences of DA and ACh, the two pathways are modulated synergically.

ROLE OF GLUTAMATE

Glutamate (or a related compound) might be the third actor in the interplay between DA, ACh and striatal medium-size spiny neurons. Glutamate is thought to be the excitatory transmitter of cortical input onto medium size spiny neurons and ACh neurons (Di Chiara and Gessa, 1981; Lehmann and Scatton, 1982; Fagg and Foster, 1983; Herrling, 1985; Cherubini et al., 1988; Consolo et al., 1990).

Recent studies favor a thalamic rather than a cortical input onto ACh neurons (Meredith and Wouterlood, 1990; Lapper and Bolam, 1992). This input is excitatory and recent studies are consistent with the possibility that this pathway utilizes glutamate as a transmitter (Z.P.Me, M.Palmi and J.P.Bolam, personal communication).

Given the importance of glutamate for the activity of striatal neurons, it is not surprising that drugs interfering with glutamate transmission are able to modify the motor effects of DA-receptor agonists and the function of striatal ACh neurons.

Some years ago various group independently reported that antagonists of NMDA receptors potentiate the motor effects of DA-receptor agonists in different models of DA deficiency including a unilaterally lesioned rat model of Parkinsonism (Carlsson and

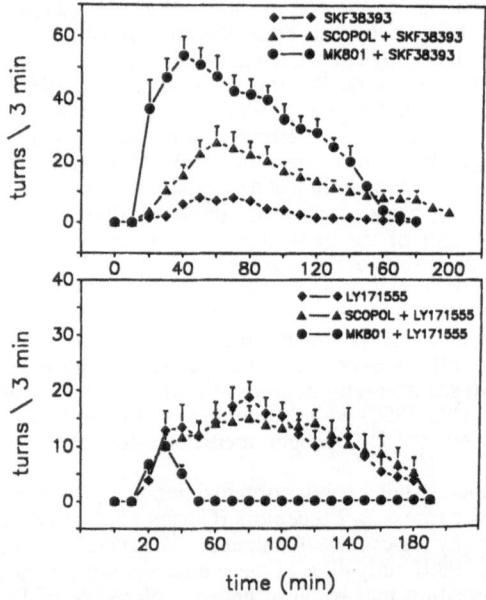

Figure 2. Contralateral turning in response to 1.5 mg/kg s.c. of the D-1 agonist SKF38393 given alone and in combination with 0.1 mg/kg i.p. of the NMDA antagonist (+)MK801 or 5 mg/kg i.p. of the muscarinic antagonist scopolamine (**upper panel**). In the **lower panel** is indicated the contralateral turning induced by 0.05 mg/kg s.c. of the D-2 agonist LY171555 alone and in combination with (+)MK801 or scopolamine at the same doses indicated above.

Carlsson, 1989; Klockgether and Turski, 1990; Morelli and Di Chiaara, 1990; Morelli et al., 1990 and 1992). However, as shown by Morelli and Di Chiaara (1990) and Morelli et al. (1990 and 1992), the ability of NMDA antagonists to potentiate DA responses does not refer to DA receptor mediated responses in general but specifically to D-1 receptor mediated responses; in fact NMDA-receptor blockade reduces D-2 mediated motor responses (Fig. 2). Recently, this opposite influence of NMDA receptor blockade on D-1 and D-2 mediated effects has been confirmed in reserpine-depleted mice (Goodwin et al., 1992; Svensson et al., 1992).

In addition to potentiate contralateral turning behavior, NMDA antagonists potentiate D-1 agonist mediated c-fos expression in the dorsolateral sensorimotor part of the caudate (Morelli et al., 1992) (Fig. 3) and also metabolic activation, measured by 2-deoxyglucose autoradiography, in striatal output areas such as the substantia nigra pars reticulata and the entopeduncular nucleus (Pontieri et al., 1992). These areas are the same that are activated by doses of D-1 agonist that are fully effective in eliciting contralateral turning (Trugman and Wooten, 1987).

Thus, blockade of NMDA receptors seems to eliminate an influence of glutamate that tends to antagonize the effect of DA on striato-nigral and striato-entopeduncular neurons. Given the direct excitatory influence of glutamate on these neurons it is unlikely that NMDA antagonists potentiate D-1 responses by blocking direct excitatory input onto medium size spiny neurons.

A more satisfactory explanation is that which considers a participation of ACh neurons. Stimulation of denervated supersensitive D-1 receptors, might stimulate a polysynaptic

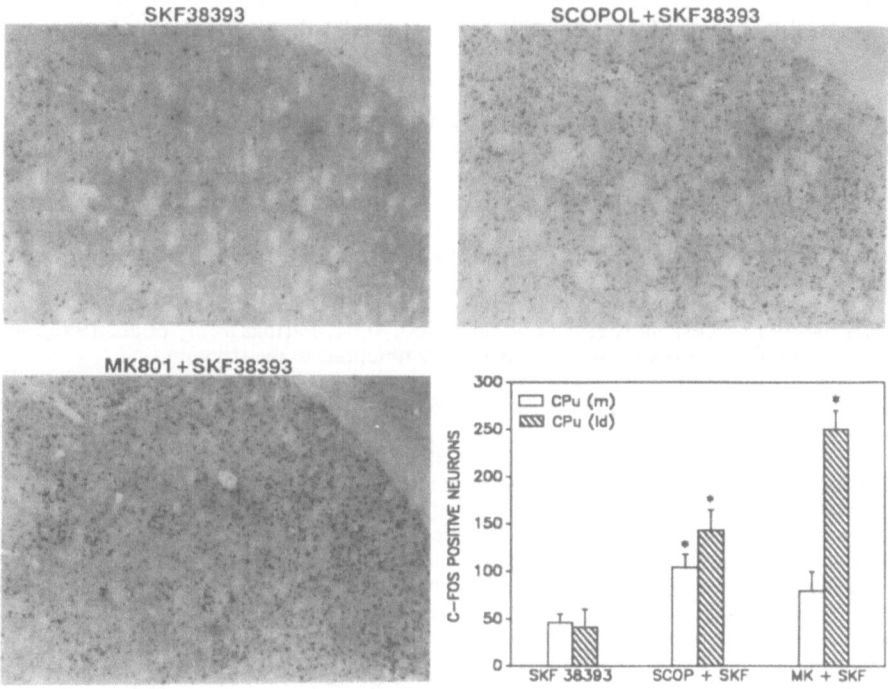

Figure 3. Photomicrographs showing the pattern of Fos positive neurons in a coronal section of the CPu correspondent to the 6-OHDA lesioned side and the number of Fos positive nuclei. Rats were treated with the D-1 agonist SKF38393 (1.5 mg/kg) alone and in combination with 0.1 mg/kg of the NMDA antagonist (+)MK801 or 5 mg/kg of the muscarinic antagonist scopolamine. In the diagram the number of Fos positive neurons in a 1 x 1 mm grid in the medial (m) and dorso-lateral (dl) part of the CPu is reported.

497

circuit reverberating from striatal spiny neurons, through thalamus and cortical excitatory projections, back to striatal ACh neurons; the resulting release of ACh onto striatal spiny neurons (Scatton and Lehmann, 1982) would act in opposition with D-1 stimulation. Consistent with this scheme is the observation that basal ACh release appears to depend upon a tonic facilitatory influence mediated by the cortex as decortication reduces basal ACh release (Consolo et al., 1990), NMDA-receptor blockade prevents D-1 stimulated release of ACh (Damsma et al., 1991) and scopolamine, an antagonist of muscarinic receptors, similarly to NMDA-blockade, potentiates D-1 (but not D-2) mediated contralateral turning behavior and striatal c-fos expression in the 6-OHDA model of denervation (Morelli et al., 1993a) (Fig. 2 and 3).

This hypothesis is consistent with the fact that D-2 receptor stimulation which reduces ACh release, potentiates D-1 mediated contralateral turning and striatal c-fos expression (Paul et al., 1992). This effect can be observed also with selective D-1 and D-2 antagonists given in combination with L-dopa (Morelli et al., 1993b) (Fig. 4 and Table 1).

It appears therefore that stimulation of supersensitive D-1 receptors activates ACh release probably through an activation of NMDA transmission in the caudate. This effect, that might be regarded as a negative feedback mechanism, has the final consequence of dampening the direct facilitatory effect of D-1 receptor stimulation on striato-nigral / striato-entopeduncular neurons (Di Chiara and Morelli, 1993).

The hypothesis that stimulation of D-2 receptors, by inhibiting ACh release can inactivate the negative feedback mechanism triggered by D-1 receptor stimulation, provides a novel interpretation for the well-known D-1/D-2 synergism i.e. for the ability of D-2 receptor stimulation to potentiate the behavioural effects of D-1 receptor stimulation and viceversa (Gershanik et al., 1983; Barone et al., 1986; Mashurano and Waddington, 1986; Longoni et al., 1987). Apparently, the Basal Ganglia offer a high degree of redundancy also for such interaction since D-1/D-2 synergism can also take place, as already pointed out, through their reciprocal influence on striato-nigral and striato-pallidal neurons (Gerfen, 1992).

ACH, DA AND NMDA TRANSMISSION AS SUBSTRATES OF A STRIATAL MODULATORY SYSTEM

The organization of Basal Ganglia has been depicted as a series of somatotopically organized channels running parallel from specific cortical areas to discrete subdivisions of the striatum and, through the globus pallidus and substantia nigra, to specific thalamic nuclei (Alexander et al., 1986).

This model of Basal Ganglia organization, already debated with respect to its specific issue of parallel processing (Percheron and Filion, 1991; Parthasarathy et al., 1992), also fails to account for the organization of modulatory functions in the striatum.

Thus, the dopaminergic projections from the substantia nigra pars compacta to the striatum appear to bridge two largely segregated domains such as the dorsal and the ventral striatum by way of the projections these neurons receive from the ventral striatum (Nauta et al., 1978). Moreover, in spite of the existence of a rough topography, nigro-striatal input to the caudate-putamen does not appear to be somatotopically organized (Parent, 1990).

A similar consideration applies to thalamic projections from the parafascicular nucleus (PF), traditionally part of the intralaminar nuclei of the thalamus. Although PF receives projections from the GPi (entopeduncular nucleus of the rat), it also receives widespread cortical input from layer V as well as from a large variety of brain areas including limbic-related structures and in turn projects to the cortex via collaterals of axons terminating in the striatum. PF excitatory projections thus provide the striatum with a template of the widespread, non-somatotopic input it provides to cerebral cortex (Jones and Leavitt, 1974; Jayaraman, 1984; Royce and Bromley, 1984; Peercheron et al., 1991). As pointed out by Lapper and Bolam (1992) the possibility that PF nuclei tonically stimulate ACh neurons is consistent with the observation that, in spite of the non-cholinergic nature of PF-striatal projections, lesions of PF nuclei reduces cholinergic markers in the striatum (Samuel et al., 1990).

Another origin of non-specific excitatory input to the striatum might be the cortex itself as cortical deep layer V projects to the striosome/patch compartment and this projection does not appear somatotopically organized (Gerfen, 1992).

Table 1. Effect of the D-2 antagonist raclopride and the D-1 antagonist SCH 23390 on L-dopa-mediated turning behavior and on the number of Fos-positive neurons in the 6-OHDA lesioned CPu.

Drug	Fos nuclei (1mm^2 grid)		Turns (2hrs)
	medial	dorso-lateral	
L-dopa(4)	124±30	293±54$^+$	480±90
+Raclo(3)	63±14	49±16*	11±2*
+SCH(0.1)	0*	0*	6±1*
L-dopa(12.5)	253±44	492±70$^+$	700±90
+Raclo(3)	196±29	317±45^{*+}	720±80
+SCH(0.1)	18±5*	10±1*	400±5*
Raclo+SCH	0*	0*	12±2*

Analysis of Variance $^*p<0.05$ L-dopa (4 or 12.5 mg/kg) vs raclopride (3 mg/kg) or/and SCH 23390 (0.1 mg/kg) + L-dopa; $^+p<0.05$ medial CPu vs dorso-lateral CPu

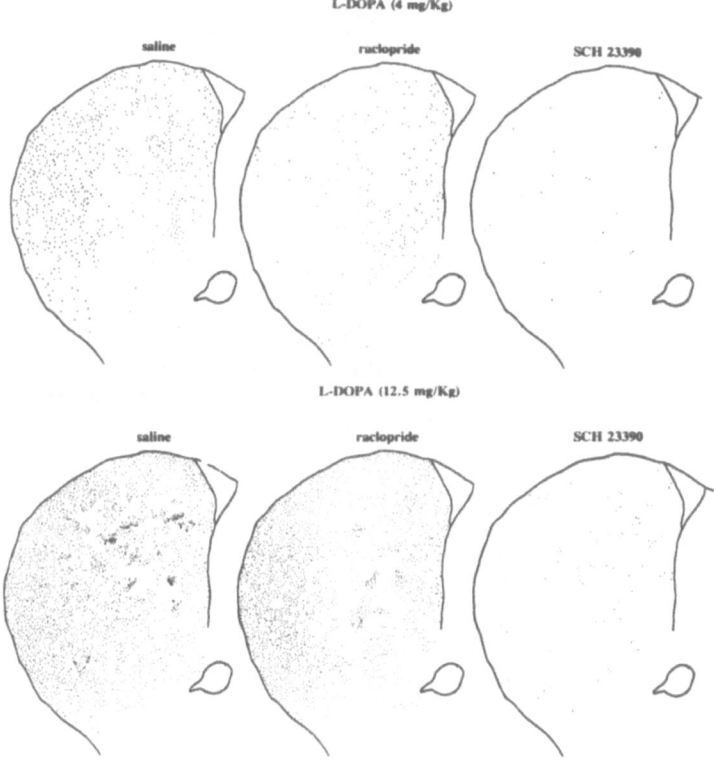

Figure 4. Representative drawing of the distribution of Fos-positive neurons in the 6-OHDA lesioned CPu of rats treated with two different doses of L-dopa. The D-2 antagonist raclopride (3 mg/kg s.c.) and the D-1 antagonist SCH23390 (0.1 mg/kg s.c.) were administered 30 min before L-dopa.

Another neuronal system that does not seem to conform to the currently held view of segregated striatal channels are the Ach interneurons. These neurons appear to integrate (atleast in the rat) wide territories of the striatum by way of their large dendritic and axonal fields.

In fact ACh neurons might be the substrate of an intrastriatal integrative network superimposed to the parallel anatomo-functional channels. Such network might provide a means for uniformly and contemporarily influence functionally different modules within the striatum.

Excitatory projections arising from intralaminar thalamus (PF) and from cortex (deep layers), DA projections arising from substantia nigra and ACh interneurons might form an interconnected system that we call the **striatal modulatory system** (see diagram in Figure 5). This system has at least three major characteristics: that of being organized non-somatotopically, to act tonically and to exert modulatory actions on its final target, the medium size spiny neurons.

These characteristics are differential with respect to the "specific" cortico-striatal system that is somatotopically organized, phasically active and exerts "fast" transmitter-like effects on medium size spiny neurons.

The striatal modulatory system might perform reward-related (incentive) modulation of the responsiveness of striatal medium size spiny neurons to specific, somatotopically organized cortical input. It is interesting in this respect, that the tonically firing striatal neurons that respond non-somatotopically to sensory stimuli that predict reward (Kimura et al., 1984; Kimura, 1986; Apicella et al., 1991) are suggested to be ACh neurons (Wilson et al., 1990); it is also notable that the somatotopically organized burst discharge of striatal spiny neurons appears to coincide with, rather than precede, movement (Kimura, 1986).

Non-ACh medium size aspiny neurons using somatostatin-NPY-neurotensin as transmitters might participate into this network being strategically placed to bridge the striosome and matrix compartment (Smith and Parent, 1986; Chesselet and Graybiel, 1986; Chesselet et al., 1987). These neurons are innervated by ACh neurons and in turn innervate medium size spiny neurons (Vuillet et al., 1992; Chang and Kita, 1992); moreover, similarly to ACh neurons, are activated by low amplitude excitatory input (Wilson et al., 1989).

FUNCTIONAL OPERATIONS OF THE STRIATAL MODULATORY SYSTEM

The actions of the striatal modulatory system are the result of the action of ACh, DA and glutamate (or a related compound) on muscarinic, D-1 and D-2 (D-3, D-4 etc...) and NMDA receptors. Activation of these modulatory receptors, rather than directly triggering action potentials or inhibiting their generation has the effect of setting the excitability of the neuron to incoming phasic inputs mediated by "fast" neurotransmitter receptors (AMPA, nicotinic, GABA-A). Modulatory actions have long kinetics of activation/desensitization being related to modulation of voltage-dependent channels (e.g. muscarinic and dopaminergic receptors) or to opening of voltage-sensitive receptor-operated cationic channels (NMDA receptors). The sign of these modulatory influences (facilitatory or inhibitory) might actually depend from the voltage of the post-synaptic membrane (Kitai and Surmeier, 1993).

Under basal conditions striatal spiny neurons are silent being in a hyperpolarized state; in this condition NMDA transmission is inoperative due to Mg^{++}-dependent blockade of NMDA-operated channels (Herrling, 1985; Cherubini et al., 1988). DA by acting on D-1 receptors can modulate a slow potassium current in a voltage dependent manner (Kitai and Surmeier, 1993). At high voltage (the state of the spiny neurons) D-1 receptor stimulation inactivates the current (Kitai and Surmeier, 1993). Since this current contributes to maintain the spiny neurons in a hyperpolarized resting state, D-1 receptor stimulation would have the effect of making the neuron more easily depolarized by "fast" EPSP's. This in turn makes NMDA transmission operative; as a result of the combined action of these modulatory influences the neuron can fire in bursts in response to incoming excitatory input (Calabresi et al., 1992). D-1 receptors can also inhibit a slowly depolarizing Na^+ current (Calabresi et al. 1987) but this mechanism is likely to have a consistent influence only in the activated state (partially depolarized). In this condition D-1 stimulation is expected to stabilize the neuron. It appears therefore that D-1 receptor stimulation has the effect of setting the excitability of the spiny neuron within a range more suitable for responding to excitatory phasic input.

ACh under resting conditions (-75 mV, hyperpolarized state) activates the inward potassium A-current thus stabilizing the neurons. Instead, under active firing conditions (55mV, depolarized state) (Wilson, 1990) ACh attenuates the potassium A-current thus contributing to maintain the neurons in the active state (Kitai and Surmeier, 1993). Therefore, ACh rather than simply facilitate or inhibit neuronal excitability, tends to stabilize the shift into a given state (resting or active) of the neuron following the influence of other inputs.

Similar receptor mechanisms probably operate on the ACh neuron but the final influences they produce on neuronal excitability might be quite different as a result of the different functional state and intrinsic properties of these neurons in comparison with medium size spiny neurons (Wilson et al., 1990). In fact ACh neurons are tonically active and their voltage is at a near threshold value. In this condition NMDA transmission can be fully operative in facilitating the generation of burst firing in response to low amplitude excitatory input (Wilson et al., 1990). Blockade of NMDA receptors is expected to have no influence on the activity of medium size spiny neurons but might conceivably depress the response of ACh neurons to afferent excitatory input. This is indeed what we have postulated to be the mechanism by which low doses of NMDA antagonists potentiate behavioural and biochemical D-1 responses in the striatum (see above). Consistent with the

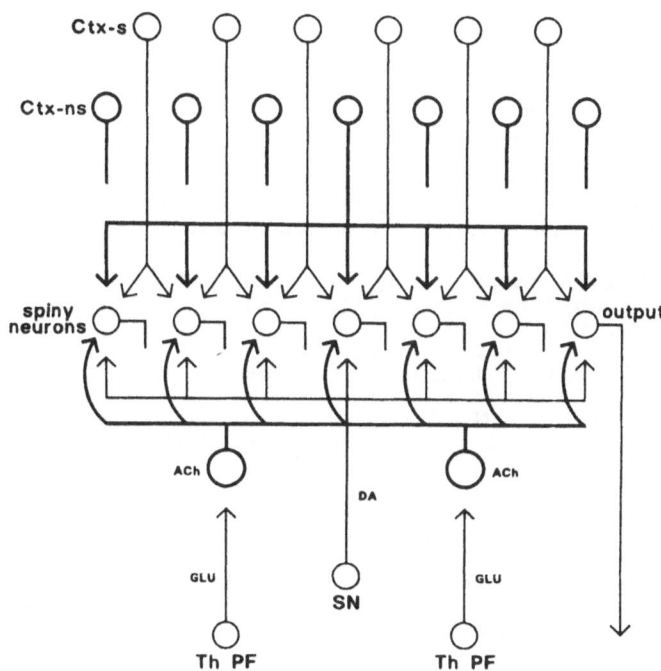

Figure 5. Diagramm of the striatal modulatory system.

The striatal modulatory system is depicted as composed of a striatal modulatory network and by three non-somatotopic inputs. The striatal modulatory network is made by striatal cholinergic interneurons (ACh) and NPY-somatostatin interneurons (not depicted). Afferent non-somatotopic input is provided by the cortex (Ctx-ns) through projections from deep layer V onto spiny neurons. Somatotopic input (Ctx-s) to spiny neurons arises instead from superficial layer V. The second non-somatotopic input arises from the intralaminar thalamus (Th), and in particular from parafascicular nucleus (PF) and terminates through glutamatergic (GLU) projections onto ACh neurons. The third non somatotopic input is the dopaminergic (DA) projection from substantia nigra pars compacta (SN) that terminates in part on ACh neurons but mostly converges with ACh input and cortical and thalamic non-somatotopic input onto medium size spiny neurons projecting outside the striatum (out-put).

above model is also the observation that high doses of NMDA antagonists reduce D-1 mediated effects, probably by an action on medium size spiny neurons activated by D-1 dependent modulation.

Acknowledgements

The authors wish to thank Paolo Calabresi, Steve Kitai and Gérard Percheron for the discussion over specific issues of this article. This work was supported by a grant from CNR, progetto finalizzato "FATMA" sottoprogetto "Stress". The authors wish also to thank Mrs A. Marchioni for typing of the manuscript.

REFERENCES

Albin, R.L., Young, A.B., and Penney, J.B., 1989, The functional anatomy of basal ganglia disorders, *Trends Neurosci.*, 12:366-375.

Alexander, G.E., DeLong, M.R., and Strick, P.L., 1986, Parallel organization of functionally segregated circuits linking basal ganglia and cortex, *Annu. Rev. Neurosci.* 9:357-381.

Apicella, P., Scarnati, E., and Schultz, W., 1991, Tonically discharging neurons of monkey striatum respond to preparatory and rewarding stimuli, *Exp. Brain Res.* 84:672-675.

Arenas, E., Alberch, J., Perez-Navarro, E., Solsona, C., and Marsal, J., 1991, Neurokinin receptors differentially mediate endogenous acetylcholine release evoked by tachykinins in the neostriatum, *J.Neurosci.*11,8:2332-2338.

Barbeau, A., 1962, The pathogenesis of Parkinson's disease: a new hypothesis, *Can. med. Ass.J.* 87:802-807.

Barone, P., Davis, T.A., Braun, A.R., and Chase, T.N., 1986, Dopaminergic mechanisms and motor function: characterization of D-1 and D-2 dopamine receptor interactions, *Eur.J.Pharmacol.* 123:109-114.

Bernard, V., Normand, E., and Bloch, B., 1992, Phenotypical characterization of the rat striatal neurons expressing muscarinic receptor genes, *J. Neurosci.* 12,9:3591-3601.

Bertorelli, R., and Consolo, S., 1990, D_1 and D_2 dopaminergic regulation of acetylcholine release from striata of freely moving rats, *J.Neurochem.* 54:2145-2148.

Bertorelli, R., Zambelli, M., Di Chiara, G., and Consolo, S., 1992, Dopamine depletion preferentially impairs D_1-over D_2-receptor regulation of striatal in vivo acetylcholine release,*J.Neurochem.*59:353-357.

Bolam, J.P., Ingham, C.A., Izzo, P.N., Levey, A.I., Rye, D.B., Smith, A.D., and Wainer, B.H.,1986, Substance P-containing terminals and synaptic contact with cholinergic neurons in the neostriatum and basal forebrain: a double immunocytochemical study in the rat, *Brain Res.* 397:279-289.

Calabresi, P., Mercuri, N., Stanzione, P., Stefani, A., and Bernardi, G., 1987, Intracellular studies on the dopamine-induced firing inhibition of neostriatal neurons in vivo: evidence for D-1 involvement, *Neurosci.* 20:757-765.

Calabresi, P., Pisani, A., Mercuri, N.B., and Bernardi, G., 1992, Long-term potentiation in the striatum is unmasked by removing the voltage-dependent magnesium block of NMDA receptor channels, *Eur.J.Neurosci.* 4:929-935.

Carlsson, M., and Carlsson, A., 1989, Dramatic synergism between MK-801 and clonidine with respect to locomotor stimulatory effect in monoamine-depleted mice, *J.Neural Transm.* 77:65-71.

Chang, H.T., 1988, Dopamine-acetylcholine interaction in the rat striatum: a dual-labelling immunocytochemical study, *Brain Res.* 21:295-304.

Chang, H.T., and Kita, H., 1992, Interneurons in the rat striatum: relationships between parvalbumin neurons and cholinergic neurons, *Brain Res.* 574:307-311.

Cherubini, E., Herrling, P.L., Lanfumey, L., and Stanzione, P., 1988, Excitatory amino acids in synaptic excitation of striatal neurones in vitro, J.Physiol., 400:677-690.

Chesselet, M.F., and Graybiel, A.M., 1986, Striatal neurons expressing somatostatin-like immunoreactivity: evidence for a peptidergic interneuronal system in the cat, *Neurosci.*17,3:547-557.

Chesselet, M.F., Weiss, L., Wuenschell, C., Tobin,m A.J., and Affolter, H.-V., 1987, Comparative distribution of mRNAs for glutamic acid decarboxylase, tyrosine hydroxylase, and tachykinins in the basal ganglia: an **in situ** hybridization study in the rodent brain, *J.Comp. Neurol.* 262:125-140.

Consolo, S., Girotti, P., Russi, G., and Di Chiara, G., 1992, Endogenous dopamine facilitatestriatal in vivo acetylcholine release by acting on D_1 receptors localized in the striatum, *J.Neurochem.* 59,4:1555-1557.

Consolo, S., Salmoiraghi, P., Amoroso, D., and Kolasa, K., 1990, Treatment with oxitracetam or choline restores cholinergic biochemical and pharmacological activities in striata of decorticated rats, *J.Neurochem.* 54:571-577

Consolo, S., Wu, C.F., Fiorentini, F., Ladinsky, H., and Vezzani, A., 1987, Determination of endogenous acetylcholine release in freely moving rats by transstriatal dialysis coupled to a radioenzymatic assay: effect of drugs, *J.Neurochem.* 48:1459-1465.

Damsma, G., Robertson, G.S., Tham, C.S., and Fibiger, H.C., 1991, Dopaminergic regulaation of striatal acetylcholine-release: importance of D_1 and N-methyl-D-aspartate receptors, *J. Pharmacol. Exp. Ther.* 259:1064-1072.

Damsma, G., Tham, C.S., Robertson, G.S., and Fibiger, H.C., 1990, Dopamine D_1 receptor stimulation increases striatal acetylcholine release in the rat, *Eur.J.Pharmacol.* 186:335-338.

Di Chiara, G., and Gessa, G.L., 1981, " Glutamate as a Neurotransmitter, "Vol. 27, Raven Press, New York.

Di Chiara, G., and Morelli, M., 1993, Dopamine-Acetylcholine-Glutamate interactions in the striatum. A Working Hypothesis, *Adv. in Neurol.* 60:102-106.

Di Chiara, G., Carboni, E., Morelli, M., Cozzolino, A., Tanda, G.L., Pinna, A., Russi, G., andConsolo, S., 1993a, Stimulation of dopamine transmission in the dorsal caudate by pargyline as demonstrated by dopamine and acetylcholine microdialysis and Fos immunohistochemistry, Neurosci. in press.

Dimova, R., Vuillet, J., Nieoullon, A., and Kerkebian-Le Goff, L., 1993, Ultrastructural features of the choline acetyltransferase-containing neurons and relationships with nigral dopaminergic and cortical afferent pathways in the rat, *Neurosci.* in press.

Fage, D., and Scatton, B., 1986, Opposing effects of D-1 and D-2 receptor antagonists onacetylcholine levels in the rat striatum, *Eur.J.Pharmacol.* 129:359-364.

Fagg, G.E., and Foster, A.C., 1983, Amino acid neurotransmitters and their pathways in the mammalian central nervous system, *Neurosci.* 9:701-709.

Freund, T.F., Powell, J.F., and Smith, A.D., 1984, Tyrosine hydroxylase-immunoreactive boutons in synaptic contact with identified striatonigral neurons, with particular reference to dendritic spines, *Neurosci.* 13:1189-1215.

Gerfen, C.R., 1991, Substance P (neurokinin-1) receptor mRNA is selectively expressed in cholinergic neurons in the striatum and basal forebrain, *Brain Res.* 556:165-170.

Gerfen, C.R., 1992, The neostriatal mosaic: multiple levels of compartmental organization in the basal ganglia, *Ann.Rev.Neurosci.* 15:285-320.

Gerfen, C.R., Engber, T.M., Mahan, L.C., Susel, Z., Chase, T.N., Monsma, F.J., and Sibley, D.R., 1990, D_1 and D_2 dopamine receptor-regulated gene expression of striatonigral and striatopallidal neurons, *Science* 250:1429-1432.

Gershanik, O., Heikkila, R.E., and Duvoisin, R.C., 1983, Behavioural correlation of dopamine receptor activation, *Neurol.* 33:1489-1494.

Goodwin, P., Starr, B.S., and Starr, M.S., 1992, Motor responses to dopamine D_1 and D_2 agonists in the reserpine-treated mouse are affected differentially by the NMDA receptor antagonist MK 801, *J. Neural Transm.* 4:15-26.

Guennoun, R., and Bloch., B., 1992, Ontogeny of D_1 and DARPP-32 gene expression in rat striatum: an in situ hybridization study, *Molec. Brain Res.* 12:131-139.

Guo, N., Robertson, G.S., and Fibiger, H.C., 1992, Scopolamine attenuates haloperidol-induced c-fos expression in the striatum, *Brain Res.* 588:164-167.

Hattori, T., Singh, V.K., McGeer, E.G., and McGeer, P.L., 1976, Immunohistochemical localization of choline acetyltransferase containing neostriatal neurons and their relationship with dopaminergic synapses, *Brain Res.* 102:164-173.

Herrling, P.L., 1985, Pharmacology of the corticocaudate excitatory postsynaptic potential in the cat: evidence for its mediation by quisqualate or kainate receptors, *Neurosci.* 14:417-426.

Hulme,E.C., Birdsall, N.J.M., Buckley, N.J., 1990, Muscarinic receptor subtypes, *Annu. Rev Pharmacol. Toxicol.* 30:633-639.

Hyttel, J., 1983, SCH 23390: the first selective dopamine D-1 antagonist, *Eur.J.Pharmacol.* 91:153-158.

Iorio, L.C., Barnett, A., Leitz, F.H., Houser, V.P., and Korduba, C.A., 1983, SCH 23390, a potential benzazepine antipsychotic with unique interactions on dopaminergic systems, *J.Pharmacol.Exp.Ther.* 226:462-468.

Izzo, P.N., and Bolam, J., 1988, Cholinergic synaptic input to different parts of spiny striatonigral neurons in the rat, *J.Comp.Neurol.* 269:219-234.

Jarvie, K.R., and Caron, M.G., 1993, Heterogeneity of dopamine receptors, *in*: "Parkinson's disease, Adv. in Neurol., 60," H.Narabayashi, T.Nagatsu, N.Yanagisawa, and Y.Mizuno, eds., Raven Press, New York, pp.325-333.

Jayaraman, A., 1984, Thalamostriate projections - An overview, *in*: "The Basal Ganglia", J.S.McKenzie, R.E.Kemm, and L.N.Wilcock, eds., Plenum Press, New York, pp. 69-86.

Jones, E.G., and Leavitt, R.Y., 1974, Retrograde axonal transport and the demonstration of non-specific projections to the cerebral cortex and striatum from thalamic intralaminar nuclei in the rat, cat and monkey, *J.Comp.Neurol.* 154:349-378.

Kimura, M., 1986, The role of primate putamen neurons in the association of sensory stimuli with movement, *Neurosci.Res.*3:436-443.

Kimura, M., Raijkowski, J., and Evarts, E., 1984, Tonically discharging putamen neurons exhibit set-dependent responses, *Proc.Natl.Acad.Sci.USA* 81:4998-5001.

Kitai,S.T., and Surmeier, D.J., 1993, Cholinergic and dopaminergic modulation of potassium conductances in neostriatal neurons, *Advances in Neurol.* 60:40-52.

Klockgether, T., and Turski, L., 1990, NMDA antagonists potentiate antiparkinsonian actions of L-DOPA in monoamine-depleted rats, *Ann.Neurol.* 28:539-546.

Kubota, Y., Inagaki, S., Shimada, S., Kito, S., Eckenstein, F., and Tohyama, M., 1987, Neostriatal cholinergic neurons receive direct synaptic inputs from dopaminergic axons, *Brain Res.* 413:179-184.

Lacey, M.G., Mercuri, N.B., and North, R.A., 1987, Dopamine acts on D-2 receptors to increase potassium conductance in neurones of the rat substantia nigra zona compacta, *J.Physiol.* 392:397-416.

Laduron, P., 1983, Dopamine - sensitive adenylate cyclase as a receptor site, in: "Dopamine Receptors," C.Kaiser, J.W.Kebabian, eds., American Chemical Society, Washington DC, pp. 46-52.

Lapper, S.R., and Bolam, J.P., 1992, Input from the frontal cortex and the parafascicular nucleus to cholinergic interneurons in the dorsal striatum of the rat, *Neurosci.* 51,3:533-545.

Lehmann, J., and Langer, S.Z., 1983, The striatal cholinergic interneuron: synaptic target of dopaminergic terminals?, *Neurosci.* 10,4:1105-1120.

Lehmann, J., and Scatton, B., 1982, Characterization of the excitatory amino acid receptor- mediated release of [3H]acetylcholine from rat striatal slices, *Brain Res*.252:77-83.

Le Moine, C., Normand, E., and Bloch, B., 1991, Phenotypical characterization of the rat striatal neurons expressing the D_1 dopamine receptor gene, *Proc. Natl.Acad.Sci.* 88:4205-4209.

Le Moine, C., Normand, E., Guitteny, A.F., Fouque, B., Teoule, R., and Bloch, B., 1990, Dopamine receptor gene expression by enkephalin neurons in rat forebrain, *Proc.Natl.Acad.Sci.USA* 87:230-234.

Longoni, R., Spina, L., and Di Chiara, G., 1987, Permissive role of D-1 receptor stimulation for the expression of D-2 mediated behavioral responses: A quantitative phenomenological study in rats, *Life Sci.*41:2135-2145.

Martone, M.E., Armstrong, D.M., Young, S.J., and Groves, P.M., 1992, Ultrastructural examination of enkephalin and substance P input to cholinergic neurons within the rat neostriatum, *Brain Res.* 594:253-262.

Mashurano, M., and Waddington, J.L., 1986, Stereotyped behaviour in response to the selective D-2 dopamine receptors agonist RU 24213 is enhanced by pretreatment with the selective D-1 agonist SK&F 38393, *Neuropharmacol.* 25:947-949.

Maura, G., Giardi, A., and Raiteri, M., 1988, Release-regulating D-2 dopamine receptors are located on striatal glutamatergic nerve terminals, *J.Pharmacol.Exp.Ther.* 247:680-684.

McGeer, P.L., Boulding, J.E., Gibson, W.C., and Foulkes, R.G., 1961, Drug-induced extrapyramidal reactions, *J.Am.med.Ass.* 177:665-670.

Meredith, G.E., and Wouterlood, F.G., 1990, Hippocampal and midline thalamic fibers and terminals in relation to the choline acetyltransferase-immunoreactive neurons in nucleus accumbens of the rat: a light and electron microscopic study, *J.Comp.Neurol.* 296:204-221.

Morelli, M., and Di Chiara, G., 1990, MK-801 potentiates dopaminergic D-1 but reduces D-2 responses in the 6-hydroxydopamine model of Parkinson's disease, *Eur.J.Pharmacol.* 182:611-612.

Morelli, M., Cozzolino, A., Pinna, A., Fenu, S., Carta, A., and Di Chiara, G., 1993b, L-Dopa stimulates c-fos expression in dopamine denervated striatum by combined activation of D-1 and D-2 receptors, *Brain Res.* in press.

Morelli, M., Fenu, S., Cozzolino, A., Pinna, A., Carta, A., and Di Chiara, G., 1993a, Blockade of muscarinic receptors potentiates D_1 dependent turning behavior and c-fos expression in 6-hydroxydopamine-lesioned rats but does not influence D_2 mediated responses, *Neurosci.* 3:673-678.

Morelli, M., Fenu, S., Pinna, A., and Di Chiara, G., 1992, Opposite effects of NMDA receptor blockade on dopaminergic D_1- and D_2-mediated behavior in the 6-hydroxydopamine model of turning: relationship with c-fos expression, *J.Pharmacol.Exp.Ther.* 260,1:402-408.

Nauta, W.J.H., Smith, G.P., Faull, R.L.M., and Domesick, V.B., 1978, Efferent connections and nigral afferents of the nucleus accumbens septi in the rat, *Neurosci.* 3:385-401.

Onali, P., Olianas, M.C., and Gessa, G.L., 1985, Characterization of dopamine receptors mediating inhibition of adenylate cyclase activity in rat striatum, *Mol. Pharmacol.* 28:138-142.

Paul, M.L., Graybiel, J.-C., and Robertson, H.A., 1992, D1-like and D2-like dopamine receptors synergistically activate rotation and c-fos expression in the dopamine-depleted striatum in a rat model of Parkinson's disease, *J. Neurosci.* 12,10:3729-3742.

Parent, A., 1990, Extrinsic connections of the basal ganglia, *Trends Neurosci.*13,7:254-258.

Parthasarathy, H.B., Schall, J.D., and Graybiel, A.M., 1992, Distributed but convergent ordering of corticostriatal projections: analysis of the frontal eye field and the supplementary eye field in the macaque monkey, *J.Neurosci.* 12,11:4468-4488.

Percheron, G., and Filion, M., 1991, Parallel processing in the basal ganglia: Up to a point, *Trends Neurosci.*14:55-56.

Percheron, G., François, C., Parent, A., Sadikot, A.F., Fénelon, G., and Yelnik, J., 1991, The primate central complex as one of the basal ganglia, *Adv. in Behav. Biol.* 39:177-186.

Petitet, F., Glowinski, J., and Beaujouan, J.-C., 1991, Evoked release of acetylcholine in the rat striatum by stimulation of tachykinin NK-1 receptors, *Eur.J.Pharmacol.* 192:203-204.

Phelps, P.E., Houser, C.R., and Vaughn, J.E., 1985, Immunocytochemical localization of choline acetyltransferase within the rat neostriatum: A correlated light and electron microscopic study of cholinergic neurons and synapses, *J. Comparative Neurol.* 238:286-294.

Pickel, V.M., and Chan, J., 1990, Spiny neurons lacking choline acetyltransferase immunoreactivity are the major targets of cholinergic and catecholaminergic terminals in the rat striatum, *J.Neurosci.Res.* 25:263-280.

Pontieri, F.E., Morelli, M., Orzi, F., Terenzi, R., and Di Chiara, G., 1992, Metabolic mapping of the synergism between MK-801 and SKF 38393 in rats with unilateral lesions of the dopaminergic nigrostriatal pathway, *Synapse* 12:255260.

Robertson, G.S., Vincent, S.R., and Fibiger, H.C., 1990, Striatonigral projection neurons contain D_1 dopamine receptor-activated c-fos, *Brain Res.* 523:288-290.

Robertson, G.S., Vincent, S.R., and Fibiger, H.C., 1992, D_1 and D_2 dopamine receptors differentially regulate c-fos expression in striatonigral and striatopallidal neurons, *Neurosci.* 49,2:285-296.

Royce, G.J., and Bromley, S., 1984, Fluorescent double labeling studies of thalamostriatal and corticostriatal neurons, *Adv. in Behav. Biol.* 27:131-137.

Samuel, D., Kerkerian Le Goff, L., Kumar, U., Errami, M., Scarfone, E., and Nieuollon, A.,1990, Changes in striatal cholinergic, GABAergic, dopaminergic and serotoninergic biochemical markers after kainic acid-induced thalamic lesions in the rat, *J.Neural Transm.(Parkinson's Disease Section)* 2:193-203.

Scatton, B., and Lehmann, J., 1982, N-Methyl-D-aspartate-type receptors mediate striatal ^{3}H-acetylcholine release evoked by excitatory amino acids, *Nature* 297:422-424.

Setler, P.E., Sarau, H.M., Zirkle, C.L., and Saunders, H.L., 1978, The central effects of a novel dopamine agonist, *Eur.J.Pharmacol.* 50:419-425.

Smith, Y., and Parent, A., 1986, Neuropeptide Y-immunoreactive neurons in the striatum of cat and monkey: morphological characteristics, intrinsic organization and Co-localization with somatostatin, *Brain Res.* 372:241249.

Svensson, A., Carlsson, A., and Carlsson, M.L., 1992, Differential locomotor interactions between dopamine D1/D2 receptor agonists and the NMDA antagonist dizocilpine in monoamine- depleted mice, *J.Neural Transm.* 90:199217.

Trugman, J.M., and Wooten, G.F., 1987, Selective D_1 and D_2 dopamine agonists differentially alter basal ganglia glucose utilization in rats with unilateral 6-hydroxydopamine substantia nigra lesions, *J.Neurosci.* 7,9:2927-2935.

Vuillet, J., Dimova, R., Nieoullon, A., and Kerkerian-Le Goff, L., 1992, Ultrastructural relationships between choline acetyltransferase and neuropeptide Y-containing neurons in the rat striatum, *Neurosci.* 46:351-360.

Wilson, C.J., 1990, Basal Ganglia, *in*: "The Synaptic Organization of the Brain", G.M.Shepherd, ed., Oxford University Press, New York, pp. 64-71.

Wilson, C.J., Chang, H.T., and Kitai, S.T., 1990, Firing patterns and synaptic potentials of identified giant aspiny interneurons in the rat neostriatum, *J.Neurosci.* 10,2:508-519.

Wilson, C.J., Kita, H., and Kawaguchi, Y., 1989, GABAergic interneurons, rather than spiny cell axon collaterals, are responsible for the IPSP responses to afferent stimulation in neostriatal spiny neurons, *Soc. Neurosci. Abst.*15:907.

APOMORPHINE-SUSCEPTIBLE AND APOMORPHINE-UNSUSCEPTIBLE WISTAR RATS: A NEW TOOL IN THE SEARCH FOR THE FUNCTION OF THE STRIATUM IN SWITCHING BEHAVIOURAL STRATEGIES

Alexander R. Cools[1], Nynke Y. Rots[1,2], Ronald De Kloet[2]
and Bart Ellenbroek[1]

[1]Department of Psycho- and Neuropharmacology, University of Nijmegen,
  The Netherlands
[2]Center for BioPharmaceutical Sciences, University of Leiden,
  The Netherlands

INTRODUCTION

Individual variation in vulnerability for basal ganglia disorders such as Parkinson's disease is a well known phenomenon in man. In addition, signs and symptoms of basal ganglia disorders vary from patient to patient. Furthermore, patients suffering from basal ganglia disorders show an individual variaton in vulnerability for unwanted side-effects of drugs such as L-DOPA and bromocriptine, i.e. agents known to affect the dopaminergic transmission within the basal ganglia. Apparently, the make-up and/or the reactivity of the basal ganglia vary between individuals. It is unknown whether genetic, postnatal and/or environmental factors determine this individual variation. Given the fact that most human studies are descriptive and correlative by nature, basic knowledge has to be collected in animal studies.

Individual variation in behaviour is a well known phenomenon in the animal kingdom. In fact, unselected outbred populations of rodents are marked by a bimodal shape of variation in behaviour (winners vs losers; high vs low responders to novelty; etc.; Masur et al., 1975; Benus, 1988; Piazza et al., 1991). The bimodal shape of variation in behaviour shown by Wistar rats in the so-called defeat-test (freezing displayed by the winners vs fleeing displayed by the losers) forms part of an overall bimodal variation in behaviour and physiology which is normally present in unselected outbred populations of Wistar rats (Fokkema et al., 1986; Koolhaas et al., 1986; Cools et al., 1990 and 1993a). A few years ago, we have discovered that each of these fundamentally distinct types of individuals is marked by its own make-up and reactivity of the brain, endocrinological and immunological system (Cools et al., 1990; Cools et al., 1993a). We have also discovered that these two distinct types of individuals are differentially susceptible for apomorphine-induced gnawing: the winners (freezing rats) are apomorphine-unsusceptible, whereas the losers (fleeing rats) are apomorphine-susceptible (Cools, 1988; Cools et al., 1990). Since apomorphine-induced gnawing behaviour is primarily due to stimulation of dopaminergic receptors in the striatum, this finding has provided the first piece of evidence that the dopaminergic mechanisms within the striatum significantly differ between the two distinct types of individuals which normally co-exist in unselected outbred populations of Wistar rats.

Today, animal research has provided evidence that striatal regions which are innervated by dopaminergic fibers arising in the substantia nigra, pars compacata (A$_9$ fibers), allow the organism to switch arbitrarily behaviour, and that striatal regions which are innervated by dopaminergic fibers arising in the ventral tegmental area (A_{10} fibers), allow the organism to switch behaviour with the help of cues (Cools, 1980; Van den Bos and Cools, 1989; for review: Cools, 1990; Van den Bos et al., 1991). Given the above-mentioned difference in the hardware and/or software of the striatum between apomorphine-susceptible and apomorphine-unsusceptible rats, we have postulated that these two distinct types of rats are valid tools in the search for the function of the striatum in switching behavioural strategies. As discussed in the final paragraph of this article, this is indeed the case.

ORIGIN OF THE TWO DISTINCT TYPES OF INDIVIDUALS WHICH NORMALLY CO-EXIST IN OUTBRED POPULATIONS OF WISTAR RATS

Starting from the notion that normal outbred populations of Wistar rats are marked by a bimodal shape of variation in behaviour and/or physiology (see above), we have reasoned that there should be a bimodal shape of variation in the physiological state of the brain in these animals. For, it is the physiological state of the brain that directs the display of behaviour. By chance, we have discovered that there exists such a causal relationship between the susceptibility of α-noradrenergic receptors in the ventral striatum, especially the nucleus accumbens, and the display of freezing behaviour in the so-called defeat-test, in which an intruder has to enter the home cage of a resident (Cools, 1988). Intruders which have susceptible α-noradrenergic receptors normally flee and lose ultimately the fight, when the resident is allowed to attack them, whereas intruders which have unsusceptible α-noradrenergic receptors in this brain region, stop the fight by adopting sooner or later a freezing posture. When the (un)susceptibility of these α-noradrenergic receptors is experimentally reversed by a pharmacological manipulation, the fleeing rats become freezers, and vice versa.

These findings have provided evidence that freezing in the defeat-test is not just an intervening variable but, instead, a basic feature being the consequence of a particular physiological state in a circumscribed part of the brain. The causal relationship itself has formed the startingpoint for studies tracing additional, individual differences in the neural network in which the nucleus accumbens is embedded. As a result, we have been able to show that the bimodal variation in "freezing behaviour" is consistently coupled to a bimodal variation in a great variety of neurobiological variables of this network (Cools et al., 1990). Given this individual consistency in behaviour and brain physiology, any of the neurobiological variables of which its bimodal variation is coupled to the bimodal variation in "freezing behaviour", can be used as criterion for selecting the two distinct types of individuals. For practical reasons, we have commonly used the bimodal variation in the susceptibility to the dopaminergic agonist apomorphine. Systemic administration of this drug (1.5 mg/kg, s.c.) produces very high gnawing scores in "non-freezing" animals, but very low gnawing scores in "freezing" animals. Since this behaviour can be easily quantified in an objective manner, this criterion has been used to select female and male rats for breeding two distinct lines of a normal outbred Wistar strain of rats (Cools et al., 1990): Apomorphine-susceptible (APO-SUS) and Apomorphine-unsusceptible (APO-UNSUS) Wistar rat lines. We have used a particular breeding procedure to maintain the genotypic heterogeneity which was present in the initial outbred strain, apart from the alleles at the loci (or locus) involved in the determination of the chosen trait. The chosen trait is in part genetically determined. However, recent observations have suggested that its phenotypic expression can be modified during the early postnatal period, especially in APO-UNSUS rats (Willemen et al., unpublished).

The advantage of these lines is evident. First, each of the two lines consist of individuals which represent one of the two fundamentally distinct types of individuals which normally co-exist in unselected outbred populations of Wistar rats. Second, the striatal, dopaminergic mechanisms within the basal ganglia markedly differ between the two distinct lines (see also below). Accordingly, the individuals belonging to these lines appear to to be excellent tools in the search for the function of the striatum in switching behaviour. Moreover, these lines appear to be valid animal models in the search for mechanisms and factors (genetic, postnatal and/or environmental) giving rise to the individual variation in vulnerability for basal ganglia disorders and unwanted side-effects of dopaminergic agents (see: Cools et al., 1993a).

LIMBIC-STRIATAL SYSTEM

Using the susceptibility of circumscribed brain regions to administration of specific agonists and/or antagonists of various receptors as dependent variable, it has become possible to establish the functional activity of the transmitter systems under study. The results which are described elsewhere in detail (Cools et al., 1990, 1993a) are given in Table 1.

Table 1. APO-SUS and APO-UNSUS rats: line-specific differences in the functional activity of some transmitters in structures belonging to the limbic-striatal system

	APO-SUS		APO-UNSUS
α-noradrenaline, accumbens		<<	
dopamine (D_2), accumbens		<<	
dopamine (DA_i), accumbens		>>	
dopamine (D_2)), neostriatum		<<	
GABA, substantia nigra, pars reticulata		<<	
GABA, deeper layers of colliculus superior		>>	
apomorphine-induced gnawing		>>	

APO-SUS rats fully differ from APO-UNSUS rats as far as it concerns the functional activity of various transmitter systems forming part of the circuitry in which the nucleus accumbens is embedded: noradrenaline at the level of α-noradrenergic receptors in the nucleus accumbens (accumbens), dopamine at the level of so-called DAi receptors located on terminals of amygdaloid-accumbens fibers in the nucleus accumbens (accumbens), dopamine at the level of D2 receptors located on terminals of hippocampal-accumbens fibers in the nucleus accumbens (accumbens), dopamine at the level of postsynaptic D2 receptors of nigro-striatal fibers terminating in the striatum (striatum), GABA released from striato-nigral fibers in the substantia nigra, pars reticulata (SNR), and GABA released from nigro-collicular fibers in the deeper layers of the colliculus superior (CS). For understanding the relationship between the various variables (Table 1), it is useful to recall that the hippocampus-accumbens-SNC-striatum-SNR-CS are connected in series (Scheel-Krüger and Willner, 1991), and that the relationship between the various transmitter activities fits in with the currently available knowledge about these systems (Cools et al., 1991; Scheel-Krüger and Willner, 1991).

The line-specific differences in the hitherto known neurochemical features of structures belonging to the limbic-striatal system are summarized in Table 2. In general, the data presented in Table 2 are consistent with those presented in Table 1. The line-specific difference in the functional activity of noradrenaline in the accumbens (Table 1) is in line with the line-specific difference in the amount of noradrenaline in this nucleus (Table 2; Cools et al., 1990). The line-specific difference in the functional activity of dopamine in the neostriatum (Table 1) is in line with the line-specific difference in the affinity of dopamine for D_2 receptors in the neostriatum (Table 2; Cools et al., 1993a; Rots et al., in prep.): this difference cannot be attributed to a line-specific difference in the firing rate of the nigro-striatal dopaminergic neurons, since the metabolic activity of the substantia nigra, pars compacta does not differ between both lines as shown in (^{14}C)-2-D-deoxyglucose (DEOX) studies (Cools et al., 1993a; Cools et al., in prep.). The line-specific difference in the susceptibility for apomorphine-induced gnawing (Table 1) is also in line with the line-specific differences in the metabolic activity of various brain structures of apomorphine-treated rats as found in DEOX studies (Table 2; Cools et al., 1993a; Cools et al., in prep.). Finally, the line-specific difference in the metabolic activity of the CA1 - CA3 neurons in the hippocampus of untreated rats (Table 2; Cools et al., 1993a; Cools et al., in prep.) is understandable in view of the line-specific difference in the baseline release of dynorphin (Cools et al., 1993b): for, dynorphin is known to inhibit these CA neurons (Moises and Walker, 1985).

In sum, both the make-up and the reactivity of the limbic-striatal system significantly differ between APO-SUS and APOI-UNSUS rats.

Table 2. APO-SUS and APO-UNSUS rats: line-specific differences in some neurochemical features of structures belonging to the limbic-striatal system

	APO-SUS	APO-UNSUS
hippocampus		
dynorphin B, baseline release	<<	
metabolic activity, CA_1 - CA_3, naive rats	>>	
nucleus accumbens		
noradrenaline, amount	<<	
neostriatum		
dopamine (D_2) receptors, affinity	<<	
substantia nigra, pars compacta		
mRNA, tyrosine hydroxylase	>>	
metabolic activity, apomorphine-treated rats		
substantia nigra, pars compacta	>>	
substantia nigra, pars reticulata	>>	
ventral tegmental area	>>	
globus pallidus	>>	
subthalamus	>>	

note: dopamine has been found to have a high affinity for sulpiride binding sites in APO-SUS rats, implying that it has a high affinity for D2 receptors in the "antagonist state", but a low affinity for D2 receptors in the "agonist state".

LIMBIC-HYPOTHALAMIC-PITUITARY-ADRENAL SYSTEM

Given the fact that the hippocampus forms part of both the limbic-striatal system and the hippocampal-hypothalamic-pituitary-adrenal system, we have studied a number of features of the latter system. The data summarized in Table 3 clearly show that both the make-up and the reactivity of the corticosteroid controlled hippocampal-hypothalamic-pituitary-adrenal system significantly differ between APO-SUS and APO-UNSUS rats. Considering, for instance, the release of ACTH and corticosteroids in response to novelty-induced stress (Van Eekelen et al., 1992; Rots et al., 1992; Rots et al., in prep.), it clearly appears that APO-SUS rats are far more stress-sensitive than APO-UNSUS rats. Another important line-specific difference concerns the changes in the dynorphin B content of the hippocampus in response to novelty-induced stress: novelty-induced stress significantly reduces the dynorphin B content in APO-SUS rats, but significantly increases it in APO-UNSUS rats (Cools et al., 1993b), providing additional evidence that APO-SUS and APO-UNSUS rats differ in their stress response. This finding together with the line-specific differences mentioned in Table 3 (Sutanto et al., 1989; Sutanto et al., 1992; Rots et al., 1992; Rots et al., in prep.) show that the limbic-hypothalamic-pituitary-adrenal system is differentially active in these lines.

Table 3. APO-SUS and APO-UNSUS rats: line-specific differences in some features of the limbic-hypothalamic-pituitary-adrenal system

	APO-SUS		APO-UNSUS
hippocampus			
mineralocorticoid receptors, number		>	
dynorphin B, novelty-induced release	>		<
plasma			
ACTH, baseline release		>	
ACTH, novelty-induced release		>	
free corticosteroids, baseline release		>	
free + bound corticosteroids, baseline release		=	
free + bound corticosteroids, novelty-induced release		>	

IMMUNE SYSTEM

Given our initial hypothesis that the interline differences can be considered as long-term consequences of differences in levels of circulating glucocorticoids during the critical period of development, i.e. the stress hyposensitive period from 4 to 12 days postnatally (Cools et al., 1990), we have reasoned that all systems which are controlled by corticosteroids will differ between both lines.

Given the fact that glucocorticoids play a crucial role in immune responses, we have started to investigate the immune system itself. One of the most intriguying findings concerns the enormous difference in susceptibility to experimental allergic encephalomyelitis (EAE) which is a paralytic disease that can be induced by evoking immune responses to antigens in central nervous system myelin, and has been studied as a model for multiple sclerosis in man (Sternberg et al., 1989): male APO-UNSUS rats are far more susceptible than male APO-SUS rats (Huitinga et al., 1991; Cools et al., 1993a). This finding is understandable in view of the fact that the plasma levels of free corticosteroids which are known to control the symptoms of EAE, are higher in APO-SUS rats than in APO-UNSUS rats (Table 3; Rots et al., in prep.).

Table 4. APO-SUS and APO-UNSUS rats: line-specific differences in features related to the immunological system

	APO-SUS	APO-UNSUS
susceptibility, experimental allergic encephalomyelitis	<<	
number of pan-macrophages, thymus (medulla)	>>	
weight, spleen (stress-hyposensitive period)	>>	
weight, thymus (stress-hyposensitive period)	>>	

Given the fact that the growth of the thymus and the spleen is controlled by corticosteroids, we have made growth-curves of these organs. Indeed, the growth-curves of the weights of these organs are significantly different between both lines during the stress hyposensitive period (Cools et al., 1993a).

Given the fact that the thymus contains macrophages and T-lymphocytes which play a critical role in the immune response which differs between both lines (see above), we have started to investigate line-specific diffences in the basic features of these cells. As expected, we have found remarkable line-specific differences in this respect. For instance, the medulla of the thymus in APO-SUS rats contains a significantly higher amount of macrophages than the medulla of APO-UNSUS rats at postnatal days 30 and 60 (Cools et al., 1993a).

In sum (Table 4), it is evident that the immune system also significantly differs between APO-SUS and APO-UNSUS rats.

SWITCHING BEHAVIOURAL STRATEGIES

The available data show that each of the two fundamentally distinct types of individuals which normally co-exist in unselected outbred populations of Wistar rats, has its own make-up and reactivity of the brain, endocrinological and immune system. As discussed elsewhere (Cools et al., 1993a), these features allow them to cope fully different, but nevertheless normally, with challenges from the external and internal environment: "flourishing" respectively "perishing" occurs, when there is a "match" respectively "mismatch" between (a) the genetically predisposed make-up and reactivity of the brain and the body of a particular individual and (b) the actual nature of the factors challenging this individual. Future research must show to which degree the present rodent findings may help us to understand the mechanisms and factors underlying the individual variation in vulnerability for basal ganglia disorders and unwanted side-effects of dopaminergic agents in man.

Apart from this, the line-specific differences in the hardware of the brain and the body can be used to predict the functional consequences of these differences for the two distinct types of individuals. Below, this is elaborated for the limbic-striatal system.

Given the fact that the dopaminergic activity in the accumbens regions which are innervated by dopaminergic, A_{10} fibers, allows the organism to display cue-bound switching (Van den Bos and Cools, 1989; Van den Bos et al., 1991), we have predicted that APO-SUS rats will have a greater capacity to display cue-bound switching than APO-UNSUS rats: for, the functional activity at the level of DA_i receptors is greater in APO-SUS rats than in APO-UNSUS rats (Cools et al., 1990). Since the dopaminergic activity at the level of D_2 receptors in the neostriatal regions which are innervated by dopaminergic, A_9 fibers, allows the organism to switch arbitrarily behaviour (Cools, 1980; Cools, 1990), we have predicted that APO-SUS rats will have a smaller capacity to switch arbitrarily behaviour than APO-UNSUS rats: for, the functional activity at the level of D_2 receptors in the neostriatum is smaller in APO-SUS rats than in APO-UNSUS rats (Cools et al., 1990). Evidence is available that APO-SUS and APO-UNSUS rats are indeed showing these features in the open field test and the defeat-test (Cools et al., 1990). At the time of the latter publication, however, we were unacquainted with the fact that the aminergic mechanisms in the nucleus accumbens are extremely sensitive for challenges. As elaborated elsewhere in detail (Cools, 1988; Cools et al., 1991), there is now evidence that challenges like novelty temporarily reverse the neurochemical state of the nucleus accumbens of APO-SUS rats into that of unchallenged APO-UNSUS rats, and vice versa. At that time, we had also overlooked the fact that dopamine which is released within the nucleus accumbens, activates not only DA_i receptors, but also D_2 receptors. Accordingly, the ability to display cue-bound switching was erroneously ascribed to the DA_i activity in the nucleus accumbens at that time (Cools et al., 1990). Today, it is clear that the ability to display cue-bound switching is mediated by the D_2 activity within the nucleus accumbens (Van den Bos et al., 1991). These two sets of data explain why APO-SUS rats display more cue-bound behaviour in the open field test and the defeat test than APO-UNSUS rats do. In both tests the rats are challenged by stressors ("novelty" in the open field test and "attacking resident" in the defeat test). Accordingly, the challenged APO-SUS rats are marked by a high D_2 activity in the nucleus accumbens and, thus, an enhanced ability to display cue-bound switching, whereas the challenged APO-UNSUS rats are marked by a low D_2 activity in this nucleus and, thus, a reduced ability to display cue-bound switching. There is no evidence that challenges affect the neurochemical state of neostriatal, dopaminergic receptors in the terminal areas of the A_9 fibers: accordingly, the original explanation for the line-specific difference in switching arbitrarily behaviour holds still true (Cools et al., 1990).

As mentioned above, the D_2 activity in the nucleus accumbens allows the organism to display cue-bound switching. As discussed elsewhere in detail (Cools et al., 1991), a high D_2 activity in the nucleus accumbens inhibits the arrival of hippocampal signals in the nucleus accumbens (the hippocampal gate is closed), whereas a low D_2 activity in the nucleus accumbens allows the arrival of these signals (the hippocampal gate is open). Thus, the hippocampal gate is open both in unchallenged APO-SUS rats and in challenged APO-UNSUS rats; in contrast, it is closed both in challenged APO-SUS rats and in unchallenged APO-UNSUS rats (Table 5). As elaborated elsewhere (Cools et al., 1991), a high DA_i activity in the nucleus accumbens inhibits the arrival of amygdaloid signals (the amygdaloid gate is closed), whereas a low DA_i activity in the nucleus accumbens allows the arrival of these signals (the amygdaloid gate is open). Thus, the amygdaloid gate is closed both in unchallenged APO-SUS rats and in challenged APO-UNSUS rats; in contrast, it is open both in challenged APO-SUS rats and in unchallenged APO-UNSUS rats (Table 5).

Today, evidence is available that the amygdala is concerned with the distinct individual attributes of a single stimulus: it encodes and computes the distinct elements of each stimulus (for review: Scheel-Krüger and Willner, 1991). Moreover, the basolateral amygdaloid nucleus and its projection to the nucleus accumbens are critically involved in the mechanisms by which conditioned, or secondary, reinforcers come to control behaviour by association with a primary reinforcer (Cador et al., 1989; Everitt et al., 1989). Thus, an open amygdaloid gate allows the organsism to acquire incentive motivational learning, whereas a closed amygdaloid gate prevents the organsim to acquire this type of learning. On the other hand, the hippocampus is concerned with the attribution of an overall significance of the relationships among multiple stimuli ("configural association"; Sutherland and McDonald, 1990). Moreover, the hippocampus is able to mediate the storage and retrieval of contextual/spatial information (Hirsh, 1974). Thus, an open hippocampal gate allows the organism (a) to attribute a significance to the relationship among multiple stimuli, and (b) to store and retrieve contextual/spatial information, whereas a closed hippocampus prevents the organism (a) to acquire this so-called configural association, and (b) to store and retrieve contextual/spatial information. Considering this information in view of the data summarized in Table 5, it can be precisely predicted how unchallenged respectively challenged APO-SUS and APO-UNSUS rats will perform in various learning tasks. Below, three of these predictions are elaborated.

Table 5. Overview of the functional consequences of the neurochemical state of noradrenergic and dopaminergic activities in the nucleus accumbens of APO-SUS and APO-UNSUS rats under unchallenged (U) and challenged (C) conditions (Cools et al., 1991)

	APO-SUS		APO-UNSUS	
	U	C	U	C
A. ß-noradrenaline activity, accumbens	low	high	high	low
D_2 activity, accumbens	low	high	high	low
consequence: access, hippocampal input	yes	no	no	yes
B. α-noradrenaline activity, accumbens	low	high	high	low
DA_i activity, accumbens	high	low	low	high
consequence: access, amygdaloid input	no	yes	yes	no
C. D_2 activity, neostriatum	low	low	high	high
consequence: arbitrarily switching	no	no	yes	yes

First, it can be predicted that challenged APO-SUS rats which are marked by an open amygdaloid gate, will have a better capacity to acquire the coupling between stimuli and a response than challenged APO-UNSUS rats which are marked by a closed amygdaloid gate. Indeed, recent experiments have shown that challenged APO-SUS rats learn faster than challenged APO-UNSUS rats in a task, in which the association of a single stimulus with a primary reinforcer (food) is required for correct performance: they had to acquire learning in a four-arm maze with a fixed position in a room, of which each wall contains a single stimulus in a direct line with one of the four arms of the maze (Cools et al., 1993b).

In a learning task, in which a configural association has to be coupled with a primary reinforcer, the outcome must be different. For, challenged APO-SUS rats which are marked by a closed hippocampal gate, first need to acquire the configural association itself: they need to adapt to the challenge in order to get an open hippocampus. In other words, challenged APO-SUS rats need to adapt to the challenge before they can perform correct responses. In such a task, challenged APO-UNSUS rats will easily acquire a configural association because of their open hippocamal gate, but will have difficulties with incentive learning per se because of their closed amygdaloid gate. Unchallenged APO-UNSUS rats need to adapt to the challenge in order to get an open amygdaloid gate. In other words, challenged APO-UNSUS rats also need to adapt to the challenge before they can perform correct responses. Accordingly, it can be predicted that the acquisition of a task, in which a configural association has to be acquired for correct performance, will not significantly differ between unchallenged APO-SUS rats and unchallenged APO-UNSUS rats, assumming that the initial adaptation to the challenge does not differ between both lines. Indeed, recent experiments have shown that the acquisition of spatial learning in the Morris water maze task does not differ between unchallenged APO-SUS rats and unchallenged APO-UNSUS rats (Cools et al., in prep.).

It can also be predicted that unchallenged APO-SUS rats which are marked by an open hippocampal gate, will have a better capacity to store and retrieve contextual/spatial information than unchallenged APO-UNSUS rats which are marked by a closed hippocampal gate. Furthermore, it can be predicted that unchallenged APO-SUS rats will have a better capacity to acquire a configural association than unchallenged APO-UNSUS rats. Indeed, recent experiments have shown that unchallenged APO-SUS rats show a much better performance in the place-retention trial of the Morris water maze task than unchallenged APO-UNSUS rats: unchallenged APO-SUS rats have a far better retrieval than unchallenged APO-UNSUS rats (Cools et al., in prep.). In the same series of experiments it has also been found that the acquisition of learning to locate the replaced platform following the probe trial is much better in unchallenged APO-SUS rats than in unchallenged APO-UNSUS rats, showing that the acquisition of a configural association is indeed faster in unchallenged APO-SUS rats than in unchallenged APO-UNSUS rats. Although the majority of the remaining predictions will have to come to light in our ongoing experiments, it is evident that both lines are excellent tools in the search for the function of the striatum in switching behavioural strategies.

In sum, the available data show that the fundamentally different make-up and reactivity of the noradrenergic and dopaminergic mechanisms in the ventral and dorsal striatum of APO-SUS and APO-UNSUS rats give rise to the display of fundamentally different strategies to acquire, store and retrieve information. Furthermore, the available data give a completely new insight into the function of the nucleus accumbens. The above-mentioned data reveal that the nucleus accumbens is concerned with at least two types of cue-bound switching: (a) switching with the help of a cue being a single, conditioned stimulus, when the amygdaloid gate is open; and (b) switching with the help of a cue being the relationship between two or more conditioned stimuli, when the hippocampal gate is open.

CONCLUSION

The present article shortly summarizes the hitherto known differences in the make-up and the reactivity of the limbic-striatal system, the limbic-hypothalamic-pituitary system, and the immunological system between the two fundamentally distinct types of individuals which normally co-exist in unselected populations of outbred Wistar rats. In addition, the present article summarizes evidence that the individual-specific differences in the make-up and reactivity of the noradrenergic and dopaminergic mechanisms in the ventral and dorsal striatum between these two types of individuals give rise to the display of fundamentally different strategies to acquire, store and retrieve information. Finally, the available data reveal that the nucleus accumbens is concerned with at least two types of cue-bound switching: (a) switching with the help of a cue being a single, conditioned stimulus, when the nucleus accumbens has access to information from the basolateral amygdala, and (b) switching with the help of a cue being the relationship between two or more conditioned stimuli, when the nucleus accumbens has access to information from the hippocampus. The data are derived from studies in which two distinct lines of Wistar rats have been used: the line-specific individuals represent the two distinct types of individuals which normally co-exist in unselected outbred populations of Wistar rats. These lines appear to be excellent tools in the search for the function of the striatum in switching behaviour. Moreover, these lines provide valid animal models in the search for the individual variation in vulnerability for basal ganglia disorders and unwanted side-effects of dopaminergic agents. Since the latter aspect is discussed elsewhere (Cools et al., 1993c), it will be not included in the present article.

REFERENCES

Benus, I., 1988, "Agression and Coping", Drukkerij van Genderen B.V., Groningen.

Bos van den, R., and Cools, A.R., 1989, The involvement of the nucleus accumbens in the ability of rats to switch to cue-directed behaviours, *Life Sci.* 44:1697- 1704.

Bos van den, R., Charria Ortiz, G.A., Bergmans, A.C., and Cools, A.R., 1991, Evidence that dopamine in the nucleus accumbens is involved in the ability of rats to switch to cue-directed behaviours, *Behav. Brain Res.* 42:107-114.

Cador, M., Robbins, T.W., and Everitt, B.J., 1989, Involvement of the amygdala in stimulus-reward associations: interaction with the ventral striatum, *Neuroscience* 30:77-86.

Cools, A.R., 1980, Role of the neostriatal dopaminergic activity in sequencing and seleting behavioural strategies: facilitation of processes involved in selecting the best strategy in a stressful situation, *Behav. Brain Res.* 1:361-378.

Cools, A.R., 1987, Transformation of emotion into motion: role of mesolimbic noradrenaline and neostriatal dopamine, *in:* "Neurobiological Approaches to Human Disease," D. Hellhamer, I. Florin and H. Weiner, eds., Hans Huber Publishers, Toronto, pp 15-28.

Cools, A.R., 1990, Role of neostriatal and mesostriatal or mesolimbic dopaminergic fibers in Parkinson's disease with and without dementia: prospects, concepts and facts, *Jpn. J. Psychopharmacol.* 10:15-34.

Cools, A.R., Brachten, R., Heeren, D., Willemen, A., and Ellenbroek, B., 1990, Search after neurobiological profile of individual-specific features of Wistar rats, *Brain Res. Bull.* 24:49-69.

Cools, A.R., Bos van den, R., Ploeger, G., and Ellenbroek, B., 1991, Gating function of noradrenaline in the ventral striatum, *in:* "The Mesolimbic System: from Motivation to Action," P. Willner and J. Scheel-Krüger, eds., John Wiley and Sons, New York, pp 141-173.

Cools, A.R., Dierx, J., Coenders, C., Heeren. D., Ried, S., Jenks, B., and Ellenbroek, B., 1993a, Apomorphine-susceptible and apomorphine-unsusceptible Wistar rats differ in novelty-induced changes in hippocampal dynorphin B expression and two-way active avoidance: a new key in the search for the role of the hippocampal-accumbens axis, *Beh. Brain Res.* in press.

Cools, A.R., Ellenbroek, B., Heeren, D., and Lubbers L., 1993b, Use of high and low responders to novelty in rat studies on the role of the ventral striatum in radial maze performance: effects of intra-accumbens injections of sulpiride, *Can. J. Physiol. Pharmacol.* in press.

Cools, A.R., Rots, N.Y., Ellenbroek, B., and De Kloet, E.R. 1993c, Bimodal shape of individual variability in behaviour of Wistar rats: the overall outcome of a fundamentally different make-up and reactivity of the brain, endocrinological and immunological system, *Neuropsychobiol.* in press.

Eekelen van, J.A.M., Rots, N.Y., De Kloet, E.R., and Cools, A.R., 1992, Central corticosteroid receptors and stress responsiveness in two pharmacogenetically selected rat lines, *Soc. Neurosci. Abstracts* 18:1514.

Everitt, B.J., Cador, M., and Robbins, T.W., 1989, Interactions between the amygdala and ventral striatum in stimulus-reward associations: studies using a second-order schedule of sexual reinforcement, *Neuroscience* 30:63-75.

Fokkema, D.S., Smit, K., Van der Gugten, J., and Koolhaas, J.M., 1988, A coherent pattern among social behaviour, blood pressure, corticosterone and catecholamine measures in individual male rats, *Physiol. Behav.* 42:485-489.

Hirsh, R., 1974, The hippocampus and contextul retrieval of information from memory: a theory, *Behav. Biol.* 12:421-444.

Huitinga, I., Dijkstra, C.D., Kraal, G., Ried, S., and Cools, A.R., 1991, Susceptibility to experimental allergic encephalomyelitis linked to defined behavioural responses, *J. Neuroimmunol.*, Suppl. 1:161.

Koolhaas, J.M., Fokkema, D.S., Bohus, B., and Van Oortmerssen, G.A., 1986, Individual differences in blood pressure reactivity and behavior of male rats, *in:* "Biological and Psychological Factors in Cardiovascular Disease," T.H. Schmidt, T.M. Dembroski and G. Blümchen, eds., Springer Verlag, Berlin, pp 517-526.

Masur, J., Maroni, J.B., and Benedito, M.A.C., 1975, Genetically selected winner and loser rats in the tunnel competition: influence of apomorphine and DOPA, *Behav. Biol.* 14:21-30.

Moises, H.C. and Walker, J.M., 1985, Electrophysiological effects of dynorphin peptides on hippocampal pyramidal cells in the rat, *Europ. J. Pharmacol.* 108:85-98.

Piazza, P.V., Deminière, J.M., Maccari, S., Le Moal, M., Mormède, P., and Simon, H., 1991, Individual vulnerability to drug self-administration: action of corticosterone on dopaminergic systems as a possible pathophysiological mechanism, *in :* "The Mesolimbic Dopamine System: From Motivation to Action," P. Willner and J. Scheel-Krüger, eds., John Wiley and Sons, Toronto, pp 473-495.

Rots, N.Y., De Kloet, E.R., Rostene, W.H., and Cools, A.R., 1992, A rat line selected for enhanced apomorphine susceptibility displays increased neuroendocrine stress responsiveness, Proceedings of the 7th Int. Catecholamine Symposium, Amsterdam, pp 277.

Scheel-Krüger, J., and Willner, P., 1991, The mesolimbic system: principles of operation, *in:* "The Mesolimbic System: from Motivation to Action", P. Willner and J. Scheel-Krüger, eds., John Wiley and Sons, Toronto, pp 559-597.

Sternberg, E.M., Hill, J.M., Chrousos, G.P., Kamilaris, T., Listwak, S.J., Gold, P.W., and Wilder, R.L., 1989, Inflammatory mediator-induced hypothalamic-pituitary-adrenal axis activation is defective in streptococcal cell wall arthritis-susceptible Lewis rats, *Proc. Natl. Acad. Sci.* 86:2374-2378.

Sutanto, W.E., De Kloet, E.R., De Bree, F., and Cools, A.R., 1989, Differential corticosteroid binding characteristics to the mineralocorticoid (type I) and glucocorticoid (type II) receptors in the brain of pharmacogenetically-selected apomorphine-susceptible and apomorphine-unsusceptible Wistar rats, *Neuroscience Res. Comm.* 5:19-27.

Sutanto, W.E., Oitzl, M.S., Rots, N.Y., Schöbitz, B., Van den Berg, D.T., Van Dijken, H.H., Mos, J., Cools, A.R., Tilders, F.J., Koolhaas, J.M., and De Kloet, E.R., 1992, Corticosteroid receptor plasticity in the central nervous system of various rat models, *Endocrine Regulations* (199) 26:111-118.

Sutherland, R.J., and McDonald, R.J., 1990, Hippocampus, amygdala, and memory deficits in rats, *Behav. Brain Res.* 37:57-79.

CLINICAL DISORDERS AND PHARMACOLOGY

OF THE BASAL GANGLIA

MOVEMENT DISORDERS IN PATIENTS WITH STRIATAL AND/OR PALLIDAL LESIONS

Pierre Rondot and Nguyen Bathien

Service de Neurologie
Centre Raymond Garcin
Hôpital Sainte Anne
1 rue Cabanis
7674 Paris Cedex 14

It is well known that lesions of the basal ganglia induce serious motor disorders, particularly in children: abnormal movements, dystonia, rigidity, akinesia. Thanks to the development of non-invasive diagnostic techniques, it is now possible to visualize the site of the lesions. However the newly developed methods have their limitations. Even MRI does not allow the visualization of all the lesions and it gives only a partial view of those involved. In spite of these imperfections, it is tempting for the clinician to compare clinical observations with radiological data.

In the basal ganglia, the striatum is particularly vulnerable as it is supplied by the deep branches of the middle cerebral artery, the lateral and medial lenticulo-striate arteries which originate from the upper rim of the first segment of the middle cerebral artery. This position facilitates the migration of embolism and also makes a traumatism more serious. In fact, Maki et al (1980) have shown in monkey that a cranial traumatism can provoke the bending and thrombosis of these arteries at their origin. As in the case of head injuries in man, the cerebral mass has a tendency to move forward while the arteries stay with the base of the skull. In circumstances such as embolism or traumatism, ischemia can result in the area supplied by the lenticulo-striate arteries, i.e. putamen and caudate nucleus. The lesion is rarely selective. It is thus difficult to link the symptoms to one or the other structure.

Another cause of the vulnerability of these formations is related to their intense metabolism, clearly demonstrated by PET-scan (Roland et al., 1982), which explains frequency of the lesions after cerebral anoxia. Under such conditions, the pallidum appears more exposed than the striatum: it is known that carbon monoxide poisoning involves the pallidum as well as the cerebral cortex. Infections can also provoke necrosis of these structures (Goutières et Aicardi, 1982).

POPULATION

Twenty-two cases of striatal and/or pallidal lesions have been collected thanks to CT-scan and M.R.I. data. Two types of criteria were chosen in the selection of our cases: the first criterion concerns the site of the lesion in the striatum or the pallidum. The second concerns the symptoms of motor disorder: abnormal movement, dystonia, akinesia, rigidity. Cases provoked by tumours were discarded because it is very difficult, even after gadolinium injection, to determine the precise extent of certain cerebral tumours especially when they are infiltrative. According to these criteria, 22 cases remained: 13 men, 9 women,

The Basal Ganglia IV, Edited by
G. Percheron *et al.*, Plenum Press, New York, 1994

out of which 8 were bilateral. The initial symptoms appeared at an early age: 9 years and 4 months on average. Six patients were older than 15 years at the onset but one of them was affected by a congenital lesion, an arterio-venous angioma.

With regards to the site of the lesion, we established three groups (Table 1): striatal lesions, lesions of the putamen and pallidum, pallidal lesions. In 20 of these cases, the extent of the lesion was evaluated quantitatively according to the method of Harris et al (1992).

Table 1. Site of lesions and frequencies of motor disorders.

Site of lesions	Number of cases	Abnormal movements	Dystonia
Striatum	14		
Putamen + Pallidum	2	10	16
Pallidum	6	1	3
Total	22	11	19

The striatal lesion group was the largest, 14 of which 2 were bilateral. In one case the lesion was located in the caudate nucleus (case 6), in 2 cases in the putamen (cases 8 and 16). Most often when both the caudate nucleus and putamen were simultaneously involved, the dominant lesion was located either in caudate nucleus either in putamen. In 2 other cases, the lesions were associated in putamen and pallidum and in 6 cases (third group) the pallidum was affected, bilaterally in five cases.

In spite of the precision with which these subdivisions were established, our methods had their limits. This is probably the reason why we did not find focal lesions in many cases of abnormal movements due to neonatal anoxia. The same notations can be found elsewhere (Marsden et al, 1985). In any case, it seems certain that the lesions go beyond the limits established according to the radiological data.

ABNORMAL MOVEMENTS

Abnormal movements were observed in 10 out of 16 cases of striatal (14) and pallido-striatal lesions (2). The movements were slow, confined to the extremities, crawling, corresponding to the definition of athetotic movements. In one patient, the only case in which the lesion was localized in the head of the caudate nucleus, myoclonia was observed. The 6 cases without abnormal movements were older than 14 when the illness started. Among the group of patients older than 14 years who had extrapyramidal symptoms, only one, older that 17 at the beginning of the disease, was affected by abnormal movements which presented as synkinesia. It thus seems evident that young age is a factor facilitating the occurence of abnormal movements after striatal lesions whether isolated or accompanied by pallidal lesions.

We did not analyse the relative importance of lesions to the caudate nucleus or putamen in the onset of abnormal movements, because the sub-groups were too small numerically.

The 6 patients with pallidal lesions were free of all abnormal movement when we examined them. In one of them, slow movements of the extremities of the upper limb, and in another action myoclonia, were temporarely observed during the month following the initial acute episode.

OTHER MOTOR DISORDERS

Akinesia or dystonia were observed in the following conditions: in 16 patients with striatal lesions, either isolated or associated with a pallidal lesion, dystonia was observed. It affected the 4 limbs when lesion of the putamen was bilateral. In one case, only the opposite side of the largest lesion was dystonic. No difference was observed whether the lesion was striatal or pallidal. However in putamen lesions dystonia was more severe when measured using the scale of Burke et al. (1986) according to the size of the lesion evaluated by the method of Harris et al. (1992). Such a relation was not found when the severity of dystonia was compared to the size of lesion of the caudate nucleus (Fig. 1) which brings out the importance of putaminal lesion in the pathophysiology of dystonia.

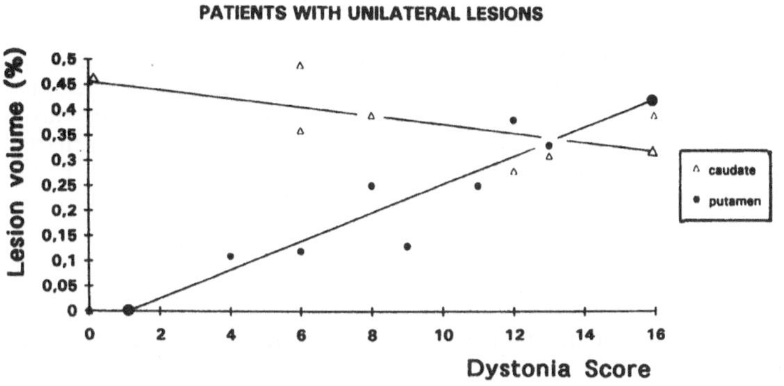

Figure 1. Plots of putamen and caudate lesion volumes (evaluated as a percentage of the total volume of each structure) with dystonia scores of patients with unilateral striatal lesions. Putamen volume is significantly correlated with the dystonia scores.

When the pallidum alone was involved according to MRI data, in three cases we observed a severe dystonia of the four limbs and of the face. In three other cases following carbon monoxide poisoning, there was no dystonia but slight rigidity and akinesia. In a sixth case, the dystonia showed up after an ytrium implantation at the tip of the medial segment of the globus pallidus in order to improve a traumatic action tremor: the tremor was markedly improved, but twelve years later the upper limb opposite to the implantation became dystonic. The dystonia was not accompanied by abnormal movement. In no case did we note either perseveration or obsessive-compulsive behaviour.

DISCUSSION

The comments will be limited to three questions: - the relation between the site of the lesion and the existence of abnormal movements, - the relation between striatal and/or pallidal lesions and the initiation of movement, - the pathophysiology of striatal or pallidal dystonia.

Striatal and/or Pallidal Lesions and Abnormal Movements

Striatal lesions most frequently provoke abnormal movements: choreoathetotic movements or myoclonia: 10 cases of abnormal movements out of 16 cases with uni or bilateral striatal lesions. On the other hand, in the case of pallidal lesions, abnormal movements were observed in only 1 case out of 6, and only in the first months after the onset of the disease. These clinical data do not confirm the current hypothesis about the function of the internal pallidal segment which would be activated by the subthalamo-pallidal pathway under the influence of glutamate (Parent et al., 1989). Activation of medial pallidal neurons would be inhibitory on the ventro-lateral nucleus of the thalamus. The disappearence of this inhibition after a subthalamic or medial pallidal lesion would provoke hemiballismus. In fact, in previous cases medial pallidal lesions did not provoke abnormal movements, but rather akinesia. What is more, internal pallidotomy, performed to improve experimental hemiballismus, stopped the abnormal movement (Carpenter et al., 1950). It is therefore necessary to reconsider the role of the medial pallidal segment with these data. The function of the medial pallidum with regards to the thalamus should be considered not as inhibitory, but as reinforcing the activity of the striato-pallido-thalamo-cortical circuit. This interpretation is not very far removed from that of Hallet et al. (1980) for whom the basal ganglia have an energizing effect upon the motor system.

Striatal/Pallidal Lesions and the Initiation of Movement

Movements are perturbed by striatal and pallidal lesions. Simple reaction time is prolonged and if the amplitude of the movement is not perturbed, its speed is slowed down (Anderson and Horak, 1985) particularly in the case of dystonia. The movement-related potentials (Kornhuber and Deecke, 1965) which accompany the preparatory phase of movement are profoundly modified by basal ganglia lesions (Fève et al., 1993). At the same time, the first BP and the second negative NS' components preceding the movement have an amplitude inferior to normal. Moreover, while the BP potential is maximal on medial derivations and NS' on the opposite motor cortex, this predominance disappears after uni or bilateral striatal or pallidal lesions.

This significant diminution of movement-related potentials which has been observed many years after the lesion inducing a motor disorder, makes it possible to explain the delay in the execution of movement but not akinesia. In fact, Shibasaki et al. (1978) showed that the amplitude of these movement-related potentials is also diminished after a cerebellar lesion which does not provoke akinesia but a delay in the movement, as in the case of striatal or pallidal lesions. It is to be noted that in the case of unilateral striatal lesion the movement-related potentials are diminished on both hemispheres. Here we come back to what Hallett and Khosbin (1980) qualified as an "inappropriate energizing of muscles". This interpretation seems to us more likely than the one which evokes a dysfunction of pallidal or striatal neurons altering the preparation of movement. In fact, the activity of most of the striatal or pallidal neurons is modified only after, but not before, the onset of the movement, as shown by Alexander and DeLong (1985) and Liles (1985).

In such conditions, it is difficult to suppose that a lesion of these structures can damage the preparation of movement. However, a decrease of the basic stimulation which comes from these structures can deprive the primary and supplementary motor areas, from which the potentials NS' are generated, of activation necessary in order to facilitate the motor command.

Pathophysiology of Dystonia

The frequency of dystonia due to striatal or pallidal lesions is relevant to two factors: first the enhancement of the shortening reaction which freezes the limb at the end of a movement and blocks or defers an opposite movement (Rondot and Scherrer, 1966; Bathien et al., 1981), and secondly the perturbation of postural patterns which leads to a return to archaic patterns (Rondot, 1991).

CONCLUSION

This clinical study of 22 cases with basal ganglia lesions underlines the frequency of dystonia in children following a striatal or pallidal lesion. It confirms that abnormal movements are related to striatal but not pallidal lesions. It opens the discussion about the role of the medial pallidal segment which should be considered not as inhibiting but as somehow activating the ventral lateral nucleus of thalamus.

REFERENCES

Alexander, G.E., and DeLong, M.R., 1985, Microstimulation of the primate neostriatum. II. Somatotopic organization of striatal microexcitable zones and their relation to neuronal response properties, *J. Neurophysiol.* 63:1417-1430.

Anderson, M.E., and Horak, F.B., 1985, Influence of the globus pallidus on arm movements in monkeys. III Timing of movement-related information, *J. Neurophysiol.* 64:433-448.

Bathien, N., Rondot, P., and Toma, S., 1981, Etude de la réaction de raccourcissement presente chez l'homme dans diverses affections neurologiques. Rôle des afférences articulaires, *E.E.G. and Clin. Neurophysiol.* 61:156-164.

Burke, R.E., Fahn, S., Marsden, D., Bressman, S.B., Moskowitz, C., and Friedman, J., 1985, Validity and reliabity of a rating scale for the primary torsion dystonias, *Neurology* 35:73-77.

Carpenter, M.B., Whittier, J.R., and Mettler, F.A., 1950, Analysis of choreoid hyperkinesia in the rhesus monkey, *J. Comp. Neurol.* 92:293-3Z2.

Fève, A., Bathien, N., and Rondot, P., 1993, Movement-related potentials in patients with lesions of basal ganglia and anterior thalamus, *J. Neurol. Neurosurg. Psychiat.* in press.

Goutieres, F., and Aicardi, J., 1982, Acute neurological dysfunction associated with destructive lesions of the basal ganglia in children, *Ann. Neurol.* 12:322-332.

Hallett, M., and Khoshbin, S., 1980, A physiological mechanism of bradykinesia, *Brain* 103:301-314.

Harris, G.J., Pearlson, G.D., Peyser, C.E., Aylward, E.H., Roberts, J., Barta, P.E., Chase, G.A., and Folstein, S.E., 1992, Putamen volume reduction on magnetic resonance imaging exceeds caudate changes in mild Huntington's disease, *Ann. Neurol.* 31:69-73.

Kornhuber, H.H., and Deecke, L., 1966, Hirnpotentialanderungen bei Willkurbewegungen und pasiven Bewegungen des Menschen: Bereitschaftpopotential und reafferente Potentiale, *Pflügers Archiv.* 284:1-17.

Liles, S.L., 1986, Activity of neurons in putamen during active and passive movements of wrist, *J. Neurophysiol.* 63:217-236.

Maki, Y., Akimoto, H., Enomoto, T., 1980, Injuries of basal ganglia following trauma in children, *Childs Brain* 7:113-123.

Marsden, C.D., Obeso, J.A., Zarranz, J.J., and Lang, A.E., 1986, The anatomical basis of symptomatic hemidystonia, Brain 108:463-483.

Parent, A., Hazrati, L.N., and Smith, Y., 1989, The subthalamic nucleus in primates. A neuroanatomical and immunohistochemical study. Neural mechanisms in disorders of movement, *in*: "Current Problems in Neurology," A.R. Crossman and M.A. Sambrook, eds., John Libbey, London, Paris, pp. 29-36.

Roland, P.E., Meyer, E., Shibasaki, H., Yamamoto, Y.L., and Thompson, C.J., 1982, Regional cerebral blood flow change in cortex and basal ganglia during voluntary movements in normal human volunteers, *J. Neurophysiol.* 48:467-480.

Rondot, P., 1991, The shadow of the movement, *J. Neurol.* 238:411-419.

Rondot, P., and Scherrer, J., 1966, Contraction réflexe provoquée par le raccourcissement passif du muscle dans l'athétose et les dystonies d'attitude, *Rev. Neurol.* 114:329-337.

Shibasaki, H., Shima, F., and Kuroiwa, Y., 1978, Clinical studies of the movement-related cortical potential (MP) and the relationship between the dentato-rubro-thalamic pathway and readiness potential (RP), *J. Neurol.* 219:15-26.

DEFICITS IN SACCADIC EYE MOVEMENTS IN BASAL GANGLIA DISORDERS

Masaya Segawa[1], Okihide Hikosaka[2], Hideki Fukuda[1, 3], Kimiaki Uetake [1] and Yoshiko Nomura[1]

[1]Segawa Neurological Clinic for Children, Tokyo Japan
[2]National Institute for Physiological Sciences, Okazaki, Japan
[3]National Institute of Industrial Health, Kawasakai, Japan

INTRODUCTION

Saccadic eye movements are controlled by the basal ganglia with two serial inhibitory connection systems, both GABAergic, which initiate in the caudate nucleus and terminate in the superior colliculus, changing the synapse at the pars reticulata of the substantia nigra (SNr) (Hikosaka et al., 1989). The nigrostriatal dopamine (NS-DA) neurons were verified to modulate this system in monkeys exposed to 1-methyl-4-phenyl-1,2,3,6-tetrahydropyridine (MPTP) (Miyashita et al., 1990). Abnormalities in voluntary saccades were also shown clinically in cases with Dopa responsive disorders, such as Parkinson disease (PD) (Hikosaka et al., 1987) and hereditary progressive dystonia with marked diurnal fluctuation (HPD) (Nomura et al., 1987). It was also revealed that the mode of abnormalities of the saccades differed between PD and HPD (Hikosaka et al., 1993).

These evidences suggest that in humans, voluntary saccades are also modulated by the direct projection of the striatum and particular components of the basal ganglia involve particular parameters of voluntary saccades. In this study we analyzed the parameters of voluntary saccade in various kinds of basal ganglia disorders and correlated them to the particular lesion(s) in the basal ganglia detected by neuroimaging studies.

MATERIALS AND METHODS

Voluntary saccades were evaluated by Hikosaka's method (Hikosaka et al., 1993). Horizontal eye movements were recorded with a pair of electrodes placed on the lateral sides of the orbits. Vertical eye movements were recorded for the right eye. Two kinds of voluntary saccades were evaluated by two kinds of behavioral paradigms, saccade task and delayed saccade task.

The saccade task was designed to induce visually guided saccade or visual saccade. The eye movement made in the saccade task was a saccade guided by visual information derived from the target light which appeared in 48 locations with 5, 10, 15, 20, 30 and 40 degrees in 8 directions. The delayed saccade task was designed to induce memory guided saccade. While the subject was fixating the central spot, another spot of light was flashed (duration: 50 msec) to indicate the future location of the saccade target. The subject was required to maintain fixation for another random period of time (3 sec) until the fixation spot went off. The subject was asked to shift herlhis gaze immediately after the turning off of the fixation point toward the predicted location of the target. The target came on 0.6 sec

The Basal Ganglia IV, Edited by
G. Percheron *et al.*, Plenum Press, New York, 1994

after the fixation spot went off. It was natural for the subject to make a saccade before the target actually came on. The resultant saccade was therefore guided by visual spatial memory, that is, memory guided saccade or memory saccade.

Besides 16 normal individuals with ages ranging from 10 to 60 years, 7 patients with HPD, 2 with dopa-responsive dystonia (DRD) other than HPD, 4 with PD and 5 with symptomatic basal ganglia disorders (Bgl D) were subjected to this study. The two cases with dopa responsive dystonia (DRD) other than HPD were a 28-year-old female (M.N.) with action retrocollis and a 25-year-old male (S.I.) with levodopa induced dyskinesia. Of 4 cases with Parkinson's disease (PD) one was a 59-year-old female hemi-PD with left side predominance and two with a rigid-akinetic type: a 65-year-old female and a 71-year-old male. The other case with PD was a 71-year-old male with pure akinesia. Of the 5 patients with symptomatic Bgl D, 4 were symptomatic dystonia with lesion in the basal ganglia detected by neuroimaging, and one, a 55-year-old male, with parkinsonism due to a lacnar vascular lesion with additional infarction in the left striatum. All cases of dopa responsive disorders, except for one HPD were under levodopa therapy. In one 12-year-old female case with HPD, voluntary saccades were examined twice at the initial period of treatment, before and after attaining optimal dosis of levodopa. In two cases with HPD and case M.N. with DRD other than HPD, the examinations were performed under withdrawal of levodopa for one day, but clinically they had no obvious motor disability. In the patient with hemi-PD, examinations were performed twice in the state of off and on period.

RESULTS

Voluntary Saccades in Normal Subjects

The average values of latency of each individual ranged from 170 to 240 msec (206.2 ± 16.1 msec) for visual saccade and from 230 to 370 msec (281.8 ± 46.4 msec) for memory saccade. There was no apparent age variation in these values of either saccade.

The amplitudes of both saccades estimated by the degree of the initial saccade to the target lighted in the eccentricity of 20 degrees were analyzed. The average degrees of visual saccade of each subject were larger than 17 degrees (18.9 ± 1.1) without age variation. On the other hand, those of memory saccade were of similar values of 18.5 ± 2.2 degrees as visual saccade in those under the age of 60 years. But two subjects older than 60 years showed smaller amplitudes of 12.3 and 14.4 degrees, respectively, where the initial saccade tended to be hypometric to the target.

Frequencies of memory saccade, that is, rates of saccade task occurring before the peripheral predicted target comes on, were not less than 90% in normal subjects (95.2 ± 3.3%). These values showed no apparent age variation, though an individual aged 60 years showed fairly low values of 64.0%.

Voluntary Saccades in Movement Disorders

Parkinson's Disease (PD). In figure 1 electro-oculogram (EOG) of voluntary saccades representative for PD and HPD are shown in comparison with those of a normal subject. As shown in this figure of a 59-year-old female case with hemi-PD with left side predominance examined in the off period, abnormalities were observed in the delayed saccade task, showing delay at the start of the saccade, that is, prolongation in latency, smallness in amplitude of the initial saccad or hypometric to the target, and the occurrence of a fair number of saccades after the peripheral predicted target came on, that is, decrease in the frequency of the memory saccade. These tendencies were more marked in the saccade toward the left, which is shown as a downward deflection in this figure. On the contrary, the visual saccade was rather well preserved in both latency and amplitude.

Two cases with PD of the rigid-akinetic type showed similar abnormalities in the parameters of memory saccade as hemi-PD mentioned above. Although both of them showed hypometria in the visual saccade the latencies of the visual saccade were variable.

In a 71-year-old male with pure akinesia, all components of the memory saccade were also affected. However, the mode of involvement in these parameters differed from that of the rigid-akinetic type (Fig. 2). That is, in the pure akinesia type the frequency of memory saccade was markedly decreased, while the amplitude of memory saccade was less affected showing a reverse tendency to the rigid akinetic type. The latencies of both saccades and amplitude of the visual saccade were also affected in pure akinesia.

Hereditary Progressive Dystonia with Marked Diurnal Fluctuation (HPD). In an ll-year-old female case with HPD the memory guided saccade before treatment showed similar characteristics as PD, but in contrast to PD delay in latency and hypometria were also observed in e visual saccade.

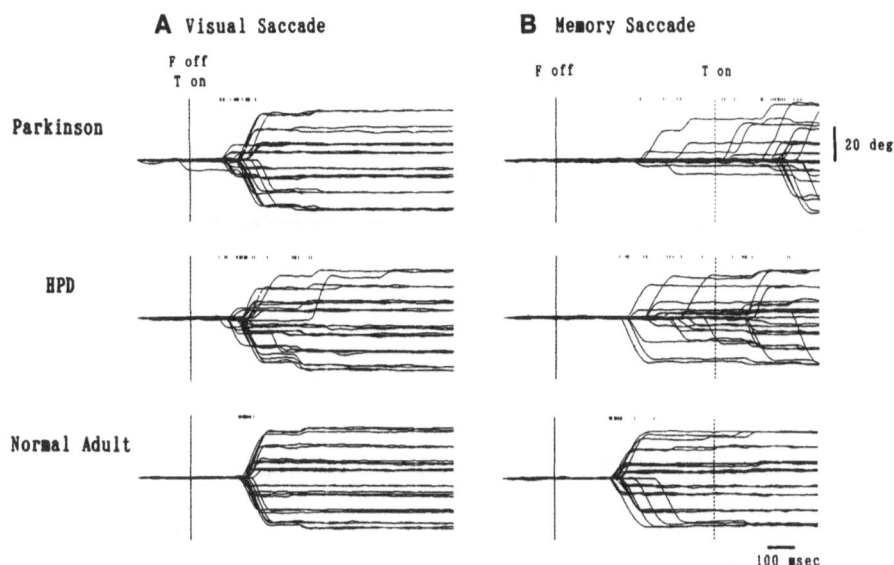

Figure l. Visual (A) and memory (B) saccades of Parkinson's disease (PD), hereditary progressive dystonia with marked diumal fluctuation (HPD) and a 44 year-old normal male. PD was a S9-year-old female case with left hemi-Parkinson's disease examined in the off phase of levodopa therapy. HPD was a 1 l-year-old girl examined before treatrnent. Each record shows superimposed traces of eye position for saccades to the targets on the horizontal meridian (5, 10, 20, and 30 degrees; right side upward, left side downward). The vertical lines indicate the turning off of the central fixation point (F:off) and the onset of the peripheral target points (T:on). Tick marks at the top and bottom indicate the onsets of saccades detected by the computer. Changes in the DC level, which were present in the original records, were removed by assuming that the subject was fixating at the central spot at the beginning of the records.

Of the other six cases with HPD, one with sub-optimal dosis of levodopa and two under one day withdrawal of levodopa showed a decrease in frequency and prolongation of the latency in the memory saccade. As for the parameters of the visual saccade, the former also showed a delay in latency while in the latter hypometria was observed.

All of the other 3 cases under levodopa showed to be normal in all parameters of voluntary saccades or showed slight hypometria in the memory saccades. The oldest case aged 44 years in whom the start of levodopa was delayed as 30 years after the clinical onset showed abnormalities in the latency and frequency of the memory saccade.

Effects of Levodopa on Voluntary Saccades. After levodopa all parameters were improved in PD (Fig. 3), where the improvement of the amplitude was more marked than that of the frequency. In HPD the abnormalities in the components of both memory and visual saccades observed before treatment improved after levodopa. Similar improvements were also observed in cases examined under suboptimal dosis or withdrawal of levodopa, after taking optimal dosis or re-administration of levodopa.

DRD other than HPD. Of the two cases with DRD other than HPD, one 29 year-old female case (M.N.) with action retrocollis, examined under one day withdrawal of levodopa, showed abnormalities in all components except the amplitude of visual saccade. The other case was a 25 year-old male (S.I.) with dyskinetic movement showed a prolonged latency of visual saccade, and hypometria of both saccades towards the right, however, those towards the left and other parameters were within normal range.

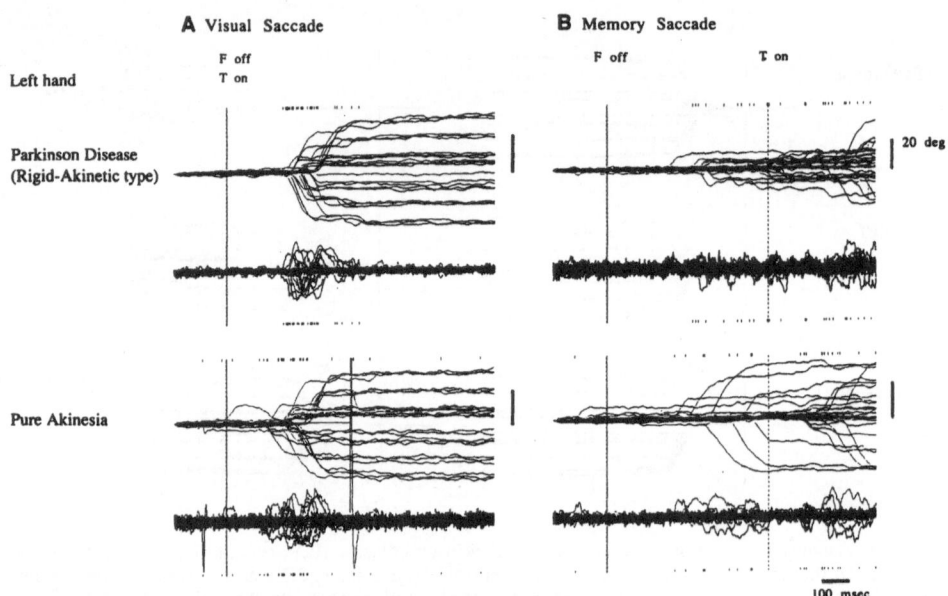

Figure 2. Visual (A) and memory (B) saccades of Parkinson's disease. Top is a 71-year-old case with a rigidakinetic type. Bottom is a 71-year-old case with pure akinesia. Lower traces are the velocity records, smoothed using a median after method based on three successive date points. For further explanation, see Figure 1.

Symptomatic basal ganglia diseases. Patients in this group showed a difference in the mode of abnormalities in voluntary saccades, depending on the etiologies or locus of the lesion in the basal ganglia.

The case with CT lesion in the medial segment of both globus pallidus, showed hypometria and a slight reduction in frequency of the memory saccade while others were preserved normally. The female case with atrophy of the caudate nucleus bilaterally showed abnormalities in all components of the memory saccade, predominantly in its frequency. The visual saccade showed delay in latency but only slight hypometria. A case with vascular parkinsonism with additional left putaminal lesion showed abnormalities of all components except for the latency of the memory saccade. Of them, the amplitude of both saccades and frequency of the memory saccade were predominantly affected. A case with torsion dystonia with bilateral putaminal lesions also revealed to have similar abnormalities.

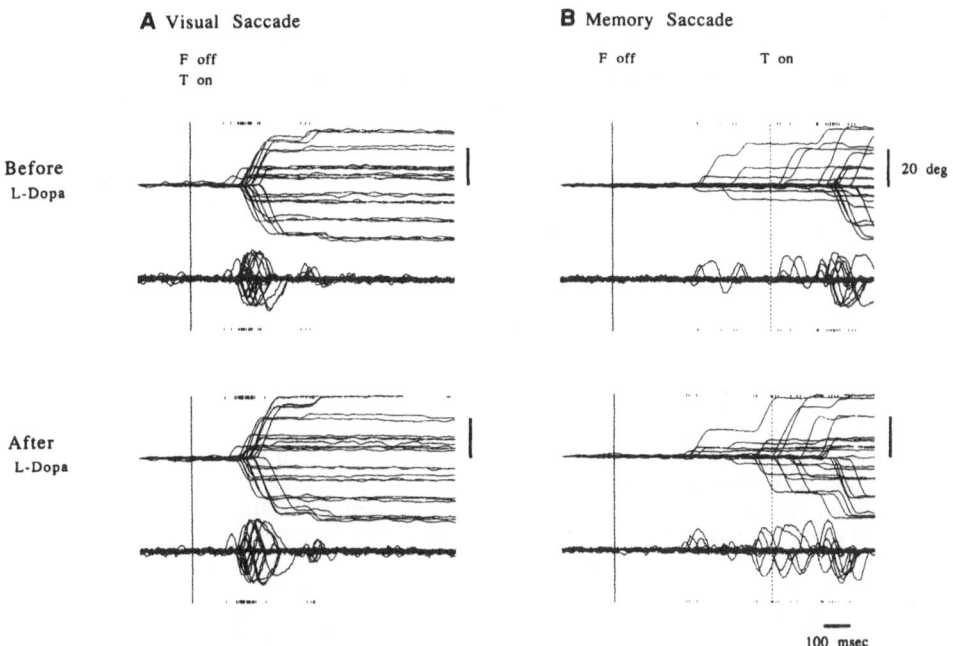

A Visual Saccade

F off
T on

B Memory Saccade

F off   T on

Before
L-Dopa

After
L-Dopa

20 deg

100 msec

Figure 3. Visual (A) and memoy (B) saccades of hemi-Parkinson's disease examined in the off phase (top) and on phase (bottom) of levodopa. The same case is shown in Figure 1. For further explanation, see Figures 1 and 2.

DISCUSSION

These results revealed that cases with lesions in the basal ganglia or the NS-DA neuron showed abnormalities in voluntary saccades. However, the mode of involvement of their parameters differed among patients or depended on the etiologies of the diseases.

The basal ganglia mediate voluntary saccades with their direct striatal projection to the SNr, which connects to the GABAergic neurons to the superior colliculus, and this pathway is also modulated by the NS-DA neurons. Furthermore, studies in MPTP monkeys revealed a difference in the mode of abnormalities of voluntary saccades depending on whether the toxin was injected into the caudate nucleus or in the rostral part of the putamen, that is, in the former, the memory saccade was affected, while in the latter the visual saccade was involved (Hikosaka et al., 1993). So the differences observed in the pattern of abnormalities of voluntary saccades among basal ganglia disorders are thought to be dependent on the differences of the loci and character of the lesion in the basal ganglia.

HPD is thought to have functional lesions restricted to the NA-DA neurons (Segawa et al., 1986) of which those connecting to the direct projections of the striatum particularly those innervating to the SNr are involved predominantly with preservation of those connecting to the indirect projection (Segawa and Nomura, 1993). There are no evidences of the existence of a receptor supersensidvity (Segawa et al, 1986).

So the abnormalities observed in HPD patients before levodopa are considered as the results of hypofunction of the NS-DA neurons, which mediate via the direct projections to the SNr. They are prolonged latency, hypometria and decreased frequency of the memory saccade and the prolongation of latency and hypometria of the visual saccade. The involvement of parameters of both saccades suggests that in HPD, the direct projections to

529

the SNr from the putamen as well as those from the caudate nucleus, are affected. Although similar abnormalities of the memory saccade were observed in cases with HPD treated with suboptimal dosis or under one day withdrawal of L-Dopa, the results of parameters of the visual saccade were variable and showed normal values in most cases. These results assume that in HPD the caudato-fugal projections to the SNr are affected predominantly.

In a 59 year old patient with Parkinson's disease, examined during off phase, similar abnormalities were observed in the memory saccade but parameters of the visual saccade were preserved normally. However, cases with Parkinson's disease over 70 years showed longer latency and hypometria in the visual saccade, even under levodopa. So it is suggested that in PD, of the two putative direct projections to the SNr, the caudato-fugal projection is affected selectively. In advanced stages or at senile ages, another direct projection from the putamen may be involved, but the pathophysiology may be different.

Polysomnographies of cases with action dystonia revealed abnormalities suggesting supersensitivity of the DA receptors that mediate the pallidofugal pathway to the thalamus (Segawa et al., 1988). The target of the stereotaxic operation for action dystonia is proposed as the nucleus ventralis oralis posterior (Sakata et al.; Lenz et al., 1992). For the pathophysiology of dopa induced dyskinesia the involvement of the indirect projection of the striatum with dopamine receptor supersensitivity which consequently connects to the pallidofugal thalamic pathway is implicated (Crossman, 1990a) and clinically, the abnormal movements are alleviated by thalamotomy with the target in the nucleus ventralis lateralis thalamus (Narabayashi et al., 1984). However, the results observed in the two cases with DRD other than HPD are explained by the involvements of the direct projections. So it is suggested that the components of the basal ganglia connected to the pallidofugal thalamic pathways are not involved in the modulation of voluntary saccades.

The involvement of the pallidal projection to the pedunculo pontine nucleus was shown in MPTP parkinsonian monkeys (Crossman, 1990b). The stereotaxic rejection to the pallidofugal descending pathways was shown to alleviate parkinsonian rigidity and L-Dopa non-responsive akinesia, but the loci of the operation differed from each other (Shima and Fukui, 1993). The difference in the involvement of the parameters of voluntary saccades between the rigid-akinetic type and pure akinesia may be due to the difference in the involvement of the pallidofugal descending pathways, which modulate differently the mechanism of voluntary saccades, though the difference in the involvement of the direct projections of the striatum or of cortical input is not ruled out.

Involvement of the striatum or the globus pallidus is suggested as the pathophysiology of the symptomatic torsion dystonia by studies on neuropathology and neuroimaging and the involvement of the direct projection of the striatum was implicated for the pathophysiology of dystonia peak dose dystonia in MPTP monkeys (Mitchell et al., 1990). So it is natural that dystonia shows abnormalities in voluntary saccades.

The results observed in cases with lesion in the caudate nucleus or the putamen neurophysiologically confirms the involvement of the direct projection to the SNr in dystonia. Furthermore, it is suggested that the medial segment of the globus pallidus is involved in the modulation of certain components of the memory saccade.

CONCLUSION

Studies on voluntary saccades make it possible to evaluate the functions of the direct projections of the striatum to the SNr, one from the caudate nucleus and the other from the putamen. So these studies are useful to clarify the pathophysiology of basal ganglia diseases.

REFERENCES

Crossman, A.R., 1990a, Animal models of movement disorders, " First International Congress on Movement Disorders," Washington, D.C.
Crossman, A.R., 1990b, A hypothesis on the pathophysiological mechanisms that underlie levodopa or dopamine against induced dyskinesia in Parkinson's disease. Implications for future strategies in treatment, *Mov. Disord.* 5:100-108.

Hikosaka, O., Fukuda, H., Kato, M., Uetake, K., Nomura, Y., and Segawa, M., 1993, Deficits in saccadic eye movements in hereditary progressive dystonia with marked diurnal fluctuation, *in*: "Hereditary Progressive Dystonia with Marked Diurnal Fluctuation," M. Segawa, ed., Parthenon, Carnforth, U.K., pp. 159-177.

Hikosaka, O., Sakamoto, M., and Usui, S., 1989, Functional properties of monkey caudate neurons. I. Activities related to saccadic eye movements, *J. Neurophysiol.* 61:780-798.

Hikosaka, O., Segawa, M., and Imai, H., 1987, Voluntary saccadic eye movement: application to analyze basal ganglia disease, *in*: "Highlights in Neuro-Ophthalmology," S. Ishikawa, eds., Aeolus Press, Amsterdam, pp. 133-138.

Lenz, F.A., Jaeger, C.J., Seike, M.S., Lin, Y.C., DeLong, M.R., Tasker, R.R., and Vitek, J., 1992, Cross-correlation analysis of thalamic neuronal and EMG signals in patients with dystonia, *Mov. Disord.* 7 (Suppl. 1):126.

Mitchell, I.J., Luquin, R., Boyce, S., Clarke, C.E., Robertson, R.G., Sambrook, M.A., and Crossman, A.R., 1990, Evidence from a 2-deoxyglucose uptake study in a primate model of dopamine agonist-induced dystonia, *Mov. Disord.* 5:49-54.

Miyashita, N., Matsumura, M., Usui, S., Kato, M., Kori, A., Gardiner, T.W., and Hikosaka, O., 1990, Deficits in task-related eye movements induced by unilateral infusion of MPTP in the monkey caudate nucleus, *Soc. Neurosci. Abstr.* 16:235.

Narabayashi, H., Yokochi, F., and Nakajima, Y., 1984, Levodopa-induced dyskinesia and thalamotomy, *J. Neurol. Neurosurg. Psychiatr.* 47:831-839.

Nomura, Y., Segawa, M., Soda, M., and Hikosaka, O., 1987, Voluntary saccadic eye movements in basal ganglia disorders, *in*: "Highlights in Neuro-Ophthalmology," S. Ishikawa, ed., Aeolus Press, Amsterdam, pp. 139-145.

Sakata, S., Shima F., and Fukui, M., 1993, Vop thalamotomy. *Functional Neurosurgery* (Tokyo), in press.

Segawa, M., and Nomura, Y., 1993, Hereditary progressive dystonia with marked diurnal fluctuation, *in*: "Hereditary progressive dystonia with marked diurnal fluctuation," M. Segawa, ed., Parthenon, Carnforth, UK, pp. 3-19.

Segawa, M., Nomura, Y., and Kase, M., 1986, Diurnally fluctuating hereditary progressive dystonia, *in*: "Handbook of Clinical Neurology Vol. 5 (49): Extrapyramidal Disorders," P.J. Vinken, G.W. Bruyn, H.L. Klawans, eds., Elsevier, Amsterdam, pp. 529-539.

Segawa, M., Nomura, Y., Tanaka, S., Hakamada, S., Nagata, E., Soda, M., and Kase, M., 1988, Hereditary progressive dystonia with marked diurnal fluctuation: Consideration on its pathophysiology based on the characteristics of clinical and polysomnographical findings, *in*: "Advances in Neurology, Vol. 50: Dystonia 2," S. Fahn, C. D. Marsden, D. B. Calne, eds., Raven Press, New York, pp. 367-376.

Shima, F., and Fukui, M., 1993, to be published by *Igaku no yumi* (Tokyo).

EXPLORING THE ROLE OF THE BASAL GANGLIA
IN MOTOR CONTROL USING EXPERIMENTAL RESULTS
AND MATHEMATICAL MODELING

Anne Beuter

Centre Nonlinear Dynamics Physiology and Medicine (McGill University)
Department of Kinanthropology
Neurokinetics Laboratory (N-8280)
University of Quebec at Montreal
Montreal, Qc, H3C 3P8, Canada

INTRODUCTION

The basal ganglia form one of the main reentrant subcortical loops which modulate the output of the cerebral cortex via thalamo-cortical projections. These neural structures play a major role in sensorimotor integration which is evident to anyone who has observed the disturbances of movement, posture, and muscle tone that result when these nuclei are damaged through injury or disease (DeLong and Georgopoulos, 1979). The role of sensorimotor integration is also evident in some dystonic patients whose movements can be improved by cutaneous stimulation of the affected body part (Albin et al., 1989). However, little is known about the nature of the sensorimotor integration that takes place in these subcortical structures (Mitchell et al., 1991) and the complex relationships existing between basal ganglia lesions and behavioral functions remains largely to be understood.

Today, it is generally agreed that organized movements depend neither entirely on feedback from the peripheral senses, nor on the programming of preset patterns of muscle activation (Jeannerod, 1988), but no definitive synthesis has been offered by the advocates of the socalled peripheralist or centralist theories (Abbs and Winstein, 1990). To understand how peripheral feedback and preprogrammed commands are processed, or the rules that govern this processing in the basal ganglia, descriptive studies serve a useful but limited function, and we are now ready to combine theory-driven and data-driven investigations.

To develop a theoretical model from experimental data, a choice about the complexity of each modelling element and the number of elements to be used in the model must be made. In general the number of elements used is inversely related to the level of complexity of each element. Thus, on the one hand, the networks developed in computer models known at parallel distributed processors (Mitchell et al., 1991) interconnect several simple elements and each element is only capable of performing trivial mathematical operations such as summing the values of inputs. However, the highly interconnected nature of the network enables the system to solve extremely complex problems. Parallel distributed processors contain a large number of simple elements and represent one of the promising computational models of the basal ganglia (Graybiel, 1991).

On the other hand, models can be based on a few complex elements such as those used in nonlinear dynamics and framed, for example, by nonlinear differential delay equation(s). These complex elements may be coupled together and used to describe qualitative changes in dynamics of normal and abnormal oscillatory phenomena occurring as relevant control

The Basal Ganglia IV, Edited by
G. Percheron *et al.*, Plenum Press, New York, 1994

parameters (such as neural conduction time or the number of membrane receptors and their affinity) are systematically manipulated. These qualitative changes, also called bifurcations, range from stable equilibria to simple and complex periodic oscillations, to aperiodic (or chaotic) fluctuations (Mackey and Glass, 1977).

Nonlinear dynamicists attempt to understand the physiology of complex systems by developing parcimonious mathematical representations of the systems behavior. A model is good if it can successfully predict and reproduce qualitative changes (i.e., steady state, beating, oscillating) observed experimentally over a wide range of parameter values. In a way nonlinear dynamicists emphasize more the prediction of changes than the explanation of the mechanisms causing these changes. This approach is especially important in a clinical setting in which it is often difficult to identify the mechanism(s) causing the dynamical abnormalities observed. This is why recent experimental work has focused on studying bifurcations induced by transient alterations in neural dynamics. In humans, parameter manipulations have typically included changing the time delay (Lee, 1950; Smith, 1962; Merton et al., 1967; Glass et al., 1988) or the gain (Stern, 1944, Stark, 1962; Scotto and Oliva, 1984; Longtin and Milton, 1988) in sensory feedback loop(s).

In our laboratory, we collect experimental data from patients with basal ganglia diseases, especially Parkinson's disease (PD), performing simple motor tasks in which control parameters, such as the time delay in visual feedback are manipulated, and we simulate experimental results using mathematical models based on nonlinear dynamics. Therefore, this chapter will attempt to address the following two points: the first point is that sensorimotor integration should be explored by modifying progressively relevant control parameters, rather than using drastic manipulations (such as removing vision or proprioception) which usually lead to a reorganization of the CNS and mask the very effect we want to see. Indeed in Parkinson's disease, the first symptoms appear after 80 to 90% of the neurons of the substantia nigra are destroyed (Hornykiewicz and Kish, 1987) and the effects of this neurodegeneration are hidden for a long time, suggesting that the central nervous system continuously reorganizes itself in order to maintain the highest possible level of functional integrity. Clearly, it is impossible to eliminate one structure or one function without deeply affecting the physiology of the other ones; the second point is that nonlinear dynamics may be an appropriate theoretical framework to explore the principles of sensorimotor integration in the basal ganglia. By modifying progressively control parameters relevant to the system studied, it is possible to detect when qualitative changes in behavior occur and gently force the system to reveal its underlying dynamics or the simple rules governing it. This approach can be used to determine whether Parkinson's disease corresponds to what Mackey and Glass (1977) and Glass and Mackey (1979) call a dynamical disease, that is, a disease occurring in an intact physiological control system operating in a range of parameter values that lead to abnormal dynamics. The temporary reversibility of hypokinetic symptoms made possible through pharmacological replacement therapy (Albin et al., 1989) appears to suggest that PD is indeed a dynamical disease.

EXPERIMENTAL APPROACH

We use a motor task which consists of maintaining a given position of the index finger at the metacarpo-phalangeal joint using visual feedback presented to the subject on a screen. Although extremely simple, this task has interesting characteristics which are mentioned below:
* it requires no learning (after an initial calibration period of a few seconds);
* it requires no specific cognitive abilities and can be performed equally well by most subjects;
* it is executed by the index metacarpo-phalangeal joint, a system well developed in most subjects;
* it involves most neural structures specialized in visual tracking and voluntary control of movement;
* it can be performed for several minutes without excessive fatigue;
* it is recorded in a noninvasive way in human subjects constrained to work in a closed loop control mode;
* it is sensitive to parameter manipulations through visual feedback (delay, gain, noise) and proprioceptive feedback (stimulation, vibration);

* it generates different responses when performed by patients with lesions of the central nervous system; and

* it offers a way to study simultaneously the voluntary and involuntary components of motor control in a continuously responding system (i.e., voluntary displacements are superimposed on involuntary microdisplacements).

Data are recorded using position lasers working by triangulation. The resolution of these lasers is between 20 and 30 μm. There is no physical contact between the sensing unit and the subject. A small piece of white thin cardboard is placed on the subject's fingernail to reflect the laser beam. Interphalangeal joints are immobilized by a light splint. Visual feedback is provided by an oscilloscope screen on which the subject can see a line moving vertically, corresponding to the position of his/her index finger. Data are recorded for 50 s at a sampling frequency of 100 Hz. When experimental delays are inserted, they vary between 300 and 1500 ms. Data are collected on patients with PD, stages I to III (Hoehn and Yahr, 1967) and control subjects matched for age, sex and handedness. Transitions and intermittent phenomena not seen in control subjects are frequently noticed in patients. Introduction of a delay induces low frequency oscillations in performance of control subjects and in about 50% of the patients with PD (Beuter et al., 1990).

MODELLING A "SIMPLE" NONLINEAR SYSTEM CONTAINING DELAYS AND NOISE

The apparently simple system formed by the subject's index finger maintained in a given position using visual feedback can be described by three characteristics: "the system contains delays, noise and nonlinearities".

The system studied contains actual neural transport delays associated with afferent and efferent information transmission, delays necessary for processing the incoming information, and delays existing between muscle contraction and movement (Abbs and Winstein, 1990). Delays depend on the distance covered, the neural circuit used and the nature of the task. It takes between 200 and 300 ms to influence the accuracy of an ongoing movement and since the work of Warrick (1948), we know that increasing a time delay in a negative feedback loop tends to destabilize the system and to impair accuracy. However, the presence of multiple loops each containing different delays seems to reduce the tendency for oscillations of the system containing them (Oguztoreli and Stein, 1976).

The system studied is noisy. The noisy dynamics observed in this and other physiological systems may reveal the presence of combined deterministic and stochastic influences. Physiological tremor contributes in part to the stochastic (i.e., noisy) influences of peripheral and central origins. In this system, tremor corresponds to a more or less regular oscillation of the finger around the metacarpo-phalangeal joint at a typical frequency between 6 and 12 Hz. Several factors are known to contribute to tremor, including natural limb resonance, inherent firing properties of the motoneurones, levels of circulating catecholamines, ballistocardiac impulses and supraspinal input to motoneurones (such as vision), etc... Although tremor is often described as an unwanted noise source present in the system, it may serve a functional role in movement initiation and control.

The system studied is nonlinear in many ways. From the muscle spindle gain and synaptic transmission, to the corrective movements executed by the subject (smaller in the neighborhood of the reference line, larger when the finger is far from the reference line, and absent when the finger position is outside the screen monitored by the subject) the behavior of the system is nonlinear.

Working backwards from the oscillations to the system dynamics, a simple model composed of two negative feedback loops was developed. For simplicity, these two loops may be associated with vision and proprioception respectively. They correspond to the minimum number of loops necessary to include in a model capable of reproducing the dynamics observed experimentally. Simulations with one loop tend to produce oscillations which have a period that is too long and intermittencies observed experimentally cannot be reproduced.

The system is modeled by two first order differential equations with delay. This is, in our opinion, the simplest mathematical framework in which we can develop a model sufficiently dynamically rich to account for the observed behavior and yet, amenable to an understanding of the role played by physiological parameters (Beuter et al., 1993). The nonlinear feedback function is a hyperbolic tangent. Numerical simulations are run with a fourth order Runge Kutta algorithm with time steps of 0.001 and observations are made after a large number of iterations, so transients have died away. A Hopf bifurcation (super critical) is noted for $A\tau = \pi/2$ with A representing the gain and τ the delay in the feedback loop. At this point the dynamics go from a stable point to a stable limit cycle. A bifurcation analysis completed the numerical simulations (Bélair, 1993). This bifurcation analysis makes it possible to visualize the behavior of the system as control parameters are manipulated.

In our studies, the delays and gains contained in the two negative feedback loops are systematically varied and stability and instability zones are identified analytically. Details dealing with the mathematics of the model can be found in Bélair (1993) and Beuter et al. (1993). Data from control subjects performing the motor task with an added experimental delay reveal the presence of an irregular low frequency oscillation with a period of approximately 2-4 times the inserted delay. The amplitude of the oscillations increases as the delay in the visual feedback increases, and a regular oscillation is observable intermittently. Superimposed on the low frequency oscillations is a low amplitude oscillation with a frequency between 8 and 12 Hz which corresponds to physiological tremor (Beuter et al, 1990). The simulated data of these experimental results show that with proper parameter selection, intermittent oscillations can be generated which look similar to those recorded experimentally. Data from a patient with PD who performs the delayed visual feedback motor task while having a large amplitude pathological tremor, and simulated results for this same patient, suggest that by increasing the noise term and the gain value in one of the feedback loops, it is possible to generate times series with morphologies similar to that observed in this particular patient (Beuter et al., 1993). The oscillatory behaviors observed experimentally when a delay is inserted are also present in the modelled data. The noisy fluctuations superimposed on the signal reproduce the general aspect of physiological and pathological tremor, suggesting that motor control can be effectively modelled in control subjects and in patients with PD using a combination of simple deterministic and stochastic elements.

CONCLUDING REMARKS

Qualitative changes or transitions in behavior are commonly observed in motor control. Going from walking to running, for example, is a complex transition which depends on biomechanical, energetic, psychological and neural parameters (Beuter and Lefebvre, 1988). Transitions are rarely studied in motor control because of their complexity. However, they probably contain rich, relevant and unique information to understand motor control and they can now be modelled and explored using nonlinear dynamics.

Mackey and Milton (1987) have recently proposed to classify transitions into three categories:
  * Oscillations may arise in a non-oscillating system. For example, in Parkinson's Disease (PD) the cogwheeling phenomenon is a rythmical resistance at a frequency of about 6-9 Hz in a mobilized joint;
  * Oscillations may disappear in an oscillating system. For example, in PD, the rhythmical motion of the swinging arms during locomotion disappears as the disease progresses;
  * Oscillations may change in amplitude or frequency. For example, in some patients with PD tremor tends to increase in amplitude and decreases in frequency.

As indicated above, qualitative changes caused by Parkinson's disease are temporarily hidden by powerful compensatory mechanisms. When symptoms appear clinically, they are partially reversible through pharmacalogical replacement therapy. However, the effects of this therapy are limited (Divac, 1992) and symptoms tend to fluctuate in complicated ways. Thus, other ways to alleviate Parkinsonian symptoms must be found and eventually combined with pharmacological treatments. These ways may include early detection and massive mobilization of compensatory mechanisms (Divac et al., 1987). Detecting subtle changes in the dynamics may indicate the presymptomatic presence of Parkinson's disease and suggest the need for an early intervention. The challenge becomes to reposition the system from a region of parameter space in which it exhibits pathological behaviors into a more healthy region. Repositioning the system could be done by compensatory mechanisms based on sensorimotor processes through electrical, mechanical and/or visual stimulation.

To explore these sensorimotor compensatory mechanisms, nonlinear modelling of neural systems will have to be further developed. Specifically, in the example presented here and dealing with motor control, the following questions will have to be addressed before meaningful answers can be proposed:

* Where does noise come from in the basal ganglia? Papers by Rolls et al., (1984) Filion et al., (1988) and Bloxham et al., (1987) suggest that a dysfunction in the basal ganglia caused by a lack of dopamine may change the signal to noise ratio and may affect the motor periphery by favoring the onset of oscillations. Further work will help determine how and where this noise is injected in the basal ganglia-thalamocortical circuits;

* What are the characteristics of the noise present in the basal ganglia? Injecting white or colored noise has different effects on the dynamics. Vasilakos and Beuter (1993) have suggested that parkinsonian tremor can be modelled with colored noise having a Gaussian distribution and a peak frequency around 5 Hz. Further work should help determine the color of "normal and pathological" noise.

* How can qualitative changes or bifurcations be detected in the presence of noise? Noise may advance or postpone the occurrence of the bifurcation and there is no easy way to answer this question (Longtin et al., 1990). Further work should focus on developing valid and reliable tools to identify the occurrence of bifurcation;

* Should we talk about delay or delays when modelling basal ganglia physiology? Feedback loops are composed of one or several populations of neurones. Mathematical models of motor control must take into account the presence of time delays arising because of the finite velocity at which graded and actions potentials are transmitted between these populations. Delays used in models have fixed values, but further work will be needed to decide if a distribution of delays rather than a discrete delay is more appropriate to describe the dynamics of this motor system;

* How many loops should be used to model the system? Our results, examining systems with one or two negative feedback loops, have generated surprisingly complex dynamics. We expect however, that further work will have to consider multi-looped negative feedback systems in order to reproduce the variety of oscillatory fluctuations observed in normal and pathological motor control.

The basal ganglia reveal their function very gradually, but there is no doubt that nonlinear modelling of experimental results offers a promising way to explore their secrets.

Acknowledgements

This work was supported by NSERC of Canada and FCAR of Quebec. The author would like to thank Dr. Anne de Geoffroy for helping in the preparation of this paper.

REFERENCES

Abbs, J.H., and Winstein, C.J., 1990, Functional contributions of rapid and automatic sensory-based adjustments to motor output, in: "Attention and Performance XIII," M. Jeannerod, ed., Lawrence Erlbaum Associates, Hillsdale (New Jersey), pp. 627-652.

Albin, R.L., Young, A. B., and Penney, J.B., 1989, The functional anatomy of basal ganglia disorders, Trends Neurosci.12, 10:366-375.

Bélair, J., 1993, in review.

Beuter, A., Bélair, J., and Labrie, C., 1993, Feedback and delays in neurological diseases: a modelling study using dynamical systems, Bull. Math. Biol. 55, 3:525-541.

Beuter, A., and Lefebvre, R., 1988, Un modèle théorique de transition de phase dans la locomotion humaine, J. Canad. Sci. Sport 13:247-253.

Beuter, A., Milton, J., Labrie, C., Glass, L., and Gauthier, S., 1990, Delayed visual feedback and movement control in Parkinson's disease, Exp. Neurol. 110:228-235.

Bloxham, C.J., Dick, D.J., and Moore, M., 1987, Reaction times and attention in Parkinson's disease, J. Neurol. Neurosurg. Psychiat. 50:1178-1183.

DeLong, M.R., and Georgopoulos, A.D., 1979, Motor functions of the basal ganglia as revealed by studies of single cell activity in the behaving primate, in: "The Extrapyramidal System and its Disorders," L.J. Poirier, T.L. Sourkes, and P.J. Bédard, eds., Adv. Neurol. 24:131-140.

Divac, I., Oberg, G.E.., and Rosenkilde, C.E.., 1987, Patterned neural activity: Implications for neurology and neuropharmacology, *in*: "Basal Ganglia and Behavior: Sensory Aspects of Motor Functioning," J.S. Schneider and T.I. Lidsky, eds., Hans Huber Publishers, Toronto.

Divac, I., 1992, Is the brain too complicated for simple replacement therapy?, *Trends Neurosci.* 15, 9:339.

Filion, M., Tremblay, L., and Bédard, P.J., 1988, Abnormal influences of passive limb movement on the activity of globus pallidus neurons in parkinsonian monkeys, *Brain Res.* 444:165-176.

Glass, L., Beuter, A., and Larocque, D., 1988, Time delays, oscillations and chaos in physiological control systems, *Math. Biosci.* 90:111-125.

Glass, L., and Mackey M.C., 1979, Pathological conditions resulting from instabilities in physiological control systems, *Ann. New York Acad. Sci.* 316:214-235.

Graybiel, A.M., 1991, Basal ganglia: Input, neural activity, and relation to cortex, *Current Opinion Neurobiol.* 1:644-651.

Hoehn, M.M., and Yahr, M.D., 1967, Parkinsonism: Onset, progression and mortality. *Neurology* 17:427-442.

Hornykiewicz, O., and Kish, S.J., 1987, Biochemical pathophysiology of Parkinson's disease, *Adv. Neurol.* 45:19-34.

Jeannerod, M., 1988, "The Neural and Behavioral Organization of Goal-directed Movements," Clarendon Press, Oxford.

Lee, B.S., 1950, Effects of delayed speech feedback, *J. Acoustical Soc. Amer.* 22:824-826.

Longtin, A., and Milton, J.G., 1988, Complex oscillations in the human pupil light reflex with "mixed" and delayed feedback, *Math. Biosci.* 90:183-199.

Longtin, A., Milton, J.G., Bos, J.E. and Mackey, M.C., 1990, Noise and critical behavior of the pupil light reflex at oscillation onset, *Phys. Rev. A* 41, 12:6992-7005.

Mackey, M.C. and Glass, L., 1977, Oscillation and chaos in physiological control systems, *Science* 197:287-289.

Mackey, M.C., and Milton, J.G., 1987, Dynamical diseases, *Ann. New York Acad.Sci.* 504:16-32.

Merton, P.A., Morton, H.B., and Rashbass, C., 1967, Visual feedback in hand tremor, *Nature* 216:583-584.

Mitchell, I.J., Brotchie, J.M., Brown, G.D.A, and Crossman, A.R., 1991, Modelling the functional organization of the basal ganglia, *Mov. Dis.* 6, 3:189-204.

Rolls, E.T., Thorpe, S.J., Boytim, M., Szabo, I., and Perrett, D.I., 1984, Responses of striatal neurons in the behaving monkey. 3. Effects of iontophoretically applied dopamine on normal responsiveness, *Neurosci.* 12, 4:1201-1212.

Oguztoreli, M.N., and Stein, R.B., 1976, The effects of multiple reflex pathways on the oscillations in neuromuscular systems, *J. Math. Biol.* 3:87-101.

Scotto, M., and Oliva, G.A., 1984, Limit cycle oscillations in the human eye, *Biol. Cybern.* 51:33-41.

Smith, K.U., 1962, "Delayed Sensory Feedback and Behavior," Saunders, Philadelphia.

Stern, H.J., 1944, A simple method for the early diagnosis of abnormality of the pupillary reaction, *Brit. J. Ophthalmol.* 28:275-276.

Stark, L., 1962, Environmental clamping of biological systems: pupil servomechanism, *J. Optic. Soc. Amer.* 52:925-930.

Vasilakos, K., and Beuter, A., 1993, Effects of noise on a delayed visual feedback system, *J. Theoret. Biol.* in press.

Warrick, M.J., 1948, "Effect of transmission type control lags on tracking accuracy," USAF Air Material Command Technical Report number 5918.

PATHOPHYSIOLOGY OF LEVODOPA-INDUCED
DYSKINESIA: CHANGING CONCEPTS

Pierre Blanchette, Paul J. Bédard, Masaru Matsumura, Hélène Richard
and Michel Filion

Centre de recherche en neurobiologie, Hôpital de l'Enfant-Jésus
1401 18e rue, Québec, Canada, G1J 1Z4

INTRODUCTION

Levodopa-induced dyskinesia (LID) is a common side effect in patients treated chronically for idiopathic Parkinson's disease (PD). First described by Cotzias et al.(1967), it occurs in 30-80% of patients treated with levodopa, the incidence in our own patient population reaching 58% (Blanchette et al., unpublished observations). Although mild in intensity and even not recognized by many patients, LID may interfere with volitional movements and may become more disabling than parkinsonism itself. LID may be of different types: choreic, ballic, dystonic or myoclonic. It may occur at the peak of the effect of levodopa (so-called peak-dose dyskinesia), or towards the beginning and/or end of the effect (labeled as diphasic dyskinesia). The following discussion will focus on the most frequent type of LID which is choreic or choreoathetoid in nature and occurs at the peak of improvement of motor symptoms.

As pointed out by Nutt (1990), LID requires chronic levodopa stimulation of a dopamine (DA)-denervated but otherwise intact striatum. From a clinical standpoint, the duration of levodopa treatment (Fahn and Bressman, 1984; Diamond et al., 1989; Horstink et al., 1990; Blanchette et al., unpublished observations) and the cumulative levodopa dose administered (Blanchette et al., unpublished observations) do not seem to be primarily related to the timing of onset of LID. On the other hand, patients younger at disease onset (Lesser et al., 1979; Pederzoli et al., 1983; Cedarbaum et al., 1991; Blanchette et al., unpublished observations) and the more severely affected (Lang et al., 1982; Langston and Ballard, 1984; Blin et al., 1988; Horstink et al., 1990) appear particularly prone to LID. Therefore, the degree of dopaminergic nigrostriatal neuronal degeneration appears to contribute to the induction process of LID. Secondary postsynaptic changes related to chronic levodopa therapy are suspected to play a role as well (Fabbrini et al., 1988; Mouradian et al., 1988).

Even though the pathophysiologic substrate of LID remains elusive, experimental models of parkinsonism have recently provided interesting clues which could be of therapeutic significance. The availability of newer selective DAergic D-1 and D-2 agonists enables us to shift the balance between different populations of striatal output neurons to understand the contribution of each of the 2 pathways transmitting neural inputs directly and indirectly to the main output stations of the basal ganglia. In this chapter, we review past and current evidence which help understanding the neural mechanisms of LID. We also present new pharmacological and neurobiological data to support the importance of the hypothetical balance of inputs ultimately integrated by the output stations of the basal ganglia to limit the emergence of LID.

The Basal Ganglia IV, Edited by
G. Percheron *et al.*, Plenum Press, New York, 1994

EVIDENCE IMPLICATING D-1 RECEPTOR ACTIVATION

Previous clinical observations comparing levodopa (a non-selective D-1 and D-2 indirect DA agonist) and bromocriptine (a direct D-2 agonist with antagonistic D-1 properties) revealed that while the latter agent is less efficacious in relieving parkinsonism, it is also much less dyskinetogenic when given alone to PD patients (Lees and Stern, 1981) or 1-methyl-4-phenyl-1,2,3,6-tetrahydropyridine (MPTP)-exposed parkinsonian primates (Bédard et al., 1986), compared with levodopa. The early combination of these 2 drugs in the treatment of PD is also less dyskinetogenic after 5 years than levodopa alone (Rinne, 1987). These results were interpreted as reflecting the necessity to stimulate both D-1 and D-2 receptors for maximal antiparkinson efficacy, and incriminated the stimulation of D-1 receptors to explain the higher incidence of dyskinesia seen with levodopa. This hypothesis was supported by the postmortem demonstration of increased [^{3}H]flupenthixol D-1 binding in the putamen of PD patients treated with levodopa, especially in those with LID (Rinne et al., 1985). In rats bearing a unilateral nigrostriatal lesion caused by 6-hydroxydopamine, intermittent treatment with levodopa reversed the increased density of DA D-2 receptors but further enhanced the sensitivity of adenylate cyclase to DA (Parenti et al., 1986), suggesting supersensitivity of striatal DA D-1 receptors and supporting the data obtained in humans. The best strategy was therefore to develop more selective DA D-2 agonists to relieve parkinsonism without dyskinesia.

Unfortunately, this strategy has not improved the therapeutic profile in experimental parkinsonism. The nonergot selective DA D-2 agonist (+)-4-propyl-9-hydroxynaphtoxazine [(+)-PHNO] was shown to be highly efficacious to alleviate parkinsonism in monkeys but induced dyskinesia much as levodopa (Gomez-Mancilla and Bédard, 1992; Luquin et al., 1992). Therefore, stimulation of D-1 receptors is not necessary to explain the dyskinetogenic potential of levodopa. In addition, further postmortem DA receptor studies on human parkinsonian striatal tissues have revealed no change in D-1 receptor binding in the putamen of levodopa-treated patients (Raisman et al., 1985), whereas another group documented a small but significant **decrease** in D-1 and D-2 receptors in the caudate nucleus only (Rinne et al., 1991). Pierot et al. (1988) reported that the densities of both types of receptors were unchanged. More importantly, the last 2 studies failed to demonstrate a correlation between receptor densities and clinical variables including LID. These results are also in accordance with data obtained in dyskinetic MPTP-exposed parkinsonian primates (Gagnon et al., 1990) and suggest that the upregulation of DA receptors seen in early PD is reversed by treatment with levodopa. This would downplay the role of denervation supersensitivity as a crucial mechanism in LID.

EVIDENCE SUPPORTING DYSFUNCTION ALONG THE INDIRECT STRIATOPALLIDAL PATHWAY IN LID

Experimental stereotaxic injections of GABA antagonists in and around the lateral segment of the globus pallidus (referred to as the GPl, see Crossman et al., 1988) and 2-deoxyglucose uptake studies performed on dyskinetic MPTP-exposed primates (Mitchell et al., 1992), together with clinical observations that lesions of the subthalamic nucleus (STN) can produce severe dyskinesia led Crossman (1990) to propose that hyperkinetic disorders originate from relatively excessive inhibition of putaminal cells projecting to the GPl, leading to overactivity of the GPl and physiological inhibition of the STN. The abnormal patterns of neural activity in LID would therefore be primarily recruited within the so-called 'indirect' striatopallidal pathway that connects the putamen with the medial segment of the globus pallidus (GPm) through the GPl and the STN. The putaminal neurons of origin of this pathway bear D-2 receptors (vide infra), thus putting more weight on the preferential activation of this type of DA receptors as the underlying mechanism in LID.

The electrophysiological studies of Filion and Tremblay (1991) confirmed the occurrence of abnormally high firing rates in the GPl and low firing rates in the GPm upon dosing with a DA agonist in LID-primed parkinsonian monkeys (Fig.1A, B). This imbalance between the activities of the pallidal segments is exactly the opposite to that observed in untreated parkinsonian monkeys and at variance with the nearly equal firing rates in intact monkeys (Miller and DeLong, 1987; Filion and Tremblay, 1991). In order to find out whether this imbalance observed in LID is a general feature in other experimental models of dyskinesia, Matsumura et al. (unpublished observations) recorded the extracellular GP activity in 2 intact monkeys before, during and after stereotaxic administration of the GABA$_A$ antagonist bicuculline (0.5 µl, 15 µg/µl) in the GPl along the paradigm followed by Crossman et al. (1988). In fact, during bicuculline-induced dyskinesia, firing rates increased in the GPl near the injection site, and GPm neurons decreased their activity (Fig.1C, D). The changes in rates coincided with the occurrence of abnormal firing patterns consisting of long bursts and pauses. The spike potentials preceding the pauses became more and more widely spaced and their amplitude did not decrease, suggesting that the pauses were not due to excessive depolarization but rather to hyperpolarization (GABA$_B$?) or dysfacilitation. At

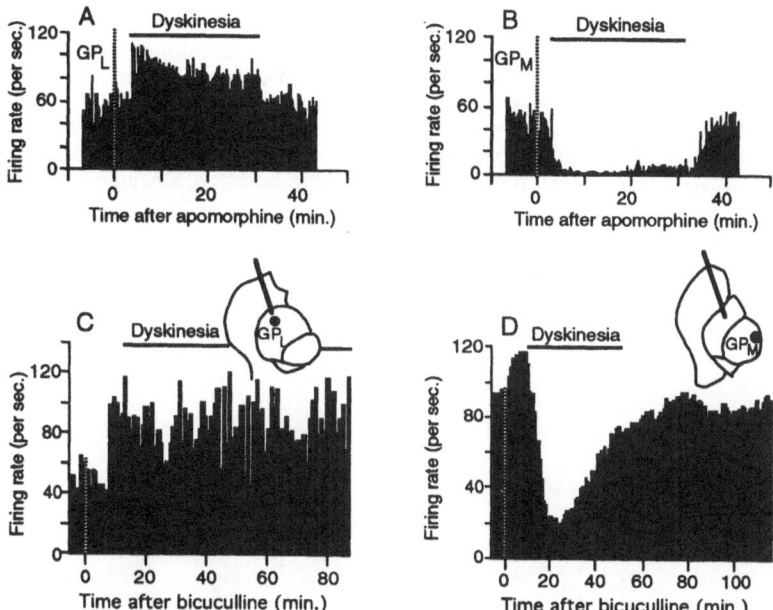

Figure 1. Histograms showing changes in the firing rate of single GP neurons (their locations in (C) and (D) are indicated by the black dot in the insets). Systemic injections of apomorphine (40 µg/kg, s.c.) increase GPl (A) and decrease GPm activity (B), concomitantly inducing dyskinesia in a monkey with MPTP-induced parkinsonism. Similarly in an intact monkey, local injections of bicuculline in the GPl (locations indicated by the tip of the black bar in the insets) also increase GPl (C) and decrease GPm activity (D) while inducing dyskinesia. In (D), the main effect in the GPm is a decrease from 10 to 70 minutes after the injection. It is however preceded by an increase which may be explained by lateral inhibition within GPl, which results in disinhibition in GPm. The increase reverses to a decrease as the drug diffuses in GPl to the site that was first submitted to lateral inhibition.

distance from the injection site, neuronal activity, surprisingly, decreased in the GPl. Inversely in the GPm, neurons with decreased activity were surrounded by neurons with increased activity. Thus, as in the case of dyskinesia induced by DA agonists in MPTP-exposed monkeys, those induced by injection of bicuculline in the GPl of normal monkeys also correspond to increased GPl and decreased GPm activity. However, the foregoing observations reveal that the abnormalities of GP activity during dyskinesia are more complex then heretofore reported and include abnormal patterns of activity both in the temporal and spatial domains. Abnormal bursting activities and centre-surround organization (inhibited neurons surrounded by excited neurons or vice-versa, see Tremblay and Filion, 1989) may also contribute to inappropriate occurrence and focusing of movements and postures.

STRATEGIES TO IMPROVE THE NEURAL BALANCE IN STRIATO-PALLIDAL FUNCTION IN LID

The relative contribution of the stimulation of D-1 and D-2 receptors to the antiparkinson and dyskinetic responses obtained with levodopa is not well established. In addition, there is no general agreement about the precise localization of these receptors within the striatum. In vitro autoradiography techniques (labeling DA receptors binding sites) combined with in situ hybridization histochemistry (to detect the messenger RNA encoding for such receptors) have been used to differentiate striatal output neurons. Gerfen et al. (1990) have documented the presence of DA D-1 receptors on striatonigral neurons in rats, the cells of origin of the so-called **direct pathway** that project directly to the GPm in primates, while D-2 receptors are expressed on striatal neurons of origin of the **indirect pathway** projecting to the GP proper in rats or to the GPl in primates. The same conclusions were drawn from selective suicide transport experimental lesions in rats (Harrison et al.,1990; 1992), and from 2-deoxyglucose autoradiography experiments that allow to examine the differential functional metabolic consequences mediated by selective DA receptor stimulation on various brain structures (Trugman and Wooten, 1987). However, other studies suggest that D-2 receptors are more widely distributed in the rat striatum and can be found on a significant proportion of striatonigral neurons (Ariano et al., 1992) and cholinergic interneurons as well (Le Moine et al., 1990). Selective DA agonists can therefore

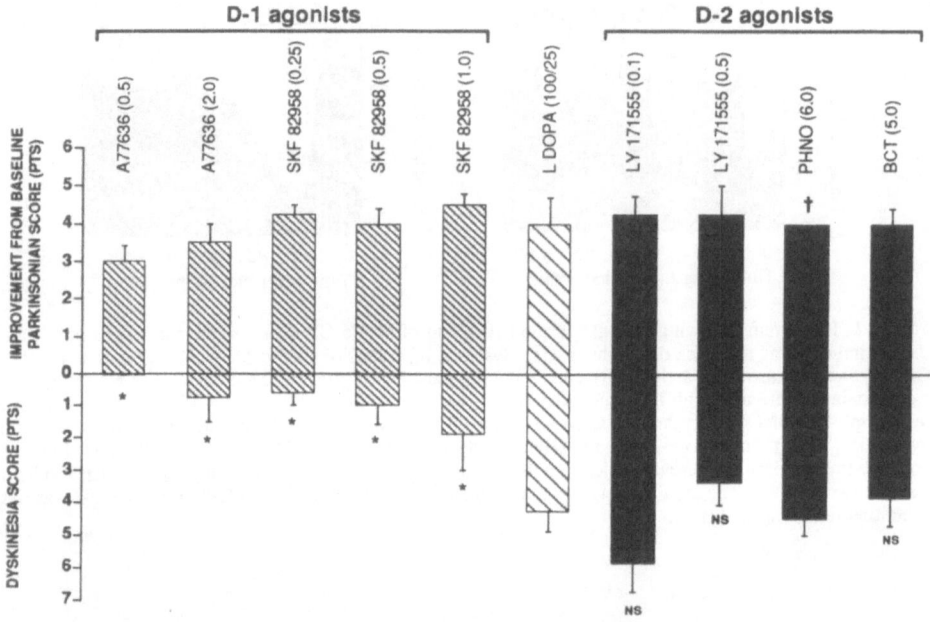

Figure 2. Improvement of parkinsonism reflected by the difference (in points) between baseline score and best motor function (top figure), and absolute dyskinesia scores (bottom figure) following acute drug administration. Results on rating scales are pooled for all 4 animals for a given drug. Doses given are in milligrams per kilogram, except for L-DOPA/Benserazide administered as a single 100 mg/25 mg capsule, and (+)-PHNO given in micrograms per kilogram. Drugs were all administered subcutaneously, except L-DOPA/Benserazide and Bromocriptine (BCT) given orally. Scores + S.E.M. (*p<0.01 by comparison with L-DOPA); NS: not significant compared to L-DOPA; Results for 2 monkeys only.

activate fairly specifically striatal output neurons that differentially modulate the activity of the basal ganglia.

We analyzed the differential behavioral effects produced by selective DA receptor stimulation in 4 cynomolgous (Macaca fascicularis) monkeys with a chronic and stable MPTP-induced parkinsonian state (Blanchette et al., in press). Animals were already dyskinetic following chronic daily treatment for 4 weeks with (+)-PHNO (2 animals), levodopa/benserazide (1 animal), and CY 208-243 (1 animal) which is a non-selective D-1 agonist. Dyskinesia could be consistently reproduced upon dosing with levodopa, and consisted of a combination of chorea and dystonia predominating in the lower limbs. Subsequently, animals were administered selective D-1 agonists (SKF 82958, A-77636) and D-2 agonists (LY 171555, (+)-PHNO, Bromocriptine) that were compared with levodopa. Motor effects were evaluated with rating scales to quantitate both the relief of parkinsonism and severity of dyskinesia obtained with each drug. Each experiment was conducted after a washout period of 1-4 weeks.

As observed previously by another group (Luquin et al., 1990), we have found similar antiparkinson efficacy between the selective DA D-1 and D-2 agonists used (Fig. 2, top histogram). Surprisingly, D-1 agonists were overall much less prone to reproduce dyskinesia for a comparable antiparkinson benefit (Fig. 2, bottom histogram). Only this class of agents could sometimes alleviate parkinsonism without concomitant dyskinesia in individual monkeys. However, the phenotypic expression of dyskinesia was essentially similar following D-1 or D-2 receptor stimulation. A slight dose-dependent increase in dyskinesia scores was apparent for A-77636 and SKF 82958. D-2 agonists always produced higher dyskinesia scores with improvement in motor function and more behavioral stereotypies (chewing, licking). The better therapeutic profile seen in our dyskinetic animals with selective D-1 agonists could theoretically come from the pharmacological bypass of the pathological events leading to underactivity of the STN in LID by tilting the balance of

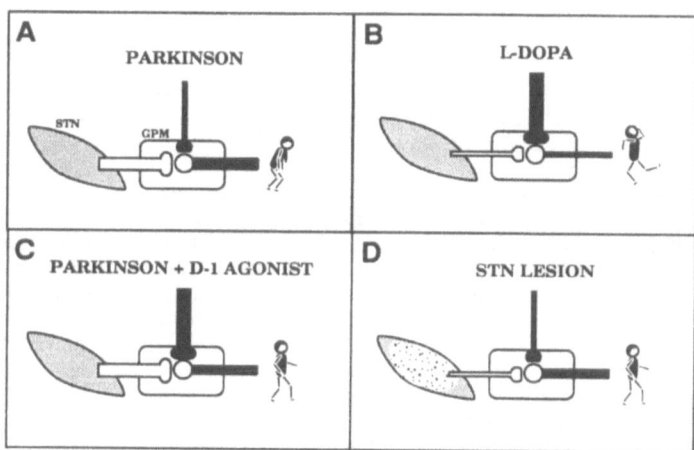

Figure 3. Highly simplified schematic diagram of the hypothetical balance of basal ganglia inputs transmitted to the medial segment of the globus pallidus along the direct inhibitory striatopallidal pathway (vertical dark connection), and the excitatory subthalamopallidal pathway (horizontal white connection) which is under the inhibitory control of a distinct indirect striato-(lateral) pallidal pathway. The size of the connections indicates the relative contribution of each pathway. The abnormal patterns of activity underlying parkinsonian symptoms (A) and levodopa-induced dyskinesia (B) are represented. A new balance of inputs is re-established following the administration of a selective D-1 agonist (C), or following selective lesioning of the STN (D) which is overactive in parkinsonism (Bergman ct al. 1990). Abbreviations: GPm, globus pallidus, medial segment; STN, subthalamic nucleus.

striatal output towards the D-1-linked direct striatopallidal pathway, which connects the putamen monosynaptically with the GPm in primates.

The strategy using selective D-1 agonists could also lead to a more harmonious integration of inputs transmitted to the GPm from the striatum and the STN (the last relay station along the indirect pathway). Figure 3 illustrates the relative contribution of each pathway in different states, as proposed by Crossman (1990) and supported by Filion et al. (1991). Compared with the pattern of neural activity obtained with levodopa (panel B, Fig.3), D-1 agonists presumably enhance transmission along the direct pathway only, leaving the heightened excitation from the STN in parkinsonian subjects untouched (panel C). This excitation is probably important to limit temporally and spatially the main inhibitory signal from the striatum, creating a better balance of inputs. Lesioning the STN is another way to reestablish this balance and correct parkinsonian signs (Bergman et al., 1990) by removing the heightened excitation from this structure to match the low inhibitory signal provided by the DA-depleted striatum (panel D). Therefore, preferential activation of the D-2-linked indirect striatopallidal pathway appears to be more closely involved in the genesis and perpetuation of LID. The poor dyskinetogenic potential of bromocriptine in drug-naive subjects could stem in its greater potential to desensitize D-2 receptors in MPTP-exposed parkinsonian primates compared to other agonists (Gagnon et al., 1990), rather than in its D-1 antagonistic properties.

Because the inhibitory loop between the GPl and the STN becomes overactive in LID, we selected 2 dyskinetic MPTP-exposed parkinsonian monkeys (Macaca fascicularis) and performed a unilateral stereotaxic lesion of the GPl using the excitotoxin ibotenic acid in an attempt to ameliorate contralateral LID (P.J. Blanchette, R. Boucher and P.J. Bédard, unpublished observations). To our surprise, this lesion produced opposite effects and worsened chorea in the contralateral leg upon challenge with levodopa or selective D-2 agonists (LY 171555 and (+)-PHNO). Histology disclosed nearly complete ablation of the GPl in one animal, and ablation of the mid-portion only in the other. The putamen and GPm were left largely intact in both cases. This negative result suggests that complex disturbances take place in LID-primed animals, and that the GPl is not singly involved in the pathogenesis of chorea in this model.

INTERMITTENT LEVODOPA ADMINISTRATION: EVIDENCE FOR STRIATONIGRAL PATHWAY DESENSITIZATION AND EFFECTS OF CONTINUOUS TREATMENT

We have seen that levodopa and selective D-2 agonists are prone to induce dyskinesia. Several reports are now indicating that the intermittent schedule of levodopa administration used clinically is susceptible to be involved in the genesis of the functional state underlying motor response fluctuations and LID in PD. The mode of pharmacologic stimulation can differentially affect the subsequent responsiveness to other drugs (Post, 1980). Chronic **intermittent** levodopa administration can produce behavioral sensitization with enhancement of the rotational response normally seen with apomorphine (Juncos et al., 1989), a non-selective D-1 and D-2 agonist. Engber et al. (1991) also demonstrated that the same treatment markedly increased the rotational response to a D-2 agonist but greatly diminished the response to a selective partial D-1 agonist. On the other hand, **continuous** levodopa administration slightly enhanced the rotational response to the D-2 agonist, while it had no effect on the response to the D-1 agonist. This was supported electrophysiologically by Weick et al. (1990) who found that the physiological inhibition of the neuronal activity in the substantia nigra pars reticulata to a selective partial D-1 agonist or to iontophoresed GABA was decreased by the intermittent administration of levodopa, unlike continuous levodopa.The intermittent administration of levodopa (akin to the clinical situation) appears then to shift the balance of striatal output to strongly favor D-2-mediated responses. Such an abnormal pharmacological behavior could with time promote the emergence of LID.

The schedule of drug administration can also differentially affect the enzymatic activity and the expression of DA receptors and neuropeptide content in the basal ganglia. Juncos et al. (1989) first reported that rats with unilateral lesions of the nigrostriatal pathway and treated intermittently with levodopa showed no DA receptors supersensitivity in spite of behavioral sensitization, whereas Ariano et al. (1991) have seen noticeable downregulation of D-1 receptors in such animals treated intermittently compared with those treated continuously. In the same rat model, extensive biochemical analyses were conducted following intermittent or continuous levodopa replacement therapy (Engber et al., 1991). In comparison with untreated lesioned rats with unilateral DA deafferentation, intermittent levodopa therapy induced several changes in neurochemical markers and altered peptide levels in striatonigral neurons but not in striatopallidal neurons. These changes may be linked with the desensitization in D-1-mediated responses alluded to earlier. In contrast, continuous levodopa therapy induced fewer perturbations and further elevated enkephalin levels in the GP compared to control values. The authors concluded that continuous delivery of levodopa may be more physiological and offered that pharmacological manipulation of striatal neuropeptides may be useful alternatives in the management of PD. According to Gerfen et al. (1990) who studied the mRNA expression for DA receptors and different peptides, continuous replacement therapy would not be universally beneficial to correct the anomalies brought about by DA depletion. In their rats with unilateral nigrostriatal pathway lesions, only a selective D-1 agonist administered **intermittently** reversed the reduction in substance P and D-1 receptor mRNA expression seen in the lesioned striatum, whereas only a selective D-2 agonist administered in a **continuous** fashion reversed the elevation in striatal mRNA expression for enkephalin and D-2 receptors resulting from DA deafferentation.

These data provided support for the evaluation of more sustained forms of DA receptor stimulation. Preliminary observations in our laboratory on MPTP-exposed primates treated chronically for 4 weeks with a highly selective full D-2 agonist revealed behavioral sensitization in the group receiving the drug intermittently, all 3 animals exhibiting dyskinesia within 7 days. The same drug given continuously through minipumps resulted in partial desensitization in all 3 animals, only 1 showing transient dyskinesia. Similarly, continuous subcutaneous administration of lisuride (a D-2 agonist) for 3 months greatly ameliorated both motor response fluctuations and LID in levodopa-treated PD patients (Baronti et al., 1992). On the other hand, the administration of a sustained-release form of (+)-PHNO (another potent selective D-2 agonist) to fluctuating levodopa-treated patients led to early pharmacological tolerance with loss of antiparkinson efficacy, suggesting DA D-2 (and D-1?) receptor desensitization (Cedarbaum et al., 1990). However, several patients noted marked reductions in LID in the immediate post-study period, and 1 patient had practically no dyskinesia for 1 month in spite of a significant temporary increase in levodopa requirements. This last observation would also support the idea that D-2 receptor-mediated effects of DAergic drugs are primarily involved in LID.

CONCLUSION

Although the induction process of levodopa-induced dyskinesia (LID) in parkinsonian patients is not well established, recent reports have provided deeper knowledge of the regulatory mechanisms of D-1 and D-2 receptors and downplayed the role of dopamine receptors supersensitivity and D-1-mediated events as the underlying cause for this frequent complication. We presented experimental data suggesting that an abnormal striatal balance created by exogenous dopaminergic drugs in the sense of a preferential activation of D-2-linked neural mechanisms is more tightly associated with LID.

Different techniques have allowed to detect the presence of D-1 receptors on the 'direct' striatopallidal (striatonigral) pathway, while D-2 receptors are preferentially (but not exclusively) located on the 'indirect' striatopallidal pathway that would be chiefly recruited in LID. The neural patterns activated in LID involve profound and opposite changes in the 2 segments of the globus pallidus (GP), with neuronal hyperactivity seen in the lateral segment (GPl) and hypoactivity recorded in the medial segment (GPm). These changes appear rather similar to those seen in another model of drug-induced dyskinesia resulting from the stereotaxic injections of the GABA antagonist bicuculline in the GPl. Therefore, LID originates from a detrimental imbalance in striatal output with preferential inhibition by exogenous dopamine of putaminal neurons bearing D-2 receptors, resulting in inhibition of the subthalamic nucleus (STN) by an overactive GPl. This in turn leads to a very asymmetric and complex balance of inputs transmitted to the GPm which could also be detrimental to the final spatial and temporal integration of the signals involved before they reach the thalamus. The oversimplistic attempt to correct these abnormal inputs by lesioning selectively the GPl in dyskinetic primates worsened LID contralaterally to the lesion. We offer the hypothesis that the preferential activation of D-1 receptors can rearrange this balance, thereby benefiting patients with PD. The role of D-1 agonism in PD deserves further clarification.

We also reviewed recent evidence that intermittent levodopa replacement therapy sensitizes D-2-mediated behavioral responses, and leads to D-1 receptors desensitization in rats with unilateral dopamine deafferentation. Continuous replacement therapy with levodopa or a selective D-2 agonist appears more physiological but may lead to partial pharmacological tolerance. Synergistic intermittent administration of a selective D-1 agonist might provide additional benefit. Pharmacological manipulation of other neurotransmitter systems in the striatum, including neuropeptides, could also provide original ways of targeting specific neuronal populations to limit the emergence of LID.

Acknowledgments

This work was supported by the MRC of Canada and the Parkinson Foundation of Canada. P.J. Blanchette holds a fellowship grant from the FRSQ. The authors wish to thank René Boucher, Gaston Côté, Laurent Grégoire and François Tardif for their technical assistance.

REFERENCES

Ariano, M.A., Engber, T.M., Susel, Z., and Chase, T.N., 1991, Striatal D_1 dopamine receptor morphochemistry following continuous or intermittent L-DOPA replacement therapy, *Exp. Neurol.* 112:112-118.

Ariano, M.A., Stromski, C.J., Smyk-Randall, E.M., and Sibley, D.R., 1992, D_2 dopamine receptor localization on striatonigral neurons, *Neurosci. Lett.* 144:215-220.

Baronti, F., Mouradian, M.M., Davis, T.L., Giuffra, M., Brughitta, G., Conant, K.E. and Chase, T.N., 1992, Continuous lisuride effects on central dopaminergic mechanisms in Parkinson's disease, *Ann. Neurol.* 32:776-781.

Bédard, P.J., DiPaolo, T., Falardeau, P., and Boucher, R., 1986, Chronic treatment with L-DOPA, but not bromocriptine induces dyskinesia in MPTP-parkinsonian monkeys. Correlation with [^{3}H]spiperone binding, *Brain Res.* 379:294-299.

Bergman, H., Wichmann, T., and DeLong, M.R., 1990, Reversal of experimental parkinsonism by lesions of the subthalamic nucleus, *Science* 249:1436-1438.

Blanchette, P.J., Bédard, P.J., Britton, D.R., and Kebabian, J.W., Differential effect of selective D-1 and D-2 dopamine receptor agonists on levodopa-induced dyskinesia in MPTP monkeys, *J. Pharmacol. Exp. Therap.* (in press).

Blin, J., Bonnet, A.M., and Agid, Y., 1988, Does levodopa aggravate Parkinson's disease?, *Neurology* 38:1410-1416.

Cedarbaum, J.M., Clark, M., Toy, L.H., and Green-Parsons, A., 1990, Sustained-release (+)-PHNO [MK-458 (HPMC)] in the treatment of Parkinson's disease: evidence for tolerance to a selective D_2-receptor agonist administered as a long-acting formulation, *Movement Disorders* 5:298-303.

Cedarbaum, J.M., Gandy, S.E., and McDowell, F.H., 1991, "Early" initiation of levodopa treatment does not promote the development of motor response fluctuations, dyskinesias, or dementia in Parkinson's disease, *Neurology* 41:622-629.

Cotzias, G.C., Van Woert, M.H., and Schiffer, L.M., 1967, Aromatic amino acids and modification of parkinsonism, *N. Engl. J. Med.* 276:374-379.

Crossman, A.R., 1990, A hypothesis on the pathophysiological mechanisms that underlie levodopa- or dopamine agonist-induced dyskinesia in Parkinson's disease: implications for future strategies in treatment, *Movement Disorders* 5:100-108.

Crossman, A.R., Mitchell, I.J., Sambrook, M.A., and Jackson, A., 1988, Chorea and myoclonus in the monkey induced by gamma-aminobutyric acid antagonism in the lentiform complex, *Brain* 111:1211-1233.

Diamond, S.G., Markham, C.H., Hoehn, M.M., McDowell, F.H., and Muenter, M.D., 1989, Effect of age at onset on progression and mortality in Parkinson's disease, *Neurology* 39:1187-1190.

Engber, T.M., Susel, Z., Kuo, S., Gerfen, C.R., and Chase, T.N., 1991, Levodopa replacement therapy alters enzyme activities in striatum and neuropeptide content in striatal output regions of 6-hydroxydopamine lesioned rats, *Brain Res.* 552:113-118.

Fabbrini, G., Mouradian, M.M., Juncos, J.L., Schlegel, J., Mohr, E., and Chase, T.N., 1988, Motor fluctuations in Parkinson's disease: central pathophysiological mechanisms, part I, *Ann. Neurol.* 24:366-371.

Fahn, S., and Bressman, S.B., 1984, Should levodopa therapy for parkinsonism be started early or late? Evidence against early treatment, *Can. J. Neurol. Sci.* 11 (suppl.):200-206.

Filion, M., and Tremblay, L., 1991, Abnormal spontaneous activity of globus pallidus neurons in monkeys with MPTP-induced parkinsonism, *Brain Res.* 547:142-151.

Filion, M., Tremblay, L., and Bédard, P.J., 1991, Effects of dopamine agonists on the spontaneous activity of globus pallidus neurons in monkeys with MPTP-induced parkinsonism, *Brain Res.* 547:152-161.

Gagnon, C., Bédard, P.J., and DiPaolo, T., 1990, Effect of chronic treatment of MPTP monkeys with dopamine D-1 and/or D-2 receptor agonists, *Eur. J. Pharmacol.* 178:115-120.

Gerfen, C.R., Engber, T.M., Mahan, L.C., Susel, Z., Chase, T.N., Monsma, Jr., F.J., and Sibley, D.R., 1990, D_1 and D_2 dopamine receptor-regulated gene expression of striatonigral and striatopallidal neurons, *Science* 250:1429-1432.

Gomez-Mancilla, B., and Bédard, P.J., 1992, Effect of chronic treatment with (+)-PHNO, a D2 agonist in MPTP-treated monkeys, *Exp. Neurol.* 117:185-188.

Harrison, M.B., Wiley, R.G., and Wooten, G.F., 1990, Selective localization of striatal D_1 receptors to striatonigral neurons, *Brain Res.* 528:317-322.

Harrison, M.B., Wiley, R.G., and Wooten, G.F., 1992, Changes in D2 but not D1 receptor binding in the striatum following a selective lesion of striatopallidal neurons, *Brain Res.* 590:305-310.

Horstink, M.W.I.M., Zijlmans, J.C.M., Pasman, J.W., Berger, H.J.C., and van't Hof, M.A., 1990, Severity of Parkinson's disease is a risk factor for peak-dose dyskinesia, *J. Neurol. Neurosurg. Psych.* 53:224-226.

Juncos, J.L., Engber, T.M., Raisman, R., Susel, Z., Thibaut, F., Ploska, A., Agid, Y., and Chase, T.N., 1989, Continuous and intermittent levodopa differentially affect basal ganglia function, *Ann. Neurol.* 25:473-478.

Lang, A.E., Meadows, J.C., Parkes, J.D., and Marsden, C.D., 1982, Early onset of the "on-off" phenomenon in children with symptomatic parkinsonism, *J. Neurol. Neurosurg. Psych.* 45:823-825.

Langston, J.W., and Ballard, P., 1984, Parkinsonism induced by 1-methyl-4-phenyl-1,2,3,6-tetrahydropyridine (MPTP): implications for treatment and the pathogenesis of Parkinson's disease, *Can. J. Neurol. Sci.* 11 (suppl.):160-165.

Le Moine, C., Tison, F., and Bloch, B., 1990, D_2 dopamine receptor gene expression by cholinergic neurons in the rat striatum, *Neurosci. Lett.* 117:248-252.

Lees, A.J., and Stern, G.M., 1981, Sustained bromocriptine therapy in previously untreated patients with Parkinson's disease, *J. Neurol. Neurosurg. Psych.* 44:1020-1023.

Lesser, R.P., Fahn, S., Snider, S.R., Cote, L.J., Isgreen, W.P., and Barrett, R.E., 1979, Analysis of the clinical problems in parkinsonism and the complications of long-term levodopa therapy, *Neurology* 29:1253-1260.

Luquin, M.R., Herrero, T., Tubia, M., and Obeso, J.A., 1990, Drugs with selective D-1 or D-2 receptor activity exhibit similar antiparkinsonian potency in the MPTP-treated monkey [Abstract], *Neurology* 40 (suppl.1):291.

Luquin, M.R., Laguna, J., and Obeso, J.A., 1992, Selective D2 receptor stimulation induces dyskinesia in parkinsonian monkeys, *Ann. Neurol.* 31:551-554.

Miller, W.C., and DeLong, M.R., 1987, Altered tonic activity of neurons in the globus pallidus and subthalamic nucleus in the primate MPTP model of parkinsonism, *in:* "The Basal Ganglia II, (Advances in Behavioral Biology, Vol. 32)," M.B. Carpenter and A. Jayaraman, eds., Plenum, New York, pp. 415-427.

Mitchell, I.J., Boyce, S., Sambrook, M.A., and Crossman, A.R., 1992, A 2-deoxyglucose study of the effects of dopamine agonists on the parkinsonian primate brain, *Brain* 115:809-824.

Mouradian, M.M., Juncos, J.L., Fabbrini, G., Schlegel, J., Bartko, J.J., and Chase, T.N., 1988, Motor fluctuations in Parkinson's disease: central pathophysiological mechanisms, part II, *Ann. Neurol.* 24:372-378.

Nutt, J.G., 1990, Levodopa-induced dyskinesia: review, observations, and speculations, *Neurology* 40:340-345.

Parenti, M., Flauto, C., Parati, E., Vescovi, A., and Groppetti, A., 1986, Differential effect of repeated treatment with L-DOPA on dopamine-D1 or -D2 receptors, *Neuropharmacol.* 25:331-334.

Pederzoli, M., Girotti, F., Scigliano, G., Aiello, G., Carella, F., and Caraceni, T., 1983, L-Dopa long-term treatment in Parkinson's disease: age-related side effects, *Neurology* 33:1518-1522.

Pierot, L., Desnos, C., Blin, J., Raisman, R., Scherman, D., Javoy-Agid, F., Ruberg, M., and Agid, Y., 1988, D1 and D2-type dopamine receptors in patients with Parkinson's disease and progressive supranuclear palsy, *J. Neurol. Sci.* 86:291-306.

Post, R.M., 1980, Intermittent versus continuous stimulation: effect of time interval on the development of sensitization or tolerance, *Life Sci.* 26:1275-1282.

Raisman, R., Cash, R., Ruberg, M., Javoy-Agid, F., and Agid, Y., 1985, Binding of [^{3}H]SCH 23390 to D-1 receptors in the putamen of control and parkinsonian subjects, *Eur. J. Pharmacol.* 113:467-468.

Rinne, U.K., 1987, Early combination of bromocriptine and levodopa in the treatment of Parkinson's disease: a 5-year follow-up, *Neurology* 37:826-828.

Rinne, J.O., Laihinen, A., Lönnberg, P., Marjamäki, P., and Rinne, U.K., 1991, A post-mortem study on striatal dopamine receptors in Parkinson's disease, *Brain Res.* 556:117-122.

Rinne, J.O., Rinne, J.K., Laakso, K., Lönnberg, P., and Rinne, U.K., 1985, Dopamine D-1 receptors in the parkinsonian brain, *Brain Res.* 359:306-310.

Tremblay, L., and Filion, M., 1989, Responses of pallidal neurons to striatal stimulation in intact waking monkeys, *Brain Res.* 498:1-16.

Trugman, J.M., and Wooten, G.F., 1987, Selective D$_1$ and D$_2$ dopamine agonists differentially alter basal ganglia glucose utilization in rats with unilateral 6-hydroxydopamine substantia nigra lesions, *J. Neurosci.* 7:2927-2935.

Weick, B.G., Engber, T.M., Susel, Z., Chase, T.N., and Walters, J.R., 1990, Responses of substantia nigra pars reticulata neurons to GABA and SKF-38393 in 6-hydroxydopamine-lesioned rats are differentially affected by continuous and intermittent levodopa administration, *Brain Res.* 523:16-22.

NON-DOPAMINERGIC APPROACHES TO THE TREATMENT
OF PARKINSON'S DISEASE

Jonathan Brotchie, Yannick Maneuf, Neill Hughes, Susan Duty
and Alan Crossman

School of Biological Sciences
University of Manchester
Manchester, M13 9PT, U.K.

INTRODUCTION

It is widely appreciated that the primary pathology underlying the symptoms of Parkinson's disease is degeneration of the dopaminergic nigrostriatal pathway. Over a quarter of a century ago, Cotzias et al. (1967) demonstrated the potential of dopamine replacement by L-DOPA as a treatment for Parkinson's disease. This realisation led to the widespread use of such dopamine-replacement therapies. L-DOPA has undoubtedly been of considerable importance in improving the quality of life of patients suffering from parkinsonism. However, dopamine-replacement therapies are associated with side-effects following chronic administration. These can range from a wearing-off of efficacy of the drug to severely debilitating hyperkinetic movement disorders, the L-DOPA-induced dyskinesias (Marsden and Parkes, 1977). Dyskinetic side-effects are generally considered to result from a change in the response to dopamine of cells in the striatum (Mitchell et al., 1992). The possibility therefore arises that anti-parkinsonian therapies, based neither on dopaminergic actions nor targeting the striatum, might have great clinical value. Here we present data, from both *in vitro* neurochemical studies and *in vivo* studies in experimental models of parkinsonism, that describe the actions of three classes of non-dopaminergic compounds in the pallidal complex and substantia nigra pars reticulata. These findings may have potential in developing novel therapeutic approaches to Parkinson's disease.

In the last five years work in this and other laboratories has provided rapid advances in our understanding of the neural mechanisms underlying parkinsonian symptoms. A wide range of molecular, metabolic, neurochemical, electrophysiological and behavioural pharmacological approaches have been applied to animal models of parkinsonism to provide data which have been synthesised into a simple theoretical model of the neural mechanisms that underlie parkinsonism (Penney and Young, 1986; Mitchell et al., 1989). The initial degeneration of dopaminergic neurons of the substantia nigra pars compacta initiates a sequence of changes in neural activity throughout the basal ganglia. These changes are ultimately made manifest in overactivity of basal ganglia outputs, from the internal segment of the globus pallidus (GPi) and the substantia nigra pars reticulata (SNr), to other non-basal ganglia motor regions. This abnormal overactivity appears to be of great importance in the generation of parkinsonian symptoms as procedures which reduce the activity of basal ganglia outputs produce marked alleviation of parkinsonian symptoms.

Two abnormalities in neural activity within the basal ganglia appear critical to the generation of this overactive basal ganglia output, and hence symptoms,
1) overactivity of the excitatory amino acid (EAA)-utilising afferents to the GPi and SNr from the subthalamic nucleus, and,

2) underactivity of the inhibitory GABA/dynorphin/substance P-utilising afferents to the GPi and SNr from the striatum.

We, and others, have previously shown that reversal of the first of these abnormalities, namely attenuation of the overactive subthalamic excitation of the GPi and SNr, can alleviate parkinsonian symptoms in rat and primate models. This can be achieved either by surgical lesion of the subthalamic nucleus (Bergman et al., 1990) or by antagonism of EAA transmission in the GPi (Klockgether and Turski, 1990; Klockgether et al., 1991; Brotchie et al., 1991). However, with regard to the clinical relevance of these findings, it is necessary to consider other factors which might limit the utility of such novel therapeutic approaches. For instance a pharmaceutical manipulation would probably be preferable to neurosurgery in most patients in the early stages of the disease. While, EAA antagonist approaches appear promising, the possibility of additional pharmacological therapies ought not to be overlooked as all theoretical problems with EAA-related treatments have yet to be fully investigated. For example, EAA antagonists might lack anatomical selectivity for basal ganglia EAA receptors when administered systemically. EAA receptors in other, non-basal ganglia, regions of the brain may be blocked which could compromise important cerebral functions such as memory and may also induce dissociative anaesthesia and ataxia. Such reasons probably underlie the failure of systemic administration of some EAA antagonists, e.g. the NMDA antagonist MK-801, to alleviate symptoms in the MPTP-treated primate (Crossman et al., 1989).

In this chapter we propose three alternative pharmacological means of attenuating the overactivity in the outputs of the basal ganglia seen in parkinsonism. The first of these strategies is based upon kappa opioid agonists and targets the overactive EAA input to the GPi and SNr. The other approaches rely on ATP-sensitive potassium channel blockers and cannabinoids to increase the levels of GABAergic transmission, and thus inhibit neuronal excitability, in these regions. In all three cases we suggest that the regional distribution of the receptors involved would confer selectivity for the output regions of the basal ganglia leaving neurotransmission in other areas of the basal ganglia and CNS relatively unaffected.

KAPPA OPIOID RECEPTOR AGONISTS

Dynorphin, co-transmitted with GABA by the striatal efferents to the GPi and SNr, exerts its actions primarily by interactions with kappa opioid receptors (Chavkin et al., 1982). The level of dynorphin transmission is thought to be decreased in parkinsonism (Jiang et al., 1990; Engber et al., 1991). Though the role of GABA-dynorphin co-transmission remains to be fully elucidated, evidence of an interaction at a functional level does exist. A pattern of similar functional roles for dynorphin and GABA emerges from electrophysiological (Lavin and Garcia-Munoz, 1985), behavioural pharmacological (Herrera-Marschitz et al., 1984) and in vivo microdialysis studies (Reid et al., 1990). Recent experiments in this laboratory may have begun to shed light on one possible neuronal mechanism by which this apparent GABA-dynorphin interaction occurs in the basal ganglia.

In slices of SNr, the selective kappa opioid agonist CI-977 reduces depolarisation-evoked release of EAAs. Slices of rat SNr (400µm) were loaded with [^{3}H]-glutamate and then incubated in aerated artificial cerebrospinal fluid in the presence of 30µM dihydrokainic acid to prevent reuptake. After equilibration the release of [^{3}H]-glutamate was measured in each 5 minute time bin over 50 minutes. A 5 minute pulse of KCl (50mM) was added to the medium to evoke release of [^{3}H]-glutamate. K$^+$-evoked release was measured in the presence or the absence of CI-977 at various concentrations (10=200µM). A dose-dependent decrease of the K$^+$-evoked release of [^{3}H]-glutamate from nigral slices was observed, the effect being maximal (83%) at 200µM CI-977. We interpret this result as suggesting that kappa opioids act to decrease EAA release from terminals of afferents from the subthalamic nucleus. If this is the case, then the functional interaction between co-released GABA and dynorphin in the GPi and SNr is indirect, however, both transmitters have the net effect of reducing the firing rate of GPi and SNr output neurons. This finding is consistent with a mounting body of evidence that kappa opioid agonists can reduce EAA release and EAA-mediated effects in other areas of the brain (Lambert et al., 1991) . Thus, recently CI-977 has been shown to have anti-epileptic and neuroprotective effects similar to those seen with EAA antagonists (Singh et al., 1990; Mackay et al., 1991). However, our studies to date cannot rule out additional mechanisms of action of dynorphin which may be occurring at the post-synaptic membrane for instance.

Given this action of dynorphin and CI-977 we predicted that intracerebral injection of kappa opioid agonists into the GPi would alleviate parkinsonian symptoms, by reducing the excessive EAA release. CI-977 was injected into the entopeduncular nucleus (rodent homologue of the GPi) in rats rendered parkinsonian by administration of reserpine (4mg/kg, s.c.). Injections of CI-977 (0.5 µl, 5-500 mM) were made unilaterally and locomotion measured (as described in Brotchie et al., 1991). Briefly, the locomotion of the rats was monitored in an open field arena divided into squares of 5cm. Locomotion scores (in locomotion units, LU) were attained by enumerating the number of squares entered during the 25 minute period of assessment. This score provides a quantitative measure that is related to distance moved by the animal. Following vehicle injections, rats remained parkinsonian (3 ± 0.6 LU). In contrast, a marked, dose-dependent alleviation of akinesia was observed following intracerebral CI-977 injections. Maximal effects were observed following injection of 100mM CI-977, locomotor score 143 ± 19 LU.

In contrast to EAA receptors, kappa opioid receptors have a discrete distribution in the brain. Amongst the areas with highest levels of kappa receptors are the output regions of the basal ganglia (Nock et al., 1988; Unterwald et al., 1991). Considering this distribution, it is tempting to speculate that systemic administration of kappa opioid agonists would manipulate EAA transmission specifically in those areas of the basal ganglia that are overactive in parkinsonism. Thus kappa opioids might not attenuate EAA transmission in other regions of the CNS and non-basal ganglia-related side-effects would be avoided. In the reserpine-treated rat model of parkinsonism we have therefore assessed parkinsonism following systemic administration of CI-977. CI-977 resulted in a significant dose-dependent reversal of parkinsonian symptoms when given intraperitoneally. Following injection of 43µg/kg CI-977 the locomotor score was 66.8 ± 8 LU. Saline injections had no anti-parkinsonian effects, locomotor score 2 ± 0.6 LU.

ATP-SENSITIVE POTASSIUM CHANNEL BLOCKERS

In the pancreas the ATP-sensitive potassium channel (K_{ATP}) plays a critical role in the modulation of insulin release. The sulphonylurea class of compounds, typified by glibenclamide and tolbutamide, have been shown to block K_{ATP}, and have been used treat diabetes mellitus since the 1950s. Sulphonylureas act to increase insulin release from the pancreatic ß-cell. Blockade of K_{ATP} causes depolarisation and consequently increases intracellular calcium via the opening of voltage-sensitive calcium channels. This rise in intracellular calcium elicits secretion of insulin by exocytosis. Within the CNS, a sulphonylurea-binding site has been described in homogenate preparations of whole pig brain using [^{3}H]-glibenclamide binding assays. Two groups of studies have demonstrated that this sulphonylurea binding site is indeed K_{ATP}. Firstly, direct measurements ionic flow through potassium channels, using ^{86}Rb$^+$ efflux, show the presence, in the SNr, of a potassium channel that is opened by conditions that reduce ATP concentrations (Amoroso et al., 1990). This channel is blocked by sulphonylureas. Secondly, electrophysiological techniques have been employed to demonstrate that in the SNr sulphonylureas have a profound effect on changes in neuronal excitability in response to anoxia or alterations in glucose concentration (Murphy and Greenfield, 1991).

Following lesion of the striatum, the levels of K_{ATP}, measured by [^{3}H]-glibenclamide binding fall markedly in the globus pallidus, entopeduncular nucleus and SNr of the rat (decreases of 40%, 11% and 24% respectively, p<.05, paired t-test with Bonferroni correction, n = 6) (Brotchie et al., in press). No significant changes were seen in other brain regions. We interpret this finding as suggesting that, in the basal ganglia, K_{ATP}s are located, in part at least, on the terminals of GABAergic striatal efferents. This location suggests the possibility that K_{ATP} may have a role to play in modulating GABA transmission in the basal ganglia. Indeed, we and others have demonstrated that in the basal ganglia K_{ATP} modulates GABA release. It is assumed that mechanisms underlying this modulation are similar to those seen in the modulation of insulin release in the pancreas. Thus, K_{ATP} blockers increase GABA release in the SNr (Amoroso et al., 1990) whilst the K_{ATP} opener diazoxide reduces GABA release in the pallidal complex (Brotchie et al., in press). Given that GABAergic afferents to the GPi and SNr are underactive in parkinsonism the potential for an anti-parkinsonian therapy based on K_{ATP} blockade has been identified. Injections of tolbutamide into the entopeduncular nucleus of the reserpine-treated rat model of parkinsonism alleviate akinesia.

In the rodent, the regional distribution of K_{ATP} has been studied using [3H]-glibenclamide and [125I]-glyburide receptor autoradiography. K_{ATP} was found to be distributed heterogeneously throughout the CNS (Mourre et al., 1990; Treherne and Ashford, 1991; Gehlert et al., 1991). We have recently investigaated the regional distribution of K_{ATP} in the brain of the primate, *Macaca fascicularis*. In both the rodent and primate, the basal ganglia were found to contain by far the highest concentrations of K_{ATP} in the brain. Thus it seems likely that treatments based upon the idea of increasing GABA transmission by blockade of K_{ATP} might be developed to act selectively in the basal ganglia, without modulating GABA transmission in other areas of the CNS. Additionally, an important qualitative difference has emerged between the rodent and primate distribution. In the primate the levels of K_{ATP} in the GPi and SNr are much higher than those in the external segment of the globus pallidus (GPe). In the rat the homologous regions, entopeduncular nucleus, SNr and globus pallidus, have similar levels of sulphonylurea binding. This finding has important implications for potential therapeutic application of sulphonylureas to Parkinson's disease. The neural mechanisms mediating parkinsonian symptoms are thought to be characterised by abnormally low levels of GABA release in the GPi and SNr, but abnormally high levels of GABA release in the GPe. The selective concentration of K_{ATP} in the primate GPi and SNr, compared to GPe, raises the possibility not only of targeting the basal ganglia, but of targeting those areas of the basal ganglia where GABA transmission is depressed.

CANNABINOID RECEPTOR AGONISTS

Preparations of the plant *Cannabis sativa*, such as marijuana and hemp, were amongst the first psychoactive agents and have long history of usage as therapeutic agents (Mechoulam, 1986). There have been several reports of cannabis preparations affecting symptoms associated with basal ganglia-related disorders. The major active component of these plant preparations is now known to be Δ9-tetrahydrocannabinol (Δ9-THC), the prototypical cannabinoid (Razdan, 1986). At the molecular level, the mechanism of action of cannabinoids has been shown to involve interaction with a G-protein-coupled receptor (Howlett et al., 1986; Little and Martin, 1991).

The regional distribution of cannabinoid receptors within the brain has been studied using [3H]-CP 55,940 receptor autoradiography (Herkenham et al., 1990). Cannabinoid receptors are selectively concentrated in the basal ganglia, specifically in the pallidal complex and SNr. It has been shown that, in the rat at least, these cannabinoid receptors are located pre-synaptically on the GABAergic terminals of striatal efferents to the entopeduncular nucleus and SNr (Herkenham et al., 1991). As with K_{ATP}, in the primate and human, receptor levels are higher in GPi and SNr than in GPe. Functional interaction between the cannabinoid and GABA$_A$ receptor complex in the globus pallidus has been suggested by behavioural experiments where Δ9-THC was found to mimic the locomotor effects of intrapallidal injections of benzodiazepines. Additionally, strong synergism between the two compounds in producing these basal ganglia-mediated locomotor effects was reported (Pertwee and Wickens, 1991).

Given these findings, we predicted that, in the output regions of the basal ganglia, the cannabinoid receptor exerts pre-synaptic modulation of GABAergic transmission such that cannabinoid agonists have effects that enhance GABAergic function. Two possibilities were apparent that could account for such effects,
i) an increase in GABA release, and/or,
ii) a decrease in GABA uptake.

We thus investigated the effects of the cannabinoid receptor agonist WIN 55,212-2 on both GABA release and uptake in the pallidal complex. In slices of globus pallidus WIN 55,212-2 had no effect on either depolarisation-induced or basal release of pre-loaded [3H]-GABA. However, cannabinoid modulation of GABA uptake into pallidal slices was apparent. Pallidal slices (400μm) were dissected out, weighed and then incubated in aerated artificial cerebrospinal fluid. [3H]-GABA (0.5μM) was then added to the medium and the uptake of radioactivity measured after 30 minutes. Non-specific uptake was defined as that observed in sodium-free conditions. The cannabinoid agonist WIN 55,212-2 (6-100 μM) decreased the specific uptake of [3H]-GABA in a dose-dependent manner ($IC_{50} = 30μM$).

The maximal effects of WIN 55,212-2 were seen at a concentration of 100μM, at this concentration GABA uptake was reduced by 84%.

This reduction of GABA uptake by cannabinoid agonists suggested the possibility that cannabinoids might potentially be capable of increasing the levels of GABA transmission in the output regions of the basal ganglia. Such an action would reverse the overactivity characteristic of parkinsonism. We thus injected WIN 55,212-2 into the entopeduncular nucleus in the reserpinised rat model of parkinsonism. Injections were made unilaterally and locomotion monitored as described above. Following vehicle injection rats remained parkinsonian (3.5 ± 1.2 LU, n=6). Following injection of WIN 55,212-2 (30mM) a marked alleviation of akinesia was observed (102 + 8.5 LU). Well-co-ordinated movements of the limbs contralateral to the injection were observed.

Given that cannabinoid receptors are selectively concentrated on the terminals of striatal efferents we proposed that systemic administration of WIN 55,212-2 might act preferentially in the basal ganglia and reverse parkinsonian symptoms. In order to test this hypothesis we have quantified descent latency in the bar test as described by Klockgether et al. (1986). This test provides a very sensitive measure of locomotion in reserpine-treated animals. Descent latency was defined as the time, in seconds, for which the rat maintained its position after its forelimbs were placed on a block 9cm high. Following injection of vehicle, rats remained parkinsonian, descent latency was unchanged at any time point. However, following intramuscular injection of WIN 55,212-2 a marked dose and time-dependent decrease in descent latency was observed, as parkinsonian rigidity and akinesia were alleviated. At 90 minutes post-injection the descent latency for the vehicle treated animals was 590 s, this was significantly higher than that in animals treated with WIN 55,212-2 at either 1 mg/kg or 5mg/kg (258s and 170s respectively). These anti-parkinsonian effects of WIN 55,212-2 lasted for up to 3 hours.

CONCLUSIONS

In this chapter we have presented data to suggest that kappa opioid agonists, ATP-sensitive potassium channel blockers and/or cannabinoids might have useful roles to play in developing novel therapeutic approaches to Parkinson's disease. It is interesting to note that anecdotal evidence for at least two of these approaches has been in the public domain for many years. In the classic neurological text of the nineteenth century, Gowers described "a great improvement" in parkinsonian symptoms following treatment with hemp. No mechanism of action of this cannabis preparation was apparent at that time (Gowers, 1888). More recently, Gates and Hyman (1960) reported "a considerable reduction of tremor or rigidity or both" in eleven out of fifteen patients suffering from Parkinson's disease following treatment with tolbutamide. In 1960 tolbutamide was not known to act by blocking ATP-sensitive potassium channels. Indeed, the presence of such channels in the basal ganglia was not to be appreciated until thirty years later.

The data reported in this chapter provide the scientific basis to explain these clinical observations of anti-parkinsonian actions of cannabinoids and sulphonylureas. Additionally, we suggest that treatments incorporating these agents or kappa opioid agonists may provide a useful alternative to current dopamine replacement strategies. Alternatively by serving as an adjunct to dopamine therapy, they may reduce the amounts of L-DOPA required for treatment of symptoms. This approach would lead to a reduction in the incidence of iatrogenic side-effects accompanying treatment of patients with Parkinson's disease.

REFERENCES

Amoroso, S., Schmid-Antomarchi, H., Fosset, and M., Lazdunski, M., 1990, Glucose, sulphonulureas and neurotransmitter release: role of ATP-sensitive K^+ channel, *Science* 247:852-854.

Bergman, H., Wichmann, T., and Delong, M.R., 1990, Amelioration of parkinsonian symptoms by inactivation of the subthalamic nucleus in MPTP-treated monkeys, *Science* 249:1436-1438.

Brotchie, J.M., Crossman, A.R., Mitchell, I.J., Duty, S., Carroll, C., Cooper, A.C., Henry, B., Hughes, N.R., and Maneuf, Y., Chemical signalling in the globus pallidus, *in*: "Chemical Signalling in the Basal Ganglia," G. Arbuthnott, ed., Elsevier, Amsterdam, in press.

Brotchie, J.M., Mitchell, I.J., Sambrook, M.A., and Crossman, A.R., 1991, Alleviation of parkinsonism by antagonism of excitatory amino acid transmission in the medial segment of the globus pallidus in rat and primate, *Mov. Disord.* 6:133-138.

Chavkin, C., James, I.F., and Goldstein, A., 1982, Dynorphin is a specific endogenous ligand of the kappa opioid receptor, *Science* 215:413-415.

Cotzias, G.C., Van Woert, M.H., and Schiffer, L.M., 1967, Aromatic amino acids and modification of parkinsonism, *New Eng. J. Med.* 276:374-379.

Crossman, A.R., Peggs, D., Boyce, S., Luquin, M.R., and Sambrook, M.A., 1989, Effect of the NMDA antagonist MK-801 on MPTP-induced parkinsonism in the monkey, *Neuropharmacol.* 28:1271-1273.

Engber, T.M., Susel, Z., Kuo, S., Gerfen, C.R., and Chase, T.N., 1991, Levodopa replacement therapy alters enzyme activities in striatum and neuropeptide content in striatal output regions of 6-hydroxydopamine lesioned rats, *Brain Res.* 552:113-118.

Gates, E.W., and Hyman, I., 1960, Use of tolbutamide in paralysis agitans, *J.A.M.A.* 172:1351-1354.

Gehlert, D.R., Gackenheimer, S.L., Mais, D.E., and Robertson, D.W., 1991, Quantitative autoradiography of the binding sites for [^{125}I]-iodoglyburide, a novel high affinity ligand for ATP-sensitive potassium channels in rat brain, *J. Pharmacol. Exp. Ther.* 257:901-907.

Gowers, W.R., 1888, "A Manual of Diseases of The Nervous System," Blackiston, Son and Co., Philadelphia.

Herkenham, M., Lynn, A.B., Little, M.D., De Costa, B.R., and Richfield, E.K., 1991, Neuronal localisation of cannabinoid receptors in the basal ganglia of the rat, *Brain Res.* 547:267-274.

Herkenham, M., Lynn, A.B., Little, M.D., Ross-Johnson, M., Melvin, L.S., De Costa, B.R., and Rice, K.C., 1990, Cannabinoid receptor localisation in the brain, *Proc. Natl. Acad. Sci.* 87:1932-1936.

Herrera-Marschitz M., Hokfelt, T., Ungerstedt, U., Terenius, L., and Goldstein, M., 1984, Effect of intranigral injections of dynorphin fragments and a-neoendorphin on rotational behaviour in the rat, *Eur. J. Pharmacol.* 102:213-227.

Howlett, A.C., Qualy, J.M., and Khachatarian, L.L., 1986, Cannabinoid inhibition of Gi in the inhibition of adenylate cyclase by cannabimimetic drugs, *Mol. Pharmacol.* 29:307-313.

Jiang, H.K., McGinty, J.F., and Hong, J.S., 1990, Differential modulation of striatonigral dynorphin and enkephalin by dopamine receptor subtypes, *Brain Res.* 507:57-64.

Klockgether T., and Turski, L., 1990, NMDA antagonists potentiate anti-parkinsonian actions of L-dopa in monoamine-depleted rats, *Ann. Neurol.* 28:539-546.

Klockgether, T., Turski, L., Honore, T., Zhang, Z., Gask, G.M., Kurlan, R., and Greenamyre, J.T., 1991, The AMPA receptor antagonist NBQX has antiparkinsonian effects in monoamine-depleted rats and MPTP-treated monkeys, *Ann. Neurol.* 30:715-723.

Klockgether, T., Turski, L., Schwarz, M., and Sontag, K.-H., 1986, Motor actions of excitatory amino acids and their antagonists within the rat ventromedial thalamic nucleus, *Brain Res.* 399:1-9.

Lambert, P.D., Woodruff, G.N., Hughes, J., and Hunter, J.C., 1991, Inhibition of L-glutamate release: a possible mechanism of action for the neuroprotective effects of the kappa-selective agonist CI-977, *Mol. Neuropharmacol.* 1:77-82.

Lavin, A., and Garcia-Munoz, M., 1985, Electrophysiological changes in substantia nigra after dynorphin administration, *Brain Res.* 369:298-302.

Little, P.J., and Martin, B.R., 1991, The effects of Δ9-THC and other cannabinoids on cAMP accumulation in synaptosomes, *Life Sciences* 48:1133-1141.

Mackay, K.B., Kusuinoto, K., and McCulloch, J., 1991, The neuroprotective effect of the kappa agonist CI-977 in a rat model of focal cerebral ischemia, *Br. J. Pharmacol.* 104:303P.

Marsden, C.D., and Parkes, J.D., 1977, Success and problems of long term therapy in Parkinson's disease, *The Lancet* i:345-349.

Mechoulam, R., 1986, The pharmacology of Cannabis sativa, *in:* "Cannabinoids as Therapeutic Agents," R. Mechoulam, ed., CRC Press, Boca Raton, pp. 1-19.

Mitchell, I.J., Boyce, S., Sambrook, M.A., and Crossman, A.R., 1992, A 2-deoxyglucose study of the effects of dopamine agonists on the parkinsonian brain, *Brain* 115:809-824.

Mitchell, I.J., Clarke, C.E., Boyce, S., Robertson, R.G., Peggs, D., Sambrook, M.A., and Crossman, A.R., 1989, Neural mechanisms underlying parkinsonian symptoms based upon regional uptake of 2-deoxyglucose in monkeys exposed to 1-methyl-4-phenyl-1,2,3,6-tetrahydropyridine (MPTP), *Neuroscience* 32:213-226.

Mourre, C., Widmann, C., and Lazdunski, M., 1990, Sulphonylurea bonding sites associated with ATP-regulated K$^+$ channels in the central nervous system: an autoradiographic analysis of their distribution and ontogenesis, and their localisation in mutant mice cerebellum, *Brain Res.* 519:29-43.

Murphy, K.P.S.J., and Greenfield, S.A., 1991, ATP-sensitive potassium channels counteract anoxia in neurons of the substantia nigra, *Exp. Brain Res.* 84:355-358.

Nock, B., Pajpara, A., O'Connor, L.H., and Cicero, T.J., 1988, Autoradiography of [^{3}H]-U-69,593 binding sites in rat brain. Evidence for kappa opioid subtypes, *Eur. J. Pharmacol.* 154:27-37.

Penney, J.B., and Young, A.B., 1986, Striatal inhomogeneities and basal ganglia function, *Mov. Disord.* 1:3-15.

Pertwee, R.G., and Wickens, A.P., 1991, Enhancement by chlodiazepoxide of catalepsy induced in rats by intravenous or intrapallidal injections of enantiomeric cannabinoids, *Neuropharmacology* 30:237-244.

Razdan, R.K., 1986, Structure-activity relationships in cannabinoids, *Pharmacol. Rev.* 38:75-149.

Reid M.S., Herrera-Marschitz M., Hokfelt T., Lindefors N., Persson H., and Ungerstedt, U., 1990, Striatonigral GABA, dynorphin, substance P and neurokinin A modulation of nigrostriatal dopamine release: evidence for direct regulatory mechanisms, *Exp. Brain Res.* 82: 293-303.

Singh, L., Vass, C.A., and Hunter, J.C., 1990, The anti-convulsant action of CI-977, a selective kappa-opioid receptor agonist: a possible involvement of the glycine/NMDA receptor complex, *Eur. J. Pharmacol.* 191:477-480.

Treherne, J.M., and Ashford, M.L., 1991, The regional distribution of sulphonylurea binding sites in the rat brain, *Neuroscience* 40:523-531.

Unterwald, E.M., Knapp, C., and Zukin, R.S., 1991, Neuroanatomical localization of kappa1 and kappa2 opioid receptors in rat and guinea pig brain, *Brain Res.* 562:57-65.

ANTAGONISTS OF EXCITATORY AMINO ACID RECEPTORS IN THE TREATMENT OF PARKINSON'S DISEASE

Ubaldo Bonuccelli and Paolo Del Dotto

Institute of Clinical Neurology, University of Pisa
Pisa, Italy

INTRODUCTION

Recent progress in the connection anatomy of basal ganglia-thalamocortical circuits and in excitatory amino acid (EAA) research has provided new insights into possible pathophysiological mechanisms responsible for the appearance of the main symptoms in Parkinson's disease (PD).

It is now well known that there are two output pathways originating in separate cell populations of the striatum directed to the internal globus pallidus (GPI): a direct pathway and an indirect pathway via the external pallidal segment (GPE) and the subthalamic nucleus (STN). The direct striato-pallidal pathway is GABAergic and its activation tends to inhibit GPI neurons; activation of the indirect pathway exerts an opposite effect on GPI neurons; in fact, this circuit passes first to the GPE via striato-pallidal GABAergic projections, then from GPE to STN via another GABAergic pathway, and finally to GPI via an excitatory glutamatergic projection from STN.

The basal ganglia output nuclei, GPI and substantia nigra pars reticulata (SNR), send GABAergic fibers to the motor thalamus, thus exerting a tonic inhibitory effect on this target which in turn projects to the cortex.

According to this model, Albin et al. (1989) proposed that in PD the degeneration of dopaminergic neurons projecting from the substantia nigra pars compacta (SNC) to the striatum decreases the activity of striatal projection to GPI and SNR (direct pathway), and increases the activity of the striatal projection to GPE (indirect pathway). Consequently the final result is a significant increase in tonic neural discharge in STN and GPI, leading to excessive inhibition of thalamo-cortical neurons.

OVERACTIVITY OF STN IN PD: EXPERIMENTAL DATA

There is a growing body of evidence to support the view that dopamine loss in PD models leads to overactivity of subthalamic neurons. Miller and De Long (1987) have reported a tonic increase in the discharge rate of subthalamic neurons in monkeys treated with 1-methyl-4-phenyl-1,2,3,6-tetrahydropyridine (MPTP); moreover, studies on the regional brain uptake of 2-deoxyglucose in MPTP-exposed monkeys showed that the subthalamic and pallidal projection neurons are abnormally active (Mitchell et al., 1989).

Recently, based on the assumption that excessive subthalamo-pallidal drive might play a critical role in the pathophysiology of PD symptoms, surgical lesions of STN have been performed in experimental models of PD. Bergman et al. (1990) evaluated the effects of lesions of the subthalamic nucleus (obtained by injections of ibotenic acid using a combined injection-recording device) in two monkeys rendered parkinsonian by treatment with MPTP. Immediately after lesioning all of the major motor disturbances, such as akinesia, rigidity and tremor, were markedly ameliorated in the contralateral limb. Similar results have been obtained by Aziz et al. (1991) in four MPTP-treated primates using non traditional neurosurgical techniques such as stereotactic thermocoagulation of the STN: dramatic improvement of the whole spectrum of parkinsonian symptoms was seen in both sides of the body even after unilateral subthalamotomy. Finally, improvement in motor performance was obtained in the same model of PD using a non-destructive method, i.e. high frequency stimulation of the STN with a stimulating electrode implanted stereotactically (Benazzouz et al., 1993).

OVERACTIVITY OF STN IN PD: CLINICAL DATA

Clinical data also confirm the critical role of STN in the pathophysiology of parkinsonian symptoms. Recently, contralateral disappearance of parkinsonian signs has been described in a PD patient who developed a unilateral subthalamic hematoma (Sellal et al., 1992). In this patient, a unilateral hemiballism suddenly developed after the lesion; when the ballistic movements disappeared, the parkinsonian symptomatology (including akinesia) did not reappear. A similar clinical course had already been observed following surgical treatment of tremor in PD patients due to a STN lesion as a complication of stereotactic maneuvers (Brion et al., 1965; Buruma and Lakke, 1986). Very recently, Pollak et al. (1993) have obtained a slight improvement of bradykinesia in a PD patient following electrical stimulation of STN.

NEUROPHARMACOLOGY OF THE GLUTAMATERGIC SYSTEM IN PD

The validity of the model of an imbalance between the glutamatergic system projecting from the cortex to the basal ganglia output nuclei (via the STN) and the dopaminergic system (and the GABAergic striato-pallidal and striato-nigral projections to these nuclei) is underscored by recent data from rodents and primates. First, Carlsson and Carlsson (1989) showed that the non-competitive N-methyl-D-aspartate (NMDA) antagonist MK-801 administered systemically caused a pronounced locomotor stimulation in mice depleted of monoaminergic stores by means of pretreatment with reserpine and a-methyl-p-tyrosine. This has been subsequently confirmed in the same model using other i. p. administered NMDA antagonists such as PCP and D-CPPene (Carlsson and Svenson, 1990). Klockgether and Turski (1990) have found that when CPP is injected locally into STN, SNR or entopeduncular nucleus (EPN) (the rat homologous of the internal pallidal segment) of monoamine-depleted rats, a marked stimulation of locomotor activity is visible, while systemic administration of this drug or MK-801 is able to potentiate the effect of L-dopa. Moreover, in monoamine-depleted mice, i.p. administration of D-CPPene and CGP37849, in combination with dopamine agonists in doses which alone produce no or very weak effects, results in clearly stronger motor behavior than with each treatment alone (Kannari and Markstein, 1991).

Further evidence supporting the view of antiparkinsonian activity of NMDA antagonists comes from other non primate models of PD. For example, MK 801 is able to exert an anticataleptic effect in haloperidol-treated rats (Schmidt and Bubser, 1989; Metha and Ticku, 1990). Moreover, in rats with unilateral lesions of the nigro-striatal DA pathway with 6- hydroxydopamine (6-OHDA), MK-801 potentiates the acute contralateral turning effects of L-Dopa, SKF 38393, a selective agonist of D1 receptors (Morelli and Di Chiara, 1990a) and apomorphine, a mixed D1/D2 agonist (Morelli and Di Chiara, 1990b) (Table 1).

Studies using glutamatergic antagonists in primate models of PD are still controversial. Crossman et al. (1989) showed that MK-801 had no antiparkinsonian effect in a single MPTP-treated cynomolgus monkey when systemically administered, while microinjections of MK-801 and kynurenic acid, a broad spectrum EAA antagonist (Graham et al., 1990; Brotchie et al., 1991), as well as 7-chlorokynurenate, an antagonist at the NMDA receptor-associated glycine site (Brotchie et al., 1992), into the medial pallidal segment of MPTP-treated primates, reverse motor symptoms of parkinsonism. Recently, in rhesus monkeys with stable, bilateral parkinsonism induced by MPTP, the AMPA receptor

Table 1. Antiparkinsonian effects of EAA antagonists: experimental data in rodents

Authors	Drugs	Models	Results
Carlsson and Carlsson (1989)	MK-801	monoamine-depleted mice	+
Carlsson and Svensson (1990)	PCP, CPP		
Schmidt and Bubser (1989)	MK-801	haloperidol-induced catalepsy in rats	+
Metha and Ticku (1990)			potentiates DA agonists
Klockgether and Turski 1990)	MK-801 CPP (sistemically or microinjected into STN, EPN, SNR)	monoamine-depleted rats	potentiate L-dopa
Morelli and Di Chiara (1990a,b)	MK-801	6-OH-dopamine-unilaterally lesioned rats	potentiates turning induced by L-dopa, D1 agonists and apomorphine
Kannari and Markstein (1991)	CPP CGP 37849	monoamine-depleted mice	potentiate DA agonists
Klockgether et al. (1991)	NBQX (systemically or microinjected into STN, EPN, SNR)	monoamine-depleted rats	+ potentiates L-dopa

STN = subthalamic nucleus; EPN= entopeduncular nucleus; SNR= substantia nigra pars reticulata
+ = stimulation of locomotor activity

antagonist NBQX, injected intramuscularly, has been shown to produce a pronounced reduction of tremor, akinesia, posture and gross motor skills. When a small dose of L-Dopa was combined with a dose of NBQX that had a modest antiparkinsonian effects, a dramatic synergistic interaction was apparent (Klockgether et al, 1991). These observations have been confirmed by Loschmann et al. (1991) in MPTP-treated marmosets (Table 2).

Another rationale for the use of NMDA antagonists in PD comes from the discovery that some classic antiparkinsonian drugs have NMDA receptor antagonistic properties. Olney et al. (1987) have demonstrated that all anticholinergic agents used in PD therapy (especially procyclidine and ethopropazine) block NMDA neurotoxicity in vitro and inhibit specific binding of some phencyclidine ligands, with inhibition constants in the low micromolar range. These drugs, therefore, are NMDA antagonists by a non-competitive mechanism at the ion channel. Greenamyre and O'Brien (1991) noted that serum levels and central nervous system concentrations reached by anticholinergic drugs after commonly employed doses appear to be excessive relative to the number and affinity of muscarinic receptors in the brain. On the other hand, at such brain concentrations, anticholinergic drugs may interact significantly with NMDA receptors. Other currently used antiparkinsonian medications that possess NMDA antagonistic properties are memantine and amantadine. Memantine at brain concentrations achieved during PD treatment is able to bind to the MK-801 binding site of the NMDA receptor in the postmortem human frontal cortex (Kornhuber et al., 1991) and blocks NMDA receptor channels in cultured neurons in a manner similar to that of MK-801 (Borman, 1989).

Similar binding results have been obtained with amantadine. Although the mechanism of action of this drug is thought to be an enhancement of dopamine release, recently Jackish et al. (1992) have found that only high concentrations of amantadine increase the spontaneous release of dopamine in cultured rabbit brain slices.

Moreover, memantine and amantadine have an inhibitory effect on NMDA-evoked acetylcholine release in the rabbit caudate nucleus in vitro (Lupp et al., 1992). From these observations, it seems rather unlikely that dopaminomimetic properties alone are responsible for the antiparkinsonian effects of memantine and amantadine.

Finally, budipine, an antiparkinsonian agent under clinical studies whose mechanism of action is unknown, has recently shown non-competitive NMDA antagonist properties in receptor binding experiments (Klockgether et al., 1993).

Table 2. Antiparkinsonian effects of EAA antagonists: experimental data in primates

Authors	Drugs	Models	Results
Graham et al. (1990)	MK-801 kynurenate (microinjected into medial pallidum)	MPTP-treated monkeys	+
Klockgether et al. (1991)	NBQX (sistemically)	MPTP-treated monkeys	+ potentiates L-dopa
Brotchie et al. (1991, 1992)	kynurenate 7-chlorokynurenate (microinjected into medial pallidum)	MPTP-treated marmosets	+
Loschmann et al. (1991)	NBQX CPP (sistemically)	MPTP-treated marmosets	+ potentiate L-dopa

+ = improvement of parkinsonian symptoms

DEXTROMETHORPHAN: A CLINICALLY TESTABLE EAA IN PD

Dextromethorphan (D-3-methoxy-17-methylmorphinan, DXM) is a widely used non-opioid drug with low toxicity and low potential for producing drug dependence. There are many experimental studies in vivo and in vitro which demonstrate the noncompetitive NMDA antagonistic properties of DXM and its main metabolite, dextrorphan (D-3-hydroxy-N-methylmorphinan, DOR) (Church et al.,1989). In fact, DXM and DOR antagonize the excitatory effects of NMDA on rat spinal cord neurons in vivo (Church et al., 1985), and in neocortical neurons in vitro (Choi et al., 1987; Wong et al.,1988), inhibit NMDA-induced convulsions in mice (Leander et al., 1988; Ferkany et al., 1988), and exert anticonvulsant activity in guinea pig (Wong et al., 1988) and rat (Aram et al., 1989) neocortical slices in vitro.

In addition, both drugs have been found to attenuate glutamate-induced neurotoxicity in murine neocortical cell cultures (Choi et al., 1987), and to reduce ischemic brain damage in rabbits and rats in vivo by virtue of their NMDA-antagonistic effect (Steinberg et al., 1988; Prince and Feeser, 1988).

Following oral administration, DXM is rapidly metabolized; two major metabolic pathways, namely O-demethylation to DOR and N-demethylation to methoxymorphinan, are known. These metabolites may be further demethylated to hydroxymorphinan and then excreted with urine in the form of glucuronide and sulfate ester conjugates.

The peak plasma level of DOR (the blood levels of unmetabolized DXM is very low) is reached at 2 hours after oral administration (Ramachander et al., 1977). This metabolite is able to bind to the NMDA receptor-channel with an affinity higher than that of DXM (Jaffe et al., 1989; Franklin and Murray, 1991); thus, although at present there are no published data on the antitussive activity of DOR, the clinical effects of DXM are probably due to its conversion to this metabolite.

Large intersubject differences have been reported in O-demethylation to DOR (Pfaff et al., 1983; Woodworth et al., 1987). On the basis of plasma DXM and urinary DOR excretion, fast, intermediate, and poor DXM metabolizers have been identified. Poor metabolizers (about 10% of human subjects) also exhibit an inherited deficiency of debrisoquine hydroxylation.

Recently, DXM has been used as a neuroprotective agent in many disease conditions in which experimental evidence suggest a physiopathological role of glutamate-induced neurotoxicity, such as Huntington's chorea (Walker and Hunt, 1989), amyotrophic lateral sclerosis (Appelbaum et al., 1991; Askmark et al., 1993), and brain ischemia (Albers et al., 1991). Fisher et al. (1990) have also employed a sustained-release formulation of DXM in patients with complex partial seizures. None of these studies has shown any significant symptomatic effect of DXM; the side effects were usually not severe despite the high doses of DXM administered (twice or more the usual antitussive dose).

To determine whether reducing glutamatergic transmission affects extrapyramidal function, an open trial of DXM was conducted in patients with PD, and results in abbreviated form have been already published (Bonuccelli et al., 1992).

Twelve PD patients (8 males and 4 females) with a mean age ($\pm$ SD) of 57.3 $\pm$ 6.5 years and mean disease duration of 2.5 ($\pm$ 1.4) years were included in the study. Disease severity, evaluated via the Hoehn and Yahr staging, was 1.8 $\pm$ 0.5. Six patients had never been treated (de novo); the remaining 6 took L-Dopa/carbidopa or L-Dopa/carbidopa plus bromocriptine with a mean therapy duration of 3.2 $\pm$ 1.2 years. None of these patients showed dementia, depression or fluctuations in motor symptoms DXM was administered orally following an open-label dose escalation at a starting dose of 45 mg per day (in three doses) for a week.

Subsequently, the dose was increased to 90, 120, 180 mg per day (in four doses) for a period of one week at each dose. Motor function was assessed at baseline, at the end of each week of treatment and one week after drug withdrawal using the motor examination subitem of the Unified Parkinson's Disease Rating Scale (UPDRS) and the tapping test.

Statistical analysis was performed using ANOVA with repeated measures and multiple comparisons followed by Scheffé F-test and Friedman test with multiple comparisons, as appropriate.

Eleven patients completed the study; one de novo patient withdrew at the end of the 90 mg dose period because of lightheadedness, drowsiness and mild ataxia. The lowest DXM doses of 45 and 90 mg per day did not modify parkinsonian symptoms, while dose-dependent, significant improvement in UPDRS (motor examination) and tapping test scores were observed for the doses of 120 and 180 mg per day (Table 3).

Table 3. Clinical effects of DXM as evaluated by means of the UPDRS (motor examination) and tapping test

	Baseline	45 mg	90 mg	120 mg	180 mg	Wash-out
	(n=12)	(n=12)	(n=12)	(n=11)§	(n=11)§	(n=11)§
UPDRS (motor examination)	19,7+7,4	19,3+6,9	19,0+7,5	17,1+7,4*	16,0+6,9*	19,1+7,6
TAPPING (Tap number/15 sec)	36,2+6,8	36,0+6,9	37,5+7,2	40,5+6,7°	42,6+6,7°	36,7+7,0

§ At the end of the 90 mg dose period, one de novo patient withdrew due to side effects.
* Significant compared to baseline and wash-out scores (Friedman test with multiple comparisons).
° Significant compared to baseline and wash-out scores (ANOVA with repeated measures and multiple comparisons, Scheffé F-test).
Differences have been considered statistically significant when p values were < 0,05.

One week after drug withdrawal, motor performance returned to baseline levels. Examining each subitem of UPDRS , the greatest improvement was obtained in rigidity, tremor and in those subscores measuring manual dexterity.

The response to DXM was more impressive for the de novo patients than for the previously treated ones. In fact, significant improvement in UPDRS and tapping test scores was achieved at the dose of 120 mg per day in the de novo, and at the dose of 180 mg per day in treated patients (Figure 1).

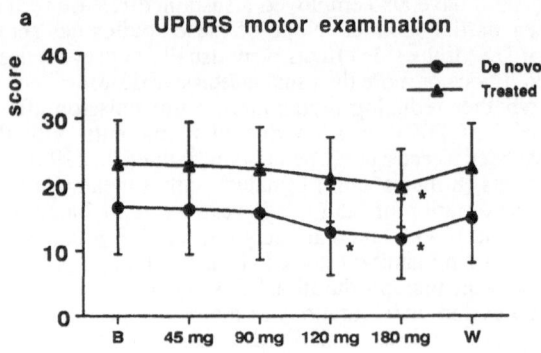

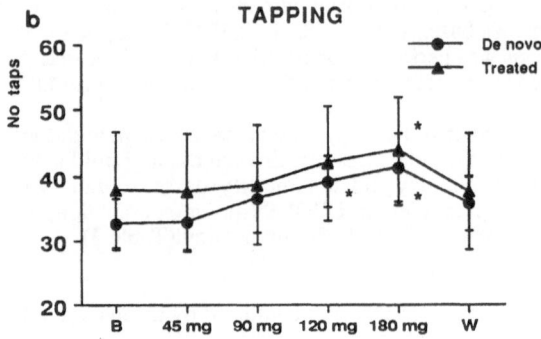

Figure 1. UPDRS (motor examination) (a) and tapping test (b)scores in de novo and treated PD patients. * p<0.05 compared to baseline (B) and wash-out (W) scores.

DISCUSSION

Drugs directed at antagonizing the effect of glutamate transmission could form the basis for novel strategies for symptomatic therapy of PD. In our study a more marked antiparkinsonian effect of DXM was observed in the de novo PD patients; a negative pharmacokinetic interference between the DXM and L-dopa in treated patients cannot be excluded. On the other hand, it is possible that some patients under L-dopa therapy were "poor metabolizers" of DXM, thus producing lower plasma levels of DOR which seems to possess more potent NMDA antagonistic properties than DXM. The results of this study have been recently confirmed by Saenz et al. (1993).

Until now few and contrasting studies concerning NMDA receptor antagonists in the treatment of PD have been available. Riederer et al. (1991) found no improvement in three out of four PD patients in advanced stages of disease after the addition of memantine to L-dopa therapy. On the other hand, two of them developed pharmacotoxic psychosis. In a previous study, the efficacy of memantine had been reported in PD patients in early stages of disease (Schneider et al., 1984); recently, Rabey et al. (1992) have added memantine to 14 parkinsonian patients with motor fluctuations taking L-dopa, with consequent improvement of parkinsonian features and "off" episodes. In the above-mentioned study of Riederer et al. (1991) the more advanced disease of the patients (IV-V Hoehn-Yahr staging) and the drugs they were taking (amantadine or anticholinergics) before and during the trial could have hampered an effect of memantine at the NMDA receptor level.

Montastruc et al. (1992) have employed ifenprodil, a non-competitive antagonist at the polyamine modulatory site of the NMDA receptor, in two groups of PD patients with and without motor fluctuations. This drug, which also possesses alpha-adrenoceptor blocking properties, is in clinical use as a cerebral and peripheral anti-ischemic agent. Ifenprodil did not modify the parkinsonian symptoms in any group. It is possible that the negative results could be due to the low dose, or the weak NMDA antagonistic properties of this drug which interacts with a modulatory site of the NMDA receptor complex as mentioned.

More recently, lamotrigine, an anti-epileptic drug, has been successfully administered to five PD patients (Zipp et al., 1993); this drug is not an NMDA receptor antagonist but is able to inhibit the release of glutamate (Leache et al., 1991).

All these studies have been conducted in open label fashion, but an indirect argument in favour of the potential efficacy of drugs that antagonize glutamatergic transmission in PD treatment, comes from the study of Giuffra et al. (1993) in which milacemide has been administered to six patients with advanced PD, under double-blind, placebo-controlled conditions. Acute administration of milacemide, a glycine prodrug that positively modulates NMDA transmission, (having the patients withdrawn from other antiparkinsonian medications twelve hours before the study), increased overall parkinsonian severity transiently, mostly due to an increase in rigidity.

The most serious criticism of using NMDA antagonists in the treatment of PD, is that massive blockade of NMDA receptors other than those localized at the level of subthalamo-pallidal pathway (e.g. cortico-striatal pathway), might cause side effects, such as learning impairment, ataxia and confusion. Antagonism of these glutamatergic synapses may also account for the failure of MK-801 to have an antiparkinsonian effect when systemically administered to a single monkey rendered parkinsonian by MPTP (Crossman et al., 1989). It is however possible that systemic admministration of EAA receptor antagonists could determine a selective pharmacological blockade of the glutamatergic subthalamic output (which is the only glutamatergic pathway that is abnormally increased in PD) due to the use-dependence phenomenon (Honey et al., 1985), i.e. an increase in antagonistic activity in the presence of the agonist.

Besides the symptomatic effect, another possible role for antagonists of EAA in PD treatment is their clinical use as neuroprotective agents. In fact, Sonsalla et al. (1989, 1991) have shown that non-competitive antagonists at NMDA receptors (such as MK-801, ketamine, PCP, ifenprodil and SL 82.0715) as well as competitive antagonists (CGS 19755 and NCP 12626), provide high protection against methamphetamine-induced toxicity in mice, a model of PD. Other experiments carried out in rats, a species insensitive to systemic administration of MPTP, showed that MK-801 and some non-competitive NMDA antagonists are able to prevent the dopamine neuronal degeneration induced by intranigral or systemic administration of MPP+ (Turski et al., 1991). Similar results have been obtained in mice treated with MPTP and acetaldehyde, an experimental model of PD characterized by specific degeneration of dopaminergic cells of the SNC (Zuddas et al., 1989); in this model MK-801, parenterally administered 30 minutes before MPTP and acetaldehyde, completely protects against cell-body damage (Corsini et al., 1992). These data were not confirmed by Sonsalla et al. (1989, 1992), Michel and Agid (1992), and Finiels-Martin et al. (1993), but were in agreement with those of Storey et al. (1992). In the latter study, the MPP+ toxicity in rat striatum was not only completely blocked by MK-801, but also markedly attenuated by prior decortication, which removes the cortico-striatal glutamatergic input.

More recently, Zuddas et al. (1992) have observed that systemic administration of MK-801 prevented the development of the parkinsonian syndrome induced by MPTP in three cynomolgus monkeys; morphological and biochemical data showed that MK-801 also prevented the extensive degeneration of nigral dopaminergic perikarya induced by MPTP and attenuated the dopaminergic depletion at the levels of striatal terminals. Moreover, MK-801 shows neuroprotective properties in other models of extrapyramidal disorders; in fact, it prevents degeneration of striatal neurons caused by intrastriatal injection of quinolinic acid in rats, which produces an animal model of Huntington's disease (Foster et al., 1988). A neuroprotective effect of DXM and DOR was finally observed in rats following administration of some neurotoxins which act at the striatum, hippocampus and cortex, such as 3,4-methylendioxymethamphetamine (Finnegan et al., 1990) and p-cloroamphetamine (Finnegan et al., 1991; Henderson and Fuller, 1992). These substances exert a direct toxic effect on serotoninergic neurons which is antagonized by DXM and DOR, probably via their effects at the NMDA receptor-channel complex. Recently we observed a significant protection against MPTP plus diethyldithiocarbamate-induced nigro-striatal lesions in mice pretreated with DXM (Bonuccelli et al., in preparation).

The exact role of EAA antagonists in preventing or reducing MPTP, MPP^+ or methamphetamine neurotoxicity in animal models remains to be determined. According to Albin and Greenamyre (1992), it is possible that an impairment of energy metabolism leads to a membrane depolarization with persistent activation of NMDA receptors and excitotoxic neuronal degeneration (at the normal resting membrane potential, this receptor ion channel is blocked by normal concentrations of extracellular magnesium). In the MPTP model of PD, an impairment of oxidative metabolism has been demonstrated, due to MPP^+ which is generated by metabolism of MPTP. MPP^+ accumulates inside mitochondria and inhibits complex I of the electron transport chain. Some authors have reported a diminished activity of complex I in substantia nigra and peripheral tissues of PD patients (Parker et al., 1989; Schapira et al., 1990).

Disruption of ATP synthesis may lead to partial neuronal depolarization with activation of NMDA receptors; thus, excitatory amino acids might be the final mediator of nigral neuronal death in PD.

CONCLUSIONS

Clinical and experimental data suggest that NMDA receptor antagonists might have beneficial symptomatic effects on PD, alone or in combination with L-Dopa or dopamine agonists. If oxidation products of dopamine and L-dopa exert toxic effects on nigral cells (Olanow, 1990; Mena et al., 1992), NMDA antagonists may allow reduction of L-dopa doses. Furthermore, it has been recently reported that 2,4,5-trihydroxyphenylalanine (6-OH-Dopa), a 6-hydroxylate derivate of L-dopa, is a powerful excitotoxin (Olney et al., 1990) displaying toxic activity in rat cortical neurons (Rosenberg et al., 1991) and in rat mesencephalic neurons (Skaper et al., 1992); this toxicity might be blocked by the non-NMDA antagonist CQNX (Rosenberg et al., 1991). The neuroprotective action of NMDA antagonists in PD models may provide a further rationale for the use of such drugs in PD, if enviromental toxins, current drug therapy, or bioenergetic defects are shown to play a role in causing and/or aggravating PD.

REFERENCES

Albers, G.W., Saenz, R.E., Moses, J.A., and Choi, D.W., 1991, Safety and tolerance of oral dextromethorphan in patients at risk for brain ischemia, *Stroke* 22:1075-1077.
Albin, R.L., and Greenamyre, J.T., 1992, Alternative excitotoxic hypotheses, *Neurology* 42:733-738.
Albin, R.L., and Young, A.B., Penney, J.B., 1989, The functional anatomy of basal ganglia disorders, *Trends Neurosci.* 12:366-375.
Appelbaum, J.S., Salazar-Grueso, E.F., Richman, J.G., Shanahan, M., and Roos, R.P., 1991, Dextromethorphan in the treatment of ALS: a pilot study, *Neurology* 41 (Suppl. 1):393.

Aram, J.A., Martin, D., Tomczyk, M., Zeman, S., Millar, J., Pohler, G., and Lodge, D., 1989, Neocortical epileptogenesis in vitro: studies with N-methyl-D-aspartate, phencyclidine, sigma and dextromethorphan receptors ligands, *J. Pharmacol. Exp. Ther.* 248:320-328.

Askmark, H., Aquilonius, S.M., Gillberg, P.G. Liedholm, L.J., Stalberg, E., and Wuopio, R., 1993, A pilot trial of dextromethorphan in amyotrophic lateral sclerosis, *J. Neurol. Neurosurg. Psychiatry* 56:197-200.

Aziz, T.Z., Peggs, D., Sambrook, M.A., and Crossman, A.R., 1991, Lesion of the subthalamic nucleus for the alleviation of 1-methyl-4-phenyl-1,2,3,6-tetrahydropyridine (MPTP)-induced parkinsonism in the primate, *Mov. Disorders* 6:288-292.

Benazzouz, A., Gross, C., Féger, J., Boraud, T., and Bioulac, B., 1993, Reversal of rigidity and improvement in motor performance by subthalamic high frequency stimulation in MPTP-treated monkeys, *Eur. J. Neurosci.* .5:382-389.

Bergman, H., Wichmann, T., and DeLong, M.R., 1990, Reversal of experimental parkinsonism by lesions of the subthalamic nucleus, *Science* 249:1436-1438.

Bonuccelli, U., Del Dotto, P., Piccini, P., Beghé, F., Corsini, G.U., and Muratorio, A., 1992, Dextromethorphan and parkinsonism, *Lancet* 340:53.

Bormann, J., 1989, Memantine is a potent blocker of N-methyl-D-aspartate (NMDA) receptor channels, *Eur. J. Pharmacol.* 166:591-592.

Brion, S., Guiot, G., Derome, P., and Comoy, C., 1965, Hémiballismes postopératoires au cours de la chirurgie stéréotaxique: à propos de 12 observations dont 2 anatomocliniques dans une série de 850 interventions, *Rev. Neurol.* 112:410-443.

Brotchie, J.M., Carrol, C., Cooper, A.J., Crossman, A.R., and Mitchell, I.J., 1992, Role of NMDA associated glycine sites in the mediation of parkinsonism, *Fourth Intern. Meet.Basal Ganglia Soc. Abstr.*:16.

Brotchie, J.M., Mitchell, I.J., Sambrook, M.A., and Crossman, A.R., 1991, Alleviation of parkinsonism by antagonism of excitatory amino acid transmission in the medial segment of the globus pallidus in rat and primate, *Mov. Disorders*, 6:133-138.

Buruma, O.J.S., and Lakke, J.P.W.F., 1986, Ballism, *in*: "Handbook of Clinical Neurology, Vol. 5 (49)," P.J. Vinken, G.W. Bruyn, and H. L. Klawans, eds., Elsevier Science, Amsterdam, pp. 369-380.

Carlsson, M., and Carlsson, A., 1989, The NMDA antagonist MK-801 causes marked locomotor stimulation in monoamine-depleted mice, *J. Neural. Transm.* 75:221-226.

Carlsson, M., and Svensson, A., 1990, The non-competitive NMDA antagonists MK-801 and PCP, as well as the competitive NMDA antagonist SDZ EAA494 (D-CPPene), interact synergistically with clonidine to promote locomotion in monoamine-depleted mice, *Life Sci.* 47:1729-1736.

Choi, D.W., Peters, S., and Viseskul, V., 1987, Dextrorphan and levorphanol selectively block N-methyl-D-aspartate receptor-mediated neurotoxicity on cortical neurons, *J. Pharmacol. Exp. Ther.* 242:713-720.

Church, J., Jones, M.G., Davies, S.N., and Lodge, D., 1989, Antitussive agents as N-methylaspartate antagonists: further studies, *Can. J. Physiol. Pharmacol.* 67:561-567.

Church, J., Lodge, D., and Berry, S.C., 1985, Differential effects of dextrorphan and levorphanol on the excitation of rat spinal neurons by amino acids, *Eur. J. Pharmacol.* 111:185-190.

Corsini, G.U., Vaglini, F., Zuddas, A., Fornai, F., Saginario, A., and Scalori, V., 1992, Excitatory amino acids, MPTP toxicity, and dopaminergic nerve cell death in Parkinson's disease, *in*: "Current Trends in the Treatment of Parkinson's Disease," Y. Agid, ed., John Libbey, London, pp. 5-17.

Crossman, A.R., Peggs, D., Boyce, S., Luquin, M.R., and Sambrook, M.A., 1989, Effects of the NMDA antagonist MK-801 on MPTP-induced parkinsonism in the monkey, *Neuropharmacol.* 28:1271-1273.

Ferkany, J.W., Borosky, S.A., Clissold, D.B., and Pontecorvo, M.J., 1988, Dextromethorphan inhibits NMDA-induced convulsions, *Eur. J. Pharmacol.* 151:151-154.

Finiels-Marlier, F., Marini, A.M., Williams, P., and Paul, S.M., 1993, The N-methyl-D-aspartate antagonist MK-801 fails to protect dopaminergic neurons from 1-methyl-4-phenylpyridinium toxicity in vitro, *J. Neurochem.* 60:1968-1971.

Finnegan, K.T., Kerr, J.T., and Langston, J.W., 1991, Dextromethorphan protects against the neurotoxic effects of p-chloroamphetamine in rats, *Brain Res.* 558:109-111.

Finnegan, K.T., Skratt, J.J., Irwin, I., and Langston, J.W., 1990, The N-methyl-D-aspartate (NMDA) receptor antagonist, dextrorphan, prevents the neurotoxic effects of 3,4-methylenedioxymeth-amphetamine (MDMA) in rats, *Neurosci. Lett.* 105:300-306.

Fisher, R.S., Cysyk, B.J., Lesser, R.P., Pontecorvo, M.J., Ferkany, J.T., Schwerdt, P.R., Hart, J., and Gordon, B., 1990, Dextromethorphan for treatment of complex partial seizures, *Neurology* 40:547-549.

Foster, A.C., Gill, R., and Woodruff, G.N., 1988, Neuroprotective effects of MK-801 in vivo: selectivity and evidence for delayed degeneration mediated by NMDA receptor activation, *J. Neurosci.* 8:4745-4754.

Franklin, P.H., and Murray, T.F., 1991, High affinity 3H-dextrorphan binding in rat brain is localized to a noncompetitive antagonist site of the activated N-methyl-D-aspartate receptor-cation channel, *Mol. Pharmacol.* 41:134-146.

Giuffra, M.E., Sethy, V.H., Davis, T.L., Mouradian, M.M., and Chase, T.N., 1993, Milacemide therapy for Parkinson's disease, *Mov. Disorders* 8:47-50.

Graham, W.C., Robertson, R.G., Sambrook, M.A., and Crossman, A.R., 1990, Injection of excitatory amino acid antagonists into the medial pallidal segment of a 1-methyl-4-phenyl-1,2,3,6-tetrahydropyridine (MPTP)-treated primate reverses motor symptoms of parkinsonism, *Life Sci.* 47:PL91-PL97.

Greenamyre, T.J., and O'Brien, C.F., 1991, N-methyl-D-aspartate antagonists in the treatment of Parkinson's disease, *Arch. Neurol.* 48:977-981.

Henderson, M.G., and Fuller, R.W., 1992, Dextromethorphan antagonizes the acute depletion of brain serotonin by p-chloroamphetamine and H75/12 in rats, *Brain Res.* 594:323-326.

Honey, C.R., Miljkovic, Z., and Mac Donald, J.F., 1985, Ketamine and phencyclidine cause a voltage dependent block of responses to L-aspartic acid, *Neurosci. Lett.* 61:135-139.

Jackisch, R., Link, T., Neufang, B., and Koch, R., 1992, Studies on the mechanism of action of the antiparkinsonian drugs memantine and amantadine: no evidence for direct dopaminomimetic or antimuscarinic properties, *Arch. Int. Pharmacodyn. Ther.* 231:456-464.

Jaffe, D.B., Marks, S.S., and Greenberg, D.A., 1989, Antagonist drug selectivity binding sites on voltage-gated and N-methyl-D-aspartate receptor-gated Ca^{2+} channels, *Neurosci. Lett.* 105:227-231.

Kannari, K., and Markstein, R., 1991, Dopamine agonists potentiate antiakinetic effects of competitive NMDA-antagonists in monoamine-depleted mice, *J. Neural. Transm. (Gen. Sect.)* 84:211-220.

Klockgether, T., Jacobsen, P., Loschmann, P.A., and Turski, L., 1993, The antiparkinsonian agent budipine is an N-methyl-D-aspartate antagonist, *J. Neural. Transm. (P-D Sect.)* 5:101-106.

Klockgether, T., Turski, L., 1990, NMDA antagonists potentiate antiparkinsonian actions of L-Dopa in monoamine-depleted rats, *Ann. Neurol.* 28:539-546.

Klockgether, T., Turski, L., Honoré, T., Zhang, Z., Gash, D.M., Kurlan, R., and Greenamyre, J.T., 1991, The AMPA receptor antagonist NBQX has antiparkinsonian effects in monoamine-depleted rats and MPTP-treated monkeys, *Ann. Neurol.* 30:717-723.

Kornhuber, J., Bormann, J., Hubers, M., Rusche, K., and Riederer, P., 1991, Effects of the 1-amine-adamantanes at the MK-801 binding site of the NMDA receptor gated ion channel: a human postmortem study, *Eur. J. Pharmacol. (Mol. Pharmacol. Sect.)* 206:297-300.

Leache, M.J., Baxter, M.G., and Critchley, M.A.E., 1991, Neurochemical and behavioral aspects of lamotrigine, *Epilepsia* 32 (Suppl. 2):S4-S8.

Leander, J.D., Rathbun, R.C., and Zimmerman, D.M., 1988, Anticonvulsant effects of phencyclidine-like drugs: relation to N-methyl-D-aspartic acid antagonism, *Brain Res.* 454:368-372.

Loschmann, P.A., Lange, K.W., Kunow, M., Rettig, K.J., Jahnig, P., Honoré, T., Turski, L., Wachtel, H., Jenner, P., and Marsden, C.D., 1991, Synergism of the AMPA-antagonist NBQX and the NMDA-antagonist CPP with L-Dopa in models of Parkinson's disease, *J. Neural. Transm. (P-D Sect.)* 3:203-213.

Lupp, A., Lucking, C.H., Koch, R., Jackisch, R., and Feuerstein, T.J., 1992, Inhibitory effects of the antiparkinsonian drug memantine and amantadine on N-methyl-D-aspartate-evoked acetylcholine release in the rabbit caudate nucleus in vitro, *J. Pharmacol. Exp. Ther.* 263:717-724.

Mena, M.A., Pardo, B., Casarejos, M.J., Fahn, S., and de Yébenes, J.G., 1992, Neurotoxicity of levodopa on catecholamine-rich neurons, *Mov. Disorders* 7:23-31.

Metha, A.K., and Ticku, M.K., 1990, Role of the N-methyl-D-apartate (NMDA) receptors in experimental catalepsy in rats, *Life Sci.* 46:37-42.

Michel, P.P., and Agid, Y., 1992, The glutamate antagonist, MK-801, does not prevents dopaminergic cell death induced by the 1-methyl-4-phenylpyridinium ion (MPP+) in rat dissociated mesencephalic cultures, *Brain Res.* 597:233-240.

Miller, W.C., and DeLong, M.R., 1987, Altered tonic activity of neurons in the globus pallidus and subthalamic nucleus in the primate MPTP model of parkinsonism, *in*: "The Basal Ganglia II. Structure and Function-Current Concepts," M.B. Carpenter and A. Jayaraman, eds., Plenum Press, New York, pp. 395-403.

Mitchell, I.J., Clarke, C.E., Boyce, S., Robertson, R.G., Peggs, D., Sambrook, M.A., and Crossman, A.R., 1989, Neural mechanisms underlying parkinsonian symptoms based upon regional uptake of 2-deoxyglucose in monkeys exposed to 1-methyl-4-phenyl-1,2,3,6-tetrahydropyridine, *Neuroscience* 32:213-226.

Montastruc, J.L., Rascol, O., Senard, J.M., and Rascol, A., 1992, A pilot study of N-methyl-D-aspartate (NMDA) antagonist in Parkinson's disease, *J. Neurol. Neurosurg. Psychiatry* 55:630-631.

Morelli, M., and Di Chiara, G., 1990a, MK-801 potentiates dopaminergic D1 but reduces D2 responses in the 6-hydroxydopamine model of Parkinson's disease, *Eur. J. Pharmacol.* 182:611-612.

Morelli, M., and Di Chiara, G., 1990b, Stereospecific blockade of N-methyl-D-aspartate transmission by MK-801 prevents priming of SKF 38393-induced turning, *Neuropharmacol.* 101:287-288.

Olanov, C.W., 1990, Oxidation reactions in Parkinson's disease, *Neurology* 40:32-37.

Olney, J.W., Price, M.T., Labruyere, J., Salles, K.S., Frierdich, G., Mueller, M., and Silverman, E., 1987, Anti-parkinsonian agents are phencyclidine agonists and N-methyl-aspartate antagonists, *Eur. J. Pharmacol.* 142:319-320.

Olney, J.W., Zorumski, C.F., Stewart, G.R., Price, M.T., Wang G., and Labruyere, J., 1990, Excitotoxicity of L-DOPA and 6-OH-DOPA: implications for Parkinson's and Huntington's diseases, *Exp. Neurol.*, 108:269-272.

Parker, W.D., Boyson, S.J., and Parks, J.K., 1989, Abnormalities of the electron transport chain in idiopathic Parkinson's disease, *Ann. Neurol.* 36:719-723.

Pfaff, G., Briegel, P., and Lamprecht, I., 1983, Inter-individual variation in the metabolism of dextromethorphan, *Int. J. Pharm.* 14:173-189.

Pollak, P., Benabid, A.L., Gross, C., Gao, D.M., Laurent, A., Benazzouz, A., Hoffmann, D., Gentil, M., and Perret, J., 1993, Effets de la stimulation du noyau sous-thalamique dans la maladie de Parkinson, *Rev. Neurol.* 149:175-176.

Prince, D.A., and Feeser, H.R., 1988, Dextromethorphan protects against cerebral infarction in rat model of hypoxia-ischemia, *Neurosci. Lett.* 85:291-296.

Rabey, J.M., Nissipeanu, P., and Korczyn, A.D., 1992, Efficacy of memantine, an NMDA receptor antagonist, in the treatment of Parkinson's disease, *J. Neural. Transm. (P-D Sect.)* 4:277-282.

Ramachander, G., Williams, F.D., and Emele, J.F., 1977, Determination of dextrorphan in plasma and evaluation of bioavailability of dextromethorphan hydrobromide in humans, *J. Pharmac. Sci.* 66:1047-1048.

Riederer, P., Lange, K.W., Kornhuber, J., and Danielczyk, W., 1991, Pharmacotoxic psychosis after memantine in Parkinson's disease, *Lancet* 338:1022-1023.

Rosenberg, P.A., Loring, R., Xie, Y., Zaleskas, V., and Aizenman, E., 1991, 2,4,5-Trihydroxyphenylalanine in solution forms a non-N-methyl-D-aspartate glutamatergic agonist and neurotoxin, *Proc. Nat. Acad. Sci. USA*, 88:4865-4869.

Saenz, R., Tanner, C.M., Albers, G., Kurth, M., and Tetrud, J., 1993, A preliminary study of dextromethorphan (DM) as adjunctive therapy in Parkinson's disease (PD), *Neurology* 43 (Suppl. 2):155.

Schapira, A.H.V., Cooper, J.M., Dexter, D., Jenner P., Clark, J.B., and Marsden, C.D., 1990, Mitochondrial complex I deficiency in Parkinson's disease, *J. Neurochem.* 54:820-827.

Schmidt, W.J., and Bubser, M., 1989, Anticataleptic effects of the N-methyl-D-aspartate antagonist MK-801 in rats, *Pharmacol. Biochem. Behav.* 32:621-623.

Schneider, E., Fischer, P.A., Clemens, R., Balzereit, F., Funfgeld, E.W., and Haase, H.J., 1984, Effects of oral memantine on symptoms of Parkinson's disease, *Dtsch. Med. Wochenschr.* 109:987-990.

Sellal, F., Hirsch, E., Lisovoski, F., Mutschler, V., Collard, M., and Marescaux, C., 1992, Contralateral disappearance of parkinsonian signs after subthalamic hematoma, *Neurology* 42:255-256.

Skaper, S.D., Facci, L., Schiavo, N., Vantini, G., Moroni, F., Dal Toso, R., and Leon, A., 1992, Characterization of 2,4,5-trihydroxyphenylalanine neurotoxicity in vitro and protective effects of ganglioside GM1: implications for Parkinson's disease, *J. Pharmacol. Exp. Ther.* 263:1040-1046.

Sonsalla, P.K., Nicklas, W.J., and Heikkila, R.E., 1989, Role for excitatory amino acids in methamphetamine-induced nigrostriatal dopaminergic toxicity, *Science* 243:398-400.

Sonsalla, P.K., Riordan, D.E., and Heikkila, R.E., 1991, Competitive and noncompetitive antagonists at N-methyl-D-aspartate receptors protect against methamphetamine-induced dopaminergic damage in mice, *J. Pharmacol. Exp. Ther.* 256:506-512.

Sonsalla, P.K., Zeevalc, G.D., Manzino, L., Giovanni, A., and Nicklas, W.J., 1992, MK-801 fails to protect against the dopaminergic neuropathology produced by systemic 1-methyl-4-phenyl-1,2,3,6-tetrahydropyridine in mice or intranigral 1-methyl-4-phenylpyridinium in rats, *J. Neurochem.* 58:1979-1982.

Steinberg, G.K., George, C.P., De LaPaz, R., Shibata, D.K., and Gross, T., 1988, Dextromethorphan protects against cerebral injury following transient focal ischemia in rabbits, *Stroke* 19:1112-1118.

Storey, E., Hyman, B.T., Jenkins, B., Brouillet, E., Miller, J.M., Rosen, B.R., and Beal, M.F., 1992, 1-Methyl-4-phenylpyridinium produces excitotoxic lesions in rat striatum as a result of impairment of oxidative metabolism, *J. Neurochem.* 58:1975-1978.

Turski, L., Bressler, K., Rettig, K.J., Loschmann, P.A., and Wachtel, H., 1991, Protection of substantia nigra from MPP+ neurotoxicity by N-methyl-D-aspartate antagonists, *Nature* 349:414-418.

Walker, F.O., and Hunt, V.P., 1989, An open label trial of dextromethorphan in Huntington's disease, *Clin. Neuropharmacol.* 12:322-330.

Wong, B.Y., Coulter, D.A., Choi, D.W., and Prince, D.A., 1988, Dextrorphan and dextromethorphan, common antitussives, are antiepileptic and antagonize N-methyl-D-aspartate in brain slices, *Neurosci. Lett.* 85:261-266.

Woodworth, J.R., Dennis, S.R.K., Moore, L., and Rotenberg, K.S., 1987, The polymorphic metabolism of dextromethorphan, *J. Clin. Pharmacol.* 27:139-143.

Zipp, F., Baas, H., and Fischer, P.A., 1993, Lamotrigine - antiparkinsonian activity by blockade of glutamate release?, *J. Neural. Transm. (P-D Sect.)* 5:67-75.

Zuddas, A., Corsini, G.U., Schinelli, S., Johannessen, J.N., Di Porzio, U., and Kopin, I.J., 1989, MPTP treatment combined with ethanol or acetaldehyde selectively destroys dopaminergic neurons in mouse substantia nigra, *Brain Res.* 501:1-10.

Zuddas, A., Oberto, G., Vaglini, F., Fascetti, F., Fornai, F., and Corsini, G.U., 1992, MK-801 prevents 1-methyl-4-phenyl-1,2,3,6-tetrahydropyridine-induced parkinsonism in primates, *J. Neurochem.* 59:733-735.

FUNCTION OF THE BASAL GANGLIA IN MENTAL ACTIVITY

Dominique Laplane

Service de Neurologie
Groupe Hospitalier Pitié-Salpêtrière
47 Boulevard de l'Hôpital
75651 Paris Cedex 13

The hypothesis that the basal ganglia play a role in mental activity is not a new concept. The major epidemic of encephalitis lethargica (von Economo's disease) during the 1920s provided many striking examples of mental disturbances. Those disturbances which, from this standpoint, made the greatest impression on investigators at that time were referred to as "bradyphrenia" (Naville, 1922), "abulia", i.e. lack of initiative or drive that appeared to be similar to certain aspects of schizophrenia, hebephrenia and catatonia (Farran-Ridge, 1926), and manifestations of obsessive-compulsive disorder (Lewis, 1936). Implication of the basal ganglia had been widely proposed because of the predominance of lesions in subcortical structures. Unfortunately, this period of history was also marked by the expansion of psychoanalysis and misunderstanding between the two levels of the study (Jeliffe, 1929) cast a certain amount of discredit on this research. In addition, the basic sciences at that time could not provide the concepts necessary to integrate these ideas. Lastly, methods of investigation available were too rudimentary to obtain observations that would be convincing and flawless. Since the 1950s, Hassler has vigorously defended the role of the basal ganglia in mental activity and in the area of attention in particular (Hassler, 1980). His observations, especially involving behavioral patterns following lesions produced in laboratory animals by stereotaxis and in humans lacked sufficient precision from a psychological standpoint to be absolutely convincing, but his intuition however was excellent.

The lesson of epidemic encephalitis had almost been forgotten when Marsden published an article on "The mysterious motor function of the basal ganglia", in 1982, without postulating their possible role in mental activity. However, when in 1981, I reported the case of a patient with obsessional symptoms who had bilateral lesions in the lenticular nuclei, the reference to epidemic encephalitis immediately came to mind (Laplane et al., 1981). This patient also presented with inertia, mental akinesia or more accurately a disturbance which I preferred to call loss of psychic self-activation when I published a second similar case the following year (Laplane et al., 1982). Since then, I have collected twelve cases, mainly subsequent to carbon monoxide poisoning or anoxia, but also after a wasp sting (my first case, in fact), disulfiram intoxication, and trauma. Other cases of bilateral lesions in the globus pallidus causing the same disturbances as those which we presented have been reported by Ali Cherif et al. (1984), and also involve two cases of carbon monoxide poisoning. The clinical presentation was identical in all aspects, including obsessive-compulsive behavior, thereby confirming the existence of new cases and by other investigators. A similar case was published quite independently by Strub (1989) in the USA under the heading "frontal lobe syndrome". In all these patients, lesions involved primarily the lenticular nuclei. This old term has the advantage of not being more precise than the diagnostic tests at our disposal (CT-scan, MRI) can reasonably confirm. The globus pallidus was certainly involved in all cases, but the putamen appeared to be implicated in several of them. These lesions were the only ones which were common to all patients; some patients, in addition, had a certain amount of frontal lobe atrophy. These data were obtained solely from CT-scans or MRI; none of our patients underwent autopsy.

All these patients presented various degrees of psychological akinesia and loss of self-activation and, in many patients, stereotyped movements and disturbances observed sometimes appeared to be very similar to symptoms of obsessive-compulsive disorder.

Patients who participated in this study did not present with neurological disturbances or only a few minor disturbances such as parkinsonian syndrome in a few instances. One patient, in addition, had movements resembling tics. Motor impairment was never an obstacle in performing a psychological examination.

LOSS OF PSYCHIC SELF-ACTIVATION AND AKINESIA

Our first case already provided a complete description of this disorder: the clinical picture was very particular. It consisted primarily of major apragmatism but did not include any evidence of anxiety, depression, melancholia or any delusional symptom. This man who had led a very active life up until then, now did practically nothing and, outwardly at least, was no longer interested in anything or anyone even though he had been a very caring husband and father.

This apraxia resulted in a presentation of "quiet dementia" a term used by the physician who was treating him at that time. Yet, this patient could play bridge which he was still skilled at playing. Four or five months after the initial episode, the patient attempted to return to work. While up until this time he had been effectively directing a major public works project, now he no longer did anything, he slept during project meetings, and he no longer worked on his plans. After a few weeks, he was fired from his job and he went home in tears. Mental emptiness was a basic element in the clinical presentation. It was very striking that this man whose intellectual abilites were in large part preserved (IQ of 114 with a MP of 38) could spend entire days practically without doing anything. He read newspapers very little (but nonetheless was still abreast of current events), he looked at television a little but most often remained unoccupied. While this lack of activity appeared surprising to others, the patient for his part was surprised that he was not bored and did not suffer from his inactivity. When asked to describe what he was thinking of, he replied :"It's hard to say, not much; yes, my mind just seems to be blank". This expression of mental void will be used spontaneously by other patients. This patient tried to explain his unsuccessful attempt to return to work by implicating his difficulty in writing. He did however, have to recognize that he had difficulty in recognizing a given problem in its overall presentation and in being able to concentrate.

Events in this patient's affective life did not have the same impact as they would have had on most other people. He did not show a lack of emotion; he was able to empathize when the misfortunes of his existence were mentioned. Nonetheless, his affect appeared to be shallow, being more intellectual than truly emotional. When asked to comment on the consequences of the true misfortunes of his life such as the condition of his health, he denied being depressed. Unhappy ? This wasn't the right word. When asked to find the right word he stated: "It's as if something is missing (comme un manque)".

Everything happened in a certain fashion as if the patient's psychological life, both intellectual and emotional, was dependent entirely on external stimuli and that in contrast, his internal dynamism, and his spontaneity had to a wide extent disappeared.

Generally speaking, these patients suffer from a massive loss of activity and motivation, which corresponds to what is commonly called loss of drive. The disturbance most often is major, with the patient spending hours without doing anything. In less pronounced cases, patients may have some stereotyped activities that can pass for initiatives, such as going to buy a newspaper and running through it or performing a routine household chore. Indeed, the impossibility of acting on his own initiative is one of the most striking features of this condition. The patient in our series who was least affected from this standpoint appeared to be capable of small initiatives such as participating in activities of caring for his neighbors or enrolling in a pilgrimage to Lourdes. Upon closer examination, this patient was responding simply by acquiescing to proposals made or to requests by staff and which had become routine or by visitors to the hospital. Conversely, this man had been living in a joyless hospice for years and he had not made any initiative to attempt to leave even though a superficial examination would have led to think that he was perfectly capable of doing so.

This underactivity was, on the contrary, dramatically reversible by the stimulation of his immediate friends and family, although, here too, to variable extents depending on patients. One patient plays bridge if asked to play, and indeed plays very well, another, an athletic supervisor prior to his illness, was capable of very good performances, provided that his wife encouraged him in this activity. Once the activity had been started, it sometimes took on a stubborn, almost compulsive aspect, liable to last weeks or months. This patient whose most remarkable characteristic was the acceptance of his life in the hospice, prior to his illness, was an amateur artist; since then, if someone insisted that he resume his painting, he always painted the same landscape of a morose, flat plain which recalled the patient's childhood in Russia A more persevering effort by a new team of moderators enabled this patient to resume this activity on a continuing basis, but since someone had suggested to him that he change his landscape and paint a postcard depicting a castle, he then began to paint a series of castles and invaribly to place them on the banks of a body of water, which most often was imaginary; then, again in response to suggestions, he started to paint a series of historical portraits. In most other cases, the works produced in this fashion were even more sterotyped and the impetus initiated was of much shorter duration.

Verbal fluency is consistently disturbed. These subjects rarely or never are the first ones to speak or spontaneously, but if questioned they can give an entirely adequate answer. Their speech is short, limited to the bare essentials, often only a few words, but well-suited to the context. Formal tests of verbal fluency demonstrate, as in frontal lobe syndromes, a difference between tests which use a definite semantic context (e.g. names of domestic animals, flowers, etc.) from those which do not involve such a context ~name beginning with a letter of the alphabet). It may be suggested that this context serves to illustrate the particular type of difficulty these patients have in using purely internal data, contrasting with at least a relative preservation of their ability to use external data.

The disorder affects not only the behavioral aspect of the patient's life but also his internal psychological life. There is a striking similarity in statements made by patients: all of them refer to a mental "void" or "blank". It is absolutely remarkable, as has been emphasized, to see these subject who have normal intelligence spend hours and hours of doing nothing and can still say that they are not bored. One of the patients suffering from such a disorder and to whom I insisted by saying that no one can ever remain without thinking of something replied to me "maybe that is true for you, Doctor, but not for me".

Typically, from the emotional aspect, these patients appear as being completely disinterested. They have a "flat" affect. Manifestly, they are not concerned by their health, their sorry condition, or how heavy a burden they are for their families. But, if they are asked specifically about these items, they not only are able to describe their situation accurately and with correct intellectual judgment, but also, at least this is true for some patients, to appear to feel an emotional response on this subject. What is true for what concerns them is also true for what affects their families. They rarely take the initiative to inquire about one person or another, but appear to react normally to news that is given to them; in fact, the emotion does not last and these subjects quickly return to their state of mental void.

Contrary to the classical description of frontal lobe syndromes, these patients are not deprived either of insight or of foresight. When asked about their condition, they recognize their sorry and even worrisome state, but this impression does not last, and soon after they make such statements, they resume their usual attitude of indifference. When asked about their plans, they often admit that they do not have any, but if asked to do so, they are entirely capable of giving a serious, unexaggerated account of their plans, but also, without realizing it, apparently of their inability to start carry out them out.

This lack of stable inner life is obviously incompatible with the pondering anxiety of depression. However, the disorder does not resemble the one observed in patients who have undergone lobotomy and who appear to be entirely unable to feel true emotion. The degree of disturbance also is variable, depending on patients. One of our patients, with a history of depression, continued to suffer from his past depression until he finally committed suicide, as if his disorder was not pronounced enough to "cure" him and another patient appeared to be depressed since an anoxic event during surgery which had caused his condition.

This loss of innner mental activity, accompanied obviously by behavioral inertia, contrasting with the quality of cognitive and emotional performances when a patient is stimulated, deserves to be called loss of psychic self-activation but is not a matter of "all or nothing" response.

It is remarkable that if DSM-III criteria are applied in a mechanical manner, our patients can be classified in the diagnostic categories of major depression or schizophrenia. Of course, the phrase "absence of lesion" must not be taken into account.

It is true that in the most serious forms of these two disorders, the influence of the patient's friends and family are much lesser than in loss of self-activation as we have described it. However, in many cases, this influence exists to a certain extent.

DISTURBANCES RESEMBLING OBSESSIVE-COMPULSIVE DISORDER

Most but not all of our patients have been suffering from another much more spectacular trouble resembling an obsessional disorder. The first case is again very characteristic. The patient who was presented to me as suffering from "quiet dementia" appeared to be severely mentally ill because of his total inactivity. But according to his wife, the most convincing evidence was the fact that the patient had been found on all fours in the street. When asked about this event, he explained that he was kicking a stone with his foot,)but due to his motor impairment (at that time he had a parkinsonian syndrome as a result of neuroleptic drug treatment) he was no longer able to do this and since he had not reached a multiple of the number nine, he felt quite obliged to push the stone by hand. This patient had other "manias": he would sometimes spend 15 minutes manipulating a light switch on and off, he would continuously count numbers without stopping, setting the pace with his fingers. He could only stop at a multiple of three or nine. However, someone who had authority over this patient, for example, his wife or his physician, could make him stop this activity. In such a case he felt irritated but not anxious strictly speaking. He understood the absurd nature of these activities, perfectly well but he could not help himself from performing them: "It's stronger than I am", he would say. Thus, the compulsive feature of his disorder was not explained by anxiety as is usually the case in obsessive-compulsive neurosis. It is important to specify that prior to encephalopathy subsequent to a wasp sting, this patient did not have any mental disturbance, in particular no obsessional type disturbance and this was also true for other patients in this series.

In another patient, the disorder was even more spectacular because the loss of self-activation was much more moderate, so much so that the obsessional activity appeared as the main abnormality after the patient experienced carbon monoxide poisoning. But this woman also suffered from chronic depression (in the context of antisocial personality), which had not been "cured" by these lesions. She answered questions put to her only after a long period of latency. Indeed, before answering, she had to count all the words spoken and if this number was not a multiple of three, she would have to add words. She also had to pronounce stereotyped ritual words under certain circumstances (when smoking or especially drinking), which caused her to have certain guilt feelings: "that's past history", before she would drink or "that's always good" after she had drunk, etc. The rare attempts to break away from these rites led to anxiety. Here too, our patients, or at least some of them satisfied the conditions for the diagnosis of obsessive-compulsive disorder according to DSM-III criteria, except, of course for absence of an organic lesion.

Overall, six of our patients had the same type of behavior corresponding to DSM-III requirements. Some types of compulsive behavior are less compelling as demonstrated by the first patient. Three patients presented simply with counting to themselves, with no outward manifestation. This seems to be the more common type of compulsion and it is useful to ask this question of every patient suspected of suffering from this type of disorder. One patient also exhibited the action of sucking his fingers but without any specific thoughts. In summary, there is a range of disturbances, from true obsessions such as those observed in recognized neurosis to very simple sterotyped movements as may be seen in many subnormal persons.

Tests did not demonstrate any significant mental deterioration in most cases; verbal ability was normal, as was gestual activity, ability to calculate and to draw. Orientation in space and time were normal. Memory and learning were normal in all cases except two; slowing of intellectual processes, considered when tests were taken varied from one patient to another and from one moment to the next. Fluctuations in attention span accounted for variations in performance of certain tasks, but overall the disorder was moderate;

reproduction of Rey's figure from memory showed a difficulty in synthesis. Development of new strategies, passage from one attitude to another in view of adjusting to modifications in a changing situation and programmation of sequential activity were disturbed according to the Wisconsin card-sorting test, problems and series based on Luria's method. On the MMPI, personality appeared to be well-preserved but affectivity was still severely altered and a depressive state was observed in two cases.

ANALOGIES WITH THE FRONTAL LOBE SYNDROME

It is difficult to find a neuropsychological definition for the frontal lobe syndrome that is acceptable to everyone. Several elements, however, are considered as being characteristic. Among the most significant is the loss of motivation or loss of drive which was very marked in all our patients except for two, although no patient was entirely free from this disturbance. Decrease in verbal fluency was also very suggestive, including the dissociation already pointed out between semantic framework tests and those tests which ask the patient to make up his own list from a letter (Stuss and Benson, 1986). Decreased attention, another frontal lobe disturbance, was present albeit moderately in our patients. Conversely, memory disturbances were slight except in two patients.

The Wisconsin card-sorting test is often considered as the most sensitive test of frontal lobe dysfunction. It was disturbed in all patients, except for one, but qualitative abnormalities were different from one case to another.

Above all, it should be kept in mind that apathy, disinterest, and apragmatism are classical features of the frontal lobe syndrome (Blumer and Benson, 1975). Eslinger and Damasio (1985) have reported the case of a patient who was a victim of postsurgical frontal lobe damage, that was remarkable because of the normality of all functions investigated by neuropsychological tests, whose results were not only normal but especially brilliant. Nonetheless, this patient suffered from an inability to apply his intellectual abilities in any field at all, and from behavioral disturbances (disorganization, instability).

Improvement under the influence of external stimulation, so striking in our patients, has also been described in patients with frontal lobe syndrome (Stuss and Benson, 1986). It follows that this disorder may not develop under examination conditions, with the observer serving truly as a frontal lobe for the patient. As emphasized by Eslinger and Damasio (1985) also in regard to patients with frontal lobe syndrome, this disorder is much more pronounced in everyday life than in the artificial situation of the neuropsychological examination.

We ourselves have reported (1988) the case of a woman who was the victim of head injury responsible for diffuse frontal lobe lesions observed on MRI. She presented with exactly the same disorder as we reported in our patients with lesions of the globus pallidus: the same frontal lobe-like syndrome, the same "loss of psychic self-activation", and the same stereotyped activity of counting to herself, although this was not a truly compulsive element.

Obsessional type behavior has also been reported in patients with frontal lobe lesions. Eslinger and Damasio's patient previously mentioned presented with typical such activities, even though this fact is not discussed in the article. We are currently following a patient who most probably has Pick's disease, according to clinical, MRI and PET findings which reveal massive underactivity in the frontal lobes but also in the basal ganglia. This patient also has presented very pronounced obsessional activity since the onset of her illness: she gargles seventeen times after each brushing of her teeth, wipes her feet thirty or forty times on the door mat and counts anything around her that can be counted.

In summary, many elements in the neuropsychological presentation are the same, whether the patient has a frontal lesion or if there is basal ganglia involvement. This fact is already known. Indeed, frontal lobe syndromes have been described in disorders of basal ganglia such as supranuclear palsy; (Cambier et al., 1985) and in Parkinson's disease (Taylor et al., 1986; Gotham et al., 1988). However, the syndrome described here is especially pure because motor disturbances which sometimes make it difficult to carry out the neuropsychological examination were practically absent in the cases we observed.

Deoxyglucose metabolism investigated by positron emission tomography (PET) in seven of our patients (Laplane et al., 1989) did not on first examination reveal any abnormality in comparison to controls. On the other hand, there was a notable, statistically significant decrease in the frontal metabolic index in comparison to posterior areas.

Thus, neuropsychological tests and metabolic tests with PET confirmed the impression of frontal lobe syndrome suggested by the physical examination, despite the localization of the main lesions in the lenticular nuclei. Naturally, decreased metabolism in these two structures was also observed with PET. In both cases, a certain amount of cerebral atrophy was observed in addition, but did not predominate in the frontal lobes.

Obviously, it cannot be ruled out with certainty that more diffuse lesions may be present in addition to those involving the lenticular nuclei but it should be emphasized that the white matter appeared normal in MRI and that in patients having experienced carbon monoxide poisoning (as did the majority of our patients) cortical lesions are not a predominant feature nor frontal lobes lesions (Lapresle and Fardeau, 1967).

To date we have observed a loss in psychic self-activation and stereotyped activities or obsessive- compulsive behavior in patients with lesions of the frontal lobe and the basal ganglia. Two patients with "loss of drive, interest and affect" reported by Habib and Poncet (1988) are extremely similar to the cases we observed, except for the absence of obsessive-compulsive behavior or stereotyped activity. In both cases, lesions consisted of multiple lacunae distributed bilaterally in the striatum, the caudate nucleus and putamen, but not in the globus pallidus. Similar findings concerning loss of drive, interest, apathy or abulia have also been reported in cases of vascular dementia (Ishi et al.,1986; Loeb 1985). They might also be due to the frequency of lacunae in the striato-pallidal region (Katz et al. 1987).

Bilateral lesions in the dorsomedian nuclei of the thalamus consistently result in apathy, loss of drive and flattening of affect . But in such cases, disturbances in cognitive function are usually very serious in all respects. Recently, Bogousslavsky et al (1991) observed two patients suffering from a bilateral paramedian thalamo-mesencephalic infarct involving the dorsomedian nuclei. The presentation was the same as that which we observed, both from the clinical standpoint as well as the neuropsychological standpoint, with answers to appropriate diagnostic tests that are characteristic of a "frontal lobe" syndrome. They did not have obsessive-compulsive or stereotyped behaviors but this was true for two of our patients. Furthermore, SPECT imaging revealed a very pronounced decrease in blood flow in the diencephalon and frontal lobes, especially the parasagittal areas.

From these current data in the literature, a complete loop - from frontal lobe to striatum to globus pallidus to thalamus to frontal lobe - can be traced, alterations of which could cause loss of psychic self-activation and stereotyped and obsessional behavior.

CONCLUSIONS

One of the exciting aspects of these cases lies in their resemblance to certain manifestations of the major psychoses (major depression, "negative" schizophrenia, ex-hebephrenia, and obsessional psychoneurosis now known as obsessive-compulsive trouble. Clinicians have long recognized pseudopsychiatric presentations of some cerebral disorders, notably frontal lobe lesions, but their theoretical interpretation remained unknown. The resemblances, however, are such that it seems reasonable to propose the hypothesis that damaged neuronal circuits in our patients could very well be the same as those which are altered (by different mechanisms) in major psychoses.

In any event, these mainly clinical data suggest that the basal ganglia play a key role in mental activity which for over a hundred years was believed to occur solely in the cortex. These clinical findings appear to agree with data from anatomic studies. Indeed, as specifically emphasized by Percheron et al. (1984), the concentration in the globus pallidus of information from the entire cortex, especially the associative area of the cerebral cortex, suggests that it might play a role in mental activity. The normaly of the globus pallidus and the fibers which cross through it appears to be vital for the spontaneous activation of cognitive and affective functions; bilateral lesions in these structures appear to cause a particular type of disorder of consciousness, heretofore not described. Hetero-activation would, of course, pass via the sensorimotor cortex, while self-activation would occur via an associative neuronal circuitry calling into play previous experiences which feed this mental activity that is maintained, more or less vague or precise, even in the absence of external stimuli during the period of wakefulness.

In contrast, the majority of classical globus pallidus syndromes correspond to motor damage and classically it is held that the striatum and the globus pallidus are motor nuclei

(Marsden, 1982). These motor disturbances are commonly so massive that they do not allow the clinician to evaluate cognitive function.

The anatomic distinction between two parallel systems, sensorimotor and associative, within the striatum and the globus pallidus, furthermore, appears to provide a possible explanation since some pallidal syndromes are motor disorders, others are purely psychic troubles and other are both.

It is perhaps of value to compare the elementary stereotyped behavior of some of our patients (counting, snapping their fingers, continuously pushing a light switch, etc...) with the inability of one patient to interrupt a circular movement (e.g. stirring a soup) once she had started to do it: a disorder previously described by Luria in patients with frontal lobe tumors extending to the basal ganglia (1985). This then would involve a defect in inhibition of an action undertaken, regardless of whether the latter was purely a motor action, purely psychic or a combination of the two.

At the same time, these models of disease provide us with an overview of the function of the basal ganglia in the physiology of mental activity. It is amusing to observe that, once again, interest is focused on the base of the brain as the site of a vital part of mental activity which, since the time of Gall (i.e. the end of the 18th century) and especially Broca, was thought to be confined to the cortex.

REFERENCES

Ali-Cherif, A., Royere, M.L., Gosset, A., Poncet, M., Salamon, G., and Khalil, R., 1984, Troubles du comportement et de l'activité mentale après intoxication oxycarbonée. Lésions pallidales bilatérales, *Rev. Neurol.* 140:401-405.

Blumer, D., and Benson, D.F., 1975, Personality changes with frontal and temporal lobe lesions, *in*: "Psychiatric aspects of Neurologic disease, Vol. 1," D.F. Benson and D. Blumer, eds., Grune and Stratton, New York, pp. 151-170.

Bogousslavsky, J., Regli, F., Delaloye, B., Felaloye-Bischoff, A., Assal, G., and Uske, A., 1991, Loss of psychic self-activation with bithalamic infarction. Neurobehavioural, CT, MRI and SPECT correlates, *Acta Neurol. Scand.* 83:309-316.

Cambier, J., Masson, M., Viader, F., Limosin, J., and Strube, A., 1985, Le syndrome frontal de la paralysie supra-nucléaire progressive, *Rev. Neurol.* 141:528-536.

Eslinger, P.J., and Damasio, A.R., 1985, Severe disturbance of higher cognition after bilateral frontal lobe ablation: Patient E.V.R., *Neurology* 35:1731-1741.

Farran-Ridge, C., 1926, Some symptoms referable to the basal ganglia occuring in dementia praecox and epidemic encephalitis, *J. Mental Sci.* 72:513-523.

Gotham, A.M., Brown, R.G., and Marsden, C.D., 1988, Frontal cognitive functions in patients with Parkinson disease on and off levodopa, *Brain* 111:239-321.

Habib, M., and Poncet, M., 1988, Perte de l'élan vital, de l'intérêt et de l'affectivité (syndrome athymhormique) au cours de lésions lacunaires des corps striés, *Rev. Neurol.* 144:571-577.

Hassler, R., 1980, Brain mechanisms of intention and attention with introductory remarks on other volitional processes, *in*: "Motivation, motor and sensory processes of the brain; electrical potentials, behaviour and clinical use," H.H. Kornhuber and L. Deecke, eds., Elsevier, Holland, pp. 584-614.

Ishii, N, Nishiharay, Y., and Imamura, T., 1986, Why do frontal lobe symptoms predominate in vascular dementia with lacunes, *Neurology* 36:340-345.

Jeliffe, S.F., 1929, Psychologic components in postencephalitic oculogyric crises, *Arch. Neurol. Psychiat.* 21:491-532.

Katz, D.L., Alexander, M.P., and Mandell, A.M., 1987, Dementia following strokes in the mesencephalon and diencephalon, *Arch. Neurol.* 44:1127-1133.

Laplane D, Widlocher D, Pillon B, Baulac M, and Binoux F 1981 Comportement compulsif d'allure obsessionnelle par nécrose circonscrite bilatérale pallido-striatale. Encéphalopathie par piqûre de guêpe, *Rev. Neurol.* 137:269-276.

Laplane, D., Baulac, M., Pillon, B., and Panayatopoulos-Achimestos, J., 1982, Perte de l'auto-activation psychique. Activité compulsive d'allure obsessionnelle. Lésion lenticulaire bilatérale, *Rev. Neurol.* 138:137-141.

Laplane, D., Dubois, B., Pillon, B., and Baulac, M., 1988, Perte de l'auto-activation psychique et activité mentale stéréotypée par lésion frontale, *Rev. Neurol.* 144:564-570.

Laplane, D., Levasseur, M., Pillon, B., Dubois, B., Baulac, M., Mazoyer, B., Tran Dinh, S., Sette, G., Danze, F., and Baron, J.C., 1989, Obsessive-compulsive and other behavioural changes with bilateral basal ganglia lesions. A neuropsychological, magnetic resonance imaging and positron tomography study, *Brain* 112:699-725.

Lapresle, J., and Fardeau, M., 1967, The central nervous system and carbon monoxide poisoning II. Anatomical study of brain lesions following intoxication with carbon monoxide (22 cases), *Prog. Brain Res.* 24:31-74.

Lewis, A., 1936, Problems of obsessional illness, *Proc. R. Soc. Med.* 29:325-336.

Loeb, C., 1985, Vascular dementia, *in*: "Handbook of Clinical Neurology, Vol 2 (46) Neurobehavioural disorders," J.A.M. Frederiks, ed., Elsevier, Amsterdam, pp. 353-369

Luria, A.R., 1985, Two cases of motor perseveration in massive lesions of frontal lobes, *Brain* 88:1-10.

Marsden, C.D., 1982, The mysterious motor function of the basal ganglia: The Robert Wartenberg lecture, *Neurology* 32:514-539.

Naville, F., 1922, Etude sur les complications et les séquelles mentales de l'encéphalite épidémique, *Encéphale* 2:371-375.

Percheron, G., Yelnik, J., François, C., 1984, The primate striato-pallido-nigral system: an integrative system for cortical information, *in*: "The Basal Ganglia Structure and Function," J. S. McKenzie, R.E. Kemm and L.N. Wilcock, Advances in Behavioral Biology, Vol 27, Plenum Press, New York, pp. 87-105.

Strub, R.L., 1989, Frontal lobe syndrome in a patient with bilateral globus pallidus lesions, *Arch. Neurol.* 46:1024-1027.

Stuss, D.T., and Benson, D.F., 1986, "The frontal lobes," New York, Raven Press.

Taylor, AE., Saint-Cyr, A., and Lang, A.E., 1986, Frontal lobe dysfunction in Parkinson's disease, *Brain* 109:845-883.

THEORIES ON BASAL GANGLIA FUNCTIONING

THE BASAL GANGLIA: "MINIMAL COHERENCE DETECTION" IN CORTICAL ACTIVITY DISTRIBUTIONS

Dietmar Plenz[1] and Ad Aertsen[2]

[1]MPI für biologische Kybernetik, Spemannstrasse 38, D-7400
Tübingen, Germany
[2]Institut für Neuroinformatik, Ruhr-Universität, P.O. Box 102148,
D-4630 Bochum, Germany

INTRODUCTION

The basal ganglia are strongly involved in specific, motivation- and reward-mediated aspects of cortical processing: perception, cognition and action (Hikosaka et al., 1989). These aspects particularly concern "changes" in cortical activity modes (Mink and Thach, 1991), to which the basal ganglia contribute by "gating" of cortical and premotor activity via disinhibition (Chevalier and Deniau, 1990). The interactions between cortex and basal ganglia seem to be organized in multiple parallel loops (Alexander et al., 1986), with lateral competition between these loops as a possible mechanism for cortical cell assembly selection (Wickens et al., 1991).

The concept of "parallel processing" however, is problematic in view of the highly divergent-convergent projections from the first level of the basal ganglia circuit - the neostriatum - to its output nuclei, both anatomically (Percheron et al., 1984) and functionally (Tremblay and Filion, 1989). Given these strongly integrative properties of the basal ganglia, both within and between loops, it is not clear how the basal ganglia could contribute in any specific way to cortical function.

Any theory on basal ganglia function has to be adequately connected to a theory of cortical coding and function. Thus, we will ask, which solution the basal ganglia can provide to the problem of sequential cell assembly activation (Hebb's "phase sequences", 1949). In particular, we will present a theoretical framework which explains the local operations that are performed in the neostriatum, and why these are followed by a sequence of nuclei with strong integrations in the intervening projections.

We propose that the basal ganglia are optimally suited, both anatomically and functionally, to perform a "minimal coherence" analysis of changes in the cortical activity distribution in order to organize reward-mediated sequential behaviour. This evaluation of "coherence" is performed in a similar way the visual system evaluates the optical flow field in order to separate "figure" from "ground". We will develop this idea for the particular case of the motor loop (Alexander et al., 1986), but we emphasize this to be the general principle according to which the basal ganglia contribute to any kind of cortical processing.

CELL ASSEMBLIES AS THE BASIC SUBSTRATE FOR CORTICAL CODING AND FUNCTION

The cortex can be viewed as a memory structure with a large number of similar elements, showing a high degree of convergence/divergence in their input/output connectivity (Braitenberg and Schüz, 1992). In particular the connectivity patterns of cortical pyramidal cells involve many different cortical areas (Selemon and Goldman-Rakic, 1988) and are assumed to be modifiable by local learning rules. In such types of neural networks, information theory favours the principle of "sparse coding" (Palm, 1982), stating that a functional group of only a very limited number of elements is activated in unison in the coding, storage and recall of a feature or constellation of features. One way how the activation of such an identifiable group can be envisaged is by synchronization: a group of neurons is temporarily separated from the entire neuronal population, when their spike activities become synchronized over some period of time. Such a group of neurons makes up a cell assembly (Gerstein et al., 1989; Erb and Aertsen, 1992; Neven and Aertsen, 1992). Recent experimental studies in the visual cortex suggest that such synchronized activity in groups of neurons may serve perceptual integration by means of feature linking (for a review see Engel et al., 1991).

The mechanism for generating such synchrony - which is not necessarily time-locked to an external stimulus or behavioral event - might at first sight be thought to reside in strong anatomical connectivity. Several arguments, however, speak against this scheme. First, strong anatomic coupling is not a flexible means to organize cells into groups and would, in fact, lead to dedicated groups of neurons, very much like a "grandmother assembly". Moreover, experimental evaluations of multiple-neuron recordings have revealed that strong synaptic coupling is, in fact, quite rare in the cortex (e.g. Abeles, 1982, 1991). Instead, the "effective connectivity" (Aertsen et al., 1989, 1991) between cortical neurons turns out to be generally weak, and highly dependent on stimulus and behavioral context. Moreover, it is remarkably variable over time, with time constants of modulation as low as tens of milliseconds (for a review see Vaadia and Aertsen, 1992). Such context-dependent, rapid modifications of effective coupling - sometimes referred to as dynamic linking - were established in several different cortical areas (e.g. Aertsen and Gerstein, 1991; Vaadia et al., 1991). Theoretical considerations based on these findings have suggested an alternative and much more flexible scheme for generating synchrony. It was shown that the influence of a presynaptic neuron onto the spike activity of its postsynaptic target neuron (i.e. their effective coupling) is critically dependent on the population background activity and only partially reflects the underlying anatomical connectivity (Aertsen and Preissl, 1991). Thus, rapid modulations of coupling can be induced by "dynamic convergence" of activity from the entire network onto the observed neurons, in particular by temporal variations of the rates (Boven and Aertsen, 1989) and the internal coherence (Bedenbaugh et al., 1988, 1990) of background firing (see also Bernander et al., 1991).

The picture which emerges from these considerations on neuronal assemblies is a highly dynamic one: the formation of synchrony is based on rapidly changing synaptic couplings. Thus, a pyramidal neuron is able to rapidly change allegiance and thereby take part in various different cell assemblies, either switching from one to the other or joining a number of them simultaneously, depending on the immediate computational demands.

CHANGE OF CORTICAL ACTIVITY DISTRIBUTIONS: CODING OF TRANSITIONAL PROBABILITIES

A cell assembly may code for a certain stimulus feature or motor program, and can be active in various behavioural contexts (Vaadia and Aertsen, 1992). At the next higher level of organization, a "set" of cell assemblies is conceived to code for a certain constellation of features and/or actions in some behavioural context; activation of such a set constitutes the cortical activity distribution. Clearly, such a set of cell assemblies should not be considered as a cell assembly itself: otherwise one would lose combinatory power to represent different contexts having common features (von der Malsburg, 1986).

A related problem arises when considering the ordered activation of (sets of) cell assemblies in time. According to the theory on cortical "synfire chains" (Abeles, 1982,

1991), once a cell assembly is activated it is able - given the appropriate cross-connectivity - to activate a new assembly. Thus, "synfire chains" or "reverberating synfire chains" (Abeles et al., 1993) seem to be a promising new approach to organize a highly ordered time structure in cell assembly activation (Bienenstock, 1991). One important aspect of time structured activity, however, is missing so far. Consider, for instance, a change in an animal's environment. The animal will respond to this change with a certain motor behaviour, which itself is governed by sequential activation of a certain set of cell assemblies. How can we imagine that the composition of this set and/or its activation sequence is modified by the outcome of the behaviour? The answer is "reward-mediated learning", leading to the formation of a new scheme for cell assembly activation which is more appropriate under the new circumstances (Miller et al., 1990). Note that this reward-mediated plasticity, modifying the transitional probabilities between (sets of) cell assemblies, should be distinguished from the plasticity involved in the establishment of the cell assemblies themselves; the two operate at different levels of organization.

In summary, the problem of organizing sequentially ordered cell assembly activation turns out to be related to the task of tuning the transitional probabilities between (sets of) cell assemblies, guided by experience via a reward signal. We propose the basal ganglia constitute a very elegant implementation of this task.

THE CORTICO-STRIATAL PROJECTION: THE STRIATUM AS A "RETINA" FOR CORTICAL TOPOLOGY

In this part we will show that the particular features of the cortico-striatal projection provide a simple strategy in order to map cortical cell assembly dynamics. Cortico-striatal projection cells are densely scattered throughout cortical layer III and upper layer V (Goldman-Rakic and Selemon, 1984). Hence, it is reasonable to assume that the cortico-striatal projection can transmit detailed information about the entire cortical activity distribution to the striatum.

Moreover, this projection is highly divergent. Local dye injections into the cortex revealed widely distributed "patchy" axonal termination areas in the neostriatum (see Percheron et al., 1984; Goldman-Rakic and Selemon, 1984; Flaherty and Graybiel, 1991; Parthasarathy et al., 1992). However, already at this level of the basal ganglia loop there is also strong convergence. If we think of multiple cortical areas projecting in multiple patches to the neostriatum, there will be considerable overlap within the neostriatal volume occupied by the dendrites of a medium-sized spiny projection neuron (principal cell).

This topographical overlap on the neo-striatum "surface" partly preserves functional neighbourhood in cortical topology, even across different cortical areas. This is demonstrated by detailed mappings of projections from well defined cortical areas: in the caudate nucleus, for example, projection fields from the supplementary and the frontal eye fields closely overlap with one another (Parthasarathy et al., 1992). This overlap, however, is not complete: the two functionally related cortical areas have a "core" where their projection areas onto the neostriatum overlap, but this is surrounded by "shells" with no overlap. We will see that this arrangement is essential to allow the organization of new combinations of cortical cell assemblies.

Let us consider the consequences of these particular features of the cortico-striatal projections in a simple example. We assume three cortical pyramidal neurons a, b and c, each one in a different cortical area A, B and C. The neurons project to a certain striatal area (Fig. 1, λ) having multiple termination fields. Each of the neurons takes part in cortical cell assembly formation. We start with neurons a and b taking part in one cell assembly. Thus, their discharge will be synchronized for a brief period of time. Since neurons within an assembly are functionally related, their termination fields are likely to overlap. Striatal principal neurons in the regions of overlap receive synchronized excitatory input and, hence, are likely to discharge (Fig. 1) and to establish intersection domains with "winner dynamics" in the local lateral inhibition network (Wickens et al., 1991). A prominent feature of such lateral inhibition networks is the phenomenon of spatial disinhibition. Thus, the intersection domains with "winner dynamics" will support the excitation of principal cells, located at a certain mean disinhibitory distance outside these intersection domains. We have indicated one such "new" domain with disinhibition-induced activity in Figure 1 (asterisk). Let us consider now three possible changes of the temporal association of these cortical neurons, and their consequences for the activity dynamics in the striatum:

1. Neuron *b* switches in synchronization from neuron *a* to neuron *c*. This change of "cortical" liaison leads to a stabilization of the new striatal intersection domain *bc* (Fig. 1, asterisk) and a decrease in activity of the former intersection domains *ab*.

2. Neuron *a* switches in synchronization from neuron *b* to neuron *c*. This cortical change leads to a destabilization of the previously favoured striatal intersection domains *ab* and, in addition, will establish the new intersection domain *ac* (Fig. 1, dotted area).

3. Neuron *c* becomes synchronized with both neurons *a and b*. At first sight this would seem to lead (as in the first case) to a stabilization of the new intersection domain *bc*, however, without weakening the original ones (*ab*). In addition, however, yet another new intersection domain will form (*ac*, dotted area), which will engage in a competition with the previous intersection domains *ab*. This, in turn, will also weaken the intersection domain *bc*, due to a reduction of the spatial disinhibition. Consequently, this last change in cortical association leads to a complete re-adjustment of the spatio-temporal activity pattern in the striatum.

Thus, we conclude that the patchiness of the cortico-striatal projection can be viewed as a strategy to map the dynamic linking of neurons in the cortex to a unique change in spatio-temporal activity patterns in the striatum.

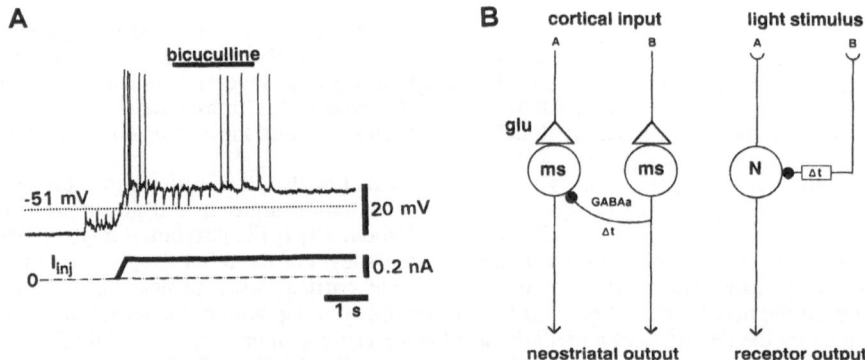

Figure 1. Dynamic linking of cortical neurons is accompanied by a change in the striatal activity patterns. Three cortico-striatal projection neurons have multiple projection fields in the striatum. The spatio-temporal activity pattern in the striatum critically depends on the dynamic linking among the temporal association of these cortical neurons (for further details see text).

THE LATERAL INHIBITION AMONG PRINCIPAL NEURONS CONSTITUTE A "MOVEMENT" DETECTOR

What kind of striatal operation does occur upon a change of cortical input activity? The extensive recombination in the cortico-striatal projection is accompanied by a striking morphological homogeneity of the main receiving neuronal substrate: the principal cells. The majority of these neurons possess - apart from their projection out of the nucleus - only locally restricted axon collaterals, mainly to cells in their direct neighbourhood, i.e. within their own dendritic field (Chang et al., 1981). Thus, in sharp contrast to the situation in the cortex, the absence of long intrinsic axonal projections restricts the neostriatal activity transformation to a locally operating mechanism.

The local operation of this strongly recurrent, striatal inhibitory network will be exemplified using case number 2, considered in the previous paragraph. When neuron *a* switches in synchronization from neuron *b* to neuron *c*, the dotted area *ac* (Fig. 1) will become a newly favoured intersection domain. Such a change of temporal coherence in the cortical activity distribution can also be interpreted as a "movement" in the input, which is accompanied ("detectet") by a local change in the activity distribution in the striatum: from the intersection(s) *ab* to the new intersection(s) *ac*. This notion of "movement detection" as a local operation occurring in the striatum helps to gain new insight into striatal function from a rather different field of study: the mechanism of movement detection in the mammalian retina. To this end, we will briefly summarize the conditions for movement detection or - more generally - for correlation detection, and will then demonstrate that these conditions are fullfilled by the striatal network.

The fundamental physiological proposal for local motion detection was made by Barlow and Levick (1965), using two input cells which were asymmetrically coupled by a delay between the input lines. In a theoretical analysis it was demonstrated that the local interaction had to be second order nonlinear (multiplication: Poggio and Reichardt, 1973). Moreover, it was shown (Torre and Poggio, 1978) that shunting inhibition approximates such multiplicative interaction, provided that two conditions are fullfilled: (1) the reversal potential of the inhibitory synapse has to overlap with the working range of the intracellular membrane potential dynamics, and (2) the shunting synapse must be close (in terms of the electrotonic length) to the excitatory synapse. Recent experiments, comparing the shunting inhibition model to other models of nonlinear interaction, demonstrated that the movement and directional sensitivity of retinal ganglion cells over a wide range of different input contrasts can indeed best be approximated by a shunting inhibition model (Amthor and Grzywacz, 1991).

As was already observed, the striatal network of principal cells is characterized by strong recurrent inhibition. Hence, the conceptual framework of "movement detection" provides quite a natural interpretation for the activity dynamics observed in striatal principal cells. Intracellular recordings in organotypic cortex-striatum co-cultures (Plenz and Aertsen, submitted) revealed *in vitro* that upon cortical, activation principal neurons stay depolarized just a few millivolts below threshold, while showing an irregular spike discharge. Such membrane potential dynamics were recently observed in principal neurons *in vivo*, where it was referred to as the "enabled state" (Wilson, 1993). It is important to note that principal neurons possess a GABAa-system with a reversal potential just below threshold, and overlapping with the membrane potential during the enabled state (Plenz and Aertsen, submitted). Thus, during this state the recurrent inhibition of principal neurons is predominantly of the shunting type. Moreover, principal neurons show a strong increase in membrane resistance upon depolarization, due to an anomalous rectifier (Wilson, 1993). This leads to an electrotonically compact dendritic tree of principal neurons during the enabled state, which strengthens the shunting operation. Upon local removal of inhibitory inputs, principal neurons start firing during the enabled state. This indicates the presence of strong inhibitory inputs stabilizing the enabled state just below threshold (Plenz and Aertsen, submitted).

In summary, we conclude that the enabled state in striatal principal neurons represents the dynamic configuration of the neostriatum to detect local correlations in the cortical activity distribution, much like the retinal network analyses the visual scene for local movement features.

"FIGURE-GROUND" DISCRIMINATION IN CORTICAL ACTVITY CHANGES

A higly simplified scheme of the local correlation detection by principal cells is outlined in Fig. 2A. This arrangement of principal cell interaction will result in output activity only if the cortical activity pattern switches from the association *ab* to the association *ac*, i.e. the input activity "moves" from A to B. However, in order to evaluate a change in the cortical activity distribution, local correlation detection is a necessary but not sufficient step. This becomes more clear when we turn once more to the metaphor of movement detection in the visual system.

Consider the situation of light points moving randomly on a screen (Fig. 2B). A single movement detector comparing two locations within the area indicated by a circle will give a certain response. This response will be high, if the movement direction of the light point within that area matches the movement orientation of the local detector (Fig. 2B, broken line). In Figure 2C, the situation for the single "local" movement detector is the same, however, a new "global" feature introduced: most of the light points on the screen move into the same direction. Clearly, local inspection of the screen does not allow to judge whether the observed local correlation results from a random distribution of moving points (Fig. 2B) or from a global - i.e. spatially coherent - change of activity on the screen (Fig. 2C). If all movement detectors have their preferred direction of movement in parallel to the one within the circled area, this coherent pattern of moving dots will give a much larger overall response of the movement detector network than the random pattern. Obviously, the example of coherence in Figure 2C is a very simple one. Such a pattern may arise e.g. from a simple structured "figure" moving against a "ground". However, from these considerations, it is also evident that "coherence" can only be defined with respect to the underlying coordinate system of the network of movement detectors. Hence, from this point of view, single-unit studies in the neostriatum, would be of limited value to address this particular type of operation.

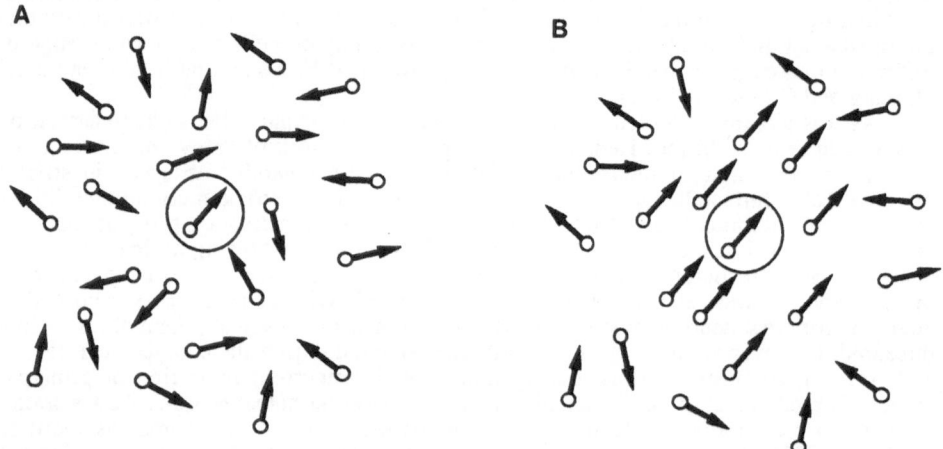

Figure 2. A. The recurrent lateral inhibition among striatal principal neurons constitutes a correlation detector. When the cortical input activity changes from *A* to *B*, this detector will give a response. If the cortical input activity would have changed from *B* to *A*, the delayed GABAa-ergic signal (Δt) from the right principal cell (ms) would have vetoed the excitatory input to the left principal cell. **glu** excitatory glutamatergic synapse from cortical afferents; **black synapse** GABAa-ergic synapse. **B.** Coherence analysis in the visual system. Light points on a retina (open circles) moving in random directions (arrows). An elementary movement detector evaluating a local region (circle) will respond to a particular movement vector (broken line). **C.** Similar situation as in B, but now there is a global coherence feature: light points over an extended region of space all move in the same direction. Observe that the response of the local movement detector is the same in both cases.

In summary, we conclude that in order to detect "global" coherence in the visual flow field (i.e. to extract "figure-movement" from "ground-noise"), spatial integration over the response of multiple elementary movement detectors has to be performed (Reichardt, 1987; Egelhaaf et al., 1988). Although the striatum has the anatomical appearance of a 3-dimensional volume, it functionally behaves like a 2-dimensional sheet: there is only one synapse between the main input and output structure (Wickens, personal communication). Hence, the visual analogy translates directly to the case of the neo-striatum: in order to detect "global" coherence in the "cortical flow field", spatial integration over the response of multiple elementary striatal "movement detectors" has to be performed. We propose that this extraction of global coherence in the changing cortical activity distribution is performed by spatial integration in the pallidal nuclei.

THE PALLIDAL NUCLEI AND THE SUBTHALAMIC NUCLEUS: "MINIMAL COHERENCE DETECTION"

It was shown in a quantitative anatomical study that neurons from the two pallidal nuclei integrate the activity of neostriatal principal neurons over different striatal volumes (Percheron et al., 1984). Thus, neurons from the globus pallidus externum integrate over a more restricted striatal volume than neurons from the globus pallidum internum. This view is supported by the higher burst index of neurons from the external pallidal segment as compared to that of neurons from the internal pallidal segment (Aldridge and Gilman, 1991). This arrangement of different spatial integration windows in the two pallidal nuclei not only makes it possible to detect a globally coherent change in the striatal activity, but also to extract the "spatial extent" of this coherence. We propose this to be realized in the basal ganglia by gain-controlled coupling of neurons from the globus pallidus externum to neurons from the globus pallidus internum.

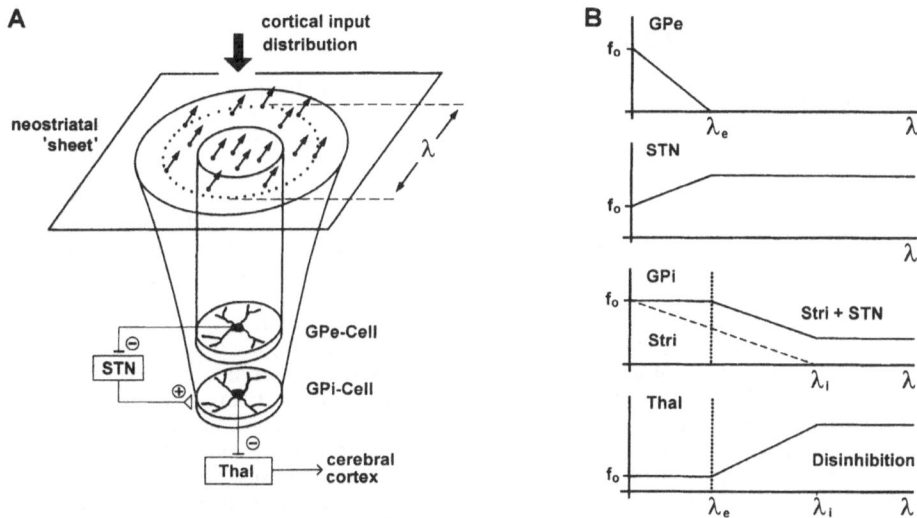

Figure 3. Minimal coherence detection. **A**. A change in the cortical activity distribution induces a certain change in the spatial pattern of activity in the neostriatal "sheet" (arrows, cf. Fig. 2). For convenience, the spatial extent of coherence is indicated as a single compact region with radius λ. Because of the smaller integration window of cells in the pallidum externum (**GPe-cell**) as compared to cells in the pallidum internum (**GPi-cell**), the inhibition exerted by the neostriatum onto a GPe-cell will saturate earlier (λe) than in a GPi cell (λi) as λ increases. **B**. The resulting decrease of GPe-cell spike activity leads to a disinhibition of the subthalamic nucleus (**STN**), which in turn increases its excitatory drive onto the GPi-cells. Consequently, the direct inhibition exerted by the neostriatum onto GPi-cells (dashed line, **Stri**) will be counterbalanced by excitation from the STN, until the extent of spatial coherence in the neostriatum exceeds the area of spatial integration of the GPe-cell (breakpoint at λe in Stri + STN curve). From here on, the activity in the GPi starts to decrease and, as a result, the disinhibitory action onto thalamic cells (**Thal**) increases. This disinhibitory action gives thalamic cells the possibility to raise cortical population activity. Thus, the STN plays a crucial role in determining the "threshold of minimal coherence" (breakpoint at λe in Thal-curve). fo spontaneous discharge activity.

It is known that the inhibitory and topographic pallido-subthalamic projection arises exclusively from the globus pallidus externum. The subthalamic nucleus (STN), besides having a recurrent feedback loop with neurons from globus pallidus externum, projects in an excitatory fashion to the internal pallidal segment (Parent, 1990). The sign inversion in the projection of the subthalamic nucleus to the internal pallidal segment is necessary. We propose that this arrangement serves to evaluate whether a certain "minimal extent of coherence" is achieved, at which the basal ganglia provide a disinhibitory signal to allow the next motor program to be performed (Fig. 3). In this conceptual framework, the interaction between the two pallidal nuclei is much more than only a "balance" between two output streams of the neostriatum. It also provides a complexity judgement on the change in cortical activity distribution, the mechanism of which can be understood from the analogy of the visual systems.

The subthalamic nucleus plays a crucial role in this framework. Along the excitatory pathway from the motor cortex, a direct gain control of the action between the two pallidal nuclei can be performed (Parent, 1990). As can be easily derived from Figure 3, an increased (decreased) activity of the STN will further increase (decrease) the minimal amount of coherence necessary for exerting a certain disinhibitory action onto thalamic neurons (Bergman et al., 1990).

SELECTION OF CORTICAL CELL ASSEMBLIES BY GENERAL CONTROL OF CORTICAL POPULATION DYNAMICS

The theory developed so far requires specificity and plasticity mainly at the level of the neostriatum. There, the cortico-striatal synaptic strength, which is under dopaminergic control (Miller et al., 1990; Calabresi et al., 1992), determines the "contrast" of two cortical inputs which have to be correlated. The high plasticity of this projection is reflected by the high spine density on principal neurons. We conclude that the dopaminergic control of this plasticity can only be adequately interpreted in terms of changes in correlated input activity. The actual extraction of coherence can be reduced to a simple spatial integration, as is reflected in the uniformity of strio-pallidal synapses (DiFiglia et al., 1982).

The overall anatomy of the neostriatal/pallidal complex strongly suggests that spatial integration over local correlation detectors is used to pass a highly complex judgement on the change of the cortical activity distribution ("minimal coherence"). Nevertheless, the information carried by the coherence signal itself (e.g. in the firing rate) is low. Its real information content resides in the time at which this signal raises the general population activity in - spatially not necessarily restricted - cortical premotor and motor areas.

The specific action of an intrinsically global population dynamics signal can be understood in the framework of cortical cell assemblies. It was shown that the efficacy of synaptic coupling of a neuron onto another critically depends on the background population activity level (Aertsen and Preissl, 1991; Bernander et al., 1991). Moreover, spike activity in neurons critically depends in an expansive nonlinear way on the level of population activity (Abeles, 1982; Eeckman and Freeman, 1991). Hence, it is reasonable to conclude that pyramdidal neurons do profit differentially from a change in background population activity: upon a certain rise in population activity, neurons already receiving some synchronized excitatory inputs are more likely to discharge than neurons receiving less synchronized inputs (Abeles, 1982; Aertsen and Preissl, 1991). Thus, the selection of the motor program depends on the active linking of neurons in (pre)-motor cortices to cell assemblies currently active in other cortical areas. The selective function exerted by this activity-dependent, dynamic linking architecture mainly defines the specificity of the gating mechanism of the basal ganglia.

SUMMARY

The view of basal ganglia function developed in this paper can be described in a very short form: the formation of cell assemblies in the cortex is accompanied by spatio-temporal changes of input activity to the neo-striatum. The correlations of this input activity are evaluated under dopaminergic control in a way similar to local "movement" detection in the visual system. The very moment a certain minimal amount of coherence is detected - compare the "pop out" of "figure" from "ground" - the basal ganglia output results in a general rise of activity in cortical (pre)-motor areas, leading to a motor action. The specificity of this general rise of activity mainly emerges from the dynamic linking of neurons to currently active cell assemblies.

Acknowledgements

We thank Valentino Braitenberg, Jeff Wickens, Robert Miller, Martin Egelhaaf, Johannes Zanker and Christian Wehrhahn for helpful discussions and Volker Staiger for expert assistance in preparing the figures.

REFERENCES

Abeles, M., 1982, " Local Cortical Circuits. An Electrophysiological Study," Springer-Verlag, Berlin.

Abeles, M., 1991, "Corticonics," Cambridge University Press, New York.

Abeles, M., Prut, Y., Bergman, H., Vaadia, E., and Aertsen, A., 1993, Integration, synchronicity and periodicity, in: "Brain Theory: Spatio-Temporal Aspects of Brain Function," A. Aertsen, ed., Elsevier Science, Amsterdam, in press.

Aertsen, A., Gerstein, G. L., Habib, M. K., and Palm, G., 1989, Dynamics of neuronal firing correlation: modulation of 'effective connectivity', J. Neurophysiol. 61:900-917.

Aertsen, A. and Gerstein, G. L., 1991, Dynamic aspects of neuronal cooperativity: fast stimulus-locked modulations of `effective connectivity', in: "Neuronal Cooperativity," J. Krüger, ed., Springer-Verlag, Berlin, pp. 52-67.

Aertsen, A., Vaadia, E., Abeles, M., Ahissar, E., Bergman, H., Karmon, B., Lavner, Y., Margalit, E., Nelken, I., and Rotter, S., 1991, Neural interactions in the frontal cortex of a behaving monkey: signs of dependence on stimulus context and behavioral state, J. Hirnforsch. 32:735-743.

Aertsen, A., and Preissl, H., 1991, Dynamics of activity and connectivity in physiological neuronal networks, in: "Nonlinear Dynamics and Neuronal Networks," H.G. Schuster, ed., VCH, Weinheim, pp. 281-301.

Alexander, G.E., DeLong, M.R., and Strick, P.L., 1986, Parallel organization of functionally segregated circuits linking basal ganglia and cortex, Ann. Rev. Neurosci. 9:357-381.

Aldrige, J.W. and Gilman, S., 1991, The temporal structure of spike trains in the primate basal ganglia: afferent regulation of bursting demonstrated with precentral cerebral cortical ablation, Brain Res. 543:123-138.

Amthor, F.R., and Grzywacz, N., 1991, Nonlinearity of the inhibition underlying retinal directional selectivity, Vis. Neurosci. 6:197-206.

Barlow, H.B. and Levick, W.R., 1965, The mechanism of directionally selective units in the rabbit's retina, J. Physiol. 178:477-504.

Bedenbaugh, P.H., Gerstein, G.L., Boven, K.-H., and Aertsen, A., 1988, The meaning of stimulus dependent changes in cross correlation between neuronal spike trains, Soc. Neurosci. Abstr. 14:651.

Bedenbaugh, P.H., Gerstein, G.L., and Aertsen, A., 1990, Dynamic convergence in neural assemblies, Soc. Neurosci. Abstr. 16:1224.

Bergman, H., Wichmann, T., and DeLong, M.R., 1990, Reversal of experimental parkinsonism by lesions of the subthalamic nucleus, Science 249:1436-1438.

Bernander, Ö., Douglas, R.J., Martin, K.A., and Koch, C., 1991, Synaptic background activity influences spatiotemporal integration in single pyramidal cells, Proc. Natl. Acad. Sci. USA 88:11569-11573.

Bienenstock, E., 1991, Notes on the growth of a 'composition machine', in: "Contributions to Interdisciplinary Workshop on Compositionality in Cognition and Neural Models," Abbaye de Royaumont (Fr), pp. 1-19.

Boven, K.-H., and Aertsen, A., 1989, Dynamics of activity in neuronal networks give rise to fast modulations of functional connectivity, in: "Parallel Processing in Neural Systems and Computers," R. Eckmiller, G. Hartmann and Hauske, G., eds., Elsevier Science Publishers, Amsterdam, pp. 53-56.

Braitenberg, V. and Schüz, A., 1992, "Studies of Brain Function: Anatomy of the Cortex," Springer, Berlin.

Calabresi, P., Maj, R., Pisani, A. Mercuri, N.B., and Bernardi, G., 1992, Long-term synaptic depression in the striatum - physiological and pharmacological characterization, J. Neurosci. 12:4224-4233.

Chang, H.T., Wilson, C.J., and Kitai, S.T., 1981, Single neostriatal efferent axons in the globus pallidus: a light and electron microscopic study, Science 213:915-918.

Chevalier, G., and Deniau, J.M., 1990, Disinhibition as a basic process in the expression of striatal functions, TINS 13:277-280.

DiFiglia, M., Pasik, P., and Pasik, T., 1982, A Golgi and ultrastructural study of the monkey globus pallidus, J. Comp. Neurol. 212:53-75.

Eeckman, F.H., and Freeman, W.J., 1991, Asymmetric sigmoid non-linearity in the rat olfactory system, Brain Res. 557:13-21.

Egelhaaf, M., Hausen, K., Reichardt, W., and Wehrhahn, C., 1988, Visual control in flies relies on neuronal computation of object and background motion, *TINS* 11:351-358.

Engel, A.K., König, P., and Singer, W., 1991, Direct physiological evidence for scene segmentation by temporal coding, *Proc. Natl. Acad. Sci. USA* 88:9136-9140.

Erb, M., and Aertsen, A., 1992, Dynamics of activity in biology-oriented neural network models: stability at low firing rates, *in*: "Information Processing in the Cortex: Experiments and Theory", A. Aertsen and V. Braitenberg, eds., Springer-Verlag, Berlin.

Flaherty, A.W., and Graybiel, A.M., 1991, Corticostriatal transformations in the primate somatosensory system. Projections from physiologically mapped body-part representations, *J. Neurophysiol.* 66:1251-1262.

Gerstein, G.L., and Aertsen, A., 1985, Representation of cooperative firing activity among simultaneously recorded neurons, *J. Neurophysiol.* 54:1513-1528.

Gerstein, G.L., Bedenbaugh P., and Aertsen, A., 1989, Neuronal assemblies, *IEEE Trans. Biomed. Engineering* 36:4-14.

Goldman-Rakic, P.S., and Selemon, L.D., 1984, Topography of corticostriatal projections in nonhuman primates and implications for functional parcellation of the neostriatum, *in*: "Cerebral Cortex: V. Sensory-Motor Areas and Aspects of Cortical Connectivity," G. Jones and A. Peters, eds., Plenum Press, New York, pp. 447-466.

Hebb, D., 1949, "The Organization of Behavior. A Neuropsychological Theory," J. Wiley, New York.

Hikosaka, O., Sakamoto, M., and Usui, S., 1989, Functional properties of monkey caudate neurons III. Activities related to expectation of target and reward, *J. Neurophysiol.* 61:814-832.

Mink, J. W., and Thach, W. T., 1991, Basal ganglia motor control. I. Nonexclusive relation of pallidal discharge to five movement modes, *J. Neurophysiol.* 64:273-300.

Miller, R., Wickens, J.R., and Beninger, R.J., 1990, Dopamine D-1 and D-2 receptors in relation to reward and performance: a case for the D-1 receptor as a primary site of therapeutic action of neuroleptic drugs, *Prog. Neuro.* 34:143-183.

Neven, H., and Aertsen, A., 1992, Rate coherence and event coherence in the visual cortex: a neuronal model of object recognition, *Biol. Cybern.* 67:309-322.

Palm, G., 1982, "Neural Assemblies, Studies of Brain Function, Vol. 7," Springer, Berlin.

Parent, A., 1990, Extrinsic connections of the basal ganglia, *TINS* 13:254-271.

Parthasarathy, H.B., Schall, J.D., and Graybiel, A.M., 1992, Distributed but convergent ordering of corticostriatal projections: analysis of the frontal eye field and the supplementary eye field in the macaque monkey, *J. Neurosci.* 12:4468-4488.

Percheron, G., Yelnik, J., and François, C., 1984, A Golgi analysis of the primate globus pallidus. III. Spatial organization of the strio-pallidal complex, *J. Comp. Neurol.* 227:214-227.

Plenz, D., and Aertsen, A., 1993, Striatal dynamics in cortex-striatum cultures, submitted.

Poggio, T., and Reichardt, W., 1973, Considerations on models of movement detection, *Kybernetik* 13:223-227.

Reichardt, W., 1987, Evaluation of optical motion information by movement detectors, *J. Comp. Physiol.* A 161:533-547.

Selemon, L. D., and Goldman-Rakic, P. S., 1988, Common cortical and subcortical target areas of the dorsolateral prefrontal and posterior parietal cortices in the rhesus monkey: a double label study of distributed neural networks, *J. Neurosci.* 8:4049-4068.

Torre, V., and Poggio, T., 1978, A synaptic mechanism possibly underlying directional selectivity to motion, *Proc. R. Soc. Lond.* 202:409-416.

Tremblay, L., and Filion, M., 1989, Responses of pallidal neurons to striatal stimulation in intact waking monkeys, *Brain Res.* 498:1-16.

Vaadia, E., and Aertsen, A., 1992, Coding and computation in the cortex: single-neuron activity and cooperative phenomena, *in*: "Information Processing in the Cortex," A. Aertsen and V. Braitenberg, eds., Springer-Verlag, Berlin, pp. .81-122.

Vaadia, E., Ahissar, E., Bergman, H., and Lavner, Y., 1991, Correlated activity of neurons: a neural code for higher brain functions? *in*: "Neuronal Cooperativity," J. Krüger, ed., Springer, Berlin, pp. 249-279.

von der Malsburg, C., 1986, 'Am I thinking assemblies,' *in*: "Brain Theory," G. Palm and A. Aertsen, eds., Springer-Verlag, Berlin, pp. 161-176.

Wickens, J.R., Alexander, M.E., and Miller, R., 1991, Two dynamic modes of striatal function under dopaminergic - cholinergic control: simulation and analysis of a model, *Synapse* 8:1-12.

Wilson, C.J., 1993, The generation of natural firing patterns in neostriatal neurons, *in*: "Chemical Signalling in the Basal Ganglia," G.W. Arbuthnott and P.C. Emson, eds., *Prog. Brain Res.* in press.

ROLE OF BASAL GANGLIA IN CONTROL OF INNATE MOVEMENTS, LEARNED BEHAVIOR AND COGNITION - A HYPOTHESIS

Okihide Hikosaka

Laboratory of Neural Control
National Institute for Physiological Sciences
Myodaijicho, Okazaki 444, Japan

INTRODUCTION

The progress of basal ganglia research has been rapid and fruitful. As is clear from articles in this book, we now know detailed neural circuits in the basal ganglia, heterogeneous distribution of chemically defined neuron groups, and the presence of various transmitter candidates and their intracellular actions. This accumulated knowledge allows us to provide reasonable interpretations for clinical symptoms of basal ganglia disorders and even to suggest novel surgical or drug treatments.

However, we still know little about how the basal ganglia work when we do something or think something. Such normal functions of the basal ganglia must be dynamic processes in which different parts of the basal ganglia work in an unique yet coordinated manner. Moreover, the basal ganglia system is only a part of the neural system that controls our behavior. Obviously we need to characterize the function of the basal ganglia in relation to such a global system. After a laborious task of investigating single cell activities, we now know that motor and non-motor signals are coded by basal ganglia neurons in various ways, frequently depending on the behavioral contexts. However, such signals are also found in other brain areas, especially in the frontal cortices.

What do all these data tell us after all? Where do such signals originate and how are they used? Are the signals merely circulating through the loop circuits formed by the cerebral cortex, basal ganglia and thalamus? What then is unique about the basal ganglia?

I think, in order to answer these questions or to figure out appropriate experimental paradigms to guide the answer, it is critical and urgent to set forth a working hypothesis. This paper is one such attempt.

DUAL MODE OF BASAL GANGLIA FUNCTION: DISINHIBITION AND ENHANCED SUPPRESSION

The two major outputs of the the basal ganglia, via the internal segment of the globus pallidus and the substantia nigra pars reticulata, are both GABAergic inhibitory and their levels are set extremely high by their tonic background discharges (DeLong and Georgopoulos, 1981) (Figure 1). Major inputs to these output areas originate from the striatum (putamen and caudate nucleus) which are also GABAergic and inhibitory (Carpenter, 1981). Here emerges a first mode of basal ganglia operation: an excitatory input from the cerebral cortex or the thalamus would lead to a disinhibition in the target · structures of the basal ganglia (Penney and Young, 1983; Chevalier and Deniau, 1990).

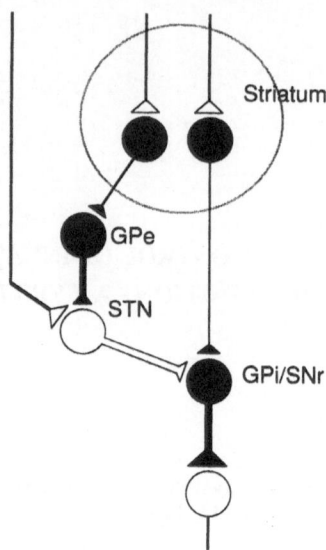

Figure 1. Schematic neural circuits in the basal ganglia. Excitatory and inhibitory neurons are shown by filled and open symbols, respectively. GPe: globus pallidus, external segment. GPi: globus pallidus, internal segment. SNr: substantia nigra pars reticulata. STN: subthalamic nucleus.

This scheme has been demonstrated for the saccadic oculomotor function in the monkey which is mediated by the caudate and substantia nigra pars reticulata whose target is the superior colliculus (Hikosaka and Wurtz, 1989). A group of caudate neurons show spike activity before saccades to task-specific targets (Hikosaka et al., 1989). The transient signal would inhibit the tonic spike activity of substantia nigra neurons (Hikosaka and Wurtz, 1983a), thereby removing the inhibition on saccade burst neurons in the superior colliculus (Hikosaka and Wurtz, 1983b). These observations led to the conclusion that the basal ganglia contribute to the initiation of movement through the mechanism of disinhibition.

For the disinhibition to work properly, however, the level of tonic inhibition must be set at an appropriate level. Furthermore, an active change of the level of the inhibition would add another perspective to basal ganglia function; that is, suppression of movements. The subthalamic nucleus is now thought to provide the output elements of the basal ganglia with an excitatory drive (Hammond et al., 1978; Kitai and Kita, 1987). In its ventral part are found a cluster of visuo-oculomotor neurons which are activated when saccades are to be suppressed or terminated (Matsumura et al., 1992). The subthalamic nucleus may thus be viewed as a tool for motor suppression. With this tool, the cerebral cortices can now actively suppress movements through either direct connections to the subthalamic nucleus or indirect connections via the striatum and the external segment of the globus pallidus.

BASAL GANGLIA CONTROL INNATE MOVEMENTS AND LEARNED MOVEMENTS

Some major targets of the basal ganglia are found in the midbrain, in addition to the superior colliculus. They include the pedunculopontine nucleus, cuneiform nucleus, periaqueductal gray and the surrounding reticular formation (Graybiel and Ragsdale, 1979; Niijima and Yoshida, 1982; Noda and Oka, 1984). These midbrain areas have crucial control over a variety of movements, such as locomotion, vocalization, mastication, respiration, vomiting, eye blinks, which are thought to be genetically determined in a species-specific manner (Grillner and Shik, 1973; Mori, 1987; Garcia-Rill, 1991). These areas also control autonomic functions, such as blood pressure and pupil dilation, and sensory functions, such as nociception (Bandler et al., 1991). These elementary behaviors, when grouped together, would constitute a purposive action, such as threat, aggression, and flight (Holstege, 1991).

The primary function of the basal ganglia would thus be schematized as follows. The lower brainstem and spinal cord contain generators of elementary movement patterns (Hepp et al., 1989; Pearson and Rossignol, 1991; Sillar, 1991). Each of the pattern generators is an intricate neural machine but would not function independently. Only when they are selectively and collectively controlled by the midbrain motor regions do the movement patterns become purposive.

During phylogenetic development animals obtained a variety of innate actions so that they could adapt to different conditions of environment. Here arose a new question: 'How can an appropriate action be selected?'. I propose that the basal ganglia evolved to play such a role of selection. The disinhibition and enhanced inhibition, which I mentioned above, are superb mechanisms for such a function. Selection is always based on some kind of evaluation and motivation. Such signals could be provided by the inputs from the limbic system (Ragsdale and Graybiel, 1988; Haber et al., 1990).

Clearly this is a very primitive form of motor control system. The actions and the constituent movement patterns are still basically innate. With the projections to the midbrain/pontine motor regions the basal ganglia can select innate actions (Figure 2). Thus, without the cerebral cortex, the animal might be able to walk, run, orient, feed, and vocalize. It would be amazing how integrated such an animal might look.

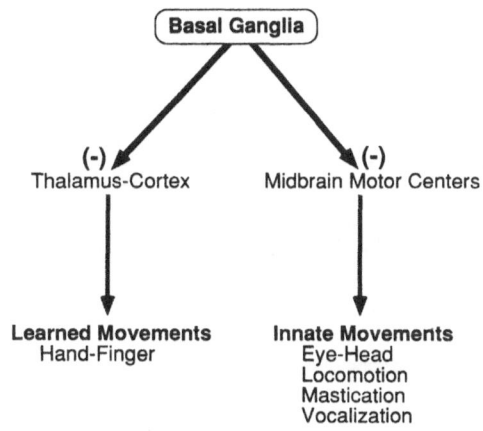

Figure 2. Basal ganglia control innate and learned movements.

But we do have learned types of movement. Here also the basal ganglia may play a role of selection. Underlying such a function may be the connection to the thalamus which is mutually connected with frontal cortical areas (Carpenter, 1981). In contrast to the midbrain projections, this pathway would control learned movements (Figure 2).

The motor control system can be viewed as a hierarchical organization (Figure 3). Central pattern generators in the spinal cord and the lower brainstem might be called 'fixed pattern generators', which imply that they generate innate movement patterns. They are coordinated by the midbrain motor regions such that synergistic, purposive actions are generated. I would like to call them 'fixed action generators', implying that action is a synergistic complex of movement patterns. They are 'fixed' because their products are still innate. True adaptability to new environment was obtained by the evolution of the cerebral cortex. The basic function of the cerebral cortex would be to combine the available actions spatially and sequentially to create new action programs ('learned action generator').

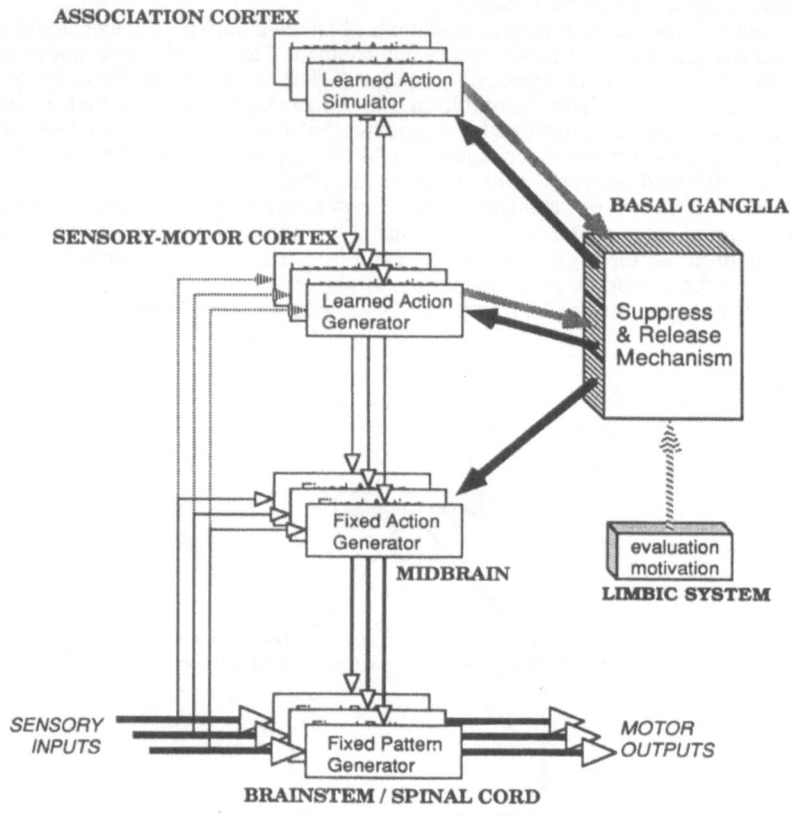

Figure 3. Hierarchical organization of motor control system and its relation to basal ganglia.

By the time an animal has grown up, many such action programs will have been accumulated in the motor-premotor cortices. Here arises the same question: 'Which action program should be selected?'. Again the basal ganglia should play a key role in the selection. This would be done by the connection through the thalamus (Schell and Strick, 1984; Nambu et al., 1991). But unlike in the case of innate actions, there is no fixed rule to be applied in terms of the algorithm of the selection. Perhaps for this reason, the basal ganglia need the information from the cerebral cortex (Kemp and Powell, 1970). In consequence, the relationship between the basal ganglia and the cerebral cortex is bound to be mutual.

The same relationship is seen between the basal ganglia and the association cortices (Ilinsky et al., 1985), which I characterized as 'learned action simulator' as opposed to 'learned action generator'. The role of the basal ganglia here would be to select non-motor signals, such as memory, attention, and expectation.

So far I have suggested that the basal ganglia select available action programs in the midbrain or in the cerebral cortex. But the function of the basal ganglia system might go beyond this: it might play an important role in formation of the action programs.

FEATURE INTEGRATION VS PROCEDURE FORMATION

Perhaps the most important feature in learning is to associate independent signals. A notable example is classical conditioning in which a previously unrelated stimulus is associated with an innate reflex. I would suggest here that there are two conceptually different modes of association (Figure 4). In either case, neurons A and B at first are supposed to function independently; then there appears another neuron that associates A and B, but in two different ways. In the first case the new element receives signals from A and B; in the second case the element sends signals to A and B.

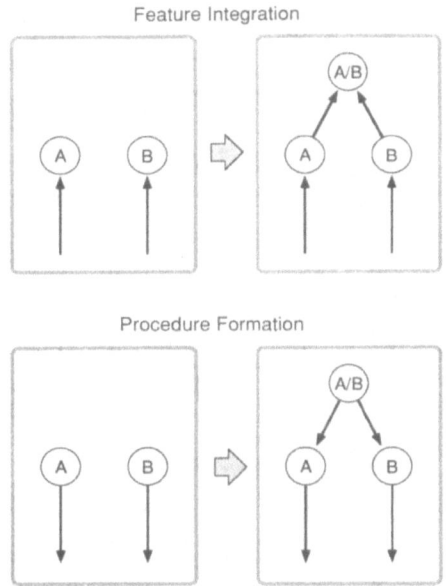

Figure 4. Two modes of association of neural signals.

The first mode (Figure 4, top) would be utilized mostly by the sensory systems. In the process that might be called 'feature integration', for example, we learn to recognize faces by combining signals of elementary features such as eye and mouth (Fujita et al., 1992). It is plausible to configure a neural mechanism underlying this process, mostly based on Hebbian or modified Hebbian type learning (Brown et al., 1990; Fregnac et al., 1992).

What seems important in the motor system is the second mode of learning (Figure 4, bottom). When we were babies we had a limited number of simple action programs or procedures. When we learned to do something, say, write our own names, the first thing we did was probably to combine the then available procedures (Newell, 1978). Such a process, which I call here 'procedure formation', may correspond to this type of neural organization. Unlike 'feature integration', it is difficult to imagine an underlying neural mechanism.

I would like to suggest that the basal ganglia may play an instructive role in procedure formation or procedural learning (Figure 5). Suppose there are neurons A and B in the cerebral cortex which send outputs independently (Figure 5, top). After mutual interplay with the basal ganglia a new set of cortical neurons that control both A and B emerges (Figure 5, bottom). The process would proceed as follows.

As already mentioned, the basic mode of basal ganglia action is inhibition and disinhibition. Before signals are fed into the basal ganglia, the outputs of the basal ganglia would continually suppress their targets, in this case, the thalamo-cortical circuits (Figure 5, top).

In the new environment, the animal may attempt to perform a motor act A simultaneously or sequentially with B, for example (Figure 5, top). It is known that there is

593

an extensive convergence in cortico-striatal connections (Parthasarathy et al., 1992). Thus it is conceivable that the signals A and B converge onto single neurons in the striatum. The neurons may at first not respond, because the striatum is probably one of the most quiet areas in the brain (Hikosaka et al., 1989). But if the combination of A & B is repeated and if the action produces reward, the combined signals may become able to activate the striatal neurons. The signals of the reward value may be transmitted by dopamine neurons in the substantia nigra pars compacta (SNc) (Romo and Schultz, 1990; Schultz and Romo, 1990; Ljungberg et al., 1992) which exert modulatory effects on striatal neurons (Gerfen et al., 1990; Garcia-Munoz et al., 1991; Calabresi et al., 1992).

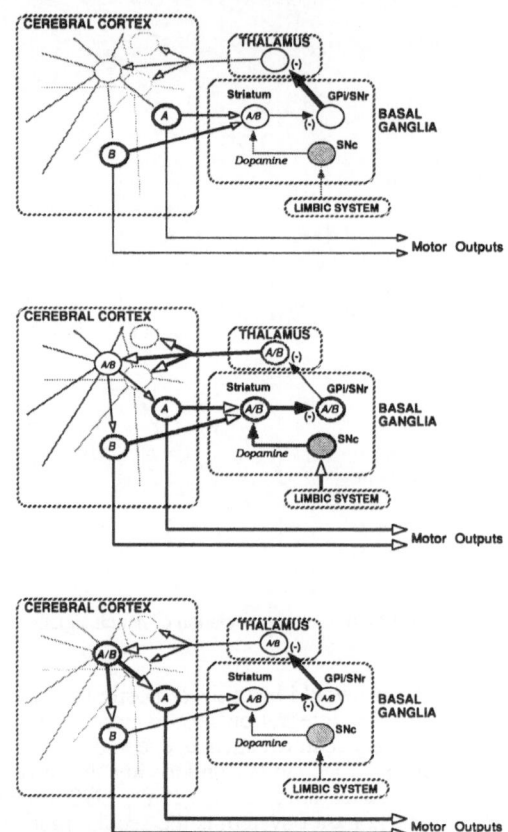

Figure 5. Hypothetical process of procedural learning.

Once the combined signal is put through the gateway of the basal ganglia, it would allow a limited portion of the thalamo-cortical circuits to be active (Figure 5, center). Note, however, that this selected set of cortical neurons are free to be active only under the condition of A/B. Let us assume that some (perhaps only a small fraction) of these neurons have already had weak and insignificant connections to the output neurons A and B. If the attempt of A/B is repeated, the efferent connections would be enhanced, perhaps based on the Hebbian rule. It may be possible that the combined cortical signals are again fed into the basal ganglia so that further complex combinations are created.

In short, the basal ganglia would temporarily retain the memory of behavioral procedures. The cerebral cortex would create motor or procedural memory based on such a neural template. An important feature is that the basal ganglia can not only combine different cortical signals but also test the validity of the combination through their outputs and the evaluating signals.

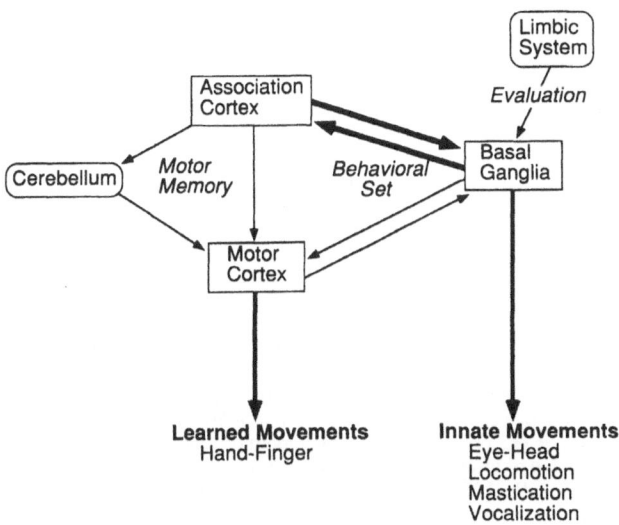

Figure 6. Differential roles of motor areas in learning.

CONCLUSION

Finally I would like to propose the hypothetical framework of the motor mechanisms (Figure 6). The basal ganglia system is a dominant structure in the lower species of animals. It would act to facilitate motor programs based on the reward-contingent inputs from the limbic system. The motor programs are still innate and thus their patterns are largely fixed. Such animals must learn, however, to associate particular environmental signals with particular motor programs. The attempted behavioral sets are first formed in the basal ganglia, and the cerebral cortex learns to create procedural memory based on the template. As the animal's behavior becomes more complex, the motor program itself must also be learned. Here again, the basal ganglia may play an instructive role so that motor memory is created efficiently in the motor cortical areas and probably also in the cerebellum.

REFERENCES

Bandler, R., Carrive, P., and Zhang, S.P., 1991, Integration of somatic and autonomic reactions within the midbrain periaqueductal grey: Viscerotopic, somatotopic and functional organization, *in*: "Role of the Forebrain in Sensation and Behavior," G. Holstege, ed., Elsevier, Amsterdam, pp. 269-305.

Brown, T.H., Kairiss, E.W., and Keenan, C.L., 1990, Hebbian synapses: Biophysical mechanisms and algorithms, *Annu. Rev. Neurosci.* 13:475.-511.

Calabresi, P., Maj, R., Pisani, A., Mercuri, N.B., and Bernardi, G., 1992, Long-term synaptic depression in the striatum: physiological and pharmacological characterization, *J. Neurosci.* 12:4224-4233.

Carpenter, M.B., 1981, Anatomy of the corpus striatum and brain stem integrating systems, *in*: "The Nervous System," V.B. Brooks, ed., American Physiological Society, Bethesda, MD, pp. 957-995.

Chevalier, G., and Deniau, J.M., 1990, Disinhibition as a basic process in the expression of striatal functions, *Trends Neurosci.* 13:277-280.

DeLong, M.R., and Georgopoulos, A.P., 1981, Motor functions of the basal ganglia, *in*: "The Nervous System," V.B. Brooks, ed., American Physiological Society, Bethesda, MD, pp. 1017-1061.

Fregnac, Y., Shulz, D., Thorpe, S., and Bienenstock, E., 1992, Cellular analogs of visual cortical epigenesis. I. Plasticity of orientation selectivity, *J. Neurosci.* 12:1280-1300.

Fujita, I., Tanaka, K., Ito, M., and Cheng, K., 1992, Columns for visual features of objects in monkey inferotemporal cortex, *Nature* 360:343-346.

Garcia-Munoz, M., Young, S.J., and Groves, P.M., 1991, Terminal excitability of the corticostriatal pathway. I. Regulation by dopamine receptor stimulation, *Brain Res.* 551:195-206.

Garcia-Rill, E.,1991, The pedunculopontine nucleus, *Progress in Neurobiology* 36:363-389.

Gerfen, C.R., Engber, T.M., Mahan, L.C., Susel, Z., Chase, T.N., Monsma, J.F.J., and Sibley, D.R., 1990, D1 and D2 dopamine receptor-regulated gene expression of striatonigral and striatopallidal neurons, *Science* 250:1429-1432.

Graybiel, A.M., and Ragsdale, C.W., 1979, Fiber connections of the basal ganglia, *in*: "Development of Chemical Specificity of Neurons," M. Cuenod, G.W. Kreutzberg, F.E. Bloom, eds., Elsevier, Amsterdam, pp. 239-283.

Grillner, S., and Shik, M.L., 1973, On the descending control of the lumbosacral spinal cord from the 'mesencephalic locomotor region', *Acta Physiol. Scand.* 87:320-333.

Haber, S.N., Lynd, E., Klein, C., and Groenewegen, H.J., 1990, Topographic organization of the ventral striatal efferent projections in the rhesus monkey: an anterograde tracing study, *J. Comp. Neurol.* 293:282-298.

Hammond, C., Denieu, J.M., Rizk, A., and Féger, J., 1978, Electrophysiological demonstration of an excitatory subthalamonigral pathway in the rat, *Brain Res.* 151:235.-244

Hepp, K., Henn, V., Vilis, T., and Cohen, B., 1989, Brainstem regions related to saccadic generation, *in*: "The Neurobiology of Saccadic Eye Movements," R.H. Wurtz, M.E. Goldberg, eds., Elsevier, Amsterdam, pp. 105-212.

Hikosaka, O., Sakamoto, M., and Usui, S., 1989, Functional properties of monkey caudate neurons. I. Activities related to saccadic eye movements, *J. Neurophysiol.* 61:780-798.

Hikosaka, O., and Wurtz, R.H., 1983a, Visual and oculomotor functions of monkey substantia nigra pars reticulata. I. Relation of visual and auditory responses to saccades, *J. Neurophysiol.* 49:1230-1253.

Hikosaka, O., and Wurtz, R.H., 1983b, Visual and oculomotor functions of monkey substantia nigra pars reticulata. IV. Relation of substantia nigra to superior colliculus, *J. Neurophysiol.* 49:1285-1301.

Hikosaka, O., and Wurtz, R.H., 1989, The basal ganglia, *in*: "The Neurobiology of Saccadic Eye Movements," R.H. Wurtz, M.E. Goldberg, eds., Elsevier, Amsterdam, pp. 257-281.

Holstege, G., 1991, Descending motor pathways and the spinal motor system: Limbic and non-limbic components, *in*: "Role of the Forebrain in Sensation and Behavior," G. Holstege, ed., Elsevier, Amsterdam, pp. 307-421.

Ilinsky, I.A., Jouandet, M.L., and Goldman-Rakic, P.S., 1985, Organization of the nigrothalamocortical systems in the rhesus monkey, *J. Comp. Neurol.* 236:315-330.

Kemp, J.M., and Powell, T.P.S., 1970, The corticostriate projection in the monkey, *Brain* 93:525-546.

Kitai, S.T., and Kita, H., 1987, Anatomy and physiology of the subthalamic nucleus: a driving force of the basal ganglia, *in*: "The Basal Ganglia II. Structure and Function-Current Concepts," M.B. Carpenter, A. Jayaraman, eds., Plenum, New York, pp. 357-376.

Ljungberg, T., Apicella, P., and Schultz, W., 1992, Responses of monkey dopamine neurons during learning of behavioral reactions, *J. Neurophysiol.* 67:145-163.

Matsumura, M., Kojima, J., Gardiner, T.W., and Hikosaka, O., 1992, Visual and oculomotor functions of monkey suhthalamic nucleus, *J. Neurophysiol.* 67:1615-1632.

Mori, S., 1987, Integration of posture and locomotion in acute decerebrate cats and in awake, freely moving cats, *Prog. Neurobiol. (Oxf)* 28:161-196.

Nambu, A., Yoshida, S., and Jinnai, K., 1991, Movement-related activity of thalamic neurons with input from the globus pallidus and projection to the motor cortex in the monkey, *Exp. Brain Res.* 84:279-284.

Newell, K.M., 1978, Some issues on action plans, *in*: "Information Processing in Motor Control and Learning," G.E. Stelmach, ed., Academic Press, New York, pp. 41-54.

Niijima, K., and Yoshida, M., 1982, Electrophysiological evidence for branching nigral projections to pontine reticular formation, superior colliculus and thalamus, *Brain Res.* 29:279-282.

Noda, T., and Oka, H., 1984, Nigral inputs to the pedunculopontine region: intracellular analysis, *Brain Res.* 322:332-336.

Parthasarathy, H.B., Schall, J.D., and Graybiel, A.M., 1992, Distributed but convergent ordering of corticostriatal projections: analysis of the frontal eye field and the supplementary eye field in the macaque monkey, *J. Neurosci.* 12:4468-4488.

Pearson, K.G., and Rossignol, S., 1991, Fictive motor patterns in chronic spinal cats, *J. Neurophysiol.* 66:1874-1887.

Penney, J.B., and Young, A.B., 1983, Speculations on the functional anatomy of basal ganglia disorders, *Annu. Rev. Neurosci.* 6:73-94.

Ragsdale, C.W., and Graybiel, A.M., 1988, Fibers from the basolateral nucleus of the amygdala selectively innervate striosomes in the caudate nucleus of the cat, *J. Comp. Neurol.* 269:506-522.

Romo, R., and Schultz, W., 1990, Dopamine neurons of the monkey midbrain: contingencies of responses to active touch during self-initiated arm movements, *J. Neurophysiol.* 63:592.-606

Schell, G.R., and Strick, P.L., 1984, The origin of thalamic inputs to the arcuate premotor and supplementary motor areas, *J. Neurosci.* 4:539-560.

Schultz, W., and Romo, R., 1990, Dopamine neurons of the monkey midbrain: contingencies of responses to stimuli eliciting immediate behavioral reactions, *J. Neurophysiol.* 63:607-624.

Sillar, K.T., 1991, Spinal pattern generation and sensory gating mechanisms, *Current Opinion in Neurobiology* 1:583-589.

PARTICIPANTS AND CONTRIBUTORS

Dr Elizabeth D. Abercrombie
Center Mol & Behavior Neurosci
Rutgers University
197 University Ave.
Newark, NJ 07102, USA

Dr Jorge Aceves
Centro de Investigacion
Departamento de Fisiologia
Apartado Postal 14-740;
Mexico 14, D.F. Mexico

Dr Ad Aertsen
Institut für Neuroinformatik
Ruhr-Universität
P.O. Box 102148
D-4630 Bochum
Germany

Dr Meryem Alamy
Dept de Neurophysiologie Générale
LNF3-CNRS
31 chemin Joseph Aiguier
13402 Marseille Cedex 9, France

Dr George Alheid
Department Behavioral Med
Box 396, Health Sciences Center
University of Virginia
Charlottesville, VA 22908, USA

Dr Marianne Amalric
Dept de Neurophysiologie Générale
LNF3-CNRS
31 chemin Joseph Aiguier
13402 Marseille Cedex 9, France

Dr Gordon Arbuthnott
Dept Preclinical Vet. Sciences
University of Edinburgh
Summerhall, Edinburgh
EH9 1QH, U.K.

Dr Sarah Augood
Department of Neurobiology
MRC Group
Institute Animal Physiol & Gen Res
Babraham, CB2 4AT Cambridge U.K.

Dr Rebecca L.M. Aylward
University Dept Pharmacology
Mansfield Road
Oxford OW1 3QT
U. K.

Dr Tipu Zahed Aziz
Dept Cell and Structural Biology
University of Manchester
Oxford Road
Manchester M13 9PT U.K.

Dr Anna Bal
Inserm U318 Equipe Lapsen
Pav Neurologie CHU Grenoble
BP 217
38043 Grenoble Cedex

Dr Mark S. Baron
Dept Neurology, WMB
6th Floor, P.O. Drawer V
Emory Univ School Med
Atlanta, GA 30322, USA

Dr Nguyen Bathien
Service de Neurologie
Hôpital Sainte Anne
1 rue Cabanis
75674 Paris Cedex 14

Dr Marielle Baudrimont
Inserm U 106 Pav Cl Bernard
Hôpital de la Salpêtrière
47 Bd de l'Hôpital
75651 Paris Cedex 13 France

Dr M. Flint Beal
Neurology Service
Massachusetts General Hospital
Boston, Massachusetts
USA

Dr Paul J. Bédard
Laboratoire de Neurobiologie
Hôpital de l'Enfant Jésus
1401 18ème rue
G1J 1Z4 Québec Canada

Dr Carlos Beltramino
Depaartment of Otolaryngology
Health Sciences Center
University of Virginia
Charlottesville, VA 22908, USA

Dr Abdelhamid Benazzouz
Laboratoire de Neurophysiologie
CNRS URA 1200
Université de Bordeaux II
146 rue Léo Saignat
33076 Bordeaux

Dr Ben D. Bennett
MRC Anatomical
Neuropharmacological Unit
Mansfield Road
Oxford OX1 3TH, U.K.

Dr Henk W. Berendse
Dept of Anatomy & Embryology
Vrije Universiteit
Van der Boechorststraat 7
1081 BT Amsterdam The Netherlands

Dr Hagai Bergman
Dept Physiology Hebrew Univ
Hadassah Med School
P.O.B. 1172, Jerusalem
Israel 91010

Dr Véronique Bernard
Lab Histologie & Embryologie
Université Bordeaux II
146 rue Léo Saignat
33076 Bordeaux Cedex

Pr Giorgio Bernardi
Clinica Neurologica Univ Tor Vergata
Via O. Raimondo 8
00173 Roma, Italy

Dr Sabina Berretta
Dept Brain Science & Cognition
M.I.T. E25-618
45 Carleton Street
Cambridge, MA 02139, USA

Pr Marie-Jo Besson
Laboratoire de Neurochimie
& Anatomie - Bâtiment B
9 Quai Saint Bernard
75005 Paris, France

Pr Anne Beuter
Department of Kinanthropologie
UQAM CP 8888
Suc A Montréal
H3C 3P8 Canada

Dr Mark Bevan
Dept Cell & Struct Biol
University of Manchester
Oxford Road
Manchester, M13 9PT, UK

Dr Pierre Blanchette
Laboratoire de Neurobiologie
Hôpital de l'Enfant Jésus
1401 18ème rue
G1J 1Z4 Québec Canada

Dr Alain Bloc
Dept de Neurophysiologie Générale
LNF-CNRS
31 chemin Joseph Aiguier
13402 Marseille Cedex 9, France

Pr Bertrand Bloch
Lab Histologie & Embryologie
Université Bordeaux II
146 rue Léo Saignat
33076 Bordeaux Cedex

Dr Frank Block
Department of Neurology
University of Essen
Hufeland str. 55
D - 4300 Essen, Germany

Dr J. Paul Bolam
MRC Anatomical
Neuropharmacological Unit
Mansfield Road
Oxford OX1 3TH, U.K.

Dr Antonello Bonci
Clinica Neurologica
Universita Tor Vergata
Via O. Raimondo 8
00173 Roma, Italy

Dr Ubaldo Bonuccelli
Institute of Clinical Neurology
University of Pisa
Via Roma 67
56100 Pisa Italy

Dr Thomas Boraud
Laboratoire de Neurophysiologie
CNRS URA 1200
Université de Bordeaux II
146 rue Léo Saignat
33076 Bordeaux

Dr Alex Braun
Department of Pathology
VA Medical Center
Northport, New York 11768
U.S.A.

Dr Serge Brique
Université de Lille II - CHRU
Sce Clinique Neurologique A
Hôpital B
59037 Lille Cedex

Dr Alexander A. Britain
University of Wales
Department of Psychology
Bangor, Gwynedd LL57 2DG
UK

Dr David Brooks
MRC Cyclotron Unit
Hammersmith Hospital
Du Cane Road
London W12 OHS, UK

Dr Jonathan Brotchie
Dept Cell & Struct Biol
University of Manchester
Oxford Road
Manchester, M13 9PT, U.K.

Dr Emmanuel Broussole
Service de Neurologie
Hôpital de l'Antiquaille
69321 Lyon Cedex 05
France

Dr Gordon Brown
University of Wales
Department of Psychology
Bangor, Gwynedd LL57 2DG UK

Dr Lucy L. Brown
Albert Einstein College of
Medicine, Bldg K, Room 601
1300 Morris Park Avenue
Bronx, NY10461 USA

Dr Pierre Burbaud
Laboratoire de Neurophysiologie
CNRS URA 1200
Université de Bordeaux II
146 rue Léo Saignat
33076 Bordeaux

Dr Paolo Calabresi
Clinica Neurologica
Universita Tor Vergata
Via O. Raimondo 8
00173 Roma, Italy

Dr Dominique Caparros-Lefebvre
Clinique Neurologique
Hôpital B-CHRU Lille
59037 Lille Cedex

Pr Malcolm B Carpenter
380 E. Chocolate Avenue
Hershey, Pa 17033
USA

Dr Franca Cerito
Dept Experimental Medicine
University of L'Aquila
School of Medicine
Collemaggio-67100 L'Aquila, Italy

Dr Gilles Chevalier
Lab Neurosci Vision
Université P. et M. Curie
4 Place Jussieu
75230 Paris Cedex 05, France

Dr Deborah Joy Clarke
Oxford University
Dept Human Anatomy
South Parks Road
Oxford OX1 3QX, U.K.

Dr Henri Condé
Univ Paris-Sud Bât 440
Laboratoire de Neurobiologie
91405 Orsay France

Pr Alexander R. Cools
Department of Neuropharmacology
University of Nijmegen
P.O. Box 9101
6500HB Nijmegen, The Netherlands

Dr Pierre-Yves Côté
Laboratoire de Neurobiologie
Hôpital de l'Enfant Jésus
1401 18ème rue
G1J 1Z4 Québec Canada

Dr Giuseppe Crescimanno
Institute Human Physiology
University Palermo
Corso Tukory n° 129
Palermo

Pr Alan Crossman
Dept Cell & Struct Biol
University of Manchester
Oxford Road
Manchester, M13 9PT, UK

Dr András Czurkó
Institute of Physiology
Univ Med School
Pecs, H-7643 Pecs
Szigeti str. 12, Hungary

Dr Annie Daszuta
LNCF - CNRS
31 Chemin Joseph Aiguier
13402 Marseille
France

Dr Peter DeBoer
Center Mol & Behavior Neurosci
Rutgers University
197 University Ave.
Newark, NJ 07102, USA

Dr Ronald De Kloet
Center Bio-Pharmaceutical Sci
University of Leiden
The Netherlands

Dr Silvano de las Heras
Departamento de Morfologia
Facultad de Medicina (U.A.M.)
C/Arzobispo Morcillo S/N
28029 - Madrid, Spain

Dr Mahlon R. DeLong
Dept Neurology, WMB
6th Floor, P.O. Drawer V
Emory Univ School Med
Atlanta, GA 30322, USA

Dr Paolo Del Dotto
Institute of Clinical Neurology
University of Pisa
Via Roma 67
56100 Pisa Italy

Dr J.M. Deniau
Lab Neurosci Vision
Université P. et M. Curie
4 Place Jussieu
75230 Paris, Cedex 05, France

Dr Marcel Desban
Service de Neuropharmacologie
Collège de France
11, Place Marcelin Berthelot
75231 Paris, France

Pr Gaetano Di Chiara
Ist Farmacologia e Tossicologia
Viale Diaz 182
09 100 Cagliari
Italy

Dr Silvia Di Loreto
Tissue Typing Institute CNR
University of L'Aquila
School of Medicine
Collemaggio-67100 L'Aquila, Italy

Dr Ivan Divac
Neurofysiologisk Institut
Panum Institute
Blegdamsvej 3C
2200 Copenhagen N, Denmark

Pr Bruno Dubois
Service de Neurologie
Hôpital de la Salpêtrière
47 Boulevard de l'Hôpital
75651 Paris Cedex 13 France

Dr Stephen B. Dunnett
Dept Experimental Psychology
University of Cambridge
Cambridge, UK

Dr Susan Duty
Dept Cell & Struct Biol
University of Manchester
Oxford Road
Manchester, M13 9PT, U.K.

Dr Bettina Eberth
Hospice Civil
68 Avenue Jean Jaurès
67100 Strasbourg Cedex
France

Dr Bart Ellenbroek
Department of Neuropharmacology
University of Nijmegen
P.O. Box 9101
6500HB Nijmegen, The Netherlands

Dr Adriana Emmi
Institute Human Physiology
University Palermo
Corso Tukory n° 129
Palermo

Dr Piers Emson
Department of Neurobiology
MRC Group
Institute Animal Physiol & Gen Res
Babraham, CB2 4AT Cambridge U.K.

Dr Michèle Fabre-Thorpe
Faculté de Médecine, Bât A3
Recherches Cerveau & Cognition
133 route de Narbonne
31062 Toulouse - France

Dr James H. Fallon
Dept Anatomy & Neurobiology
University of California
Irvine, CA 92717
USA

Dr Béla Faludi
Institute of Physiology
Univ Med School
Pecs, H-7643 Pecs
Szigeti str. 12, Hungary

Dr Richard L. M. Faull
Department of Anatomy
School Med University of Auckland
Private Bag Auckland
New Zealand

Pr Jean Féger
Laboratoire de Pharmacologie
Faculté de Pharmacie
4 Avenue de l'Observatoire
75270 Paris Cedex 06, France

Dr Samuel M. Feldman
New York University
New York, NY 10003
USA

Dr Gilles Fénelon
Inserm U 106 Pav Cl Bernard
Hôpital de la Salpêtrière
47 Bd de l'Hôpital
75651 Paris Cedex 13 France

Dr Robert Ferrante
Ger Res Clin Center 182B
Bedford Vet Medical Center
200 Springs Road
Bedford, MA 01730 USA

Dr Michel Filion
Laboratoire de Neurobiologie
Hôpital de l'Enfant Jésus
1401 18ème rue
G1J 1Z4 Québec Canada

Dr Benjamin Floran
Centro de Investigacion,
Departamento de Fisiologia
Apartado Postal 14-740
México 14, D.F. México

Dr Tiziana Florio
Dept Biomedical Technology
University of L'Aquila
School of Medicine
Collemaggio-67100 L'Aquila, Italy

Dr Jack Foucher
Hospice Civil
68 Avenue Jean Jaurès
67100 Strasbourg Cedex
France

Dr Chantal François
Inserm U 106 Pav Cl Bernard
Hôpital de la Salpêtrière
47 Bd de l'Hôpital
75651 Paris Cedex 13 France

Dr Ken-ichi Fujimoto
Department of Neurology
Jichi Medical School
Minamikawachi-machi
Tochigi-ken, 329-04, Japan

Dr Hideki Fukuda
Segawa Neurol Clinic for Children
2-8 Surugadai Kanda
Chiyoda-ku, Tokyo 101
Japan

Dr Lesley Furmidge
Dept Preclinical Vet Sciences
University of Edinburgh
Summerhall, Edinburgh
EH9 1QH, U.K.

Dr José Antonio Gandia
Departamento de Morfologia
Facultad de Medicina (U.A.M.)
C/Arzobispo Morcillo S/N
28029 - Madrid, Spain

Dr Martha Garcia
Escuela Nacional
de Ciencias Biol del IPN
México D.F., México

Dr Marianela Garcia-Munoz
Dept of Psychiatry 0603
School of Medicine, UCSD
9500 Gliman Drive
La Jolla, CA 92093-0603 USA

Dr Christian Gauchy
Service de Neuropharmacologie
Collège de France
11, Place Marcelin Berthelot
75231 Paris, France

Dr Jochen Gehrmann
Department of Neuromorphology
Max-Planck-Institute Psychiatry
8033 Martinsried
Germany

Dr Anne de Geoffroy
Department of Kinanthropologie
UQAM CP 8888
Suc A Montréal
H3C 3P8 Canada

Dr Charles R. Gerfen
National Inst Mental Health
Building 36, Room 2D-10
Bethesda, MD 20892
USA

Pr Jose Manuel Giménez-Amaya
Departamento de Morfologia
Facultad de Medicina (U.A.M.)
C/Arzobispo Morcillo S/N
28029 - Madrid, Spain

Dr Henri Gioanni
Institut des Neurosciences
9 Quai Saint Bernard, Bât B, 3ème ét
75005 Paris, France

Pr Jacques Glowinski
Chaire de Neuropharmacologie
Collège de France
11, Place Marcelin Berthelot
75231 Paris, France

Dr Wendy Graham
Dept Cell and Structural Biology
University of Manchester
Oxford Road
Manchester M13 9PT U.K.

Pr Ann M. Graybiel
Dept Brain Science & Cognition
M.I.T. E25-618
45 Carleton Street
Cambridge, MA 02139, USA

Dr Karen E. Griffith
Program in Neural Science
Department of Psychology
Indiana University
Bloomington, IN 47405 USA

Dr Henk J. Groenewegen
Dept of Anatomy & Embryology
Vrije Universiteit
Van der Boechorststraat 7
1081 BT Amsterdam The Netherlands

Dr Irena Grofova
Department of Anatomy
Michigan State University
East Lansing, MI 48824-1316
USA

Dr Christian Gross
Laboratoire de Neurophysiologie
CNRS URA 1200
Université de Bordeaux II
146 rue Léo Saignat
33076 Bordeaux

Dr Philip M. Groves
Dept Psychiatry, 0603
University of California, San Diego
9500 Gilman Dr.
La Jolla, CA 92093-0603, USA

Dr Rosalinda Guevara Guzman
Department of Neurobiology
MRC Group
Institute Animal Physiol & Gen Res
Babraham, CB2 4AT Cambridge U.K.

Dr Bernard Guibert
CNRS Inst Alfred Fessard
Avenue de la Terrasse
91198 Gif sur Yvette Cedex
France

Dr Jorge Guridi
Experimental Neurology group and
Movement Disorders Unit
Dpto Neurol, Clin Universitaria
Apdo. 192,
31080 Pamplona, Spain

Pr Suzanne N. Haber
Dept of Neurobiology & Anatomy
Univ Rochester School of Medicine
601 Elmwood Avenue, Box 603
Rochester, NY 14642, USA

Dr Constance Hammond
Inserm U 29
123 Boulevard Port Royal
75014 Paris France

Dr Peter J. Hand
University of Pennsylvania
School of Veterinary Medicine
Philadelphia, PA 19104
USA

Dr John L. Haracz
Program in Neural Science
Department of Psychology
Indiana University
Bloomington, IN 47405 USA

Dr Yasuo Hasegawa
Dept Anatomy, Fac Medicine
Mie University
2-174 Edobashi, Tsu City
Mie 514, Japan

Dr Toshi Hattori
Dept Anatomy & Cell Biol
Medical Sciences Building
Toronto, Ontario M5S 1A8
Canada

Dr Nobuaki Hayase
Department of Neurosurgery
Gunma University
School of Medicine
3-39 Showa-machi
Maebashi, Gunma, 371-Japan

Dr Christine Heim
Dept of Psychiatric Clinic
University of Göttingen
Von Sieboldt Str.
D-3400 Göttingen

Dr Maria-Trinidad Herrero
Experimental Neurology group and
Movement Disorders Unit
Dpto Neurol, Clin Universitaria
Apdo. 192,
31080 Pamplona, Spain

Dr Okihide Hikosaka
National Institut for
Physiological Sciences
Myodaiji, Okazaki 444, Japan

Dr Masafumi Hirato
Department of Neurosurgery
Gunma University
School of Medicine
3-39 Showa-machi
Maebashi, Gunma, 371-Japan

Dr Noboru Hiroi
Dept Brain Science & Cognition
M.I.T. E25-618
45 Carleton Street
Cambridge, MA 02139, USA

Dr Neill Hughes
Dept Cell & Struct Biol
University of Manchester
Oxford Road
Manchester, M13 9PT, U.K.

Dr Yoshiaki Ikai
Dept Morphological Brain Science
Faculty of Medicine
Kyoto University
Kyoto 606-01, Japan

Dr Hisamasa Imai
Dept Neurology Juntendo Univ
School of Medicine
1-1, Hongo 2
Bunkyo-ku, Tokyo 113, Japan

Dr Carolyn A. Ingham
Preclinical Veterinary Sciences
University of Edinburgh
Summerhall, Edinburgh
EH9 1QH, Scotland UK

Dr A. Jayaraman
Department of Neurology
LSU Medical Center
1542 Tulane Avenue
New Orleans
Louisiana 70112-2822, USA

Dr Luke Johnson
University Dept Pharmacology
Mansfield Road
Oxford OW1 3QT
U. K.

Dr Steven W. Johnson
Vollum Institute
Oregon Health Sci Univ L-474
3181 S.W. Sam Jackson Park Road
Portland, OR 97201, USA

Dr Jeffrey Joyce
127 Clinical Research Building
Psychiatry Res Department
422 Currie Boulevard
Univ Penn School Medicine
Philadelphia, PA 19104-6141, USA

Dr Zoltán Karádi
Institute of Physiology
Univ Med School
Pecs, H-7643 Pecs
Szigeti str. 12
Hungary

Dr Benny Karmon
Dept Physiology Hebrew Univ
Hadassah Med School
P.O.B. 1172, Jerusalem
Israel 91010

Dr Makoto Kato
National Institute for
Physiological Sciences
Okazaki 444
Japan

Dr Yasuhiro Kawashima
Department of Neurosurgery
Gunma University
School of Medicine
3-39 Showa-machi
Maebashi, Gunma, 371-Japan

Dr Nathalie Kayadjanian
Laboratoire de Neurochimie
& Anatomie - Bâtiment B
9 Quai Saint Bernard
75005 Paris, France

Dr Tetsuro Kayahara
Department of Anatomy
Sch Medicine, Mie University
Tsu, Mie, 514, Japan

Dr Paul A.T. Kelly
Dept Clinical Neurosciences
University of Edinburgh
Summerhall, Edinburgh
EH9 1QH, U.K.

Dr Marie-Lou Kemel
Service de Neuropharmacologie
Collège de France
11, Place Marcelin Berthelot
75231 Paris, France

Dr Keith M. Kendrick
Department of Neurobiology
MRC Group
Institute Animal Physiol & Gen Res
Babraham, CB2 4AT Cambridge U.K.

Dr Hideo Kiba
Department of Neurology
LSU Medical Center
1542 Tulane Avenue
New Orleans
Louisiana 70112-2822, USA

Dr Minoru Kimura
Department of Physiology
Jichi Medical School
Minamikawachi-machi
Tochigi-ken, Japan 329-04

Dr Jiro Kishimoto
Department of Neurobiology
MRC Group
Institute Animal Physiol & Gen Res
Babraham, CB2 4AT Cambridge U.K

Dr Hitoshi Kita
Dept Anatomy & Neurobiology
The University of Tennessee
875 Monroe Avenue
Memphis, TN 38163, USA

Pr Stephen T. Kitai
The University of Tennessee
Center for the Health Sciences
875 Monroe Avenue, Room 402
Memphis, TN 38163, USA

Dr Neil W. Kowall
Ger Res Clin Center 182B
Bedford Vet Medical Center
200 Springs Road
Bedford, MA 01730 USA

Pr George Krauthamer
Dept Neurosci & Cell Biology
UMDNJ Robert Wood Johnson Med Sch
675 Hoes Lane
Piscataway, NJ 08854-5635, USA

Dr Marie-Odile Krebs
Service de Neuropharmacologie
Collège de France
11, Place Marcelin Berthelot
75231 Paris, France

Dr Georg W. Kreutzberg
Department of Neuromorphology
Max-Planck-Institute Psychiatry
8033 Martinsried
Germany

Dr Wolfgang A.A. Kunze
The University of Melbourne
Department of Physiology
Parkville 3052
Victoria, Australia

Dr Béatrice Lannes
Institut de Neurologie
11 rue Humann
67085 Strasbourg Cedex
France

Pr Dominique Laplane
Service de Neurologie
Hôpital de la Salpêtrière
47 Boulevard de l'Hôpital
75651 Paris Cedex 13

Dr Brigitte Lavoie
Centre de Neurobiologie
Hôpital de l'Enfant Jésus
1401, 18ème Rue
Québec G1J 1Z4, Canada

Dr Tiffany Lee
Dept Anatomy & Neurobiology
University of California
Irvine, CA 92717 USA

Mr Eric Legallet
Dept de Neurophysiologie Générale
LNF3-CNRS
31 chemin Joseph Aiguier
13402 Marseille Cedex 9, France

Pr László Lénárd
Institute of Physiology
Univ Med School
Pecs, H-7643 Pecs
Szigeti str. 12, Hungary

Dr Vincent Leviel
CNRS Inst Alfred Fessard
Avenue de la Terrasse
91198 Gif sur Yvette Cedex
France

Dr Theodore I. Lidsky
New York Inst Basic Research
Staten Island, NY 10314
USA

Dr Liu Lizhi
Department of Neurobiology
MRC Group
Institute Animal Physiol & Gen Res
Babraham, CB2 4AT Cambridge U.K.

Dr Manuel Oscar Lopez-Figueroa
Neurofysiologisk Institut
Panum Institute
Blegdamsvej 3C
2200 Copenhagen N, Denmark

Dr Sandra Loughlin
Dept Anatomy & Neurobiology
University of California
Irvine, CA 92717
USA

Dr Tomas Ljungberg
Institute of Physiology
University of Fribourg
CH-1700 Fribourg
Switzerland

Dr Eileen Lynd-Balta
Dept of Neurobiology & Anatomy
Univ Rochester School of Medicine
601 Elmwood Avenue, Box 603
Rochester, NY 14642, USA

Pr. J.S. McKenzie
The University of Melbourne
Department of Physiology
Parkville 3052
Victoria, Australia

Dr Deborah McRitchie
Pathology Department
Neuropathology Unit
Sydney University
N.S.W. 2006 Australia

Dr Yannick Maneuf
Experimental Neurology Group
Dept Cell & Struct Biology
Stopford Building, Oxford Road
Manchester M13 9PT, UK

Dr Monique Manier
Inserm U318 Equipe Lapsen
Pav Neurologie CHU Grenoble
BP 217
38043 Grenoble Cedex

Dr Jean-Louis Martiel
Department of Mathematics
Faculty of Medicine, Grenoble
38700 La Tronche
France

Dr Guillaume Masson
URA CNRS 1166
IBHOP, Traverse C. Susini
13388 Marseille Cedex 13
France

Dr Masaru Matsumura
Laboratoire de Neurobiologie
Hôpital de l'Enfant Jésus
1401 18ème rue
G1J 1Z4 Québec Canada

Dr Mihalis Mavridis
Laboratoire de Neurochimie
& Anatomie - Bâtiment B
9 Quai Saint Bernard
75005 Paris, France

Dr Françoise Mennicken
Inserm U318 Equipe Lapsen
Pav Neurologie CHU Grenoble
BP 217
38043 Grenoble Cedex

Dr Nicola B. Mercuri
Clinica Neurologica
Universita Tor Vergata
Via O. Raimondo 8
00173 Roma, Italy

Dr Gloria Meredith
Dept Anatomy & Embryology
Fac Medicine, Vrije Universiteit
P.O. Box 7161, 1007 MC Amsterdam
The Netherlands

Dr Gabriel Micheletti
Institut de Neurologie
11 rue Humann
67085 Strasbourg Cedex
France

Dr Jacques Mirenowicz
Institute of Physiology
University of Fribourg
CH-1700 Fribourg
Switzerland

Dr Richard R. Miselis
Dept Animal Biology
School Veterinary Medicine
University of Pennsylvania
Philadelphia, Pennsylvania, 1104
USA

Dr Ian J. Mitchell
Dept Cell and Structural Biology
University of Manchester
Oxford Road
Manchester M13 9PT U.K.

Dr Nobuo Miyashita
National Institute for
Physiological Sciences
Okasaki 444
Japan

Dr Noboru Mizuno
Dept Morphological Brain Science
Faculty of Medicine
Kyoto University
Kyoto 606-01, Japan

Dr Diana Mogoseanu
MRC Anatomical
Neuropharmacological Unit
Mansfield Road
Oxford OX1 3TH, U.K.

Dr Marie-Françoise Montaron
Institut des Neurosciences
Dept Neusosci Vision active
9 Quai Saint Bernard, Bât B 6è ét.
75005 Paris, France

Dr Francisco Mora
Department of Human Physiology
Faculty of Medicine
University Complutense Madrid
28040 Madrid, Spain

Dr Rosario Moratalla
Dept Brain Science & Cognition
M.I.T. E25-618
45 Carleton Street
Cambridge, MA 02139, USA

Dr Micaela Morelli
University of Cagliari
Ist Farmacologia e Tossicologia
Viale Diaz 182
09 100 Cagliari Italy

Dr Patrick Mouchet
Inserm U318 Equipe Lapsen
Pav Neurologie CHU Grenoble
BP 217
38043 Grenoble Cedex

Dr Hakima Moukhles
LNCF - CNRS
31 Chemin Joseph Aiguier
13402 Marseille
France

Dr Mireille Mouroux
Laboratoire de Pharmacologie
Faculté de Pharmacie
4 Avenue de l'Observatoire
75270 Paris Cedex 06, France

Dr Shinichi Muramatsu
Department of Neurology
Jichi Medical School
Minamikawachi-machi
Tochigi-ken, 329-04, Japan

Dr Eiji Nakamura
Department of Neurology
Jichi Medical School
Minamikawachi-machi
Tochigi-ken, 329-04, Japan

Dr Satoshi Nakamura
Department of Neurology
Jichi Medical School
Minamikawachi-machi
Tochigi-ken, 329-04, Japan

Dr Katsuma Nakano
Department of Anatomy
Sch Medicine, Mie University
Tsu, Mie, 514
Japan

Dr Bethany Neal-Baliveau
Department of Psychology
Indiana University
Indianapolis, IN 46205-2810
USA

Dr Csaba Niedetzky
Institute of Physiology
Univ Med School
Pecs, H-7643 Pecs
Szigeti str. 12, Hungary

Pr André Nieoullon
Unité de Neurochimie - LNF
31, Chemin Joseph Aiguier
13402 Marseille Cedex 9 France

Dr Yoshiko Nomura
Segawa Neurol Clinic for Children
2-8 Surugadai Kanda
Chiyoda-ku, Tokyo 101
Japan

Dr Paula Norris
Department of Neurobiology
MRC Group
Institute Animal Physiol & Gen Res
Babraham, CB2 4AT Cambridge U.K.

Dr R. Alan North
Vollum Institute
Oregon Health Sci Univ L-474
3181 S.W. Sam Jackson Park Road
Portland, OR 97201, USA

Dr Johannes Noth
Department of Neurology
University of Aachen
Pauwelsstr. 30
5100 Aachen, Germany

Pr Jose A. Obeso
Experimental Neurology group and
Movement Disorders Unit
Dpto Neurol, Clin Universitaria
Apdo. 192,
31080 Pamplona, Spain

Pr Chihiro Ohye
Department of Neurosurgery
Gunma University
School of Medicine
3-39 Showa-machi
Maebashi, Gunma, 371-Japan

Dr Valérie Olivier
CNRS Inst Alfred Fessard
Avenue de la Terrasse
91198 Gif sur Yvette Cedex
France

Dr Jose de Olmos
Inst Investigacion Medica
Mercedes y Marin Ferrerya
Cordoba, Argentina

Dr Antonio G. Paolini
The University of Melbourne
Department of Physiology
Parkville 3052
Victoria, Australia

Dr Patricia Patino
Dept of Psychiatry 0603
School of Medicine, UCSD
9500 Gliman Drive
La Jolla, CA 92093-0603 USA

Dr Jacqueline Penit-Soria
Laboratoire de Neurochimie Anat.
Institut des Neurosciences
9 Quai Saint Bernard
75005 Paris, France

Dr David Peggs
Dept Cell and Structural Biology
University of Manchester
Oxford Road
Manchester M13 9PT U.K.

Docteur Gérard Percheron
Inserm U 106 Pav Cl Bernard
Hôpital de la Salpêtrière
47 Boulevard de l'Hôpital
75651 Paris Cedex 13 France

Dr Antonio Pisani
Clinica Neurologica
Universita Tor Vergata
Via O. Raimondo 8
00173 Roma, Italy

Dr Dietmar Plenz
Max-Planck Institute for
Biological Cybernetics
Spemannstr. 38
7400 Tubingen, Germany

Mr Jean-Claude Pons
Dept de Neurophysiologie Générale
LNF3-CNRS
31 chemin Joseph Aiguier
13402 Marseille Cedex 9, France

Dr Alberto Porras
Department of Human Physiology
Faculty of Medicine
University Complutense Madrid
28040 Madrid, Spain

Dr Emmanuelle Pourcher
Laboratoire de Neurobiologie
Hôpital de l'Enfant Jésus
1401 18ème rue
G1J 1Z4 Québec Canada

Dr Jerome K. Puotz
Program in Neural Science
Department of Psychology
Indiana University
Bloomington, IN 47405 USA

Dr Bruce Quinn
Dept Pathology, UCLA Med Sch
10833 Le Conte Ave
Los Angeles
CA 90024-1732 USA

Dr George V. Rebec
Program in Neural Science
Department of Psychology
Indiana University
Bloomington, IN 47405 USA

Dr Hélène Richard
Laboratoire de Neurobiologie
Hôpital de l'Enfant Jésus
1401 18ème rue
G1J 1Z4 Québec Canada

Dr Christopher D. Richards
Oxford University
Department of Pharmacology
Mansfield Road
Oxford OX1 3TH, U.K.

Dr Rosalinda Roberts
School of Medicine
Maryland Psychiatric Res Ctr
P.O. Box 21247
Baltimore, Maryland 21228 USA

Dr Robert G. Robertson
Dept Cell and Structural Biology
University of Manchester
Oxford Road
Manchester M13 9PT U.K.

Dr Ranulfo Romo
Institute of Physiology
University of Fribourg
CH-1700 Fribourg
Switzerland

Pr Pierre Rondot
Service de Neurologie
Hôpital Sainte Anne
1 rue Cabanis
75674 Paris Cedex 14

Dr Ariane Rosa-Kenig
Program in Neural Science
Department of Psychology
Indiana University
Bloomington, IN 47405 USA

Dr Nynke Rots
Department of Neuropharmacology
University of Nijmegen
P.O. Box 9101
6500HB Nijmegen, The Netherlands

Dr Sharleen T. Sakai
Michigan State University
Department of Anatomy
East Lansing, MI 48824-1316
USA

Dr Michael A. Sambrook
Dept Cell and Structural Biology
University of Manchester
Oxford Road
Manchester M13 9PT U.K.

Dr Kenji Satake
Department of Neurosurgery
Gunma University
School of Medicine
3-39 Showa-machi
Maebashi, Gunma, 371-Japan

Dr Eugenio Scarnati
Dept Biomedical Technology
University of L'Aquila
School of Medicine
Collemaggio-67100 L'Aquila, Italy

Dr Wolfram Schultz
Institute of Physiology
University of Fribourg
CH-1700 Fribourg
Switzerland

Dr Michael Schwarz
Department of Neurology
University of Aachen
Pauwelsstr. 30
5100 Aachen, Germany

Dr Masaya Segawa
Segawa Neurol Clinic for Children
2-8 Surugadai Kanda
Chiyoda-ku, Tokyo 101
Japan

Dr Rosa Senaris
Department of Neurobiology
MRC Group
Institute Animal Physiol & Gen Res
Babraham, CB2 4AT Cambridge U.K.

Dr Vincent Seutin
Vollum Institute
Oregon Health Sci Univ L-474
3181 S.W. Sam Jackson Park Road
Portland, OR 97201, USA

Dr Tohru Shibazaki
Department of Neurosurgery
Gunma University School Med
3-39 Showa-machi
Maebashi, Gunma, 371-Japan

Dr Yasuhide Shinonaga
Dept Morphological Brain Science
Faculty of Medicine
Kyoto University
Kyoto 606-01, Japan

Dr Maria Sieklucka
Department of Pharmacology
University of Lublin
Lublin, Poland

Dr Yoland Smith
Laboratoire de Neurobiologie
Hôpital de l'Enfant Jésus
1401 18ème Rue
Québec Canada G1J 1Z4

Dr Marylou Solbrig
University of California
College of Medicine
Department of Neurology
Irvine, CA 92717
USA

Pr Karl-Heinz Sontag
Max-Planck-Institut für Exper Med.
Hermann-Rein Strasse 3
D-3400 Göttingen
Germany

Dr Will P.J.M. Spooren
Dept of Neurobiology & Anatomy
Univ Rochester School of Medicine
601 Elmwood Avenue, Box 603
Rochester, NY 14642, USA

Doctor Alessandro Stefani
Clinica Neurologica
Universita Tor Vergata
Via O. Raimondo 8
00173 Roma, Italy

Dr Francesca Stratta
Clinica Neurologica
Universita Tor Vergata
Via O. Raimondo 8
00173 Roma, Italy

Dr Masahiko Takada
Dept Morphological Brain Science
Faculty of Medicine
Kyoto University
Kyoto 606-01, Japan

Dr Akio Takahashi
Department of Neurosurgery
Gunma University
School of Medicine
3-39 Showa-machi
Maebashi, Gunma, 371-Japan

Dr Boualam Talbi
Inserm U 106 Pav Cl Bernard
Hôpital de la Salpêtrière
47 Bd de l'Hôpital
75651 Paris Cedex 13 France

Dr Alain Teinturier
Service de Neurologie
Hôpital Sainte Anne
1 rue Cabanis
75674 Paris Cedex 14

Dr Rudolf Töpper
Department of Neurology
University of Aachen
Pauwelsstr. 30
5100 Aachen, Germany

Dr Susan Totterdell
University Dept Pharmacology
Mansfield Road
Oxford OW1 3QT
U. K.

Dr Léon Tremblay
Service de Neuropharmacologie
Collège de France
11, Place Marcelin Berthelot
75231 Paris, France

Dr Elisabeth Trouche
Dept de Neurophysiologie Générale
LNF3-CNRS
31 chemin Joseph Aiguier
13402 Marseille Cedex 9, France

Dr JoAnn T. Tschanz
Program in Neural Science
Department of Psychology
Indiana University
Bloomington, IN 47405 USA

Dr Kimiaki Uetake
Segawa Neurol Clinic for Children
2-8 Surugadai Kanda
Chiyoda-ku, Tokyo 101
Japan

Dr Derek J. Van Der Kooy
Dept Anatomy Univ Toronto
Medical Sci Bldg
Toronto, Ontario
Canada M5S 1A8

Dr François Viallet
Chef du Service de Neurologie
Centre Hospitalier Général
14 Avenue des Tamaris
F - 13616 Aix en Provence

Dr Imre Vida
Institute of Physiology
Univ Med School
Pecs, H-7643 Pecs
Szigeti str. 12, Hungary

Dr Jacqueline Vuillet
LNCF - CNRS
31 Chemin Joseph Aiguier
13402 Marseille
France

Dr Stephen R. Wachtel
Center Mol & Behavior Neurosci
Rutgers University
197 University Ave.
Newark, NJ 07102, USA

Dr Zhongrui Wang
Program in Neural Science
Department of Psychology
Indiana University
Bloomington, IN 47405 USA

Dr Thomas Wichmann
Dept Neurology, WMB
6th Floor, P.O. Drawer V
Emory Univ School Med
Atlanta, GA 30322, USA

Dr Jeffrey R. Wickens
Dept Anatomy
Univ of Otago Medical School
P.O. Box 913
Dunedin, New Zealand

Dr Klas Wictorin
Dept of Medical Cell Research
University of Lund
Lund, Sweden

Dr Sidney I. Wiener
CNRS Physiologie Neurosensorielle
15 rue de l'Ecole de Médecine
75270 Paris Cedex 06
France

Dr Charles Wilson
Dept Anatomy & Neurobiology
University of Tennessee
875 Monroe Avenue
Memphis, TN 38163 USA

Dr Floris G. Wouterlood
Dept of Anatomy & Embryology
Vrije Universiteit
Van der Boechorststraat 7
1081 BT Amsterdam The Netherlands

Dr Ann K. Wright
Dept Preclinical Vet. Sciences
University of Edinburgh
Summerhall, Edinburgh
EH9 1QH, U.K.

Dr Weining Xu
Department of Neurobiology
MRC Group
Institute Animal Physiol & Gen Res
Babraham, CB2 4AT Cambridge U.K.

Dr Hiroshi Yamada
Department of Neurology
Jichi Medical School
Minamikawachi-machi
Tochigi-ken, 329-04, Japan

Dr Yukikhiko Yasui
Department of Anatomy
Sch Medicine, Mie University
Tsu, Mie, 514
Japan

Dr Jérôme Yelnik
Inserm U 106 Pav Cl Bernard
Hôpital de la Salpêtrière
47 Bd de l'Hôpital
75651 Paris Cedex 13 France

Dr Fusako Yokochi
Tokyo Metropolitan Neurol Hospital
2-6-1, Musashidai
Fuchu, Tokyo 183
Japan

Pr Mitsuo Yoshida
Department of Neurology
Jichi Medical School
Minamikawachi-machi
Tochigi-ken, 329-04, Japan

Dr Stephen J. Young
Dept of Psychiatry 0603
School of Medicine, UCSD
9500 Gliman Drive
La Jolla, CA 92093-0603 USA

Dr Marc Ziegler
Service de Neurologie
Centre Raymond Garcin
Hôpital Sainte Anne
1 rue Cabanis
75674 Paris Cedex 14

INDEX